Integrated CAD by Optimization

B K Chakrabarty

Integrated CAD by Optimization

Architecture, Engineering, Construction, Urban Development and Management

 Springer

B K Chakrabarty
(Formerly Executive Director, HUDCO)
Independent Researcher
New Delhi, India

ISBN 978-3-030-99308-5   ISBN 978-3-030-99306-1   (eBook)
https://doi.org/10.1007/978-3-030-99306-1

This Springer imprint is published by the registered company Springer Nature Switzerland AG
The registered company address is: Gewerbestrasse 11, 6330 Cham, Switzerland

Acknowledgement

The seed of this book was laid many years back when I was deputed by my organization, HUDCO to a few reputed Universities of the USA for interaction under a USAID-funded research grant on 'Affordable Housing Design and Analysis'. During that time, I was fortunate to attend a presentation by Professor Charles Eastman at Carnegie Mellon University on 'Application of Management Science on Urban Issues'. This inspired me to work further on this topic resulting in this book for which I owe an enormous debt of gratitude to late Professor Charles Eastman.

I want to thank my daughter Piali, my son-in-law Debasish, and my loving granddaughter Urmika, all of whom put up with me throughout the arduous process of writing this book and provided me with inspiration and emotional support that made this book possible.

Contents

Chapter 1
Integrated Computer-Aided Design by Optimization: An Overview

1.1 Introduction

Computer-Aided Design (CAD), quantitative analysis, and optimization are important application areas and essential elements of emerging and all-pervasive information technology. Today, Information Technology (IT) and computer applications are everywhere shrinking the effects of time and distance, and altering the very nature of work in an organization and in all sectors of society in almost all countries in the world, irrespective of their income or development statuses. It is viewed as the second Industrial Revolution by many writers and is generating a wide range of fundamental questions in all sectors of society needing urgent answers (Piercy N. 1984). Many sectors of society and knowledge disciplines have already initiated actions for adequate response to this new challenge. This is equally applicable in the knowledge disciplines of Architecture (including Urban Planning), Engineering-Civil-Building Engineering, and Construction (usually classified as AEC sector), in various countries. Hence, it is desirable that the AEC sector in various countries initiates vigorous and comprehensive actions for adequate response to this new challenge.

In recent years, there is tremendous increase in capabilities of computer hardwares and graphics. As a result, there are unparalleled transformations in computers in terms of their capacity to capture data, to analyse, manipulate, and store them, to disseminate and communicate information, and to apply these capabilities in different areas and fields of specialization including optimization. Such transformations are occurring with increasing rapidity and frequency in many areas. The rate at which IT and Computer Applications are spreading to various sectors is truly extraordinary and has few, if any, parallels in the technological history. The pervasiveness and cheapness of IT are influencing the way we live, work, and enjoy ourselves. AEC sector and its professionals cannot remain aloof from such influences to ensure that they remain modernized and relevant.

The adoption of IT and CAD in various fields has accelerated with recent advances in microcomputer technology and graphics. As a result, computer-assisted design analysis, optimization, and graphics no longer a luxury only for a few large companies that can afford mainframe computers. Engineers, Architects, Planners, and Urban Managers everywhere have now access to better and cheaper computer hardware and software to produce sophisticated computer analysis, optimization, graphics, and drafting. The capabilities of computer hardware and graphics are increasing so rapidly that software engineers in various fields are constantly finding new applications of this technology in various aspects of design work. Information technology and computers are also playing a significant role in addressing various social, environmental, and development issues for a just society with informed public participation in the decision-making process on such issues, which is very relevant in AEC. For example, even a lay user citizen can use the software CAUB, described in later chapters, to find suitable (even optimum) values of built-form-element in a site-specific-case, to have an informed and meaningful dialogue with the Planning Authority (fixing planning norms) to arrive at a consensus in such matters, resolving conflicting viewpoints of diverse stakeholders. This makes it imperative for the Planning Authority and AEC professionals to develop and use such IT tools in their operations to make them more efficient and effective avoiding any redundancy.

At present, in AEC, especially in Architecture, CAD is primarily used in producing neater, more attractive drafting-drawing packages only to lessen the services of architectural draftsmen, utilizing the commercially available drafting-drawing software. Since, drafting-drawing is just one of the five components of integrated CAD; in this approach, the AEC professionals are rarely able to derive full benefits of integrated CAD, quantitative analysis, and optimization techniques cited above, creating many inefficiencies and inequities and the consequent urban problems. Computer-Aided Design (CAD) is no longer a fad, and its real utility is not confined to just producing sophisticated solid geometry and more attractive drawing packages only, but can cover many other aspects of design work including optimization to improve productivity and derive other benefits at various operation levels as outlined in various chapters later.

To gain such advantages, the pace at which information technology and CAD are being adopted by various sectors of industry and professional disciplines, using ever increasing computing power, is truly extraordinary compared to any other technology developed in the past. This is equally applicable in AEC sector, where even now more stress is given to cosmetic such as producing neater, more attractive drawing packages only as cited above. Instead of this, adoption of an integrated computer-aided design approach (covering all its components including optimization), involving hundreds of variables, will improve productivity, i.e. efficiency, effectiveness and derive other benefits in AEC operation, particularly in Urban Development and Management.

As discussed later, optimization is a human trait that is as old as antiquity and even the ancients applied it. Tedious and voluminous computations involved, in utilizing optimization and operations research techniques and to solve optimizing models, have prevented their application in the past. The IT revolution cited above has made it very easy to apply integrated CAD, quantitative analysis, and optimization techniques in all sectors including AEC sector, to derive many benefits. AEC professionals commit huge resources (like money, materials, land, and so on) in planning and design decisions in their operations. Generally, resources are always inadequate compared to the need, whether in low- or high-income countries. No doubt low-income countries are usually subjected to severe resource-constraints necessitating scaling-down of needs in the context of available resources. But, even in affluent countries, such resource-constraints are applicable because of scaling-up of needs to have higher standard of living consuming more resources. Moreover, all of us *live in a finite earth with finite resources* requiring conservation of resources by optimization by all responsible citizens in society. In view of large variations in quantitative design efficiency indicators in various AEC operations (as earlier reported in published articles in different journals and shown in various chapters later) irrespective of the context of income or development statuses of countries, it is desirable that such variations are presented by AEC professionals in a transparent manner (by application of integrated CAD, quantitative analysis, and optimization techniques) to all stakeholders including user citizens in all countries, for appropriate planning and design decisions, and thus, help derive the benefit of achieving a resource-efficient urban problem solution with social equity.

This chapter introduces the fundamental principles of major CAD elements covering all its five components, i.e. (1) geometric modeling, (2) cost engineering and database management, (3) planning and design analysis, (4) design optimization, (5) architectural and engineering drafting. It provides the reader an overview of Integrated CAD emphasizing the importance of computerizing the design process in AEC sector and its utility in urban planning, development, and management. The chapter stresses the need of viewing AEC planning and design as one of the five urban management functions (i.e. planning function) for improving productivity, social equity, and social responsiveness in urban operations. Such an approach is essential for resource-efficient solution of urban development problems, viewing planning and design as a problem-solving process involving search through numerous potential solutions to select a particular solution that meets a specified criteria or goal. This makes computer use and CAD by optimization indispensable in such activities involving civil engineers, architects, and planners (i.e. all AEC professionals) who commit huge resources (like money, materials, land, and so on) in their planning and design decisions in such operations, enjoining on them an accountability and ethical responsibility to society to ensure efficient use of resources so committed. The book emphasizes utility of CAD as an integrating tool enabling above professionals to adopt such an integrated problem-solving approach to derive above benefits.

Since AEC professionals commit huge resources in their planning and design decisions, particularly related to urban development and management to improve the *quality of life for all*, it is desirable to view AEC planning and design as one of the five urban management functions as stated above. Therefore, a separate Chap. 2 on urban management is added to enable AEC professionals to comprehend the '*big picture*' of urban operations to solve problems resolving conflicting viewpoints in the process, and use of integrated CAD by optimization as a planning and management tool. This is followed by nine chapters, namely on (1) elements of numerical techniques and operations research (OR), (2) elements of cost engineering and database management, (3) elements of computer graphics and architectural/engineering drafting-drawing, (4) planning and design analysis, (5) design optimization, (6) integrated computer-aided structural design by optimization, (7) integrated computer-aided building-housing design by optimization, (8) integrated computer-aided layout design by optimization, and (9) integrated computer-aided dwelling-layout system design by optimization, respectively. Several software for urban development design analysis and drafting-drawing (CAUB/CADUB), urban development design optimization and drafting-drawing (HudCAD), and for structural design optimization and drafting-drawing (RCCAD) are presented with source code listings. In the above chapters, design analysis, optimization, and drafting-drawing are demonstrated using software's incorporating geometric programming techniques, highlighting the possibility of using available commercial participatory-collaborative IT tools in AEC sector for managing the dynamics of urbanization to arrive at more satisfactory urban problem solutions with citizen participation achieving *resource-efficiency* and *social equity*.

As CAD is an important application area of IT, introduction of compulsory and comprehensive CAD education, which is rare in AEC sector in many countries, can be one of the important actions towards such efforts. Such an action has attained added urgency for AEC sector, as Operations Research (OR) and Computer-Aided Design (CAD) are essential integrative techniques to improve productivity, i.e. efficiency & effectiveness (Koontz et al., 1976, 1990). Since AEC professionals commit huge resources in their planning-design-decisions, it is basic that they use such basic tools to discharge their accountability to the society to achieve productivity and social equity in the use of huge (but inadequate compared to need) resources committed by them in their plans and designs to produce buildings and urban services.

The purpose of this chapter is to provide to students (future AEC professionals) and to practicing professionals in these fields an overview of integrated CAD by optimization in AEC sector in the context of ongoing IT revolution, before discussing the individual topics of CAD included in the book.

1.2   Information Technology Revolution: Impact on Society and Effect on the Working of Organizations and Professions in Various Sectors

Information technology is a vast subject beyond the scope of this book. Here, some salient aspects of the current information technology revolution are discussed only to give an idea about their possible impact in society affecting the working of organizations and professions in various sectors including AEC sector. Information Technology (IT), which is the coming together of the existing computing, communication, and control engineering technologies, affects the way we live, work, and enjoy ourselves, and also our competitive ability in the 'information society'. Many writers view Information Technology (IT) as the second Industrial Revolution and see diffusion of IT as a continuous process of rationalization as part of this New Industrial Revolution, which can lead to major economic growth, wealth creation, and the overall expansion of employment, as was the case during past technological breakthroughs (invention of steam power, internal combustion engine, electric motor, and so on), i.e. first Industrial Revolution.

At one level, information technology might be regarded as no more than simply a way of upgrading existing processes in organizations, and the products and services they provide; as computerized data processing is faster, cheaper, and more widely available; word processors replace typewriters; microcomputers and PCs replace filing systems; and so on. But, information technology has potentially a somewhat more fundamental impact within organizations, and in the relationships between organizations and people. IT entails two factors of vital importance, namely the convergences of previously separate technologies and its pervasive impact on society, questioning many assumptions like those relating to what causes efficiency, what constitutes work and the value of organizational hierarchy, and so on. In this context, the possibility of the organization as a community, based on consultation and democratic principles in decision-making, is often suggested, where information technology is an enormously powerful catalyst enabling workers-users-planners-managers in an entity-field-sector-country to work in geographically diverse locations, linked by terminals accessing a computer system and a shared database. Such an approach is very relevant in AEC and urban operations including that in urban organizations, as it would permit decision-making in urban sector (e.g. planning norms) in the context of individual, local, national, and even global constraints, based on consultations and democratic principles, thus helping a resource-efficient, equitable, and more satisfying urbanization and development process fulfilling people's need.

Computers and IT have been applied for many years in problem-solving in manufacturing industry, since effective management of manufacturing engineering required the collection, generation, communication, recording, manipulation, and exploitation of information. Such IT application areas in engineering production and management were in Computer-Aided Planning (CAP), Computer-Aided Design (CAD), and Computer-Aided Manufacture (CAM) together forming Computer-

Aided Engineering (CAE). The history of computer use in problem-solving has different stages. In the initial stage, computers were used only by specialist technicians, who later allowed a much wider range of users to access computers through them. But the current IT revolution permits computers to be 'user friendly' from outset allowing all professionals and even the lay user citizens (again very relevant in respect of AEC-urban operations) to access computers directly in 'user' language, in which dialogues relate directly to the relevant knowledge base or 'user' priority and concerns (e.g. urban sector). One of the characteristics of IT is the enormous flexibility of the underlying technical systems and ideas and the possibilities seem to be almost endless.

Application of information technology to the workplace is having major and even revolutionary impacts on the work organization and its members. This is because the new IT permits the principles, practice, and terminology of information handling to be treated in a united, systematic way. In view of above and the fact that in all problem solution, collection, generation, communication, recording, manipulation, computation, and exploitation of information are essential activities, use of computers and IT is also essential in such activities (including AEC and urban operations) to help effective problem solution in respective area.

IT, which is the modern name for the means by which we collect, store, process, and use information, is built around the scientific, technological, engineering, and management disciplines used in information handling and processing. This, in turn, is achieved by the proper selection and implementation of policies and plans involving many factors, which interact with each other creating the operational complexity. IT assists the introduction of an integrated systems approach to such complex problem-solving, resolving conflicting viewpoints and demands. As a result, IT is an essential part in most manufacturing-service organizations, in product design processes, in delivery of services, and in inter-organizational relations. It is shrinking the effects of time and distance, thereby altering the very nature of work in an organization. These are equally applicable in AEC and urban organizations and in their operations.

One fundamental characteristic of IT is the rapidity of technological convergence and shared technological change. Its impact is far greater than simply adding peripherals like word processors, AutoCAD, and so on and leaving it at that. At least two distinct phases may be considered in the information technology revolution in various societal operations including operations in the AEC sector. At the initial phase-1 stage, the main impact of IT is simply to facilitate operations as they are now being carried out in conventional process. For example, the word processor replaces the typewriter, the microcomputer and PCs replace the filing cabinet, and draftsman is partly replaced by AutoCAD. While significant, these innovations simply increase the speed and quality of work as it is now done in conventional process, but does not derive many other benefits which can accrue using IT. The second phase of IT application portends revolutionary changes to the organization, in relationships between organizations, and in the interaction between organizations, people, and the society as a whole. It may involve outright elimination of many intermediary functions, adding new functions and is generally brought about by the convergence of many technologies like electronic data processing, telecommunications, computerizing the planning and design process, and so on. Thus, diffusion of information technology can be seen as a continuous process of rationalization in all activities of an organization or entity.

In fact, implications and impact of IT on organizations-disciplines-sector-entity are inseparable from its broader effects on the society, and may impinge on the 'social legitimacy' of disciplines-organizations-entity-sector, unless these entities-constituents develop proper responses to this revolution. To support such transformation, research is promoted to develop different IT tools including CAD tools and individual courses are offered at various levels to produce IT specialists. High-level computer languages have also been developed in artificial intelligence research demonstrating reasoning and drawing conclusions and ultimately helping decision-making, creating exciting possibility of using information technology and computers for solving difficult intellectual problems of a non-numerical character, like simulation of cognitive processes such as perception, learning, concept formation, and induction (Aguiler, 1973; Alexander, 1964).

In the AEC sector, many decisions are of non-numerical character and have to be taken on subjective considerations resulting to dissatisfaction of stakeholders. Similar information technology and artificial intelligence research to develop artificial intelligence systems for use in the AEC sector may help more satisfying problem solution in this sector. Again, knowledge disciplines like mechanical and electrical engineering have introduced individual courses on CAD, a vital application area of IT, long time back and are taking appropriate actions in adapting themselves to cope with this fundamentally different working situation with responses more strategic and proactive, so that they become more efficient and survive in the competitive market and also retain social legitimacy. Similarly, it is desirable that AEC sector should also develop proper responses to this IT-revolution and focus attention on issues involved in adapting disciplines-organizations-entity in this sector. Research to develop integrated CAD tools and introduction of individual courses on integrated CAD can be one such response to cope with this fundamentally different working situation with responses more strategic and proactive, and not merely an ad hoc reaction to implementation problem of IT by simply adding peripherals and leaving it at that, as cited above, so that they become more useful to the society in resolving resource-efficient problem solution and retain their social legitimacy.

Information Technology (IT), which is the coming together of the existing computing, communication, and control engineering technologies, affects the way we live, work, and enjoy ourselves, and also our competitive ability in the 'information society'. Many writers view Information Technology (IT) as the second Industrial Revolution and see diffusion of IT as a continuous process of rationalization as part of this New Industrial Revolution, which can lead to major economic growth, wealth creation, and the overall expansion of employment, as was the case during past technological breakthroughs (invention of steam power, internal combustion engine, electric motor, and so on), i.e. first Industrial Revolution.

At one level, information technology might be regarded as no more than simply a way of upgrading existing processes in organizations, and the products and services they provide; as computerized data processing is faster, cheaper, and more widely available; word processors replace typewriters; microcomputers/PCs replace filing systems; and so on. But, information technology has potentially a somewhat more fundamental impact within organizations, and in the relationships between organizations. IT entails two factors of vital importance, namely, the convergences of previously separate technologies and its pervasive impact on society, questioning many assumptions like those relating to what causes efficiency, what constitutes work, and the value of organizational hierarchy and so on. In this context, the possibility of the *organization as a community*, based on consultation and democratic principles in *decision-making*, is often suggested, where information technology is an enormously powerful catalyst enabling workers/users/planners/managers in an entity/field/sector/country to work in geographically diverse locations, linked by terminals accessing a computer system and a shared database. Such an approach is very relevant in respect of **Urban Organizations**, as it would permit *decision-making* in urban sector (e.g. planning norms) in the context of individual, local, national, and even global constraints, based on consultation and democratic principles, thus helping a *resource-efficient*, *equitable*, and more satisfying urbanization and development process.

Computers and IT have been applied for many years in problem-solving in manufacturing industry, since effective management of manufacturing engineering required the collection, generation, communication, recording, manipulation, and exploitation of information. Such IT application areas in engineering production and production management were in Computer-Aided Planning (CAP), Computer-Aided Design (CAD), and Computer-Aided Manufacture (CAM) together forming Computer-Aided Engineering (CAE) with the ultimate goal of achieving the automatic factory covering activities from automatic parts ordering right through to automatic warehousing of finished goods (p. 17). Many of the basic elements needed within IT also predate microelectronic revolution. However, the history of computer use in problem-solving has different stages. In the initial stage, computers were used only by specialist technicians, who later allowed a much wider range of users to access computers through them. But the current IT revolution permits computers to be 'user friendly' from outset allowing all professionals and even lay user citizens (again very relevant in respect of **Urban Operations**) to access computers directly in 'user' language, in which dialogues relate directly to the relevant knowledge base or 'user' priority and concerns (e.g. urban sector). One of the characteristics of IT is the enormous flexibility of the underlying technical systems and ideas, and hence the tremendous variety of hardware and diversity of software designed to accompany them. The possibilities seem to be almost endless and has progressed a huge distance beyond the achievements embodied in the early computers.

As a result, application of information technology to the workplace is having major and even revolutionary impacts on the work organization and its members. This is because the new IT permits the principles, practice, and terminology of information handling to be treated in a united, systematic way. In view of above and the fact that in all problem solution, collection, generation, communication, recording, manipulation, and exploitation of information are essential activities, use of computers and IT is also essential in such activities (including AEC activities) to help effective problem solution in respective area. IT, which is the modern name for the means by which we collect, store, process, and use information, is built around the scientific, technological, engineering, and management disciplines used in information handling and processing. Engineering production and production management are concerned with the effective use of people, products, plant, and processes in modern manufacturing industry. This, in turn, is achieved by the proper selection and implementation of policies and plans involving many factors, which interact with each other creating the operational complexity. IT assists the introduction of an integrated systems approach to such complex problem-solving in modern manufacturing industry, which makes IT such a potentially powerful production management tool. As a result, IT is an essential part of the processes internal to most manufacturing/service organizations, to product design processes, to delivery of services, and to inter-organizational relations. It is shrinking the effects of time and distance and altering the very nature of work in an organization, and becoming the lifeblood of most industrial and service organizations.

One fundamental characteristic of IT is the rapidity of technological convergence/shared technological change, and its impact is far greater than simply adding peripherals like word processors/AutoCAD and so on and leaving it at that. At least two distinct phases may be considered in the information technology revolution in various societal operations including operations in the AEC sector. At the initial phase-1 stage, the main impact of IT is simply to facilitate operations as they are now being carried out in conventional process. For example, the word processor replaces the typewriter, the microcomputer replaces the filing cabinet, and draftsman is partly replaced by AutoCAD. While significant, these innovations simply increase the speed and quality of work as it is now done in conventional process. However, it is becoming increasingly apparent that IT **Revolution** has more fundamental impact within organizations, in relationships between organizations, and in the interaction between organizations. These implications at the phase-2 stage of IT **Revolution** can be classified under four categories—communication and decision-making, the hierarchy and managers, flexibility of work time and place, and effects on lower level personnel. This second phase of IT application portends revolutionary changes to the organization, in relationships between organizations, and in the interaction between organizations, people, and the society as a whole. It may involve outright elimination of many intermediary functions, adding new functions and is generally brought about by the convergence of many technologies like electronic data processing, telecommunications, office machines, computerizing the planning and design process and so on, seeing diffusion of information technology as a continuous process of rationalization.

In fact, implications/impact of IT on organizations/disciplines/sector/entity is inseparable from its broader effects on the society, and may impinge on the 'social legitimacy' of disciplines/organizations/entity/sector, unless these entities/constituents develop proper responses to this revolution. Hence, there is need to focus attention on issues involved in adapting organizations/disciplines/sector/entity to cope with this fundamentally different working situation with responses more strategic and proactive and not merely an ad hoc reaction to implementation problem of IT, or simply adding peripherals and leaving it at that, as cited above, so that they become more useful to the society.

To support such transformation, research is promoted to develop different IT tools including CAD tools and individual courses are offered at various levels to produce IT specialists. High-level computer languages have also been developed in artificial intelligence research demonstrating reasoning and drawing conclusions and ultimately helping decision-making, creating exciting possibility of using information technology and computers for solving difficult intellectual problems of a non-numerical character, like simulation of cognitive processes such as perception, learning, concept formation, and induction (Aguiler, 1973; Alexander, 1964). Many such artificial intelligence systems are developed and in use in robotics and the automatic factory improving productivity and giving other benefits. In the AEC sector, many decisions are of non-numerical character and have to be taken on subjective considerations resulting to dissatisfaction of stakeholders. Similar information technology and artificial intelligence research to develop artificial intelligence systems for use in the AEC sector may help more satisfying problem solution in this sector. Again, knowledge disciplines like mechanical and electrical engineering have introduced individual courses on CAD, a vital application area of IT, long time back and are taking appropriate actions in adapting themselves to cope with this fundamentally different working situation with responses more strategic and proactive (e.g. automating the process industry using IT), so that they become more efficient, and retain their social legitimacy and become more useful to the society. Similarly, it is desirable that AEC sector should also develop proper responses to this IT-revolution and focus attention on issues involved in adapting disciplines/organizations/entity in this sector—research to develop CAD tools and introduction of individual courses on CAD can be one such response—to cope with this fundamentally different working situation with responses more strategic and proactive and not merely an ad hoc reaction to implementation problem of IT, so that they become more useful to the society and retain their *social legitimacy*.

1.3   CAD as Information Technology Application Area

There are many application areas of IT, which is expanding continuously with new developments in IT. Computer-Aided Design (CAD) is an important application area of Information Technology (**IT**) as mentioned above. Collection, generation, communication, recording, manipulation, and exploitation of information are essential activities in solution of most problems. IT can be considered as the means by which we collect, store, process, and use information. Hence, use of computers and IT is essential in all activities (including AEC activities) to help effective problem solution in the respective area. It is built around the scientific, technological, engineering, and management disciplines used in information handling and processing. Engineering production and production management are concerned with the effective use of people, products, plant, and processes in modern manufacturing industry. This, in turn, is achieved by the proper selection and implementation of policies and plans involving many factors, which interact with each other. In view of this operational complexity involving information handling and processing, IT is in use in engineering production and production management for many years, and several of the basic ingredients of IT predate the present microelectronic revolution. IT also assists the introduction of an *Integrated Systems Approach* to such complex problem-solving in modern manufacturing industry, which makes IT such a potentially powerful production management tool. As a result, IT and computer application is an essential part of the processes internal to most manufacturing/service organizations, to product design processes, to delivery of services, and to inter-organizational relations.

History of computer use in problem-solving has three stages. In the initial stage, only computer specialists were the user of computers. In the second stage, computer user base became much wider via the computer specialists. Now in the third stage, even lay users can use computers as the current IT revolution has made computers extremely 'user-friendly' from the outset where computer is accessed directly in 'user' language, in which dialogue relates directly to the relevant knowledge base. Oldest IT application area was at the level of Management Information System (MIS) in Computer-Aided Planning (CAP) within manufacturing industry. If proper data collection discipline is applied, computers make it much easier to keep production managers aware of the status of orders, finished stocks, work-in-progress, and costs. Such information are of immediate tactical use in production planning, production scheduling, raw material purchasing, delivery date forecasting, and so on. Such software packages at the tactical level can relieve the workload on clerks, buyers, progress chasers, and foremen, once the strategic policy is decided, thus improving efficiency. Besides CAP as the application area of IT at tactical level, there are a number of ways in which IT is being used via interactive simulation at the strategic (policy making) level developing *static* simulation models and using *operations research techniques* to explore and compare feasibility plans for short term and medium term. Similarly IT is applied in management dynamics (also called system dynamics, industrial dynamics) by establishing dynamic simulation models of the firm to explore the improvement in performance which can result from particular management policies relating to marketing, production, storage, and distribution functions of a company. All these IT application areas form part of computer-aided strategic planning within manufacturing industry. Attempts were also made for computer-aided strategic planning in urban development using the concept of *urban dynamics*. Such techniques if properly developed can also be very useful tools in urban development and management.

The next IT application area is the computer-aided design (CAD). The manufacturing and process industry has a long history of CAD use. In computer-aided design, which is in use in manufacturing industry for many years, computers plus graphics are used as tools in both conceptual and detail design phases of a new product. Since the CAD output is of the form needed for Computer-Aided Manufacturing (CAM), there is now much stress to develop CAD/CAM system in which a product can be manufactured in computer-controlled machines directly from the CAD database. IT is making huge improvement in production engineering and production management performance via CAD and CAM (i.e. Computer-Aided Engineering or CAE), as the boundaries between CAD and CAM become more blurred, via improved data capture and database design, and the automatic factory is becoming an increasingly available option for industrialists to consider as a way of operating expensive machinery.

CAD allows designers to spend maximum time on creative work leaving much drudgery of the designers and draughtsman to the computer. If proper software-hardware is available, the outcome of CAD is a set of working drawings produced much quickly and consistently compared to the manual means and drawing office productivity increases as much as 1:1 have been reported. Since speed is an important factor in the project feasibility stage, CAD can play a big role at project tendering phase to get orders for new projects especially when three-dimensional sketches are required. In some application areas in manufacturing/process industry, CAD is now incorporating *Iterative Optimization Techniques* so that a drawing, once created by the designer, will be the 'best' at satisfying the particular functional specifications, instead of merely being adequate being the first satisfactory solution to the problem to emerge from the creative process p. 26. There is lot of scope to apply such *Iterative Optimization Techniques* using computers (i.e. CAD) in *planning and design process* in urban development and management, structural engineering, and so on (i.e. in AEC) to derive *efficiency* and *effectiveness* as discussed and illustrated later in the book.

Information Technology (IT) is making huge improvement in performance in the production engineering and in production management, via CAD and CAM (i.e. Computer-Aided Engineering or CAE), as the boundaries between CAD and CAM are becoming more blurred, via improved data capture and database design. As a result, automatic factory is becoming an increasingly available option for industrialists to consider as a way of operating expensive machinery. IT-CAD is also making profound impact on manufacturing methods and manufacturing management. The traditional boundaries between functional designs, design for manufacture, software development, planning control, and production processes are fast disappearing. The result will be the generation of a new breed of 'manufacturing professionals'—the system experts who can link all these specialists together to reduce lead time, reduce costs, and improve product quality simultaneously. Similarly, IT can make huge improvement in performance, in the architecture, engineering, construction, (i.e. AEC sector) and urban management, via CAD and CAP if proper data collection discipline is applied, and research and development activities to develop and apply various IT tools and techniques relevant to the sector are promoted. In this background, the AEC sector and the various disciplines comprising it cannot remain insulated from this **IT-CAD Revolution** changing the way we live and work. Hence, there is need to develop IT tools and techniques relevant to the AEC sector and use such modern tools and techniques, and thereby, contribute to more *Efficient*, *Effective*, and *Equitable* solution of various problems faced by the society in the AEC sector. Such contributions by using such *Modern Tools* are essential if the sector and its professionals are to remain relevant in the society and retain their *social* *legitimacy* in the above changed context, and thus, help with *Social Responsiveness* a *Resource-Efficient, Effective,* and *Equitable* solution of problems faced by the society in the AEC sector.

1.4 Information Technology Revolution and CAD in the AEC Sector

As mentioned above, **Information Technology Revolution** has more fundamental impact within organizations, in relationships between organizations, and in the interaction between the organizations, people, and the society as a whole in different sectors. The pervasiveness and cheapness of IT are generating a wide range of fundamental questions in all sectors of society needing urgent answers. The AEC sector is also facing such fundamental questions for effective response to this Information Technology Revolution.

At least two distinct phases may be considered in the Information Technology Revolution in various societal operations. This is equally applicable for operations in the AEC sector. Initially the main impact of this new technology has been simply to facilitate operations as they are now being carried out. For example, the word processor replaces the typewriter, the microcomputer replaces the filing cabinet, and draftsman is partly replaced by AutoCAD. While significant, these innovations simply increase the speed of work and also improve the quality of work (say improved quality of drawing using AutoCAD) as it is now done. However, a second phase of application, which portends revolutionary changes to the organization, in relationships between organizations, and in the interaction between the organizations, people, and the society as a whole, involves outright elimination of many intermediary functions, adding many new functions and involvement of many concerned entities in decision-making (say involvement of people who use and are affected by such decisions particularly in the AEC sector) changing the very way of working. This has been brought about by the convergence of the many technologies such as electronic data processing, telecommunications including Internet, office machines, AutoCAD, and similar tools. To illustrate, take the example of a construction/consultancy firm where field personnel needs to frequently telephone the central office for drawing/data of ongoing projects or for pricing and product information for bidding new projects. Phase-1 automation would be illustrated by the adoption of microcomputers for storing this information, and provision of video display terminals to the clerks/assistants answering the phone calls from the field force. In contrast, phase-2 would involve provision of portable terminals to the field force, or even, conceivably, to the customer/user citizens, thus eliminating the clerks and even some of the field/design personnel. Similarly, managers/designers would communicate directly with one another or with customers/clients/user citizens through 'electronic mail' and interconnected data banks eliminating many intermediary functions and involving many *stakeholders* in decision-making for better problem solution and achieving the goals.

There are two basic types of workers mainly related to office and IT application. The first is the 'information worker' who is involved in the routine entry, recording, storage, and transmission of information. These include typists, secretaries, clerks, and data entry personnel. The second group consists of those who analyse and utilize this information. These are known as 'knowledge workers'. Phase-1 of the information technology or microelectronic revolution mainly affects the 'information workers', while Phase-2 will affect the 'knowledge workers' as well. Organizational and human implications take place in a more substantial way in the phase-2 of microelectronic revolution, particularly in the medium to longer term. These implications will fall under four categories-communication and decision-making, the hierarchy and managers, flexibility of work time and place, and effects on lower level personnel. It is necessary to focus attention on issues involved in adapting each sector (including AEC sector)/organizations/entities/disciplines covering all the above four categories to cope with these very different working situations.

Present discussion is concerned neither with the technology itself—in its accelerating rate of change or the convergence of its components—nor even directly with its broad impact on society at large. Rather the interest here is with the implications of new information technology for one major sector of society—those planning and designing in Architecture, Civil/Building Engineering and Construction (usually classified as AEC field) and those managing at various levels in both commercial and public urban organizations. Clearly, such organizations are dependent on their various environments—for resources, markets, *social legitimacy*, and so on. So the implications of Information Technology (IT) for AEC sector, and its organizations, disciplines (e.g. Architecture, Civil/Building Engineering and Construction management) and other constituents, are to some extent inseparable from its broader effects on the society, and may impinge on their 'social legitimacy', unless these entities/constituents/disciplines develop proper/effective responses to this revolution. At the level of enterprise strategy, the managerial concern is essentially with the *social legitimacy* of the organization and its role in society, in the light of the uncontrollable and sociopolitical aspects of environment. One authority has noted that 'during past twenty years, a major escalation of environmental turbulence has taken place. For the firm it has meant a change from a familiar world of marketing and production to an unfamiliar world of strange technologies, strange competitors, new consumer attitudes, new dimensions of social control and, above all, a questioning of the firm's role in society'. Although this view is expressed with regard to strategic management related to business policy and planning in general, it is equally applicable in respect of the disciplines of Architecture, Civil/Building Engineering and Construction management and both commercial and public urban organizations.

Moreover, as earlier mentioned, the pervasiveness and cheapness of **IT** are influencing the way we live, work, and enjoy ourselves with deep impact on society being transformed into an 'information society'. In view of this and the rapidity of technological convergence/shared technological change, attempts to address the problem in AEC sector merely by an ad hoc reaction to implementation problem of IT, say, simply by adding peripherals like word processors/AutoCAD (mostly improving drafting) and leaving it at that, would not provide adequate/effective response to this new challenge. Hence, there is need to focus attention on issues involved in adapting AEC organizations/disciplines/constituents to cope with this fundamentally different working situation by initiating actions and developing responses more strategic and proactive so that they become more useful to the society in this changed context and the competitive ability of such entities is also improved. This is also essential if the AEC sector is to prevent adverse effects on its '*social legitimacy*' in the context of this new challenge and IT revolution.

Many sectors of society and knowledge disciplines in developed/-industrialized countries have already initiated actions for adequate response to this new challenge. To support such transformation, individual courses are offered at various levels to produce IT specialists. As mentioned above, many knowledge disciplines like mechanical and electrical engineering have introduced individual courses on CAD, a vital application area of IT, long time back (even the knowledge disciplines of Architecture and Civil/Building Engineering in many developed/industrialized countries have initiated actions for adequate response to this new challenge/revolution by introduction of individual courses on CAD, which is the first step towards such action) and are taking appropriate actions in adapting themselves to cope with this fundamentally different working situation with responses more strategic and proactive (e.g. automating the process industry using IT), so that they become more efficient, and retain their social legitimacy and become more useful to the society. This is equally applicable in Architecture, Civil/Building Engineering and Construction (usually classified as **AEC** sector) dealing with many public issues and concerns where **IT** can influence the way of working with deep impact on society being transformed into an 'information society'. Hence, it is desirable that AEC sector and knowledge disciplines of Architectural (including Planning) Engineering-Civil/Building Engineering and Construction, in various countries initiate vigorous and comprehensive actions to develop proper responses to this revolution and new challenge, focusing attention on issues involved in adapting various disciplines/organizations in this sector to cope with this fundamentally different working situation with responses more strategic and proactive and not merely an ad hoc reaction to implementation problem of IT, so as to become more useful to the society and retain their '*social legitimacy*'. There is also need to analyse the implications of IT for the AEC sector, an important sector of society, and for the fields of Architecture (including Planning), Building/Civil Engineering and Construction (forming part of the AEC sector), and Managing the Urban Sector as a whole to get better performance of the sector and its disciplines to improve the *quality of life* for all people with their participation within the given constraints, which is greatly facilitated by IT-CAD and other IT-techniques.

For example, Eastman et al. have pointed out that sharp and bitter conflict has become an ingredient of urban life and the public policy affecting that life and conflict emerges around decisions dealing with our human and natural resources. This conflict has many manifestations like caste and racial rebellions, street demonstrations, increases in violent crimes, protest marches, public interest litigations, and escalating demands for improved housing, employment, education, social, and public services by citizen's groups, leaving 'cities' in a more or less perpetual state of foment subject to the urban dynamics. In fact, in recent times, the situation has aggravating further in a lot of cities in many countries. Current conventional urban planning and urban management processes and techniques have not been able to address effectively this serious social problem leading to parallel activities by people (e.g. proliferation of slums, squatter settlements, and so on) and by urban professionals (preparing master/development plans and so on which are not accepted by people), making the activities of the latter (i.e. master/development plans and so on) almost redundant in many cases, and consequent loss of *social legitimacy* of such urban professionals. This is often erroneously seen as a pure legal problem, although intervention even by the national apex courts (in Delhi for example) has failed to address the problem. Corrective actions, such as Application of IT and **Integrated Urban Management Approach**, ensuring *Efficiency, Effectiveness, and Equity* within the given *Resource and Environmental Constraints*, seizing advantage of **IT Revolution** including **IT-CAD**, may help to cope with the situation with *people's participation*, assisting urban professionals to find *Efficient, Effective,* and *Equitable* solution of such problems with *Social Responsiveness*, and thus retain their *social legitimacy*.

As mentioned above, IT is shrinking the effects of time and distance and altering the very nature of work in an organization, becoming the lifeblood of most industrial and service organizations, which are equally applicable in AEC organizations. A novel characteristic of IT is its ability to create participative system in so far as all levels in the organization and outside it will be able to interact with others in networks, and thus instant feedback may be facilitated. This is very relevant in respect of the AEC/urban sector. It is to be anticipated that the computer will impose its logic on the organization. There is clear potential for better two-way communication, consultation and even co-determination both directly and indirectly. For example, the user citizen/individual will be able to be involved directly, or his representative's access to the system will be facilitated, for more satisfying *participative* design-decision-making. If a system is created, say a *planning/design proposal*, various participative subunits and the interested user citizens could be online with each other and give inputs for a satisfying optimal solution of an urban problem in a given context.

Creating environments that are more responsive to human needs will require changes in traditional practice and role of the professionals, who will need to be multifaceted. Designers will require new skills and knowledge as an enabler, technical advisor, social worker, and bureaucratic trouble-shooter. Learning how to listen, not only to the paying client, but also to people who use and are affected by the environment, within the social and historic context, can produce professional with an expanded capacity for shaping the future, and to help physical environment to become better through design intervention. If planners and designers in the AEC sector add capacities like IT-CAD expertise to their conventional knowledge, the usual mismatch between planning/plans and people's need/affordability and consequent squatter development/slums referred above could be minimized. Moreover, IT revolution help application of the concept of *organization as a community*, based on consultation and democratic principles in decision-making, which can have profound and revolutionary effect on AEC organizations and AEC professional activities, since such activities are mostly concerned with people and the user citizens. Such a concept is being applied even in industrial organizations deriving benefits of IT revolution. Since, AEC organizations and AEC professional activities are mostly concerned with people and the user citizens, it is desirable that such a concept is applied in AEC sector organizations/professional activities creating capacity for the same taking full advantage of the current IT revolution. In the past, designers have thought of the built environment as dependent upon them, but it is becoming increasingly evident that their decision domain is shrinking and the role of the professional is continually being questioned, as is the issue of human *accountability* as it has been practiced. Therefore, AEC professionals need to be equipped with appropriate skills such as IT-CAD, for effective response to this new challenge.

In recent times, there has been a considerable movement towards the direct involvement of the public in the definition of their physical environment with increased sense of social responsibility when a new feeling of community consciousness started prevailing in many low-income urban neighbourhoods. People need to participate in the creation of their environment with a feeling of control, which is the only way that their needs and values can be taken into consideration. Moreover, participation is a means of protecting the interests of groups of people as well as of individuals, because it satisfies their needs, but these are very often totally or partially ignored by organizations, institutions, bureaucracies, and the 'expert' planners and designers. To facilitate introduction of the users in the decision-making process (i.e. *participation*), designers have to add new capacities to their conventional experience and skills. It is well known that all problem solution in the AEC sector involves collection, generation, communication, recording, manipulation, and exploitation of information. Therefore, use of computers and IT is essential in AEC activities to help effective problem solution. The current IT revolution permits computers to be 'user friendly' from outset allowing all professionals and even lay people/citizens to access computers directly in 'user' language, in which dialogues relate directly to the relevant knowledge base or 'user' priority and concerns (e.g. urban sector). This is of special importance in problem solution in AEC sector, where IT application areas of Computer-Aided Planning (CAP) and Computer-Aided Design (CAD) can play a very important role. Hence, it is essential that designers in AEC sector add capacities like IT-CAD expertise to their conventional knowledge to help the above process.

In the above context, AEC/urban sector by not applying IT-CAD and other IT innovations in their various activities cannot remain immune to the revolutionary effects of IT on the work organizations and its members, if the sector/disciplines have to make useful contributions to the all-round development of a country/society, and also remain relevant in the society without adversely affecting their 'social legitimacy'. As cited above, implications/impact of IT on sector/organizations/entities/disciplines is inseparable from the broader effects on society and social legitimacy of sector/organizations/entities/disciplines, and therefore, by not applying

IT-CAD and other IT innovations AEC sector and its disciplines may be instrumental in adversely affecting their 'social legitimacy'. One of the important steps towards this direction (and one of the urgent actions towards such efforts) could be introduction of individual courses for compulsory and comprehensive Education on CAD (a vital application area of IT but rare in **AEC** sector in many countries), in various disciplines in AEC sector, and also promoting Research and Development activities in this emerging area, to help equip them to develop and apply IT techniques and tools in their operations.

As mentioned above, there are unparalleled transformations in computers in terms of their capacity to capture data, to analyse, manipulate, and store them, to disseminate and communicate information, and to apply these capabilities in different areas and fields of specialization. Such transformations are occurring with increasing rapidity and frequency in many areas and the rate at which IT and computer applications are spreading to various sectors is truly extraordinary. Today, information technology and computer application are everywhere shrinking the effects of time and distance and this second Industrial Revolution is altering the very nature of work in an organization including AEC organizations. Again, pervasiveness and cheapness of **IT** are influencing the way we live, work, and enjoy ourselves and is generating a wide range of fundamental questions in all sectors of society needing urgent answers. As outlined above, many sectors of society and knowledge disciplines have already initiated actions for adequate response to this new challenge. However, knowledge disciplines of Architectural (including Planning) Engineering-Civil/Building Engineering and Construction (usually classified as AEC) in many countries are slow to initiate vigorous and comprehensive actions for adequate response to this new challenge, at the least by introducing compulsory and comprehensive CAD (a vital application area of **Information Technology**) instructions covering all its components including Optimization in Architectural/Civil-Building Engineering Schools particularly in the developing countries.

The author feels that such actions are essential if the above *disciplines* have to remain relevant in the society by contributing to the *Effective* and *Efficient* solution of societal problems in the AEC sector, and thus retain their *social legitimacy* in the above changed context. For example, application of IT-CAD in *Building Economics* covering development of economic performance measures of interest to developers, owners, contractors, and users; linking cost engineering including cost indices, elemental estimating, computerized information systems, and model building could be very useful tool for economic analyses of projects both for single buildings and building components leading to much more *Effective* and *Efficient* solution of such societal problems. Similarly, *Computer-Aided Building Design* including ***Optimization*** covering building engineering design process, methodology, identification of objectives, building codes, formulation of design problems, development and evaluation of design alternatives; conceptual building design; spatial requirements, design of space layout; preliminary building design and synthesis, and performance evaluation using modeling, sensitivity analysis, and cost estimation using computer-aided design tools could lead to much more *Effective* and *Efficient* solution of such societal problems. Lack of necessary hardware, software, and appropriate courses of instruction and textbooks, covering different topics/components of CAD relevant to Architectural and Civil/Building Engineering Design, are some of the inhibiting factors affecting above CAD instruction/application. Steady decline in the cost of computers has now made such hardware's affordable even to the smallest Architectural and Civil/Building Engineering Schools/Offices. However, most of the other deficiencies are still to be adequately addressed.

1.5 Conventional Design and CAD

There is no unanimity regarding the precise description and the scope of CAD. However, in essence, CAD can be viewed as an effective tool made available to the designers to deal with all aspects of design work, particularly because CAD offers many exclusive capabilities and advantages over the conventional design methods, and can be applied in any field such as electrical, mechanical, architectural, civil, and even fashion design. In fact, creative designers can develop many innovative use of CAD in their field of specialization by proper research and development studies in this evolving discipline, which needs to be encouraged. Keeping this background, CAD should be considered neither simply as exotic computer graphics nor as just sophisticated computer analysis, but a combination of all such novel features as aids to the designer in the design process for improved performance, and not a passing fad. Actually, if the designers look at the powerful computers and peripherals, such as graphics display terminals/workstations, simply as design aids, in the same way they have looked in the past on the design-aid tools such as the slide rule and handheld calculators, any confusion regarding description and scope of CAD would disappear. Recognizing CAD as a new tool for designers for the above purpose, the designers must be aware of its strength and capabilities and learn the special characteristics of this tool so as to make its most effective use to derive maximum benefits in the design process. Accordingly, some unique capabilities, characteristics, and distinctiveness of CAD are discussed below:

1.5.1 Some Unique Capabilities and Advantages of CAD and Graphics

CAD has many unique capabilities and advantages such as graphical depiction and solid geometric modeling, animation, quick mathematical analysis/computation including design optimization, to name a few, which are not available in the traditional design methods. In fact, increasing and cheaper availability of computers with tremendous computing capability and other functions (graphic representation and so on) and also development of different programming languages led in the 1960s to a revolution, in scientific, engineering, and operations research problem-solving, which is still evolving. Such capabilities and advantages of CAD over the traditional design methods, leading to saving of money and time and improved quality of drafting products, are the main reasons for its unquestionable popularity and irresistible adoption by the industry and to some extent in the AEC sector. There are many areas and applications of graphics such as modeling, animation, virtual reality, visualization, image processing, computational photography, video games, visual effects, simulation, medical imaging, information visualization, CAD-CAM, and so on (Shirley et al., 2011). Some unique capabilities of CAD and graphics are enumerated below:

1. **Interactive Computer Graphics and Geometric Modeling**

   Graphics has been an increasingly extensive and important area of computer application and is an important component of CAD to generate graphic portrayal of objects with complex geometries for better appreciation of geometric properties and arrangement of spaces in the designed objects. By means of interactive graphics, the geometry of the object could be changed by simple commands to the computer to find an acceptable solution. Geometric modeling defines an object in terms of geometric properties/arrangement of spaces. Thus, characteristic applications of computer graphics techniques include engineering and architectural drafting, production of perspectives, and mapping, while computer graphics and geometric modeling are essential components of computer-aided design in various fields, which is equally applicable for the AEC sector.

   Interactive Computer Graphics in CAD can provide real-time or 'instantaneous' communication between the computer and the designer or operator, allowing sketching and drawing to be both input to and output from the computer, offering the operator much more freedom over the traditional input methods that use the keyboard. It enhances the designer's visual appreciation of the component's geometry and also permits the designer to identify obvious errors prompting immediate action by the designer to rectify these errors by adjusting proper design parameters. At the initial or conceptual design stage, the shape of a facility could be represented graphically on a monitor screen even using a computer mouse. This can be converted into a proper two-dimensional or three-dimensional geometric model and linked to appropriate numerical analysis models for planning/design analysis and optimization to derive its final shape fulfilling the prescribed requirements. The geometric model can then be used as input to an automated drafting system to produce architectural, planning, and engineering drawings.

   CAD can demonstrate graphically the complex geometries of a designed object and thus facilitating better appreciation of the object geometry. In this method, the designer and other stakeholders can observe the designed object from selected views with perspectives (e.g. isometric views) to judge its acceptability, and any modification of the geometry (whether two- or three-dimensional) can be achieved in an interactive manner by simple commands to the computer (by all stakeholders) taking hardly any time. Thus, use of CAD can obviate the need of making prototype solid three-dimensional design models (taking lot of time and resources), which are generally prepared in conventional design methods to help better understanding of the object geometry and its geometric properties.

As earlier mentioned, the IT revolution permits better two-way communication, consultation, and even co-determination both directly and indirectly, and the user citizens/individual stakeholders can be involved directly, or his representative's access to the system can be facilitated, for more satisfying *participative design-decision-making*. Thus, if a system is created, say a *planning/design proposal* (both in alphanumeric and graphic version), various participative sub-units and the interested user citizens could be online with each other to give input for a satisfying optimal solution of an urban problem in a given context. This is not possible by conventional manual design methods, but using CAD technique such an objective can be easily achieved.

2. Image Processing

An application closely related to computer graphics is *image processing* that is computer analysis of photographs and other visual materials. Typical applications of *image processing* technology include remote sensing and analysis of earth satellite photographs to extract data about crops, atmospheric and water pollution, mineral deposits, land use, etc. which are very useful data for efficient and effective urban planning and management in the context of urban dynamics.

3. Computer Animation and Virtual Reality

In CAD, computer animation of scenes in perspectives allows sophisticated simulations of movements through an environment to be produced. Real-time visual simulation using a colour television screen has been applied in the simulation of aircraft and spacecraft landing and ship docking. CAD has the unique capability to animate sequences of events on a graphical screen, which is of great value to the designer for evaluating moving objects in simulation studies and in urban traffic control. For example, before a decision is taken on final design and implementation, a real life operation of an urban facility could be simulated using such CAD and virtual reality technique (which is not possible in conventional planning and design methods), for better appreciation of the design by all stakeholders. Virtual reality attempts to submerge the user of a CAD system into a 3D virtual reality world. Thus, using this technique, an urban development plan could be simulated in live form for better appreciation of the advantages and disadvantages of a facility plan and carrying out suitable amendments, before it is actually implemented.

4. Artificial Intelligence and Expert Systems

High-level computer languages have been developed in *artificial intelligence* research demonstrating reasoning and drawing conclusions and ultimately helping decision-making, creating exciting possibility of using information technology and computers for solving difficult intellectual problems of a non-numerical character. Expert systems can be viewed as a subset of artificial intelligence. There is rapid progress in artificial intelligence research resulting in the design and implementation of systems possessing a bewildering variety of capabilities such as simulation of cognitive processes such as perception, learning, concept formation, and induction, performing complex physical manipulation tasks, and solving spatial problems such as path finding in complex physical environments (Aguiler, 1973; Alexander, 1964). Many such artificial intelligence systems are developed and in use in robotics and the automatic factory. In the AEC sector, many decisions are of non-numerical character and have to be taken on subjective considerations resulting to dissatisfaction of stakeholders. Similar information technology and artificial intelligence research linked with CAD to develop artificial intelligence systems for use in planning and design in the AEC sector may help more satisfying problem solution in this sector.

5. **Speedy and Sophisticated Design and Engineering Analysis**

   CAD is capable to perform not only planning and design analysis using simple algebraic formulations, but also sophisticated engineering analysis using powerful techniques such as finite element method. Such analysis in conjunction with interactive graphics makes the planning and design process much more meaningful to all the stakeholders. CAD permits presentation of the results in graphic form in a comprehensive manner showing critical areas in components with high stress or strain concentrations, or too much deformations, which can be shown with a distinct colour code, enabling a designer to detect such critical areas instantly and at a glance. Thus, more efficient engineering design is facilitated by such information. Conventional manual design methods do not perform such functions.

6. **Design Optimization**

   Engineering, architecture, and urban professionals commit huge resources in their *planning and design decisions* making them ethically responsible and accountable to society for *efficiency* and *equity* in use of resources so committed (which are scarce in many cases). Design optimization is essential to attain such *efficiency* and *equity* to produce various products and services. In fact, *no rational decision is really complete without optimization.* Since all such operations involve numerous variables and many solutions, with large variations in efficiency indicators, search of optimal solution becomes quite complex, making it difficult to attain by conventional manual methods. CAD is of particular importance for application of optimization involving hundreds of design variables and constraints although difficult to achieve by conventional manual design methods. Importance of optimization in planning and design process and special capability of CAD to apply optimization is discussed later.

7. **Participatory Design Using Interactive Graphics and Collaborative IT Tools**

   As mentioned above, the current IT revolution permits computers to be 'user friendly' from outset allowing all professionals and even lay user citizens (very relevant in respect of **Urban Operations**) to access computers directly in 'user' language, in which dialogues relate directly to the relevant knowledge base or 'user' priority and concerns (e.g. urban sector). Thus, CAD permits a *participatory design* as many CAD systems can be interfaced with construction system and participatory design systems using collaborative IT tools, resulting to more satisfying, efficient, and equitable design solutions. Such *participatory planning* *and design*, involving all *stakeholders* even if in geographically diverse locations but linked by terminals accessing a computer system and a shared database, is not possible in conventional manual design methods.

8. **A Broad List of Benefits of CAD Compared to Conventional Design Methods**

   To sum up, CAD offers many exclusive capabilities and advantages over the conventional design methods, and some of these benefits of CAD, compared to traditional design methods, could be listed as follows:

   (a) CAD improves the quality of design with excellent graphical representation of facility geometries and design drawings.

   (b) CAD enables designers to visually inspect the geometry of the facility being created/designed from every possible viewing angle. This obviates the need to build prototypes/physical models of facilities as usually adopted in conventional/traditional design processes in architecture, civil/building, engineering, and urban development.

   (c) Graphic simulation and animation of a CAD system creating virtual reality can minimize human error and planning and design failure to achieve the planning/design objectives particularly in the architecture, civil/building engineering, and urban development and in the AEC sector as a whole.

   (d) Using CAD, geometric definition of a facility can be exhaustively studied and modified as required, as CAD permits faster data creation and manipulation including graphic data, taking little time in correcting and revising geometric designs and documentation of those designs.

   (e) As already mentioned above, ever improving and increasingly available computing power has further enlarged the scope of CAD in solving design problems involving hundreds of variables providing designers with optimal solutions of complex problems in many cases.

   (f) CAD provides great flexibility and economy in the management of design data incorporating all relevant design information and facility geometries and specifications including cost engineering data, faster data output including graphics, and more convenient data storage.

   (g) CAD provides designers the means for faster production of design, and for easy modification of existing design files and construction drawings and thus saves storage space and also time at all decision levels improving productivity.

   (h) Many CAD systems can be interfaced with construction system and *participatory design systems* involving all stakeholders including design specifications and cost engineering approaches (see Chap. 4) providing live cost engineering data. For example, CAD and graphic simulation permit online interactive design and *participative design* where lay-user-stakeholder can meaningfully interact and participate in the design process, by remaining online and giving input, using collaborative **IT** (Information Technology)-**Tools**, such as e-mail/Internet. Adoption of such a method would give more satisfying, *efficient* and *equitable* design problem solutions, particularly in the AEC sector.

(i) CAD can provide real-time or 'instantaneous' communication between the computer and the designer or operator in the form of *interactive computer graphics*, allowing sketching and drawing to be both input and output from the computer, offering the operator much more freedom over the traditional input methods that use the keyboard. It enhances the designer's visual appreciation of the component's geometry and also permits the designer to identify obvious errors prompting immediate action by the designer to rectify these errors by adjusting proper design parameters.

(j) CAD provides planners and designers the means for instant linkage and conversion of *optimal* design in numerical terms into a graphic output, thus facilitating application of *optimization techniques* for *resource-efficient* and *equitable* solution of planning and design problems involving hundreds of variables, particularly in the architecture, civil/building engineering, and urban development, and in the AEC sector as a whole.

The above is only a broad list of advantages of CAD. In fact, Information Technology Revolution and CAD have removed many inhibiting factors which hampered application of optimization in the past, and thus, opening up a vast possibility of applying design optimization in different sectors to improve efficiency in various activities, particularly in the AEC sector. Some of these issues are discussed below.

1.6   Computerizing the Design Process: Need of Hardwares and Software

Although fundamental design principles of conventional design remain intact in CAD, it is quite distinct from traditional design methods in many respects because of use of computers. The human being is creative but slow, whereas computer is uncreative, fast and never gets tired. To take advantage of both, the objective in CAD is to allow a human being and the computer to work together as a team to control the CADD (computer-aided drafting and design) functions. Moreover, the languages of human being and computer are very different. Human being thinks in terms of symbols and pictures, whereas computers understand only simple electrical impulses. Therefore, to computerize the design process, the design language structure for a computer-aided design-display needs to be a series of commands (in the form of software or keyboard strokes) that move the VDT (video display tube) beam, plotter pen, and so on, such that the end result is a properly drawn design and working drawings. Hence, the CAD process requires both hardwares and software based on computer algorithms, which is one of its distinctive features.

1.6.1   CAD and Hardwares

The graphics workstation (at the minimum, consisting of hardware items such as system unit, display monitor, keyboard, and a printer or plotter) is an example of such human-computer control complementing each other. Hence, to computerize the design process, hardware items such as system unit (containing Central Processing Unit or CPU having microprocessor and memory board subcomponents along with hard discs/drives with adequate RAM, i.e. random access memory, and ROM, i.e. read-only memory capacity), Video Display Tube or VDT with adequate resolution and other attributes, operating system selected, display monitor, keyboard, printer/plotter, and other peripheral equipment are obvious need, which is not required in traditional design methods. Personal computers/laptop computers can serve as graphic workstations since they do not depend on an outside processing source. When the computer is switched on, it boots up in text mode and its graphic capabilities depend on the type of monitor and the video display adapter such as CGA, EGA, VGA, and so on, attached to it.

As most commercially available CAD hardware components are constantly being replaced by upgraded versions, detailed description of CAD hardwares (which may become outdated in no time) at this point would not serve any purpose, and hence, not attempted, but only some broad items are indicated here. A CAD hardware system may consist of a CPU (Central Processing Unit) with adequate memory disks and disk drives, keyboard, monitor, plotter, printer, and digitizer. The CPU is the most essential component of a digital computer with two subcomponents, i.e. microprocessor controlling all system operation related activities and performing all arithmetic and logic functions, and the memory boards performing both read-only memory (ROM) and the random access memory (RAM) functions. The main function of ROM is to store the system software, while RAM incorporates instructions pertaining to a job to be performed by the computer as per the application software supplied by the user. Memory storage required for most application software programs is more than that available in RAM, and the additional memory is provided by memory discs. The keyboard is the principal input device in a computer containing alphanumeric and function keys. Monitor displays the user's input and the computer's output (soft copy) in either alphanumeric or graphic form. Plotter and

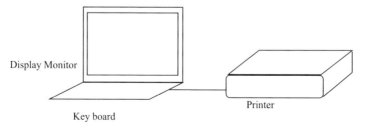

Fig. 1.1 Schematic diagram showing some minimal hardware components for a PC-based CAD system

printer produce permanent record (hard copy) of the CAD output. Digitizer tablet or graphic tablet is used for graphic input to the computer and is essential for interactive graphical operation of CAD. There are many textbooks, which describe the working principles of the above hardware components, and therefore, no rigorous treatment of these materials is attempted here. However, a diagram indicating some minimal hardware components for a PC-based CAD system is shown in Fig. 1.1.

1.6.2   Software in CAD

Although fundamental design principles of conventional design remain intact in CAD, there are two distinct characteristics of CAD, namely its strong dependence on software programs and its strong reliance on computer algorithms, which are quite different from conventional way of design. CAD primarily depends on (1) system software—supplied with the hardware and (2) application software—available commercially or to be developed in-house. System software is used for the internal data management and for operation of the components of the system. Even if the application software is purchased from commercial outlets as a 'black box', designers must be knowledgeable about the principles of the application software used for the job so that they can prepare the input and interpret the output result correctly. In many cases, particularly in the AEC sector, designers may have to develop application software in house for their specific requirements, as commercial availability of application software in this sector is comparatively less.

1.6.3   Computer Algorithms in CAD

All computations performed by computers are based on numerical approximation techniques. Computer algorithms with iterative procedures for numerical approximation are standard practice in engineering analysis and numerical techniques for CAD. The graphics algorithms also incorporate such concepts for computerization of the design process. Thus, graphic display of a circle on a monitor screen may be construed as a polygon composed of many miniature straight-line segments. Sound knowledge of engineering and numerical techniques is essential while adopting the most effective algorithm for formulation of software program and code development of a specific problem for CAD.

1.6.4   Principles of Computer Graphics

Any use of computers to create and manipulate images is denoted by the term 'computer graphics' (Shirley et al., 2011). In general, a graphic image is made of tiny dots known as picture elements or pixels.

Computer-aided graphics (i.e. computer-aided drafting and design or CADD) are a fundamental form of architectural and engineering graphics, and the concept is applicable in any form of display that must be done in any activity. A raster screen contains a large number of picture elements (i.e. pixels) or points, which may be more than a quarter of a million in an average quality raster screen. Creation of images is done by lighting some of these pixels and leaving the rest unlit or by lighting some in one colour and others in other colours. A bit map contains the information about the condition (on/off) or the colour of each of these pixels, and requires a large buffer memory and fast processors for *raster display* in graphic terminals. A variety of colours can be produced by varying the combinations of intensities of the emission from each pixel at a specific optical spectrum (e.g. in the three basic colours of red, blue, and green), and the computer memory bit planes (i.e. an assembly of bit maps) built-in for each pixel controls such emission. Working principles of *liquid crystal display* or LCDs are similar to that of raster display on a refresh CRT. Here LCD cells replace the pixels and LCD cells are installed in the matrix form with a number of horizontal and vertical electrodes that apply the necessary driving voltage through the crystal cells. The change of optical properties of the electrified crystal cells controls the amount of light to be passed through these cells. Visual images can thus be created on the screen and viewed by the operator. Rapid advances in the LC layer structures and materials, and strong market demand for flat-panel monitors, are leading to replacement of bulky CRT monitors by LC monitors not only in television industries but also in the computer systems.

1.6.4.1   Mathematical Formulations for Computer Graphics

Much of the graphics translates mathematics into codes. In computer graphics, the straight lines form the basis for display of all types of shapes, whether two-dimensional or three-dimensional. For example, joining a large number of small line segments, a circle may be drawn. To draw a line segment, it is possible to assign its two endpoints to specific locations on a computer screen. Most graphic packages contain a line drawing command that takes two end points in screen coordinates and draws a line between them. The routine should draw some reasonable set of pixels that approximate a line between general screen coordinate end points (x_0, y_0) and (x_1, y_1). On a raster display, the location of these points is defined in terms of the sequence number or address of a pixel, and the row number to which the pixel belongs. In order to draw a continuous-looking line, the computer must be able to pick up a number of other intermediate pixels that should also be lighted in addition to the two end points, say, by drawing line using implicit line equations.

One of the popular methods to identify intermediate pixels is symmetric *digital differential analyzer* (DDA) expressing the equation of a line by a pair of parametric equations:

$$x = x_1 + (x_2 - x_1)^* s$$
$$y = y_1 + (y_2 - y_1)^* s$$

Here, the parameter s varies from 0 to 1. If s is incremented by, say 0.1, then the DDA will generate 9 intermediate points between the two ends of the line. For each value of s, the real numbers are found for the x and the y coordinates of the intervening pixels from the above equations. Since a pixel can have only integer coordinates, a hardware device usually converts each fractional coordinate into the nearest integer number. On determination of the addresses of all pixels, these are stored in a frame buffer, and the display driver reads the array of addresses and illuminates the corresponding pixels. For a line to look continuous, it is necessary to increment the parameter s in a manner consistent with the resolution of the display device. Thus, if a larger increment is selected, the line may appear to be discontinuous, while choice of a too small increment may cause a large number of pixels illuminated near each other, making the line appear thicker and brighter, and the drawing process slower.

1.6.5   Principles of CADD

CADD is a way of converting computer impulses/electronic data into documents in various fields such as engineering, architecture, and so on, and also conversely, a way to translate design operator's instructions into electronic data. A powerful CADD software package can be a two-dimensional (or three-dimensional) computer-aided design and drafting tool that is as at home with initial sketches and conceptual designs as it is with final presentation drawings and full-scale engineering or architecture projects. Whether one creates electrical schematics, mechanical assembly drawings, or architectural floor plans, CADD offers a way to get the needed documents, which is the reason for its unmatched popularity in the industry and for its wide use by many design and engineering professionals in various disciplines although use by AEC professionals is comparatively less. Human mind tends to solve a problem by heuristics (that is by trial and error), whereas a CADD system solves it by the use of algorithms (an error-free sequence of logic), and thus, letting both human mind and machine work to their best capacity. This results to a new and better method, which can be automated, leading to improvement or elimination of certain or all parts of manual labour involved in doing a job. However, this does not mean the elimination of human beings from the scene, since one has to start and stop the process (either directly by pushing a button or indirectly by a software programming the design process), and also has to refine the input data in response to the current output until the final acceptable output is arrived at, particularly in the AEC design process. Plotters, the first automatic drafting machines, eliminated the need for a designer to push and pull a triangle and T-squire around a drafting table, but the designer has to be there to run a CADD operation much as was done before, although the plotter is only one of several input-output (I/O) operations available with the same central processing unit.

A CADD workstation accepts symbolic values from a processor (a central processing unit or CPU) and converts this input into physical values through electrical signals (which are usually digital pulses, but analog varying voltages can also be used) that are translated into VDT (video display tube) line segments. The generation of line motions, required in most applications, is achieved by suitable programmed LISP routines inside the CADD software called device drivers (e.g. device drives by AutoDesk, Inc), and it is the device driver that enables the VDT to produce movable line segments. When unique CADD graphics, requirements cannot be satisfied using existing LISP routines, the designer can write special purpose AutoLISP routines that provide direct control of line segments and generation of graphic shapes as desired. Commercial CADD software such as AutoSketch is a product for CADD users who can gain experience without drowning in a sea of CADD commands and options with emphasis on basic principles rather than on learning how a graphics system work. Just as word processors have changed the ease and speed with which reports are written, AutoSketch is changing how designers gather information, and using this approach, a designer can create clean, precise images, and revise notes and drawings without having to redraw. One is able to use common forms and shapes, duplicate these automatically, and make instant copies of base drawings and detailed notes. Similarly, there are a number of other commercial software such as GenCADD Architectural/Civil Series and so on, which covers concepts basic to computer-aided design including preparing the design layout so that computer can automatically generate detail drawings, location, and clearance fit checks and communicate design intent allowing inclusion of all information's for design analysis (testing), prototype construction (experimentation), and production of a product. GenCADD Architectural/Civil Series is for use by architects, space planners, homebuilders, facility planners, interior designers, landscape architects, and civil engineers. From field book to finished drawings, finished detail drawings, floor plans, elevations, door and window schedules, automatic stair generation, and roof lines are only a small portion of the work done in this package. AutoCAD is the largest single and most expensive commercial software package, but is only one of several available.

Computerized descriptive geometry (CDG) is used for image construction, image maintenance, and the creation of project CAD to produce drawings of all types and descriptions. CDG in design process, consisting of several stages such as conceptualization, engineering and planning analysis, engineering, architectural and planning description, starting from development of an idea and converting it into a finished product, plays an important role in communicating design information. Computerized Vector Geometry (CVG) is used for study of graphic statics, with graphics workstation assistance for solution of structural problems. Computerized Solid

Geometry (CSG), available in commercial software like AutoDesk's AME (Advanced Modeling Extension) can be used for the study of computer-generated modeling with wireframe models, geometric models, two-axis models, and so on.

In essence, the most fundamental difference between traditional design and the CAD process is not in the design methodology, but in the implementation of the design method. Thus, computerization of the design process produces speedy and accurate design analysis and imposing easily modifiable graphic displays, but the fundamental design principles as applicable in traditional design method remain unchanged in the CAD process. In fact, it is rarely possible for a designer, using CAD process, to produce a good design without adequate knowledge of traditional design principles of respective discipline, say in the AEC sector.

1.7 Planning and Design Process

Different persons can construe the term planning and design differently. According to some simple definition to design is to formulate a plan for the satisfaction of a human need. Some views design as a goal directed problem-solving activity causing all the properties, incorporated in the goal, to be present in the end result. In fact, Planning and Design, particularly in Architecture, Planning, and Civil Engineering (say, Building/Housing/Urban Services), i.e. AEC sector, consuming huge *Resources* and affecting the *quality of life* of people, needs to be viewed as a *problem-solving process* in a resource-efficient manner, adopting cost engineering principles and techniques as may be relevant. In this context, planning and design problem can be considered as a system, with decision variables as input and the properties as output to solve problems. Many facility planning and design problems incorporate all the characteristics of such a system and can be designed accordingly. The laws determining the ways in which given properties vary under the influence of different decisions constitute the conditions within which the problem must be solved. The properties, which are required to be exhibited by the proposed system, are defined by the goals or objectives of the problem. In many cases, a goal-decision system can be constructed so that it contains one and only one objective, so that the system can be identified and named in terms of the objective it contains. For example, in an optimizing model, the single-objective may be put as either the least-cost or most-benefit, and the goal-decision system model can be constructed accordingly. Goal-decision systems can also be constructed with the aim to select a set of states for the decision variables such that the resulting set of states of the properties satisfy their respective objectives, or that the degree of fulfillment of objectives is optimized. For example, in an urban-built-form (this is further elaborated later in the book) a set of states selected for the decision variables (say values of Building Rise, Semi-Public Open-Space per Built-Unit, and so on) in a particular case constitutes a design proposal or design, and if the proposal is implemented, the consequences would be a set of states of the properties (say values of built-unit-space, built-unit-density, built-unit-cost, and so on) which can be called the outcome or performance of the proposal arising from the set of states of decision variables selected. Similarly, a goal-decision system, causing some of the above properties to be present to a satisfactory degree in the end result, could be constructed in the form of an optimizing model, as discussed and presented later in the book.

Creating built environments that are more responsive to human needs (all classes) would require changes in traditional practice, making professionals multifaceted requiring new skills and knowledge as an enabler, technical advisor, social worker, and a trouble-shooter, and thus producing professionals with an expanded capacity for shaping the future with *accountability*. In this context, Planning and Design in Architecture, Planning, Civil Engineering (say, Building/Housing/Urban Services consuming huge *Resources* and affecting the *quality of life* of people) and Construction (i.e. AEC) needs to be viewed as a *problem-solving process* involving *search* through *numerous* potential solutions (function of *numerous variables* with large variations in *resource-efficiency* indicators) to select a particular solution that meets a specified criteria or *goal* (say, *Resource-Efficient* and affordable shelter design), making computer use indispensable (to help *search* through *numerous* potential solutions) for problem solution with *Efficiency* and *Effectiveness* in all branches of AEC. This view for urban facility planning and design in AEC sector is quite relevant especially in developing countries to achieve the national *goals* of providing adequate shelter and urban services for all, within the affordability and *Resource-Constraints* at the individual, local, and national level. With increasing population and dwindling per capita *Resources*, this view is equally applicable even in the developed countries, as *we live in a Finite Earth with Finite Resources*.

In the above background, acquiring skills in computer-aided design (CAD) by *optimization* is one of the most important prerequisites for designers, particularly in the AEC sector to attain design objective(s) subject to various constraints and solve problems. A designer in the AEC sector needs to study the physical environment and help to become better by design interventions in respect of various facilities being created. In this context planning and design need to be viewed as a problem-solving activity, where all activities are directed towards the provision of certain properties, or certain states of certain properties in the end result so as to fulfill design objective (s). Generally, in this process, the decision variables (which are normally within the control of designer) are the input, the properties are output, and the laws determining the ways in which the given properties of a facility vary, under the influence of different decisions, constitute the conditions within which the problem must be solved. However, whether an input variable is a decision variable or a context variable, and whether or not a decision variable is also a property, are all part of the definition of the problem. In many facilities, planning and design problems cost is paramount and, therefore, the planning and design process is converted into production of a system with the lowest possible cost (Least-Cost Optimizing Model) for a given level of effectiveness, or producing a system with maximum possible effectiveness (Most-Benefit Optimizing Model) for a given cost, and thus fulfill the given objective(s).

1.7.1   Stages in Planning and Design Process

One can visualize many stages in the planning and design process, depending on the emphasis placed on particular aspect(s) of design by a designer. If design is conceived as a problem-solving process to achieve some objective(s), generally, there can be six types of operations/stages in a facility planning and design process, namely,

1. Agreeing on Planning and Design Objectives:
    A planning and design problem may be concerned with many objectives depending on the requirements and priority of different stakeholders who may have conflicting interests, viewpoints, and demands. Hence, agreeing on Planning and Design Objectives resolving conflicting interests of *multiple stakeholders* is the first step in which all stakeholders in the facility planning and design need to participate. This is of special importance in the AEC sector, as one of the reasons for many failures in the large-scale urban design and planning (e.g. mass housing) is that the product does not fit the way people live (p. 239). Such participation would enable identification of social patterns necessary for the multiple stakeholders and incorporate them into the planning and design process, so that a planner/designer, unable to intuit the needs of multiple stakeholders with which he/she is unfamiliar, do not impose his/her own values and needs on those for whom he/she builds. At this stage, the design objectives are tentatively decided which would be subjected to a thorough design synthesis (i.e. putting together the parts or elements or components to produce new effects creating an overall order, i.e. the facility) study as discussed later.
2. Setting Up the Facility Design Specifications:
    On deciding the objective, the Facility Design Specifications need to be set up in broad term based on the identification of the properties or conditions required by the objectives. A facility design specification defines the nature of the facility to be designed in terms of *data structures*, intended use, material, environmental and other *constraints, and objectives, cost limits, quality, and other relevant conditions.* A typical specification may include the intended use of the facility, material specifications for different components of the facility, space, cost and environmental requirements/constraints, quality, durability, and other relevant conditions, which may be applicable in a specific case. After the facility specification is defined, a designer need to start the process of synthesis and optimization.
3. Conducting a Facility Design Synthesis:
    Synthesis can be defined as the process of putting together the parts or elements to produce new effects and to demonstrate that these effects create an overall order, i.e. the facility. Once the design objectives and specifications are finalized in broad terms, a process of Facility Design Synthesis needs to be started by logical search of all potential solutions based on analysis of various design conditions and constraints as stipulated in the facility design specifications above, and by determining the relationships/interactions between varying states of

properties and the consequent varying degrees of fulfillment of respective objectives (say interaction between built-space, open-space, and cost).

This process of synthesis and optimization is greatly facilitated by adopting a systems approach in the design process. A system is defined as a collection of entities or components interrelated in some ways (defining the structure of the system) to accomplish a specific purpose fulfilling the constraints that restricts the behaviour of the system. Many facility planning and design problems, particularly in the AEC sector, incorporate all the characteristics of a system and, therefore, it should be designed as a system to fulfill the given objective(s). For this purpose, the system will first have to be defined in the form of a descriptive problem model identifying the purpose, components, structure, and constraints of the system, thus giving a representations of the real situation and reflecting the important factors that can be evaluated, manipulated, and prescribed so that the system designer could propose a specific system solution in descriptive form that met with the approval and the value of the society, multiple-stakeholder, or decision maker. In many cases, the descriptive problem model could be converted into a mathematical model presenting the situation in mathematical symbolic terms, thus permitting the designers/decision makers to adopt an analytical procedure for providing a system that will optimally satisfy the stated purpose or goal, fulfilling the given constraints. Thus, the design synthesis process may consist of the following four broad steps:

(i) Itemize the major expected performance capabilities of the facility design and agree on the objectives

Listing, quantifying, and prioritizing the major requirements of the facility as incorporated in the specification are to be done in this step. This should lead to an agreement on the objectives by all stakeholders in the design of the facility to be created by the design process. For example, major performance requirements of the facility of affordable housing can be itemized as built-space, open-space, cost, and so on, and all stakeholders could agree either on built-space, open-space, or cost as the priority. Alternatively, one of these criteria can be the main design objective and the limits of acceptability for the other criteria can be prescribed. This could be based on the determination of the interdependence between the above requirements and a compromise of these competing requirements could be arrived at by design analysis and optimization, and suitably incorporated in the final design. Accordingly, the determination of the interdependence between the above requirements can be the next step in the design process as discussed below.

(ii) Itemize the major design constraints

There may be many design constraints that tell the designer what not to do. Many of these constraints may be prescribed in the design specification and cost considerations. Some constraints may be quantitative/specific/definite, which can be translated into technical terms and incorporated into the design process. On the other hand, many constraints may be qualitative/subjective judgment making it difficult to incorporate into the design process in technical terms (e.g. type of urban development—low-rise, medium-rise, high-rise). This process should lead to identification of the laws controlling the interdependence, if any, of the properties.

(iii) Set the design criteria

Designer can identify the relevant design criteria for the facility by reviewing the requirements detailed in the specification and also based on the assessments in steps 1 and 2. However, it is desirable that all those requirements are translated into technical terms and that they be met based on design analysis and optimization discussed later. For example, in affordable housing design in an urban development, the common design criteria may be either least-cost design or maximum-built-unit-space design of the urban development system. In this context, cost engineering needs to be considered while setting the design criteria, and should be an important part of design analysis and optimization. While setting the design criteria, it should be ensured that the interdependence of the properties constitutes a realm of feasibility and that this lies at least in part in the domain of acceptability.

(iv) Itemize the major elements in the design consideration

After identifying the critical requirements and constraints, the designer can put together a package of all major design considerations relevant to the facility design and the options that are available indicating the advantages and disadvantages of each option. A logical optimal solution may then be derived from such analysis. A number of potential or alternative solutions may emerge from this design synthesis process.

The objective of this facility design synthesis is to conduct a thorough logical search and analysis of all conditions and constraints defined in the facility specification establishing the limiting and ideal states of the properties and hence an arena or the domain of acceptability implied by the objectives, leading to deriving possible solutions that satisfy the design specifications. An optimized, but not necessarily the best solution, as well as a number of alternative approaches, may emerge from this process.

4. Establishing Conceptual Geometry of the Facility

As mentioned above, through the design synthesis process, a number of potential or alternative solutions may be generated which may be in a descriptive form or descriptive model, say, in the form of module. In this operation, based on the descriptive model of facility design generated above, a designer can develop sketch designs giving provisional geometry and dimensions of the facility. But, such sketch designs may be only indicative geometric patterns and dimensionless, in case optimization of the design to find optimal dimensions of the facility design is the objective. In this case, the optimizing model, as per the design objective chosen (say least-cost or most-benefit objective), will give the actual dimensions of the facility. Alternatively, sketches with tentative dimensions can be developed. In both cases, provisional geometry of the facility design will have to be tested for its feasibility and acceptability as per design criteria specified in steps (i) to (iii) above. Such provisional geometries can be created in computer screen using interactive graphic capability of CAD, where all stakeholders can participate to give input in an interactive input-output session, as modification of the geometry can be done with great ease at any instant, using such interactive graphic capability of CAD.

5. Performing Facility Design Analysis and Optimization

This operation may include strength analysis, cost analysis, specific performance analysis, and so on, using the provisional geometry of the facility as above to ensure that the tentative facility design not only satisfy the specified objectives, conditions, and constraints but also is feasible in terms of technical (say structural soundness, based on relevant codes of practice), financial (say cost-effectiveness based on local/national context), and social conditions if applicable (say planning regulations and cost-affordability by the intended target group for an urban facility, based on local/national policy context and policy analysis). For example, using design analysis technique an urban planning policy analysis can be carried out to derive appropriate planning norms and local planning regulations based on the local and national context and resource-constraints as illustrated later in the book. In design analysis task, a designer can develop and use descriptive models representing the problem in mathematical terms (i.e. mathematical modeling technique), construct geometric models giving shape of the conceptual design of a component (i.e. geometric modeling technique), and also employ interactive graphics techniques providing to the designer the visual appreciation of the input data and the consequent output. At this stage, the designer will develop one possible solution based on the adopted options, and by repeating this design analysis process using different options/conditions, many more possible solutions, i.e. alternative solutions can be developed which may appear equally acceptable. However, if a unique optimized design solution is desired, an optimization technique would have to be used as explained and illustrated later in the book. If cost is a consideration—in fact, cost may be paramount in many facility design cases—cost engineering would be an important component of such design analysis and optimization process. Importance of Design Analysis and Optimization in the Planning and Design Process and the utility of CAD techniques in carrying out these tasks are also discussed below.

6. Finalizing the Facility Design Geometry

When the above five operations/stages in a facility planning and design process confirm that all the design criteria is met, the final geometry of the facility is decided with suitable modification, if any, as may be required in a specific situation.

The above operations/stages may not be sequential and may overlap, and a designer may have to perform much lengthy iteration among various functions in each of the above operations/stages.

1.7.2   Planning and Design in Architecture, Planning, Civil Engineering and Urban Development, i.e. AEC Sector, as Problem-Solving Process: Need of CAD, Planning, and Design Analysis and Optimization for Problem Solution

Planning and Design in Architecture, Planning, Civil Engineering (say, Building/ Housing/Urban Services consuming huge *Resources* and affecting the *quality of life* of people) needs to be viewed as a *problem-solving process* involving *search* through *numerous* potential solutions (because of *numerous* design variables involved) to select a particular solution that meets a specified criteria or *goal* (say, *Resource-Efficient* and affordable shelter design). This makes computer use indispensable for problem solution in all branches of AEC including architecture and civil engineering, particularly in the context of *search* through *numerous* potential solutions with large variations in *resource-efficiency* indicators. This view for planning and design in Architecture, Planning, Civil Engineering/Urban Development is quite relevant especially in developing countries to achieve the national *goals* of providing adequate shelter and urban services for all, within the affordability and *Resource-Constraints* at the individual, local, and national level. With increasing population and dwindling per capita *Resources*, this view is equally applicable even in the developed countries, as *we live in a Finite Earth with Finite Resources*.

    Utility of computer-oriented methods for solving civil engineering problems was accepted long time back emphasizing the urgency for recognition of this necessity by the researchers, analysts, and practicing engineers in their service to the civil engineering profession. But, the concept of integrated CAD by optimization is still not part of civil engineering education curricula in most of our educational institutions inhibiting development and application of such modern techniques in actual civil engineering practice. Some institutions have started CAD subject related to individual fields like structural engineering. But, resource use in such individual parts is much less compared to the larger systems like housing-urban-development systems, formed by such parts. Moreover, planning and designing in housing/urban development are a *problem-solving process* to achieve the national *goals* of providing adequate shelter and urban services to all, within the affordability and resource-constraints. A problem-*solving process* involves *search* through *numerous* potential solutions to select a particular solution that meets a specified *goal*, requiring a comprehensive planning and design analysis. Civil engineers, who are also quite strong in comprehensive planning and design analysis, need to play a vital role to solve urban development problems. To emphasize this role, American Society of Civil Engineers has a separate division of urban planning and development. However, civil engineers in most other countries generally play a passive role in urban development, where a key role being played by architects, planners, and administrators, who generally *avoid intricate planning and design analysis* using *mathematical models*. It is desirable that civil engineers in India and other countries correct this situation by taking more active role in the housing/urban development process, and enhance their professional contribution by mastering such techniques & applying integrated CAD by optimization in such larger systems to help attain a *resource-efficient* and *equitable* problem solution in the respective countries.

1.8   Design Analysis and CAD

Design analysis is the most important part of the whole planning and design process, and generally has two components, namely (1) planning, architectural, engineering, and costing analysis and (2) interactive graphics. Planning, architectural, engineering, and costing analysis is performed on the conceptual or initial geometry of the facility to ensure that not only the facility will satisfy the specified conditions and criteria (say compliance with space requirements, resource-constraints, and cost-affordability) but also satisfy the planning, architectural, and engineering norms and regulations. Analytical results can be derived from planning and designing codes/regulations, mathematical modeling, and model analysis covering descriptive and optimizing models, and so on. For example, using a descriptive model of an Urban-Built-Form in mathematical terms (mathematical modeling), a planner/designer could determine the **Interaction** between the selected elements of Urban-Built-Form, and also carry out a **sensitivity analysis**, selecting appropriate set of independent/dependent design variables and the *Iteration Choice*, and develop an **insight** into their interrelationships including cost. Based on such **sensitivity analysis** and insight, the desired value(s) of specified Built-Form-Elements (BFEs) could be selected, which is acceptable to *multiple stakeholders*. Geometric consistency of such value(s) of BFEs can be tested using geometric modeling techniques. Similarly, strength and performance analysis could be performed on the conceptual or initial geometry of the facility/component to confirm its compliance with the specified design conditions and criteria including structural soundness of the component.

Interactive graphics will give to the planner/designer and other stakeholders the visual appreciation of the input data and the corresponding output in numeric and graphic form, to help arrive at a planning and design solution acceptable to all stakeholders. Using the design analysis techniques at each stage, a designer can develop one of many possible design solutions based on the adopted options. The design analysis process would indicate to the designer that adopting other options, which may be equally relevant, he/she could get many alternative solutions which need to considered before a final decision is taken. CAD including interactive graphics can be very useful tool in this design analysis process, giving to the planner/designer and other stakeholders the visual appreciation of the input-output data to help arrive at a planning and design solution acceptable to all stakeholders, permitting participation of even lay user-stakeholder of the facility in the planning and design process. To find the unique optimal solution of the facility design for the specified/chosen design conditions, a detailed design optimization procedure would have to be adopted in a particular case. Importance of design optimization especially in the AEC sector and the special capability of CAD to apply optimization are discussed below.

1.9   Design Optimization and CAD

The human mind or a facility designer confronting a task or problem can recognize more than one course of action or alternative designs which have to be synthesized involving selection of an alternative out of a number of options some of which may perform better than the others, based on certain concepts. Thus, the designer tries to optimize the design followed by the selection (*decision*) of what is considered the desirable action, often involving a value judgment or qualitative analysis of the facility design. For example, in urban built-form, one can adopt low-rise or medium-rise. Considering the land price and built-unit-cost, the medium-rise may be a better option, but in terms of *quality of life* some people may prefer (value judgment) a low-rise built-form even if built-unit-cost is higher, on the other hand, some people may prefer medium-rise built-form if the built-unit-cost is lower. In such qualitative analysis, it is often difficult to develop mathematical relationships directly quantifying such value judgments, but such mathematical relationships could be developed and used for an indirect measure of such value judgment as illustrated later in the book. However, once a facility design concept is formulated, a designer can give definite optimal shape to the design by using relevant *Optimization Techniques* and using a measure of effectiveness. This is quite relevant in AEC sector and particularly in housing and urban development at each system level, where building/*urban professionals/user citizens* have to select an alternative from among many alternatives, frequently with different *efficiency indicators*. Tedious and voluminous computations involved, in utilizing *Optimization* and Operations Research Techniques and to solve and use *Optimizing Models*, have prevented their applications in the AEC sector specially in traditional building/urban *planning* and *design* in the past. But with the information technology revolution and advances made in CAD techniques, application of optimization in different sectors has become much easier with many advantages as discussed below.

1.9.1   Special Capability of CAD to Apply Optimization

It is difficult to apply *Optimization* using conventional manual methods while CAD greatly facilitates it. In fact, tedious and voluminous computations involved, in utilizing *optimization* and operations research techniques and to solve and use *optimizing models* like above, have prevented their applications in the traditional building/urban *planning* and *design* in the past. The continued development of cheap and powerful computers with interactive graphics systems has made it possible to perform many planning and design functions integrating graphics with complex computations, design analysis, cost engineering, and *optimization* techniques on a computer. Now, there are many commercial CAD packages, which perform various degrees of design optimization involving hundreds of variables and constraints. Conventional manual design cannot deal with such optimization process so effectively as compared to CAD method. Development of MICROCADD or 'PC-CADD' has revolutionized the fields of design, engineering, and drafting because of many advantages over the traditional manual methods, leading to substantial improvement in **Productivity** (Goetch 90). PC-CADD is a more productive approach to design and drafting than the conventional manual design because of:

1. Faster production of design and faster data creation.
2. Faster data manipulation including graphic data, taking little time in correcting and revising designs and documentation of those designs.
3. Faster, more convenient data storage.
4. Faster data output including graphic output.
5. Instant linkage and conversion of optimal design in numerical terms into a graphic output, thus facilitating application of optimization techniques for resource-efficiency.

Design optimization needs to be considered as part of the planning and design process. Crucial need of optimization in AEC sector and the utility of CAD as a design optimization tool to derive benefits of optimization, avoiding difficulties mentioned earlier, are discussed below.

1.9.2 Optimization: A Crucial Need in AEC Sector

Optimization is a human trait that is as old as antiquity. Even the *ancients*, who had little *technology* and *resource-constraints/population pressure*, applied *Optimization* and perceived that a circle has the greatest ratio of area to perimeter, which is the reason that many ancient cities are circular to minimize the length of city walls to enclose the city's fixed area (Brotchie et al., 1973; Brown, 1978). It is unfortunate that in spite of availability of modern highly powerful computers/graphic systems, *optimization* is rarely applied in our modern building/housing/urban operations, i.e. AEC sector, creating many *inequities* and *inefficiencies* and the consequent urban problems (e.g. proliferation of slums, inadequate housing/urban services, environmental degradation, and so on), especially in the context of rising population and dwindling *resources in many countries. Optimization* in AEC sector, i.e. Building, Housing and Urban Development is a Crucial Need, since *we live in a Finite Earth with Finite Resources*, and it is obligatory to apply *Optimization* in all such operations consuming *resources*, especially because such housing and urban operations involve many *variables* giving large variations in *resource-efficiency indicators* depending on the *planning-design-decisions*, creating large scope for *Resource-Optimization*. As the Urban Professionals *Commit* huge *Resources in their planning-design-decisions*, it is also *Basic* that they use *Operations Research*

(OR)/*Optimization* (*Basic Productivity Tools*) to *discharge their <u>Accountability</u> to the Society to achieve Productivity and Equity* in the use of *Resources* so *committed* by them in their *plans/designs* to produce *affordable* building/urban services for the people.

Optimization is a *three-step rational decision-making process,* i.e. knowledge of the system, finding a measure of system effectiveness and choosing those values of the system variables that yield *optimum* effectiveness. A problem can be *maximized* for benefit/profit or *minimized* for cost/loss, and which is desired is a matter of statement. Urban development is both a physical and a socio-economic system and may have conflicting goals/viewpoints, which need to be resolved using *optimization* both in qualitative (judged by human preference) and quantitative (using exact mathematical means) terms. Economic *optimization* has to consider human values expressible in quantitative form even if indirectly. For example, the cost difference of given built-space between chosen development types, say between low-rise and medium-rise urban development, could be an indirect measure for qualitative evaluation of these two types of development, while selecting the preferred alternative, with the *participation* of *user citizen* using collaborative **IT** (Information Technology)-**Tools**.

Essential characteristics of OR/*Optimization* are its emphasis on *models* (indicating physical depiction of a problem), *goals* (measure of effectiveness), *variables*, *constraints*, and putting them in *mathematical terms,* i.e. a *generative system* discussed earlier. *Optimization* theory is more than a set of numerical recipes for finding optima. By studying various optimization techniques, each suitable for different quantitative situations (even if idealized), one often discerns general *decision rules* appropriate to problems, and can develop an *insight* in recognizing the proper form of an optimal solution even when a problem is not completely formulated in mathematical terms. One advantage of such optimization study is that it can foster recognition of information valuable enough to be gathered to depict a system well enough for it to be optimized (bringing agreement on the measure of effectiveness and expressing its quantitative dependence on system variables), thus, help apply *optimization* in problem solution. Such an approach is *crucial* in different urban development operations to achieve a *resource-efficient* and *equitable* solution of urban problems.

Basic tool of Operations Research/*Optimization* in traditional management is the development and use of conceptual models, which could be 'descriptive', if they are designed only to describe the relationships between *variables* in a problem, or '*optimizing*' designed to lead to the selection of a best course of action among available alternatives (Anderson, 1973; Aoshima & Kawakami, 1979). These are equally applicable in urban management to produce efficiently the building/housing/urban services. Developing the descriptive model could also be the first step of the *optimization* process, i.e. acquiring knowledge of the system. Design optimization is a special subject of its own, and many disciplines have developed and adapted optimization techniques for their respective design problems. As application of integrated CAD incorporating optimization is rare in the AEC sector, this book gives emphasis on application of optimization techniques in design problems in AEC sector, adopting an integrated Computer-Aided Design (CAD) procedure including optimization and interactive graphic analysis, as illustrated in subsequent chapters.

Application of such concepts in housing, urban development, and urban management is illustrated in subsequent chapters of the book. Design optimization is a special subject of its own, and many disciplines have developed and adapted optimization techniques for their respective design problems. As application of integrated CAD incorporating optimization is rare in the AEC sector, this book gives emphasis on application of optimization techniques in design problems in AEC sector, adopting an integrated Computer-Aided Design (CAD) procedure including optimization and interactive graphic analysis, as illustrated in subsequent chapters. The elements of such Integrated Computer-Aided Design (CAD) procedure incorporating optimization and interactive graphic analysis are outlined below.

1.10 Integrated Computer-Aided Design by Optimization

As outlined above, CAD has many components, and the principles and mathematical formulations of computer graphics, geometric modeling, interactive graphics, architectural and engineering drafting, design analysis, and design optimizations are essential elements of a modern CAD system for any sector including the AEC sector. Similarly, cost engineering would be an important component of a CAD system where cost is a consideration for design, as in the AEC sector, in which cost is paramount in many cases to attain a *resource-efficient* and effective problem solution. Data management, storage, and transfer should also form part of a CAD system, to permit efficient data management for speedy information handling and decision-making resulting to higher productivity (i.e. efficiency and effectiveness), which is one of the basic objectives of CAD. As earlier mentioned, planning and design needs to be viewed as a *problem-solving process* in a resource-efficient manner. This view is especially relevant in Architecture, Planning, and Civil Engineering (say, Building/Housing/Urban Services), i.e. AEC sector, consuming huge *Resources* and affecting the *quality of life* of people in many countries. In view of above, all the components of CAD need to be integrated in the form of a comprehensive planning and design tool, i.e. as an integrated CAD system and applied in a specific problem situation to maximize the benefits of CAD in solving a problem in a resource-efficient and effective manner.

Ideally, an integrated CAD system should cover the entire portion of the professional planning and design practices in different sectors. To build such a comprehensive tool suitable for specific design needs and for a specific sector, a lot of research and development is necessary, consuming a lot of time and money. There are many commercial CAD software packages named as turnkey systems for some sectors, developed by integrating some major components of CAD such as geometric modeling, design analysis, engineering drafting, and data management, storage, and transfer; but these are very costly. These are created as 'black boxes', for practical application requiring no further in-house development or system modification by thc user in most cases. Turnkey natures of these systems make it possible to effectively use the systems by the users, having little knowledge or specialized training in software programming. However, many such systems are so costly that they are beyond the reach of most small sized industries, offices, professional organizations, or individual professionals. In view of above, there is need for research and development to develop integrated CAD software for AEC sector which should be less costly ensuring liberal availability of research results to all concerned including users of integrated CAD for further development to meet their specific needs as they arise while using such systems by the user.

As building a comprehensive integrated CAD tool covering all components is an uphill task and very costly, author feels that initially it should include major components such as (1) geometric modeling, (2) cost engineering, (3) planning and design analysis, (4) design optimization, (5) architectural and engineering drafting, and (6) data management, storage, and transfer. Salient details of each of above components are outlined below.

1.10.1   Geometric Modeling

Geometric modeling defines an object/facility in terms of its geometric properties and arrangement of spaces. Geometric properties such as geometric patterns (and size of independent variables if no size optimization is desired) indicating the relationship between different components of the object/facility is essential input for subsequent design analysis and optimization operations. A geometric model can be created by a designer on the monitor screen using the mouse or can be constructed using graphic primitives such as dots, lines, arcs, and curves in case of two-dimensional models. Three-dimensional models can be produced by turnkey systems.

1.10.2   Cost Engineering Data Management, Storage, and Transfer

Cost engineering is that area of engineering principles where engineering judgment, and experience are utilized in the application of scientific principles and techniques to problems of cost estimating, cost control, profitability analysis, planning, and scheduling. Cost is a major consideration in all human activities and is an important parameter to remain competitive in all sectors of activities. In fact, cost is paramount in AEC sector and in urban management where *resource-efficient* (and also *cost-effective*) and *equitable* supply of urban services is a prime objective. In view of importance of cost engineering in planning and design process, a separate Chap. 4 is devoted to the topic of cost engineering and database management, giving more details to enable readers to apply such techniques in their problem situations.

Effective management, storage, and transfer of data/information's related to planning, architectural, and engineering design process (including cost engineering data) should be an essential feature of an integrated CAD system (especially for AEC sector), for speedy information handling and decision-making resulting to higher productivity (i.e. efficiency and effectiveness), which is one of the basic objectives of CAD. A database is defined as a logical collection of related information, which should be (a) dynamic in information handling, (b) multi-user oriented, and (c) multi-layer structured. A database may contain millions of different sets of data, necessitating development of a system that can accept, organize, store, and retrieve information's quickly. Main thrust in the construction of a planning and design database should be to include efficient description of geometries, specifications, and unit cost details related to different components of a facility. A design database may consist of six groups: (1) general data, (2) technical data, (3) geometric data, (4) facility specification data, (5) facility cost data, and (6) data on previous projects. A database management system is the software systems developed for efficient operation of the above functions related to a database. A suitable system for planning and design decision support and for database management, stressing the use of information technology to secure *user citizen participation* in *planning and design process,* is essential in AEC sector. In view of importance of data management, storage, and transfer in planning and design process, a separate section in this chapter is devoted to this topic to give a broad understanding to the reader.

1.10.3   Planning and Design Analysis

As earlier mentioned, planning and design analysis is the most important part of the whole planning and design process, and generally has two components, namely (1) planning, architectural, and engineering, and may include strength analysis, cost analysis, specific performance analysis, and so on, (2) interactive graphics. Such analysis is performed on the conceptual or initial geometry of the facility to ensure that not only the facility will satisfy the specified objectives, conditions, and criteria (say compliance with space requirements, resource-constraints, and cost-affordability) but also is feasible in terms of technical (say satisfaction of planning, architectural, and engineering norms and regulations, satisfaction of structural soundness, based on relevant codes of practice, and so on), financial (say cost-effectiveness based on local/national context), and social conditions as may be applicable (say cost-affordability by the intended target group for an urban facility, based on local/national context and policy analysis). Analytical results can be derived from planning and design codes/regulations, mathematical modeling, and model analysis. In view of importance of planning and design analysis in planning and design process, a separate Chap. 6 is devoted to this topic giving its broad details to enable readers to apply such techniques in their problem situations.

In planning and design analysis, designer will develop one possible solution based on the adopted options, and by repeating this design analysis process, using different options/conditions, many more possible solutions, i.e. alternative solutions can be developed which may appear equally acceptable. However, if a unique optimized design solution was desired, an optimization technique would have to be used as explained below.

1.10.4 Design Optimization

As mentioned above, once a facility design concept is formulated, a designer can give definite optimal shape to the design by using relevant *optimization techniques* and using a measure of effectiveness. This is quite relevant in AEC sector and particularly in housing and urban development at each system level, where building/ *urban professionals/user citizens* have to select an alternative from among many alternatives, frequently with different *efficiency indicator*, making *optimization* a crucial need for *resource-efficient* and *equitable* problem solution. Tedious and voluminous computations involved, in utilizing *optimization* and operations research techniques, have prevented their applications in the AEC sector especially in traditional building/urban *planning* and *design* in the past. But with the Information Technology Revolution and technical advances made, CAD has special capability to apply optimization with many advantages as discussed earlier and also illustrated with examples in relevant chapters later. In view of importance of design optimization in planning and design process, a separate Chap. 8 is devoted to this topic giving the broad details to enable readers to apply such techniques.

1.10.5 Architectural and Engineering Drafting

Drafting is the documentation of the computer-aided design process from the initial sketches and conceptual designs in geometric modeling and analysis to the final presentation or working drawings of full-scale architectural and engineering projects. A fully integrated CAD system should be able to convert all end results of a planning and design process (incorporating design analysis and optimization) into final presentation or working drawings in architectural and engineering projects in AEC sector. However, this will need a lot of research and development in the AEC sector. With the ongoing Information Technology Revolution and advances made in CAD techniques in different sectors, drafting can be paperless with computer monitors displaying drawings as required, which can be instantly transmitted to the sites/other desired locations through the electronic means.

As mentioned above, there are many commercial CAD software packages named as turnkey systems for some sectors, developed by integrating some major components of CAD. But, many such systems are so costly that they are beyond the reach of most users. Moreover, such turnkey systems which can only be used as 'black boxes' make such systems inaccessible due to their proprietary nature, hindering their maximum possible effective use, and making the users fully dependent on suppliers of such 'black box' software for this purpose. Since, these are in the form of 'black boxes', the users are not able to develop any insight in their operations, which is essential for any problem solution. Moreover, because of their very nature users are not permitted to modify these suitably to meet their specific design needs in specific situation, or to develop the systems further for such needs as they arise (which is very usual in most design situations) while operating these systems. Such freedom is the usual phenomena in all scientific and engineering activities, giving stress on free scientific knowledge, which is the driving force resulting to the modern developed world from a primitive world. The above situation in integrated CAD may not change even if the costs of such 'black box' software are drastically reduced.

The ongoing Information Technology Revolution gives an opportunity to correct this situation in integrated CAD field in different sectors, if our academic and research institutions, professional bodies, and professionals in different sectors give stress to carry out research and development in this evolving field, ensuring liberal availability of their research results to all concerned including users of integrated CAD. Towards this objective, the author is carrying research to develop integrated CAD software for AEC sector. Since cost is a paramount factor for problem solution in AEC sector, cost engineering is an important component of the integrated CAD software being developed. Most microcomputer-based commercial CAD systems such as AutoCAD are constructed primarily to perform only geometric modeling and drafting, and their performance is effectively limited to what is included in them. Integrated CAD tool being developed in modular form attempts to remove this limitation, and covers all the six major components mentioned above, and permits planning and design analysis and optimization to help attain a resource-efficient, equitable, and optimal housing development design solution, incorporating cost engineering. Promoting further research and development to cover all components can enlarge the scope of such integrated CAD tool in AEC sector incorporating different modules as required by different users. Application of this integrated CAD software for problem solution is illustrated in subsequent chapters.

1.11 Planning and Design as Management Function: Utility of IT-CAD

Management knowledge is generally organized around the five essential managerial *functions* such as *planning*, *organizing*, *staffing*, *leading,* and *controlling* (Koontz & Weihrich, 1990; Koontz & O'Donnel, 1976). This is equally applicable in urban management. As per management theory, *planning* involves selecting missions and objectives and deciding the courses of action (from among alternatives) to achieve them. *Controlling* is the measuring and correcting activities to ensure that events conform to plans, which are not self-achieving. No real plan exists, until a decision, in terms of commitment of money, human, and material resources, is made (before such a decision it is only a planning study, analysis, or a proposal). Hence, urban planning (including land-use planning) should be viewed as a *planning function* of *Integrated Urban Management* incorporating the other four *functions*, and also the decisions regarding commitment of *resources*, to get the things done and achieve objectives. Design should be part of this planning function. Also, development control, building bulk regulations, and similar activities should be part of *control function* to achieve results or goals *resource-efficiently* and *equitably* within the given constraints at each level (say, at the level of individual, local, or national stakeholder), serving the multiple stakeholders and resolving their conflicting interests, viewpoints, and demands at each level as may be applicable.

Systems approach is an accepted practice in traditional management to help integration of all five managerial functions identifying the critical variables, constraints, and their interactions at each level, and forcing managers to be constantly aware that one single element or problem should not be treated without regard for its interacting consequences with other elements (Koontz & O'Donnel, 1976). Such an approach is essential in urban management, involving a vast number of interacting elements, for integration of all five urban management functions and coordination of efforts at each level, to achieve a *resource-efficient* and *equitable* solution of urban problems. This is all the more a necessity, since, in urban operations a number of actors and disciplines (urban planning, architecture, engineering, and so on), often with conflicting viewpoints, have to work together at each system level. An integrated urban management approach helps integrate such activities at each system level (e.g. project), and to get the things done, i.e. achieve objectives and solve urban problems using efficiently the *resources* (physical, financial, and so on) at hand, within control of an urban sector entity, availing the opportunities and respecting the constraints beyond its control (e.g. resource-constraints). Thus, the external constraints and systems environment, including the given macro and micro policy constraints, should be accepted as given conditions not amenable to change, while taking decision at a point of time, until more effective policies are evolved based on parallel policy-performance analysis, as part of *integrated urban management*. Accordingly, results have to be achieved within the given conditions by discharging efficiently all the five urban management *functions* at each level, using various management techniques including computer-based tools such as IT and integrated CAD, so that the individual, disciplinary, and the organizational objectives get translated into social attainments in, say, improved urban services and *quality* of life for all in a resource-efficient and equitable manner.

1.11.1 IT-CAD as Management Techniques to Improve Productivity, Social Equity, and Social Responsiveness in AEC Operations

Productivity, defined as the output-input ratio, which implies effectiveness (achievement of objectives) and efficiency (achievement of ends with the least amount of resources) in individual and organizational performance, is the essence of management in any sector (Koontz & Weihrich, 1990; Koontz & O'Donnel, 1976). Although inadequate, huge amounts of resources (money, materials, labour, land, and so on, i.e. inputs) are spent or committed by AEC/Urban Professionals (planners, architects, civil/building engineers) and planning authorities at different AEC/urban operation levels, while taking planning and design decisions at each system level (project, 'spatial unit', etc.) to produce AEC/urban services (i.e. outputs). This makes them accountable to the society or people to achieve productivity (i.e. efficiency and effectiveness) and also equity, particularly in urban operations, in the use of scarce resources so committed. AEC/urban professionals can use various planning and management techniques (e.g. IT, CAD, OR, and so on discussed below) to discharge this professional accountability to achieve efficiency and equity in the use of scarce resources committed, and thus, help achieve a resource-efficient and equitable solution of our mounting problems in the AEC/urban sector, in the context of urban dynamics and uncertainties, affordability, and resource-constraints at various levels applicable in many countries.

1.11.2 Urban Management Education

There are many other traditional management principles and techniques, which are equally applicable in urban management and are stressed in the holistic course design for urban management degree education (133 credit points) developed by the author under a research and consultancy study (Chakrabarty, 1997a, b). It is desirable that professional education and training institutions start such a field of study to produce professional urban managers, and also urban organizations (UOs) use such urban managers to enable them to adopt an integrated urban management approach, applying various management principles and techniques (IT, OR, optimization, integrated CAD, and so on) in their urban operations. This is central for improved performance of our UOs as effective instruments for resource-efficient and equitable solution of our mounting urban problems, in the context of urban dynamics and the individual, local, and national resource-constraints, and the conflicting interests, viewpoints, and demands of multiple stakeholders. It would help convert cities as efficient engines for economic growth with environmental sustainability, maximizing beneficial impacts of urbanization and, improving resource-efficiently the quality of life for all. This is discussed in more derails in the next chapter where abstract of the full-length course design for urban management degree education

(133 credit points) is presented. This course design includes a full-length subject of study on integrated computer-aided design for AEC sector. The proposed content of this subject of study is also presented. Hopefully, this book could be an important study material for this subject.

1.12 Closure

In recent years, there are unparalleled transformations in computers in terms of their capacity to capture data, to analyse, manipulate, and store them, to disseminate and communicate information, and to apply these capabilities in different areas. This is resulting to an all-pervasive Information Technology (IT) Revolution, viewed by many as the second Industrial Revolution, influencing the very way we live and work in society. Computer-Aided Design (CAD) is an essential part of this Information Technology (IT).

CAD can provide designers with many unique capabilities, as outlined in Sects. 1.5, 1.7–1.9, many of which would never have been thought possible with the traditional method. With these additional capabilities, fully developed for AEC sector, architects, planners, and engineers can handle projects involving a great many more variables and constraints as well as complexities and varieties in terms of planning and design conditions, and integrating the concerns of each discipline. Use of such planning-design analysis, optimization, and interactive computer graphics techniques is of special importance in the AEC sector, as this would allow multiple stakeholders in the sector, to participate in the facility planning and design as part of integrated urban management, to help resolve their conflicting interests, viewpoints, and demands, since one of the reasons for many failures in the large-scale urban planning and design (e.g. mass housing) is that the product does not fit the way people live, their constraints, and priorities.

The rate at which IT, CAD, and computer applications are spreading in different sectors is truly extraordinary and has few parallels in technological history. Today, information technology and computer application are everywhere shrinking the effects of time and distance and altering the very nature of work in organizations and professions in different sectors. This is equally applicable in Architecture, Civil/Building Engineering and Construction, i.e. AEC sector, which need to initiate vigorous and comprehensive actions for adequate response to this new challenge. Introduction of comprehensive education on urban management and IT-CAD, giving emphasis on integrated CAD and optimization as an urban management tool, can be a very important step to equip AEC professionals to effectively use these new tools and face this new challenge, and thus, become more relevant, efficient, effective, and useful in the society, and retain their social legitimacy in the above changed context.

This chapter provides the reader with an overview of integrated CAD and its utility in AEC sector, urban planning, development, and management. It stresses the need of viewing AEC planning and design as one of the five urban management functions (i.e. planning function) for improving productivity, social equity, and social responsiveness in urban operations. Such an approach is essential for resource-efficient solution of urban development problems, viewing planning and design as a problem-solving process involving search through numerous potential solutions to select a particular solution that meets a specified criteria or goal. This makes computer use and CAD by optimization indispensable in such activities involving civil engineers, architects, and planners (i.e. all AEC-urban professionals) who commit huge resources (like money, materials, land, and so on) in their planning and design decisions in such operations, enjoining on them an account-ability and ethical responsibility to society to ensure efficient use of resources so committed. The book aims to emphasize utility of CAD as an integrating tool enabling above professionals to adopt such an integrated problem-solving approach to derive above benefits.

Since AEC professionals are mostly involved in committing huge resources in planning and design decisions, particularly related to urban development and man-agement to improve the *quality of life for all*, a separate Chap. 2 on urban manage-ment is added to enable them to comprehend the '*big picture*' of urban operations to solve problems resolving conflicting viewpoints in process, and use of integrated CAD by optimization as a planning and management tool. This is followed by Chap. 3 on elements of numerical techniques and operations research (OR), Chap. 4 on elements of cost engineering and database management, and Chap. 5 on elements of computer graphics and architectural/engineering drafting-drawing, respectively. Several software for urban development design analysis and drafting-drawing (CAUB/CADUB), urban development design optimization and drafting-drawing (HudCAD), and for structural design optimization and drafting-drawing (RCCAD) are presented with source code listings. Separate chapter is added on planning and design analysis in Chap. 6, where application of quantitative analysis-optimization and computer graphics using above software are also demonstrated. Chapter 7 on design optimization is added where the principles of optimization are outlined and elements of optimization techniques such as geometric programming technique are presented with examples in AEC sector. Three chapters are devoted on integrated computer-aided structural design by optimization in Chap. 8, integrated computer-aided building-housing design by optimization in Chap. 9, and integrated computer-aided layout design by optimization in Chap. 10. The last chapter is on integrated computer-aided dwelling-layout system design by optimization in Chap. 11. In the above chapters, design analysis, optimization, and drafting-drawing are demonstrated using above software incorporating geometric programming tech-niques. Such advanced topics are presented at a level that is suitable for undergrad-uate students, highlighting the possibility of using available commercial participatory-collaborative IT tools in AEC sector for managing the dynamics of urbanization to arrive at more satisfactory urban problem solutions with citizen participation getting *resource-efficiency* and *social equity*.

Chapter 2
Urbanization, Urban Management, and Computer-Aided Planning and Design

2.1 Introduction

In Chap. 1, it is mentioned that in recent years, there is a revolution in the general capacity to process information and to apply these capabilities in quite novel areas. Today information technology and computer application are everywhere altering the very nature of work in organizations in different sectors with profound impact on management, managers, and in various professions in such sectors. This is equally applicable for AEC professionals (i.e. Architects-including Urban Planning, Engineers-Civil-Building Engineering and Construction) and in the urbanization, urban management, and the urban sector as a whole, where Information Technology (IT), including computer-aided planning and design as an important application area of IT, will also play a vital role, particularly in the context of conflicting interests, viewpoints, and demands of *multiple stakeholders* in the urbanization and development operations. Hence, AEC/urban sector needs to initiate vigorous and comprehensive actions for adequate response to this new challenge, and thus, become more relevant, efficient, effective, and useful in the society. It was stressed that introduction of comprehensive education on urban management and IT-CAD, giving emphasis on integrated CAD and optimization as an urban management tool, can be a very important step to equip AEC and urban professionals to effectively use these new tools and face this new challenge.

This chapter gives an outline of urbanization and its beneficial impacts, and the need of *Integrated Urban Management Approach* to derive the beneficial impact of urbanization on society, and to help attain a *Resource-Efficient* and *Equitable* solution of our mounting housing and urban problems so as to improve the *quality of life* for all. The limitations of the conventional approaches in achieving this goal and the utility of modern and traditional management principles, techniques, and tools to overcome these limitations are highlighted with illustrative examples and case studies. It is stressed that urban professionals like architects, engineers, and planners can make significant improvement in the urban environment by combining their base-disciplinary knowledge with the application of modern planning and management principles and techniques in such urban operations.

Operations Research, *Optimization*, and *Computer-Aided Design* (CAD) are *Basic Techniques* to improve *Productivity* in Management (Koontz et al., 1990) including urban management. AEC professionals commit huge resources (money, materials, labour, land, and so on) while taking *Planning* and *Design Decisions,* making them *Accountable* to the Society or People to achieve *Productivity* (i.e. *Efficiency* and *Effectiveness*) and also *Equity* in the use of scarce *resources* so committed by them. Hence, it is *Basic* that AEC/urban professionals apply such planning and management *Techniques* (e.g. IT, CAD, OR, and so on) to discharge their above professional *Accountability* to achieve *Efficiency* and *Equity* in the use of scarce *resources* committed by them. However, such *Planning* and *Design Decisions*, using integrated CAD and optimization (as urban management tools), should be part of *Integrated Urban Management,* and not taken in isolation as a fad, if these are to be effective to help achieve a *resource-efficient* and *equitable* solution of our mounting problems in the AEC/urban sector.

Therefore, before illustrating various elements of integrated CAD and optimization, this chapter presents a concept of *Urban Management* and some basic management principles, which are very relevant in the urban operations. This will enable students to comprehend the above 'big picture' during learning and also while applying integrated CAD and optimization as urban management tools in the role of urban professionals, and thus become more effective. An outline of a course design of urban management education of 133 credits is also given, where Computer-Aided Design (CAD) forms a full-length *subject of study* as part of the core course group: 'Quantitative Methods, Optimization and Computers', which is one of five core course groups (as per national technical education norm) of this urban management education program. This will give a holistic understanding to the students while learning and applying Computer-*Aided Design* (CAD) as a problem-solving urban planning and management tool to derive the beneficial impact of urbanization on society, and to help attain a *Resource-Efficient* and *Equitable* solution of our mounting housing and urban problems so as to improve the *quality of life* for all.

2.2   Need of Effective Urban Management to Derive Beneficial Impacts of Urbanization and Improve the Quality of Life for All

Urbanization is the most dominant phenomenon in all developing countries. Many World Development Reports and other studies show that urbanization level and economic development status of a country is intimately linked, i.e. higher the economic development status reached by a country, the higher is its urbanization level, indicating the inevitability of the urbanization process if a country is to develop (WB, 1990, 1991; Smith, 1975). However, urbanization tends to accentuate a number of problems such as inadequate housing and urban services (water, sanitation, transport, and so on), spiraling land prices and construction costs, proliferation of slums, pollution, and deterioration of urban environment, and so on, leading to a sort of urban crisis in many countries. This is mainly because of nonexistence of integrated urban management approach to overcome the limitations of the conventional approaches in urban planning and development operations. There are many beneficial impacts of urbanization as outlined below which can be derived and augmented if an *integrated management approach*, incorporating modern management principles and techniques, is adopted, which would help remove the limitations of the conventional approaches in urban operations.

Environmental movement often involves a debate between several artificial competing factions stressing forests, rivers, industries, and so on, which often obscures the fact that development of sustainable cities can be one of the most important factors in creating workable solutions to world environmental problems. This is because cities are one of the humankind's most efficient institutions for mobilizing resources (often providing more than 60% of GDP), increasing the number of people and economic activities that a region can support for economic uplift and improving the quality of life of people. Development of sustainable cities as 'engines for economic growth' can also provide environmental protection beyond the city's boundaries. For example, cities generate large, new, off-farm employment absorbing surplus farmers, decreasing pressure to cultivate marginal lands, and helping to consolidate most suitable land parcels for more sustainable and efficient farming. More than 60% of deforestation in Asia is reported to be caused by land-hungry farmers lacking alternative employment and, countries with higher urbanization level have lower rates of deforestation. Since urbanization generally leads to decline in birth rates, it could arrest the rising population in many countries, which is a serious threat to the environment. Studies (on the basis of equivalent value of goods and services produced) indicate that even in terms of pollution control and energy efficiency, performance of urban industrial economy is better than rural economy and, energy requirements per unit of GNP reduce with urbanization. No doubt, because of all these factors, urbanization is the most dominant phenomenon in all developing countries, and consequently, the world has crossed a demographic milestone, i.e. for the first time in the history of humankind urban population is more than rural population (United Nations, 1992).

In view of above, the goal of 'raising income while minimizing adverse impacts' to the environment can be achieved more efficiently in urban areas. But, most of our cities have major environmental problems mainly because of poorly managed urbanization and ineffective environmental management. This is because, in conventional practice instead of professional urban managers, usually, the general administrators are employed as urban managers. However, rarely such general administrators are imparted the intimate knowledge of modern management concepts, principles, and techniques, including modern systems management techniques, tools, and information technology. Because of this background, frequently they take a superficial generalist view, making them ineffective urban managers unable to take a holistic problem-solving managerial view with *accountability* to *efficiently* accomplish the selected *aims* resolving the conflicting interests, viewpoints, and demands of *multiple stakeholders* to attain *equity*, keeping in view the *urban dynamics and uncertainties*; need and demand of housing and urban services; cost of land and other *inputs*; and the *resource*, affordability, and environmental *constraints*, as may be applicable in a given context. Similarly, there is a tendency of each *base-discipline* working in the urban sector to look at the urban problems only from their base-disciplinary angle, instead of an integrated problem-solving managerial approach. Consequently, there are many limitations of conventional practice to manage urbanization and development process to attain *efficiency* and *equity* resolving conflicting viewpoints and demands, which results to many urban problems as illustrated below with example problems.

2.3 Conventional Practice in the Urban Sector: Limitations and Need for an Integrated Management Approach Using Modern IT Tools

As mentioned above, urbanization is the most dominant phenomenon in all developing countries and tends to create a number of urban problems such as inadequate housing and urban services, spiraling land prices and construction costs, proliferation of slums, and deterioration of urban environment. In many countries, the above problems are getting accentuated due to rising population, low affordability of a large section of population, and lack of resources. Therefore, it is imperative that professionals working in the urban sector should ensure *efficiency*, *equity*, and *effectiveness* in the utilization of scarce resources in the urban development and operations process, keeping in view the urban dynamics, economic growth potentials, and the interests of multiple stakeholders in the sector. However, it is difficult to achieve these objectives by the conventional approaches because of a number of limitations as discussed below.

2.3.1 Problem of Multiple Actors and Stakeholders in Context of Urban Dynamics

In the urban sector, a number of urban organizations have to work together at each system level, and in each of the urban organizations, a number of individuals have to work together in groups to efficiently accomplish the selected aims at each system level. Moreover, in the urbanization, urban operations and development process a number of *base-disciplines* such as architecture, engineering, physical planning, geography, sociology, economics, finance, and so on, have also to work together at each system level to achieve the selected aims at each system level.

Thus, in urban operations and development process, there are multiple actors and stakeholders including the people and user citizens, for whom the facilities are generally planned, designed, implemented, and managed by the professionals. Moreover, there can be *large variations* in the *efficiency indicators* of design of urban-built-form and urban facilities (depending on the rules such as *planning regulations*, the *dynamics* of input-prices, and other relevant factors). Urban dynamics and changes in the economic base structure of a city accentuate such variations. User *Citizens, Builders,* other *Stakeholders* cannot ignore such variations as they would like to maximize '*productivity*' (input-output ratio) and see that each of their money produces maximum benefit along with the desired environmental standards. This highlights the need to apply *dynamic* approaches in a *dynamic* city instead of frequent *static* master planning approach discussed below. In this context, it is imperative to integrate activities of various entities and multiple problem dimensions at each system level, resolving the conflicting interests, viewpoints, and demands of multiple stakeholders to achieve efficiently the selected aims and solve urban problems keeping in view the urban dynamics and uncertainties. This requires an integrated urban management approach covering all the five essential urban managerial functions, i.e. *planning, organizing, staffing, leading,* and *controlling,* outlined in more details later.

2.3.2 Difficulty to Adopt an Integrated Approach for Urban Problem Solution

In the conventional approach, generally the urban planners attempt to integrate key urban problem dimensions in their plans. However, usually, they have a little role in the other four urban managerial functions to be effective (particularly in the context of *urban dynamics* and *uncertainties*) and be *accountable* to get the things done and achieve *efficiently* the selected *aims* as per the plans. In the absence of professional urban managers, usually, the general administrators are employed as urban managers with decisive role in all the five urban managerial functions. However, rarely such administrators are imparted the intimate knowledge of urban issues, their complexities and interactions and of the modern systems management techniques, tools, and information technology to enable them to take an *Integrated Management Approach* for urban problem solution. Because of this background, frequently they take a superficial generalist view, instead of a holistic problem-solving managerial view with *accountability* to achieve aims and objectives and solve urban problems, ensuring efficient use of resources within control, availing the opportunities and respecting constraints beyond control at each system level, in the context of urban dynamics and uncertainties.

Similarly, there is a tendency of each *base-discipline* working in the urban sector (i.e. architects, engineers, and planners) to look at the urban problems only from their base-disciplinary angle, instead of an integrated problem-solving managerial approach respecting the constraints beyond control (such as scarcity of resources, low affordability of substantial section of population who are important *stakeholders*, and so on) and availing the opportunities (say, deriving an *optimal-urban-built-form* to *minimize resource* consumption) in a given context. Moreover, each urban professional (belonging to various *base-disciplines* particularly architecture, engineering, and physical planning) commits substantial resources (money, materials, land, and so on) while taking decisions in their respective area of activities making them *Accountable* to achieve *Efficiency* and *Equity* in the use of scarce *resources* committed by them. Since the resources are scarce quantities, society will expect, and also professional ethics demand that urban professionals re-iterate their decisions in terms of the *efficiency*, *equity,* and *public welfare* achieved (using a common 'value scale' to resolve conflicting interests, viewpoints, and demands) commensurate with the resources committed at each system level. This requires knowledge of integrated urban management and also skills for using appropriate management tools (e.g. IT-CAD Tools) and quantitative methods, which knowledge is rarely imparted to such urban professionals.

In this situation, no professional group is equipped to be *accountable* to adopt an *Integrated Management Approach* for *resource-efficient* solution of urban problems achieving *efficiency* and *equity*, and availing the opportunities and taking into account the urban dynamics, uncertainties, and the resource-constraints at the national, local, and individual user levels. As a result, such an *Integrated Management Approach* is rather rare in the conventional practice and needs correction. This is further discussed and illustrated below with examples and case studies.

2.3.3   Urban Planning, Development Control, Efficiency, and Social Equity in Urbanization: A Case Study

In conventional practice, one of the objectives of urban planning and development controls is to take care of the deficiency of the market economy and of inadequate knowledge of people who may not be the best judges of their own welfare (Harvey, 1987; Smith, 1975). This may be in terms of making allowances for externalities, making inadequate provision for open-space and for future demand affecting the efficient allocation of land resources, and so on (Harvey, 1987; Smith, 1975). For example, a prescriptive principle of Floor Space Index, that is also called Floor Area Ratio (FAR), is generally followed as part of urban planning and development control mechanism to achieve some of the above objectives (Harvey, 1987; Hoch, 1969). FAR is generally defined as the ratio between the total built-space on all floors of an urban development project and its land area. In its purest form, a principle incorporates a dependent and an independent variable. The scientific principle of floor area ratio could be developed in terms of the dependent and independent variables as elaborated in Chap. 6. However, for ready reference, in this case study, some relevant basic built-form variables are discussed, such as Built-

Unit-Space (BUS), i.e. the average built-space per built-unit (i.e. dwelling unit, commercial unit, and so on) of an urban development project; Building-Rise (BR i.e. number of floors), and the Semi-Public Open-Space (SPO) per Built-Unit (BU) which is the average open-to-sky-space per BU (i.e. built-unit) in the project layout. As outlined in Chap. 6, the scientific principle of FAR could be expressed, in terms of the above basic variables such as BUS (i.e. the space per built-unit), BR (i.e. number of floors), and SPO (i.e. the semi-public-open-space per BU), as follows:

$$FAR = \frac{BUS \times BR}{BUS + BR \times SPO} \qquad (2.1)$$

The relationship between FAR and built-unit-density (BUD i.e. number of BUs per unit area), and between FAR and Ground Coverage Ratio (GCR) could be expressed as follows:

$$BUD = \frac{FAR}{BUS} \qquad (2.2)$$

$$GCR = \frac{FAR}{BR} \qquad (2.3)$$

The above expressions can be used for any type of urban development, i.e. residential, office, and commercial development, and for any size of the planning and development area.

In conventional practice, urban planners and local bodies usually specify city *planning regulations* such as a limiting FAR (frequently without establishing the scientific relationships between variables as above). We may take as an example, a limiting FAR = 1 specified by a local-body, and a medium-rise walk-up residential development (say BR = 4) with built-unit-space (BUS) of 60 m^2 chosen by a low-income community. The above scientific relationship will show that in this case, a density of about 167 BUs per Hectare is achieved, and to comply with the above limiting value of FAR, the community will have to provide an Open-Space (SPO) of 45 m^2 per BU (i.e. 75% of the BUS giving a GCR of only 25%) which is much higher than the standard (usually 15–20 m^2 per BU specified by the Bureau of Indian Standards for such type of development). This may increase the cost per BU beyond their affordability (particularly because of spiraling land prices in many urban areas) pricing them out from shelter supply, adversely affecting the very public welfare on which consideration this limiting value of FAR is specified. If an open-space standard of 20 m^2 per BU is adopted, a higher value of FAR = 1.71 will have to be allowed, achieving a higher density of 286 BUs per Hectare, thereby reducing the incidence of land cost per BU and improving the affordability. As a *Policy*, there should be no objection to such *flexibility* in *planning regulations,* i.e. enhancing FAR here, as the open-space standard has been fulfilled and the low-income households may give more priority to the built-space per dwelling unit (BUS) with a reasonable open-space per BU. In conventional practice in many cases, a pre-conceived generalized prescription regarding density and Building-Rise (BR) such as low-rise is made as a National Policy (NCU, 1988) based on some misconceptions. In this case, if a low-rise of BR = 2 is insisted—as per the above pre-conceived generalized prescription in conventional practice—it can be shown, using the Software CAUB discussed later, that to achieve the same density (i.e. 286 BU/Hectare here), the open-space (i.e. SPO) will reduce to only 5 m^2 per BU giving a very high Ground Coverage Ratio (GCR) of 85.5%, making it an infeasible and unacceptable solution. Thus, in this case, adoption of the conventional approach leading to the application of *rigid* development control regulations in isolation, without adopting *Integrated Management Approach* incorporating *management principles* such as *flexibility principles*, instead of achieving the objective of *welfare of the low-income people*, may hinder the same, creating *inefficiency* and *inequity* in urbanization and development process.

To help apply such *Integrated Management Approach* incorporating *management principles* such as *flexibility principles*, the above mathematical model, i.e. **Operations Research Model** has been converted into Application **Software** 'CAUB/CADUB' for use as an urban planning and management tool where multiple *stakeholders* can interactively participate in planning and design analysis and effective decision-making to attain *efficiency* and *equity* in a given context. Details of these software are given in Chap. 7. However, here a few application results of the software highlighting some misconceptions in conventional practice (creating *inefficiency* and *inequity* in many situations), with regard to two planning issues even at the highest *Policy Level* in some countries, are presented below.

2.3.4   Some Application Results of Operations Research in Urban Development Planning and Design Analysis for Urban Policy Evaluation

Application of the concept of Operations Research (OR) in urban development planning and design policy analysis is illustrated in Chaps. 6 and 7 using the 'descriptive' and 'optimizing models' for urban built-form incorporated in the **Application Software 'CAUB/CADUB'** and '**HudCAD**' which are also presented there. Here, two example case studies, namely, FAR & built-unit-cost interaction and building-rise-density interaction are presented. These examples are selected because there are lots of misconceptions with regard to these issues even at the highest *Policy Level* in some countries like India (e.g. National Commission on Urbanization formed by the Government of India). These interactions are analysed using the **Operations Research Models** and the above Application Software. Application results highlight the utility of the integrated CAD Application Software like above to help better *design-decisions* to achieve *efficiency* and *equity* in building, housing, and urban development as part of *Integrated Urban Management*.

2.3.4.1 FAR & Built-Unit-Cost Interaction: Urban Management and Housing and Urban Development Planning and Design Policy Implications

The above descriptive models could also incorporate easily the unit land price and building construction costs, to derive the total cost per built-unit (i.e. Building-Unit-Cost/Built-Unit-Cost, or BUC) considering the prevailing market land price, and thus the effect of a prescribed planning regulation, like FAR, on the BUC could be determined instantly to confirm feasibility/acceptability of such planning regulations with people's (user citizens) participation (bottom-up approach), rather that imposing such regulations from top (usual top-down approach) which may not work. Application results of Module '**CAUB**' (of Application CAD Software: '**CAUB/CADUB**'), which incorporates the unit land price and building construction costs to derive the total cost per built-unit and the *scientific relationships* between built-form-elements, land price, and building construction costs, is shown in Fig. 2.1 (adopting land price of 10,000 Monetary Units (MU) per m^2, BUS = 110 m^2, SPO = 20 m^2). Figure 2.1 shows the relationship (*nonlinear*) between the '**FAR**' and '**Building-Unit Cost**', indicating that the respective *BU-Costs* at **FAR** of **0.846** (minimum feasible 'FAR' keeping above values of BUS and SPO, unchanged), and **0.5** (frequently specified by a Planning Authority ('PLA'), requiring an increased SPO = 110 m^2, even for a single-storey development), are **2.14 times** and **3.2 times the *Minimum-BU-Cost* occurring** *at the Optimum FAR* = **4. 018** (derived by '**CAUB**'). Such cost-variations only due to *Planning Regulations* may be even higher at higher land prices. Urban Planning Authorities (PLAs) in many countries like India continue to keep the *Planning Regulations* unchanged, say since the first Master Plan prepared in 1960s (thus, insisting and trying to enforce a low FAR prescribed 50–60 years back) irrespective of the land price increase to astronomical figures since then (even by 1000 times in some cities, and consequent astronomical increase in *BU-Cost*), creating huge *inefficiency* and *inequity* and the consequent housing and urban problems leading to the Public Interest Litigations (PIL), which in many cases receives attention even at the **National Apex Court**.

It is high time that such Urban Planning & Development problems are seen as part of Integrated Urban Management & Civil Engineers/Urban Professionals START using Management Techniques like OR/CAD as above to attain a Resource-Efficient & Equitable solution.

2.3.4.2 Building-Rise-Density Interaction: Urban Management and Building and Urban Development Planning and Design Policy Implications

The variations of density (i.e. BUD or Built-Unit-Density)/FAR/GCR with the variation in Building Rise (i.e. BR) are shown in Fig. 2.2 (produced by '**CAUB**' using *scientific relationships* between built-form-elements). Figure 2.2 indicates that in none of the three possible cases (i.e. adopting constant values of independent

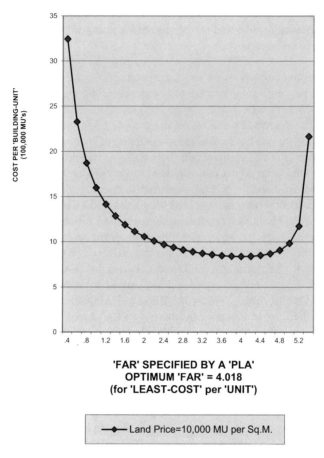

Fig. 2.1 Variation in cost per 'building-unit' with 'far' (same built & open-space per 'unit')

variables of: (1) BUS-SPO, (2) SPO-FAR, and (3) FAR-BUS, while BR is changed)
a low-rise leads to high density. This is *quite contrary to the frequent generalized prescription of 'Low Rise High Density (LRHD) Built-Form'* in the *Conventional Planning Practice and Policy* even at the highest national level. Density can be increased with 'Low-Rise' (which also increases the BU-cost creating *inefficiency* due to less sharing of *horizontal-diaphragms*, foundation costs, etc., compared to 'Medium-Rise'), only by reducing BUS or the SPO (and by increasing the GCR). This may not be acceptable to people and user citizens who are important **stake-holders** in the urban operations, if they are informed (for an *informed-decision-making with transparency*) of such BU-cost increase and BUS/SPO reduction (creating *inequity*), while giving such *pre-conceived prescriptions*. To overcome such limitations of conventional practice to achieve *efficiency* and *equity*, it is imperative for a *Policy* to use computer-aided planning & design techniques like **Software 'CAUB/CADUB'** and **'HudCAD'** discussed earlier.

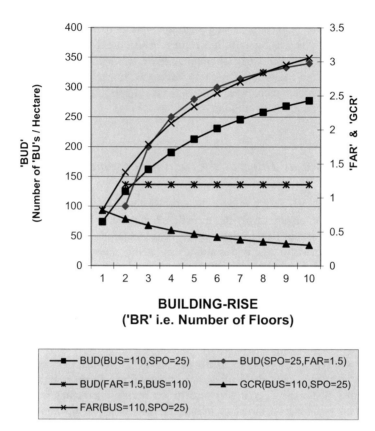

Fig. 2.2 Variation of density ('BUD'), 'FAR' & 'GCR' with building rise ('BR')

The above results show that *only due to the Planning-Regulation, i.e. FAR specified by a Planning-Authority (entirely within its control) there is very large increase in Cost per BU, even though adequate open-space—a basic objective of such Regulations—is achieved even at the Optimum FAR minimizing the Cost/Building-Unit.* Such **large cost-variations** are difficult to be ignored by user **Citizens**/other **Stakeholders**, who would like to *maximize 'productivity' (output-input ratio)* along with the desired environmental standards. *This may be one of the reasons for frequent illegal-constructions which may* be minimal if **User Citizens** are convinced through **Informed-Participation** (using Computer Software such as '**CAUB/CADUB**' & collaborative information technology tools for **transparency**) that the planning regulations do not require any change to achieve their above objectives, including *the desired environmental standards*.

To secure such **_Informed-Participation_** of **User Citizens**, Software '**CADUB**' cited above can be an useful tool, as even lay **Citizens** can give *Interactive Inputs* to get both *Numeric* and *Graphic Outputs* for a meaningful participation of such *stakeholders* in the *decision-making* regarding *building-layout-systems* taking into account their *constraints* and *priority*. Such results in more details along with typical graphic outputs of '**CADUB**', using such an input-output-session, are shown in Chap. 6. But, this is feasible only if a *Policy Decision* exists to include *in professional education* such *Modern Computer-Aided-Planning and Design Techniques* and use it, as part of *Integrated Urban Management*, to overcome the limitations of conventional practice to achieve *Efficiency*, *Effectiveness*, *Equity*, and *Optimization* in our building and urban operations.

2.3.5   Need for an Integrated Urban Management Approach

An urban area is a highly complex dynamic system of interacting activities and interrelated components involving growth, change, and uncertainties requiring application of various management principles, techniques, and tools to manage effectively the interacting activities and interrelated components, recognizing various technological, economic, and social issues to promote a resource-efficient urban operations and development to improve the quality of life for all. One of the distinctive features of an urban area, besides its higher population density and non-agricultural land use, is that its population is characterized by specialization of occupations and of labour with interdependence, requiring extensive managerial organization to provide services deemed essential for civilized survival sustaining productive activities. Urban areas are also subjected to a number of external dynamic influences creating many difficulties and opportunities. Inadequate attention and absence of holistic managerial response to the above interacting factors lead to many urban problems in terms of inadequate access to shelter and urban basic services, proliferation of slums and squatter settlements, inadequate urban infrastructure and transport systems and their high costs, congestion, pollution, and other environmental degradations.

As urbanization has many positive impacts outlined above and benefits individuals, communities, businesses, and generates economic wealth of the nation as a whole, going against urbanization may amount to going against the tide of national development and the economic, social, and political forces that spur it (Smith, 1975). Therefore, efficient urban management is the only way by which urban problems could be minimized while maximizing the above beneficial impacts of urbanization enhancing the role of cities as 'engines for economic growth'. Governments, urban organizations (both in the public and private sector), and urban professionals play a key role in the above urbanization process, and a resource-efficient and equitable solution of urban problems and cost-effective urban operations depends very much on the efficiency and effectiveness of the above entities. It is universally recognized that management expertise is very crucial to improve efficiency and effectiveness of any organization to achieve optimal results and goals. Even for the not-for-profit organizations with the goal of public service, management expertise is necessary to accomplish them with minimum of resources or to accomplish as much as possible with a given available resources. However, such integrated urban management expertise is very rare in the urban organizations in many countries, which no doubt is one of the crucial factors contributing to the deterioration of urban environment in many countries.

As mentioned earlier, in the absence of professional urban managers, usually, the general administrators are employed as urban managers in the conventional practice. However, rarely such administrators are imparted the required knowledge and skills for managing effectively the interacting activities and interrelated components, recognizing various technological, economic, and social issues to promote a resource-efficient urban operations and development to improve the quality of life for all. In this situation, frequently they try a superficial generalist view while carrying out various urban operations creating many urban problems as above. Intimate knowledge of urban issues, and of the modern systems management techniques, tools, and information technology is essential to help determine preferred plans, to predict the outcome of any decision, and the actual pursuit of a course of action to achieve the desired outcomes, taking into account the uncertainties and the dynamics of the urban processes, and thus enabling them to take an integrated problem-solving managerial approach in various urban operations. In this background, often they are unable to take a holistic problem-solving managerial view with *accountability* to achieve a *resource-efficient* and *equitable* solution of our mounting urban problems, ensuring efficient use of resources within control, availing the opportunities and respecting the constraints beyond control at each system level, in the context of urban dynamics and uncertainties.

Similarly, in the urbanization, urban operations and development a number of *base-disciplines* such as architecture, engineering, physical planning, economics, geography, sociology, and so on, have to work together at each system level (project, urban organization, 'spatial unit', and so on), to *efficiently* accomplish the selected aims and solve urban problems respecting the constraints beyond control (e.g. *resource-scarcity* and *low affordability* of a large section of population who are important *stakeholders*, and so on) and availing the opportunities (e.g. deriving an *optimal-urban-built-form* to *minimize* cost and *resource* consumption) in a given context. Moreover, urban professionals (particularly belonging to architecture, engineering, and physical planning discipline) commit substantial resources (money, land, materials, and so on) while taking decisions in their respective area of activities making them *Accountable* to the society to achieve *Efficiency* and *Equity* in the use of scarce *resources* **so** committed by them. In general, their creative ideas can take a real shape only by consuming appropriate resources. Since the resources are scarce quantities, society will expect that urban professionals re-iterate their decisions in terms of the *efficiency*, *equity,* and *public welfare* achieved at each system level (resolving the conflicting viewpoints and demands of multiple stakeholders using a common 'value scale') commensurate with the *resources* committed by them. All these require knowledge of integrated urban management and use of appropriate integrative management tools (including IT-CAD Tools), which are rarely emphasized in the education of such urban professionals.

As a result, in the conventional practice, each *base-discipline* tends to look at the urban problems only from its own base-disciplinary angle, instead of an integrated problem-solving managerial approach with emphasis on availing the opportunities and respecting the constraints beyond control at the national/organizational/individual level (e.g. land price and the affordability of multiple stakeholders) in the context of urban dynamics and uncertainties. AEC professionals such as architects, engineers, and planners, working in the urban operations can make significant improvement in the urban environment by combining their base-disciplinary knowledge with application of modern management principles and techniques using IT-CAD tools in such operations. But, this is rarely attempted in conventional practices, since hardly ever such AEC professionals are exposed to the modern management principles, techniques, quantitative methods, and IT-CAD tools to give them the knowledge and skills to adopt an integrated problem-solving managerial approach in the urban operations and development process to attain a *resource-efficient* and *equitable* urban problem solution.

The example problems above highlights how, adoption of the conventional approaches, without adopting *Integrated Management Approach* incorporating *management principles* such as *flexibility principles*, instead of achieving the objective of *welfare of the low-income people*, may hinder the same, creating *inefficiency* and *inequity* in urbanization and development process. The above results also show that *only due to the Planning Regulations, entirely within the control of a Planning Authority, there is very large increase in Cost per BU, even though adequate open-space—a basic objective of such Regulations—is achieved at the Optimum Planning Solution Minimizing the Cost per Building-Unit.* Such **large cost-variations** is difficult to be ignored by user **Citizens and** other **Stakeholders**, who would like to *maximize 'productivity' of their resources while achieving* the desired environmental standards, which *may be one of the reasons for frequent illegal-constructions.* These *may be minimal if* **User Citizens** *are convinced, through* **Informed-Participation** *using* modern IT tools like *Computer Software* presented in Chap. 7 along with collaborative *Information Technology Tools for* **Transparency**, *that the Planning Regulations do not require any change to achieve their above objectives.* This is hardly attempted in conventional practices using IT-CAD tools.

The above results also highlight how great *inefficiencies* and *inequities* are created in many situations due to misconceptions in conventional practice with regard to some planning issues even at the higher *Policy Level* in some countries, which need to be corrected using appropriate modern IT tools as part of *Integrated Urban Management.* The above results are only in respect of a few urban issues and case studies. In actual urban operations, there are numerous urban planning, development, and management issues which have to be addressed to remove the limitations of the conventional practice, which create a number of urban problems, *inefficiencies* and *inequities* in the urban operations and development process jeopardizing the role of cities as engines for economic growth and public welfare. This requires adoption of an *Integrated Urban Management* approach (covering all managerial functions as per the management theory) in all urban operations for *efficient* and *equitable* solution of urban problems. A concept of such *Integrated Urban Management* based on *Management Theory* is presented below.

2.4 Management Theories and a Concept of Integrated Urban Management

Management theory defines management as the process of designing and maintaining an environment in which individuals, working together in groups, efficiently accomplish selected aims (Koontz & Weihrich, 1990; Koontz & O'Donnel, 1976). In all sectors of activity of society, people form groups to accomplish aims that they could not achieve as individuals. This is equally applicable in the urban sector where

1. A number of urban organizations have to work together in an integrated manner at each system level (say the city/'urban spatial units' at various levels) to efficiently accomplish the aims at that system level, in consonance with the aims of the higher system level in the hierarchy (say the district, state, region, nation), as may be applicable.
2. In each of the urban organizations, a number of individuals have to work together in an integrated manner in groups to efficiently accomplish the selected aims at each group/system level (say in departments, projects, and so on) in consonance with the aims of the higher system level, i.e. the urban organization as a whole.

Thus, urban management should not only encompass management of the urban enterprises and organizations (such as urban local bodies, private/public developers, private/public urban infrastructure organizations, and so on) somewhat similar to the conventional business management (provided adequate stress is given to the urban issues and the urban dynamics), but also, the additional tasks of the management of the process of urbanization, development, and urban operations. Again, in this process, a number of urban enterprises have to work together (with organized cooperation and social responsiveness for effective response to the *urban dynamics* and uncertainties) to *efficiently* accomplish the selected *aims* at various system levels, availing opportunities and respecting the *constraints* beyond control, in the 'urban spatial units' and the urban sector as a whole involving *multiple stakeholders*. This is a much more difficult management task compared to managing a business for which conventional management courses are offered, which are mostly oriented towards manufacturing and production organizations and activities. Therefore, such conventional management courses require reorientation towards urban issues with a separate Urban Management Body of Knowledge (UMBOK) with stress on urban concerns (including social, economic, environmental, financial factors, and so on), to serve the urban enterprises and the urban sector as a whole.

Management knowledge is generally organized around the five essential managerial functions: *planning*, *organizing*, *staffing*, *leading,* and *controlling* (Koontz & Weihrich, 1990; Koontz & O'Donnel, 1976). This is equally applicable in urban management. The task of urban managers is to transform the inputs from external environment (capital, people, skills, and so on), into outputs (housing and urban services, profits, satisfaction, integration of the goals of various claimants and stakeholders to the urban organization, and so on) by discharging all the five essential managerial functions, i.e. *planning, organizing, staffing, leading,* and *controlling*, in an *effective* and *efficient* manner, taking due to consideration of the external environment and variables. As mentioned above, in urban operations, a number of actors, stakeholders, and *disciplines* (urban planning, architecture, engineering, and so on), often with conflicting interests and viewpoints, have to work together at each system level. An *integrated urban management approach* helps integrate such activities and viewpoints at each system level (e.g. project), and to get the things done, i.e. achieve *objectives* and solve urban problems using *efficiently* the *resources* (physical, financial, and so on) at hand, within control of an urban sector entity, availing the opportunities and *respecting* the *constraints* beyond its control (e.g. resource-constraints). As cited above, urban managers have to consider and respond to the external environment while discharging managerial functions. In this context, the external systems environment, opportunities, and constraints, including the given macro and micro policy constraints, should be accepted as given conditions not amenable to change (at that point of time, at various decision and action levels, say at the level of project, urban organization, and so on), until more effective policies are evolved (Chakrabarty, 1998). The new policies should be evolved based on parallel and continuous process of *policy-performance analysis* in the urban sector, forming part of urban management process establishing linkages with the urban management functions at appropriate system level and at appropriate time using IT for corrective actions. Accordingly, result(s) and goal(s) in the area of concern have to be achieved by an urban sector entity (say an urban organization in public or private sector, various *disciplines,* and so on), within the given conditions and constraints beyond control, and using the available resources and opportunities at each system level, discharging *efficiently* all the five urban management functions at each system level, using various *Management Techniques* and computer-based tools including IT tools, so that the individual, *disciplinary*, and the organizational *objectives* get translated into *social attainments* in the urban sector, say, in the improved urban services and *quality of life* for all in a *resource-efficient* and *equitable* manner, as established by information's generated through suitable feedback loop(s) in the integrated urban management process using IT (Koontz & Weihrich, 1990; Koontz & O'Donnel, 1976; Chakrabarty, 1997a, b).

The conceptual structure of integrated urban management on the above lines is shown diagrammatically in Fig. 2.3. The diagram indicates a feedback loop showing outputs of an urban sector entity, analysis of the outputs with respect to the degree of its goals/objectives achievement, and if not satisfactory, feedback for corrective actions through managerial functions. The numbers indicated in the diagram could be used to describe classification of urban organizations based on the degree of integration achieved in its operations, between its *goals* in the area of *concern* and the effective and efficient utilization of *resources* to get the things done, discharging efficiently all five managerial functions. For example, an urban organization operating in the area designated by 1 is achieving its goals resource-efficiently attaining full integration of the above three factors in its operations, a very ideal case and need to be strived for by all urban organizations although difficult to attain (Koontz & Weihrich, 1990; Koontz & O'Donnel, 1976; Chakrabarty, 1997a, b).

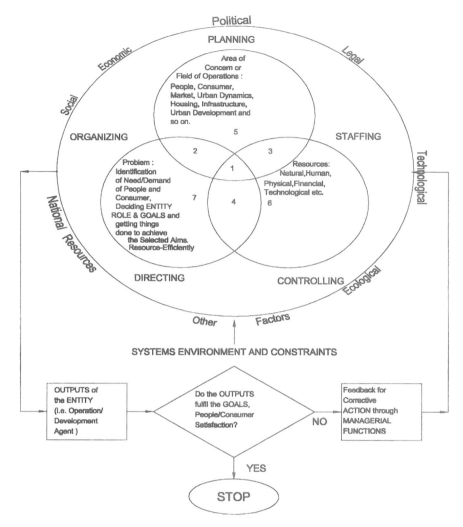

Fig. 2.3 Conceptual structure of integrated urban management

Just as in traditional management, the operational approach to management theory and science, which attempts to bring together the pertinent management knowledge to the managerial job and functions and tries to integrate the concepts, principles, and techniques, needs to be applied in urban management. The operational approach to management theory and science incorporates the systems theory, management science, decision theory, and so on, as deemed useful and relevant to managing. Some of these concepts, principles, and techniques, which are equally applicable to urban management, are discussed below.

2.4.1 Systems Approach to Urban Management

Systems approach, which is one of many different approaches to the analysis of management, recognizes importance of studying interrelatedness of planning, organizing, and controlling in an organization and its interaction with external environment (Koontz & Weihrich, 1990). Systems approach is an accepted practice in traditional management enabling managers to identify the critical variables, *constraints* and their interactions with one another at each system level, i.e. manager's organization, department, section, and so on (Koontz & O'Donnel, 1976). Such an approach forces managers to be constantly aware that one single element, phenomenon, or problem should not be treated without regard for its interacting consequences with other elements. (Koontz & O'Donnel, 1976).

The above approach is very relevant and useful in urban operations and management, particularly because urban operations are extremely dependent and interrelated with the external and internal environment of urban organizations which are part of larger systems such as the sector to which it belongs, economic system, and the society as a whole. The urban organization receives inputs, transforms them through the managerial functions of *planning, organizing, staffing, leading,* and *controlling*, into outputs and exports the outputs to the environment. The inputs from the external environment may include people, fund/capital, knowledge, skills, and so on; and outputs may include housing and urban services, profits, satisfaction, integration of the goals of various claimants and stakeholders to the urban organization, and so on. Various interest groups will make demands on the organization. For example, consumers will demand safe, reliable, and adequate housing and urban services at affordable prices, which need to be very important outputs for urban organizations. Governments will expect taxes/revenues to be paid by the organization, which is also required to fulfill the relevant social and economic objectives complying the pertinent laws and regulations. Investors and financial institutions will want not only a high return on investment but also security for their money. Professional bodies and individual professions will want respective professional excellence, i.e. architectural excellence, engineering excellence, and so on, which may contradict the affordability criteria. Community may demand maximizing job creation with minimization of pollution. Thus, there may be many claimants to the organization with conflicting interests, viewpoints, and demands, and it is the urban manager's job to integrate the legitimate objectives of the claimants by recognizing the systems nature of the problem.

The task of managers is to transform the inputs into outputs in an effective and efficient manner. This managerial transformation process can be comprehensively analysed considering systems approach to management as the foundation for organizing managerial knowledge dealing with various aspects of managerial functions of *planning, organizing, staffing, leading,* and *controlling*, and also covering the basis of management, and interaction between the organization and its environment highlighting social responsibility and ethics cutting across all the five *managerial functions* (Koontz). A communication system using modern IT is essential in all phases of managerial process to integrate managerial functions and to link the organization with its external environment. For example, objectives set in *planning* need to be communicated to devise suitable organization structure, staffing, and effective leadership; and whether performance conform to plans can be determined only through communication. Thus, communication makes managing possible. Moreover, customers or user citizens in respect of urban organizations, supplying urban services, are outside an organization, although they are the reason for the very existence of an organization (including urban organizations), It is only through a communication system using modern IT that the needs of customers or user citizens could be identified, enabling organizations to provide the services in a resource-efficient and equitable manner satisfying the needs of customers or user citizens. An effective communication system using modern IT is also essential to find out the potential problems and constraining factors in the operations of an organization.

By adopting a systems approach to urban management, all the five *urban management functions* could be integrated at various system levels, while linking an urban organization and a spatial unit, with its environment, and with the urban sector as a whole to achieve the missions and objectives. As mentioned earlier, the task of urban managers is to utilize the inputs to the urban organization and transform them to outputs through *effective* and *efficient* managerial functions, taking due to consideration of the external variables, i.e. the elements and forces of their external environment. Although urban managers may have little or no power to control or change the external environment, they have no alternative but to respond to it, and discharge their *social responsibility* to help solve social problems, being a part of the social system. Since urban operations involve a vast number of interacting elements at each level, *systems approach* to urban management is essential to help coordination of efforts at each system level, so that individual and organizational objectives become translated into social attainments in the urban sector.

This is all the more a necessity, since, in urban operations, a number of actors, stakeholders, and *disciplines* (urban planning, architecture, engineering, and so on), often with conflicting interests, demands, and viewpoints, have to work together at each system level. An *integrated urban management approach* helps integrate such activities and viewpoints at each system level (e.g. project), and to get the things done, i.e. achieve *objectives* and solve urban problems using *efficiently* the *resources* (physical, financial, and so on) at hand, within control of an urban sector entity, availing the opportunities and *respecting* the *constraints* beyond its control (e.g. resource-constraints). Finally, it is necessary to recognize that in this systems model of urban management operations, some of the outputs may become important inputs (say, satisfaction of employees, reinvestment of surplus of income over costs, corrective actions based on feedback and feed forward analysis of outputs, and so on) to the system (i.e. reenergizing the system), through suitable feedback and feed forward loops as depicted in the conceptual structure of integrated urban management shown in Fig. 2.3, to ensure satisfactory performance of an organization, discharging efficiently and effectively all five managerial functions outlined below (and also shown in Fig. 2.3).

2.4.2   Managerial Functions

In the literature, functions of managers are used as a framework for organizing managerial knowledge, and it is also opined that there are no new ideas, research findings, or techniques that cannot be placed in the classifications of planning, organizing, staffing, leading, and controlling (Koontz & Weihrich, 1990). Therefore, all five managerial functions are outlined below, to give an understanding of management.

2.4.2.1   Planning

In the context of management, *planning* involves selecting *aims,* i.e. missions and objectives and the actions to achieve them, requiring decision-making, i.e. choosing future courses of action from among alternatives adopting a rational approach (Koontz & Weihrich, 1990). Any organized operation must have a mission or purpose if it is to be meaningful. The missions or purpose signifies the basic function or task of an agency (or of part of it), which is assigned to it by the society in every social system. Thus, the purpose of the state housing and urban development department or housing board is to plan, design, build, and promote or facilitate housing and urban development operations in the state. Similarly, purpose of the city level housing and urban development authority is to plan, design, build and promote, or facilitate housing and urban development operations in the city. The objectives or goals not only represent the end point of planning, but also the ends toward which activities of organizing, staffing, leading, and controlling are aimed. The goal of all managers in an organization should be a 'surplus', that is accomplishing objectives (say supply of given level of housing and urban services) using least amount of resources, i.e. with the least amount of time, money, materials, and so on; or to achieve as much as possible of a desired objective (say supply of housing and urban services) using available resources. All managerial functions should be aimed to attain this goal. There are various types of plans, ranging from overall purposes and objectives to the most detailed actions to be taken. A plan does not really exist until a decision to commit resources (capital, material, human, and so on) has been made. Before such a decision is made outlining also a planning horizon, all that exists is a planning study, an analysis, or a proposal; but no real plan. This tends to be the case in urban planning discussed later.

2.4.2.2   Organizing

As defined above management is a process where individuals, working together in groups, efficiently accomplish selected aims or goals. In any such group effort to achieve the goals, the role of individuals needs to be defined preferably in a structured manner by someone who intends people to contribute to the group effort in a specific way. Organizing is that part of managing that entails establishing an intentional structure of roles for people in an organization, implying that their job has a definite objective which fits into group effort, and for this purpose, they have the necessary authority, tools, and information to accomplish the task. Thus, the job-objectives of architects, engineers, and planners working in an urban organization should fit into the group effort or organization objective (say supply of given level of housing and urban services using least amount of resources), and for this purpose, they should have necessary tools (such as CAD tools) and authority in the form of an organization structure. The purpose of an organization structure, as a management tool, is to help in creating an environment for human performance (Koontz & Weihrich, 1990). It is intended to ensure that all tasks necessary to accomplish goals are assigned, and also allotted to people who can do them best. Designing an effective organizational structure is quite an intricate task encountering many problems in terms of making structures fit situations, defining jobs to be done, and finding people to do them. This is equally applicable in respect of urban organizations and urban management.

2.4.2.3   Staffing

Staffing is that part of managing which involves filling, and keeping filled, the positions in the organization structure by suitable candidates who can accomplish their tasks effectively and efficiently. This is done by identifying work-force requirement; recruiting, placing, promoting, appraising, career planning, and training fresh candidates and existing jobholders to create capacity to accomplish their tasks effectively and efficiently. Thus, architects, engineers, and planners recruited and placed in an urban organization should be capable to accomplish their tasks effectively and efficiently to fulfill the group effort or organization objective (say supply of given level of housing and urban services using least amount of resources). Suitable career planning and training (say on use of CAD tools in housing and urban planning and design) of such professionals could be organized to create such capacity.

2.4.2.4   Leading

Leading is the influencing of people so that they will contribute to organization and group goals, and involves predominantly interpersonal aspects of managing such as motivation, leadership styles and approaches, and communication. An effective urban manager needs to be an effective leader, should be conversant with behavioural science, and should understand the human factor in their operations to produce desired results, harmonizing individual and urban organizational objectives.

2.4.2.5 Controlling

Controlling is the measuring and correcting activities to ensure that events conform to plans, which are not self-achieving. Thus, *controlling* is necessary to compare plans with results, and for corrective actions to achieve results, making *planning* and *controlling* functions inseparable, as people must know first where they want to go (the result of the task of *planning*), if they are to know whether they are going where they want to go (the result of the task of *controlling*) (Koontz & Weihrich, 1990). As discussed earlier, in the conventional urban planning and control operations, generally *rigid* development control regulations are applied in isolation, without adopting *Integrated Urban Management Approach* embracing other *managerial functions* and *management principles*. Although the *goal* of conventional urban planning and control operations is *welfare of the people*, such conventional approach with stipulation for application of *rigid* development control regulations in isolation, without adopting *Integrated Urban Management Approach* incorporating other *managerial functions* and *management principles* such as *flexibility principles* using information technology and CAD techniques for effective response to the *urban dynamics* and uncertainties, rarely achieve the above *goal*, and instead creates many *inefficiencies* and *inequities* in the urbanization and development process, as illustrated above with examples. Hence, it is imperative to consider conventional urban planning as the *planning function*, and conventional urban planning control as *control function* of urban management, incorporating other *managerial functions*, *management principles*, and *management techniques* including information technology and CAD techniques, to help achieve the desired *goals* in such urban operations. This is further elaborated below.

2.4.3 *Urban Planning as the Planning Function in Urban Management*

In the urbanization and development process, cities are '*engines for economic growth*' (particularly in the context of a developing country) and, places for *people to live and work*. Although, in conventional practice, urban planners prepare plans with the *aim* of making a city a viable and desirable place in which to live and work, it may be noted that traditional urban planning incorporates only planning function (that too mainly physical planning) in isolation, without linking it to the goal oriented urban management incorporating all the five urban managerial functions. As a result stated *aims* are often not achieved. Moreover, urban planning should accomplish a *physical planning structure* matching with the *economic-base-structure* (creating work) and the consequent *demographic-distribution-structure*, covering the *multiple stakeholders* including the *urban poor*, to achieve *Efficiency* and *Equity*, and create a balanced community. However, in conventional planning practice, this is often neglected, and *urban dynamics*, uncertainties (difficult to foresee and plan) and conflicting interests and viewpoints of *multiple stakeholders* also make it difficult to achieve such a match creating many urban problems like proliferation of slums in a city.

This conventional physical planning approach, without linking it to the goal oriented urban management incorporating all the five urban managerial functions of *planning, organizing, staffing, leading*, and *controlling* to achieve efficiency and equity, often leads to many urban planning failures (particularly because of the urban dynamics and uncertainties), inefficiencies, and inequities as cited above with examples. On the other hand, a goal oriented urban management would include all the above five urban managerial functions (in addition to physical planning), incorporating *flexibility principles* (as explained later) for effective response to the urban dynamics and uncertainties, for promoting economic growth, wholesome urban development, improved urban operations, and improved urban living environment at each system level, in the environment of change (physical change, value change, and so on), so as to improve the *quality of life* for all.

As mentioned above, *planning* involves selecting *aims* and the actions to achieve them. A real plan should indicate the missions and objectives and the actions, including commitment of resources, with *accountability* to achieve them adopting an integrated management approach covering all *managerial functions*, so as to ensure that events conform to plans, which are not self-achieving. In fact, a plan does not really exist until a decision to commit resources (capital, material, human, and so on) has been made, outlining also a planning horizon and linkages with other *managerial functions* to get the things done as per plan. Before such a decision is made all that exists is a planning study, an analysis, or a proposal; but no real plan. If operations of an urban organization are to be meaningful, it must have a mission or purpose, which signifies the basic function or task of an agency (or of part of it), which is assigned to it by the society in every social system. Therefore, an urban planning and development authority preparing plans for housing and urban development operations in a city should not only indicate the objectives or goals representing the end point of *planning*, but also highlight the ends towards which its activities of *organizing, staffing, leading*, and *controlling* are aimed, to get the things done and attain goals as per plan, i.e. adopting an integrated management approach covering all *managerial functions*. However, conventional *urban planning* and master planning approach at the city level tends to be a planning study, an analysis, or a proposal, as these are not linked with commitment of resources and with other four managerial functions of *organizing, staffing, leading*, and *controlling*, as part of *Integrated Urban Management* to get the things done, so as to attain with *accountability* the missions and objectives incorporated in the urban plan, achieving both *Equity* and *Efficiency*.

Therefore, it is imperative that urban planning (including land-use planning at 'urban spatial unit' level) should be viewed as a part of the '*planning function*' of the integrated urban management, incorporating the other four *urban managerial functions*; and also comprising the decisions regarding commitment of *resources* mentioned above, so as to efficiently accomplish the selected aims, missions, and objectives at each system level. This is more so because plans are not self-achieving. Design should be part of this *planning function*. As mentioned above, *controlling* is necessary to compare plans with results, and for corrective actions to achieve results, making *planning* and *controlling* functions inseparable. There are similar linkages between the other urban managerial functions. In this context, development control and building bulk regulations should be part of '*control function*' to achieve results and goals *resource-efficiently* within given *constraints* at each system level, serving the *multiple stakeholders*. In this process information technology, CAD techniques including *Optimization*, and other *Management Techniques* and *Principles*, including *Flexibility Principles* discussed below, should be extensively used to resolve conflicting interests, demands, and viewpoints of *multiple stakeholders*, and for effective response to the *urban dynamics* and uncertainties, so as to achieve the goal of *resource-efficient* and *equitable* solution of our mounting urban problems.

2.5 Applicability and Usefulness of Modern Management Principles and Techniques in Urban Planning and Management

A scientific approach requires not only concepts but also the principles and techniques. In modern management, there are a number of principles and techniques to efficiently accomplish the selected aims. Many of these are equally applicable in urban management to derive the advantages of efficient and effective management and also the beneficial impacts of urbanization discussed earlier. However, this is possible only if appropriate *Policy* exists to apply such principles and techniques in various urban operations. Some of these modern management *Principles* and *Techniques* are outlined below only as illustrative examples showing their applicability and usefulness in urban planning and management.

2.5.1 Modern Management Principles in Urban Operations: Some Illustrative Examples

As mentioned above, cities are for people to live and work. It is desirable that urban planning should accomplish a *physical planning structure* matching with the *economic-base-structure* (creating work) and the consequent *demographic-distribution-structure* (which should be accepted as a given condition at appropriate decision level, and should cover the *multiple stakeholders* including the *urban poor*). However, *urban dynamics*, uncertainties (difficult to foresee and plan) and conflicting interests and viewpoints of *multiple stakeholders* make it difficult to achieve such a match, creating many urban problems such as proliferation of slums and so on. In this context, application of modern management principles of flexibility and navigational change is essential to cope with such *urban dynamics*, uncertainties (difficult to foresee and plan) and conflicting interests, and pre-empt such urban problems. Similarly, application of the '*principle of social responsiveness*' in urban management is imperative as urban problems have significant social and environmental dimension. Even though inadequate, huge amounts of *resources* (i.e. inputs) are committed by the urban professionals and spent by the society in the urban operations and development processes, to achieve selected *aims* (i.e. outputs). The society would expect that urban organizations and professionals apply the modern management '*principle of productivity*' to ensure *efficiency* and *effectiveness* in utilization of *resources* in their activities to attain the selected *aims*, particularly because the resources are *scarce quantities* compared to the total need. Although there are many other modern *management principles*, which are equally applicable in urban management, and can be learnt through a full-length course of education, applicability and usefulness of above *principles* in urban operations are explained below as illustrative examples only.

2.5.1.1 Principle of Productivity

Productivity, defined as the input-output ratio within a time period with due consideration for quality, is the *essence* of management in any sector, and implies *effectiveness* and *efficiency* in individual and organizational performance (Koontz & Weihrich, 1990; Koontz & O'Donnel, 1976). *Effectiveness* is the achievement of objective(s), while *efficiency* is the achievement of ends with the least amount of *resources*. This definition of *productivity* can be applied not only to the productivity of organizations, managers, professionals, workers, and so on, but also to the urban development projects undertaken by urban organizations. Although inadequate, huge amounts of *resources* (i.e. inputs such as money, land, materials, labour, and so on) are committed by the urban professionals and spent by the urban organization and society in the urban sector, including urban operations and development processes, to achieve the selected *aims* (i.e. outputs such as housing and urban services, profits, satisfaction, integration of the goals of various claimants and stakeholders to the urban organization, and so on). The society would expect that urban professionals engaged in the process transform the inputs into outputs, ensuring *productivity, efficiency,* and *effectiveness* of utilization of *resources* in their activities, particularly because the resources are *scarce quantities*. The '*principle of productivity*' can be applied in different urban operations, including urban development projects, and at different levels of an urban organization. For example, in the context of urban development projects undertaken by an urban organization, *productivity* can be improved by:

1. Increasing outputs (say, the number of built-units and the area of saleable built-space/plot-spaces produced) with the same inputs (say, capital investment, land, materials, and labour).

2. Decreasing inputs (say, capital investment, land, materials, and labour) while maintaining the same outputs (say, the number of built-units and the area of saleable built-space/plot-spaces produced),
3. Increasing outputs and decreasing inputs to change the above ratio favourably.

Since the *resources* are always inadequate compared to the need, a *Policy* commitment to the application of the '*principle of productivity*' in various urban operations is an imperative need, if the urban problems are to be solved within the *resources* at each system level, including that at the individual, 'spatial-unit', and national level, and serving the multiple stakeholders to ensure *equity*. There are many techniques in traditional management to improve productivity. For example, in traditional management **Operations Research**, **Management Science**, and **Computer-Aided Design** are employed in planning and managing operations to improve **Productivity** (Koontz & Weihrich, 1990).

Application of above techniques are equally applicable in urban management, particularly because in urban operations, urban professionals (e.g. engineers, architects, planners, and so on) tend to look for their respective professional sophistication rather than ways of producing the product or urban facility at a reasonable cost. Such divergent interests of functionally oriented professionals need to be integrated using appropriate management techniques. Besides, operations research, which is one of the modern techniques for planning, controlling, and improving productivity, there are other techniques such as value engineering and computer-aided design which are employed in traditional management to improve productivity, and equally applicable in urban management. Through use of such techniques, the management planning and control have been given more rigorous treatment by clear-cut goals, development of measures of effectiveness in achieving goals and by emphasizing the logical physical representation of a problem in the form of mathematical models, enhancing the capability of managers to develop quantified answers to many problems (Koontz & Weihrich, 1990; Koontz & O'Donnel, 1976). However, application of the '*principle of productivity*' in various urban operations is feasible only if a *Policy* exists to develop and use such techniques in urban planning and management, and thus, removing the limitations of conventional practice discussed above.

2.5.1.2   Principle of Social Responsiveness

Urban problems have significant social and environmental dimension requiring application of the '*principle of social responsiveness*' in urban management. Social responsiveness means 'the ability of a corporation to relate its operations and policies to the social environment in ways that are mutually beneficial to the company and to society'. This is receiving increased attention even in conventional business operations, due to the *interdependencies* of the many groups in our society (Koontz & Weihrich, 1990). *Such interdependencies are more crucial in respect of urban organizations, as urban development is a joint public-private enterprise where decisions about the use of resources are made by the market place and by the governments (at various levels) in an interactive way*. Therefore, there is need for a *Policy* to apply the '*principle of social responsiveness*' in the operations of urban organizations, to indicate that they are discharging their social responsibilities. The concept of '*social audit*' defined as 'a commitment to systematic assessment of and reporting on some meaningful, definable domain of the company's activities that have social impact', is also applied in traditional management (Koontz & Weihrich, 1990). Social responsiveness in urban operations implies actions and urban enterprise responses to achieve the social goals in the urban sector, and to solve urban problems. Therefore, to help evaluate social performance of urban organization, the concept of '*social audit*' should also be applied in their operations.

A *Policy* commitment to the application of the principles of '*social responsiveness and social audit*' in urban operations, may discourage additional government regulations and interventions, giving greater freedom and *flexibility* in decision-making, bettering *efficiency* of urban organizations and achieving organized cooperation between such organizations, improving performance of the urban sector as a whole. Urban research is necessary to develop computer-aided techniques (including computer-aided planning and computer-aided design techniques), to facilitate application of such principles in various urban operations, and help develop more effective urban policies based on systematic assessment, evaluation, and analysis of such urban policies (including *social-cost-benefit* analysis), adopting a *systems approach*.

2.5.1.3 Principles of Flexibility and Navigational Change

As cited above, it is desirable that urban planning should accomplish a *physical planning structure* matching with the *economic-base-structure* (creating work) and the consequent *demographic-distribution-structure* (which should be accepted as a given condition at appropriate decision level, and should cover the *multiple stakeholders* including the *urban poor*) in a city, to achieve *Efficiency* and *Equity*, and to create a balanced community. However, *urban dynamics*, uncertainties (difficult to foresee and plan), and conflicting interests and viewpoints of *multiple stakeholders* make it difficult to achieve such a match, creating many urban problems such as proliferation of slums and frequent illegal construction of habitats and land-use change (mostly by informal entrepreneur's creating workplaces and jobs) in violation of conventional rigid statutory city-master-plans and *planning regulations* in many countries, subjected to rapid urbanization, demographic change, soaring land prices and construction costs, and acute affordability and resource-constraints. For example, in India, the per capita land resource is one of the lowest in the world, and dwindling fast from 1.42 in 1901 to only 0.32 hectare in 2000, but city-master-plans generally continue to prescribe a density range of about 60–600 persons per hectare, while density in some city-slums is as high as 5000–10,000 persons per hectare, which is highly **iniquitous**. In conventional practice, stress is given on periodic *mass-scale* dismantling and demolition of such informal industries, workplaces, and habitats, resulting to loss of jobs and shelters of millions of people, and also to a huge wastage of individual and national resources in a country like India, which has nearly one third of its population below the poverty line with gigantic unemployment problems. It is imperative to modify the conventional practice using modern management principles of flexibility and navigational change, to avoid such urban planning and management failures, adversely affecting the interests of people, and often creating an urban crisis.

The 'principles of flexibility and navigational change' providing for reviewing plans from time to time and redrawing them if this is required by the changed events and expectations is an accepted planning practice in traditional management to cope with uncertainties (Koontz & Weihrich, 1990). It is also suggested that because of too numerous pay-off matrices and uncertainties in the urbanization and development process, an incremental planning approach with 'flexibility principles' should be adopted for guiding the development of cities, providing for constant feedback about the response of the urban system, and for changing again and again the rules (such as zoning, augmenting or new provision of infrastructure, and so on), based on such response so that things do not work out in an undesirable way (Smith, 1975). Continuous planning with flexibility and short range forecast for land use in selected areas is also suggested to cope with the size and complexity of growth of cities, which are becoming more dynamic (Branch, 1970). Moreover, there can be *large variations* in the *efficiency indicators* of urban-built-form (depending on the rules such as *planning regulations* and the *dynamics* of input-prices). *User Citizens, Builders,* other *Stakeholders* in general would not like to ignore such *large variations*, and would be keen to maximize '*productivity*' (input-output ratio) and see that each of their money produces maximum benefit along with the desired environmental standards. Therefore, it is necessary to convince *User Citizens*, who are important stakeholders, invoking their *participation* in the planning and design process, using *Computer-Aided Planning and Design Techniques* (presented in later chapters) and information technology for *transparency*, that the *Planning Regulations* do not require any change to attain their above aims, and thus, minimize their violations. The above urban problems highlight the need to apply such *dynamic* approaches (modern IT tools make it very easy to apply such approaches) in a *dynamic* city instead of frequent *static* master planning approach discussed above with case studies.

Availability of even PCs having tremendous computing power and graphic capability, and rapid development in the information technology (including GIS and Internet), makes it easier now to apply such '*flexibility principles*' for timely corrective actions, with *transparency* and *informed-participation of multiple stakeholders*, and thus, help develop *effective response* to the *urban dynamics/*uncertainties. Such an approach will pre-empt above urban problems at the *nascent-stage* before they get out of control, achieving *efficiency* and *equity in the context of urban dynamics/uncertainties*, fulfilling *appropriate environmental requirements in the urbanization and economic development context*. However, this will require a *Policy*, particularly in developing countries subjected to rapid economic and demographic change, to review the conventional practice to treat the City Master Plan as a *rigid* legal document, and instead, view it as a *live* city development plan *responsive* to *People's need*, and subject to '*navigational change*' (including change of land use and *planning regulations* if needed, thus, avoid *freezing* the evolving *options* for 20 long years, the usual Master Plan period, which further aggravates the problems) as part of '*integrated urban management*'. Urban research is necessary to develop various techniques to facilitate application of such management principles, in various urban operations. For example, use of modern management techniques such as **Management Science, Operations Research,** Information Technology, and Computer-Aided Planning and Design, as applied in the traditional management, is equally applicable in urban operations and is feasible only if urban research is carried out to develop such techniques with urban application orientation, and also a *policy* exists to apply such techniques in the urban planning and management process. The author is carrying out research to develop such techniques and some computer software, developed as urban planning, design and management tools, are presented in subsequent chapters of the book.

2.5.2   Modern Management Techniques in Urban Operations: Some Clarifying Examples

Techniques are ways of doing things to accomplish results, reflect theory, and are a means of helping managers undertaking activities most effectively (Koontz & Weihrich, 1990). Operational approach draws on knowledge from other fields like systems theory, decision theory, and so on, and from the application of mathematical analysis and concepts, for effective operations incorporating activities necessary to deliver a service as well as a product (Koontz & Weihrich, 1990, p. 17). To facilitate traditional management, and particularly operations management, there are many tools and techniques such as operations research, linear programming, time-event networks, value engineering, **Computer-Aided Design**, and so on, which are employed in planning and managing operations to improve **Productivity** (Koontz & O'Donnel, 1976, p. 206; Koontz & Weihrich, 1990).

In production operations, engineers often look for engineering sophistication rather than ways of producing a product at a reasonable cost (Koontz & Weihrich, 1990, p. 443). Similarly, in urban operations concerned with provision of urban facilities and services, urban professionals (e.g. engineers, architects, planners, and so on) tend to look for their respective professional sophistication rather than ways of producing and delivering urban facilities and services at a reasonable cost. Such divergent interests of functionally oriented professionals need to be integrated using above tools, techniques, and other appropriate management techniques for efficient and effective operations.

Through use of above techniques, the management planning and control have been given more rigorous treatment by clear-cut goals, development of measures of effectiveness in achieving goals and by emphasizing the logical physical representation of a problem in the form of mathematical models, enhancing the capability of managers to develop quantified answers to many problems (Koontz & Weihrich,

1990; Koontz & O'Donnel, 1976). Computer-aided analysis and information technology also facilitate application of the *flexibility principles* in urban operations. Some of these modern management techniques, which are very relevant to urban operations, are outlined below just as illustrative examples.

2.5.2.1 Management Science

Management science sees managing as mathematical processes, concepts, symbols, and models; looking at management as a logical process, expressed in mathematical symbols and relationships. Management science is focused on the problems of developing and converting management theory to practice without losing sight of behavioural and economic realities, reflecting mutuality of interest of managers and management scientists in the total exercise of the management functions (management science-v-16). This is a very useful tool to make management as a logical process to attain desired objectives.

2.5.2.2 Operations Research

Most acceptable definition of operations research, which can be called 'quantitative common sense', is that '**operations research** is the application of scientific methods to the study of alternatives in a problem situation, with a view to obtaining a quantitative basis for arriving at the best solution'. Operations research is sometimes also called '**Management Science**', making these two terms interchangeable. In operations research, the emphasis is on scientific method, on use of quantitative data, on goals, and on the identification of the best means to reach the goals. All these tools are very relevant and useful in urban operations to *efficiently accomplish the selected aims* in such operations. Elements of operations research are elaborated in more details in next chapter.

2.5.2.3 Cost Engineering

Cost is paramount in urban operations and in the supply of urban services. It is a major consideration in all human activities, and as the technology and society advances it becomes necessary to estimate costs more closely and accurately to remain competitive. Broadly cost engineering is that area where engineering judgement and experience is utilized in the application of scientific principles and techniques to problems of cost estimating, cost control, profitability analysis, planning, and scheduling (Jelen, p. 4). In the urban management context, in addition to profitability analysis, cost-affordability analysis for the intended target groups for supply of various urban services is very important, to attain equitable and socially responsive operations of urban organizations. Application and utility of cost engineering in urban operations are shown in more details in Chap. 4.

2.5.2.4 Information Technology

As mentioned in Chap. 1, Information Technology (IT) is the modern name for the means by which we collect, store, process, and use information. Communication is essential in all phases of managerial process for two reasons, namely for integration of managerial functions, and for linking the organization with its environment. It is communication that makes managing possible. It may be noted that claimants, customers, and user citizens, who is the reason for the existence of virtually all organizations, are outside the organization. Hence, it is only through the communication system that the needs, priorities, and constraints of customers and user citizens can be identified to enable organizations to provide the goods and services to them accordingly, serving the multiple stakeholders and claimants to the organization. In recent years, there is a revolution in the capacity to capture, manipulate, and store data, and to disseminate and communicate information, affecting the way we live and work in the growing 'information society', making IT a powerful planning and management tool (Piercy, 1984). This is equally applicable in urban operations and management. CAD is an important component of IT.

2.5.2.5 Computer-Aided Design

In Chap. 1, some unique capabilities and advantages of CAD along with a broad list of benefits of CAD compared to conventional design methods are presented, highlighting the special capability of CAD to apply optimization, which is a crucial need in all AEC operations consuming *Resources*, especially because such operations involve many *variables* giving large variations in *resource-efficiency-indicators* depending on the *planning-design-decisions*, creating large scope for *Resource-Optimization*. Utility of CAD as a management technique to improve productivity, equity, and social responsiveness in AEC operations is also stressed in Chap. 1.

As mentioned above, CAD is an important component of IT and can be a very good urban planning and management tool to attain *Efficiency*, *Effectiveness*, *Equity*, and *Optimization*, resolving the conflicting interests, viewpoints, and demands of *multiple stakeholders*, using an integrated systems approach in problem-solving, in the context of urban dynamics and uncertainties, and the affordability and *resource-constraints* at various system levels. But, urban application of CAD to derive above benefits including *Optimization*, and to increase public awareness and to secure and invoke effective *citizen participation* in informed decision-making in the urbanization and development process, is very rare, creating many *inefficiencies* and *inequities* and consequent urban problems as cited earlier. This needs correction by using CAD as an urban management tool in the urbanization and development process. This book attempts to help create such capacity for AEC professionals engaged in such urban planning, design, and management process.

2.6   Urban Management Education

As earlier pointed out, an urban area is a highly complex dynamic system of interacting activities and interrelated components involving economic growth, change, and uncertainties requiring application of various management principles, techniques, and tools to manage effectively the interacting activities and interrelated components, recognizing various technological, economic and social issues, difficulties, and opportunities to promote a resource-efficient urban operations and development so as to provide efficiently urban services deemed essential for civilized survival sustaining productive activities. Inadequate attention and lack of holistic managerial response to the above interacting factors often lead to many urban problems in terms of inadequate access to shelter and urban services, proliferation of slums and squatter settlements, inadequate urban infrastructure and transport systems and their high costs, congestion, urban decay, pollution, and various other environmental degradations which are visible in many countries. As urbanization benefits individuals, communities, businesses, and generates economic wealth of the nation as a whole, going against urbanization amounts to going against the tide of national development and the economic, social, and political forces that stimulate it. Therefore, efficient urban management is the only way by which urban problems could be minimized while maximizing the beneficial impacts of urbanization, enhancing the role of cities as 'engines for economic growth', and improving the *quality of life for all*. A resource-efficient and equitable solution of urban problems, and cost-effective urban operations depend very much on the urban management expertise of Governments, Urban organizations (both in the public and private sector), and urban professionals who play a key role in this process. Even for not-for-profit organizations with the goal of public service, management expertise is necessary to accomplish them with minimum of resources or to accomplish as much as possible with a given available resources. However, in many countries such integrated urban management expertise is rare in above entities, becoming a factor to the worsening of urban environment by improperly managed urbanization and ineffective environmental management.

Removing the limitations of the conventional practice in deriving the beneficial impacts of urbanization, and thus, improve the *quality of life* for all, can be greatly facilitated by adopting an *integrated management approach* (covering all managerial functions as per the management theory) in urbanization and development operations to *efficiently* accomplish the selected *aims* resolving the conflicting interests, viewpoints, and demands of *multiple stakeholders* to achieve *equity*, keeping in view the *urban dynamics and uncertainties*; need and demand of housing and urban services; cost of land and other *inputs*; and the environmental, affordability, and *resource-constraints*, as may be applicable in a given context. Such an approach will improve the capacity of our cities and urban organizations for efficient and integrated urban and environmental management to achieve high economic growth and a sustainable urbanization, development, and urban operations, applying modern management principles and techniques including flexibility principles and information technology, invoking increased public awareness and participation in this process. There are many other modern and traditional management principles and techniques, which are equally applicable in urban management and needs to be applied to derive benefits and obviate many urban problems.

However, in conventional practice, professional urban managers with capacity to get the things done taking a holistic problem-solving managerial view with *accountability* to *efficiently* accomplish the selected *aims* in the above context are rarely employed in urbanization and development operations. Again, professionals of each *base-discipline* (e.g. engineers, architects, planners, and so on) working in the urban sector, instead of adopting an integrated problem-solving managerial approach, tend to look at the urban problems only from their base-disciplinary angle, wanting respective professional excellence (i.e. architectural excellence, engineering excellence, and so on), rather than ways of producing the product or urban facility at a reasonable cost, which may contradict the problem-solving managerial approach taking into account the affordability and other relevant criteria. Such divergent interests of functionally oriented professionals also need to be integrated using appropriate management techniques incorporated in urban management education. The above situation is mainly because qualified professional urban managers are rarely available in many countries. This makes it imperative to introduce a full-length and comprehensive urban management education to produce professional urban managers, and thus, create such urban management capacity in the urban sector of various countries.

2.6.1   Urban Management Education as a Distinct Discipline: A Course Design

Education in urban management as a distinct discipline should provide the required knowledge and skills to the students for efficient and effective discharge of all urban management functions such as *planning, organizing, staffing, leading/directing,* and *controlling,* at various system levels in the urban operations and development process, subject to the given constraints at that level, and recognizing various technological, economic, and social dynamics to promote a *resource-efficient* urban growth and urban operations. It should initiate students to *logical thinking* using quantitative data and *management science* techniques, and assist students as future urban managers to improve urban management *decision-making,* and also to manage the urbanization and urban services delivery process, to achieve the desired outcomes, taking into account uncertainties and the dynamics of urban processes, using information technology and other management techniques. The urban management education course should highlight the concept of urban management *to get things done,* i.e. achieve objectives and solve problems with the available resources within the control of an entity, looking internally rather than externally particularly for resources, and subject to the external constraints beyond its control although indirectly   influenced   by   various   policy   and   strategic   interventions   and non-interventions. It should emphasize the concept of development management and local economic development for financial resource generation, stressing the role of *cities as engines of economic growth*, as well as a tool for improving human development and social systems performance, adopting various social and human development indicators, with focus on human settlement as a continuum, duly related to the concerned decision-making entity to achieve desired outcome with *accountability*. The urban management education should cover a number of foundation courses, functional courses, and management science techniques to prepare urban managers to function as coordinating and integrating members of the professional teams dealing with urban operations, including development and use of urban space and managing the urbanization process, at the local, city, state, and regional levels, to achieve *resource-efficient* and *equitable* urban operations, development, and supply of urban services, so as to improve the *quality of life* for all.

Towards this effort, a holistic course design for urban management education ('MBA-Urban Services' with 133 credits) as a *distinct discipline*, presently non-existent in most countries, has been prepared with productivity, social responsiveness and systems approach as the core, giving stress on management theory (Chakrabarty, 1997a, b, 1998, 2001). In this course design, 'Quantitative Methods, Systems Analysis, Operations Research and Optimization, Information Technology, and Computers' form one of the five Core Course Groups (as per national technical education norm in India) where Computer-Aided Design (CAD) is included as a subject of study. But, this subject of study of Computer-Aided Design (CAD) is rarely available in the AEC sector in most countries in the world, which is a *serious gap* in education of AEC professionals. The present book is designed as a textbook for this subject of study of Computer-Aided Design (CAD) relevant to AEC operations and is an attempt to help remove the above *serious gap* in education of AEC professionals.

Proposed full-length Master's Degree Curricula for **Urban Management Education** (133 credit points) as a **Distinct Discipline** are shown in Tables 2.1 and 2.2. More than 50 subject areas are incorporated in the course design. These can form the Urban Management Body of Knowledge (UMBOK) and can be modified and enlarged depending on a specific situation, and if required, research and development activities may be carried out to augment UMBOK to meet individual needs, as urban management is a new area of knowledge.

It is paradoxical that while a course on *urban planning*—which is one of the five *urban managerial functions* cited above—exists; no full-length urban management course as a distinct *discipline*, covering all the five essential *urban managerial functions*, exists (in most countries), to create urban management capacity to ensure that events conform to such plans, which are not self-achieving and may remain only as a planning study, or a proposal, but no real plan to achieve desired *objectives* and *goals* (as is the usual case in conventional practice, with consequent creation of many urban problems as discussed above), in the absence of above capacity creation, linked with commitment of resources and with other four managerial functions of *organizing*, *staffing*, *leading*, and *controlling*, as part of *Integrated Urban*

Table 2.1 Proposed master's degree curricula for urban management education as a distinct discipline (MBA-urban services)

Part-A: Core courses for all specialization groups			
Core courses	Suggested credit	Core courses	Suggested credit
Area-I: Foundation		**Area-IV: Marketing management**	
1. Urban Management Concepts, Principles, and Techniques	1.5	16. Marketing Management	3
2. Country-Specific Social Environment	3	17. Strategic Management	1.5
3. Micro and Macro Economics	3	**Area-V: Urban operations and development management**	
4. Political Economy of Urbanization	1.5	18. Urban Development Policy and Planning	1.5
5. Urban Economics	1.5	19. Urban Infrastructure Management	1.5
6. Urban Land Management	1.5	20. Urban Transport Management	1.5
7. Urban Legal Environment	3	21. Urban Social Services	1.5
8. Urban Poverty Alleviation	1.5	22. Urban Housing Development Management	1.5
Area-II: Organization and management		23. Urban Environmental Management	3
9. Behaviours in Organization	3	24. Project Management	1.5
10. Organization and Management **or** Organizational Analysis and Design	3	25. Urban Development Management	3
11. Organization and Social Change	3	**Area-VI: Quantitative methods, systems analysis, operations research and optimization, information technology, and computers**	
12. Personnel Management	3		
13. Managerial Analysis and Communications	3	26. Quantitative Methods, OR and Mathematical Optimization Techniques, Information Technology	6
Area-III: Financial management		27. Urban Systems Analysis	3
14. Financial Management	6	28. Management Information System and Decision Support	3
15. Management Accounting	3	29. Urban Research Methods **or** **Computer-Aided Design by Optimization**	3
Total credit for core courses for all groups = 75			

Management to get the things done, so as to attain with *accountability* the missions and *objectives* incorporated in the urban plan, achieving both *Equity* and *Efficiency*.

It is desirable that Universities, Professional Education, and Training Institutions start such a field of study to produce professional urban managers, and also Urban Organizations (UOs) use such urban managers to enable them to adopt an *Integrated Urban Management Approach*, applying various management principles and

Table 2.2 Proposed master's degree curricula for urban management education as a distinct discipline (MBA-urban services)

Part-B: Separate courses for each specialization group			
Courses	Suggested credit	Courses	Suggested credit
(I) Local government management group		**(III) Infrastructure and urban development management group**	
1. Local Government Management and Urban Municipal Finance	3	1. Infrastructure Planning and Economic Development	3
2. Government Business Interface, or, Development Administration, or Marketing for Local Urban Services	3	2. Infrastructure Finance	3
3. Public Utilities and Services	3	3. Public Management	3
4. Urban Operations Research	3	4. Geographic Information Systems or Marketing for Infrastructure Services	3
5. Infrastructure Planning and Economic Development	3	5. Urban Dynamics and Urban Modeling	3
6. Urban Dynamics and Urban Modeling or Public Management	3	6. Systems Analysis and Design Economics or Development Administration	3
Sub-Total Credit for the Courses of this Specialization Group	18	**Sub-Total Credit for the Courses of this Specialization Group**	18
(II) Housing development management group		**(IV) Real estate and urban business management group**	
1. Housing Development Management	3	1. Management of Private Organization	3
2. Cost-Effective Building Materials and Construction Techniques	3	2. Economics of the Firms	3
3. Housing Finance	3	3. Business Policy	3
4. Systems Analysis and Design Economics	3	4. Construction Management or Systems Analysis and Design Economics	3
5. Infrastructure Planning and Economic Development or Marketing for Housing and Estate Management	3	5. Real Estate Finance and Marketing	3
6. Urban Dynamics and Urban Modeling or Public Management	3	6. Urban Business Management	3
Sub-Total Credit for the Courses of this Specialization Group	18	**Sub-Total Credit for the Courses of this Specialization Group**	18
Part-C: Credit for the thesis and field works for all groups = 40			
Total credits for entire urban management course for all groups = 133			

techniques (IT, OR, *Optimization*, Integrated CAD, and so on) in their urban operations. This is central for improved performance of our UOs as effective instruments for *Resource-Efficient* and *Equitable* solution of our mounting urban problems, in the context of *urban dynamics* and the individual, local, and national

Resource-Constraints, and the conflicting interests, viewpoints, and demands of multiple stakeholders in this process. It would help convert cities as *efficient engines for economic growth with environmental sustainability,* maximizing beneficial impacts of urbanization and, improving *resource-efficiently* the *quality of life* for all.

2.6.2   Elements of Urban Management as a Subject of Study for AEC Professionals

As earlier stated, professionals such as Engineers, Architects, Planners, and so on (i.e. AEC professionals) tend to look for their respective professional sophistication rather than looking for ways of producing an urban facility at a reasonable cost covering all its dimensions to serve the multiple stakeholders. Such divergent interests of functionally oriented urban professionals need to be integrated incorporating technical, financial, and managerial aspects by using appropriate management techniques for better operational results. In fact, professionals such as architects, engineers, and planners, working in the urban operations and development process can make significant improvement in the urban environment by combining their base-disciplinary knowledge with application of modern management principles and techniques in specific urban problem situation, thus, help attain a *resource-efficient* and *equitable* urban operations and development process to solve urban problems. It is also necessary that such professionals belonging to various base-disciplines should be able to comprehend the above 'big picture' of urban operations and development process while applying their base-disciplinary knowledge in the role of urban professionals to solve urban problems, and thus, become more effective. In the above context, to give a managerial orientation to all professionals of such *base-disciplines* who are likely to deal with urban operations, it is desirable to include 'Elements of Urban Management' as a **Subject of Study**, covering only the basic elements of management related to urban operations, in the curricula of such base-disciplines, i.e. AEC professionals. There are many books such as '**Essentials of Management**' by Harold Koontz and Heinz Weihrich, which give introduction to managing embracing principles, concepts, and theories. Such books can be used as textbooks for this subject of study by suitable addition and modification by the course instructor to give an urban management orientation.

2.6.3   Computer-Aided Design as a Subject of Study for AEC Professionals

As already mentioned, planning and design in AEC operations, consuming huge *Resources* and affecting the *quality of life* of people, needs to be viewed as a *problem-solving process* involving *search* through *numerous* potential solutions (because of *numerous* design variables involved) to select a particular solution that meets a specified criteria or *goal* (say, *Resource-Efficient* and affordable shelter design). This makes computer use indispensable for planning and design analysis and *Optimization* for problem solution in all branches of AEC including architecture, planning, civil engineering, and construction, particularly in the context of *search* through *numerous* potential solutions with large variations in *resource-efficiency* indicators. This view for planning and design in Architecture, Planning, Civil Engineering, and Urban Development is quite relevant especially in developing countries to achieve the national *goals* of providing adequate shelter and urban services for all, within the affordability and *Resource-Constraints* at the individual, local, and national level. With increasing population and dwindling per capita *Resources*, this view is equally applicable even in the developed countries, as *we live in a Finite Earth with Finite Resources*. In this context, *Optimization* in AEC sector, i.e. building, housing and urban development is a crucial need, and it is obligatory to apply *Optimization* in all such operations consuming *Resources*, especially because such housing and urban operations involve many *variables* giving large variations in *resource-efficiency-indicators* depending on the *planning-design-decisions*, creating large scope for *Resource-Optimization*. Application of CAD in such operations is imperative, in view of special capability of CAD to apply optimization as outlined in Chap. 1. Moreover, as urban professionals <u>*Commit*</u> huge *Resources* in their *planning-design-decisions*, it is <u>*Basic*</u> that they use Operations Research (OR) and optimization (*Basic Productivity Tools*) to discharge their <u>*Accountability*</u> to the society to achieve *Productivity* and *Equity* in the use of *Resources* so committed by them in their plans and designs to achieve *resource-efficient*, *equitable*, and affordable supply of building and urban services for all sections of the people.

As AEC and urban operations involve numerous variables, pay-off matrices and uncertainties in the context of urban dynamics and resource-constraints, large variations in efficiency indicators, and multiple stakeholders with conflicting interests and viewpoints, it is difficult to achieve efficiency and equity and solve urban problems by the conventional methods as illustrated above. As discussed above, management is concerned with efficiently accomplishing the selected aims, and '*Planning*' (which includes Design) is one of the five *Management Functions*. To overcome the above limitations of conventional practices, an *Integrated Urban Management Approach* is crucial, and the Civil Engineering, Building, and Urban 'Planning and Design' Function should be seen as its Part (i.e. <u>Planning Function</u>), for a resource-efficient and equitable solution of our mounting urban problems in the context of urban dynamics and uncertainties (Chakrabarty, 1998, 2001). Towards this effort, a holistic course design for urban management, i.e. '<u>MBA-Urban Services</u>' is prepared and presented above, where a concept of *Urban Management* and some basic management principles, which are very relevant in the urban operations, are also presented.

This is done, before illustrating various elements of integrated CAD and optimization, to enable students to comprehend the above 'big picture' during learning and also while applying integrated CAD and optimization as urban management tools in the role of urban professionals, and thus, become more effective. In the course design of urban management education of 133 credits presented in Tables 2.1 and 2.2, Computer-Aided Design (CAD) forms a full-length *subject of study* for master's degree urban management students from AEC base-disciplines, as part of the Core Course Group VI: 'Quantitative Methods, Systems Analysis, Operations Research and Optimization, Information Technology, and Computers', which is one of five Core Course Groups (as per national technical education norm in India) of this urban management education program (students who had CAD as *subjects of study* in their base-discipline course, may take the alternative subject in this course). Computer-Aided Design (CAD), along with 'Elements of Urban Management' as suggested above, should also form full-length *subjects of study* in the curricula of AEC professional courses, i.e. in the undergraduate courses of *base-disciplines* such as Architecture, Planning, Civil Engineering, and so on. This will give a holistic understanding to the students while learning and applying Computer-*Aided Design* (CAD) as a problem-solving urban planning and management tool to derive the beneficial impact of urbanization, and to help attain a *Resource-Efficient* and *Equitable* solution of our mounting housing and urban problems so as to improve the *quality of life* for all.

Proposed curricula for **Computer-Aided Planning and Design by Optimization ('CAPDO')** as a **Subject of Study** for AEC Professionals are shown in Table 2.3. Hopefully, this book could be a textbook for this subject of study.

2.7 Equipping Urban Professionals as Accountable Managers for Resource-Efficient and Equitable Problem Solution in the Urban Sector

As discussed earlier, urban management can be viewed as an applied realm of knowledge with its own theoretical framework, and also drawing knowledge from the *theories* and *principles* in management and urban operation related *base-disciplines*, to formulate the rules and procedures in the context of urban operations to *efficiently* accomplish the selected *aims* and to improve the performance of urban professionals (who may belong to various *base-disciplines*), organizations, and the urban sector as a whole. In this context, all urban professionals such as engineers, architects, planners act not only as technical experts in their respective *base-disciplines* and fields of specialization but also as accountable managers to achieve the goal of *resource-efficient* and *equitable* problem solution in the urban sector where they are working. Thus, improving *productivity*, which is one of the basic principles in management, is equally applicable and vital in the area of operation of each *base-discipline* to attain the above goal. This is particularly because urban professionals *Commit* huge *Resources* in their *planning-design-decisions*, and hence, it is *Basic* that they apply the principle of *Productivity* and *Equity* in the use of *Resources* so committed by them in their plans and designs using *Basic Productivity Tools* discussed above to discharge their *Accountability* to the society to achieve

Table 2.3 Proposed curricula for 'computer-aided planning and design by optimization' ('CAPDO') as a subject of study for AEC professional courses

A Suggestive Course-Content
An Overview of Computer-Aided Planning and Design
CAD and Traditional Design
Computerizing the Planning & Design Process
Commercial CAD Systems
Components of Integrated CAD System
Numerical Techniques for CAD
Elements of Computer Graphics & Architectural/Engineering Drafting
Elements of Cost Engineering
Planning and Design Databases
Elements of *Operations Research* (OR) & *Optimization* Techniques
Planning/Design *Analysis* and *Optimization*
Towards a Theory of *Optimal* Urban Built-Form
Software for Analysis and *Optimization* of Building/Development Control—
Role of Information Technology (IT) for *Citizen Participation*
Managing the *Dynamics* of Urbanization—Applying the
Urban Management *Principles of Flexibility* with *Participatory* & Collaborative IT Tools
Models and Computer-Aided Layout Design by *Optimization*
Models and Computer-Aided Building-Housing Design by *Optimization*
Models and Computer-Aided Dwelling-Layout System Design by *Optimization*
Models and Computer-Aided Structural Design by *Optimization*—Reinforced Concrete Sections
Computer-Aided Infrastructure Design by *Optimization*
Integrated CADD by *Optimization* from AutoLISP
Interfacing Design *Analysis,* Design *Optimization* & Graphic Design using C++, AutoCAD & AutoLISP
Automating *Descriptive/Optimal* Design Drawings, Costing & Bill of Materials
Integrative, Multidisciplinary & ***Participatory Optimal Planning & Design*** with *User Interface* using Collaborative **IT Tools**

resource-efficient, *equitable*, and affordable supply of building and urban services for all sections of the people. The '*principle of flexibility and navigational change*', providing for reviewing plans from time to time and redrawing them if this is required by the changed events and expectations, is an accepted practice in traditional management (Koontz & Weihrich, 1990). This is equally applicable in urban planning and management to avoid *inefficiencies* and *inequities* and the consequent urban problems (e.g. proliferation of slums), often because of adopting an *inflexible* and *static* plan in a *dynamic* city. Urban organizations (**UOs**) and urban professionals should not wait for the urban problems to develop before preparing to face them, and instead *Pro-action* and *Flexibility* need be an essential part of the urban planning and management for *effective response* to the *urban dynamics* and *uncertainties* in the development process. Therefore, urban professionals even if belong to a *base-discipline* should be conversant with such basic management principles if they are to be more effective in urban problem solution adopting a *Pro-active* and *Flexible* approach as an accountable manager.

The '*principle of social responsiveness*' in terms of 'the ability of a corporation to relate its operations and policies to the social environment in ways that are mutually beneficial to the company and to society', is receiving increased attention even in conventional business operations, due to the interdependencies of the many groups in our society (Koontz & Weihrich, 1990). *Such interdependencies are more crucial in respect of* **UOs**, *as urban development is a joint public-private enterprise where decisions about the use of resources are made by the market place and by the governments (at various levels) in an interactive way.* Therefore, such *principles* should be applied in the operations of **UOs**, to indicate that they are discharging their *social responsibilities*, in terms of actions and enterprise responses to achieve the *social goals* in the urban sector. *Social Audit* defined as 'a commitment to systematic assessment of and reporting on some meaningful, definable domain of the company's activities that have social impact' (Koontz & Weihrich, 1990), should also be applied to evaluate the social performance of **UO's/PLAs** (i.e. Planning Authorities at the City Level devising and managing city planning regulations), particularly in terms of *productive* and *equitable* use of scarce *resources*, to serve the *multiple stakeholders* for solution of urban problems.

As discussed above, there are many management tools and techniques, which are employed in planning and managing urban operations to improve ***Productivity***. Supply of housing and urban services with *efficiency* and *equity* is not only affected by the *planning regulations* as discussed above, but also, by the architectural and engineering designs. But, it is difficult to consider the combined-effect of all such decisions by the conventional approaches, as urban professionals (e.g. Engineers, Architects, Planners and so on) tend to look at the urban problems only from their base-disciplinary angle, and for their respective professional sophistication rather than ways of producing the product or urban facility at a reasonable cost serving the multiple *stakeholders*, taking an integrated problem-solving managerial approach. In this process, integrated computer-aided design (including *Optimization*) can be a very important modern planning, design, and management technique *to efficiently accomplish the selected aims* in urban development and management operations, and thus, help urban professionals to discharge their role not only as technical experts in their respective *base-disciplines* but also as accountable managers to achieve the goal of *resource-efficient* and *equitable* problem solution in the urban sector as a whole where they are working.

Many of the above techniques, including the computer-aided analysis and opti-mizing urban development models, can be used in urban planning and management for more effectiveness and help decide appropriate urban policies and strategies and manage the *dynamic* and *incremental* process of urban development with *flexibility*, giving stress on the *productivity*, *resource-efficiency* and *people's welfare* (Chakrabarty, 1998). However, this is feasible only if an *Urban Policy* initiative is taken by the *urban professionals* and *policy makers* to (1) carry out urban research to develop different urban operations oriented management techniques for more *equitious* and *productive* urban planning and management, covering various urban issues, (2) modify the conventional practice, by incorporating the use of such urban planning and management techniques, and (3) make all *urban professionals* committing *resources*, conversant with the use of such techniques (by making these techniques *part of the curriculum of urban planning and management education*), and thus, equipping them as accountable managers for resource-efficient and equitable problem solution in the urban sector. This would help remove the limitations of the conventional practice discussed earlier, and facilitating more effective discharge of their *ethical responsibility* and *accountability* in achieving *efficiency* and *equity* in various urban operations.

In view of above, it is desirable to impart the knowledge of modern management concepts, principles, and techniques to the urban professionals belonging to various *base-disciplines* to equip them as accountable managers for resource-efficient and equitable problem solution in the urban sector. This can be at two levels. In the first level, curricula of *base-disciplines* (e.g. Civil/Building Engineering, Architecture, Planning) of all AEC professional courses may include: (1) 'Elements of Urban Management' as a subject of study, covering only the basic elements of management related to urban operations, and also (2) 'Computer-Aided Planning and Design by Optimization' as **Subjects of Study**, which will enable such professionals to take a holistic problem-solving managerial view with *accountability* to *efficiently* accomplish the selected *aims* resolving the conflicting interests, viewpoints, and demands of *multiple stakeholders* to attain *equity*. At the second level, AEC professionals may be imparted the intimate knowledge of modern management concepts, principles, and techniques, including modern systems management techniques, tools, and information technology through a full-length urban management education as a *Distinct Discipline*, and thus becoming professional urban managers with different specializations as per the course design of '**MBA-URBAN SERVICES**' presented above.

2.8   Closure

In this chapter, an outline of urbanization and its beneficial impacts is presented, and the need of *Integrated Urban Management Approach* to derive the beneficial impact of urbanization, and to help attain a *Resource-Efficient* and *Equitable* solution of the mounting housing and urban problems so as to improve the *quality of life* for all, is highlighted. The limitations of the conventional practice in attaining this goal and the utility of modern management principles, techniques, and tools including Computer-Aided Planning and Design as part of *Integrated Urban Management*, to overcome these limitations are illustrated with examples and case studies. A concept of *Urban Management* and some basic management principles and techniques, which are very relevant in the urban operations, are presented to give an understanding of basic elements of *Integrated Urban Management*. This is done, before illustrating various elements of integrated CAD and optimization, to enable students to comprehend the above 'big picture' during learning and also while applying integrated CAD and optimization as urban management tools in the role of urban professionals, and thus become more accountable and effective in solving urban problems and achieving the desired goals. This is essential as huge amounts of *Resources* are *committed* by urban professionals while taking *planning* and *design-decisions* at each *system level* to produce urban services, making them *Accountable to the Society* to achieve *Efficiency*, *Effectiveness*, and *Equity* in the use of *scarce Resources* so *committed* by them.

The imperative need to introduce a full-length and comprehensive urban management education to produce professional urban managers, and thus, create such urban management capacity in the urban sector of various countries, is stressed. To equip urban professionals belonging to various *base-disciplines* as accountable managers for resource-efficient and equitable problem solution in the urban sector, a two-level education program to impart knowledge of modern management concepts, principles, and techniques is suggested. In the first level, inclusion of 'Elements of Urban Management' and 'Computer-Aided Planning and Design by Optimization' as subjects of study in the curricula of *base-disciplines* of all AEC professional courses is suggested. At the second level, introduction of a full-length urban management education as a *Distinct Discipline*, to produce professional urban managers with different specializations is suggested. To help this process, a course design of urban management education ('**MBA-URBAN SERVICES**') of 133 credits to produce professional urban managers is presented in Tables 2.1 and 2.2. In this course design, Computer-Aided Design (CAD) forms a full-length *subject of study* as part of the Core Course Group VI: 'Quantitative Methods, Systems Analysis, Operations Research and Optimization, Information Technology, and Computers', which is one of five core course groups (as per the national norm for such education prescribed by the All India Council of Technical Education, the National Apex Body for such Education in India) of this urban management education program based on management theory. A course design for '**Computer-Aided Planning and Design by Optimization**' as a **Subject of Study** for AEC Professionals is also presented in Table 2.3, and this book is designed to be a textbook for this subject of study.

The concept of *Urban Management* and some basic management principles and techniques presented in this chapter will give a holistic understanding to the students as future urban professionals to comprehend this 'big picture' of urban operations, while applying Computer-*Aided Design* (CAD) as a problem-solving urban planning and management tool to help derive the beneficial impact of urbanization, and to help attain a *Resource-Efficient* and *Equitable* solution of our mounting housing and urban problems so as to improve the *quality of life* for all. The remainder of the book will illustrate various elements of integrated CAD and optimization.

Chapter 3
Elements of Numerical Techniques and Operations Research for CAD

3.1 Introduction

Use of modern computer techniques including computer-aided design is very diffi-cult without skilful application of approximate and numerical analysis and tech-niques. Power of computers lies mainly on their incredible speed and memory capacity mainly to perform arithmetical functions, and not on their intrinsic intelli-gence or independent judgement. Computers perform complex mathematical oper-ations from simple trigonometric functions to solving complex differential equations, by converting all these complicated mathematical functions and equations to simple arithmetical operations that can be handled by the computers in the form of numerical techniques and error minimization in the solutions. In fact in computer-aided design by optimization frequently, we have to formulate and solve linear and nonlinear simultaneous equations. In this process, matrices are very useful technique as illustrated below. Numerical techniques and recipes are a vast subject. In this chapter, elements of only a few numerical techniques relevant to computer-aided design by optimization are presented below.

3.2 Matrices

A rectangular array of numbers enclosed by a pair of parentheses, ( ), or brackets, [ ], such as

$$\text{(a)} \begin{pmatrix} 2 & 5 & 1 \\ -3 & 7 & 2 \end{pmatrix} \quad \text{or} \quad \text{(b)} \begin{vmatrix} 1 & 0 & 6 \\ 3 & 4 & 8 \\ 5 & 6 & 9 \end{vmatrix}$$

© The Author(s), under exclusive license to Springer Nature Switzerland AG 2022
B. K. Chakrabarty, *Integrated CAD by Optimization*,
https://doi.org/10.1007/978-3-030-99306-1_3

and subject to given rules of operations is called a matrix. The matrix (a) can be considered as the **coefficient matrix** of the following system of homogeneous linear equations:

$$\begin{pmatrix} 2x + 5y + z = 0 \\ -3x + 7y + 2z = 0 \end{pmatrix}$$ or, the **augmented matrix** of the system of non-homogeneous linear equations such as : $$\begin{pmatrix} 2x + 5y = 1 \\ -3x + 7y = 2 \end{pmatrix}$$

The matrix can be used to obtain solution of these systems of linear equations as shown later. The matrix (b) could be given a similar interpretation or its rows may denote coordinates of points (1, 0, 6), (3, 4, 8), and (5, 6, 9) in ordinary space.

In the matrix

$$A = \begin{pmatrix} a_{11} \; a_{12} \; a_{13} \; a_{14} \dots\dots\dots\dots\dots\dots\dots a_{1n} \\ a_{21} \; a_{22} \; a_{23} \; a_{24} \dots\dots\dots\dots\dots\dots\dots a_{2n} \\ a_{31} \; a_{32} \; a_{33} \; a_{34} \dots\dots\dots\dots\dots\dots\dots a_{3n} \\ \dots\dots\dots\dots\dots\dots\dots\dots\dots\dots\dots\dots \\ \dots\dots\dots\dots\dots\dots\dots\dots\dots\dots\dots\dots \\ a_{n1} \; a_{n2} \; a_{n3} \; a_{n4} \dots\dots\dots\dots\dots\dots\dots a_{nn} \end{pmatrix}$$

the numbers or functions a_{ij} are called the **elements** of the matrix, where the first subscript indicates the row, and the second subscript indicates the column to which the element belongs. A matrix of m rows and n columns is designated to be of order 'm by n' or $m \times n$. When the order of the matrix has been established, it can be designated simply as 'the matrix A'. There are many types of matrices such as square matrix (as shown above where $m = n$), identity matrix, and so on (Ayres, 1974).

3.2.1   Determinant of a Matrix

The determinant of a matrix A denoted by D is the sum of all the different signed products called terms of D which can be formed from the elements of A, and the determinant of a square matrix of order n is called a determinant of order n. Two example formulas for evaluating determinant D of a matrix of order two and three are shown below in Expression (3.1):

$$\begin{pmatrix} a_{11} & a_{12} \\ a_{21} & a_{22} \end{pmatrix} D = a_{11} \times a_{22} - a_{12} \times a_{21}; \quad \begin{pmatrix} a_{11} & a_{12} & a_{13} \\ a_{21} & a_{22} & a_{23} \\ a_{31} & a_{32} & a_{23} \end{pmatrix} \quad (3.1)$$

$$D = a_{11} \begin{pmatrix} a_{22} & a_{23} \\ a_{32} & a_{33} \end{pmatrix} - a_{12} \begin{pmatrix} a_{21} & a_{23} \\ a_{31} & a_{33} \end{pmatrix} + a_{13} \begin{pmatrix} a_{21} & a_{22} \\ a_{31} & a_{32} \end{pmatrix}$$

If from n-square matrix A, the elements of its ith row and jth column are removed, the determinant of the remaining $(n-1)$-square matrix is called a **first minor** of A or D and is denoted by (M_{ij}) and it is also called the **minor** of a_{ij}. The signed minor, $(-1)^{i+j} (M_{ij})$ is called the **cofactor** of a_{ij} and is denoted as: α_{ij}. The value of the determinant D of the n-square matrix A is the sum of the products obtained by multiplying each element of a row (column) by its cofactor as shown below (Ayres):

$$D = a_{i1}\alpha_{i1} + a_{i2}\alpha_{i2} + a_{i3}\alpha_{i3} + \cdots + a_{in}\alpha_{in} = \sum_{k=1}^{n} aik\alpha ik \quad (3.2)$$

$$D = a_{1j}\alpha_{1j} + a_{2j}\alpha_{2j} + a_{3j}\alpha_{3j} + \cdots + a_{nj}\alpha_{nj} = \sum_{k=1}^{n} akj\alpha kj \quad (3.3)$$

Example 1 Using the Expression (3.1), evaluate the determinant of the matrix A below:

$$A = \begin{pmatrix} 2 & 3 & 2 \\ 1 & 2 & -3 \\ 3 & 4 & 1 \end{pmatrix} \quad D = 2\begin{pmatrix} 2 & -3 \\ 4 & 1 \end{pmatrix} - 3\begin{pmatrix} 1 & -3 \\ 3 & 1 \end{pmatrix} + 2\begin{pmatrix} 1 & 2 \\ 3 & 4 \end{pmatrix}$$

or $D = 2(2 + 12) - 3(1 + 9) + 2(4 - 6) = 28 - 30 - 4 = -6$

3.2.2   Rank of a Matrix

If we have a rectangular matrix as:

$$A = \begin{pmatrix} a_{11}a_{12} \ldots \ldots \ldots .a_{1n} \\ a_{21}a_{22} \ldots \ldots \ldots .a_{2n} \\ \ldots \ldots \ldots \ldots \ldots \ldots \\ a_{m1}a_{m2} \ldots \ldots \ldots .a_{mn} \end{pmatrix}$$

and if we isolate k arbitrary rows and k arbitrary columns ($k \leq m, k \leq n$) in this matrix A, a determinant of the kth order composed from the elements of the matrix A located at the intersection of the isolated rows and columns is known as a *minor* of the kth order of the matrix A. The *rank* of the matrix A is the greatest order of the nonzero minor of that matrix. If all the elements of the matrix are equal to zero, then the rank of the matrix is taken as zero. The rank of the matrix A is designated as: $r(A)$. If $r(A) = r(B)$, the matrices A and B are said to be equivalent. The rank of a matrix does not change as a result of elementary transformations.

3.3   Numerical Solution of Systems of Linear Simultaneous Equations

Let us have a system of n linear equations in n unknowns as shown below:

$$
\left.
\begin{aligned}
a_{11}x_1 + a_{12}x_2 + a_{13}x_3 + \ldots\ldots\ldots\ldots\ldots\ldots a_{1n}x_n = b_1 \\
a_{21}x_1 + a_{22}x_2 + a_{23}x_3 + \ldots\ldots\ldots\ldots\ldots\ldots a_{2n}x_n = b_2 \\
a_{31}x_1 + a_{32}x_2 + a_{33}x_3 + \ldots\ldots\ldots\ldots\ldots\ldots a_{3n}x_n = b_3 \\
\ldots\ldots\ldots\ldots\ldots\ldots\ldots\ldots\ldots\ldots\ldots\ldots\ldots\ldots\ldots\ldots \\
a_{n1}x_1 + a_{n2}x_2 + a_{n3}x_3 + \ldots\ldots\ldots\ldots\ldots\ldots a_{nn}x_n = b_n
\end{aligned}
\right\}
\qquad (3.4)
$$

The above system of linear equations can be solved using a matrix and can be expressed more compactly as: $Ax = b$, where

$$
A = \left(a_{ij}\right) = \begin{pmatrix}
a_{11}\, a_{12}\, a_{13}\, a_{14}\ldots\ldots\ldots\ldots\ldots\ldots\ldots\ldots\ldots a_{1n} \\
a_{21}\, a_{22}\, a_{23}\, a_{24}\ldots\ldots\ldots\ldots\ldots\ldots\ldots\ldots\ldots a_{2n} \\
a_{31}\, a_{32}\, a_{33}\, a_{34}\ldots\ldots\ldots\ldots\ldots\ldots\ldots\ldots\ldots a_{3n} \\
\ldots\ldots\ldots\ldots\ldots\ldots\ldots\ldots\ldots\ldots\ldots\ldots\ldots\ldots\ldots \\
\ldots\ldots\ldots\ldots\ldots\ldots\ldots\ldots\ldots\ldots\ldots\ldots\ldots\ldots\ldots \\
a_{n1}\, a_{n2}\, a_{n3}\, a_{n4}\ldots\ldots\ldots\ldots\ldots\ldots\ldots\ldots\ldots a_{nn}
\end{pmatrix}
$$

is the matrix of coefficients, and

$$
b = \begin{pmatrix} b_1 \\ b_2 \\ b_3 \\ . \\ b_n \end{pmatrix}
\qquad
x = \begin{pmatrix} x_1 \\ x_2 \\ x_3 \\ . \\ x_n \end{pmatrix}
$$

b is the column matrix of the constant terms, while x is the column matrix of the unknowns. If the matrix A is non-singular, i.e.

$$D = \det A = \begin{pmatrix} a_{11}\ a_{12}\ a_{13}\ a_{14} \dots\dots\dots\dots\dots\dots\dots\dots a_{1n} \\ a_{21}\ a_{22}\ a_{23}\ a_{24}\dots\dots\dots\dots\dots\dots\dots a_{2n} \\ \dots\dots\dots\dots\dots\dots\dots\dots\dots\dots\dots \\ a_{n1}\ a_{n2}\ a_{n3}\ a_{n4}\dots\dots\dots\dots\dots\dots\dots a_{nn} \end{pmatrix} \neq 0$$

then the system of equations (3.4) has the unique solution.

3.3.1 Solution of Systems of Linear Equations Using Cramer's Rule

The values of the unknowns x_i ($i = 1, 2, 3, \dots n$) in the system of equations (3.4) can be found using the Cramer's rule which is given by:

$$x_i = \det A_i / \det A,$$

where the matrix A_i is obtained from the matrix A by replacing its ith column by the column of the constant terms of the matrix A. Application of the above procedure for solution of linear equations using **Cramer's rule** is illustrated below with a numerical example.

Example 2 Solve the system of linear equations below:

$$\left.\begin{array}{l} 0.04x_1 - 0.08x_2 + 4x_3 = 20 \\ 4x_1 + 0.24x_2 - 0.08x_3 = 8 \\ 0.09x_1 + 3x_2 - 0.15x_3 = 9 \end{array}\right\} \quad \text{using Cramer's Rule.}$$

We find

$$A = (a_{ij})$$

$$= \begin{pmatrix} 0.04 & -0.08 & 4 \\ 4 & 0.24 & -0.08 \\ 0.09 & 3 & -0.15 \end{pmatrix}$$

and $\det A = 0.04((0.24(-0.15) - 3(-0.08)))$
$- (-0.08)(4(-0.15) - 0.09(-0.08))$
$+ 4\,(4 \times 3 - 0.09 \times 0.24) = 47.874336$

Similarly, we find

$$A_1 = \begin{pmatrix} 20 & -0.08 & 4 \\ 8 & 0.24 & -0.08 \\ 9 & 3 & -0.15 \end{pmatrix}$$

and $\det A_1 = 20((0.24(-0.15) - 3(-0.08)))$
$- (-0.08)(8(-0.15) - 9(-0.08))$
$+ 4(8 \times 3 - 9 \times 0.24) = 91.4016,$

and

$$A_2 = \begin{pmatrix} 0.04 & 20 & 4 \\ 4 & 8 & -0.08 \\ 0.09 & 9 & -0.15 \end{pmatrix}$$

and $\det A_2 = 0.04((8(-0.15) - 9(-0.08)))$
$\qquad\qquad - 20((4(-0.15) - 0.09(-0.08)))$
$\qquad\qquad + 4(4 \times 9 - 0.09 \times 8) = 152.9568,$

and

$$A_3 = \begin{pmatrix} 0.04 & -0.08 & 20 \\ 4 & 0.24 & 8 \\ 0.09 & 3 & 9 \end{pmatrix}$$

and $\det A_3 = 0.04\,((9 \times 0.24 - (3 \times 8))$
$\qquad\qquad - (-0.08)((4 \times 9 - 0.09 \times 8)))$
$\qquad\qquad + 20(4 \times 3 - 0.09 \times 0.24) = 241.5168$

Hence, by using Cramer's rule, the values of unknowns x_1, x_2, and x_3 are as follows:

$$x_1 = A_1/A = 91.4016/47.874336 = 1.909$$
$$x_2 = A_2/A = 152.9568/47.874336 = 3.195$$
$$x_3 = A_3/A = 241.5168/47.874336 = 5.045$$

Using Cramer's rule to solve systems of linear equations in n unknowns involves computing $n + 1$ determinants of the nth order, which is a quite laborious task especially when the number n is quite large. There are many alternative methods of solving systems of linear equations belonging to two groups, i.e. exact and iterative. In exact methods, computations are carried out exactly without rounding and give exact values of the unknowns x_i although even such results are approximate because of unavoidable rounding in practical applications. The Gaussian method is one of the most popular exact methods for solving systems of linear equations in n unknowns which is outlined below.

3.3.2 Solution of Systems of Linear Equations Using Gaussian Elimination

An algorithm for successive elimination of the unknowns is the most common technique for solving a system of linear equations which is also called **Gaussian Elimination**. To illustrate the concept in a simple manner, let us convert the expression (3.4) above into a system of four equations in four unknowns as shown below which is to be solved:

$$
\left.
\begin{aligned}
a_{11}x_1 + a_{12}x_2 + a_{13}x_3 + a_{14}x_4 &= a_{15} \ \ (\text{a}) \\
a_{21}x_1 + a_{22}x_2 + a_{23}x_3 + a_{24}x_4 &= a_{25} \ \ (\text{b}) \\
a_{31}x_1 + a_{32}x_2 + a_{33}x_3 + a_{34}x_4 &= a_{35} \ \ (\text{c}) \\
a_{41}x_1 + a_{42}x_2 + a_{43}x_3 + a_{44}x_4 &= a_{45} \ \ (\text{d})
\end{aligned}
\right\}
\tag{3.5}
$$

Solution Let us assume that the leading element in the equation (a) of the system of equations (3.5), i.e. $a_{11} \neq 0$. Dividing the equation (a) of system (3.5) by this leading element a_{11}, we get:

$$
x_1 + b_{12}x_2 + b_{13}x_3 + b_{14}x_4 = b_{15}
\tag{3.6}
$$

$$
\text{where } b_{1j} = a_{1j}/a_{11} \ \ (j = 2, 3, 4, 5)
\tag{3.7}
$$

Using Eq. (3.6), we can eliminate the unknown x_1 from the equations (b), (c), and (d). This is done by multiplying Eq. (3.6) by a_{21}, a_{31}, and a_{41}, and subtracting the results so obtained from the equations (b), (c), and (d) of the system (3.5), respectively, and thus, finally we get the following three equations eliminating the unknown x_1:

$$
\left.
\begin{aligned}
a_{22}^{(1)}x_2 + a_{23}^{(1)}x_3 + a_{24}^{(1)}x_4 &= a_{25}^{(1)} & (\text{e}) \\
a_{32}^{(1)}x_2 + a_{33}^{(1)}x_3 + a_{34}^{(1)}x_4 &= a_{35}^{(1)} & (\text{f}) \\
a_{42}^{(1)}x_2 + a_{43}^{(1)}x_3 + a_{44}^{(1)}x_4 &= a_{45}^{(1)} & (\text{g})
\end{aligned}
\right\}
\tag{3.8}
$$

where the coefficients $a_{ij}^{(1)}$ are given by the following expressions:

$$
a_{ij}^{(1)} = a_{ij} - a_{i1}b_{1j} \quad (i = 2, 3, 4; j = 2, 3, 4, 5).
\tag{3.9}
$$

Now dividing the equation (e) by the leading element $a_{22}^{(1)}$, we get the equation:

$$
x_2 + b_{23}^{(1)}x_3 + b_{24}^{(1)}x_4 = b_{25}^{(1)},
\tag{3.10}
$$

where

$$
b_{2j}^{(1)} = a_{2j}^{(1)}/a_{22}^{(1)} \ \ (j = 3, 4, 5).
\tag{3.11}
$$

Eliminating x_2 in the same way, we eliminated the unknown x_1 as above, we get the following system of equations:

$$
\left.
\begin{aligned}
a_{33}^{(2)}x_3 + a_{34}^{(2)}x_4 &= a_{35}^{(2)} & (\text{h}) \\
a_{43}^{(2)}x_3 + a_{44}^{(2)}x_4 &= a_{45}^{(2)} & (\text{k}),
\end{aligned}
\right\}
\tag{3.12}
$$

where

$$a_{ij}^{(2)} = a_{ij}^{(1)} - a_{i2}^{(1)} b_{2j}^{(1)} \quad (i=3, 4 \quad j=3, 4, 5). \tag{3.13}$$

Now dividing the equation (h) by the leading element $a_{33}^{(2)}$, we get the equation:

$$x_3 + b_{34}^{(2)} x_4 = b_{35}^{(2)} \tag{3.14}$$

where

$$b_{3j}^{(2)} = a_{3j}^{(1)} / a_{33}^{(1)} (j=4, 5). \tag{3.15}$$

Using the Eq. (3.14) for value of x_3 in equation (k), we can eliminate x_3 from that equation. Thus, we get the following equation:

$$a_{44}^{(3)} x_4 = a_{45}^{(3)}, \tag{3.16}$$

where

$$a_{4j}^{(3)} = a_{4j}^{(2)} - a_{43}^{(2)} b_{3j}^{(2)} \quad (j=4, 5). \tag{3.17}$$

Thus, we have converted the system of equations (3.5) into an equivalent system with a triangular matrix as follows:

$$\left. \begin{array}{r} x_1 + b_{12}\, x_2 + b_{13}\, x_3 + b_{14}\, x_4 = b_{15} \\ x_2 + b_{23}^{(1)} x_3 + b_{24}^{(1)} x_4 = b_{25}^{(1)} \\ x_3 + b_{34}^{(2)} x_4 = b_{35}^{(2)} \\ a_{44}^{(3)} x_4 = a_{45}^{(3)} \end{array} \right\} \tag{3.18}$$

Using the system of equations (3.18), we can easily find the values of x_1, x_2, x_3, and x_4 in a consecutive order as:

$$\left. \begin{array}{l} x_4 = a_{45}^{(3)} / a_{44}^{(3)}, \\ x_3 = b_{35}^{(2)} - b_{34}^{(2)} x_4, \\ x_2 = b_{25}^{(1)} - b_{24}^{(1)} x_4 - b_{23}^{(1)} x_3, \\ x_1 = b_{15} - b_{14}\, x_4 - b_{13}\, x_3 - b_{12}\, x_2 \end{array} \right\}. \tag{3.19}$$

Thus, process of solution of linear equations using Gaussian method involves two procedures, namely,

1. Procedure to reduce the system of equations (3.5) to a triangular form as shown in (3.18) which is also called *direct procedure*.
2. Procedure to find the values of the unknowns by the formula (3.19) which is also called *reverses procedure*.

It is also to be noted that this method is applicable only if all the leading elements are nonzero. Application of the above generalized procedure for solution of linear equations using Gaussian method is illustrated below using a numerical example.

Example 3 Solve the following system of linear equations by the Gaussian Method:

$$
\left.\begin{array}{ll}
2.0x_1 - 4.0x_2 - 3.25x_3 + 1.0x_4 = 4.84 & \text{(a)} \\
3.0x_1 - 3.0x_2 - 4.3x_3 + 8.0x_4 = 8.89 & \text{(b)} \\
1.0x_1 - 5.0x_2 + 3.3x_3 - 20.0x_4 = -14.01 & \text{(c)} \\
2.5x_1 - 4.0x_2 + 2.0x_3 - 3.0x_4 = -20.29 & \text{(d)}
\end{array}\right\} \quad (3.20)
$$

Dividing the equation (a) of system (3.20) by its leading element $a_{11} = 2.0$, we get:

$$x_1 - 2.0x_2 - 1.625x_3 + 0.50x_4 = 2.42$$

Hence, $b_{12} = -2.0$, $b_{13} = -1.625$, $b_{14} = 0.50$, $b_{15} = 2.42$.

Using the formula (3.9) presented earlier, we can calculate the coefficients $a_{ij}^{(1)} = a_{ij} - a_{i1}b_{1j}$ ($i = 2, 3, 4; j = 2, 3, 4, 5$) and also form here the system (3.8) as below.

For $i = 2$

$$
\begin{aligned}
a_{22}^{(1)} &= a_{22} - a_{21}b_{12} = -3.0 - 3.0 \times (-2.0) = 3.0 \\
a_{23}^{(1)} &= a_{23} - a_{21}b_{13} = -4.3 - 3.0 \times (-1.625) = 0.575 \\
a_{24}^{(1)} &= a_{24} - a_{21}b_{14} = 8.0 - 3.0 \times (0.5) = 6.5 \\
a_{25}^{(1)} &= a_{25} - a_{21}b_{15} = 8.89 - 3.0 \times (2.42) = 1.63
\end{aligned}
$$

For $i = 3$

$$
\begin{aligned}
a_{32}^{(1)} &= a_{32} - a_{31}b_{12} = -5.0 - 1.0 \times (-2.0) = -3.0 \\
a_{33}^{(1)} &= a_{33} - a_{31}b_{13} = 3.3 - 1.0 \times (-1.625) = 4.925 \\
a_{34}^{(1)} &= a_{34} - a_{31}b_{14} = -20.0 - 1.0 \times (0.5) = -20.5 \\
a_{35}^{(1)} &= a_{35} - a_{31}b_{15} = -14.01 - 1.0 \times (2.42) = -16.43
\end{aligned}
$$

For $i = 4$

$$
\begin{aligned}
a_{42}^{(1)} &= a_{42} - a_{41}b_{12} = -4.0 - 2.5 \times (-2.0) = 1.0 \\
a_{43}^{(1)} &= a_{43} - a_{41}b_{13} = 2.0 - 2.5 \times (-1.625) = 6.0625 \\
a_{44}^{(1)} &= a_{44} - a_{41}b_{14} = -3.0 - 2.5 \times (0.5) = -4.25 \\
a_{45}^{(1)} &= a_{45} - a_{41}b_{15} = -20.29 - 2.5 \times (2.42) = -26.34
\end{aligned}
$$

Using the above results, we get the following system of three equations in three unknowns:

$$
\left.
\begin{array}{ll}
3.0x_2 + 0.575x_3 + 6.5x_4 & = 1.63 \qquad\quad \text{(e)} \\
-3.0x_2 + 4.925x_3 - 20.5x_4 & = -16.43 \quad\; \text{(f)} \\
1.0x_2 + 6.0625x_3 - 4.25x_4 & = -26.34 \quad\; \text{(g)}
\end{array}
\right\}
\qquad (3.21)
$$

Dividing the above equation (e) by $a_{22}^{(1)} = 3.0$, we get:

$$
x_2 + 0.19167x_3 + 2.16667x_4 = 0.54333 \quad \text{(h)}
$$

where $b_{23}^{1} = 0.19167$, $b_{24}^{1} = 2.16667$, $b_{25}^{1} = 0.54333$.

Using the formula (3.13), i.e. $a_{ij}^{(2)} = a_{ij}^{(1)} - a_{i2}^{(1)}b_{2j}^{(1)}$ $(i = 3, 4 \; j = 3, 4, 5)$, we compute the coefficients $a_{ij}^{(2)}$ and form the system of equations (3.12) as follows:

For $i = 3$, we get:

$$
\begin{aligned}
a_{33}^{(2)} &= a_{33}^{(1)} - a_{31}^{(1)}b_{23}^{(1)} = 4.925 - (-3.0) \times 0.19167 &&= 5.5 \\
a_{34}^{(2)} &= a_{34}^{(1)} - a_{31}^{(1)}b_{24}^{(1)} = -20.5 - (-3.0) \times 2.16667 &&= -14/0 \\
a_{35}^{(2)} &= a_{35}^{(1)} - a_{31}^{(1)}b_{25}^{(1)} = -16.43 - (-3.0) \times 0.54333 &&= -14.8
\end{aligned}
$$

For $i = 4$, we get:

$$
\begin{aligned}
a_{43}^{(2)} &= a_{43}^{(1)} - a_{41}^{(1)}b_{23}^{(1)} = 6.0625 - 1.0 \times 0.19167 &&= 5.87083 \\
a_{44}^{(2)} &= a_{44}^{(1)} - a_{41}^{(1)}b_{24}^{(1)} = -4.25 - 1.0 \times 2.16667 &&= -6.41667 \\
a_{45}^{(2)} &= a_{45}^{(1)} - a_{41}^{(1)}b_{25}^{(1)} = -26.34 - 1.0 \times 0.54333 &&= -26.88333
\end{aligned}
$$

Using the above results, we get the following system of two equations in two unknowns:

$$
\left.
\begin{array}{ll}
5.5x_3 \quad\;\;\; - 14x_4 & = -14.8 \qquad\qquad\; \text{(h)} \\
5.87083x_3 - 6.41667x_4 & = -26.88333 \quad\; \text{(k)}
\end{array}
\right\}
\qquad (3.22)
$$

Dividing the equation (h) by $a_{33}^{(2)} = 1$, we get:

$$
x_3 - 2.54545x_4 = -2.69091,
$$

where $b_{34}^{(2)} = -2.54545$, $b_{35}^{(2)} = -2.69091$.

Using the formula (3.17), we can find the coefficients $a_{4j}^{(3)}$ as:

$$
\begin{aligned}
a_{44}^{(3)} &= a_{44}^{(2)} - a_{43}^{(2)}b_{34}^{(2)} = -6.41667 - 5.87083 \times (-2.54545) = 8.52723 \\
a_{45}^{(3)} &= a_{45}^{(2)} - a_{43}^{(2)}b_{35}^{(2)} = -26.88333 - 5.87083 \times (-2.69091) = -11.08546.
\end{aligned}
$$

We can now get one equation with one unknown as follows:

$$8.52723x_4 = -11.08546.$$

Thus, we get the equivalent system in the form of a triangular matrix as follows:

$$\left.\begin{array}{ll} x_1 - 2.0x_2 - 1.625x_3 + 0.50x_4 & = 2.42 \\ x_2 + 0.19167x_3 + 2.16667x_4 & = 0.54333 \\ x_3 - 2.54545x_4 & = -2.69091 \\ 8.52723x_4 & = -11.08546 \end{array}\right\} \quad (3.23)$$

Here, the direct procedure is terminated and using the reverse procedure with the system (3.23), we successively get the values of unknowns as follows:

$$\begin{array}{ll} x_4 = -11.08546/8.52723 & = -1.30 \\ x_3 = -2.69091 + 2.54545(-1.3) & = -6.0 \\ x_2 = 0.54333 - 0.19167 \times (-6.0) - 2.16667(-1.3) & = 4.51 \\ x_1 = 2.42 + 2.0 \times 4.51 + 1.625 \times (-6.0) - 0.50(-1.3) = 2.34 \end{array}$$

3.4 Numerical Differentiation

As earlier mentioned, use of modern computer techniques including computer-aided design is very difficult without skilful application of approximate and numerical analysis and techniques. Computers cannot perform calculus such as differentiation and integration. But it can be used to perform these tasks using the concept of numerical differentiation and integration. Thus, using such techniques, even complex differential equations can be converted into linear equations and solved easily using computers. Numerical differentiation is an important tool in this process, and it can also help in automating the procedure for solution of systems of nonlinear simultaneous equations and optimization problems as available in literature. Here, only a few formulas are presented to give an idea about this numerical technique.

The formulas for numerical differentiation are derived using various interpolation formulas such as Newton's interpolation formulas, Stirling's interpolation formula, and so on which are available in the literature. Thus, the formula for numerical differentiation will vary with the specific interpolation formula adopted. Formula for numerical differentiation is used in the case when, for example, a function $y = f(x)$ is represented by a table of its values $y_i = f(x_i)$ for equally spaced points $x_i = x_0 + ih$ ($i = 0, \pm 1, \pm 2, \ldots$) where h is the spacing. Here, $f'(x)$ is $\frac{\Delta y}{\Delta x}$ designated as the derivative of $f(x)$. Choosing a set of $n + 1$ points, the function $y = f(x)$ can be replaced by an interpolation polynomial $P_n(x)$. Thereafter, the derivative of this polynomial $P'_n(x)$ is used for numerical differentiation, thus, giving:

$$f'(x) \approx P'_n(x).$$

Thus, for example, by differentiating the Stirling's interpolation polynomial, we get:

$$y_0 \approx f'(x_0)$$

$$\approx 1/h \left(\frac{\Delta y - 1 + \Delta y 0}{2} - \frac{1}{6} \frac{\Delta^3 y - 2 + \Delta^3 y - 1}{2} + \frac{1}{30} \frac{\Delta^5 y - 3 + \Delta^5 y - 2}{2} + \cdots \right).$$

$$(3.24)$$

Retaining only the first term of the formula (3.23), we get the formula in simplified form as:

$$y'_0 \approx (y_1 - y_{-1})/2h. \tag{3.25}$$

Similarly, retaining only the first two terms of the formula (3.23), we get the formula in simplified form as:

$$y'_0 \approx (y_{-2} - 8y_{-1} + 8y_1 - y_2)/2h. \tag{3.26}$$

The formula (3.23) is more convenient for computation and also has a higher degree of accuracy. Therefore, the above formula (3.25) developed from formula (3.23) is used later in the solution of nonlinear equations by means of numerical differentiation replacing the manual formulations of derivatives, thus permitting more automation in the solution process.

3.5 Numerical Solution of Systems of Nonlinear Simultaneous Equations

Frequently the classical optimization techniques require the solution of systems of nonlinear simultaneous equations. Newton–Raphson approximation is one of the numbers of numerical approximation techniques available for solution of such problems. This technique is illustrated below using a single variable nonlinear equation, two variable nonlinear simultaneous equations, and three variable nonlinear simultaneous equations, respectively. In this section, derivatives of functions are derived manually and incorporated in the solution process. However, there is no need for manual derivation, and the process can be automated if numerical differentiation technique is adopted as shown later.

Fig. 3.1 Diagrammatic representation of Newton–Raphson approximation technique

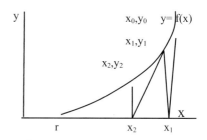

3.5.1   Solution of Single Variable Nonlinear Equation

A single variable nonlinear equation expressed as $y = f(x)$ representing a curve as shown in Fig. 3.1.

To obtain the positive root r of the equation $y = f(x)$, let us arbitrarily guess the root as x_0 and draw a tangent to the curve at the point (x_0, y_0) which cuts the x-axis at the point x_1. This abscissa value x_1 replaces x_0 as an improved value of the root. This process is repeated using the point (x_1, y_1) to draw a tangent to the curve at the point (x_1, y_1) which intersects the x-axis at the point x_2 which replaces x_1 as an improved value of the root, and so on until a satisfactory approximation of the root is achieved.

The equation of the tangent to the curve $y = f(x)$ at the point (x_0, y_0) can be written as:

$$f'(x) = (y - y_0)\,(x - x_0)^{-1}$$

The second approximation x_1 is the x-axis intercept of this tangent. Substituting $y = 0$, $y_0 = f(x_0)$, we get:

$$f'(x_0) = -\,[f(x_0)/(x_1 - x_0)]$$
$$x_1 = x_0 - f(x_0)/f'(x_0) \qquad \text{(or)}$$

Hence, the general expression for the arbitrarily chosen root x_n will be as follows:

$$x_{n+1} = x_n - f(x_n)/f'(x_n) \qquad (3.27)$$

The expression (3.23) is the Newton–Raphson formula for the $(n + 2)$ nd Approximation (Stark). In the optimization problem with the single variable x, we seek the root of the first derivative, i.e. $z'(x)$ of the objective function $z(x)$, which is equated to zero. Hence, in this case, $f(x_n)$ is replaced by $z'(x_n)$ and $f'(x_n)$ is replaced by $z''(x_n)$, i.e. second derivative of the objective function $z(x)$, and the formula becomes:

$$x_{n+1} = x_n - z'(x_n)/z''(x_n) \qquad (3.28)$$

This formula is used repeatedly until putting the current value of x_{n+1} in the expression for $z'(x)$ equates to zero with acceptable approximation as specified, giving the optimal solution, i.e. root r.

Any function infinitely differentiable in the interval $x_0 - r < x < x_0 + r$ can be expanded in that interval into infinite Taylor's series as follows (Danko et al., 1983):

$$f(x) = f(x_0) + f'(x_0)\,(x - x_0)/1! + f''(x_0)(x - x_0)^2/2! + \cdots$$

When the value of x_{n+1} is close to the root r, the Eq. (3.27) can be interpreted as the Taylor expansion of $f(x)$ about x_n incorporating only the first order terms as shown below:

$$f(x) = f(x_n) + f'(x_n)(x - x_n)$$

for $x = x_{n+1}$ and $f(r) = 0$, it yields,

$$f(x_n) + f'(x_n)(x_{n+1} - x_n) = 0 \ \text{ or, } \ f(x_n) + f'(x_n)x_{n+1} = f'(x_n)\,x_n \ \text{ or,}$$
$$f'(x_n)x_{n+1} = f'(x_n)x_n - f(x_n)$$

or, $x_{n+1} = x_n - f(x_n)/f'(x_n)$ this is same as the Eq. (3.27) above.

Application of the above procedure for solution of nonlinear equations with single variable is illustrated below using a numerical example.

Example 4 Solve the following single variable nonlinear equation using **Newton–Raphson Approximation Technique**:

$$Z(x) = x^3 - 3x + 10 = 0$$
$$f(x) = x^3 - 3x + 10 \ \text{ and } f'(x) = 3x^2 - 3, \ \text{ and}$$

for $x_0, f(x_0) = x_0^3 - 3x_0 + 10$ and $f'(x_0) = 3x_0^2 - 3$.

Thus, the Newton–Raphson formula becomes:

$$x_{n+1} = x_n - (x_0^3 - 3x_0 + 10)/(3x_0^2 - 3)$$

If we choose an arbitrary value of $x_0 = 0.5$, we get $f(0.5) = 8.625$ and $f'(0.5) = -2.25$ and therefore, we get

$x_1 = 0.5 - 8.625/(-2.25) = 4.33333$, $f(x_1) = 78.3704$, $f''(x_1) = 53.33333$ and therefore,

$x_2 = 4.33333 - 78.3704/53.33333 = 2.86389$, $f(x_2) = 24.8975$, and $f'(x_2) = 21.60551$ and therefore,

$x_3 = 2.86389 - 24.8975/21.60551 = 1.71152$, $f(x_3) = 9.87901$ and $f'(x_3) = 5.78792$ and therefore,

$x_4 = 1.71152 - 9.87901/5.78792 = 0.00469137$, $f(x_4) = 9.98593$ and $f'(x_4) = -2.99993$ and therefore,

$x_5 = 0.00469137 - 9.98593/(-2.99993) = 3.33341$, $f(x_5) = 37.0393$ and $f'(x_5) = 30.33487$ and therefore,

$x_6 = 3.33341 - 37.0393/30.33491 = 2.11239$, $f(x_6) = 13.0887$ and $f'(x_6) = 10.3866$ and therefore,

$x_7 = 2.11239 - 13.0887/10.3866 = 0.85224$, $f(x_7) = 8.06228$ and $f'(x_7) = -0.82109$ and therefore,

$x_8 = 0.85224 - 8.06228/(-0.82109) = 10.67124$, $f(x_8) = 1193.17$, $f'(x_8) = 338.625$ and therefore,

$x_9 = 10.67124 - 1193.17/338.625 = 7.14765$, $f(x_9) = 353.722$ and $f'(x_9) = 150.267$ and therefore,

$x_{10} = 7.14765 - 353.722/150.267 = 4.79368$, $f(x_{10}) = 105.775$ and $f'(x_{10}) = 65.9382$.

It is obvious from above that solving the nonlinear equations by manual calculations even for a single variable nonlinear equation is very tedious and even after 10 iterations it did not converge to a solution. But, using computers, it becomes very easy. Therefore, a C++ program NR123.CPP to solve nonlinear equations of one, two, and three variables using Newton–Raphson approximation technique is developed and included in this chapter. Running this C++ NR123.CPP Program, the above problem is solved instantly after 22 iterations giving the value of value of x as -2.61289 as shown below, and also, since putting this value of x in the function $z(x)$ gives the value of $f(x) = -1.98937e - 05\sim0$, the value of $x = -2.61289$ is the approximate value of x, correct to the five decimal places.

The results as above obtained instantly by running the C++ NR123.CPP Program, in the form of a typical output in this case is shown below.

Output of the C++ program NR123.CPP in this case:

```
PRESET VALUE OF LIMIT OF ITERATION NUMBER 'ITRL' = 50
PRESET VALUE OF ACCEPTABLE MARGIN OF ERROR 'ITOL' = 0.0001

DO YOU WANT TO CHANGE THE ABOVE PRESET VALUES (Y/N) ? n
YOU HAVE NOT CHANGED THE VALUES OF 'ITRL' & 'ITOL'

'ITERATION MODELS' AVAILABLR HERE ARE SHOWN BELOW :

ITMODEL TO SOLVE SINGLE VARIABLE NONLINEAR EQUATION (ITMODEL- 1)
ITMODEL TO SOLVE TWO VARIABLES NONLINEAR EQUATIONS (ITMODEL- 2)
ITMODEL TO SOLVE THREE VARIABLES NONLINEAR EQUATIONS (ITMODEL- 3)

PUT YOUR CURRENT CHOICE OF ITMODEL (1/2/3)
1
PUT THE CURRENT VALUE OF X .5
CURRENT VALUE OF ITERATION NUMBER IS = 0
CURRENT VALUE OF fx IS = -2.25
CURRENT VALUE OF f IS = 8.625
CURRENT VALUE OF x IS = 0.5
PUT ANY CHARACTER TO CONTINUE n
NEW VALUE OF ITERATION NUMBER IS = 1
NEW VALUE OF x IS = 4.33333
```

```
NEW VALUE OF f IS = 78.3704
NEW VALUE OF ITERATION NUMBER IS = 2
NEW VALUE OF x IS = 2.86389
NEW VALUE OF f IS = 24.8975
NEW VALUE OF ITERATION NUMBER IS = 3
NEW VALUE OF x IS = 1.71152
NEW VALUE OF f IS = 9.87901
NEW VALUE OF ITERATION NUMBER IS = 4
NEW VALUE OF x IS = 0.00469137
NEW VALUE OF f IS = 9.98593
NEW VALUE OF ITERATION NUMBER IS = 5
NEW VALUE OF x IS = 3.33341
NEW VALUE OF f IS = 37.0393
NEW VALUE OF ITERATION NUMBER IS = 6
NEW VALUE OF x IS = 2.11239
NEW VALUE OF f IS = 13.0887
NEW VALUE OF ITERATION NUMBER IS = 7
NEW VALUE OF x IS = 0.852234
NEW VALUE OF f IS = 8.06228
NEW VALUE OF ITERATION NUMBER IS = 8
NEW VALUE OF x IS = 10.6712
NEW VALUE OF f IS = 1193.17
NEW VALUE OF ITERATION NUMBER IS = 9
NEW VALUE OF x IS = 7.14765
NEW VALUE OF f IS = 353.722
NEW VALUE OF ITERATION NUMBER IS = 10
NEW VALUE OF x IS = 4.79368
NEW VALUE OF f IS = 105.775
NEW VALUE OF ITERATION NUMBER IS = 11
NEW VALUE OF x IS = 3.18953
NEW VALUE OF f IS = 32.8788
NEW VALUE OF ITERATION NUMBER IS = 12
NEW VALUE OF x IS = 1.99478
NEW VALUE OF f IS = 11.9531
NEW VALUE OF ITERATION NUMBER IS = 13
NEW VALUE OF x IS = 0.657345
NEW VALUE OF f IS = 8.31201
NEW VALUE OF ITERATION NUMBER IS = 14
NEW VALUE OF x IS = 5.53616
NEW VALUE OF f IS = 163.07
NEW VALUE OF ITERATION NUMBER IS = 15
NEW VALUE OF x IS = 3.70283
NEW VALUE OF f IS = 49.6608
NEW VALUE OF ITERATION NUMBER IS = 16
NEW VALUE OF x IS = 2.40052
NEW VALUE OF f IS = 16.6314
NEW VALUE OF ITERATION NUMBER IS = 17
NEW VALUE OF x IS = 1.23646
NEW VALUE OF f IS = 8.18097
NEW VALUE OF ITERATION NUMBER IS = 18
NEW VALUE OF x IS = -3.92008
NEW VALUE OF f IS = -38.4799
NEW VALUE OF ITERATION NUMBER IS = 19
NEW VALUE OF x IS = -3.0273
NEW VALUE OF f IS = -8.662
```

```
NEW VALUE OF ITERATION NUMBER IS = 20
NEW VALUE OF x IS = -2.67366
NEW VALUE OF f IS = -1.09158
NEW VALUE OF ITERATION NUMBER IS = 21
NEW VALUE OF x IS = -2.61448
NEW VALUE OF f IS = -0.0278836
NEW VALUE OF ITERATION NUMBER IS = 22
NEW VALUE OF x IS = -2.61289
NEW VALUE OF f IS = -1.98937e-05
CURRENT VALUE OF ITERATION NUMBER IS = 22
CURRENT  VALUE OF f IS = -1.98937e-05
CURRENT VALUE OF x  IS = -2.61289
CURRENT VALUE OF ITOL CHOSEN IS = 0.0001

SINCE THE CURRENT SOLUTION MEETS THE VALUE OF CHOSEN
MARGIN OF ERROR, THE SOLUTION ACCEPTED & PROGRAM TERMINATED
```

For functions of several variables, the procedure is similar to that shown above for single variable, and is outlined below, which are also incorporated in the C++ program NR123.CPP mentioned above.

3.5.2   Solution of a System of Two Nonlinear Simultaneous Equations

Let us assume $f(x,y) = 0$ and $g(x,y) = 0$ are the two nonlinear simultaneous equations. Using (x_n, y_n) as the point of expansion and considering first order terms we can write,

$$\left.\begin{aligned}
f(x, y) &= f(x_n, y_n) + \frac{\partial f}{\partial x}\bigg|_n (x - x_n) + \frac{\partial f}{\partial y}\bigg|_n (y - y_n) \\
g(x, y) &= g(x_n, y_n) + \frac{\partial g}{\partial x}\bigg|_n (x - x_n) + \frac{\partial g}{\partial y}\bigg|_n (y - y_n)
\end{aligned}\right\} \tag{3.29}$$

Here, the subscripted derivatives indicate evaluation at (x_n, y_n). If $x = x_{n+1}$ and $y = y_{n+1}$ and also these are close to the roots $x = r$ and $y = s$, then both $f(x_{n+1}, y_{n+1})$ ~0 and $g(x_{n+1}, y_{n+1})$~0, giving two simultaneous equations in two variables x_{n+1}, and y_{n+1} as follows:

$$\left.\begin{aligned}
f(x_n, y_n) + \frac{\partial f}{\partial x}\bigg|_n (x_{n+1} - x_n) + \frac{\partial f}{\partial y}\bigg|_n (y_{n+1} - y_n) &= 0 \qquad\qquad\text{(a)} \\
g(x_n, y_n) + \frac{\partial g}{\partial x}\bigg|_n (x_{n+1} - x_n) + \frac{\partial g}{\partial y}\bigg|_n (y_{n+1} - y_n) &= 0 \qquad\qquad\text{(b)} \\
\text{From Eq.(3.26) (b) we get } (y_{n+1} - y_n) &= -\left\{ g(x_n, y_n) - \frac{\partial g}{\partial x}\bigg|_n (x_{n+1} - x_n) \right\} \Big/ \frac{\partial g}{\partial y}\bigg|_n \quad\text{(c)}
\end{aligned}\right\}$$

$$\tag{3.30}$$

From (a) and (c), we get

$$f(x_n, y_n) + \left.\frac{\partial f}{\partial x}\right|_n (x_{n+1} - x_n) - \left.\frac{\partial f}{\partial y}\right|_n \left\{ g(x_n, y_n) + \left.\frac{\partial g}{\partial x}\right|_n (x_{n+1} - x_n) \right\} / \left.\frac{\partial g}{\partial y}\right|_n = 0$$

$$\text{or } \left.\frac{\partial f}{\partial y}\right|_n \left\{ g(x_n, y_n) + \left.\frac{\partial g}{\partial x}\right|_n (x_{n+1} - x_n) \right\} / \left.\frac{\partial g}{\partial y}\right|_n = f(x_n, y_n) + \left.\frac{\partial f}{\partial x}\right|_n (x_{n+1} - x_n)$$

$$\text{or, } \left.\frac{\partial f}{\partial y}\right|_n \left\{ \left.\frac{\partial g}{\partial x}\right|_n (x_{n+1} - x_n) \right\} / \left.\frac{\partial g}{\partial y}\right|_n - \left.\frac{\partial f}{\partial x}\right|_n (x_{n+1} - x_n) = f(x_n, y_n) - \left.\frac{\partial f}{\partial y}\right|_n \{ g(x_n, y_n) \} / \left.\frac{\partial g}{\partial y}\right|_n$$

$$\text{or, } \left.\frac{\partial f}{\partial y}\right|_n \left.\frac{\partial g}{\partial x}\right|_n (x_{n+1} - x_n) - \left.\frac{\partial g}{\partial y}\right|_n \cdot \left.\frac{\partial f}{\partial x}\right|_n (x_{n+1} - x_n) = f(x_n, y_n) \cdot \left.\frac{\partial g}{\partial y}\right|_n - \left.\frac{\partial f}{\partial y}\right|_n \cdot g(x_n, y_n)$$

$$(x_{n+1} - x_n) \left\{ \left.\frac{\partial f}{\partial y}\right|_n \left.\frac{\partial g}{\partial x}\right|_n - \left.\frac{\partial g}{\partial y}\right|_n \cdot \left.\frac{\partial f}{\partial x}\right|_n \right\} = f(x_n, y_n) \cdot \left.\frac{\partial g}{\partial y}\right|_n - g(x_n, y_n) \cdot \left.\frac{\partial f}{\partial y}\right|_n$$

$$\text{Or, } (x_{n+1} - x_n) \left\{ \left.\frac{\partial f}{\partial x}\right|_n \cdot \left.\frac{\partial g}{\partial y}\right|_n - \left.\frac{\partial f}{\partial y}\right|_n \left.\frac{\partial g}{\partial x}\right|_n \right\} = - \left\{ f(x_n, y_n) \cdot \left.\frac{\partial g}{\partial y}\right|_n - g(x_n, y_n) \cdot \left.\frac{\partial f}{\partial y}\right|_n \right\}$$

$$\text{Or, } x_{n+1} = x_n - \left\{ f(x_n, y_n) \cdot \left.\frac{\partial g}{\partial y}\right|_n - g(x_n, y_n) \cdot \left.\frac{\partial f}{\partial y}\right|_n \right\} / \left\{ \left.\frac{\partial f}{\partial x}\right|_n \cdot \left.\frac{\partial g}{\partial y}\right|_n - \left.\frac{\partial f}{\partial y}\right|_n \left.\frac{\partial g}{\partial x}\right|_n \right\}$$

Similarly, we can get the expression for y_{n+1}.

Thus, solving the above two simultaneous linear equations (i.e. Eqs. (3.30) (a) and (b) above), we finally get the Newton–Raphson formula for solving two variable nonlinear simultaneous equations as follows:

$$\left.\begin{aligned} x_{n+1} &= x_n - \left\{ f(x_n, y_n) \cdot \left.\frac{\partial g}{\partial y}\right|_n - g(x_n, y_n) \cdot \left.\frac{\partial f}{\partial y}\right|_n \right\} / \left\{ \left.\frac{\partial f}{\partial x}\right|_n \cdot \left.\frac{\partial g}{\partial y}\right|_n - \left.\frac{\partial f}{\partial y}\right|_n \left.\frac{\partial g}{\partial x}\right|_n \right\} \\ y_{n+1} &= y_n - \left\{ g(x_n, y_n) \cdot \left.\frac{\partial f}{\partial x}\right|_n - f(x_n, y_n) \cdot \left.\frac{\partial g}{\partial x}\right|_n \right\} / \left\{ \left.\frac{\partial f}{\partial x}\right|_n \cdot \left.\frac{\partial g}{\partial y}\right|_n - \left.\frac{\partial f}{\partial y}\right|_n \left.\frac{\partial g}{\partial x}\right|_n \right\} \end{aligned}\right\}$$

$$(3.31)$$

The expressions (3.31) are the Newton–Raphson formula for solving two variable nonlinear simultaneous equations. The above formula can be written in a more simplified form by expressing the partial derivatives as f_x, f_y, g_x, and g_y where x and y subscripts to f and g denote partial derivative of the functions with respect to the x and y variables, respectively. Thus the formula can be written as:

$$\left.\begin{aligned} x_{n+1} &= x_n - \{ f(x_n, y_n) \cdot g_y - g(x_n, y_n) \cdot f_y \} / \{ f_x \cdot g_y - f_y \cdot g_x \} \\ y_{n+1} &= y_n - \{ g(x_n, y_n) \cdot f_x - f(x_n, y_n) \cdot g_x \} / \{ f_x \cdot g_y - f_y \cdot g_x \} \end{aligned}\right\}$$

$$(3.32)$$

It may be noted that the above Newton–Raphson formula for $f(x,y)$ and $g(x,y)$ can be expressed in matrix form as follows, where x and y subscripts to f and g denote partial derivative as explained above:

$$\begin{pmatrix} f_x & f_y \\ g_x & g_y \end{pmatrix} \begin{pmatrix} x_{n+1} - x_n \\ y_{n+1} - y_n \end{pmatrix} = \begin{pmatrix} -f(x_n, \ y_n) \\ -g(x_n, \ y_n) \end{pmatrix} \tag{3.33}$$

This method can be generalized to functions of several variables x, y, $\ldots$, z as follows:

$$\begin{pmatrix} f_x & f_y \cdots\cdots\cdots\cdots\cdots f_z \\ g_x & g_y \cdots\cdots\cdots\cdots\cdots g_z \\ \cdots\cdots\cdots\cdots\cdots\cdots\cdots \\ h_x & h_y \cdots\cdots\cdots\cdots\cdots h_z \end{pmatrix} \begin{pmatrix} x_{n+1} - x_n \\ y_{n+1} - y_n \\ \cdots\cdots \\ h_{n+1} - h_n \end{pmatrix} = \begin{pmatrix} -f(x_n, \ y_n \cdots\cdots\cdots z_n) \\ -g(x_n, \ y_n \cdots\cdots\cdots z_n) \\ \cdots\cdots\cdots\cdots\cdots \\ -h(x_n, \ y_n \cdots\cdots\cdots z_n) \end{pmatrix}$$

Example 5 Solve the following system of two nonlinear equations using the **Newton–Raphson Approximation Technique**:

$$x^3 + y^3 - 6x + 3 = 0, x^3 - y^3 - 6y + 2 = 0.$$

Here, $f(x, y) = x^3 + y^3 - 6x + 3$ and $g(x, y) = x^3 - y^3 - 6y + 2$
Therefore, here the values of $f_x, f_y, g_x,$ and g_y are as follows:

$$f_x = 3x^2 - 6; \quad g_x = 3x^2$$
$$f_y = 3y^2; \quad g_y = -3y^2 - 6$$

The arbitrary selection of the values of $x_0 = 0.75$, $y_0 = 0.75$ starts the process. Thus, we get

$$f(0.75, 0.75) = -0.65625 \quad g(0.75, 0.75) = -2.5$$
$$f_x|_0 = -4.3125 \quad\quad g_x|_0 = 1.6875$$
$$f_y|_0 = 1.6875 \quad\quad g_y|_0 = -7.6875$$

By substituting these values into the above Newton–Raphson formula, we get:

$$x_1 = 0.75 - \{(-0.65625) \times (-7.6875) - (-2.5) \times 1.6875\}$$
$$/\{(-4.3125) \times (-7.6875) - 1.6875 \times 1.6875\} = 0.444316$$
$$y_1 = 0.75 - \{(-2.5) \times (-4.3125) - (-0.65625) \times 1.6875\}$$
$$/\{(-4.3125) \times (-7.6875) - 1.6875 \times 1.6875\} = 0.357695.$$

Substituting these new values of x and y, i.e. x_1 and y_1, we get new values of $f(x,y)$, $g(x,y)$, f_x, f_y, g_x, and g_y and applying these values into the above Newton–Raphson formula, we get new values of x and y, i.e. x_2 and y_2, and repeating this process, we finally get the values of x_3, y_3, and so on as follows:

$$x_2 = 0.530188$$
$$y_2 = 0.349336$$
$$x_3 = 0.532368$$
$$y_3 = 0.351257$$

Values of $x = 0.532368$ and $y = 0.351257$ give the values of functions $f(x,y)$ and $g(x,y)$ within the tolerance limit of error $= 0.0001$ and hence can be adopted as solution of the nonlinear equations in this case.

The results for x_2, y_2, and x_3, y_3 shown above are obtained by running the NR123. CPP Program discussed earlier, and its output after three iterations giving the solution satisfying acceptable margin of error in this case is shown below.

Output of the C++ program NR123.CPP in this case:

```
PRESET VALUE OF LIMIT OF ITERATION NUMBER 'ITRL' = 50
PRESET VALUE OF ACCEPTABLE MARGIN OF ERROR 'ITOL' = 0.0001

DO YOU WANT TO CHANGE THE ABOVE PRESET VALUES (Y/N) ? n
YOU HAVE NOT CHANGED THE VALUES OF 'ITRL' & 'ITOL'

 'ITERATION MODELS' AVAILABLR HERE ARE SHOWN BELOW :

ITMODEL TO SOLVE SINGLE VARIABLE NONLINEAR EQUATION(ITMODEL- 1)
ITMODEL TO SOLVE TWO VARIABLES NONLINEAR EQUATIONS (ITMODEL- 2)
ITMODEL TO SOLVE THREE VARIABLES NONLINEAR EQUATIONS (ITMODEL- 3)

PUT YOUR CURRENT CHOICE OF ITMODEL (1/2/3)
2
PUT THE CURRENT VALUE OF X .75
PUT THE CURRENT VALUE OF Y .75
CURRENT VALUE OF ITERATION NUMBER IS = 0
CURRENT VALUE OF fx IS = -4.3125
CURRENT VALUE OF fy IS = 1.6875
CURRENT VALUE OF gx IS = 1.6875
CURRENT VALUE OF gy IS = -7.6875
CURRENT VALUE OF f IS = -0.65625
CURRENT VALUE OF g IS = -2.5
CURRENT VALUE OF x IS = 0.75
CURRENT VALUE OF y IS = 0.75
PUT ANY CHARACTER TO CONTINUE n
NEW VALUE OF ITERATION NUMBER IS = 1
NEW VALUE OF f IS = 0.467588
NEW VALUE OF g IS = -0.104222
NEW VALUE OF x IS = 0.444316
NEW VALUE OF y IS = 0.357695
NEW VALUE OF ITERATION NUMBER IS = 2
```

```
NEW VALUE OF f IS = 0.010537
NEW VALUE OF g IS = 0.0103882
NEW VALUE OF x IS = 0.530188
NEW VALUE OF y IS = 0.349336
NEW VALUE OF ITERATION NUMBER IS = 3
NEW VALUE OF f IS = 1.14399e-05
NEW VALUE OF g IS = 3.69472e-06
NEW VALUE OF x IS = 0.532368
NEW VALUE OF y IS = 0.351257
CURRENT VALUE OF ITERATION NUMBER IS = 3
CURRENT VALUE OF f IS = 1.14399e-05
CURRENT VALUE OF g IS = 3.69472e-06
CURRENT VALUE OF x IS = 0.532368
CURRENT VALUE OF y IS = 0.351257
THE VALUE OF ITOL CHOSEN IS = 0.0001

SINCE THE CURRENT SOLUTION MEETS THE VALUE OF CHOSEN
MARGIN OF ERROR, THE SOLUTION ACCEPTED & PROGRAM TERMINATED
```

3.5.3 Solution of a System of Three Nonlinear Simultaneous Equations

Let us assume $f(x,y,z) = 0$, $g(x,y,z) = 0$, and $h(x,y,z) = 0$ are the three nonlinear simultaneous equations. Using (x_n, y_n, z_n) as the point of expansion and considering first order terms, we can write:

$$
\left.\begin{aligned}
f(x,y,z) &= f(x_n, y_n, z_n) + \frac{\partial f}{\partial x}\bigg|_n (x - x_n) + \frac{\partial f}{\partial y}\bigg|_n (y - y_n) + \frac{\partial f}{\partial z}\bigg|_n (z - z_n) \\
g(x,y,z) &= g(x_n, y_n, z_n) + \frac{\partial g}{\partial x}\bigg|_n (x - x_n) + \frac{\partial g}{\partial y}\bigg|_n (y - y_n) + \frac{\partial g}{\partial z}\bigg|_n (z - z_n) \\
h(x,y,z) &= h(x_n, y_n, z_n) + \frac{\partial h}{\partial x}\bigg|_n (x - x_n) + \frac{\partial h}{\partial y}\bigg|_n (y - y_n) + \frac{\partial h}{\partial z}\bigg|_n (z - z_n)
\end{aligned}\right\} \quad (3.34)
$$

Here, the subscripted derivatives indicate evaluation at (x_n, y_n, z_n). If $x = x_{n+1}$, $y = y_{n+1}$, $z = z_{n+1}$, and also these are close to the roots $x = r$, $y = s$, and $z = t$, then $f(x_{n+1}, y_{n+1}, z_{n+1}) \sim 0$, $g(x_{n+1}, y_{n+1}, z_{n+1}) \sim 0$, and $h(x_{n+1}, y_{n+1}, z_{n+1}) \sim 0$, giving three simultaneous linear equations in three variables x_{n+1}, y_{n+1}, and z_{n+1} as follows:

$$
\left.\begin{aligned}
f(x_n, y_n, z_n) + \frac{\partial f}{\partial x}\bigg|_n (x_{n+1} - x_n) + \frac{\partial f}{\partial y}\bigg|_n (y_{n+1} - y_n) + \frac{\partial f}{\partial z}\bigg|_n (z_{n+1} - z_n) &= 0 \quad (a) \\
g(x_n, y_n, z_n) + \frac{\partial g}{\partial x}\bigg|_n (x_{n+1} - x_n) + \frac{\partial g}{\partial y}\bigg|_n (y_{n+1} - y_n) + \frac{\partial g}{\partial z}\bigg|_n (z_{n+1} - z_n) &= 0 \quad (b) \\
h(x_n, y_n, z_n) + \frac{\partial h}{\partial x}\bigg|_n (x_{n+1} - x_n) + \frac{\partial h}{\partial y}\bigg|_n (y_{n+1} - y_n) + \frac{\partial h}{\partial z}\bigg|_n (z_{n+1} - z_n) &= 0 \quad (c)
\end{aligned}\right\}
$$

$$(3.35)$$

Using the above three simultaneous linear equations in three variables x_{n+1}, y_{n+1}, and z_{n+1}, and the elimination procedure adopted in case of two simultaneous linear equations in two variables as shown above, we can derive the Newton–Raphson formula for solving a system of three nonlinear simultaneous equations in three variables as indicated below.

To make it simpler, we can write the above equations in a more simplified form by expressing the partial derivatives as $f_x, f_y, f_z, g_x, g_y, g_z, h_x, h_y,$ and h_z where x, y, and z subscripts to f, g, and h denote partial derivative of the functions with respect to the x, y, and z variables, respectively. Similarly, we can make the following substitutions:

$$f = f(x_n, y_n, z_n)$$
$$g = g(x_n, y_n, z_n)$$
$$h = h(x_n, y_n, z_n)$$

$$F = f^*\left(g_y{}^*h_z - h_y{}^*g_z\right) + g^*\left(h_y{}^*f_z - f_y{}^*h_z\right) + h^*\left(f_y{}^*g_z - g_y{}^*f_z\right);$$
$$G = g^*\left(h_z{}^*f_x - f_z{}^*h_x\right) + h^*\left(f_z{}^*g_x - g_z{}^*f_x\right) + f^*\left(g_z{}^*h_x - h_z{}^*g_x\right);$$
$$H = h^*\left(f_x{}^*g_y - g_x{}^*f_y\right) + f^*\left(g_x{}^*h_y - h_x{}^*g_y\right) + g^*\left(h_x{}^*f_y - f_x{}^*h_y\right);$$
$$R = f_x{}^*\left(g_y{}^*h_z - h_y{}^*g_z\right) + g_x{}^*\left(h_y{}^*f_z - f_y{}^*h_z\right) + h_x{}^*\left(f_y{}^*g_z - g_y{}^*f_z\right);$$

Thus the Newton–Raphson formula for three simultaneous nonlinear equations in three variables can be written as:

$$x_{n+1} = x_n - (F/R); \quad y_{n+1} = y_n - (G/R); \quad z_{n+1} = z_n - (H/R); \qquad (3.36)$$

By this method, we get a sequence of approximate values $x_1, x_2 \ldots x_n$; $y_1, y_2 \ldots y_n$; and $z_1, z_2 \ldots z_n$; each successive term of which is closer, compared to its predecessor, to the unknown exact values of the root of the three nonlinear simultaneous equations. This is illustrated with an example below.

Example 6 Solve the following system of three nonlinear equations using the **Newton–Raphson Approximation Technique**:

$$x^2 + y^2 + z^2 - 1 = 0, \ 2x^2 + y^2 - 4z = 0, \ 3x^2 - 4y + z^2 = 0.$$

Here, $f(x, y, z) = f = x^2 + y^2 + z^2 - 1$, $g(x, y, z) = g = 2x^2 + y^2 - 4z$, and $h(x, y, z) = h = 3x^2 - 4y + z^2$.

Therefore, expressions for $f_x, f_y, f_z, g_x, g_y, g_z, h_x, h_y,$ and h_z are as follows:

$$f_x = 2x$$
$$f_y = 2y$$
$$f_z = 2z$$
$$g_x = 4x$$
$$g_y = 2y$$
$$g_z = -4$$
$$h_x = 6x$$
$$h_y = -4$$
$$h_z = 2z$$

Let us arbitrarily select values of $x_0 = 0.5$, $y_0 = 0.5$, and $z_0 = 0.5$ to start the process. Thus, we get

$$f(0.5, 0.5, 0.5) = -0.25$$
$$g(0.5, 0.5, 0.5) = -1.25$$
$$h(0.5, 0.5, 0.5) = -1.0$$

$$
\begin{array}{lll}
f_x|_0 = 1 & g_x|_0 = 2 & h_x|_0 = 3 \\
f_y|_0 = 1 & g_y|_0 = 1 & h_x|_0 = -4 \\
f_z|_0 = 1 & g_z|_0 = -4 & h_x|_0 = 1
\end{array}
$$

By substituting these values into the above Newton–Raphson formula, we get:

$$F = f^* \left(g_y{}^* h_z - h_y{}^* g_z \right) + g^* \left(h_y{}^* f_z - f_y{}^* h_z \right) + h^* \left(f_y{}^* g_z - g_y{}^* f_z \right) = 15$$
$$G = g^* \left(h_z{}^* f_x - f_z{}^* h_x \right) + h^* \left(f_z{}^* g_x - g_z{}^* f_x \right) + f^* \left(g_z{}^* h_x - h_z{}^* g_x \right) = 0$$
$$H = h^* \left(f_x{}^* g_y - g_x{}^* f_y \right) + f^* \left(g_x{}^* h_y - h_x{}^* g_y \right) + g^* \left(h_x{}^* f_y - f_x{}^* h_y \right) = -5$$
$$R = f_x{}^* \left(g_y{}^* h_z - h_y{}^* g_z \right) + g_x{}^* \left(h_y{}^* f_z - f_y{}^* h_z \right) + h_x{}^* \left(f_y{}^* g_z - g_y{}^* f_z \right) = -40$$
$$x_{n+1} = x_n - (F/R); \quad y_{n+1} = y_n - (G/R); \quad z_{n+1} = z_n - (H/R)$$

The results for various parameters shown above are obtained by running the NR123.CPP Program discussed earlier, and its output in this case is shown below.

It would be seen that the values of $x = 0.78521$, $y = 0.496611$, and $z = 0.369923$ give the values of the functions $f(x,y,z)$, $g(x,y,z)$, and $h(x,y,z)$ within the tolerance limit of error $= 0.0001$ after three iterations, and therefore can be adopted as the solution of the system of three nonlinear simultaneous equations in this case.

Output of the C++ program NR123.CPP in this case:

```
PRESET VALUE OF LIMIT OF ITERATION NUMBER 'ITRL' = 50
PRESET VALUE OF ACCEPTABLE MARGIN OF ERROR 'ITOL' = 0.0001

DO YOU WANT TO CHANGE THE ABOVE PRESET VALUES (Y/N) ? n
YOU HAVE NOT CHANGED THE VALUES OF 'ITRL' & 'ITOL'

'ITERATION MODELS' AVAILABLR HERE ARE SHOWN BELOW :

ITMODEL TO SOLVE SINGLE VARIABLE NONLINEAR EQUATION(ITMODEL- 1)
ITMODEL TO SOLVE TWO VARIABLES NONLINEAR EQUATIONS (ITMODEL- 2)
ITMODEL TO SOLVE THREE VARIABLES NONLINEAR EQUATIONS (ITMODEL- 3)

PUT YOUR CURRENT CHOICE OF ITMODEL (1/2/3)
3
PUT THE CURRENT VALUE OF X .5
PUT THE CURRENT VALUE OF Y .5
PUT THE CURRENT VALUE OF Z .5
CURRENT VALUEE OF ITERATION NUMBER IS = 0
CURRENT VALUE OF fx IS = 1
CURRENT VALUE OF fy IS = 1
```

```
CURRENT VALUE OF fz IS = 1
CURRENT VALUE OF gx IS = 2
CURRENT VALUE OF gy IS = 1
CURRENT VALUE OF gz IS = -4
CURRENT VALUE OF hx IS = 3
CURRENT VALUE OF hy IS = -4
CURRENT VALUE OF hz IS = 1
CURRENT VALUE OF F IS = 15
CURRENT VALUE OF G IS = 0
CURRENT VALUE OF H IS = -5
CURRENT VALUE OF R IS = -40
CURRENT VALUE OF f IS = -0.25
CURRENT VALUE OF g IS = -1.25
CURRENT VALUE OF h IS = -1
CURRENT VALUE OF x IS = 0.5
CURRENT VALUE OF y IS = 0.5
CURRENT VALUE OF z IS = 0.5
PUT ANY CHARACTER TO CONTINUE n
NEW VALUE OF ITERATION NUMBER IS = 1
NEW VALUE OF f IS = 0.15625
NEW VALUE OF g IS = 0.28125
NEW VALUE OF h IS = 0.4375
NEW VALUE OF x IS = 0.875
NEW VALUE OF y IS = 0.5
NEW VALUE OF z IS = 0.375
NEW VALUE OF ITERATION NUMBER IS = 2
NEW VALUE OF f IS = 0.0072933
NEW VALUE OF g IS = 0.0145238
NEW VALUE OF h IS = 0.0217943
NEW VALUE OF x IS = 0.789817
NEW VALUE OF y IS = 0.496622
NEW VALUE OF z IS = 0.369932
NEW VALUE OF ITERATION NUMBER IS = 3
NEW VALUE OF f IS = 2.12169e-05
NEW VALUE OF g IS = 4.24335e-05
NEW VALUE OF h IS = 6.36502e-05
NEW VALUE OF x IS = 0.78521
NEW VALUE OF y IS = 0.496611
NEW VALUE OF z IS = 0.369923
CURRENT VALUE OF ITERATION NUMBER IS = 3
CURRENT VALUE OF f IS = 2.12169e-05
CURRENT VALUE OF g IS = 4.24335e-05
CURRENT VALUE OF h IS = 6.36502e-05
CURRENT VALUE OF x IS = 0.78521
CURRENT VALUE OF y IS = 0.496611
CURRENT VALUE OF z IS = 0.369923
THE VALUE OF ITOL CHOSEN IS = 0.0001


 SINCE THE CURRENT SOLUTION MEETS THE VALUE OF CHOSEN
 MARGIN OF ERROR, THE SOLUTION ACCEPTED & PROGRAM TERMINATED
```

3.6   Automating Nonlinear Equation Solution Process Using Numerical Differentiation Technique

As mentioned earlier in Sect. 3.5, derivatives of functions are derived and incorporated manually in the above section. This is a very tedious and time consuming process, particularly when the nonlinear equations are quite complicated as are the cases in respect of planning and design analysis and optimization problems related to computer-aided design presented in this book. Using numerical differentiation technique, this process could be automated and the above tedious and time consuming procedure could be avoided. In fact, this tediousness and also complications is one of the reasons for inhibition in using optimization techniques in computer-aided design particularly, in the AEC sector related to urban development and management. Computers have removed these problems and even lay users can apply such techniques if suitable software are available. An attempt is made to provide such software in this book to help this process. Thus, the C++ program named as NRNDF.CPP is developed to solve system of nonlinear equations of one or two variables using the Newton–Raphson approximation technique incorporating the numerical differentiation procedure so as to automate the solution procedure.

In this section, the procedure for automating nonlinear equation solution process using numerical differentiation technique is illustrated by means of C++ program named as NRNDF.CPP cited above. Here, the same two examples of nonlinear equations—one for single variable and the other for two variables as adopted in Sect. 3.5—are used just to demonstrate that numerical differentiation technique gives equally accurate results.

Example 7   Here, the same single variable nonlinear equation, as solved in Example 4 earlier using manually derived expression for derivative, is chosen which will now be solved using the Newton–Raphson approximation technique incorporating numerical differentiation procedure:

Nonlinear equation is $x^3 - 3x + 10 = 0$.

Here, $y = x^3 - 3x + 10$ which is represented as 'f' in the Program NRNDF.CPP mentioned above.

The formula for numerical differentiation used in Program NRNDF.CPP is the same as presented in Eq. (3.26) which is reproduced below:

$$y'_0 \approx (y_{-2} - 8y_{-1} + 8y_1 - y_2)/2h. \qquad (3.26)$$

As in case of Example 4 the iteration started choosing the iteration Model (1 here) and value of $x = 0.5$ which are given as input to the Program NRNDF.CPP in response to the prompt given by the computer. The output generated *instantly* by the C++ program NRNDF.CPP in respect of single variable nonlinear equation is presented below. It would be seen from the output that the value of $x = -2.61289$ gives the values of the functions 'f' within the tolerance limit of error $= 0.0001$ after

22 iterations, as was the case in respect of Example 4 presented earlier, and can be adopted as the solution of the nonlinear equation in this case also.

Example 8 Here, also the same system of nonlinear simultaneous equations solved in Example 5 earlier using manually derived expressions for derivatives is solved using Newton–Raphson approximation technique incorporating numerical differentiation procedure. The system of two nonlinear simultaneous equations in this case is as follows:

$$x^3 + y^3 - 6x + 3 = 0, x^3 - y^3 - 6y + 2 = 0.$$

Here, $f(x, y) = x^3 + y^3 - 6x + 3$ and $g(x, y) = x^3 - y^3 - 6y + 2$ which are expressed as 'f' and 'g', respectively in the Program NRNDF.CPP.

As in case of Example 5, the process starts with the arbitrary selection of the values of $x_0 = 1$ and $y_0 = 1$ (in this case, the values are changed from 0.75 to 1 for both x_0 and y_0) which are given as input to the Program NRNDF.CPP in response to the prompt given by the computer when choice of ITMODEL is given as '2', i.e. model for solving a system of two nonlinear simultaneous equations is chosen. The input-output entries in the computer at the start of the process, i.e. when iteration number is = '0' are shown below:

```
PRESET VALUE OF LIMIT OF ITERATION NUMBER 'ITRL' = 50
PRESET VALUE OF ACCEPTABLE MARGIN OF ERROR 'ITOL'= 0.0001
PRESET VALUE OF CHOSEN SPACING FOR ITERATION 'h'= 0.02

DO YOU WANT TO CHANGE THE ABOVE PRESET VALUES (Y/N) ? n
YOU HAVE NOT CHANGED THE VALUES OF 'ITRL','ITOL'& 'h'

'ITERATION MODELS' AVAILABLR HERE ARE SHOWN BELOW :

ITMODEL TO SOLVE SINGLE VARIABLE NONLINEAR EQUATION(ITMODEL- 1)
ITMODEL TO SOLVE TWO VARIABLES NONLINEAR EQUATIONS (ITMODEL- 2)

PUT YOUR CURRENT CHOICE OF ITMODEL (1/2)
2
PUT YOUR CHOSEN VALUE OF X 1
PUT YOUR CHOSEN VALUE OF Y 1
CURRENT VALUE OF ITERATION NUMBER IS = 0
CURRENT VALUE OF fx IS = -3
CURRENT VALUE OF fy IS = 3
CURRENT VALUE OF gx IS = 3
CURRENT VALUE OF gy IS = -10.3728
CURRENT VALUE OF f IS = -1
CURRENT VALUE OF g IS = -4
CURRENT VALUE OF x IS = 1
CURRENT VALUE OF y IS = 1
```

It would be seen from above that the computer has calculated ***instantly*** the values of partial derivatives fx, fy, gx, and gy, based on the starting input values of x and y, i.e. 1 and 1 given by the user in this case, using the formula (3.26) for numerical differentiation presented above. Complete output generated ***instantly*** by the C++ program NRNDF.CPP in respect of the above system of two nonlinear simultaneous equations is given below. It would be seen from the output generated by the C++ program NRNDF.CPP and presented below, that the values of $x = 0.532371$ and $y = 0.351269$ give the values of functions $f(x,y)$ and $g(x,y)$ within the tolerance limit of error $= 0.0001$, and hence, can be adopted as solution of the nonlinear equations in this case also. It may be observed that for the same system of two nonlinear equations, the solution values of $x = 0.532368$ and $y = 0.351257$ were given by the Programs NR123.CPP in Example 5, as against the values of $x = 0.532371$, and $y = 0.351269$ given here by the Program NRNDF.CPP. Although the results are same up to four decimal places, the slight difference in other decimal places may be accounted for by the difference in arbitrary choice of initial values of x_0 and y_0 in the two cases.

Thus, it can be concluded that the solution obtained using numerical differentiation procedure is equally accurate and can be adopted in actual problem situation.

Output of the C++ Program NRNDF.CPP in respect of single variable nonlinear equation in this case:

```
PRESET VALUE OF LIMIT OF ITERATION NUMBER 'ITRL' = 50

PRESET VALUE OF ACCEPTABLE MARGIN OF ERROR 'ITOL'= 0.0001
PRESET VALUE OF CHOSEN SPACING FOR ITERATION 'h'= 0.02

DO YOU WANT TO CHANGE THE ABOVE PRESET VALUES (Y/N) ? n
YOU HAVE NOT CHANGED THE VALUES OF 'ITRL','ITOL'& 'h'

'ITERATION MODELS' AVAILABLR HERE ARE SHOWN BELOW :

ITMODEL TO SOLVE SINGLE VARIABLE NONLINEAR EQUATION(ITMODEL- 1)
ITMODEL TO SOLVE TWO VARIABLES NONLINEAR EQUATIONS (ITMODEL- 2)

PUT YOUR CURRENT CHOICE OF ITMODEL (1/2)
1
PUT YOUR CHOSEN VALUE OF X .5
CURRENT VALUE OF ITERATION NUMBER IS = 0
CURRENT VALUE OF fx IS = -2.25
CURRENT VALUE OF f IS = 8.625
CURRENT VALUE OF x IS = 0.5
PUT ANY CHARACTER TO CONTINUE n
NEW VALUE OF ITERATION NUMBER IS = 1
NEW VALUE OF x IS = 4.33333
NEW VALUE OF f IS = 78.3704
NEW VALUE OF ITERATION NUMBER IS = 2
NEW VALUE OF x IS = 2.86389
NEW VALUE OF f IS = 24.8975
```

```
NEW VALUE OF ITERATION NUMBER IS = 3
NEW VALUE OF x IS = 1.71152
NEW VALUE OF f IS = 9.87901
NEW VALUE OF ITERATION NUMBER IS = 4
NEW VALUE OF x IS = 0.00469137
NEW VALUE OF f IS = 9.98593
NEW VALUE OF ITERATION NUMBER IS = 5
NEW VALUE OF x IS = 3.33341
NEW VALUE OF f IS = 37.0393
NEW VALUE OF ITERATION NUMBER IS = 6
NEW VALUE OF x IS = 2.11239
NEW VALUE OF f IS = 13.0887
NEW VALUE OF ITERATION NUMBER IS = 7
NEW VALUE OF x IS = 0.852234
NEW VALUE OF f IS = 8.06228
NEW VALUE OF ITERATION NUMBER IS = 8
NEW VALUE OF x IS = 10.6712
NEW VALUE OF f IS = 1193.17
NEW VALUE OF ITERATION NUMBER IS = 9
NEW VALUE OF x IS = 7.14765
NEW VALUE OF f IS = 353.722
NEW VALUE OF ITERATION NUMBER IS = 10
NEW VALUE OF x IS = 4.79368
NEW VALUE OF f IS = 105.775
NEW VALUE OF ITERATION NUMBER IS = 11
NEW VALUE OF x IS = 3.18953
NEW VALUE OF f IS = 32.8788
NEW VALUE OF ITERATION NUMBER IS = 12
NEW VALUE OF x IS = 1.99478
NEW VALUE OF f IS = 11.9531
NEW VALUE OF ITERATION NUMBER IS = 13
NEW VALUE OF x IS = 0.657345
NEW VALUE OF f IS = 8.31201
NEW VALUE OF ITERATION NUMBER IS = 14
NEW VALUE OF x IS = 5.53616
NEW VALUE OF f IS = 163.07
NEW VALUE OF ITERATION NUMBER IS = 15
NEW VALUE OF x IS = 3.70283
NEW VALUE OF f IS = 49.6608
NEW VALUE OF ITERATION NUMBER IS = 16
NEW VALUE OF x IS = 2.40052
NEW VALUE OF f IS = 16.6314
NEW VALUE OF ITERATION NUMBER IS = 17
NEW VALUE OF x IS = 1.23646
NEW VALUE OF f IS = 8.18097
NEW VALUE OF ITERATION NUMBER IS = 18
NEW VALUE OF x IS = -3.92008
NEW VALUE OF f IS = -38.4799
NEW VALUE OF ITERATION NUMBER IS = 19
NEW VALUE OF x IS = -3.0273
NEW VALUE OF f IS = -8.662
NEW VALUE OF ITERATION NUMBER IS = 20
NEW VALUE OF x IS = -2.67366
```

```
NEW VALUE OF f IS = -1.09158
NEW VALUE OF ITERATION NUMBER IS = 21
NEW VALUE OF x IS = -2.61448
NEW VALUE OF f IS = -0.0278836
NEW VALUE OF ITERATION NUMBER IS = 22
NEW VALUE OF x IS = -2.61289
NEW VALUE OF f IS = -1.98937e-05
CURRENT VALUE OF ITERATION NUMBER IS = 22
CURRENT  VALUE OF f IS = -1.98937e-05
CURRENT VALUE OF x IS = -2.61289
CURRENT VALUE OF ITOL CHOSEN IS = 0.0001


 SINCE THE CURRENT SOLUTION MEETS THE VALUE OF CHOSEN
 MARGIN OF ERROR, THE SOLUTION ACCEPTED & PROGRAM TERMINATED
```

Output of the C++ Program NRNDF.CPP in respect of the System of Two Variable Nonlinear Simultaneous Equations in this case:

```
PRESET VALUE OF LIMIT OF ITERATION NUMBER 'ITRL' = 50
PRESET VALUE OF ACCEPTABLE MARGIN OF ERROR 'ITOL' = 0.0001
PRESET VALUE OF CHOSEN SPACING FOR ITERATION 'h' = 0.02

DO YOU WANT TO CHANGE THE ABOVE PRESET VALUES (Y/N) ? n
YOU HAVE NOT CHANGED THE VALUES OF 'ITRL', 'ITOL' & 'h'

'ITERATION MODELS' AVAILABLR HERE ARE SHOWN BELOW :

ITMODEL TO SOLVE SINGLE VARIABLE NONLINEAR EQUATION (ITMODEL- 1)
ITMODEL TO SOLVE TWO VARIABLES NONLINEAR EQUATIONS (ITMODEL- 2)

PUT YOUR CURRENT CHOICE OF ITMODEL (1/2)
2
PUT YOUR CHOSEN VALUE OF X 1
PUT YOUR CHOSEN VALUE OF Y 1
CURRENT VALUE OF ITERATION NUMBER IS = 0
CURRENT VALUE OF fx IS = -3
CURRENT VALUE OF fy IS = 3
CURRENT VALUE OF gx IS = 3
CURRENT VALUE OF gy IS = -10.3728
CURRENT VALUE OF f IS = -1
CURRENT VALUE OF g IS = -4
CURRENT VALUE OF x IS = 1
CURRENT VALUE OF y IS = 1
PUT ANY CHARACTER TO CONTINUE n
NEW VALUE OF ITERATION NUMBER IS = 1
NEW VALUE OF f IS = 3.10234
NEW VALUE OF g IS = 0.035675
NEW VALUE OF x IS = -0.0115017
NEW VALUE OF y IS = 0.321832
NEW VALUE OF ITERATION NUMBER IS = 2
NEW VALUE OF f IS = 0.129314
```

```
NEW VALUE OF g IS = 0.126586
NEW VALUE OF x IS = 0.505906
NEW VALUE OF y IS = 0.327938
NEW VALUE OF ITERATION NUMBER IS = 3
NEW VALUE OF f IS = 0.0016934
NEW VALUE OF g IS = -0.0107105
NEW VALUE OF x IS = 0.532161
NEW VALUE OF y IS = 0.35291
NEW VALUE OF ITERATION NUMBER IS = 4
NEW VALUE OF f IS = 3.45829e-06
NEW VALUE OF g IS = 0.000878399
NEW VALUE OF x IS = 0.53236
NEW VALUE OF y IS = 0.351118
NEW VALUE OF ITERATION NUMBER IS = 5
NEW VALUE OF f IS = 2.42407e-08
NEW VALUE OF g IS = -7.38371e-05
NEW VALUE OF x IS = 0.532371
NEW VALUE OF y IS = 0.351269
CURRENT VALUE OF ITERATION NUMBER IS = 5
CURRENT VALUE OF f IS = 2.42407e-08
CURRENT VALUE OF g IS = -7.38371e-05
CURRENT VALUE OF x IS = 0.532371
CURRENT VALUE OF y IS = 0.351269
THE VALUE OF ITOL CHOSEN IS = 0.0001

SINCE THE CURRENT SOLUTION MEETS THE VALUE OF CHOSEN
MARGIN OF ERROR, THE SOLUTION ACCEPTED & PROGRAM TERMINATED
```

3.7 Closure

In Computer-Aided Design (CAD) by Optimization, in the area of Architecture, Engineering, and Construction (AEC) particularly related to urban development and management, lot of numerical analysis has to be carried out and many linear and nonlinear and other complex equations have to be formulated and solved for efficient and equitable problem solution. This is because the scientific relationships between different elements in this area have to be explored and expressed in linear and nonlinear equations as may be appropriate, and thereafter, carry out planning and design analysis, and also optimization if feasible, so as to achieve ***efficiency*** and ***equity*** in problem solutions within the ***resource-constraints*** at various levels as may be applicable in many countries. However, this is rarely done, and in some cases, for example, generalized prescriptions are specified without linking scientifically the different built-form elements creating *inefficiencies* and *inequities* and consequent urban problems. Tedious and time consuming nature of numerical analysis may be one of the reasons for inhibition in the use of such techniques particularly in the area of urban planning, development and management leading some time to generalized prescriptions cited above. Such tediousness can be removed and speedy analysis could be carried out with the help of easily available and less costly modern computers with high speed and graphic capabilities permitting numerical and graphical analysis in no time. However this is possible only if professionals working in this area are conversant with the relevant numerical analysis and techniques, and also suitable software are available for this purpose.

In this chapter, an attempt is made to present some of the numerical techniques relevant to the Computer-Aided Design (CAD) by optimization, in the area of Architecture, Engineering, and Construction (AEC) particularly related to urban development and management. In this effort, stress is given to provide numerical examples in each topic, so as to familiarize and create interest of such professionals and students in such techniques, and thus, remove any inhibition to apply such techniques in Computer-Aided Design (CAD) by optimization, in their respective field to derive benefits. To help this process, a number of user-friendly software with complete listing are incorporated in this chapter. These techniques and tools are used in subsequent chapters related to design optimization, computer-aided structural, design layout design, building design, and dwelling-layout system design. Computer-Aided Design (CAD) by optimization does not even form a *full-length subject of study* in the curricula in the area of Architecture, Engineering, and Construction (AEC) particularly related to urban development and management, in many architecture and civil engineering colleges in many countries.

Hopefully, with the availability of tools and techniques presented above and in the subsequent chapters, concerns of AEC professionals and students will increase in the subject area of Computer-Aided Design (CAD) by optimization and its importance in achieving problem solution ***resource efficiently*** and ***equitably*** in their area of operations will be felt. This will popularize this subject leading to its introduction as a *full-length subject of study* in the *curricula* in the area of Architecture, Engineering, and Construction (AEC) particularly related to urban development and management.

In this chapter, two C++ programmes are developed and their use is demonstrated above. To help readers of this book to use such techniques to achieve their own problem solutions, complete listings of the respective C++ programs are included in an online space connected to this book.

Chapter 4
Elements of Cost Engineering and Database Management in Computer-Aided Design

4.1 Introduction

Cost is a major consideration in all human activities and is of paramount importance in all AEC activities including housing development where many disciplines have to interact together in a project, making cost as the common language for such interactions specifically in the computer-aided design process. Cost engineering involves using engineering judgement and experience in the application of scientific principles and techniques to problems of cost estimating, cost control, planning and scheduling including productivity analysis, optimization, and operations research (Jelen & Black, 1983). As stated earlier, there are unparalleled transformations in computers in terms of their capacity to capture data, to analyse, manipulate and store them, to disseminate and communicate information, and to apply these capabilities in different areas and fields of specialization including CAD. Such transformations are occurring with increasing rapidity and frequency in many areas and the rate at which IT and computer applications are spreading to various sectors is truly extraordinary and has few, if any, parallels in technological history. Apparently cost engineering using such techniques can greatly contribute in solving various societal problems such as housing and urban development in an efficient and effective manner. Even for a business entity, it is imperative to ensure accurate estimation of costs based on such scientific principles and techniques to remain competitive. An estimate based on overdesign will eliminate a business entity from bidding while an under design estimate may win the bid but may lead to disaster in terms of profitability.

There are many components of cost engineering. Some of these are cost indices, elemental estimating, building specification; elemental unit cost, economic analyses of projects, cost control, optimization, management aspects, and a wide range of other topics (Jelen & Black, 1983). Sources of finance, determinants of the cost of money, interest, and the time value of money are important components of cost engineering, which are also very relevant to real estate, urban development, and

© The Author(s), under exclusive license to Springer Nature Switzerland AG 2022
B. K. Chakrabarty, *Integrated CAD by Optimization*,
https://doi.org/10.1007/978-3-030-99306-1_4

management including housing development with stress on affordability, efficiency, and equity in such development process. Computerized information systems cost estimation is a very important element in **Computer-Aided Design (CAD)** by optimization which is the subject matter of this book. Hence, instead of covering the whole range of cost engineering, only the above relevant topics are discussed in this chapter while a separate Chap. 7 is devoted to optimization techniques relevant to AEC problems.

4.2   Time Value of Money and Affordability Analysis in Housing Development

Investments and loans are spread over a time period, and time has a monetary value which is applicable even without any real monetary or banking transactions. In housing development, a housing developer invests money to complete a project involving a number of dwelling units which are offered for purchase by the housing beneficiaries at a selling price P per unit depending on the investment made and the desired profit margins. A housing beneficiary can purchase a dwelling unit making full down payment for the selling price P if he or she can afford it. However, in most cases, a housing beneficiary cannot afford full down payment or he or she can afford a part down payment and the remaining amount can be paid in equated instalments (EI) from the salary, for which the housing developer will charge a rate of return i per period, which may be monthly, yearly, and so on. The equated instalment has to be calculated keeping in view the amount of capital cost or selling price which is allowed to be deferred for payment (which can be designated as DC), the period for which it is deferred (n) and the rate of return charged by the developer (i) for the period. The relationships between the above factors can be easily developed using the business mathematics principles as discussed below.

If a present value P earns a rate of return i per year, then its future value S_1 at the end of 1 year will be given by the relationship:

$$S_1 = P + Pi$$

At the beginning of second year, the starting value of investment becomes S_1 and at the end of second year, its value becomes:

$$S_2 = S_1 + S_1 i = (P + Pi) + (P + Pi)\, i = P(1 + i) + P(1 + i)i = P(1 + i)(1 + i)$$
$$= P(1 + i)^2$$

Similarly, at the end of third year, its value becomes $S_3 = P(1 + i)^3$, and at the end of n year, its value becomes:

$$S_n = P(1+i)^n \qquad (4.1)$$

The factor $(1+i)^n$ is called the *compound interest factor*. Generally, a year is taken as the unit period of time in engineering economics, with n in years and i as the yearly rate of return in percentage. In case, compounding occurs p times per year, the relationship becomes:

$$S_n = P(1+i/p)^{np} \qquad (4.2)$$

where i/p is the rate of return per period and np is the number of periods. Using the Eq. (4.1), the present value P can be expressed in terms of the future value S (representing the time value conversion relationship), as shown below:

$$P = S/(1+i)^n \text{ or } P = S\,(1+i)^{-n} \qquad (4.3)$$

In most cases, a housing beneficiary pay the housing cost in equated instalments EI for the decided repayment period n or RP. The equal end of the period payment of EI for n periods can be expressed diagrammatically as follows:

Moving all the terms to 0 time using the time value conversion relationship presented above, all the terms can be replaced by a single present value P as below:

$$P = \text{EI}\Big[1/(1+i) + 1/(1+i)^2 + 1/(1+i)^3 + 1/(1+i)^4 + \cdots + 1/(1+i)^{(n-1)}$$
$$+ 1/(1+i)^n\Big\}$$

The terms within the brackets represent a geometrical series of n terms with first term as $1/(1+i)$ and the ratio as $1/(1+i)$, and the sum of a geometrical series is given by:

Sum of geometric series $= \text{first term}^* \Big[\big(1 - \text{ratio}^{\text{number of terms}}\big)/(1 - \text{ratio})\Big]$

Thus,the previous present value P becomes $= \text{EI}^*[1/(1+i)]$
$* [1 - 1/(1+i)^n]/[1 - 1/(1+i)]$

This can be simplified as $P = \text{EI}^*[(1+i)^n - 1]/[i(1+i)^n] \qquad (4.4)$

Equation (4.4) can be modified to express EI in terms of the capital cost P as follows:

$$\mathrm{EI} = P/ * [i(1 + i)^{n}]/[(1 + i)^{n}-1] \tag{4.5}$$

The relationship between the future value S and the present amount or capital cost P is shown in expression (4.3) above. Placing this expression for P in terms of S in Eq. (4.4), we get:

$$S^{*}(1 + i)^{-n} = \mathrm{EI}^{*}[(1 + i)^{n}-1]/[i(1 + i)^{n}]$$

Simplifying, we get:

$$S = \mathrm{EI}^{*}[(1 + i)^{n}-1]/i \tag{4.6}$$

The factor $[(1 + i)^{n} - 1]/[i(1 + i)^{n}]$ is called *equal payment series present-value factor* or *annuity present-value factor*. Similarly, the factor $[i(1 + i)^{n}]/[(1 + i)^{n} - 1]$ is called the *capital cost recovery factor*. The capital cost recovery factor converts a single zero-time cost to an equivalent uniform end of the year cost for the n period. The factor $[(1 + i)^{n} - 1]/i$ is called the *equal payment series future value factor*. All these factors are generally presented in annuity tables; but with availability of cheap calculators or computers, these can be instantly calculated using such tools. Generally, only one series of equal instalments for housing loans as adopted is presented above. To improve affordability, it is desirable to adopt a variable equal payment series where instalment amounts are suitably increased for each loan repayment sub-period depending on the income of a housing beneficiary as shown later.

4.2.1 Housing Affordability Analysis: Constant Equal Payment Series in Entire Period

A person earns 100,000 MUs (Monetary Units) per month and he can afford 10% of his income for purchase of a housing unit. If money is worth 10% per year (i.e. interest rate is 10% per annum) and the maximum loan repayment period allowed is 10 years, what is the maximum loan amount he can afford to borrow?

Affordable equated monthly instalment (EMI) in this case $= 0.1*100,000 = 10,000$ MU. Running the C++ program EMI.CPP and putting the given interest rate of 10% per annum and repayment period of 10 years in the C++ program, and responding to the computer prompts gives the following result:

```
NOMENCLATURES USED IN THE PROGRAM ARE:

RI = RATE OF INTEREST IN % PER PERIOD.PERIODS MAY BE IN MONTHS,YEARS AND
SO ON.
MU = MONETARY UNITS WHICH MAY BE IN DOLLARS,RUPEES AND SO ON.
RP = REPAYMENT PERIOD FOR ENTIRE LOAN WHICH MAY BE IN MONTHS,YEARS AND SO
ON.
```

LA or AHC = LOAN AMOUNT OR AFFORDABLE HOUSING COST(AHC) IN MU
          WHICH MAY BE IN DOLLARS,RUPEES AND SO ON?
EI = EQUATED INSTALMENT PER PERIOD. IT IS DESIGNATED AS:
EMI i.e. EQUATED MONTHLY INSATALMENT, IF PERIOD IS TAKEN AS MONTH.
EM1 = EQUATED MONTHLY INSATALMENT FOR PERIOD 1 WHEN VARIABLE EMI's ARE
TAKEN.
EM2 = EQUATED MONTHLY INSATALMENT FOR PERIOD 2 WHEN VARIABLE EMI's ARE
TAKEN.
EM3 = EQUATED MONTHLY INSATALMENT FOR PERIOD 3 WHEN VARIABLE EMI's ARE
TAKEN.
EM4 = EQUATED MONTHLY INSATALMENT FOR PERIOD 4 WHEN VARIABLE EMI's ARE
TAKEN.
EM5 = EQUATED MONTHLY INSATALMENT FOR PERIOD 5 WHEN VARIABLE EMI's ARE
TAKEN.
RP1 = VALUE OF REPAYMENT SUB-PERIOD-1 FOR EM1.
RP2 = VALUE OF REPAYMENT SUB-PERIOD-2 FOR EM2.
RP3 = VALUE OF REPAYMENT SUB-PERIOD-3 FOR EM3.
RP4 = VALUE OF REPAYMENT SUB-PERIOD-4 FOR EM4.
RP5 = VALUE OF REPAYMENT SUB-PERIOD-5 FOR EM5.
RP0 = TOTAL PERIOD APPLICABLE FOR ENTIRE LOAN REPAYMENT WITH EMI OR EM0.

GENERALLY RP IS IN YEARS,RI IN % PER YEAR AND EI PER MONTH i.e. EMI;
HENCE REQUIRES SUITABLE CONVERSION IN CALCLATION AS SHOWN BELOW.

PUT INPUT VALUES: RI (% PER YEAR) & TOTAL REPAYMENT PERIOD RP(YEARS)IN
SAME ORDER:

10 10

CALCULATED VALUE OF INITIAL INTEREST FACTOR INTF0 =   2.7070
CALCULATED VALUE OF UNIFORM ANNUITY PRESENT VALUE FACTOR APVF =
75.6712
CALCULATED VALUE OF UNIFORM CAPITAL COST RECOVERY FACTOR CCRF =   0.0132

PUT ITERATION CHOICE 1, 2 OR 3:

1 FOR FINDING THE AFFORDABLE HOUSING COST(AHC)OR LOAN AMOUNT(LA)WITH
UNIFORM EMI
2 FOR FINDING THE AFFORDABLE HOUSING COST (AHC) OR LOAN AMOUNT (LA)
 WITH VARYING EMI's WITH DESIRED EMI-INCREMENT RATIO PER SUB-PERIOD
3 FOR FINDING UNIFORM EQUATED MONTHLY INSTALMENT (EMI) FOR A GIVEN LOAN
AMOUNT (LA)
1
PUT THE VALUE OF DESIRED AFFORDABLE UNIFORM EQUATED
MONTHLY INSTALMENT (EMI) IN MU
10000
CALCULATED VALUE OF LOAN AMOUNT I.E. AHC (IN MU) LA= 756711.6337
ASSUMED AFFORDABLE VALUE of EQUATED
MONTHLY INSTALMENT (IN MU) EMI = 10000.0000
ASSUMED VALUE OF RATE OF INTEREST (% PER YEAR) RI =   10.0000
ASSUMED VALUE OF REPAYMENT PERIOD (YEARS) RP =   10.0000

4.2.2   Housing Affordability Analysis: Variable Equal Payment Series in Each Sub-period

In case of home loans, generally only one EMI (Equated Monthly Instalment) is adopted for the entire RP (Repayment Period) which may be as long as 10–15 years, during which period the monthly income of the housing beneficiary may undergo substantial upward change increasing his or her capacity to pay a higher EMI and consequently a higher loan discharge capacity or a higher AHC (Affordable Housing Cost). In this dynamic situation, it is desirable to allow a variable EMI and accordingly a higher AHC or LA (Loan Amount), so that a housing beneficiary can get a better housing fulfilling his/her need during the life period, rather than restricting the LA as per the EMI only related to the present income housing expense ratio. The eligibility of LA adopting a variable EMI can be derived as follows:

RP0	RP1	RP2	RP3	RP4	ERP
0	1	2	3	4	5
EM0	EM1	EM2	EM3	EM4	

Adopting a time diagram of five loan sub-periods as above, the loan repayment periods for each of the EMIs of EM0, EM1, EM2, EM3, and EM4 are RP0, RP1, RP2, RP3, and RP4, respectively, which are all in months. ERP is the end of full loan repayment period and is taken as the focal date in this case. If the length of each sub-period is designated as SRP in months, total loan repayment period is designated as RP in years RPM in months, the values of RP0, RP1, RP2, RP3, and RP4 are as follows:

$$RP0 = RP^*12 = RPM, RP1 = RPM\text{-}SRP, RP2 = RPM\text{-}2^*SRP,$$
$$RP3 = RPM\text{-}3^*SRP, RP4 = RPM\text{-}4^*SRP$$

If the value of the first EMI (i.e. Equated Monthly Instalment) is given as EM0, and the desired EMI-increment ratios are designated as K1, K2, K3, and K4, respectively, the values of corresponding additional EMIs for each of the loan repayment sub-periods are as follows:

$$EM1 = K1^*EM0, EM2 = K2^*EM0, EM3 = K3^*EM0, EM4 = K4^*EM0$$

Using the *equal payment series future value factors* from Eq. (4.6), the future values of all the above EMIs on the above focal date can be easily calculated. Similarly, the future value of the LA on the above focal date can be calculated using the Eq. (4.3). Equating these two values, the value of eligible LA can be easily calculated for the case of varying EMIs as proposed above. A C++ program EMI. CPP for such calculations is presented below. One example of housing development affordability estimation using the above expressions and the C++ program is presented below.

4.3 Comparative Housing Affordability Analysis: Using Constant and Variable Equal Payment in Each Sub-period

To obviate the above difficulty, in the C++ program, an iteration Choice 2 is provided to run the program with varying EMIs with desired EMI-increment ratio per sub-period, with desired number of sub-periods. The results of a program run adopting such a procedure, deleting nomenclature, avoiding repetitions, and starting from input values, are presented below:

```
PUT INPUT VALUES: RI (% PER YEAR) & TOTAL REPAYMENT PERIOD RP (YEARS) IN
SAME ORDER:
10 10

CALCULATED VALUE OF INITIAL INTEREST FACTOR INTF0 =    2.7070
CALCULATED VALUE OF UNIFORM ANNUITY PRESENT VALUE FACTOR APVF =
75.6712
CALCULATED VALUE OF UNIFORM CAPITAL COST RECOVERY FACTOR CCRF =   0.0132

PUT ITERATION CHOICE 1,2 OR 3:

1 FOR FINDING THE AFFORDABLE HOUSING COST(AHC)OR LOAN AMOUNT(LA)WITH
UNIFORM EMI
2 FOR FINDING THE AFFORDABLE HOUSING COST(AHC)OR LOAN AMOUNT(LA)
  WITH VARYING EMI's WITH DESIRED EMI-INCREMENT RATIO PER SUB-PERIOD
3 FOR FINDING UNIFORM EQUATED MONTHLY INSTALMENT(EMI)FOR A GIVEN LOAN
AMOUNT(LA)
2

PUT DESIRED AFFORDABLE FIRST EMI VALUE (MU),LENGTH OF EACH LOAN
SUB-PERIOD DEVISION SRP(MONTHS),AND EFFECTIVE NUMBER NP OF SUB-PERIODS
IN SAME ORDER
10000 24 5

PUT DESIRED EMI-INCREMENT RATIO:K1,K2,K3,K4,K5 PER PERIOD IN SAME ORDER
0.2 0.4 0.6 0.8 1.0

CALCULATED VALUE OF AFFORDABLE HOUSING COST OR LOAN AMOUNT MU=
1217996.6506
CALCULATED VALUE OF AFFORDABLE LOAN AMOUNT i.e.LA IN MU's LA =
1217996.6506
ASSUMED VALUE OF RATE OF INTEREST (% PER YEAR) RI =   10.0000
ASSUMED VALUE OF TOTAL REPAYMENT PERIOD (YEARS) RP =   10.0000
REPAYMENT SUB-PERIODS ARE:RP1=  96,RP2=  72,RP3=  48,RP4=  24,RP5=  0
FUTURE VALUES OF EACH EMI-INCREMENTS,S0 = 2048449.8, S1 = 292362.2,S2 =
392445.3
, S3 = 352335.0, S4 = 211575.3, S5 =   0.0
FUTURE VALUE OF ALL EMI's ON THE FOCAL DATE-SP0 = 3297167.5
FUTURE VALUE OF LOAN AMOUNT ON THE FOCAL DATE-FLA = 3297167.5
```

It would be seen from above that adoption of a variable EMI with desired EMI-increment ratio per sub-period as: 0.2, 0.4, 0.6, 0.8, and 1.0 leads to the increase of affordable housing cost (AHC) or loan amount to 1,217,996.65 MU as against 756,711.63 MU with constant EMIs, i.e. an increase of 1.6 times although all terms such as first EMI, RI, and total RP remain unchanged except that variable EMIs are adopted as shown in the above computer results. Hence, it is desirable to adopt such housing development affordability analysis as part of CAD system to achieve housing development which is both affordable and satisfying to the housing beneficiaries.

4.4   Affordability Analysis and Computer-Aided Design

It would be seen from above that on putting 10,000 as the input value of EMI, computer gives instantly the calculated loan amount of 756,711.63 MU as the output, which should be taken as the given ceiling cost within which a housing unit should be designed, if providing *affordable housing* is the design objective. Adopting a Computer-Aided Design (CAD) procedure incorporating affordability analysis and optimal design techniques in an *integrated manner*, it is possible to achieve such a design as illustrated later. Thus, in this new technique, the design procedure is reversed, where affordable cost for a housing beneficiary is determined first using the cost engineering principles as illustrated above, and this cost is given as input constraint and the housing design is generated using the CAD procedure fulfilling this constraint, In case, there is possibility of increased income of the housing beneficiary during the repayment period of 10 years adopted in this case, a variable EMI with desired EMI-increment ratio per sub-period can be adopted (responding to the *income-dynamics* of the housing beneficiary) permitting a higher loan amount, i.e. AHC and consequently a better housing unit improving the satisfaction level. However, this is rarely adopted in conventional practice. One of the reasons for inhibition in adopting such a practice may be the complicated calculations involved if manual practice is adopted and non-availability of computer programs giving instant results.

4.5   Listing of the Complete Computer Program: EMI.CPP

To help the above process, the C++ Listing of the complete computer program: EMI.CPP is placed in an online space connected to this book. In all the 3 iteration choices with any input values of rate of interest RI (% per year) and total repayment period RP (years) can be chosen. In the iteration choice 2, i.e. program with varying EMIs, there are two options of sub-period lengths, i.e. 3 and 5 numbers. The length of each sub-period in terms of months can be chosen as desired by the stakeholders to meet the specified requirements. Thus, the program gives numerous options to meet a variety of situations to fulfil the affordability requirement and the satisfaction level of the housing beneficiary. In this chapter, two C++ programmes are developed and their use is demonstrated above. To help readers to use such techniques to achieve their own problem solutions, complete listings of the respective C++ programs are placed in an online space connected to the book as cited above.

4.6   Database Management in Cost Engineering

Cost estimating is a process to predict the final outcome of a future expenditure, taking into accounts the parameters and conditions concerning a project which are known when the cost estimate is prepared. In computer-aided design in the AEC sector cost is a vital parameter and in fact the optimal design and analysis presented in this book is based mainly on cost considerations. Therefore, it is imperative to incorporate suitable methods and techniques of cost estimating so that the costs predicted are realistic as per the site-specific conditions. In this context, suitable site-specific data particularly regarding labour and materials, which forms the basis for cost estimating in many cases, need to be generated and managed dynamically as costs vary with time.

Computer-aided design involves a variety of data related to cost, planning, design, and so on. Physical planning and design data are available in professional data literature which remains constant. As, for example, there is hardly any change in anthropometric data related to dwelling component design such as bath room, water closet, and so on. However, data related to costs of housing-building components varies with time and location of projects, although material and labour consumption coefficients (generally derived based on works and methods study) for various components such as concrete, brickwork, and so on, for such components remain constant. In view of above, it is desirable to keep the cost-related data in a comprehensive (covering all possible items) and dynamic form, say, in the form of a matrix amenable to change with time, and thus, ensure an efficient data management for cost estimating.

There are numerous project related data for various items of works in building and housing projects. There is need for a comprehensive, robust, and dynamic data management process to ensure realistic cost estimating at any point of time. Thus, it is desirable to generate, update, and keep all such data covering all possible items of works related to various types of AEC projects, in a comprehensive and dynamic manner as stated above, which is beyond the scope of this book. Interested readers may consider developing such comprehensive and dynamic cost engineering and data management tool for AEC operations. However, to illustrate the concept, a simple data management and cost engineering tool covering only 15 items related to building and housing project is presented below.

Effective management, storage, and transfer of data/information related to planning, architectural, and engineering design process (including cost engineering data) should be an essential feature of an integrated CAD system (especially for AEC sector), for speedy information handling and decision-making resulting to higher productivity (i.e. efficiency and effectiveness), which is one of the basic objectives of CAD. A database is defined as a logical collection of related information, which should be (a) dynamic in information handling, (b) multi-user oriented, and (c) multi-layer structured. A database may contain millions of different sets of data, necessitating development of a system that can accept, organize, store, and retrieve information quickly. Main thrust in the construction of a planning and design database should be to include efficient description of geometries, specifications, and unit cost details related to different components of a facility.

A design database may consist of seven groups: (1) general data, (2) technical data, (3) geometric data, (4) facility specification data, (5) facility cost data, (6) data on previous projects, and (7) cost data. A database management system is the software systems developed for efficient operation of the above functions related to a database. A suitable system for planning and design decision support and for database management, stressing the use of information technology to secure *user citizen participation* in *planning and design process,* is essential in AEC sector. In view of importance of data management, storage, and transfer in planning and design process and specifically, in respect of cost engineering data a separate section in this chapter is devoted to this topic to give a broad understanding to the reader.

4.7   Cost Engineering and Database Management Model in CAD

Keeping in view all the factors discussed above a cost engineering and database management model in the form of C++ program named as: COSTENG.CPP is developed and available in an online space attached to this book. In this program, only 15 items of work and 16 items of labour and material coefficients are considered in the form of a *Matrix* just for illustration, which can be extended to cover hundreds of items of works as in a standard schedule of rates in different countries, adopting the same technique presented in this example.

A smaller cost engineering database management program: DBM.CPP is shown below, where two cost database management matrices, namely Labour and Material Quantity Coefficients (Matrix named as: LMQC), and Labour and Material Rate Coefficients (Matrix named as: LMRC) are included only as illustration to indicate importance and relevance of such techniques in CAD:

4.7.1   Database Management Matrix–Labour and Material Quantity Coefficients (Matrix: LMQC)

```
{0.20,0.167,0.00,0.00,0.00,0.00,0.00,0.00,0.00,0.00,0.00,0.00,
0.00,0.00,0.00,0.00},
 {0.90,0.60,0.05,0.05,0.04,0.27,0.17,0.00,0.47,0.00,0.24,0.65,
```

```
0.00,0.00,0.00,1.667},
{0.15,1.07,0.35,0.35,0.00,0.275,0.05,0.28,0.00,0.00,0.00,0.00,
0.494,0.00,0.00,0.528},
{0.15,1.07,0.35,0.35,0.00,0.275,0.063,0.268,0.00,0.00,0.00,0.00,
0.494,0.00,0.00,0.528},
{0.9,0.6,0.45,0.45,0.04,0.27,0.32,0.00,0.445,0.89,0.00,0.00,0.00,
0.00,0.00,2.277},
{0.15,1.40,0.43,0.43,0.00,0.275,0.05,0.28,0.00,0.00,0.00,0.00,
0.494,0.00,0.00,1.184},
{0.15,1.40,0.43,0.43,0.00,0.275,0.063,0.268,0.00,0.00,0.00,0.00,
0.494,0.00,0.00,1.184},
{0.0103,0.094,0.094,0.00,0.00,0.0392,0.0043,0.0184,0.00,0.00,
0.00,0.00,0.00,0.00,0.00,0.08},
{0.0086,0.081,0.081,0.00,0.00,0.031,0.0036,0.0154,0.00,0.00,0.00,
0.00,0.00,0.00,0.00,0.08},
{0.0043,0.081,0.065,0.00,0.00,0.0292,0.0027,0.0077,0.00,0.00,
0.00,0.00,0.00,0.00,0.0,0.11},
{1.00,0.77,0.09,0.09,0.08,0.47,0.22,0.00,0.50,0.20,0.64,0.00,
0.00,0.00,0.00,1.723},
{1.00,0.77,0.09,0.09,0.08,0.47,0.32,0.00,0.445,0.22,0.67,0.00,
0.00,0.00,0.00,1.723},
{0.01,0.00,0.01,0.00,0.00,0.00,0.00,0.00,0.00,0.00,0.00,0.00,
0.00,1.05,0.00,1.111},
{0.00,0.018,0.0,0.0,0.018,0.0,0.0,0.0,0.0,0.0,0.0,0.0,0.0,0.0,
0.003,0.064 },
{2.70,2.025,0.00,3.375,0.00,0.675,0.37,0.00,0.445,0.22,0.67,0.00,
0.00,0.00,0.00,4.117}
```

4.7.2   Database Management Matrix: Labour and Material Rate Coefficients (Matrix: LMRC)

```
{70,60,120,100,90,80,2700,200,357,600,518,178,900,13.5,140,1.0}
```

4.7.3   Items of Illustrative Labour and Materials Considered

In this software-model example, the following labour and material items are considered:

```
b[0] = Labour:Un-Skilled-Category-1(US-1) (rate-each MU/day)
b[1] = labour:un-skilled-category-2(US-2) such as: coolie(rate-each
MU/day)
b[2] = labour:skilled-category-1(S-1) such as:mason-class-1(rate-
each MU/day)
b[3] = labour:skilled-category-2(S-2) such as:mason-class-2(rate-
each MU/day)
b[4] = labour:skilled-category-3(S-3) such as:mate(rate-each MU/per day)
```

```
b[5]= labour:skilled-category-4(S-4)like bhisti/other-1(rate-each
MU/day)
b[6]= material:-portland cement (PC)(rate MU per metric tonne)
b[7]= material:-fine sand(FS) (rate MU per cubic meter)
b[8]= material:-course sand(CS) (rate MU per cubic meter)
b[9]= material:-stone aggregate 10 mm nominal size(SA-S)(rate MU per
cubic meter)
b[10]= material:-stone aggregate 20 mm nominal size(SA-M)(rate MU per
cubic meter)
b[11]= material:-stone aggregate 40 mm nominal size (SA-L)(rate MU per
cubic meter)
b[12]= material:-1st class bricks(BR) (rate MU per 1000 no's)
b[13]= material:-mild steel/torsteel round bars(MS)(rate MU per kilo
grams(kg))
b[14]= white lime/other materials etc.(WL)(rate MU per quintal)
b[15]=miscellaneous/sundries/hire & running charges of plants etc.(OTH)
MU  = 'Monetary-Unit' in any Currency (Rupees, Dollars and so on).
```

4.7.4   Details of Illustrative Items of Works Considered

In this software-model example, the following items of works are considered:

```
Row-0: earthwork in foundation (EW)
Row-1: plain cement concrete in foundation (PC)
Row-2: brickwork in foundation (FBW-1)
Row-3: brickwork in foundation (FBW-2)
Row-4: damp proof course  (DPC)
Row-5: brick-work in superstructure  (SBW-1)
Row-6: brick-work in superstructure  (SBW-2)
Row-7: brick-work in superstructure  (SBW-2)
Row-8: cement plaster (CP-1)
Row-9: cement plaster (CP-2)
Row-9: cement plaster (CP-3)
Row-10 re-inforced cement concrete. in roof/floor (RCC-1)
Row-11 re-inforced cement concrete. in roof/floor (RCC-2)
Row-12 mild steel reinforcement for RCC work (MSW)
Row-13 colour washing (CW)
Row-14 cement concrete flooring (CCF)
```

4.7.5   Dynamic Database Management Matrix for Unit Cost Estimating

The cost engineering and database management model in the form of C++ program: COSTENG.CPP incorporates a dynamic matrix of input cost-engineering (Qties) coefficient and input labour-material-unit-cost(MKT.RATES) coefficient, giving choice to the user to change their values depending on the site-specific and country-specific condition as per the following broad dialogue structure:

- "Want to RUN COST ENGINEERING MODEL to Derive UNIT-COSTS?(put 'y' if yes)";
- "Do you want the individual Basic Results displayed?(put 'y' if yes)";
- "Do you want the individual Item Quantities displayed(put 'y' if yes)";
- "Want to change ALL Input Cost-Engineering(Qties)Co-Efficients?(put 'y' if yes)";
- "If 'y' Enter 240 values(Qties) 16 per row";
- "Want to change ALL Input Labour-Material-Unit-Cost(MKT.RATES)Co-Efficients(put 'y' if yes)";
- "If 'y' Enter 16 values i.e. Labour-Material-Unit-Cost(MKT.RATES)1 per column";
- "Want to change Some Input Cost-Engineering(Qties)Co-Efficients?(put 'y' if yes)";
- "If 'y' Enter Number of ROWS(0/1/../9)for which Inp.Cost-Engg.Co-eff.(Qties)-to be changed";
- "Want to change Some Input Labour-Material-Unit-Cost(MKT.RATES)Co-Efficients?(put 'y' if yes)";
- "If 'y' Enter Number of COLUMNS (0/1/../16) for which Input.Labour-Material-Unit-Cost(MKT.RATES)Co-Eff. to be changed";
- "If 'y' Enter Input values i.e. Unit-Costs(MKT.RATES)1 per column:";

Some values (i.e. quantities) of the above items designated here as: US-1, US-2, S-1, S-2, S-3, S-4, PC, FS, CS, SA-S, SA-M, SA-LBR, MS, WL, OTH, respectively, for each row designated as: EW, PC, BW-1, BW-2, DPC, BW-3, BW-4, CP-1, CP-2, CP-3, RCC-1, RCC-2, MSW, CW, CCF (details of each item are mentioned above), respectively, are shown in the Matrix: LMQC above.

Similarly, some values (i.e. Market Rates) of the above items designated here as: US-1, US-2, S-1, S-2, S-3, S-4, PC, FS, CS, SA-S, SA-M, SA-LBR, MS, WL, OTH, respectively, are shown in the Matrix: LMRC above.

It is desirable that the values (i.e. quantities) of the items designated here as: US-1, US-2, S-1, S-2, S-3, S-4, PC, FS, CS, SA-S, SA-M, SA-LBR, MS, WL, OTH, respectively, for each row designated as: EW, PC, BW-1, BW-2, DPC, BW-3, BW-4, CP-1, CP-2, CP-3, RCC-1, RCC-2, MSW, CW, CCF, shown in the Matrix: LMQC above are determined by scientific works and methods studies, in site-specific and country-specific conditions.

Similarly, the values (i.e. market rates) of the above items designated here as: US-1, US-2, S-1, S-2, S-3, S-4, PC, FS, CS, SA-S, SA-M, SA-LBR, MS, WL, OTH, respectively, shown in the Matrix: LMRC need to be determined by market assessment to make them realistic.

It may be seen from the above dialogue structure that the user can change all or some input cost-engineering (Qties) coefficient and input labour-material-unit-cost (MKT.RATES) coefficient, giving choice to the user to change their values depending on the site-specific and country-specific conditions, making it an universal model.

Illustrative examples of labour and material quantity coefficients change and labour and material rate coefficients change are presented below.

To permit only database management in a dynamic manner, i.e. changing (1) labour and material quantity coefficients and (2) labour and material rate coefficients, a separate C++ program **DBM.CPP** is developed and the listing of the same is given below.

4.8   Listing of an Illustrative Cost Engineering Database Management Program: DBM.CPP

```
/* THIS IS AN ILLUSTRATIVE PROGRAM TO DEMONSTRATE DATA BASE MANAGEMENT
   I.E. CHANGE OF DATA VALUES IN
   (1)MATERIAL AND LABOUR QUANTITY COEFFICIENTS AND
   (2)MATERIAL AND LABOUR RATE COEFFICIENTS, BASED ON LOCAL AND
   COUNTRY-SPECIFIC DATA INPUT VALUE, WHICH METHOD INCORPORATED
   IN PROGRAM COSTENG.CPP
 */

#include<iostream.h>
#include<math.h>
#include<fstream.h>
#include<stdio.h>
#include<stdlib.h>
void main(void)

{
  cout<<" WELCOME to SOFTWARE 'HudCAD-Cost Engineering : DBM Model'.
USING IT ONE CAN\n";
  cout<<"SEE THE DATA ON (1) MATERIAL AND LABOUR QUANTITIES COEFFICIENTS
AND\n";
  cout<<"(2) MATERIAL AND LABOUR RATE COEFFICIENTS PRE-STORED IN THE
COMPUTER AND CHANGE\n";
  cout<<"THE SAME BY INPUTTING THE SITE-SPECIFIC-INPUT-DATA in terms of
Labour\n";
  cout<<"and Material QUANTITIES AND RATE COEFFICIENTS, MAKING the
Design Process\n";
  cout<<"more Realistic\n Solutions in a SITE-SPECIFIC CONTEXT.\n";
  {
  float cf,cw,cr,cfl;
  int ni,nj;
  char any,inacpt,cec,schcec,schlmc;
  float sum;
  float c[16];
  float r[16];
  float q[15][16];
```

```
float a[15][16] ={

{0.20,0.167,0.00,0.00,0.00,0.00,0.00,0.00,0.00,0.00,0.00,0.00,
0.00,0.00,0.00,0.00},
{0.90,0.60,0.05,0.05,0.04,0.27,0.17,0.00,0.47,0.00,0.24,0.65,
0.00,0.00,0.00,1.667},
{0.15,1.07,0.35,0.35,0.00,0.275,0.05,0.28,0.00,0.00,0.00,0.00,
0.494,0.00,0.00,0.528},
{0.15,1.07,0.35,0.35,0.00,0.275,0.063,0.268,0.00,0.00,0.00,0.00,
0.494,0.00,0.00,0.528},
{0.9,0.6,0.45,0.45,0.04,0.27,0.32,0.00,0.445,0.89,0.00,0.00,
0.00,0.00,0.00,2.277},
{0.15,1.40,0.43,0.43,0.00,0.275,0.05,0.28,0.00,0.00,0.00,0.00,
0.494,0.00,0.00,1.184},
{0.15,1.40,0.43,0.43,0.00,0.275,0.063,0.268,0.00,0.00,0.00,0.00,
0.494,0.00,0.00,1.184},
{0.0103,0.094,0.094,0.00,0.00,0.0392,0.0043,0.0184,0.00,0.00,
0.00,0.00,0.00,0.00,0.00,0.08},
{0.0086,0.081,0.081,0.00,0.00,0.031,0.0036,0.0154,0.00,0.00,
0.00,0.00,0.00,0.00,0.00,0.08},
{0.0043,0.081,0.065,0.00,0.00,0.0292,0.0027,0.0077,0.00,0.00,
0.00,0.00,0.00,0.00,0.0,0.11},
{1.00,0.77,0.09,0.09,0.08,0.47,0.22,0.00,0.50,0.20,0.64,0.00,
0.00,0.00,0.00,1.723},
{1.00,0.77,0.09,0.09,0.08,0.47,0.32,0.00,0.445,0.22,0.67,0.00,
0.00,0.00,0.00,1.723},
{0.01,0.00,0.01,0.00,0.00,0.00,0.00,0.00,0.00,0.00,0.00,0.00,
0.00,1.05,0.00,1.111},
{0.00,0.018,0.0,0.0,0.018,0.0,0.0,0.0,0.0,0.0,0.0,0.0,0.0,0.0,
0.003,0.064 },
{2.70,2.025,0.00,3.375,0.00,0.675,0.37,0.00,0.445,0.22,0.67,
0.00,0.00,0.00,0.00,4.117}
};
float b[16]={70,60,120,100,90,80,2700,200,357,600,518,178,900,
13.5,140,1.0};
g1:  cout<<"\nDO YOU WANT TO SEE THE INPUT DATA PRE-STORED IN COMPUTER?
(put 'y' if yes)";
     cin>>cec;
     if ((cec== 'y')||(cec =='Y'))
   {
   cout<<"INPUT Basic-Cost-Engineering Unit Co-Efficients
(QUANTITIES) are:\n";
     for(int i=0;i<15;i++)
       {
       for(int j=0;j<16;j++)
       cout<<" "<<a[i][j];
       cout<<endl;
       }
       cout<<"INPUT Basic LABOUR-MATERIAL Unit-Rates(MARKET RATES)Co-
Efficients are:\n";
     for(int j=0;j<16;j++)
       {
       cout<<" "<<b[j]<<" ";
       }
     }
```

```
cout<<"\nAre the INPUT DATA as above acceptable?(put 'y' if yes)";
cin>>inacpt;
if ((inacpt== 'y')||(inacpt=='Y'))
goto g;
else
{
int i,j;
cout<<"\nWant to Change ONLY SPECIFIC Cost-Engg.Qties Co-Efficients &
Rate coeff.?\n"<<"(put 'y' if yes)\n";
cin>>cec;
if ((cec== 'y')||(cec=='Y'))
{
cout<<"Put ROW(0/1/../15)& COL.(0/1/../16) of Inp.Cost-Engg.Co-
eff. to be changed\n";
cin>>i>>j;
cout<<"Put the desired value of above Input Data to be changed\n";
cin>>a[i][j];
cout<<"a[i][j] ="<<a[i][j]<<endl;
cout<<"Put same or another ROW(0/1/../15)& COL.(0/1/../16) of Inp.
Cost-Engg.Co-eff. to be changed\n";
cin>>i>>j;
cout<<"Put the desired value of above Input Data to be changed\n";
cin>>a[i][j];
cout<<"a[i][j] ="<<a[i][j]<<endl;
cout<<"Put COL.(0/1/../16) of Inp.Cost-Engg.RATE Co-eff. to be
changed\n";
cin>>j;
cout<<"Put the desired value of above Input Data to be changed\n";
cin>>b[j];
cout<<"b[j]="<<b[j]<<endl;
cout<<"Put the same or another COL.(0/1/../16) of Inp.Cost-Engg.
RATE Co-eff. to be changed\n";
cin>>j;
cout<<"Put the desired value of above Input Data to be changed\n";
cin>>b[j];
cout<<"b[j]="<<b[j]<<endl;
cout<<"put any char"<<endl;
cin>>any;
goto g1;
}
cout<<"Want to change Some Input Cost-Engineering(Qties)Co-
Efficients?\n"<<"(put 'y' if yes)\n";
cin>>schcec;
{if ((schcec== 'y')||(schcec=='Y'))
{
cout<<"Enter Number of ROWS(0/1/../9)\nfor which Inp.Cost-Engg.
Co-eff.(Qties)to be changed\n";
cin>>ni;
cout<<"Enter Input values(Qties) 16 per row:\n";
for(int i=0; i<ni;i++)
{
cout<<"row "<<i<< ": ";
for(int j=0; j<16;j++)
cin>>a[i][j];
cout<<"put any ch"<<endl;
```

```
      cin>>any;
       }
     }
    cout<<"Want to change Some Input Labour-Material-Unit-Cost(MKT.
    RATES)Co-Efficients?\n"<<"(put 'y' if yes)\n";
    cin>>schlmc;
    if ((schlmc== 'y')||(schlmc=='Y'))
    {
      cout<<"Enter Number of COLUMNS i.e. nj(0/1/../16) for which\nInp.
     Labour-Material-Unit-Cost(MKT.RATES)Co-Eff. to be changed\n";
      cin>>nj;
      cout<<"Enter nj="<<nj<<"Input values i.e. Unit-Costs(MKT.RATES)1
    per column:\n";
      for(int j=0; j<nj;j++)
      {
      cin>>b[j];
      }
      cout<<endl;
    }
    }
   }

  g: cout <<"\nPut any character to DISCONTINUE Iteration/Display of
  RESULTS\n";
  cin>>any;
  }
 }
```

4.8.1   Illustrative Examples of Labour and Material Quantity Coefficients Change

As shown in the above listing, the computer gives the following prompt for user to response:

```
Want to Change ONLY SPECIFIC Cost-Engg.Qties Co-Efficients & Rate
coeff.?\n"<<"(put 'y' if yes)\n"
```

On putting 'y' response, the computer asks in terms of ROW & COL for specific data to be changed as below:

```
"Put ROW(0/1/../15)& COL.(0/1/../16) of Inp.Cost-Engg.Co-eff. to be
changed\n";
```

to which the user put the response '0 0', say, after which the computer asks specific value as below:

"Put the desired value of above Input Data to be changed\n"; on putting the value of '0.25', say, by the user the computer replaces the original value of '0.20' by this new value of '0.25' in **Matrix: LMQC** shown above. Similarly, more or all data can be changed.

4.8.2   Illustrative Examples of Labour and Material Rate Coefficients Change

As shown in the above listing, the computer gives the following prompt for user to response: as per the following dialogue structure:

```
"Put COL.(0/1/../16) of Inp.Cost-Engg.RATE Co-eff. to be changed\n";
```

to which the user put the response '0', say, after which the computer asks specific value as below:

"Put the desired value of above Input Data to be changed\n"; on putting the value of '90', say. by the user the computer replaces the original value of '70' by this new value of '90' in **Matrix: LMRC** shown above. Similarly, more or all data can be changed.

Here, only one data item change is shown just for illustration. In fact, the C++ program DBM.CPP and COSTENG.CPP provides for some or all Cost-Engg.Qties Co-Efficients & Rate coeff. Change as per the local and country-specific context, and accordingly, the C++ program COSTENG.CPP determines SOR and building component unit costs, permitting a realistic cost estimating.

Interested readers can modify and enlarge the program to cover more items and also increase its utility by including generation of files containing different cost data.

4.9   Cost Estimating Techniques: Unit Rate Technique

The unit rate technique is one of the most widely used estimating techniques both in preliminary and detailed estimate (Jelen & Black, 1983). Preliminary estimate of building cost can be done by multiplying the area (say square metre) or volume (cubic metre) of a building by the unit cost data based on average unit costs for major cities in a country. However, variations in labour and material costs and other local factors will cause variations in such unit costs and consequently in preliminary estimates. Using the unit costs of building components such as walls, floors, and foundations may give more accurate cost estimates. This approach may incorporate both bare unit cost and the complete unit cost. Labour wage costs per day, material cost per unit of weight or volume are bare unit costs. Installed costs of specific diameter pipeline per unit of length, completed cost of specified quality of concrete per unit volume are the complete unit costs.

In the unit rate technique, two basic factors which are to be considered are labour and material consumption and their unit prices leading to the estimation of the total cost in a project. The input data required for such cost estimates are as follows:

1. Units of measurement for various items which may be in terms of centimetre, metre, square centimetre, square metre, cubic centimetre, cubic metre, hour, day, kilogram, tonne, and so on.
2. Labour unit quantities of various skills which are consumed to produce a unit output of various items of work such as earthwork, masonry, plain cement concrete, reinforced cement concrete, and so on.
3. Rates per unit of labour of various un-skilled and skilled categories such as mason, plumber, and so on.
4. Material unit quantities of various categories such as cement, steel, bricks, sand, and so on which are consumed to produce a unit output of various items of work such as concrete, masonry, and so on of various specifications as per standard practice.
5. Rates per unit of material of various categories such as cement, steel, bricks, sand, and so on.
6. Miscellaneous and sundries such as hire and running charges of plants and so on.
7. Specifications of various building components such as masonry, concrete, and so on.

These input data can be assembled in the form of a matrix to generate a schedule of rates for various items of work. Generally, Public Works Departments of various countries collect such data and incorporate them in the form of a Schedule of Rates (SOR) for various items of works in a book form for a particular year. These rates are updated periodically or a percentage is added to the Schedule of Rates (SOR) for a particular year to make it at par with the current market rates. All such activities are generally carried out manually making it very difficult to respond correctly to the market conditions from time to time. In these days of Information Technology Revolution and cheap availability of computers, it is desirable that all such Schedule of Rates (SOR) should be computerized so that these can be constantly updated without any loss of time leading to more efficient and accurate estimating of costs. To help this process, the C++ program: COSTENG.CPP(cited above) on cost engineering, giving stress on the Schedule of Rates, is developed and available in an online space attached to this book.

4.10 Computer-Aided Cost Estimating and Database Management

Computer-Aided Estimating is a technique to reduce time and cost of preparing cost estimates and also to increase the accuracy of such estimates, and should be an integral part of a computer-aided design system integrating planning and design by optimization as attempted in this book. There are many examples of computer-aided estimating applications (Jelen & Black, 1983). To illustrate such applications and also to help achieve optimal design considering costs as presented later, the unit cost estimating both for various items of work and for various components of a building are taken as examples here. Using such concepts, the above cost engineering and database management model is developed as part of software HudCAD, and is outlined below.

4.11   Software HudCAD: Cost Engineering and Database Management Model

As cited above the C++ program: COSTENG.CPP on cost engineering, giving stress on the schedule of rates, is developed and available in an online space attached to this book. In the C++ program example (cited above), 15 items of work and 16 items of labour and material coefficients are considered here in the form of a *Matrix* just for illustration, which can be extended to cover hundreds of items of works as in a standard schedule of rates, adopting the same technique presented in this example. Similarly, only building components such as foundation, wall, roof, and floor types are considered to derive component unit costs, which can be extended to cover other components such as pipelines for water supply of various sizes and specification, sewerage networks of various sizes and specifications, roads and pavements of various dimensions and specifications and so on, adopting the same principles and technique presented in this example.

In this software-model example, the labour and material items considered are shown in Sect. 4.7.3 above and the items of works considered are shown in Sect. 4.7.4 above.

The Input Basic-Cost-Engineering (Quantities) Coefficients (Per Unit) both for labour and material for each item are usually generated by works and method study and its design specifications. These are generally constants for each item of work irrespective of the time and location of the work, and therefore, can be constant input to the cost engineering model. On the other hand, the rate or price per unit of labour (in terms of hour, day, and so on) and per unit of material (in terms of cubic metre, kilo gram, quintal, and so on) may vary with time and location of the work, and hence, may have to be given as fresh input every time a program is run in a site-specific case, if there is variation.

Once the inputs are given correctly to the C++ program, named as: COSTENG. CPP, it will instantly generate the site-specific schedule of rate for various items of work incorporated, and also the building component unit costs which are accurate and realistic, and consequently, the optimal design generated using such costs will also be realistic as illustrated in later chapters of the book.

4.11.1   Application of the Cost Engineering and Database Management Model with Actual Dialogue Structure: Generation of a Site-Specific Schedule of Rate

The application of the above program to generate a Schedule of Rates (SOR) *instantly* is illustrated below. Here, the actual *dialogue structure* when it is run is shown. On starting the program, the following message appears, along with the subsequent prompts and responses:

```
WELCOME to SOFTWARE 'HudCAD-Cost Engineering Model'. Using it one can
find UNIT-COSTS of Building Items and Components, given INSTANTLY on
selection
of SITE-SPECIFIC-INPUT-DATA in terms of Labour and Material Costs. Use
of such
Realistic UNIT-COSTS in the Design Process makes the Designs more
Realistic
Solutions in a SITE-SPECIFIC CONTEXT.

You want to see Definitions of Labour & Material Items(put 'y' if yes)?
n
Want to RUN COST ENGINEERING MODEL to Derive UNIT-COSTS?(put 'y'if yes)
y
Do you want the individual Basic Results displayed?
(put 'y' if yes)
n
Do you want the individual Item Quantities displayed?
(put 'y' if yes)
n
Want to change ALL Input Cost-Engineering(Qties)Co-Efficients?
(put 'y' if yes)
n
Want to change ALL Input Labour-Material-Unit-Cost(MKT.RATES)Co-
Efficients?
(put 'y' if yes)
n
Want to change Some Input Cost-Engineering(Qties)Co-Efficients?
(put 'y' if yes)
n
Want to change Some Input Labour-Material-Unit-Cost(MKT.RATES)Co-
Efficients?
(put 'y' if yes)
n
Do you want to see the Basic Input Data?
(put 'y' if yes)
n
```

On putting response to each of the above prompts of the computer, it generates **instantly** a *site-specific* Schedule of Rates (SOR) and also the building component unit costs based on SOR at the same time.

In the present model, four building components, namely foundation, wall, roof, and floor are considered, and each building component has three specification types (1/2/3). Thus, a planner-designer-engineer has to choose any of 1/2/3 types, for each of four building components by putting four values when prompted by the computer.

An application example of the cost engineering and database management model is presented below.

4.11.2   Application Example of the Cost Engineering and Database Management Model: Instant Generation of a Site-Specific Schedule of Rate and Building Component Unit Costs

As cited above, on putting response to the prompts of the computer, it generates **instantly** a *site-specific* Schedule of Rates (SOR) and also the building component unit costs based on the SOR generated at the same time. Thereafter, the planner-designer-engineer has to choose any of 1/2/3 types, for each of four building components by putting four values (here, the values of: 2,2,2,2 are given) when prompted by the computer. Accordingly, the unit costs of chosen foundation/wall/roof/floor types, and also the site-specific schedule of rate, given as OUTPUT by the model is shown below in the present case.

Choose foundation/wall/roof/floor types, i.e. choose any of
1/2/3 for each Component and put 4 values in same order
2 2 2 2

The unit costs of chosen foundation/wall/roof/floor types are as follows:
cf = 287.6 MU per RM, cw = 722.28 MU per RM
cr = 326.33 MU per SQM, cfl = 136.03 MU per SQM
The chosen roof parameters are as follows:

rcthk = 0.1 M, ast = 4.5 kg per sq.m

A Schedule of Rates (SOR) and also the above building component unit costs *instantly* given as OUTPUT by the cost engineering model (in the file: SOR.CPP) are reproduced below:

```
SITE-SPECIFIC SCHEDULE OF RATES (SOR) AND ALSO BUILDING COMPONENT
UNIT COSTS INSTANTLY GIVEN BY THE COST ENGINEERING MODEL, ARE:
Item Description of Item                        Unit      Rate(Monetary
                                                              Units)


1. Earthwork(ordinary soil)                     Cum            24.020
   in Wall Foundation
2. Plain Cement Concrete 1:4:8(1 Cement:4       Cum          1002.010
   Coarse Sand:8 Stone Agg.40 mm)in Foundation
```

  3. Brick-work in Foundation in 1:8   Cum   809.300
     (1 Cement:8 Fine Sand) cement mortar
  4. Brick-work in Foundation in 1:6   Cum   842.000
     (1 Cement:6 Fine Sand) cement mortar
  5. Providing/Laying DPC with Cement Conc.   Cum   1780.065
     1:2:4(1Cem:2C.Sand:4S.AG10)
  6. Brick-work in Superstructure in 1:8   Cum   846.700
     (1 Cement:8 Fine Sand) cement mortar
  7. Brick-work in Superstructure in 1:6   Cum   879.400
     (1 Cement:6 Fine Sand) cement mortar
  8. Cement plaster 15 mm thick (rough side of   Sqm   36.067
     wall) in 1:6 (1Cem.:6F.Sand) cement mortar
  9. Cement plaster 12 mm thick in wall   Sqm   30.462
     in 1:6 (1Cem.:6F.Sand) cement mortar
 10. Cement plaster 6 mm thick (roof/ceiling) in   Sqm   24.127
     1:4 (1Cem.:4F.Sand) cement mortar
 11. Re-infrcd.cem.conc. in roof/floor 1:3:5 (1Cem.   Cum   1404.820
     :3C.Sand:5S.AG20) excldng.shutrng./re-infrcmnt.
 12. Re-infrcd.cem.conc. in roof/floor 1:2:4 (1Cem.   Cum   1682.725
     :2C.Sand:4S.AG20) excldng.shutrng./re-infrcmnt.
 13. Mild Steel Reinforcement for RCC Work incl.   Kg   16.075
     bending, binding, placing in position
 14. Colour washing (green, blue, buff) on New Work   Sqm   3.120
     (2 or more coats) including a base coat of
     white washing with lime to give even shade
 15. Cement Concrete Flooring 1:2:4 (1 Cement:2   Cum   2338.925
     Coarse Sand:4 Stone Aggregate 20 mm Nom.Size)
     finished with a floating coat neat cement

Nomenclature of Basic-Cost-Engineering Items (Labour& Materials) are:
j=0-BELDAR(UNSKL1), j=1-COLIE(UNSKL2), j=2-MASON1/BLACKSMITH(SKL1),
J=3-MASON2(SKL2), j=4-MATE(SKL3), j=5-OTHL/BHISTI(SKL4),
j=6-CEMENT, j=7-F.SAND, j=8-C.SAND, j=9-STONE.AG10, j=10-STONE.AG20,
j=11-STONE.AG40, j=12-BRICK, j=13-STL, j=14-OTH1, j=15-OTH2
INPUT MARKET RATES (+CENTAGE) OF ABOVE BASIC COST ENGINEERING ITEMS ARE:
UNSKL1=70.000, UNSKL2=60.000, SKL1=120.000, SKL2=100.000,
SKL3=90.000, SKL4=80.000
CEMENT=2700.000, SAND.F=200.000, SAND.C=357.000, A10=600.000,
A20=518.000
A40=178.000, BRICK=900.000, STEEL=13.500, LIME=140.000, OTHER2=1.0
'cmb1'

BASED ON THE ABOVE SITE-SPECFIC 'SOR', THE BUILDING COMPONENT UNIT COSTS
OF THE CHOSEN FOUNDATION, WALL, ROOF AND FLOOR TYPES ARE:

cf= 287.600 MU per RM, cw= 722.277 MU per RM
cr= 326.330 MU per SQM, cfl=136.026 MU per SQM

THE CHOSEN ROOF AND WALL PARAMETERS ARE:

rcthk= 0.100 M, ast= 4.500 Kg per Sq.M bwthk= 0.230 M

4.12   Universal Nature of the Cost Engineering Model

It would be seen from the *dialogue structure* that the program provides for change of values of all or some 'Input Cost-Engineering (Qties) Co-Efficients' and, 'Input Labour-Material-Unit-Cost (MKT. RATES) Co-Efficients' as per site-specific requirement in a particular location. For example, 230 mm long brick sizes are adopted here the corresponding number of bricks of 494 per cubic metre of brick-work under item 6 and 7 in the 15×16 *Matrix* of item-labour-material coefficients. If the brick sizes are different, corresponding number of bricks per cubic metre should be calculated and given as input in response to the *dialogue* prompt, replacing the above number of bricks and accordingly changing the *Matrix* while running the program giving correct results. Thus, the *Software* is a *Universal Model* to generate item wise and building component wise unit costs permitting generation of optimal designs in any situation in any country on cost considerations.

4.13   Closure

Cost is of paramount importance in all AEC activities where many disciplines have to interact together in a project, making cost as the common language for such interactions specifically in the computer-aided design process. In this chapter, some elements of cost engineering and database management, which are very relevant to Computer-Aided Design (CAD) to help attain a *Resource-Efficient* and *Equitable* solution of housing and urban problems, are presented.

In the above context, affordable housing development is very essential. This is possible if affordability analysis giving the Affordable Housing Cost (AHC) and the Computer-Aided Design giving the physical design of housing are linked together. Thus, the conventional practice of first-design-and-then-cost should be reversed providing for first-cost-and-then-design. In this design procedure, the AHC for the target group of housing beneficiary should be determined by a detailed affordability analysis, and thereafter, a housing development should be designed within this constraint of AHC. To facilitate affordability analysis in Computer-Aided Design, a **C++** program: **EMI.CPP** with complete listing of source code is presented with a number of application examples. Using the program and the *Optimization Models* presented in later chapters, it should be possible to adopt the above design procedure to achieve *Resource-Efficient* and *Equitable* housing development solutions.

Here, the imperative need to incorporate suitable methods and techniques of cost estimating so that the costs predicted are realistic as per the site-specific conditions is stressed, and importance of database management in cost engineering is discussed. A method of dynamic database management matrix for unit cost estimating, covering both labour and material quantity coefficients and labour and material rate coefficients, is presented.

A procedure to make computer-aided estimating, which is a technique to reduce time and cost of preparing cost estimates and to increase the accuracy of such estimates, as part of computer-aided design system integrating planning and design by optimization, is offered. To illustrate such applications and also to help achieve optimal design considering costs, the unit rate technique, which is one of the most widely used estimating techniques, is taken as examples both for various items of work and for various components of a building. To facilitate this process, the **C++ program: COSTENG.CPP** with complete listing of the source code is presented. Using it one can find unit-costs of building items and components, given INSTANTLY on selection of site-specific-input-data in terms of labour and material costs. Use of such realistic unit-costs in the design process makes the designs more realistic solutions in a site-specific context.

In these examples, a limited number of items of work and items of labour and material coefficients are considered in the form of a *Matrix* just for illustration, which can be extended to cover hundreds of items of works as in a standard schedule of rates, adopting the same technique presented in these examples. The software is also used in optimization models presented in later chapters. Hopefully, this approach will lead to complete computerization of standard Schedule of Rates (SOR) covering all items of construction works in various countries, popularizing CAD in various projects and achieving more *efficiency* and *cost-effectiveness* in the AEC sector as a whole.

Chapter 5
Elements of Computer Graphics and Architectural and Engineering Drafting

5.1 Introduction

The phrase *computer graphics* describe any use of computers to produce and manipulate images (Shirley et al., 2011). As already mentioned in Chap. 1, computer graphics is an extensive and important component of CAD for generating graphic portrayal of objects with even complex geometries for better appreciation of geometric properties and arrangement of spaces in the designed objects, for geometric modeling, and for interactive graphics by means of which the geometry of an object could be changed by simple commands to the computer to find an acceptable solution. Application of computer graphics to computer-generated design drawings started a few decades back, and thereafter, there is rapid increase in various fields in the use of computer-aided drafting and design (or CADD), which is a fundamental form of architectural and engineering graphics, and the concept is applicable in any form of display that must be done in any activity. Thus, characteristic applications of computer graphics techniques include engineering and architectural drafting, production of perspectives, mapping, and so on. In fact, computer graphics and geometric modeling are an essential component of computer-aided design in various fields, which is equally applicable in architecture (including planning), engineering (both civil and building engineering) and construction, i.e. AEC field.

Human mind tends to solve a problem by heuristics (that is by trial and error), whereas a CADD system solves it by the use of algorithms (an error-free sequence of logic), and thus letting both human mind and machine work to their best capacity. This results to a new and better method, which can be automated, leading to improvement or elimination of certain or all parts of manual labour involved in doing a job. However, this does not mean the elimination of human beings from the scene, since one has to start and stop the process (either directly by pushing a button or indirectly by a software programming the design process), and also has to refine the input data in response to the current output until the final acceptable output is arrived at, particularly in the AEC design process. Plotters, the first automatic

© The Author(s), under exclusive license to Springer Nature Switzerland AG 2022
B. K. Chakrabarty, *Integrated CAD by Optimization*,
https://doi.org/10.1007/978-3-030-99306-1_5

drafting machines, eliminated the need for a designer to push and pull a triangle and T-squire around a drafting table, but the designer has to be there to run a CADD operation much as was done before.

A powerful CADD software package can be a two-dimensional (or three-dimensional) computer-aided design and drafting tool from initial sketches and conceptual designs to final presentation drawings and full-scale engineering or architectural projects, producing sophisticated solid geometry and neater, more attractive drawing packages with excellent drafting quality. Whether one creates electrical schematics, mechanical assembly drawings, civil and structural engineering drawings, or architectural floor plans, CADD offers a way to get the needed documents, which is the reason for its unmatched popularity in the industry and for its wide use by many design and engineering professionals in various disciplines. Its use by AEC professionals in architectural and engineering drafting, which was comparatively less, is gradually becoming popular. In Chap. 1, basic mathematical formulations for computer graphics and the principles of CADD as a way of converting computer impulses and electronic data into documents in various fields were discussed. In this chapter, instead of complicated mathematical formulations for computer graphics that are available in many books, the elements of computer graphics with emphasis on language-dependent applications of CADD in AEC field such as architectural and engineering drafting as part of integrated CAD are presented. There can be two methods for such language-dependent applications of CADD. One method is the use of graphics standards, just like standardization of FORTRAN, COBOL, Pascal, etc. in computer programming, which caused an explosion in computer utilization. The other method is the use of suitable programmed LISP routines inside the CADD software, or use of C++ graphics programming in CADD.

It is mentioned in Chap. 1 that CAD, characterized by its strong reliance on computer algorithms and software programs, primarily depends on: (1) system software— supplied with the hardware and (2) application software to be developed in-house in many cases, particularly in the AEC sector where designer-programmers have to develop application software for their specific requirements, as commercial availability of application software in this sector is comparatively less. In high-level language application programs, a graphic package is required to produce images of objects, and a typical graphics package may contain the following features:

1. Graphics primitive subroutines for points, lines, and so on.
2. Subroutines for Window, clipping, and so on.
3. Subroutines for entering and leaving the graphics mode, selecting colours, hard copying, etc.

Here, in a computer graphics sense, **window** is the rectangular portion which will be displayed, the rectangular area of the CRT or plotter (described later) where image will be displayed is referred as the **viewport** (Zecher, 1994). The parts of the object that are outside the viewport are not to be drawn and must be **clipped**.

Originally computer hardware manufacturers wrote the graphic packages in low-level machine language that worked only on the specific computer in question, and as a result, an application program that used the graphics package could be run only on the specific machine. As this situation was unsatisfactory for developing portable programs, i.e. programs that would work on all computers and display devices, it was felt that some method should be found to preclude machine dependence of graphics programs. One route taken towards this goal was the use of graphics standards, just like standardization of FORTRAN, COBOL, Pascal, etc. in computer programming, which caused an explosion in computer utilization. Accordingly, graphics standards in terms of set of specifications for graphical programming were established as GKS and CORE, which consist of a number of subroutines, which application programmers can incorporate within their programs for graphical images. GKS defines many different subroutines which may be about 200 (Hsu & Sinha, 1992). These subroutines are independent of any programming language. Some language bindings are also standardized, and the work is continuing.

The above approach requires independent development of programs for each case for specific language bindings to allow programmer to call the GKS subroutines from the corresponding high-level language being used by the programmer which is a time-consuming process. Therefore, in this book, emphasis is given to the language-dependent applications of CADD using AutoLISP routines or C++ Graphic routines where possible, that provide direct control of generation of graphic shapes as desired, in architectural and engineering drafting instead of complicated mathematical formulations for computer graphics, which are available in many books. As cited in Chap. 1, tedious and voluminous computations involved, in utilizing optimization and operations research techniques and to solve optimizing models, have prevented their application in the past, and the current IT revolution has made it very easy to apply integrated CAD, quantitative analysis, and optimization techniques in all sectors including AEC sector, to derive many benefits. Drawing-graphics are the language of AEC professionals, and hence, the micro-CAD systems of HUDCAD, RCCAD, and other software's emphasizing graphics, presented in this book have evolved mainly to promote application of quantitative analysis and optimization techniques in a user-friendly manner instead of presenting intricate mathematics and techniques of graphics for which many books are available, and to show the feasibility of integrated CAD by optimization in AEC field after many years of research and not as commercial software. But the basic purpose of demonstrating the feasibility and utility of such integrated CAD system (including optimization) in AEC field (which is presently very rare) is served as shown by the illustrative application examples using these micro-CAD systems of HUDCAD and RCCAD in different chapters. Any interested user, AEC professional, academician, and researcher can modify, enlarge, and update these systems to universalize and popularize quantitative analysis and optimization techniques and enhance their utility; and to facilitate this process, the source code of the software components is incorporated in the book in a transparent manner, instead of the usual black-box approach.

5.2   Short Overview of Graphics Systems and Applications

As cited in Chap. 1, tedious and voluminous computations involved, in utilizing optimization and operations research techniques and to solve optimizing models, have prevented their application in the past and the current IT revolution has made it very easy to apply integrated CAD, quantitative analysis, and optimization techniques in all sectors including AEC sector, to derive many benefits. Drawing-graphics are the language of AEC professionals, and hence, the micro-CAD systems of HUDCAD, RCCAD, and other software emphasizing the key role of graphics are presented in this book which has evolved mainly to promote application of quantitative analysis and optimization techniques in a user-friendly manner, instead of presenting intricate mathematics and techniques of graphics systems for which many books are available in literature. However, just to familiarize the readers, a very short overview of graphics systems and applications is outlined below.

5.2.1   Some Graphic Techniques and Hardwares

Most computer graphics images are presented in some kind of raster scan display which shows images as rectangular arrays of pixels or picture elements. Such a raster image is simply a 2-D array that stores the pixel value for each pixel. A common example is a flat panel computer display monitor in television having rectangular small light emitting pixels that can be set to different colours to create desired image. Different colours are achieved by mixing varying intensities of red, blue, and green (RGB) light. As rasters are very prevalent in devices, raster images are the most common way to store and process images. But, one may not like to display an image this way and may want to change the size or orientation of the image, correct the colours, and show the image pasted in a three-dimensional surface. Such considerations break the direct link between image pixels and display pixels, and it is best to think raster image as a device independent description of image to be displayed and the display device as a way of approximating that ideal image (Shirley et al., 2011).

Vector images which are resolution independent can be the other ways of storing description of shapes—areas bounded by lines and curves—with no reference to any particular pixel grid. This amounts to storing instructions for displaying the image rather than the pixels needed to display it (Shirley et al., 2011). Vector images are often used for text, diagrams, drawings, and other applications where crispness and precisions are important, and photographic images and complex shading are not needed. However, vector images must be rasterized before they can be displayed. A few familiar raster devices are: transistor liquid display (LCD), light emitting diode (LED) for output display; and 2D array sensor digital camera and 1D array sensor flatbed scanner for input.

Major hardware components in computer graphics systems include video monitors, hard copy devices, key boards, and other devices for graphics input or output.

Graphics hardwares are components necessary to quickly render objects as pixels in the computer screen using specialized rasterization-based hardware architecture. Basic features of some graphic hardwares are discussed below just to give some idea to the readers without going to the intricate details which are the subject matter of computer science and beyond the scope of this book.

5.2.1.1   Video Display Devices

The primary output device in a graphic system is the video monitor. The operation of most video monitors is based on the standard cathode ray tube or CRT, but several other technologies like solid state monitors are also available.

The video display unit in a computer has two components, i.e. the display screen and the display adapter. The CRT is the most common graphics display device which is of two basic types, one employing magnetic deflection and the other using electrostatic deflection.

Large majority of computer graphics systems use some type of CRT display and usually there are three types of CRT display technologies, namely Direct View Storage Tube (DVST), calligraphic refresh, and raster scan refresh displays. Generally, DVST and calligraphic refresh are for line drawing and the raster scan refresh displays are for point plotting. Just to illustrate the basic operation of a CRT, a schematic basic design of a magnetic deflection CRT is shown in Fig. 5.1 (Donald Hearn & Baker, 1997).

The beam of electrons known as cathode rays emitted by the electron gun passes through focussing and deflection systems that direct the electron beam towards the specified positions on the phosphor coated screen as shown in Fig. 5.1. The phosphor then emits a small spot of light at each position when the electrons hit the screen. As the light emitted by the phosphor fades exponentially with time, the entire picture must be refreshed or redrawn repeatedly by quickly directing the electron beam back over the same points. As earlier mentioned, this type of display device is called a refresh CRT. The maximum number of points or pixels that can be displayed without overlap on a CRT is called as **resolution of the monitor**. A more

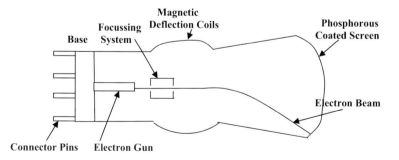

Fig. 5.1 Schematic basic design of a magnetic deflection CRT (not to scale)

precise definition of resolution is the number of points per centimetre that can be plotted horizontally and vertically. Generally, a graphic image is made of pixels. In graphics not only the resolution giving the number of pixels in horizontal and vertical direction, but also the colour of the pixels is also important. In black and white monitors, the colour of the pixels is either black or white. Some monitors support 4, 16, or more colours. Storing of a graphic image in the RAM of a computer requires memory which is much more compared to storing of characters or numbers. Higher resolution and more colours improve the image quality but increase the memory requirements. The hardware which determines the range of resolution and colours in a computer is the VDU or video display unit which consists of two devices: (1) the display screen (monitor) and (2) the video display adapter. Some of the monitors available are monochrome monitors, VGA (video graphics array) colour monitors. VGA colour monitors are generally used to generate graphic images. A screen can be in text mode or graphic mode. To produce graphic images, a screen should be in graphic mode. These are discussed in more details in Sect. 5.5 (Graphic Programming in C++) later.

5.2.1.2 Raster Scan System

The most common types of graphics monitor using a CRT is the raster scan display. In raster scan displays, the electron beam is swept across the screen one row at a time from top to bottom, and as the beam moves across each row, the beam intensity is turned on and off to create a pattern of illuminated spots. The picture definition is stored in a memory area which is called **refresh buffer** or **frame buffer**. This area holds the set of intensity values for all screen points (referred as '**picture element**' or **pixel**) which are retrieved from the refresh buffer and painted on the screen one row (**scan line**) at a time. In a simple black and white system, each screen point is either on or off, and so only one bit per pixel is needed to control the intensity of screen positions. Additional bits are needed to display colour and intensity variations. Up to 24 bits per pixel are included in high quality system. A system with 24 bits per pixel and a screen resolution of 1024 by 1024 requires 3 MB of storage for the frame buffer. In black and white system with one bit per pixel, the frame buffer is commonly called **bitmap**, and for systems with multiple bits per pixel, the frame buffer is referred to as a **pixmap**.

The refresh rate of raster scan displays is usually 60 frames per second and is independent of picture complexity, i.e. number of lines, points, and characters. The refresh rate for vector scan systems, i.e. DVST and calligraphic, depends directly on the picture complexity, greater the complexity, the longer the time taken by a single refresh cycle and the lower is the refresh rate.

Interactive raster graphic systems use a special purpose processor called the video controller or display controller in addition to the central processing unit or CPU, to control the operation of the display device. Here the video controller accesses the frame buffer to refresh the screen. Sophisticated raster systems employ other processors such as co-processors and accelerators to implement various graphic operations.

5.2.1.3   Random Scan system

Random scan monitors draws a picture one line at a time and also referred to as vector displays, or stroke-writing or calligraphic displays. This type of CRT maintains a vector representation of the displayed image in a portion of the memory called display file, which includes the endpoint coordinates of each line segment along with line type. The display file is processed continuously causing the image to be refreshed very rapidly. If the image is too complex, it will begin to fade before it is redrawn completely causing a phenomenon called flicker (Zecher, 1994). Therefore, such CRTs are most often used in applications that do not require large number of vectors to be displayed.

5.2.1.4   Input Devices

Input devices are the means by which user interface with application software allows the user to graphically select various portions of a displayed image, to modify it or to create new images. Many devices are available for data input in graphics. Keyboard is the most common input device. Additional devices are specifically designed for interactive input which may include mouse, trackball, joystick, digitizers. Mouse is a handheld cursor control device which is operated by moving it over a flat surface and mostly used for selection of graphical entities displayed on the screen or for picking specific command or menu option. Mouse is classified as relative input device and does not provide absolute positional data. Trackballs are also relative input devices and look like an inverted mouse with its ball protruding partially through the top of a rectangular box. The user rolls the ball in the same direction as the screen cursor is desired to move. Joysticks have a vertical control stick used to position screen's cursor, and a button on top of the joystick is used to send the cursor's coordinate position to the computer. Digitizer is a common device for drawing, painting, or for interactively selecting coordinate positions on an object. A digitizer is used to scan over a drawing or object and to input a set of discrete coordinate positions to approximate a curve or a surface shape. Data gloves, touch panels, image scanners, and voice systems are some input devices used for particular applications.

5.2.1.5   Hard-Copy Output Devices

Hard-copy output for graphic images can be obtained in several formats. The quality of hard-copy output of graphic images obtained from a device depends on the dot size and the number of dots per inch, or the number of lines per inch, which can be displayed. Printers are very common hard-copy output devices and produce output either by impact or by nonimpact methods. Impact printer's press formed character faces against an inked ribbon onto the paper. An example of impact printer is a line printer with the typefaces mounted on bands, chains, drums, or wheels. To get images onto paper, the nonimpact printers and plotters use laser techniques, ink-jet sprays, xerographic processes (used in photo copying machines), electrostatic methods, and electro thermal methods.

A laser beam in a laser device creates a charge distribution on a rotating drum coated with a photoelectric material such as selenium, and a toner is applied to the drum and then transferred to paper. To produce output, the ink-jet methods squirt ink in horizontal rows across a roll of paper wrapped on a drum. The electrically charged ink-stream is deflected by an electric field to produce dot-matrix patterns. Negative charge is placed on the paper by an electrostatic device, one complete row at a time along the length of the paper. Then the paper is exposed to a toner, which is positively charged, and so attracted to the negatively charged areas, where it adheres to produce the specified output. In electrothermal methods, heat is used on a dot-matrix print head giving output patterns on the heat-sensitive paper. Limited colour output can be obtained on an impact printer by using different coloured ribbons. Various techniques to combine three colour pigments (cyan, magenta, and yellow) are used by nonimpact devices to produce a range of colour patterns. The three pigments are deposited on separate passes, by the laser and xerographic devices. Ink-jet devices shoot the three colours simultaneously on a single pass along each print line on the paper (Donald Hearn & Baker, 1997).

Drafting layouts and other drawings are usually generated with ink-jet printers or pen plotters. A pen plotter can have one or more pens mounted on a carriage or cross-bar that spans a sheet of paper. Pens with varying widths and colours are used to produce a variety of line styles in the drawing. Plotters are generally used to produce quality engineering drawing (Zecher, 1994). Most plotters have built-in micropro-cessor with memory allowing them to receive commands rapidly and then buffer these commands. This allows the plotter to work offline from the computer, and thus freeing up the computer for other jobs, while plotter can work at much slower pace using the information in its buffer. Plotters are classified as vector or raster devices and can be different types like pen plotters, electrostatic plotters, and printer-plotters. Pen plotters can produce the highest quality line drawing output.

5.3   Graphics Software

There are two general classifications for graphic software, namely general program-ming packages and special purpose application packages. A general programming package usually includes an extensive set of graphics functions that can be used in a high level language such as C and C++. An example of general-purpose graphics programming package is the standard library graphic functions, interactive graphics, direct access colour graphics, and so on provided in C++. AutoLISP can also be considered as an example of general-purpose graphics programming package. Basic functions in a general package include those for generating picture components (includes graphic primitives) such as straight lines, polygons, circles, and other figures. Application graphic packages are designed for non-programmers so that users can generate displays without worrying about how the specific application graphics package works. RCCAD and HUDCAD presented in Chaps. 8 and 9, and other chapters are application graphic packages developed and designed for non-programmers and AEC professionals.

5.3.1 Some Graphic Primitives and Coordinate Systems

Basic programming techniques used to generate graphical displays are the graphic primitives like points and lines and also the higher order graphical entities such as arcs, circles, ellipses, splines, and surfaces. Most plotters and CRT's work in terms of two-dimensional Cartesian coordinate systems, where the origin is located in the lower left corner as shown in Fig. 5.2a. However, on the majority of personal computers (PCs), the origin is in the upper left corner as shown in Fig. 5.2b (Zecher, 1994).

When in graphics mode, the face of the CRT is a rectangular area divided into a grid pattern of picture elements called pixels which appear as small dots on the screen. Most CRTs are designed to operate either as a graphic terminal or as an alphanumeric or text terminal, but not both at the same time. Most CRTs are by default in text mode when powered up.

5.3.1.1 Graphic Primitives: Displaying Points

The face of the CRT when in graphics mode is an inverted (x, y) cartesian coordinate system as shown in Fig. 5.2b, the screen's layout contains the origin in the upper left-hand corner and provides 640 addressable pixels along the x-axis, and 480 addressable pixels along y-axis in a VGA monitor in screen mode-2 with 16 colours. Any pixel can be illuminated by specifying (x, y) pair of values which identifies the location of the pixel or point. Numbering of the pixels starts with 0, and therefore, in this mode values ranging from 0 to 639 can be used in the x-direction, and the values ranging from 0 to 479 in the y-direction as shown in Fig. 5.3. Thus, the address of the pixel at the top left will be $(0, 0)$, while the address of the pixel at the bottom right will be $(639, 479)$. Any point P can be drawn in the screen by specifying (x, y) pair of values in appropriate graphic program statement as shown in Fig. 5.3. For example, the C++ program statement: *putpixel (40, 40, RED)* will put a red dot at point $(40, 40)$.

Fig. 5.2 (x, y) Coordinate system on face of CRT and PC. (**a**) CRT. (**b**) PC

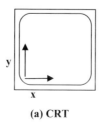 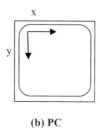

(a) CRT   (b) PC

Fig. 5.3 Coordinate system in graphic screen of VGA monitor

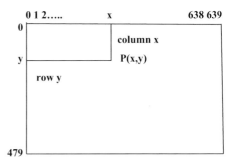

5.3.1.2   Graphic Primitives: Displaying Lines

If the CRT is initialized to be in graphic mode (more details discussed in Sect. 5.4 later), lines can be displayed by specifying numerical values for the starting pixel position $(x1, y1)$ and the ending pixel position $(x2, y2)$. For example, the C++ program statement: *line (x1, y1, x2, y2)*, when executed will draw a line joining points or pixel positions $(x1, y1)$ and $(x2, y2)$.

5.4   Graphic Programming in AutoLISP

AutoLISP is the implementation of the LISP programming language embedded within the AutoCAD, and it allows users and software program developers to write programs and functions that are well suited to graphics applications. AutoLISP supports several data types which are:

- Lists—which are an organized way of representing data—are the most versatile data type. A list is a bunch of items enclosed in parentheses, and these items are called the elements of the list.
- Symbols—these are another type of data and typically named in words, letters, phrases, or abbreviations (e.g. SQRT for 'square root'). Symbol names may contain any combination of letters and numbers.
- Strings—character strings are a type of sequence and are a subtype of vectors and not symbols. Character strings must be enclosed in double quote characters (",) using double quote key in computer keyboard.
- Numbers—there are two types of numbers: (1) integers which are whole numbers and (2) floating point numbers which are always written with a decimal point.

All AutoLISP expressions are in the form of (Function [Arguments]...). Each expression begins with a left parenthesis and ends with a right parenthesis.

5.4.1 AutoLISP Variables and Functions

There are four types of variables: real, integer, point, and strings. They can have any name that begins with an alphabetical character. Thus, 'setq' function is used to assign values to a variable and the format is

(setq variable name value)

The list function places together more than one item into a single variable. Thus, a point is a list of *x* and *y* components.

Some arithmetic expressions

(+ x y)—returns the sum of *x* and *y*
(− x y)—returns the difference of *x* and *y*
(* x y)—returns the product of *x* and *y*
(/ x y)—returns the quotient of *x* divided by *y*
(abs x)—returns the absolute value of *x*

There are many other LISP and AutoLISP functions and operators which are described and illustrated in literature (Touretzky, 2013, Schaefer & Brittain, 1989, Oliver, 1990), which is not reproduced here.

A procedure for integrated CADD (computer-aided design and drafting) by optimization using both AutoLISP and C++ is described below.

5.4.2 Integrated CADD by Optimization Using Both AutoLISP and C++

In order to only illustrate the procedure for integrated CADD by optimization using both AutoLISP and C++, a simple AutoLISP Graphic Program 'boxa.lsp' in AutoLISP for a box is developed and presented below. The program is interfaced with a C++ program 'boxc.cpp' for optimal box calculations to minimize the length of perimeter for a box of given area. To interface the C++ program 'boxc.cpp' with AutoLISP Graphic Program 'boxa.lsp', the following sentence is added in AUTOCAD's ACAD.PGP file which is a way to automate DOS and other commands.

```
BOXC, START BOXC, 1, *Application to start (Type 'BOXC'): ,
```

Each item in the line is separated by a comma. The first item 'BOXC' is what one types at the AUTOCAD's command line to bring up the program. The last item is the prompt statement and when one types it is appended and the program 'BOXC' is called up and run giving the results which are incorporated in the LISP output file 'boxout.lsp' which is generated by the 'BOXC' while it is run.

One can also run a text processor other than EDLIN by modifying the ACAD.
PGP file accordingly (Schaefer & Brittain, 1989).

5.4.2.1   A Simple Graphic Program in AutoLISP

```
(vmon)
(defun bx ()
 (lad)
 (splis1)
 (setq a1 0.0 )
 (setq uzp (getpoint "\nPut the drawing starting point (list) :  " ))
 (setq p1 uzp )
 (txt)
 (command)
 (setq p2 (polar p1 a1 lx ) )
 (setq p3 (polar p1 (+ a1 (* 0.5 pi )) ly ))
 (setq p4 (polar p2 (+ a1 (* 0.5 pi )) ly ))
 (command "line" p1 p2 p4 p3 p1"" )
 (command)
)

(defun lad ()
(load "boxout.lsp" )
(command)
)

(defun splis1 ()
 (setq lx (nth 0 dlis1 ))
 (setq ly (nth 1 dlis1 ))
 (setq r (nth 2 dlis1 ))
 (setq a (nth 3 dlis1 ))
 (setq pr (nth 4 dlis1 ))
 (command)
)

(defun txt ()
 (setq slx (rtos lx 2 2))
 (setq sly (rtos ly 2 2))
 (setq sr (rtos r 2 2))
 (setq sa (rtos a 2 2))
 (setq spr (rtos pr 2 2))
 (setq wv (* 0.55 ly) )
 (setq b (list (+ (car p1) 0.5 ) (+ (cadr p1) wv ) ) )
 (command "text"
 (list (car b) (+ (cadr b) 0.10))
 "0.2" "0" "LENGTH OF BOX (X-DIRECTION)" "text"
 (list (+ (car b) 5.50 ) (+ (cadr b) 0.10))
 "0.22" "0" slx "text"
 (list (car b) (- (cadr b) 0.35))
 "0.22" "0" "LENGTH OF BOX (Y-DIRECTION)" "text"
 (list (+ (car b) 5.50 ) (- (cadr b) 0.35))
```

```
  "0.22" "0" sly "text"
  (list (car b) (- (cadr b) 0.65))
  "0.22" "0" "LENGTH/WIDTH RATIO OF BOX" "text"
  (list (+ (car b) 5.50 ) (- (cadr b) 0.65))
  "0.22" "0" sr "text"
  (list (car b) (- (cadr b) 0.95))
  "0.22" "0" "PERIMETER OF BOX" "text"
  (list (+ (car b) 5.50 ) (- (cadr b) 0.95))
  "0.22" "0" spr "text"
  (list (car b) (- (cadr b) 1.25))
  "0.22" "0" "AREA OF BOX" "text"
  (list (+ (car b) 5.50 ) (- (cadr b) 1.25))
  "0.22" "0" sa
 )
)

(defun c:boxa ()
  (SETQ ERSC ( STRCASE
   ( GETSTRING "\nDo you want the Earlier Design(s) to be ERASED(Y/N)?
")))
    ( if (= ERSC "Y" )
     (COMMAND "ERASE" ( SSGET "X" ) "" )
     (PRINC "" )
     )         ; if
  (command "limits" (list 0.5 0.5 ) (list 12.0 9.0 ) )
  (bx)
  (command "zoom" "a" )
  (SETQ DGNCH (STRCASE (GETSTRING "\nWant to CHANGE DESIGN RE-START
Itern.(Y/N):

")))
  (IF (= DGNCH "Y" )
  (PROMPT "\nType 'boxc'& after result type'boxa'in COMMAND for REVISED
Drg.Sesn.

")
  (PRINC "" )
  )
  (command)
 )
```

5.4.2.2 A Simple C++ Program BOXC.CPP for Optimum BOX Calculations

//C++ PROGRAM BOXC.CPP FOR BOX CALCULATIONS

```
#include<math.h>
#include<fstream.h>
#include<stdio.h>
#include<stdlib.h>
```

```cpp
void main (void)
//begin main programme
{
double lx,ly,r,a,pr;
cout<< "LENGTH IN X-DIRECTION, Y-DIRECTION & RATIO OF BOX ARE:lx,ly & r
\n";
cout<< "PUT THE DESIRED AREA OF THE BOX : " ;
cin>>a;
cout<< "OPTIMUM VALUE OF RATIO r TO GIVE MINIMUM PERIMETER OF BOX = 1\n";
cout<< "PUT THE DESIRED RATIO OF THE BOX : " ;
cin>>r;
cout.precision(4);
if (r == 1 )
{
lx= sqrt (a);
ly = lx;
}
else
{
lx = sqrt ( a/r );
ly = r*lx;
}
pr = 2*( lx+ly);
cout<< "LENGTH OF THE BOX IN Y-DIRECTION IS = " << ly << "\n";
cout<< "LENGTH OF THE BOX IN X-DIRECTION IS = " << lx << "\n";
cout<< "PERIMETER OF THE BOX IS = " << pr << "\n";
cout<< "RATIO OF THE BOX IS = " << r << "\n";
//programme output in files : begin
{
   FILE*file1;
   char b =')';
   char d =')';
   if ((file1=fopen ("boxout.lsp","w"))!=NULL)
   {
    fprintf(file1,"(setq dlis1\n");
    fprintf(file1,"(list");
    fprintf(file1," %6.3f %6.3f %6.3f %6.3f %6.3f \n",lx,ly,r,a,pr);
    fprintf(file1," %c %c",b,d);
    fclose(file1);
   }
   else
   printf("unable to open file");
  }
  cout<<"\n\n THIS ITERATION SESSION IS CLOSED\n";
}

// end of main programme
```

5.4.2.3   Lisp Output File 'boxout.lsp' Generated by C++ Program for Each Cycle of Calculations

Here, two alternatives one with 'RATIO OF THE BOX'=2, and the other with 'RATIO OF THE BOX'=1 (OPTIMUM) are adopted as shown giving the perimeter of the box (last item in the list) for each alternative. The C++ program asks the user to put the desired ratio of the box by the following prompt statement.

```
"PUT THE DESIRED RATIO OF THE BOX:"
```

Lisp Output File: 'boxout.lsp' Generated in Alternative-1

```
(setq dlis1
(list 7.071 14.142 2.000 100.000 42.426
) )
```

Lisp Output File 'boxout.lsp' Generated in Alternative-2

```
(setq dlis1
(list 10.000 10.000 1.000 100.000 40.000
) )
```

5.4.2.4   Drawing Generated by the AutoLISP Graphic Program 'boxa.lsp' for Each Alternative

The drawing generated by the AutoLISP Graphic Program 'boxa.lsp' for Each Alternative is shown in Fig. 5.4.

To sum up the procedure for **Integrated CADD by Optimization using both AutoLISP and C++** as presented above can be described in the following stepwise manner:

1. OPEN the blank drawing file blank.dwg in AUTOCAD.
2. When the blank drawing file blank.dwg is opened in AUTOCAD, LOAD the **AutoLISP Program 'boxa.lsp'**.

Fig. 5.4 Output drawing generated by AutoLisp Program 'boxa.lsp' as per output results given by C++ Program 'boxc.cpp'

3. TYPE 'boxa' to run the **AutoLISP Program 'boxa.lsp'** and get the output results for various alternatives. In case the Lisp Output File: 'boxout.lsp' is already generated and available; the program 'boxa.lsp' will prompt you to give the drawing starting point, and on its supply will draw the drawing output.

4. In case Lisp Output File: 'boxout.lsp' is not available RUN the C++ program BOXC.CPP and get the Lisp Output File: 'boxout.lsp' generated and follow the procedure in step-3 to get the output drawing.

5. GET the hardcopy of the drawing using a printer.

The above procedure has been adopted in the application software and graphics packages: RCCAD and HUDCAD developed and designed for non-programmers and AEC professionals and presented in Chaps. 8 and 9, and other chapters.

5.4.3   *Integrated CADD Without Optimization Using Only Application Graphics Programming in AutoLISP: An Example*

As mentioned above, essential characteristics of OR/Optimization are its emphasis on models indicating physical depiction of a problem and describing the relationships between goals, variables, and constraints in mathematical terms. Many authors have attempted modeling the urban system, and a number of large-scale urban models are available in the literature (Helweg, 1979; Lee Jr., 1973). Development of small-scale models to address a particular policy problem is often suggested. This is very relevant in urban management, involving building and urban development, requiring both 'descriptive' and 'optimizing' models. Moreover, the relationships between many design variables connected to cost are often nonlinear, time dependent, and physical design related. Hence, it is desirable to develop small-scale urban development models, linked with physical design and adopting a shorter planning horizon, to permit planning and design analysis with more realistic costs.

To help this process, the author developed earlier a number of descriptive models in terms of simple algebraic expressions (no *optimization*) for shelter design and analysis (HUDCO, 1982). Any large-scale urban layout generally has some basic modules or planning modules, which are repeated to form the whole layout. Hence, by suitable control of the quantitative planning and design efficiency indicators (such as saleable land ratio, circulation space ratio, open-space ratio, social facility space ratio, plot density, quantity and cost of utility networks, cost per unit of saleable land, cost per plot, and so on) of such planning modules the character, land use, and cost structure of the whole layout can be controlled. Thus, the planning modules must be efficient in terms of saleable land and circulation space to ensure that the layout as a whole is also efficient in terms of these parameters. Similarly, a layout module needs to be designed in physical terms to ensure that the costs of residential plots are affordable by intended users (say urban poor) which can be ensured if the given affordable cost is taken as independent design variable in the design of a layout module to give the plot size within the above affordable cost, and selecting suitable mix of other planning and design parameter values (such as open-space per plot, utility network specification, and so on) in a number of iterations, and thus carry out a planning and design analysis before deciding the physical design of the layout module.

Descriptive models in terms of simple algebraic expressions given in HUDCO MODEL were intended to permit such planning and design analysis, which could be only in numeric terms since these models did not have any graphics programming features to make it more meaningful. By graphics programming using AutoLISP, it is possible to integrate the algebraic expressions with drafting and drawing to permit planning and design analysis both in numeric and graphic terms, and obtain cost-effective layout planning and dwelling designs in terms of physical designs, fulfilling the given cost parameters and environmental standards acceptable to different users and stakeholders as per their requirements and priority. However, generally AutoLISP can be used for simple algebraic calculations only.

HUDCO MODEL contains a typology of planning modules and dwelling options covering a number of modules and options, based on a large number of executed projects. Application of graphics programming using AutoLISP is illustrated taking the layout planning module-A, one of the many layout modules included in the HUDCO MODEL (HUDCO, 1982). The complete listing of this graphics programme, i.e. CAD application software: 'HUDCMOD' using AutoLISP is given below, which can be run by any user who can modify it as per specific requirement, and also, attempt such graphic programming for all layout modules and dwelling option incorporated in the HUDCO MODEL.

```
; HUDCMOD.LSP
; A Layout Module Design Generating Program Using Layout-Module:-A in
HUDCO-Model-1982.
; Developed as a Self-Contained Tool for Analysis & Design of Layout-
Module:-A, Considering ;Various Planning and Design Parameters
including Costs.
; Developed by Dr. B.K.Chakrabarty
(defun lis2 ()
(setq i 6.0 su 7.5 m 9.0 o 10.0 r 1.4 d 13.5)
(princ "\npre-set physical parameters are:i=6.0,su=7.5,m=9.0,
o=10.0,r=1.4,d=13.5 ")
(setq cpp ( strcase (getstring "\ndo you want to change the above values
(y/n) ? " )))
(if (= cpp "y" )
(progn
(setq i (getreal "\nenter desired width of internal road(changed) : " ))
(setq su (getreal "\nenter desired width of subsidiary road(changed) : "
))
(setq m (getreal "\nenter desired width of main road (changed) : " ))
(setq o (getreal "\nenter desired open-space(sfs) per plot (changed): "
))
(setq r (getreal "\nenter desired layout-module ratio (changed): " ))
(setq d (getreal "\nenter desired depth of plot (changed): " ))
```

```
)
(princ "" )
)
(setq ci 231.0 cs 281.0 cm 414.0 cl 10.0 pc 1592.0)
(princ "\npre-set cost parameters are:ci=231.0,cs=281.0,cm=414.0,
cl=10.0,pc=1592.0 ")
(setq ccp ( strcase (getstring "\ndo you want to change the above values
(y/n) ? " ) ) )
(if (= ccp "y" )
(progn
(setq ci (getreal "\nenter desired cost of internal road(changed)  : " ))
(setq cs (getreal "\nenter desired cost of subsidiary road(changed)  : "
))
(setq cm  (getreal "\nenter desired cost of main road (changed)  : " ))
(setq cl (getreal "\nenter desired cost of raw land (changed): " ))
(setq pc (getreal "\nenter desired affordable cost of plot (changed): "
))
)
(princ "" )
)
)
(defun lpar ()
(setq b (+ (* 6 d) (* 2 i)  su ) )
(setq l (* b r ) )
(setq a (* b l ) )
(setq x (- (* 6 l) ( + (* 4 i) (* 3 su) (* 3 m ) ) ) )
(setq q2 (+ (* (+ (* 0.5 b) l ) cs ) (* (* 0.5 b) cm ) ( * a cl) ) )
(setq q1 (* (- (+ (* 2 l) (* 4 d) ) m su ) ci ) )
(setq q (+ q1 q2) )
(setq pa (- (/ (* pc x d) q ) o ) )
(prompt "\naffordable plot size (sq.m.) = " )
(princ (eval pa) )
(getstring "\nput any character to continue : " )
(setq n (/ (* d x) (+ pa o) ) )
(princ (eval n) )
(setq rn (fix n) )
(prompt "\nrounded number of total plots in module = " )
(princ (eval rn) )
(getstring "\nput any character to continue : " )
(setq ow (/ o d) )
(setq nplw (/ pa d) )
(setq gplw (+ ow nplw) )
(setq vl (+ (* 4 d) i ) )
(setq hml (- l (+ (* 4 d) (* 2 i) (* 0.5 (+ m su) ) ) ) ) )
(setq hl1 (+ i (* 2 d) ) )
(setq hl2 (+ (* 0.5 m) (* 0.5 su) hl1 ) )
(setq hl (- l hl2 )  )
(setq tosp (* n o ) )
(setq on 0.5 os 0.5 om 0.001 oe 0.001 owp 0.001)
(princ "\npre-set open space proportions are:on=0.5,os=0.5,om=0.001,
oe=0.o01,owp=0.001")
(setq cosp ( strcase (getstring "\ndo you want to change the above values
(y/n) ? " )))
```

```
(if (= cosp "y" )
(progn
(prompt "\nput desired open space proportions(0.001 if 0)each block
(total all blocks=1) " )
(setq on (getreal "\nput desired open space proportion-north e-w block
(on-changed):" ))
(setq os (getreal "\nput desired open space proportion-south e-w block
(os-changed):" ))
(setq om (getreal "\nput desired open space proportion-middle e-w block
(om-changed):" ))
(setq oe (getreal "\nput desired open space proportion-east n-s block
(oe-changed):" ))
(setq owp (getreal "\nput desired open space proportion-west n-s block
(owp-changed):" ))
)
(princ "" )
)
(setq onl (/ (* tosp on) (* 2 d) ) )
(setq osl (/ (* tosp os) (* 2 d) ) )
(setq oml (/ (* tosp om) (* 2 d) ) )
(setq oel (/ (* tosp oe) (* 2 d) ) )
(setq owl (/ (* tosp owp) (* 2 d) ) )
(if (< hml oml )
(progn
(setq omlo oml)
(setq oml hml)
(setq owl (+ owl (- omlo hml) ) )
)
(princ "" )
)
(setq npln (/ (- hl onl) nplw ) )
(setq rnpn (fix npln) )
(setq npls (/ (- hl osl) nplw ) )
(setq rnps (fix npls) )
(setq nplm (/ (- hml oml) nplw ) )
(setq rnpm (fix nplm) )
(setq nple (/ (- vl oel) nplw ) )
(setq rnpe (fix nple) )
(setq npw (/ (- vl owl) nplw ) )
(setq rnpw (fix npw) )
)
(defun dimen ()
(setq ds 15.0)
(prompt "\npre-set dimension scale is = " )
(princ (eval ds) )
(setq cds (strcase (getstring "\ndo you want to change dimension scale ?
(y/n): " )))
( if (= cds "y")
(setq ds (getreal "\nenter desired dimension scale (changed) : " ))
(princ  "" )
)
(command "dim" "dimscale" ds )
(command)
```

```
(dim)
(command)
)
(defun dim ()
(setq w1 (list (car ip) (- (cadr ip) 12.0 ) ) )
(setq v1 (list (- (car ip) 12.0 ) (cadr ip ) ) )
(command "dim" "hor" ip f0 w1 sllyt )
(command "dim" "ver" v1 f1 v1 swlyt )
(setq pp1 (list (+ (car ps3) 0.25) (- (cadr ps3) 2) ) )
(setq d1 (list (car d) (- (cadr d) 5.5 ) ) )
(setq d2 (list (+ (car d1) 2.0) (cadr d1) ) )
(setq pp2 (list (- (car ps4) 2 ) (+ (cadr ps4) 4) ) )
(setq d3 (list (car d1) (- (cadr d1) 5 ) ) )
(setq d4 (list (+ (car d3) 2.0) (cadr d3) ) )
(setq dimm1 (list (car cornn) (+ (cadr cornn) 4 ) ) )
(setq dimm2 (list (car dmm2 ) (+ (cadr dmm2) 4 ) ) )
(setq dims1 (list (car crne) (+ (cadr crne) 4 ) ) )
(setq dims2 (list (car dms2) (+ (cadr dms2) 4 ) ) )
(setq mn1 (list (+ (car dimm1) 2.5) (+ (cadr dimm1) 4) ) )
(setq mn2 (list (+ (car dimm1) 1.0) (cadr mn1) ) )
(setq mrwtxt (strcat "main road width= " sm "m " ) )
(setq srwtxt (strcat "subsidiary road width= " ss "m " ) )
(setq irwtxt (strcat "internal road- width= " si "m " ) )
(setq pptxt (strcat "partial-size plots" ) )
(setq fptxt (strcat "full-size plots-width= " splw " m ") )
(setq inr1 (list (car ps0) (+ (cadr ps0) (* 0.5 i) ) ) )
(setq inr2 (polar inr1 pi i) )
(command "dim" "hor" dmm2 cornn mn1 mrwtxt )
(command "dim" "hor" dims1 dims2 dims2 srwtxt )
(command "dim" "hor" inr2 inr1 inr1 irwtxt )
(command "dim" "leader" pp1 d1 d2 "" pptxt )
(command "dim" "leader" pp2 d3 d4 "" fptxt )
(setq pwtxt (strcat "plot-width= " splw " m ") )
(setq pwtp (polar plm1 (* 0.5 pi) (* 0.5 i) ) )
(command "dim" "hor" pl3 pl4 pwtp pwtxt )
(command "dim" "ver" pl1 pl2 pl2 spld )
(command)
)
(defun plot ()
(command "zoom" "a" )
(setq tewpl (* (+ rnpn rnps rnpm ) 2 ) )
(setq tnspl (* (+ rnpe rnpw ) 2 ) )
(setq tnpl (+ tnspl tewpl) )
(prompt "\nrounded number (integer value) of total-plots (all blocks) =
" )
(princ (eval tnpl) )
(getstring "\nput any character to continue 1 : " )
(setq dx nplw dy d )
(setq num rnpn)
(setq ipl (polar fli (* 1.5 pi ) (* 2 d) ) )
(setq a1 0)
(setq uzp ipl)
(pldg)
```

```
(setq uzp (polar ipl (* 0.5 pi ) d ) )
(pldg)
(setq num rnps)
(setq uzp pls)
(pldg)
(setq uzp (polar pls (* 0.5 pi ) d ) )
(pldg)
(if (> hml oml )
(progn
(setq num rnpm)
(setq uzp plm)
(pldg)
(setq g1 p2)
(setq uzp (polar plm (* 0.5 pi ) d ) )
(pldg)
(setq g4 p4)
)
(princ  "" )
)
(setq dx nplw dy d a1 (* 0.5 pi) )
(setq num rnpw)
(setq uzp (polar ipi 0 d) )
(pldg)
(setq uzp (polar ipi 0 (* 2 d) ) )
(pldg)
(setq num rnpe)
(setq uzp (polar rvl1 0 d) )
(pldg)
(setq uzp rvl2 )
(pldg)
(opdg)
(command "line" opm1 opm2 opm3 opm4 opm1 "" )
(command)
)
(defun opdg ()
(setq opm1 (polar g2 pi oml) )
(setq opm4 (polar g3 pi oml) )
(setq opm2 g2 opm3 g3)
(command "line" opm1 opm4 "" )
(setq ope1 (polar rvl3 (* 1.5 pi ) oel) )
(setq ope4 (polar rvl4 (* 1.5 pi ) oel) )
(setq ope2 rvl3 ope3 rvl4)
(command "line" ope1 ope4 "" )
(setq opw1 (polar opw3 (* 1.5 pi ) owl) )
(setq opw2 (polar opw4 (* 1.5 pi ) owl) )
(command "line" opw1 opw2 "" )
(setq ops1 (polar ps1 pi osl) )
(setq ops4 (polar ps2 pi osl) )
(setq ops2 ps1 ops3 ps2)
(command "line" ops1 ops4 "" )
(setq opn1 (polar opn3 pi onl) )
(setq opn2 (polar opn4 pi onl) )
(command "line" opn1 opn2 "" )
```

```
(setq pl1 (polar ipl 0 (* 0.5 nplw) ) )
(setq pl2 (polar pl1 (* 0.5 pi) d ) )
(setq pl3 (polar ipl (* 0.5 pi) (* 1.5 d ) ) )
(setq pl4 (polar pl3 0 nplw) )
(command)
)
(defun pldg ()
(setq p1 uzp )
(repeat num
(dwof)
(setq uzp p2)
)
(command)
)
(defun dwof ()
(setq p1 uzp)
(setq p2 (polar p1 a1 dx ) )
(setq p3 (polar p1 (+ a1 (* 0.5 pi )) dy ))
(setq p4 (polar p2 (+ a1 (* 0.5 pi )) dy ))
(command "line" p1 p2 p4 p3 p1 "" )
)
(defun dwofb ()
(setq p1 uzp )
(setq p2 (polar p1 a1 dx ) )
(setq p3 (polar p1 (+ a1 (* 0.5 pi )) dy ))
(setq p4 (polar p2 (+ a1 (* 0.5 pi )) dy ))
(command "line" p1 p2 p4 p3 p1 "" )
)
(defun road ()
(setq dy 4 )
(prompt "\nroad offset length (in meter) is = " )
(princ (eval dy) )
(setq cofs (strcase (getstring "\ndo you want to change offset length ?
(y/n) : " )))
( if (= cofs "y")
(setq dy (getreal "\nenter desired offset length (changed)  : " ))
(princ  "" )
)
(setq dx (- l (* 0.5 (+ m su) ) ) )
(setq uzp (list (car f1i ) (+ (cadr f1i) su dy) ) )
(setq a1 (* 1.5 pi ) )
(drgl)
(setq cornn p2 crne p3)
(setq dx (- b su) )
(setq uzp (list (+ (car f2i) su dy) (cadr f2i ) ) )
(setq a1 pi )
(drgl)
(setq corne p2 )
(setq dx (- l (* 0.5 (+ m su) ) ) )
(setq uzp (list (car f0i ) (- (cadr f0i) (+ su dy ) ) ) )
(setq a1 (* 0.5 pi ) )
(drgl)
(setq corns p2 )
```

```
(setq dx (- b su) )
(setq uzp (list (- (car ipi) (+ m dy) ) (cadr ipi ) ) )
(setq a1 0 )
(drgl)
(setq cornw p2 )
(corner)
(rdtxt)
)
(defun drgl ()
(setq p1 uzp )
(setq p2 (polar p1 a1 dy ) )
(setq p3 (polar p2 (+ a1 (* 0.5 pi )) dx ) )
(setq p4 (polar p3 (+ a1 pi ) dy ) )
(command "line" p1 p2 p3 p4 "" )
)
(defun corner ()
(setq p2 (polar cornn pi m ) )
(setq dmm2 p2 )
(setq p1 (polar p2 (* 0.5 pi ) dy ) )
(setq p3 (polar p2 pi dy ) )
(command "line" p1 p2 p3 "" )
(setq p2 (polar corne (* 0.5 pi ) su ) )
(setq dms2 p2 )
(setq p1 (polar p2 (* 0.5 pi ) dy ) )
(setq p3 (polar p2 0 dy ) )
(command "line" p1 p2 p3 "" )
(setq p2 (polar corns 0 su ) )
(setq p1 (polar p2 (* 1.5 pi ) dy ) )
(setq p3 (polar p2 0 dy ) )
(command "line" p1 p2 p3 "" )
(setq p2 (polar cornw (* 1.5 pi ) su ) )
(setq p1 (polar p2 (* 1.5 pi ) dy ) )
(setq p3 (polar p2 pi dy ) )
(command "line" p1 p2 p3 "" )
)
(defun rdtxt ()
(setq mtx (list (- (car ipi) (* 0.1 m ) ) (+ (cadr ipi) (* 0.5 (- b su) ) ) ) )
(setq sntx (list (+ (car f1i) (* 0.5 (- l (* 0.5 (+ m su))))) (+ (cadr f1i)
(* 0.1 su))))
(setq setx (list (+ (car f0i) (* 0.9 su) ) (+ (cadr f0i) (* 0.5 (- b su) ) ) )
)
(setq sstx (list (+ (car ipi) (* 0.5 (- l (* 0.5 (+ m su))))) (- (cadr ipi)
(* 0.9 su) ) ) )
(command "text" "c" mtx tx1 "90" "main road"
"text" "c" sntx tx1 "0" "subsidiary road"
"text" "c" setx tx1 "90" "subsidiary road"
"text" "c" sstx tx1 "0" "subsidiary road" )
)
(defun txt ()
(txsze)
(command "text" "c"
(list (+ (car f2) (* 24 tx1) ) (- (cadr f2) tx ) ) tx "0"
"HUDCO MODEL:-LAYOUT MODULE-A" )
```

```
(setq cx1 (list (+ (car f2) (* 24 tx1) ) (- (cadr f2) (* 5 tx1 ) ) ) )
(command "text" "c" cx1 tx1 "0"
"SOME INPUT-DESIGN-DECISIONS GIVEN"
"text" "c" (list (car cx1) (- (cadr cx1) (* 2 tx1 ) ) ) tx1 "0"
"TO SOFTWARE & CORRESPONDING"
"text" "c" (list (car cx1) (- (cadr cx1) (* 4 tx1 ) ) ) tx1 "0"
"OUTPUT-DESIGN-PARAMETER-VALUES"
"text" "c" (list (car cx1) (- (cadr cx1) (* 6 tx1 ) ) ) tx1 "0"
"GENERATED BY THE SOFTWARE :- " )
(setq cx (list (+ (car f2) 10) (- (cadr f2) (* 16 tx2 ) ) ) )
(command "text"
(list  (car cx) (cadr cx) ) tx2 "0"
"total land area (sq.m.) in layout module   ="
"text" (list (+ (car cx) tl) (cadr cx) ) tx2 "0" stla
"text" (list (car cx ) (- (cadr cx) (* 2 tx2) ) ) tx2 "0"
"total effective number of plots in module ="
"text" (list (+ (car cx) tl) (- (cadr cx) (* 2 tx2) ) ) tx2 "0" sn
"text" (list (car cx ) (- (cadr cx) (* 4 tx2) ) ) tx2 "0"
"total blockwise number of full-size plots ="
"text" (list (+ (car cx) tl) (- (cadr cx) (* 4 tx2) ) ) tx2 "0" tnpl
"text" (list (car cx ) (- (cadr cx) (* 6 tx2) ) ) tx2 "0"
"length (e-w) of layout module (m)       ="
"text" (list (+ (car cx) tl) (- (cadr cx) (* 6 tx2) ) ) tx2 "0" sllyt
"text" (list (car cx ) (- (cadr cx) (* 8 tx2) ) ) tx2 "0"
"width (n-s) of layout module (m)       ="
"text" (list (+ (car cx) tl) (- (cadr cx) (* 8 tx2) ) ) tx2 "0" swlyt
"text" (list (car cx ) (- (cadr cx) (* 10 tx2) ) ) tx2 "0"
"given depth of plot (m)             ="
"text" (list (+ (car cx) tl) (- (cadr cx) (* 10 tx2) ) ) tx2 "0" spld
"text" (list (car cx ) (- (cadr cx) (* 12 tx2) ) ) tx2 "0"
"derived width of plot (m)          ="
"text" (list (+ (car cx) tl) (- (cadr cx) (* 12 tx2) ) ) tx2 "0" splw
"text" (list (car cx ) (- (cadr cx) (* 14 tx2) ) ) tx2 "0"
"gross cost of serviced land (mu/sq.m.)    ="
"text" (list (+ (car cx) tl) (- (cadr cx) (* 14 tx2) ) ) tx2 "0" sglc
"text" (list (car cx ) (- (cadr cx) (* 16 tx2) ) ) tx2 "0"
"market raw land price (mu/sq.m.)       ="
"text" (list (+ (car cx) tl) (- (cadr cx) (* 16 tx2) ) ) tx2 "0" slp
"text" (list (car cx ) (- (cadr cx) (* 18 tx2) ) ) tx2 "0"
"ratio of saleable land to layout area    ="
"text" (list (+ (car cx) tl) (- (cadr cx) (* 18 tx2) ) ) tx2 "0" sslr
"text" (list (car cx ) (- (cadr cx) (* 20 tx2) ) ) tx2 "0"
"ratio of circulation area to layout area  ="
"text" (list (+ (car cx) tl) (- (cadr cx) (* 20 tx2) ) ) tx2 "0" scirr
"text" (list (car cx ) (- (cadr cx) (* 22 tx2) ) ) tx2 "0"
"ratio of social-facility-area to layout area ="
"text" (list (+ (car cx) tl) (- (cadr cx) (* 22 tx2) ) ) tx2 "0" sspor
"text" (list (car cx ) (- (cadr cx) (* 24 tx2) ) ) tx2 "0"
"length of utility network per plot (m)    ="
"text" (list (+ (car cx) tl) (- (cadr cx) (* 24 tx2) ) ) tx2 "0" slut
"text" (list (car cx ) (- (cadr cx) (* 26 tx2) ) ) tx2 "0"
"plot density in layout (no.of plots/hectare) ="
"text" (list (+ (car cx) tl) (- (cadr cx) (* 26 tx2) ) ) tx2 "0" spden
```

```
"text" (list (car cx ) (- (cadr cx) (* 28 tx2) )  ) tx2 "0"
"given total affordable cost per plot (mu)  ="
"text" (list (+ (car cx) tl) (- (cadr cx) (* 28 tx2) ) ) tx2 "0" spc
"text" (list (car cx ) (- (cadr cx) (* 30 tx2) )  ) tx2 "0"
"derived affordable area of plot (sq.m.)   ="
"text" (list (+ (car cx) tl) (- (cadr cx) (* 30 tx2) ) ) tx2 "0" spla )
(setq d (list (car cx ) (- (cadr cx) (* 30 tx2) ) ) )
)
(defun paramtr ()
(setq pden (* (/ n a) 10000) )
(setq sl (/ (* n pa ) a ) )
(setq glc (/ (* n pc) a ) )
(setq li (/ (- (+ (* 2 l) (* 4 d) ) (+ m su) ) n ) )
(setq ls (/ (+ l (* 0.5 b ) ) n ) )
(setq lm (/ (* 0.5 b) n ) )
(setq lut (+ li ls lm ) )
(setq cirai (* (* (+ hl vl ) i ) 2 ) )
(setq ciras (- (+ (* l su) (* 0.5 su b ) ) (* 0.5 su su) ) )
(setq ciram (- (* 0.5 b m) (* 0.5 m su) ) )
(setq cirr (/ (+ cirai ciras ciram ) a ) )
(setq spor (/ (* n o ) a ) )
(setq sllyt (rtos l 2 2 ) )
(setq swlyt (rtos b 2 2 ) )
(setq stla (rtos a 2 2 ) )
(setq spla (rtos pa 2 2 ) )
(setq splw (rtos nplw 2 2 ) )
(setq spld (rtos d 2 2 ) )
(setq sspo (rtos o 2 2 ) )
(setq spden (rtos pden 2 2 ) )
(setq sslr (rtos sl 2 2 ) )
(setq scirr (rtos cirr 2 2 ) )
(setq sspor (rtos spor 2 2 ) )
(setq slp (rtos cl 2 2 ) )
(setq sglc (rtos glc 2 2 ) )
(setq slut (rtos lut 2 2 ) )
(setq spc (rtos pc 2 2 ) )
(setq sn (rtos n 2 2 ) )
(setq sm (rtos m 2 2 ) )
(setq ss (rtos su 2 2 ) )
(setq si (rtos i 2 2 ) )
)
(defun txsze ()
(setq tx 3.5 tx1 2.5 tx2 2.0 nx 45.0 )
(princ "\npre-set text-ht/loc. (k./nx) are:tx1=2.5,tx2=2.0,tx=3.5,
nx=45.0 ")
(setq ctx ( strcase (getstring "\ndo you want to change the above values
(y/n) ? " )))
(if (= ctx "y" )
(progn
(setq tx1 (getreal "\nenter desired text ht. (tx1-heading) (changed)  : "
))
(setq tx2 (getreal "\nenter desired text ht. (tx2-general) (changed)  : "
))
```

```
(setq tx (getreal "\nenter desired text ht.(tx-top heading) (changed)  :
" ))
(setq nx (getreal "\nenter desired text locn.(nx-mutiple data)
(changed): " ))
)
(princ "" )
)
(setq tl (* nx tx2 ) )
)
(defun view ()
(setq vwdr (strcase ( getstring "\nyou want to change drawing view
direction ? (y/n):")))
(while (= vwdr "y" )
(setq ovwdr (strcase ( getstring "\ninsert existing output-drawing view
direction i(isometric)/p(plan) ? (i/p):")))
(setq ovwdrc (strcase ( getstring "\nyou want change of output-drawing
view direction(i to p ,or, p to i ? (y/n):")))
(cond
( ( and (= ovwdr "i" ) (= ovwdrc "y" ) )
(command "vpoint" "r" "270.0" "90.0" )
)
( ( and (= ovwdr "p" ) (= ovwdrc "y" ) )
(progn
(command "vpoint" "r" "225.0" "30.0" )
(setq ciso (strcase ( getstring "\nyou want change iso-view direction
from angle 30 to 45 ? (y/n):")))
(if (= ciso "y" )
(command "vpoint" "r" "225.0" "45.0" )
(princ "" )
)
)
)
)
(setq vwdr (strcase ( getstring "\nyou want to change drawing view
direction ? (y/n):")))
)
)
(defun C:HUDCMOD ()
(command "limits" (list 0.0 0.0 ) (list 12.0 9.0 ) )
(command "zoom" "a" )
(prompt "\nWelcome to Design-Analysis-Drafting SOFTWARE for Layout-
Module-A " )
(setq ersc ( strcase ( getstring "\nDo you want the Earlier Design(s) to
be Erased(Y/N)? ")))
( if (= ersc "Y" )
(command "erase" ( ssget "x" ) "" )
(princ "" )
)
(lis2)
(lpar)
(command "zoom" "a" )
(setq ip (getpoint "\nput the drawing starting point (list) :  " ))
```

```
(setq ao 0 )
(setq f0 (list (+ (car ip) l ) (cadr ip) ) )
(setq f1 (polar ip (* 0.5 pi) b ) )
(setq f2 (polar f1 ao l) )
(setq ipi (list (+ (car ip) (* 0.5 m ) ) (+ (cadr ip) (* 0.5 su ) ) ) )
(setq f0i (list (- (car f0) (* 0.5 su ) ) (+ (cadr f0) (* 0.5 su ) ) ) )
(setq f1i (list (+ (car f1) (* 0.5 m ) ) (- (cadr f1) (* 0.5 su ) ) ) )
(setq f2i (list (- (car f2) (* 0.5 su ) ) (- (cadr f2) (* 0.5 su ) ) ) )
(setq uzp ipi)
(setq a1 0 )
(setq dx (* 2 d) )
(setq dy vl )
(dwofb)
(setq opw3 p3 opw4 p4 vb1 p1 vb2 p2)
(setq p1 (polar f0i (* 0.5 pi) (* 2 d) ) )
(setq p2 (polar p1 pi hl ) )
(setq p3 (polar f0i pi hl ) )
(setq pls p3 ps2 p1 ps1 f0i ps0 p2)
(setq ps3 (polar ps0 0 (* rnps nplw) ) )
(setq ps4 (polar pls 0 (* rnps nplw) ) )
(command "line" p1 p2 p3 f0i p1 "" )
(setq p1 (polar p2 (* 0.5 pi) i ) )
(setq plm p1 )
(setq p2 (polar p1 0 hml ) )
(setq p3 (polar p2 (* 0.5 pi) (* 2 d ) ) )
(setq p4 (polar p1 (* 0.5 pi) (* 2 d ) ) )
(setq plm1 p4 plm2 p2 plm3 p3)
(command "line" p1 p2 p3 p4 p1 "" )
(setq g2 p2 g3 p3)
(setq uzp f2i)
(setq a1 pi )
(setq dx (* 2 d) )
(setq dy vl )
(dwofb)
(setq rvl1 p4 rvl2 p3 rvl3 p2 rvl4 p1)
(setq p1 (polar f1i (* 1.5 pi) (* 2 d) ) )
(setq p2 (polar p1 0 hl ) )
(setq p3 (polar f1i 0 hl ) )
(command "line" p1 p2 p3 f1i p1 "" )
(setq opn3 p2 opn4 p3)
(plot)
(paramtr)
(command "linetype" "s" "continuous" "" )
(command "line" ip f0 f2 f1 ip "c" )
(command "linetype" "s" "continuous" "" )
(txt)
(road)
(dimen)
(command "zoom" "a" )
(view) \
)
```

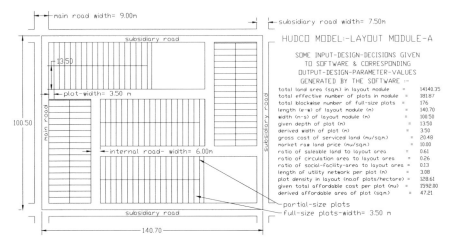

Fig. 5.5 Typical layout design drawing given instantly by 'HUDCMOD', integrating graphic designs with planning parameters and costs

Typical layout design drawing given instantly by 'HUDCMOD', integrating graphic designs with planning parameters and costs is shown in Fig. 5.5, where some input-design-decisions given to the software by a stakeholder/user/planner/designer and the corresponding output-design-parameter values instantly generated by the software are also displayed. The program calculates the affordable plot size based on the desired affordable cost of plot given as INPUT by the USER when prompted by the program.

5.5   Graphics Programming in C++

C++ language can be used not only to carry out complex, tedious, and voluminous computations involved, in utilizing optimization and operations research techniques and to solve optimizing models as shown in various chapters later, but also in a standalone full CAD package with all options. It is desirable that various professionals carry out research and development to derive full potential of the capability of C++ compilers including graphic capability, remove its current limitations, and expand its capabilities to produce neater and sophisticated images and drawings in a standalone one platform, which will be a great contribution in the field of integrated CAD by optimization. It is high time that the usual method of clicking and dragging the mouse to generate images as is done in menu driven packages, which does not permit any mathematical programming and optimization to achieve *efficiency* in resource utilization, is replaced by a program-based CAD package including *optimization* to improve *resource-efficiency*. In this section, an attempt is made to explore the possibility of utilizing C++ as a standalone CAD package which not only creates graphic images on the screen, but also solves intricate complex, tedious, and voluminous computations involved, in utilizing optimization and operations research techniques and to solve optimizing models. Here, TURBO C graphics library is used to produce graphic images on the screen. Graphics library functions require information about the video display unit (VDU) of the computer in use. Generally, a graphic image is made of tiny dots or pixels. As earlier discussed, storing a graphic image in the RAM of the computer requires memory which is much more than that required for storing numbers and characters. Higher resolution and more colour options improve the image quality but at the same time it increases the memory requirement, and graphic programs may not run because of insufficient memory. Therefore, appropriate mode (giving resolution and colours) has to be selected before a graphic program can run. When the computer is switched on, it boots up in text mode. A standard monitor in text mode has 25 rows and 80 columns displaying 2000 characters in all; there are 255 standard characters and each character is associated with an ASCII (American Standard Code of Information Interchange) number, which are available in literature (Mahapatra, 2000). As earlier discussed, the hardware that determines the range of options for resolution and colours is the VDU consisting of display screen (monitor) and video display adapter (VDA). Some monitors cannot produce graphic image or not capable of adding colour to the image, or produce poor quality image. Thus, the graphic capability of a computer depends very much on the type of monitor and the video display adapter selected. Currently, VGA colour monitors are mostly used to generate graphics images.

When an output is to be displayed on a screen, the signal from the CPU is stored in VDU which is a part of the RAM (Random Access Memory) of the system. The exact position or location in the RAM where some data is stored is defined as the address of the data. For a VGA monitor, the address of the VDU memory is 0xB8000000, and each address in the VDU memory corresponds to a specific location of the screen. The VDA repeatedly reads the information from the VDU memory, about 60–70 times in a second and the rate at which data from the VDU memory is transferred to the screen is known as the *refresh rate*.

5.5.1   Graphics Programming in C++ Using Directly the VDU Memory

If the address of the VDU memory is known, one can use the indirection operator of C (*) to store any data in the VDU memory. Accordingly, the C++ program: 'star. cpp' is developed and placed below (Mahapatra, 2000).

```
//program star.cpp to draw astrisks in screen
#include<iostream.h>
#include<math.h>
#include<fstream.h>
#include<stdlib.h>
#include<stdio.h>
#include<conio.h>
#include<graphics.h>

void main(void)
//begin main programme
{
int i;
char any;
long x=0xB8000000;
char far * v;
v= (char far *)x;
for(i=0;i<200;i=i+1)
*(v+2*i)='*';
getch();
cout<<"\nSTAR FIGURE PRODUCED ABOVE\nput any character to close the
program"<<endl;
cin>>any;
clrscr();
}
```

One output of the program is shown in Fig. 5.6. Here, the number of asterisks to be drawn in the screen is limited to 200 which can be changed as desired.

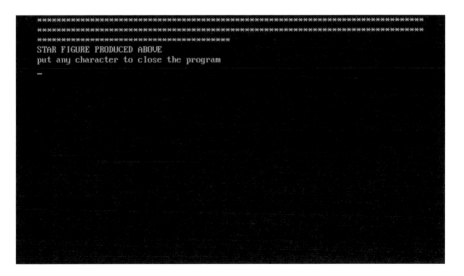

Fig. 5.6 Typical output of a C++ program where address of the VDU memory is known

5.5.2 *Screen Mode*

As earlier stated, the number of pixels in each row and column of a monitor is its resolution. The range of options for resolution and colours available in the system depends on the type of monitors and display adapter attached to the system. The C++ compiler requires information about the VDU attached to the system and it can detect it if it is not supplied. In addition to above, one has to select the appropriate screen type that is the screen mode giving the resolution and colour. The mode selected must be compatible with the VDU attached to the system; otherwise, the program may not run. The VGA monitor which is the most common works in three possible modes as shown in Table 5.1.

Header file: graphics.h of C/C++ has an enumeration named as: **graphics_modes** which defines **VGALO**, **VGAMED**, and VGAHI as 0, 1, and 2, respectively, and thus, in graphics program, the following statements are equivalent:

```
mode = 2
mode = VGAHI
```

5.5.3 *Preparing the Screen for Graphics*

It is necessary to change the mode of the screen from the text mode to graphics mode for starting the graphics work. This is done by using the C++ library graphic function: **initgraph** (). Prototype of the graphic function: **initgraph** () is as follows:

```
initgraph (int far * graphdriver, int far *graphmode,
   char far * path_bgi);
```

where

graphdriver is an integer number supplying the information about the VDU. The value of 'graphdriver' for VGA is '9', and if the 'graphdriver' is assigned the value of 0, the system checks the hardware and selects the appropriate screen mode. The

Table 5.1 Modes available in a VGA monitor

Mode	Enumeration	Resolution	Colours
0	VGALO	640 × 200	16
1	VGAMED	640 × 350	16
2	VGAHI	640 × 480	16

graphmode is the screen mode and the available screen modes for the **VGA Monitor** are: 0, 1, and 2 as shown in Table 5.1.

path_bgi is a character string which stands for the path name of the subdirectory in which the files with .bgi extension are stored. The C/C++ package comes with a set of files with .bgi extension. In the computer used for this book, these files are placed in the subdirectory c:\turboc3\bgi, and accordingly this path name of the subdirectory is supplied as an argument to the function: **initgraph** (), in C++ programs presented later.

5.5.4   Changing the Screen for Text Mode

It is possible to change the screen from graphics to text mode using the library function: **restorecrtmode** (). Again the function: **setgraphmode** (), allows to change the mode from text to graphics mode once more. The function: **closegraph** () can be used to end the graphic session in a permanent manner.

Fig. 5.7 Typical output of the C++ program DETECT.CPP giving the hardware details

5.5.5   *Detecting the Hardware*

The following program named as 'DETECT.CPP' when run detects the hardware available in the computer (Fig. 5.7).

```cpp
#include <graphics.h>
#include <stdlib.h>
#include <stdio.h>
#include <conio.h>

/* names of the various cards supported */
char *dname[] = { "requests detection",
         "a CGA",
         "an MCGA",
         "an EGA",
         "a 64K EGA",
         "a monochrome EGA",
         "an IBM 8514",
         "a Hercules monochrome",
         "an AT&T 6300 PC",
         "a VGA",
         "an IBM 3270 PC"
         };

int main(void)
{
  /* returns detected hardware info. */
  int gdriver, gmode, errorcode;

 /* detect graphics hardware available */
  detectgraph(&gdriver, &gmode);

  /* read result of detectgraph call */
  errorcode = graphresult();
  if (errorcode != grOk)  /* an error
                occurred */
  {
   printf("Graphics error: %s\n", grapherrormsg(errorcode));
   printf("Press any key to halt:");
   getch();
   exit(1); /* terminate with an error code */
  }

  /* display the information detected */
  clrscr();
  printf("You have %s video display card.\n", dname[gdriver]);
  printf("Press any key to halt:");
  getch();
  return 0;
}
```

5.6   Integrated CADD by Optimization Using Only Standalone C++

In CADD from AutoLISP presented above, only simple arithmetic calculations can be carried out although the drawing can be quite elaborate. On the other hand, using only C++ in standalone format not only complex mathematical programming including geometric programming for optimization can be carried out to generate numerical optimal results, but also these numerical optimal results can be instantly converted to drawing. Just to demonstrate the feasibility of using only C++ in standalone format for integrated CADD by optimization, an example optimization model for land subdivision design is selected as presented below.

5.6.1   A Land Subdivision Optimization Model

An optimization model for land subdivision design is presented earlier (Chakrabarty, 1986a, b), which is reproduced below (showing only objective function of the geometric programming model, the interested readers may refer to the article cited above for more details):

$$\text{Minimize } y(x) = A x_1 x_2{}^{-1} x_3 + B x_3 + C x_1 x_3 + D x_4$$

Here, model constants are A, B, C, D the cost coefficients for end circulation street, front access street, land cost, and service connection cost, respectively, for the block of lots. The model variables are:

x_1 = width of the block between the centre lines of the front access streets in metres
x_2 = length of the block between centre lines of the end circulation streets in metres
x_3 = average width of the lots allowing for the width of the end circulation streets in metres
x_4 = depth of a lot from the edge of the front access street up to the rear boundary of the lot in metres
x_5 = width of the front access street in the block of lots in metres

As shown in above reference, the solution of the geometric programming model leads to two equations—one linear and the other nonlinear—as follows:

$$w_{01} - \frac{A a_2 a_3 a_4}{(B - A a_2 a_3 a_4)} w_{04} = 0 \tag{5.1}$$

$$BD(1 - 2w_{04})^2 - a_1 a_4^2 C^2 (w_{04} - w_{01}) w_{04} = 0 \tag{5.2}$$

Substituting, $\frac{A a_2 a_3 a_4}{(B - A a_2 a_3 a_4)} = k$ in Eq. (5.1), we get $w_{01} = k\, w_{04}$ and Eq. (5.2) becomes:

BD $(1 - 2w_{04})^2 - a_1 a_4^2 C^2 (w_{04} - k\, w_{04}) w_{04} = 0$, which is a single variable nonlinear equation and can be solved easily using Newton–Raphson approximation technique as presented in Chap. 3.

Reproduced below is the source code listing of the C++ Graphic Program LANDSUB.CPP, which solves not only the nonlinear geometric programming optimization model, but also, instantly converts the *optimal* numerical OUTPUT results into drawing in the same platform, precluding the need for any other program or software for drawing.

```
/*
  C++ PROGRAM 'LANDSUB.CPP' WHICH SOLVES NOT ONLY THE
  NONLINEAR GEOMETRIC PROGRAMMING OPTIMIZATION MODEL BUT ALSO
  INSTANTLY CONVERTS THE OPTIMAL NUMERICAL RESULTS INTO
  DRAWING IN THE SAME PLATFORM OBVIATING THE NEED FOR ANY
  OTHER PROGRAM- SOFTWARE.
*/
#include<iostream.h>
#include<math.h>
#include<fstream.h>
#include<iomanip.h>
#include<stdio.h>
#include<stdlib.h>
#include<malloc.h>
#include<dos.h>
#include<iostream.h>
#include<graphics.h>
#include<conio.h>
void main(void)
//begin main programme

{
  void far restorecrtmode();
  int gdriver=DETECT;
  int gmode,i;
  char path[]="c:\crturboc3\crbgi";
{
double x,y,f,g,fx,fy,gx,gy,nr1,nr2,dr,F1,G1,H1,ITOL,x1,x2,x3,x4,
x5;
  double R,A,B,C,D,a1,a2,a3,a4,k,w01,w02,w03,w04,w,yx,n,d,la;
  double cx,cy,brx,bry,cy1,cy2,cx1,lb,wb;
  int itr,ITMODEL,ITRL;
  char cont,ANS,any;

    ITRL = 50;
    ITOL = 0.0001;
   cout<< "PRESET VALUE OF LIMIT OF ITERATION NUMBER 'ITRL' = " << ITRL <<
"\n";
    cout<< "PRESET VALUE OF ACCEPTABLE MARGIN OF ERROR 'ITOL' = " << ITOL
<< "\n";
    cout<< "DO YOU WANT TO CHANGE THE ABOVE PRESET VALUES (Y/N) ? " ;
    cin>>ANS;
    if ( ANS == 'Y' || ANS == 'y' )
```

```
      {
    cout<< "PUT NEW VALUES OF TERATION NUMBER LIMIT (ITRL) & ERROR MARGIN
(ITOL) IN SAME ORDER\n";
      cin>>ITRL>>ITOL;
      }
      else
    cout<< "YOU HAVE NOT CHANGED THE VALUES OF 'ITRL' & 'ITOL'\n";
    cout<< "PUT THE CURRENT VALUE OF X " ;
       cin>>x;
       itr=0;
       A=140.5; B=115.5; C=5.0; D=100.0; a1=80.0; a2=0.05; a3=2.0;
a4=2.5;
       k=(A*a2*a3*a4)/(B+A*a2*a3*a4);
       f = B*D*(1-2*x)*(1-2*x) - a1*a4*a4*C*C*(x-k*x)*x;
       fx = -2*B*D*(1-2*x)-B*D*(1-2*x)*2-a1*a4*a4*C*C*(1-k)*2*x;
       cout<< "CURRENT VALUE OF ITERATION NUMBER IS = " << itr << "\n";
       cout<< "CURRENT VALUE OF fx IS = " << fx << "\n";
       cout<< "CURRENT VALUE OF f IS = " << f << "\n";
       cout<< "CURRENT VALUE OF x IS = " << x << "\n";
       cout<< "PUT ANY CHARACTER TO CONTINUE ";
       cin>>cont;
       F1=fabs(f);
       if ( (F1>ITOL) ) goto s1;
    s1: {
       itr = itr+ 1;
       f = B*D*(1-2*x)*(1-2*x) - a1*a4*a4*C*C*(x-k*x)*x;
       fx = -2*B*D*(1-2*x)-B*D*(1-2*x)*2-a1*a4*a4*C*C*(1-k)*2*x;
       nr1 = f;
       dr  = fx;
       x = x - nr1/dr;
       f = B*D*(1-2*x)*(1-2*x) - a1*a4*a4*C*C*(x-k*x)*x;
       F1=fabs(f);
       cout<< "NEW VALUE OF ITERATION NUMBER IS = " << itr << "\n";
       cout<< "NEW VALUE OF x IS = " << x << "\n";
       cout<< "NEW VALUE OF f IS = " << f << "\n";
       if ( (F1 > ITOL) && (itr < ITRL) ) goto s1;
       if (itr == ITRL)
       cout<<"\nSINCE NUMBER OF ITERATIONS EXCEED LIMIT PROGRAME
TERMINATED\n";
       else
       {
       itr = itr;
       cout<< "CURRENT VALUE OF ITERATION NUMBER IS = " << itr << "\n";
       cout<< "CURRENT  VALUE OF f IS = " << f << "\n";
       cout<< "CURRENT VALUE OF x  IS = " << x << "\n";
       cout<< "CURRENT VALUE OF ITOL CHOSEN IS = " << ITOL << "\n";
        cout<<"\n SINCE THE CURRENT SOLUTION MEETS THE VALUE OF CHOSEN\n";
       cout<<" MARGIN OF ERROR, THE SOLUTION ACCEPTED & PROGRAM TERMINATED
\n";
       w04= x;
       w01= k*x;
       w03= (1-2*x);
       w02=(1-k)*x;
       x1=(B*w03)/(C*w02);
       x4=x1/a4;
       yx=(D*x4)/w04;
```

```
         x3=(yx*w02)/B;
         x2=(A*x1*x3)/(yx*w01);
         n=(2*x2)/x3;
         x5= a2*x2;
         w =x3 - (x5*x3)/x2;
         d=x4;
         la=w*d;
         printf("w01=%7.4f w02=%7.4f w03=%7.4f wo4=%7.4f\n",w01,w02,
w03,w04);
         printf("LEAST COST PER LOT= %7.2f MU \n",yx);
         printf("minimum acceptable net lot area=%7.2f sq.m\n",la);
         printf("gross width of the lot=%7.2f m\n",x3);
         printf("net width of the lot=%7.2f m\n",w);
         printf("depth of the lot =%7.2f m\n",x4);
         printf("width of the block=%7.2f m\n",x1);
         printf("length of the block=%7.2f m\n",x2);
         printf("number of lots in the block= %6.0f\n",n);
         printf("width of the front access street=%7.2f\n",x5);
         cout<<"put any character"<<endl;
         cin>>any;
         initgraph(&gdriver,&gmode,path);
         cx=125.0;
         cy=250.0;
         w=4*w; //lot width magnified 4 times for visibility
         d=4*d; //lot depth magnified 4 times for visibility
         cy1=cy+50;
         cy2=cy+75;
         moveto(cx,cy);
         for(i=0;i<15;i++)
         {
         cx=cx+w;cy=cy;brx=cx+w;bry=cy-d;
         rectangle(cx,cy,brx,bry);
         }
         cx=125; cy=250-d;
         for(i=0;i<15;i++)
         {
         cx=cx+w; cy=cy;brx=cx+w;bry=cy-d;
         rectangle(cx,cy,brx,bry);
         }
         cx1=cx-350;
         settextstyle(0,0,0.5);
      outtextxy(cx1,cy1,"DRAWING OF 30 LOTS INSTANTLY GENERATED BY C++");
         moveto(cx1,cy2);
         w=w/4;d=d/4;
         lb=0.5*n*w; wb=2*d;
         printf("net width of the lot=%7.2f m\n",w);
         printf("depth of the lot =%7.2f m\n",d);
         printf("net area of the lot =%7.2f sq.m.\n",la);
       printf("width of the block(excluding access streets)=%7.2f m\n",
wb);
         printf("length of the block(excluding end streets)=%7.2f m\n",lb);
         printf("number of lots in the block=%5.0f\n",n);
         getch();
          }
         }
        }
       }
```

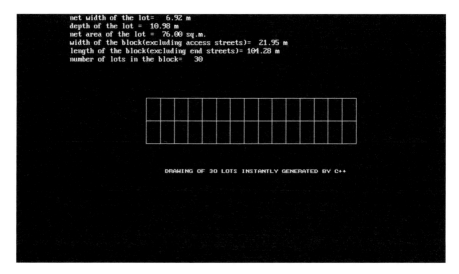

```
net width of the lot=   6.92 m
depth of the lot  =  10.98 m
net area of the lot  =  76.00 sq.m.
width of the block(excluding access streets)=  21.95 m
length of the block(excluding end streets)= 104.28 m
number of lots in the block=   30
```

DRAWING OF 30 LOTS INSTANTLY GENERATED BY C++

Fig. 5.8 Output drawing generated instantly by the C++ program using numerical *Optimal* results

The OUTPUT drawing generated by running the C++ Graphic Program for the Case-III in reference (Chakrabarty, 1986a, b) is shown in Fig. 5.8:

It may be noted that in this case, both the **Solution** of the geometric programming optimization model and the **Generation** of the optimal design drawing using the output **Numerical** *Optimal* **Results given by the C++ program** are carried out only by C++ in the same C++ platform, unlike the use of both AutoLISP and C++ in case of application graphics packages developed in AutoLISP presented earlier.

5.7   Closure

In recent years, there is tremendous increase in capabilities of computer hardwares and graphics resulting to unparalleled transformations in computers in terms of their capacity to capture data, to analyse, manipulate, and store them, to disseminate and communicate information, and to apply these capabilities in different areas and fields of specialization. Such transformations are occurring with increasing rapidity and frequency in many areas and the rate at which Information Technology (IT) and computer applications are spreading to various sectors is extraordinary and has few parallels in technological history.

As mentioned above Computer-Aided Design (CAD) is a vital application area of this ongoing all-pervasive information technology revolution shrinking the effects of time and distance, and altering the very nature of work in an organization. Modern computers can generate and process information on a scope never before possible, and if such information, generated only in alphanumeric form in hundreds of pages it may be of no worth to a human user unable to comprehend them. But these may be very meaningful and extremely useful if presented graphically. Moreover, computer graphics is of great help to make computers extremely 'user-friendly' from the outset where computer can be accessed directly in 'user' language, permitting even lay public users without any computer literacy to use it effectively and meaningfully making it feasible to adopt a participatory planning and design to address various social issues. All these are accelerating application of CAD and graphics in various fields with recent advances in microcomputer technology, steady decline in cost of computers, and tremendous increase in their computing power and graphic capabilities.

In this chapter, elements of computer graphics particularly related to computer-aided drafting and design (or CADD) are discussed, a short overview of graphics systems and applications is presented. Some graphic techniques and hardwares are outlined including **Video Display Devices**, **Raster Scan System**, **Random Scan system**, **Input Devices**, and **Hard-Copy Output Devices**. There are two general classifications for graphic software, namely general programming packages and special purpose application packages. An example of general-purpose graphics programming package is the standard library graphic functions provided in C++. AutoLISP can also be considered as an example of general-purpose graphics programming package. Application graphic packages are designed for non-programmers so that users can generate displays without worrying about how the specific application graphics package works. RCCAD and HUDCAD presented in Chaps. 8 and 9, and other chapters are application graphic packages developed and designed for non-programmers and AEC professionals. Some graphic primitives and coordinate systems are discussed.

Graphic programming in AutoLISP, which allows users and software developers to write programs and functions, is well suited to graphics applications, but it has to be supplemented by a high level language like C++, if the AutoLISP application graphics package is to be used in integrated CAD by optimization. Therefore, a general procedure for integrated CADD (Computer-Aided Design and Drafting) by optimization (which can be implemented in any application package) using both AutoLISP and C++ is developed and presented, using a simple optimization program in C++ and a graphic program in AutoLISP. A model: '**HUDCMOD**', for integrated CADD without optimization using only AutoLISP is also presented; which gives **instantly** the layout design drawing (as shown in Fig. 5.5), integrating graphic designs with planning parameters and costs, requiring to perform only simple algebraic calculations.

C++ language can be used not only to carry out complex, tedious, and voluminous computations involved, in utilizing optimization and operations research techniques and to solve optimizing models as shown in various chapters later, but also in a standalone full CAD package with all options. In this chapter, an attempt is made to explore the possibility of utilizing C++ as a standalone CAD package which not only creates graphic images on the screen, but also carries out intricate complex, tedious, and voluminous computations for utilizing optimization and operations research techniques and to solve optimizing models. In C++ graphic programming, it is also possible to use directly the VDU memory if the address of the VDU memory is known, and this process is much speedier. Accordingly, the C++ program: 'star.cpp' is developed and its output is shown in Fig. 5.6.

Here, TURBO C graphics library is used to produce graphic images on the screen. The C++ compiler requires information about the VDU attached to the system and it can detect it if it is not supplied. In addition to above, one has to select the appropriate screen type that is the screen mode giving the resolution and colour. The mode selected must be compatible with the VDU attached to the system. The VGA monitor which is most common works in three possible modes as shown in Table 5.1. It is necessary to change the mode of the screen from the text mode to graphics mode for starting the graphics work. This is done by using the C++ library graphic function: **initgraph** (). It is possible to change the screen from graphics to text mode and vice versa using the appropriate C++ library function as shown above. A program named as 'DETECT.CPP' is developed and presented above which when run detects the hardware available in the computer.

Just to demonstrate the feasibility of using only C++ in standalone format for integrated CADD by optimization, an example geometric programming optimization model for land subdivision design is selected as presented above. The **Land Subdivision GP Optimization Model** with complete listing of the C++ program: 'landsub.cpp' which solves not only the nonlinear geometric programming optimization model but also instantly converts the optimal numerical results into drawing in the same platform obviating the need for any other program- software, is presented above. The output optimal design drawing generated instantly by the C++ program using the output numerical *Optimal* results also given by the same C++ program is shown in Fig. 5.8. This demonstrates that it is feasible to use only C++ in standalone format for integrated CADD by optimization, without the help of any other software.

In this chapter, an attempt is made to explore the possibility of utilizing C++ as a standalone CAD package which not only creates graphic images on the screen, but also solves intricate complex, tedious, and voluminous computations involved, in utilizing optimization and operations research techniques and to solve optimizing models. It is desirable that various professionals carry out research and development to derive full potential of the capability of C++ compilers including graphic capability, remove its current limitations, and expand its capabilities to produce neater and more sophisticated images and drawings in a standalone single platform, which will be a great contribution in the field of integrated CAD by optimization. It is also high time that the usual method of clicking and dragging the mouse to generate images as is done in menu driven packages, which does not permit any mathematical programming and optimization to achieve *efficiency* in resource utilization, is replaced by a program-based CAD package including *optimization* to improve *resource-efficiency*, particularly in AEC operations.

Chapter 6
Planning and Design Analysis in Urban Development and Management and a Theory of Optimal Urban-Built-Form

6.1 Introduction

Planning and design in AEC sector, involving building, housing, and urban services consuming huge resources and affecting the *quality of life* of all people, need to be viewed as a problem-solving process involving search through numerous potential solutions (being function of numerous design variables involved) to select a particular solution that meets a specified criteria or goal, requiring a comprehensive analysis. This view is quite relevant especially in developing countries to achieve the national goals of providing adequate shelter and urban services for all, within the *affordability* and *resource-constraints* at the individual, local, and national level. With increasing population and dwindling per capita resources, this view is equally applicable even in the developed countries, as we live in a *Finite Earth with Finite Resources*. To facilitate problem solution, the function of planning and design in AEC sector needs to be part of holistic urban management, i.e. as *planning function* (one of the five traditional management functions as detailed in Chap. 2) to make cities viable and desirable places for people to live and work. Similarly, the principles of *efficiency* and *effectiveness* which are the essence of management as well as the principles of *flexibility* and *navigational change* and the *principles of social responsiveness* should be equally applicable in urban development and management to achieve *resource-efficient* and *equitable* urban problem solutions as earlier explained in Chap. 2. The role of planning and design analysis in the application of the above important management principles in urban operations to achieve *resource-efficient* urban problem solution with *social equity* is outlined below.

B. K. Chakrabarty, *Integrated CAD by Optimization*,
https://doi.org/10.1007/978-3-030-99306-1_6

6.2   Holistic Urban Development and Management: Principles of Efficiency and Effectiveness in Planning and Design Analysis

As explained in Chap. 2, productivity (i.e. efficiency and effectiveness) is the essence of management. Moreover, planners and designers (e.g. civil engineers, architects, and planners) in the AEC sector commit huge *resources* (e.g. money, land, materials, and so on) in their planning-design-decisions enjoining on them an *ethical responsibility* and *accountability* to the society to achieve *efficiency, effectiveness,* and *equity* in resource-use so committed for a *resource-efficient* and *equitable* housing and urban problem solutions in the context of rising population and dwindling resources. There are a number of techniques such as planning and design analysis, design optimization, and computer-aided design to help achieve these objectives. The design analysis is one of the most important parts of the whole planning and design process. Design analysis generally has two components, namely (1) planning, architectural, engineering, and costing analysis and (2) interactive graphics. Planning, architectural, engineering, and costing analysis are performed on the conceptual or initial geometry of a facility to ensure that not only the facility will satisfy the specified conditions and criteria (say compliance with space requirements, resource-constraints, and cost-affordability) but also satisfy the planning, architectural, and engineering norms and regulations for improved environment and safety. Analytical results can be derived from planning and design codes, regulations, mathematical modeling, and model analysis covering descriptive and optimizing models, and so on. For example, using a descriptive model of an urban-built-form in mathematical terms (mathematical modeling), a planner or designer could determine the **Interaction** between the selected elements of urban-built-form, and also carry out a **sensitivity analysis**, selecting appropriate set of independent and dependent design variables and also appropriate iteration choice, and thus develop an **insight** into their interrelationships including cost to achieve *efficiency* within the applicable constraints.

Based on such **sensitivity analysis** and insight, the desired values of the *Quantitative Design Efficiency Indicators* (e.g. FAR, BUD, GCR, BUC, and so on) could be selected and achieved if feasible. In this technique, planning and design process is reversed where the desired value(s) of a *Quantitative Design Efficiency Indicator* is selected first and the design fulfilling this value (if feasible) is produced later. This is rarely possible in conventional practice where such *Efficiency Indicators* are available only after the end of the complete design process and also as a fait accomplice and in this practice the planner or designer has little control on the values of such *Efficiency Indicators*. Using such **sensitivity analysis** and insight, the value(s) of specified built-form-elements (BFEs) could also be chosen, which is efficient, equitable, and acceptable to the multiple stakeholders. Geometric consistency of such value(s) of BFEs can also be tested using geometric modeling techniques. Such analysis in conjunction with interactive graphics makes the planning and design process much more meaningful to all the stakeholders. All these issues are presented below in more details using the application software for computer-aided-analysis/design-urban-built-form. In the traditional management, CAD (Computer-Aided Design) is recognized as an important technique to improve efficiency and effectiveness which is equally applicable in urban management and operations. CAD is capable to perform not only planning and design analysis using simple algebraic formulations as enumerated below with a software, but also sophisticated mathematical analysis using powerful techniques such as linear and nonlinear programming to achieve design optimization to maximize efficiency and effectiveness as illustrated in Chap. 8.

6.3 Holistic Urban Management: Principles of Social Responsiveness in Planning and Design Analysis

As urban problems have significant social and environmental dimensions, the need for application of the *'principle of social responsiveness'* in urban management is stressed. This is particularly because of the interdependencies in respect of urban organizations and many groups in our society, as urban development is a joint public-private enterprise where decisions about the use of resources are made by the marketplace and by the governments (at various levels) in an interactive way.

It is felt that *social responsiveness* in the context of urban operations should mean the ability of urban organizations to relate their operations and policies to the social environment in ways that are mutually beneficial to various urban organizations and also to the society. Such concepts are receiving increasing attention even in conventional business operations, due to the interdependencies of the many groups in our society (Koontz & Weihrich, 1990). Such interdependencies are more crucial in respect of urban organizations and urban operations, as urban development is a joint public-private enterprise where decisions about the use of resources are made by the marketplace and by the governments in an interactive way as stated above. Therefore, there is need for a policy to apply the *'principle of social responsiveness'* in the operations of urban organizations, to indicate that they are discharging their social responsibilities. The concept of *'social audit'* defined as 'a commitment to systematic assessment of and reporting on some meaningful, definable domain of the company's activities that have social impact' is also applied in traditional management (Koontz & Weihrich, 1990). Social responsiveness in urban operations implies actions and urban enterprise responses to achieve the social goals (e.g. ensuring supply of affordable housing to vulnerable groups in society) in the urban sector, and to solve urban problems confirmed by evaluation of their social performance. To help evaluate social performance of urban organization, the concept of *'social audit'* should also be applied in their operations to indicate that they are discharging their social responsibilities.

Planning and design analysis incorporating mathematical modeling and model analysis covering descriptive and optimizing models could be very useful tools to assess and to ensure compliance with such social performances. For example, ensuring adequate supply of housing and urban services at a cost affordable by various income groups should be an important indicator of social performance of urban organizations. Use of affordability analysis models as presented in Chap. 4 and planning and design models presented below (and later chapters) could be very useful tools to achieve the above social goals in the urban sector and also for 'social audit' of such performances. This is illustrated in a few examples below using the software for computer-aided-analysis and design. It is also emphasized that a policy commitment to the application of the principles of 'social responsiveness and social audit' in urban operations may discourage additional government regulations and interventions, giving greater freedom and flexibility in decision-making, bettering efficiency of urban organizations, and achieving organized cooperation between such organizations, improving performance of the urban sector as a whole. However, this will be possible only if suitable tools and techniques are applied for 'social audit' to confirm adequate social responsiveness and social performance in a transparent manner.

6.4 Holistic Urban Management: Principles of Flexibility and Navigational Change in Urban Planning, Design, and Urban Development

It is desirable that urban planning should accomplish a *physical planning structure* matching with the *economic-base-structure* (creating workplaces and jobs) and the consequent *demographic-distribution-structure*, which should be accepted as a given condition at appropriate decision level, and should cover the multiple stake-holders including the urban poor in a city, to achieve *efficiency* and *social equity*, and to create a balanced community. However, urban dynamics, uncertainties (difficult to foresee and plan) and conflicting interests, viewpoints, and demands of multiple stakeholders make it difficult to achieve such a match, creating many urban problems such as proliferation of slums and frequent illegal construction of habitats and land-use-change (mostly by informal entrepreneur's creating workplaces and jobs not foreseen and provided for in the formal master plans) in violation of the conventional rigid statutory city-master-plans and planning regulations in many countries, subjected to rapid urbanization, demographic change, soaring land prices and construction costs, and acute affordability and resource-constraints.

For example, in many countries, the per capita land-resource is dwindling fast with the increasing population while land mass of the country remains static, but, city-master-plans generally continue to prescribe a lower density range while density in some existing city-slums is 10–15 times of the above prescribed city-master plan density which is highly *iniquitous*. In conventional practice, stress is given on periodic *mass-scale* dismantling and demolition of such informal industries, work-places, and habitats, resulting to loss of jobs and shelters of millions of people, and also to a huge wastage of individual and national resources even in countries having a substantial proportion of their population living below the poverty line with gigantic unemployment problems. It is imperative to modify these conventional planning practices using modern management principles of flexibility and navigational change, to avoid such urban planning and management failures, adversely affecting the interests of people, and often creating an *urban crisis*.

The *'principles of flexibility and navigational change'* providing for reviewing plans from time to time and redrawing them if this is required by the changed events and expectations is an accepted planning practice in traditional management to cope with uncertainties (Koontz & Weihrich, 1990). These principles are equally applicable in urban planning, urban development, and management in the context of urban dynamics, uncertainties, acute affordability, and resource-constraints in many countries, and conflicting interests, viewpoints, and demands of multiple stakeholders in the urban sector. It is also suggested that, because of too numerous pay-off matrices and uncertainties in the urbanization and development process, an incremental planning approach with 'flexibility principles' should be adopted for guiding the development of cities, providing for constant feedback about the response of the urban system, and for changing again and again the rules (such as zoning, augmenting or new provision of infrastructure, and so on), based on such response so that things do not work out in an undesirable way. Continuous planning with flexibility and short range forecast for land use in selected areas is also suggested to cope with the size and complexity of growth of cities, which are becoming more dynamic. Suitable computer-based planning and design tools can help this process.

Moreover, there can be *large variations* in the *efficiency indicators* of urban-built-form (depending on the rules such as *planning regulations* and the *dynamics* of input-prices). *User Citizens*, *Builders*, other *Stakeholders* in general would not like to ignore such *large variations*, and would be keen to maximize *'productivity'* (input-output ratio) and see that each of their money produces maximum benefit along with the desired environmental standards. Therefore, it is necessary to convince *User Citizens*, who are important stakeholders in a city, invoking their *participation* in the planning and design process, using *Computer-Aided Planning and Design Techniques* (presented below and in later chapters) and information technology for *transparency*, that the *Planning Regulations* do not require any change to attain the above aims of the stakeholders, and thus minimize their violations, which often create many urban problems including slums. This requires viewing city-master-plans as 'Live' development plans (as against conventional rigid statutory city-master-plans and planning regulations) subject to navigational changes in response to the *urban dynamics* and uncertainties, so that it is responsive to the people's needs achieving social equity, efficiency, and environmental requirements in urban development.

Availability of even PC's having tremendous computing power and graphic capability, and rapid development in the information technology makes it easier now to apply such *'flexibility principles'* and *'navigational changes'* for timely corrective actions, with *transparency* and *informed-participation* of *multiple stakeholders*, and thus help develop effective response to the *urban dynamics* and uncertainties. Such an approach will pre-empt above urban problems at the nascent-stage before they get out of control, achieving *efficiency* and *social equity* in the context of urban dynamics and uncertainties, fulfilling appropriate environmental requirements in the urbanization and economic development context. However, this will require a *Policy*, particularly in developing countries subjected to rapid economic and demographic change, to review the conventional practices, to treat the city-master-plan and planning regulations as *rigid* and *static* legal-documents, and instead, view it as a *live* city development-plan *responsive to People's need* (achieving *equity and efficiency* in urban development), and subject to '***navigational changes***' (including change of land-use and *planning regulations* if needed), thus avoid *freezing* the evolving *options* for 20 long years, the usual master plan period, which further aggravates the problems. The above urban problems highlight the need to apply such *dynamic* approaches (modern IT tools make it very easy to apply such approaches) in a *dynamic* city instead of frequent *static* master planning approach. However, this has to be done in a *transparent* manner using *information technology* and computer-aided planning and design techniques permitting instant updating of the 'Live' development plans open to public, so as to avoid any corrupt practices.

The above process should be part of '***integrated urban management***' at the city and regional Level, again, requiring a *Policy* at the national and local level to recognize *Urban Management* as a *distinct full-fledged discipline* and to produce professional urban managers to take over these tasks rather than leaving these tasks to the general administrators as at present in many countries creating a sort of urban chaos. Urban research is necessary to develop various techniques to facilitate application of such management principles, in various urban operations. For example, use of modern management techniques such as ***management science, operations research***, information technology, and computer-aided planning and design, as applied in the traditional management, is equally applicable in urban operations. However, this is feasible only if a *policy* exists to apply such techniques in the urban planning and management process and urban research is carried out to develop computer-aided techniques (including computer-aided planning and computer-aided design techniques), to facilitate application of such principles in various urban operations, and help develop more effective urban policies based on systematic assessment, evaluation, and analysis of such urban policies (including *social-cost-benefit* analysis), adopting a *systems approach*.

The author is carrying out research to develop such urban planning, design and management tools and techniques some of which are presented in this book. The software 'CADUB' (Computer-Aided Analysis & Design of Urban-Built-form) presented below and other software and computer-aided planning and design tools and techniques presented in this book may be helpful in the above efforts.

6.5   A Theory of Optimal Urban-Built-Form for Resource-Efficient Urban Problem Solution

The theory of optimization studies how to describe and attain what is best (synonymous with 'most' what is good) and 'least' what is bad, (say cost), knowing to measure it by various methods, and to find them. Knowledge of optimization theory forces a professional to examine closely the description of a system, giving insight into the underlying structure of rational decisions and constraints, discerning fairly general decision rules towards *optimization*. Therefore, optimization theory in urban problem solution should be regarded as a valuable addition to the existing knowledge of AEC professionals, so that sufficient efforts are expended by them on optimization study of urban problems, realizing the value of information needed, and generating them to describe a system well enough for it to be optimized, and thus help urban problem solution *resource-efficiently*, achieving *social equity*.

As mentioned above, *urban dynamics* and uncertainties, in terms of rapid urbanization, demographic changes, soaring land prices and construction costs (often difficult to foresee and plan), and also the conflicting interests, viewpoints, and demands of multiple stakeholders, with acute affordability and resource-constraints in many countries make it imperative to apply *dynamic* urban management approaches, using optimization theory and modern IT tools, instead of a *static* approach (often flawed) in a *dynamic* situation. As cited above, there can be *large variations* in the *efficiency indicators* of urban-built-form (depending on the rules such as *planning regulations* and the *dynamics* of input-prices), which are hard to be ignored by *User Citizens, Builders*, and other *Stakeholders*. In this context, a suitable theory of three-dimensional optimal urban-built-form is required, which should incorporate many of the above factors to enable AEC professionals including planners, to develop effective response in a *transparent* manner, to the *urban dynamics*, in terms of the planning regulations, and planning and design, solutions achieving *Resource-Efficiency, Social Equity*, and the desired and affordable environmental standards in the urban development process. This is attempted below.

A schematic urban-built-form (single-building-block) illustrating some built-form-elements is shown in Fig. 6.1. This can also be extended to multiple-building-blocks as shown later.

Here, the Built-Form-Elements (BFEs), such as built-space per built-unit, i.e. Built-Unit-Space (BUS), Building Rise (BR—which is assumed as effective number of floors having continuous value as explained later), Semi-Public-Open-Space (SPO—which is the open-to-sky space per built-unit and may include recreation space, road space, and so on in a layout), and the Floor Area Ratio (FAR—which is the ratio between the total built-space covering all floors and the total land area in a layout), are related mathematically as follows:

$$FAR = \frac{BUS \cdot 3 \cdot BR}{BUS + BR \cdot 3 \cdot SPO}; \quad BUD = \frac{FAR}{BUS}; \quad GCR = \frac{FAR}{BR}; \quad (6.1)$$

Fig. 6.1 A schematic urban-built-form (single-building-block)

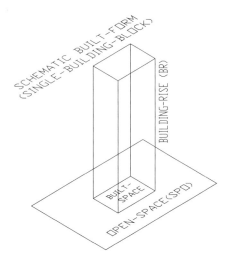

6.5.1   A Model for Optimal Design of Urban-Built-Form

We may designate BUS, SPO, FAR, and BR as decision variables: x_1, x_2, x_3, and x_4, respectively, as explained below:

$x_1 =$ optimum space per Built-Unit (BU, i.e. dwelling unit, commercial unit, so on) in a planning area,

$x_2 =$ optimum Semi-Public Open-Space (SPO), i.e. open-to-sky space per BU in the planning area,

$x_3 =$ a variable giving the optimum FAR, i.e. the ratio between the total built-space on all floors of an urban development and its land area,

$x_4 =$ variable giving optimum building rise (effective), i.e. number of effective floors in the planning area. Ideally, this should be an integer value, and consequently, nonlinear integer programming, which is quite complicated without commensurate benefit, should be applied. But, literature indicates that efficient solution methods are not available even for linear integer programming (Mark). However, the problem can be overcome by adopting the concept of 'effective building rise', and thus permitting application of nonlinear continuous geometric programming as adopted here. This concept is outlined in more details in Chaps. 8 and 10.

Similarly, we may designate the cost coefficients c_1, c_2, and c_3 as follows:

$c_{01} =$ cost coefficient relating building construction cost per unit of plinth area (monetary units/m^2),

$c_{02} =$ a cost coefficient related to building rise, in terms of the increased cost per unit of plinth area (monetary units/m^2) in the urban development. This factor can also be used as a social-cost-coefficient (say, a penalty factor for high building rise determined by the society uniformly in a transparent manner with *participation* of multiple stakeholders, instead of arbitrary imposition from top) for determining the optimal building-rise (effective) based on the social-cost-benefit-analysis principles, in a site-specific case,

c_{03} = a cost coefficient related to developed land price (including infrastructure) per unit of land area (monetary units/sq.m.).

Constraint (discussed below) function constant coefficients can be defined as follows:

c_{21} or a_1 = the desired minimum built-unit space in sq.m. per built-unit in the urban development,

c_{31} or a_2 = the desired minimum semi-public open-to-sky-space in sq.m. per built-unit in the urban development.

Using FAR, i.e. x_3 and BUS, i.e. x_1, the area of land consumed per built-unit of built-area BUS can be calculated as:

$$= BUS/FAR = x_1 x_3^{-1}$$

We may define $y(x)$ as the objective function, i.e. the total cost per built-unit (BU—residential, commercial, and so on) including, building construction cost, extra cost due to building rise, and developed land price. Using the above factors, the expression for $y(x)$, as the total cost per built-unit, in terms of the above variables and cost coefficients will be given by:

$$y(x) = c_{01} x_1 + c_{02} x_1 x_4 + c_{03} x_1 x_3^{-1}$$

The first term of the above objective function represents the building construction cost per built-unit, the second term represents the increased building construction cost per built-unit due to increase in building rise, and the third term represents the land cost per built-unit based on consumption of land as calculated below.

Using the above variables: x_1, x_2, x_3, and x_4, the expression for FAR given in Eq. (6.1) becomes:

$$x_3 = x_1 x_4 / (x_1 + x_4 x_2)$$

or,

$$x_3 x_4 x_2 + x_3 x_1 = x_1 x_4$$

Thus, to ensure compatibility between FAR (Floor Area Ratio), BUS (Built-Unit-Space), SPO (Semi-Public-Open-Space), and the BR (Building Rise) in a layout plan, the above expression can be taken as a constraint as follows:

$$x_3 x_4 x_2 + x_3 x_1 \leq x_1 x_4$$

Dividing both sides by x_1 x_4, the above constraint expression (shown as constraint-1 below) becomes as follows:

$$x_1^{-1}x_2x_3 + x_3x_4^{-1} \leq 1$$

Since the objective function is to minimize the cost per built-unit, obviously the value of BUS or x_1 tends to be reduced to even zero to reduce cost which is not a feasible or practicable solution. Hence, minimum acceptable value of x_1 (designated by the coefficient c_{21}) is included as constraint-2 as shown below. Similarly, the value of SPO or x_2 tends to be reduced to even zero to reduce cost which is not a feasible or practicable solution. Therefore, a constraint defining the minimum acceptable value of open-space per built-unit (SPO), i.e. x_2 (designated by the coefficient c_{31}) is added as constraint-3.

Using the above relationships, a model for optimal design of urban-built-form in a 'planning area' linking Built-Space (BUS), Open-Space (SPO), Building Rise (BR), land price, construction costs can be developed for optimal design of urban-built-form to give both for (1) least-cost of a built-unit of given built-space (BUS), or, for (2) most-benefit in terms of maximum built-space (BUS) within a given built-unit-cost (BUC in MU) as follows:

1. Least-Cost Optimal Design Model

Minimize Built-Unit-Cost $= y(x) = c_{01}x_1 + c_{02}x_1x_4 + c_{03}x_1x_3^{-1}$ $\qquad$ (6.2)
Subject to:

$x_1^{-1}x_2x_3 + x_3x_4^{-1}$ $\qquad\qquad \leq 1$
(FAR, Built-Unit-Space, Open Space and Building Rise compatibility) $\qquad$ (1)

$c_{21}x_1^{-1}$ $\qquad\qquad\qquad\qquad \leq 1$
(Built-Unit-Space constraint) $\qquad$ (2)

$c_{31}x_2^{-1}$ $\qquad\qquad\qquad\qquad \leq 1$
(Open Space constraint) $\qquad$ (3)

x_1, x_2, x_3, x_4 $\qquad\qquad\qquad\qquad > 0$
(Nonnegativity constraint)

Using geometric programming, a *nonlinear programming technique*, and solving the above model, we get the following two equations (7.65) and (7.66) with two unknown dual variables, namely w_{11} and w_{12} (readers interested to learn about such terms of geometric programming may refer to Chap. 7).

$$c_{03}c_{31}(1 - w_{11} - 2w_{12}) - c_{01}c_{21}w_{11} = 0 \qquad (7.65)$$

$$c_{02}c_{03}(1 - w_{11} - 2w_{12})^2 - c_{01}{}^2w_{12}{}^2 = 0 \qquad (7.66)$$

These equations could be easily solved to find the values of w_{11} and w_{12} using the Newton–Raphson method. Knowing the values of w_{11} and w_{12}, the values of other dual variables could be determined, and the problem could be completely solved.

Similarly, a model for optimal design of urban-built-form in a 'planning area' to give most-benefit in terms of maximum Built-Space (BUS) within a given Built-Unit-Cost (BUC in MU) can be developed as follows:

2. **Most-Benefit Optimal Design Model**

Maximize Built-Unit-Space $= y(x) = x_1$ $\qquad (6.3)$
Subject to:

$$x_1{}^{-1}x_2x_3 + x_3x_4{}^{-1} \qquad\qquad \leq 1 \qquad (1)$$
(FAR, Built-Unit-Space, Open Space and Building Rise compatibility)

$$c_{31}x_2{}^{-1} \qquad\qquad\qquad \leq 1 \qquad (2)$$
(Open Space constraint)

$$a_3x_1 + a_4x_1x_4 + a_5x_1x_3{}^{-1} \qquad \leq 1 \qquad (3)$$
(Affordable Built-Unit-Cost constraint)

$$x_1, x_2, x_3, x_4 \qquad\qquad\qquad > 0$$
(Nonnegativity constraint)

Here,
$a_3 = c_{01}/\text{BUC}$, $a_4 = c_{02}/\text{BUC}$, and $a_5 = c_{03}/\text{BUC}$, where BUC is the affordable built-unit-cost in MU.

Using geometric programming, a *nonlinear programming optimization technique*, the above model for optimal design of urban-built-form to give the most-benefit optimal design for maximum built-space (BUS) within a given built-unit-cost (BUC in MU) is also solved in Chap. 8, which may be referred by interested readers. It is fascinating to note that the above complex geometric programming, i.e. *nonlinear programming optimization technique* solution, also yields simple **Algebraic Formulas for Optimal Design of Urban-Built-Form** for both cases, and AEC professionals can use these simple formulas to obtain **Optimal Design of Urban-Built-Form** even without learning such complex techniques. These simplified **Algebraic Formulas for Optimal Design of Urban-Built-Form**, derived from the solution of the above *nonlinear programming optimization* models for both cases are given below.

6.5.2   Algebraic Formulas for Optimal Design of Urban-Built-Form

The complete models for optimal design of urban-built-form as presented above are solved in Chap. 8 using geometric programming, a *nonlinear programming technique*; and complete listing of the C++ program OUBF.CPP to solve the above optimization models is also presented in Chap. 8. However, AEC professionals not conversant with *nonlinear programming technique* can also obtain optimal results using the following simplified algebraic formulas derived from the solution of above optimization models in Chap. 8. In the algebraic formulas, the following substitutions have been made for simplification.

c_{01}, c_{02}, c_{03}, c_{21}, and c_{31} are replaced by c_1, c_2, c_3, a_1, and a_2, respectively.

1. **Least-Cost Optimal Design Model**

   The **Optimum Floor Area Ratio** (FAR) to give the **Least-Cost** per Built-Unit is given by the following simplified expression linked only with *constant coefficients* of the *Optimization Model*:

$$\textbf{Simplified Formula of Optimum FAR} = \frac{c_3 a_1}{\left[c_3 a_2 + \sqrt{c_3 c_2 3 a_1} \right]} \tag{6.4}$$

$$\textbf{Optimum Building Rise (BR) in Effective Number of Floors} = \sqrt{c_3/c_2} \tag{6.5}$$

   The Simplified Formula for Least-Cost (LC) per Built-Unit in Monetary Units (MU) is given by:

$$\textbf{LC} = c_3 a_2 + 23\sqrt{c_3 c_2 3 a_1} + c_1 a_1 \tag{6.6}$$

$$\textbf{Ground Coverage Ratio (GCR)} = \text{FAR}/\text{BR} \tag{6.7}$$

$$\textbf{Built-Unit-Density (BUD)} = 100003\text{FAR}/\text{BUS Built-Units/Hectare} \tag{6.8}$$

2. **Most-Benefit Optimal Design Model**

   Simplified Formula of Optimum FAR (FAR) is

$$= \frac{a_5 \times (1 - a_2 a_5)}{\sqrt{a_4 a_5} \times (1 + a_2 a_5) + a_2 a_3 a_5} \tag{6.9}$$

$$\textbf{Formula for Maximum Built-Unit-Space (BUS)} = \frac{1 - a_2 a_5}{2 \times \sqrt{a_4 a_5} + a_3} \tag{6.10}$$

$$\textbf{Optimum Building Rise (BR) in Effective Number of Floors} = \sqrt{\frac{a_5}{a_4}} \tag{6.11}$$

It may be seen that the expressions (6.4)–(6.8) give the complete solution of the model for the least-cost optimal design of urban-built-form presented in expression

(6.2) above, only in terms of the objective function constant coefficients and the constraint function constant coefficients. Similarly, the complete solution of the model for the most-benefit optimal design of urban-built-form presented in expression (6.3) above is also given by the expressions (6.9)–(6.11) only in terms of the objective function constant coefficients and the constraint function constant coefficients. Thus, the AEC professionals such as urban planners, engineers, architects, and urban managers can solve the model for optimal design of urban-built-form in their site-specific situation by applying these formulas, and even using only a handheld calculator.

However, as already stated, the C++ computer program **OUBF.CPP** incorporates one model (designated as Model-1) which uses the above algebraic equations for **Least-Cost Optimal Design**, and another model (designated as Model-3) which uses the above algebraic equations for **Most-Benefit Optimal Design**, giving instantly the complete solution of the models for optimal design of urban-built-form as outlined above. AEC professionals, even if not familiar with *nonlinear programming techniques*, can use this program to find instantly the complete solution of the model for optimal design of urban-built-form in a site-specific case.

6.5.3   Optimal Design of Urban-Built-Form Using Algebraic Formulas: A Case Study

We can take as **Case Study I**, an example to obtain the optimal design of urban-built-form, with given Construction Plinth Area Rate: CPAR = 4500 MU/sq.m., Land Price = 10,000 MU/sq.m., the increased building construction cost per built-unit due to increase in building rise $p = 0.010$, Built-Unit-Space ('BUS') = 110.00 sq.m./BU, and Open-To-Sky-Space ('SPO') = 20.00 sq.m./BU.

A typical result of the C++ computer program OUBF.CPP solving the above algebraic formulas (6.2)–(6.6), and giving complete solution of the model for optimal design of urban-built-form, taking the above input data as **Case Study 1**, is given below:

```
SUMMARY OF ''outubf.cpp'' RESULTS IN THE INSTANT CASE ARE:

INPUT MODEL CONSTANT VALUES CHOSEN BY USER ARE:
CONSTRUCTION PLINTH AREA RATE: CPAR= 4500 MU/SQ.M. LAND PRICE= 10000
MU/SQ.M. P=0.010

BUILT-UNIT-SPACE ('BUS')=110.00 SQ.M./BU  OPEN-SPACE ('SPO')= 20.00
SQ.M./BU

OUTPUT OPTIMAL MODEL VALUES, USING THE SIMPLIFIED FORMULAS, ARE:-
OPTIMAL FAR= 4.018   LEAST B-UNIT-COST (LC) = 842580 MU
 OPTIMAL BLDG.RISE (EFFECTIVE) = 14.91
 GROUND COVERAGE RATIO (GCR) = 0.260
BUILT-UNIT-DENSITY (BUD) =365.243 BUILT UNITS/HECTARE
```

Even the geometric programming weights (not indicated here, and interested readers may refer to Chap.7 for these details) are also given by the simplified algebraic formulas incorporated in the Model-I of the above C++ program OUBF.CPP. The Model-II of the above C++ program solves in no time the *nonlinear simultaneous equations* (i.e. Eqs. (7.65) and (7.66) cited above) giving the exact optimal solution of the Model. The output exact optimal solution of the same Case Study I example problem, instantly given by the Model-II of the above C++ program OUBF.CPP, is shown below:

```
SUMMARY OF ''outubf2.cpp'' RESULTS IN THE INSTANT CASE ARE:

INPUT MODEL CONSTANT VALUES CHOSEN BY USER ARE:
CONSTRUCTION PLINTH AREA RATE: CPAR= 4500 MU/SQ.M. LAND PRICE= 10000
MU/SQ.M. P=0.010
BUILT-UNIT-SPACE ('BUS')=110.00 SQ.M./BU  OPEN-SPACE ('SPO')= 20.00
SQ.M./BU

CURRENT VALUE OF ITERATION NUMBER IS = 6
CURRENT VALUE OF f IS =    0.000000000
CURRENT VALUE OF g IS =   -0.000000001
CURRENT VALUE OF x IS = 0.236366
CURRENT VALUE OF y IS = 0.086566

OUTPUT OPTIMAL MODEL VALUES GIVEN BY THE MODEL ARE:
OUTPUT OPTIMAL MODEL WEIGHTS: w11= 0.2364 w12= 0.0866 w01= 0.5865 w02=
0.0866
w03 = 0.3249 w21= 0.6626 w31=0.2364

OPTIMAL FAR= 4.018  LEAST B-UNIT-COST (LC)= 842580 MU
OPTIMAL BLDG.RISE (EFFECTIVE) = 14.91
GROUND COVERAGE RATIO (GCR) = 0.260
BUILT-UNIT-DENSITY (BUD) =365.243 BUILT UNITS/HECTARE

THE VALUE OF ITOL CHOSEN IS =0.000100

SINCE THE CURRENT SOLUTION MEETS THE VALUE OF CHOSEN
MARGIN OF ERROR, THE SOLUTION ACCEPTED & PROGRAM TERMINATED
```

It may be seen that the above output computer results (in output file: 'outubf.cpp') using the algebraic formulas/expressions (6.2)–(6.6) are exactly same as the output computer results (in output file: ''outubf2.cpp') given by solving the nonlinear equations (7.65) and (7.66) presented as **Case Study-1** above, as in both cases input model constant values chosen are same. *Thus, results of algebraic formulas are exactly same as the optimal solution represented by the solution of the system of nonlinear equations, and a user can choose any of these methods for his/her problem solution.*

Therefore, AEC professionals not having any knowledge of *Optimization* should have no inhibition in applying this model for optimal design of urban-built-form and the *optimization principles* in their planning, design, and urban management practices to help achieve *efficiency* and *social equity* in the urban development process

being managed by them, although acquiring such expert knowledge is desirable for AEC professionals to be more effective in solving our mounting urban problems.

6.6 Computer-Aided Planning and Design Analysis

All the above issues make computer use indispensable for problem solution in all branches of AEC including architecture, planning, and civil engineering, particularly in the context of search through numerous potential solutions with large variations in resource-efficiency indicators, to select a particular solution that meets a specified *goal*, requiring a comprehensive planning and design analysis. For example, planning and design in housing and urban development is a *problem-solving process* to achieve the national *goals* of providing adequate shelter and urban services to all, within the affordability and resource-constraints at various levels, and it is imperative that above management principles should be given very high priority in this process to achieve the goal of *efficiency* and *social equity* in urban development and management. However, this is possible only if suitable tools and techniques are used. Computers permit development of symbolic generative systems representing design by symbols, variables, and computer programs, producing potential solutions or goal set, help store and retrieve data and graphics describing design, and integrate quantitative analysis, optimization, and graphics into the design process, and thus help effective cost control, effective response to the constraints, and achieve design objectives and goals.

6.6.1 HUDCO Model

One of the earliest efforts of computer-based design and analysis was undertaken by HUDCO (Housing and Urban Development Corporation), resulting to a book entitled: 'Computer Based Design and Analysis of Affordable Shelter-HUDCO MODEL', which was published by HUDCO, NEW DELHI, in 1982. This generated lot of interests in housing and urban development agencies at that time. Because of non-availability of computers at affordable cost at that time, the model was based on handheld programmable calculators, which were also in short supply at that time, and now have become outdated. The book is relevant even now to clarify in a simple way many housing and urban development issues concerning *efficiency* and *social equity* in the process. The programmes in the above book, based on handheld programmable calculators, are now converted to **C++** programmes named as **hudmod.cpp, and to help readers of this book to use such techniques to achieve their own problem solutions, complete listings of this C++ programs are included in an online space connected to this book.** The **hudmod.exe** file is also provided in the **online space connected to this book**, and thus, all 14 programs of HUDCO MODEL can be run in any computer to get results. This may be of interest to AEC professionals, to carry out simple planning and design analysis.

6.6.2   Application of Traditional Management Principles in Urban Development Planning and Management to achieve Efficiency, Equity, and Social Responsiveness: A C++ Software

Software 'CADUB' (Computer-Aided-analysis & Design of Urban-Built-form) which is the Module-II of the mother Software: HudCAD described in detail in later chapters is developed as an urban planning and management IT tool, to help achieve productivity, equity, and social responsiveness in urban development. It helps apply the traditional management principles of flexibility and social responsiveness in urban operations such as in 'Planning & Design' of urban-built-form (development-control/building-bulk-regulations), with participation of multiple stakeholders and claimants (i.e. a '*bottom-up*' approach, as against the conventional '*top-down*' approach), for effective response to the urban dynamics and uncertainties (including land price and construction cost dynamics) to achieve the aim of 'adequate and affordable housing and urban services for ALL' within given constraints (e.g. resources, etc.) beyond control of various entities, and thus help attain a resource-efficient and EQUITABLE solution of our mounting urban problems.

'CADUB' has two components, namely: (1) Numeric and (2) Graphic. The numeric component is named as software: 'CAUB' (Computer-Aided-analysis of Urban-Built-form) developed in user-friendly C++ language, linking scientifically planning regulations and urban Built-Form-Elements (BFE) such as FAR, BR, BUS, BUC, K as explained below:

FAR = Floor Area Ratio, i.e. the ratio between the total built-space on all floors of an urban development project and its land-area

BU = Built-Unit, i.e. 'dwelling-unit', 'commercial-unit', etc. in urban development

BUS = Built-Unit-Space (sq.m.), i.e. average built-space per BU in development

MBUS = Maximum-Built-Unit-Space when a 'MBO' MODEL (see below) is used

BR = Building Rise, i.e. number of floors in urban development (system level)

SPO = Semi-Public-Open-Space (SPO), i.e. average open-to-sky-space (sq.m.) per BU (i.e. Built-Unit) in the project layout (at the system level)

BUD = Built-Unit-Density, i.e. number of BUs per unit-area (hectare)

GCR = Ground Coverage Ratio in the project layout (at the system level)

LP = Market Land Price in Monetary Units (MU) per unit-area (sq.m.) of land

CPAR = Base Construction-Plinth-Area-Rate in MU per unit-area(sq.m.) of BU

GPAR = Gross-Plinth-Area-Rate in MU per sq.m. of BU including land price

K = A cost-factor to allow increase in cost due to building-rise (effective),or, a PENALTY Factor, for high-rise-development, DECIDED by SOCIETY uniformly

BUC = Built-Unit-Cost, i.e. Cost per Building-Unit or 'BU', in Monetary Units

LBUC = Least-Cost per BU, when the Least-Cost-Optimizing (LCO) model is used

ABUC = Affordable Cost/BU, when Most-Benefit-Optimizing (MBO) model is used

It solves ten iteration models, including the nonlinear programming built-form optimization models (for analysis and optimization of urban-built-form in alphanumeric form), with 22 input set options (extendable), permitting analysis of numerous options by multiple stakeholders, which is rarely possible in conventional practices. The graphic component is a LISP program named 'CBDG'. Numeric component can be invoked by 'CBDG' for numeric iteration at appropriate stage(s) to generate a numeric planning solution for a given set of input conditions. On loading graphic component, i.e. 'CBDG', above numeric solution is converted instantly into graphic planning solution. In this process, even a lay user citizen (not initiated to computer) can be prompted in an interactive manner for appropriate RESPONSE at each iteration step, to derive desired solution.

6.6.3 Software Features and Benefits

Using 'CADUB' and 'CAUB', one can carry out the following tasks:

1. Determine the 'Interaction' between selected elements of urban-built-form.
2. Carry out 'sensitivity-analysis', with apt set of independent/dependent design variables and 'Iteration Choice', to develop an insight into their interrelationship including cost.
3. Select desired value(s) of BFE acceptable to multiple stakeholders, based on 'sensitivity-analysis', using 'CAUB' for numeric iteration and the graphic software: 'CBDG' (giving INSTANT VISUAL presentation of the numeric planning solution) for graphic iterations, in a site-specific case, with more informed/ meaningful stakeholder participation.
4. Determine the 'Optimal-Design' (e.g. optimum FAR) of urban-built-form in a site-specific case, taking into account the land price, with participation of multiple stakeholders.
5. Find instantly, 'Least-Cost' per 'Built-Unit' of given Plinth-Area/'Maximum-Plinth-Area' within a given investment/affordable cost, depending on the planning-parameters chosen.
6. Determine 'Planning-Regulations'/'Project Parameters' responsive to the people's need (in a site-specific-case in the context of 'urban-dynamics'), based on 'Citizen-Planning-Authority-Interactive-Participatory' 'Session' for analysis and 'informed-decision-making' by the multiple stakeholders in the urban operations, using both numeric/graphic iterations,
7. Decide the most acceptable Transfer of Development Rights Package (TDR, e.g. 'Sale of FAR' to get fund) in a transparent manner, using both numeric and graphic iterations, with participation of multiple stakeholders, to promote low-income housing even in high-land-price-areas in a city, to match with the given demographic-distribution-structure.

Thus, the software facilitates a participatory 'urban planning and policy analysis' with transparency for informed -*decision-making* to improve *equity, social responsiveness*, and *productivity* in the built-environment and urban development, in the context of urban dynamics including rising land prices and construction costs, and affordability and resource-constraints of stakeholders at various levels.

6.6.4   Operation of the Applications Software: 'CADUB'-'CAUB'-Procedure and Computer Prompts Appearing in Video Screen

On loading the software: 'CADUB' in AutoCAD and typing 'cadub' against Command line, following message appears in computer screen for RESPONSE (Fig. 6.2).

Some built-form-elements considered are illustrated in Fig. 6.1 in Sect. 6.5 above (Software can also deal with multiple-building-blocks shown later) and is reproduced below for ready reference, which is also displayed by 'CADUB' on start.

Thus, in the text part, user is also prompted for response by the following message:

Do you want to see Typical Schematic Urban-Built-Form (Y/N)?

If response is 'Y', the Schematic Urban-Built-Form shown above is displayed in the Graphic Part of Computer Screen as shown in Fig. 6.3 (to give a grasp of few BFE'S to the User), and thereafter, or, if the response is 'N', the following message appears for User response in the Text Part of screen:

You want to RUN 'Numerical Iteration SESSION' (i.e. 'CAUB')? (Y/N)

If choice is 'Y', the following message appears in computer screen for response:

Type 'CAUB' for 'Num. Itn. Session' and After Results Load 'CBDG' for 'Drawing Session'

```
WELCOME TO ::
            '' CADUB ''
A TOOL FOR 'COMPUTER-AIDED PLANNING & DESIGN'
OF URBAN 'BUILT-FORM' & 'LAYOUT'(WITH GRAPHIC),
FOR AN EFFICIENT ''URBAN MANAGEMENT'', AND TO
ACHIEVE AN URBAN DEVELOPMENT WITH :-
''EFFICIENCY AND EQUITY.''
        The SOFTWARE ''CADUB'' has a Number of MODELS.
        SELECT MODELS to RUN as per your Requirements.
```

Fig. 6.2 Computer screen message on start

Fig. 6.3 A schematic urban-built-form (single-building-block)

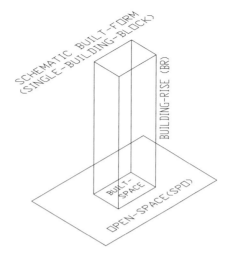

On typing: CAUB and pressing <CR>, following message appears in Computer Screen for appropriate RESPONSE by the User:

```
WELCOME TO SOFTWARE: ''CAUB''
(Computer-Aided-analysis/design-Urban-Built-form)
Using ''CAUB'', you can CARRY OUT the following TASKS :
** (1) DETERMINE Interaction between the URBAN 'BUILT-FORM-ELEMENTS'
('BFE')
** (2) CARRY OUT 'Sensitivity-Analysis'(SA), choosing an Appropriate
SET of Independent/Dependent DESIGN VARIABLES and the 'ITERATION MODEL'
** (3) SELECT Desired Values of 'BFE', using 'Sensitivity-Analysis' in
CONTEXT of 'URBAN-DYNAMICS'
** (4) DETERMINE 'Optimal Design' of URBAN-BUILT-FORM in SITE-SPECIFIC-
CASE
** (5) FIND Instantly, 'LEAST-COST' per BUILT-UNIT of GIVEN PLINTH-AREA
('BUS') or, 'MAXIMUM-PLINTH-AREA' within GIVEN INVESTMENT/ AFFORDABLE
COST('ABUC') per BUILT-UNIT
** (6) DETERMINE Acceptable 'TDRP' (Transfer of Development Rights
Package i.e. Sale of 'FAR') by a 'PARTICIPATORY-INTERACTIVE' SESSION
** (6) DETERMINE Optimal 'PLANNING REGULATIONS', Responsive to PEOPLE'S
NEED based on
'CITIZEN-Planning-Authority-PARTICIPATORY-INTERACTIVE' Session

Do you want to see Definitions of Urban-Built-Form Elements?
Type y (if 'yes') or n (if 'no')and press <CR>
```

Depending on response 'y' or 'n', Software gives following message after displaying Definitions of Urban-Built-Form Elements included in it ,or, without displaying Definitions:

```
Do you want to START the 'ITERATION SESSION'? Type y(if 'yes') or n
(if 'no')and press <CR>
```

On typing 'n' it Closes and 'y' starts 'ITERATION SESSION' with following message:

As shown below, there are ten 'ITERATION MODELS' depending on the SET OF DESIGN VARIABLES you Choose as INDEPENDENT(i.e. you want to INPUT any desired value),or, as DEPENDENT (i.e. Dependent on Values assigned by you to Independent Variables):

```
FAR - A DEPENDENT DESIGN VARIABLE('DDV' i.e. OUTPUT)(ITMODEL- 1)
BUS - A DEPENDENT DESIGN VARIABLE ('DDV' i.e. OUTPUT)(ITMODEL- 2)
SPO - A DEPENDENT DESIGN VARIABLE ('DDV' i.e. OUTPUT)(ITMODEL- 3)
BR - A DEPENDENT DESIGN VARIABLE('DDV' i.e. OUTPUT)(ITMODEL- 4)
BUD, SPO, BR - INPUTS i.e. INDEPENDENT DESIGN VARIABLES ('IDV')
(ITMODEL-5)
BUD, BUS, BR - INPUTS i.e. INDEPENDENT DESIGN VARIABLES ('IDV')
(ITMODEL-6)
BUS, FAR, SPO-INPUTS (BR-DDV): FOR 'BUC'+'TDRP'('FAR'-SALE) ANALYSIS
(ITMODEL-6)
BUC, FAR, SPO, LP, CPAR,K - INPUTS: FIND 'BUS' WITHIN A GIVEN 'BUC'
(ITMODEL- 8)
BUS, SPO, LP, CPAR, K - INPUTS: FIND 'LBUC' & OPTM.'FAR' - 'LCO MODEL'
(ITMODEL- 9)
BUC, SPO, LP, CPAR,K -INPUTS : FIND 'MBUS' & OPTM.'FAR' - 'MBO MODEL'
(ITMODEL-10)
ENTER YOUR CHOICE OF 'ITERATION MODEL' ( any of : 1/2/3/4/5/6/6/8/9/10)
```

On Iteration MODEL choice, it prompts INPUT-Set choice by the following message:

```
YOU CAN CHOOSE ANY OF FOLLOWING 22 'INPUT SET' OPTIONS:
0.When you select the pre-set computer data input
1.When you want to change values of BUS,SPO,BR
2.When you want to change values of SPO,FAR,BR
3.When you want to change values of FAR,BUS,BR
4.When you want to change values of BR,BUS
5.When you want to change values of BR,SPO
6.When you want to change values of BUS,SPO
6.When you want to change value of BR only
8.When you want to change value of BUS only
9.When you want to change value of SPO only
10.When you want to change value of FAR only
11.When you want to change value of BUD only
12.When you want to change value of CPAR only
13.When you want to change value of K only
14.When you want to change value of LP only
15.When you want to change value of BUC only
16.When you want to change values of BUD,SPO,BR
17.When you want to change values of BUD,BUS,BR
18. When you want to change values of LP,CPAR,K
19.When you want to change values of FAR,BUS,SPO,LP,CPAR,K
20.When you want to change values of BUS,SPO,LP,CPAR,K (LCO-MODEL)
21. When you want to change values of BUC,SPO,LP,CPAR,K (MBO-MODEL)
ENTER YOUR CURRENT 'INPUT SET' OPTION(PUT 0/1.../21-TDRP: 0/13/14) &
PRESS <CR>:
```

Thus, the software is very user-friendly, where a USER, even if not computer literate, is prompted in an interactive manner for appropriate RESPONSE (Selection of MODEL, INPUT-Set, Input Value(s) and so on), at each step of its run, leading him/her to the desired solution. Applications of the SOFTWARE to carry out some of the above tasks are illustrated below.

Urban Planners/Planning Authorities (PLA) prescribe 'Planning Regulations' often on ad hoc basis. Thus, a fixed FAR (e.g. 0.65) is prescribed irrespective of the land price of 100 MU/m^2, or, 20,000 MU/m^2, increasing cost per built-unit 3–4 times, creating inefficiencies, inequities, and negative-citizen-responses, as PEOPLE prefer optimization. In context of soaring land prices (in Delhi, a land price of Rs. 10,000–40,000/m^2 is reported) and construction costs and, scarcity of land-resource in India (shrinking per capita land from 1.42 ha in 1901, to only 0.3 ha in 2001), deciding Built-Form-Elements(BFE) such as FAR, becomes very critical to find a planning solution which is efficient, affordable, optimal, and acceptable to the multiple stakeholders, and thus, make it sustainable.

6.6.5 Illustrative Applications of the Software: 'CADUB'-'CAUB'

As mentioned above, there are ten models included in the software to carry out different tasks such as interaction between the urban 'built-form-elements' ('BFE'), sensitivity analysis between different built-form-elements, determining optimal urban built-form-elements, determining optimal planning regulations, responsive to the people's need based on citizen-planning-authority-participatory-interactive session, and so on. Applications of the models included in the above software are illustrated below.

6.6.5.1 Determining Interaction and Sensitivity between Urban 'Built-Form-Elements' ('BFE') and Building Rise-Density Interaction (Application of Model-1, 2, and 3)

A USER can select appropriate MODEL(S) to determine interaction between different BFEs. For example, to determine Building Rise-Density Interaction (BUS and SPO Fixed), MODEL-1 can be selected. Its illustrative INPUT-OUTPUT results and *prompting* for response in a typical RUN is shown (in *italics*) below:

```
SUMMARY OF ''CAUB'' RESULTS IN THE PRESENT CASE ARE (MODEL-1):
Iteration MODEL Chosen       : Sensitivity MODEL (Physical + Cost)
Independent Design Variables('IDV') : Building-Rise(BR), Open-Space
(SPO)
                          & Built-Unit-Space(BUS),Unit-Costs
Dependent Design Variables ('DDV'): FAR, Ground-Coverage-Ratio(GCR)&
          DENSITY, Total Built-Unit-Cost (BUC)
INPUT Parameter-Values Chosen in the present Case, are :
```

Con.Plth.-Area-Rate(CPAR)= 4500 MU/Sq.M.   LP = 10000 MU/Sq.M.   K=0.01
Building-Rise ('BR')   = 4.00   Open-Space ('SPO') = 25.00 Sq.M./BU
Built-Unit-Space ('BUS') = 110.00 Sq.M./BU   ITMODEL= 1 INCHOICE= 19
OUTPUT Parameter-Values given Instantly by the COMPUTER, are :
Total Built-Unit-Cost (BUC) = 1039800 MU, Gross-Plth.-Area-Rate
(GPAR) = 9452.6 MU/Sq.M.   FAR =
2.10   GCR = 0.524 DENSITY i.e. Built-Unit-Density ('BUD') = 190.5
BU/Hectare
One can RE-RUN the MODEL, keeping Constant the Values of any 2 IDV'S
(say-'BUS' & 'SPO') and, changing Value of the 3rd. 'IDV' (say- 'BR'), to
Determine the Sensitivity between the DDV'S (i.e. 'DENSITY', FAR & GCR
for above Choice) and the 3rd. 'IDV' (i.e. 'BR' here), as may be Chosen by
the User.
DO YOU WANT TO CHANGE 'INPUT VALUE(S)/INPUT SET/ITERATION MODEL', FOR
BETTER RESULTS/PARTICIPATORY-DECISION ? Type y (if yes), or, n (if no) &
press <CR>

The MODEL could be RE-RUN (or, a Series of ITERATIONS *automated*) with different INPUT Values by multiple *stakeholders* to find desired *Planning Solution* in a specific-case. Outputs of an *Automated* Series of ITERATIONS (changing only INPUT-BR) for 'BR-DENSITY' *Sensitivity Analysis* (MODEL-1) shown below:

MODEL-1: A Typical *Automated* Series of ITERATIONS (SPO and BUS fixed)

BR (Number of floors)	FAR	DENSITY (BU/HECTARE)	GCR	BUC (MU)	GPAR (MU/Sq. M.)
1	0.81	64.06	0.815	1849950	16818
2	1.38	125.00	0.688	1304900	11863
3	1.68	162.16	0.595	1126516	10241
4	2.10	190.48	0.524	1039800	9453
5	2.34	212.66	0.468	989650	8998
6	2.54	230.66	0.423	958033	8609
6	2.60	245.61	0.386	936693	8516
8	2.84	258.06	0.355	922100	8383
9	2.96	268.66	0.328	911662	8289
10	3.06	266.68	0.306	904500	8223

Similarly, an Illustrative INPUT-OUTPUT results of MODEL-2 are shown below:

SUMMARY OF ''CAUB'' RESULTS IN THE PRESENT CASE ARE (MODEL-2):
Iteration Model Chosen   : Interaction/Sensitivity Model (Cost
+Phy.)
Independent Design Variables ('IDV') : Building-Rise (BR), Open-Space
(SPO) &
                       Floor-Area-Ratio (FAR)
Dependent Design Variables ('DDV') : Built-Unit-Space (BUS), G.C.
Ratio, Density
INPUT Parameters with Values Chosen by User :
Con.Plth.-Area-Rate(CPAR)= 4500 MU/Sq.M. LP= 10000 MU/Sq.M. K=0.01

```
Building-Rise ('BR') = 2  Open-Space ('SPO') = 25.00  Sq.M./BU
Floor-Area-Ratio ('FAR') = 1.50   ITMODEL= 2 INCHOICE= 10
OUTPUT Parameters with Values Given by Computer :
Total BU-Cost(BUC)= 1688500 MU, Gross-P.A.Rate(GPAR)= 11256.6 MU/Sq.
M.
Built-Unit-Space(BUS)= 150.00 Sq.M./BU  Ground-Cov.Ratio(GCR)= 0.650
Density i.e. Built-Unit-Density ('BUD') = 100.00 BU/Hectare
```

Series of OUTPUT-Values of Built-Form-Elements(BFE) including DENSITY generated Running MODEL-2 (Dependent Variable BUS) with SPO and FAR fixed, are shown below for 'BR-DENSITY' *Sensitivity Analysis* in the above case.

MODEL-2: A Typical *Automated* Series of ITERATIONS (SPO and FAR fixed)

BR (Number of floors)	BUS (Sq. M.)	DENSITY (BU/HA)	GCR	BUC (MU)	GPAR (MU/Sq. M.)
1	−65.00	−200.00	1.500	−840865	11212
2	150.00	100.00	0.650	1688500	11256
3	65.00	200.00	0.500	846625	11302
4	60.00	250.00	0.365	680800	11346
5	53.56	280.00	0.300	610268	11392
6	50.00	300.00	0.250	561833	11436
6	46.63	314.29	0.214	546989	11482
8	46.15	325.00	0.188	532000	11526
9	45.00	333.33	0.166	520625	11562
10	44.12	340.00	0.150	512500	11616

Other INPUT VALUES in the above CASES are:

```
FAR=1.5 Sq.M.  SPO=25.0 Sq.M.  LP=10000 MU/Sq.M. CPAR= 4500 MU/Sq.M.
```

Negative OUTPUT-Values indicate an infeasible solution (for the INPUT Values chosen) which can be discarded (Case with BR = 1 above), or, the MODEL can be RE-RUN with changed INPUT Value (say 'FAR' above) to get a feasible solution.

An Illustrative INPUT-OUTPUT result (in *italics*) of MODEL-3 is shown below:

```
SUMMARY OF ''CAUB'' RESULTS IN THE PRESENT CASE ARE (MODEL-3):
Iteration Model Chosen          : Interaction/Sensitivity Model (Cost
+Phy.)
Independent Design Variables('IDV'): Building-Rise(BR),Built-Unit-
Space(BUS) &
                          Floor-Area-Ratio(FAR)
Dependent Design Variables('DDV'): Open-Space(SPO),GCR, Density
INPUT Parameters with Values Chosen by User :
Constn.Plth.Area-Rate(CPAR)= 5000 MU/Sq.M., LP= 5000 MU/Sq.M.,
K=0.0250,
Building-Rise ('BR')= 2 , Built-Unit-Space ('BUS') =110.00 Sq.M./BU ,
Floor-Area-Ratio ('FAR') = 1.50, ITMODEL= 3 , INCHOICE= 3
OUTPUT Parameters with Values Given by Computer :
```

Total BU-Cost (BUC)= 944166 MU Gross P.A.-Rate (GPAR)=8583.3 MU/Sq.M.
Open-Space (SPO)= 18.3333 Sq.M./BU, Ground-Coverage-Ratio (GCR)=
0.650
Density i.e. Built-Unit-Density (BUD)= 136.36 BU/Hectare

Series of OUTPUT-Values of <u>BFE</u> including DENSITY, generated by Running MODEL-3 (Dependent Variable SPO), with FAR and BUS fixed and changing only BR, would show that in this case with 'BR' change there is no change in density. Using the series of OUTPUT-Values given by software for the above three cases, the 'Building-Rise-Density' *Sensitivity* for each case could be represented graphically as shown in Fig. 6.3 (generated by using MS.Excel), where the variation of 'FAR' and 'GCR' with the variation of building-rise is also shown in case of MODEL-1.

Figure 6.4 illustrates that in NONE of the three possible cases a low-rise leads to a high density, which is quite contrary to the frequent conventional generalized urban

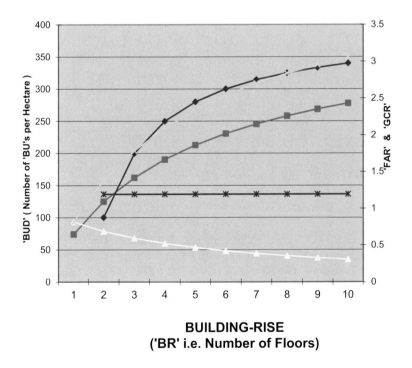

BUILDING-RISE
('BR' i.e. Number of Floors)

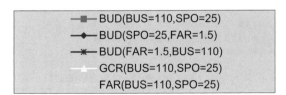

Fig. 6.4 Variation of density ('BUD'), FAR, GCR with building rise (BR)

planning policy prescription of '*Low Rise High Density (LRHD) Built-Form*' imposed from top in the *Conventional Planning-Practice and Policy* which often creates inefficiency and inequity in the urban development process, and can be corrected by **informed decision-making** using this SOFTWARE: '*CAUB*'.

6.6.5.2   Determining Interaction and Sensitivity Between Urban 'Built-Form-Elements' ('BFE'): 'FAR'- Built-Unit-Cost Sensitivity-Given Land Price (Application of Model-4)

To determine FAR-Built-Unit-Cost Interaction (BUS and SPO Fixed), MODEL-4 can be selected. Its illustrative INPUT-OUTPUT results (computer *prompting* for response similar as above) in a typical RUN is shown below:

```
SUMMARY OF ''CAUB'' RESULTS IN THE INSTANT CASE ARE:
ITERATION MODEL CHOSEN       :            INTERACTION/SENSITIVITY MODEL
(COST+PHY.)
INDEPENDENT DESIGN VARIABLES ('IDV'):   BUILT-UNIT-SPACE(BUS),OPEN-
SPACE(SPO)
                                    & FLOOR AREA RATIO(FAR)
DEPENDENT DESIGN VARIABLES ('DDV') :   BLDG.RISE(BR),GR.CV.RATIO
(GCR),DENSITY
INPUT PARAMETERS WITH VALUES CHOSEN BY USER :
CONSTN.PLTH.-AREA-RATE(CPAR)= 5000 MU/SQ.M. LAND PRICE= 5000 MU/SQ.M.
K=0.0250
BUILT-UNIT-SPACE ('BUS') =100.00  SQ.M./BU OPEN-SPACE('SPO')=33.33
SQ.M./BU
FLOOR-AREA-RATIO ('FAR') = 0.65   ITMODEL= 4  INCHOICE= 0
OUTPUT PARAMETERS WITH VALUES GIVEN BY COMPUTER:
TOTAL BU-COST (BUC)= 1169166 MU GROSS-PLINTH-AREA-RATE(GPAR)=11691.6
MU/SQ.M.
BUILDING-RISE ('BR')  = 1     GROUND-COVERAGE-RATIO (GCR)= 0.650
DENSITY i.e. BUILT-UNIT-DENSITY ('BUD') = 65.00 BU/HECTARE  br= 1.00
```

The MODEL-4 could be RE-RUN (or, a Series of ITERATIONS *automated*) with different input values by multiple *stakeholders* to find desired *Planning Solution* in a specific-case. Outputs of an *Automated* Series of Iterations (changing only Input-FAR within a given range) for 'FAR-BUC' *Sensitivity Analysis* are shown below:

MODEL-4: A Typical *Automated* Series of ITERATIONS (SPO and BUS fixed)
SENSITIVITY OF DENSITY/COST TO THE FLOOR AREA RATIO

FAR	BLDNG.RISE (BR (BRF))	DENSITY (BU/HECTARE)	GCR (MU)	BUC (MU/Sq. M.)	GPAR
0.25	1 (0.26)	22.63	0.955	4896296	44512
0.50	1 (0.55)	45.45	0.909	2696622	24525
0.65	1 (0.86)	68.18	0.864	1965965	16862

(continued)

FAR	BLDNG.RISE (BR (BRF))	DENSITY (BU/HECTARE)	GCR (MU)	BUC (MU/Sq. M.)	GPAR
1.00	1 (1.22)	90.91	0.818	1601050	14555
1.25	2 (1.62)	113.64	0.663	1383006	12563
1.50	2 (2.06)	136.36	0.626	1238543	11259
1.65	3 (2.56)	159.09	0.682	1136266	10330
2.00	3 (3.14)	181.82	0.636	1060556	9641
2.25	4 (3.81)	204.55	0.591	1002636	9116
2.50	5 (4.58)	226.26	0.545	956688	8606
2.65	6 (5.50)	250.00	0.500	922225	8384
3.00	6 (6.60)	262.63	0.455	894336	8130
3.25	8 (6.94)	295.45	0.409	862686	6934
3.50	10 (9.62)	318.18	0.364	856929	6690
3.65	12 (11.69)	340.91	0.318	846663	6696
4.00	15 (14.66)	363.64	0.263	842600	6660
4.25	19 (18.60)	386.36	0.226	846388	6694
4.50	25 (24.65)	409.09	0.182	861956	6836
4.65	35 (34.83)	431.82	0.136	899004	8163
5.00	55 (55.00)	454.55	0.091	986250	8965

Optimum FAR value = 4.018 to give Least BU-Cost = 842580 MU (see illustrative INPUT-OUTPUT results of MODEL-9 below). INPUT VALUES Chosen in the above CASE are: BUS = 110.00 Sq.M. SPO = 20.0 Sq.M. LP = 10,000 MU/Sq.M. CPAR = 4500 MU/Sq.M. $K = 0.010$

```
FRACTIONAL BLDNG.RISE IF SHOWN MEANS FULL 'BUS' IS NOT
AVAILABLE IN ONE FLOOR & NEED INPUT VALUE CHANGE IF
FULL 'BUS' IS DESIRED
```

Series of OUTPUT-Values of BFE including BUC, generated by Running MODEL-4 (Dependent Variable DENSITY, BUC), with SPO and BUS fixed and changing only FAR, would show that in this case with 'FAR' change there is change in BUC, GCR, and density. 'Sensitivity of Density/BU-Cost to the Floor Area Ratio' could be represented graphically as shown in Fig. 6.5 (generated by using MS. Excel), using the Series of OUTPUT-Values given by software for the above case. It may be seen that with the increase of FAR, the BUC gets reduced reaching a minimum and thereafter with the increase of FAR the BUC also gets increased, showing that there is an optimum value of FAR which gives minimum BUC. Model 4 does not give the exact value of optimum FAR but using Model-9 the exact value of optimum FAR could be determined which is shown above. FAR and built-unit-cost interaction and its urban management and development planning and design policy implications are discussed below, with case studies.

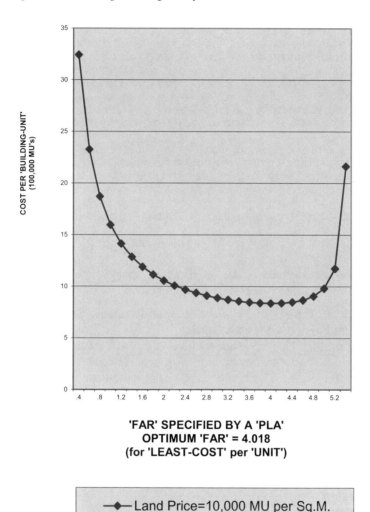

Fig. 6.5 Variation in cost per building-unit ('BU') with 'FAR' (Same built-unit-space and open-space per BU)

6.6.5.3   Determining Interaction and Sensitivity Between Urban 'Built-Form-Elements' ('BFE'): Density-Built-Space Interaction (Application of Model-5)

To determine density-built-space interaction for different combination of BFEs, MODEL-5 can be selected. Its illustrative INPUT-OUTPUT results in a typical RUN are shown below:

```
SUMMARY OF ''CAUB'' RESULTS IN THE INSTANT CASE ARE:
ITERATION MODEL CHOSEN    :    INTERACTION/SENSITIVITY MODEL (COST+PHY.)
INDEPENDENT DESIGN VARIABLES ('IDV'): OPEN-SPACE (SPO), BUILDING-RISE (BR)
                                  & BUILT-UNIT-DENSITY (BUD)
DEPENDENT DESIGN VARIABLES ('DDV') : FAR, G.C.RATIO (GCR), BUILT-UNIT-
SPACE (BUS)
INPUT PARAMETERS WITH VALUES CHOSEN BY USER :   ITMODEL= 5 INCHOICE= 0
CONSTN.PLTH.-AREA-RATE (CPAR) = 5000 MU/SQ.M. LAND PRICE= 5000 MU/SQ.M.
K=0.0250
OPEN-SPACE ('SPO')    = 20 SQ.M./BU    BUILDING-RISE ('BR') =   4
BUILT-UNIT-DENSITY ('BUD')=200.00 BU/HECTARE
OUTPUT PARAMETERS WITH VALUES GIVEN BY COMPUTER:
TOTAL BU-COST (BUC)= 910000 MU GROSS-PLINTH-AREA-RATE (GPAR)=6583.3
MU/SQ.M.
FLOOR-AREA-RATIO ('FAR') = 2.400  GROUND-COVERAGE-RATIO ('GCR')=
0.600
BUILT-UNIT-SPACE ('BUS') = 120.000 SQ.M./BU
```

Outputs of an *Automated* Series of Iterations (changing only Input-DENSITY within a given range) for 'DENSITY-BUS' *Sensitivity Analysis* are shown below:

MODEL-5: A Typical *Automated* Series of Iterations Showing Sensitivity of Built-Unit-Space (BUS) to the Density

DENSITY (BU/HA)	FAR	BUS (Sq.M./BU)	GCR	BUC (MU)	GPAR (MU/Sq.M.)
50.00	1.800	360.00	0.900	2890000	8028
65.00	1.600	226.66	0.850	1856666	8191
100.00	1.600	160.00	0.800	1340000	8365
125.00	1.500	120.00	0.650	1030000	8583
150.00	1.400	93.33	0.600	823333	8821
165.00	1.300	64.29	0.650	665614	9096
200.00	1.200	60.00	0.600	565000	9416
225.00	1.100	48.89	0.550	468889	9695
250.00	1.000	40.00	0.500	410000	10250
265.00	0.900	32.63	0.450	353636	10806
300.00	0.800	26.66	0.400	306666	11500
325.00	0.600	21.54	0.350	266923	12393
350.00	0.600	16.14	0.300	232856	13583
365.00	0.500	13.33	0.250	203333	15250
400.00	0.400	10.00	0.200	166500	16650
425.00	0.300	6.06	0.150	154606	21916
450.00	0.200	4.44	0.100	134444	30250
465.00	0.100	2.11	0.050	116316	55250

INPUT VALUES Chosen in the above CASE are:

```
SPO= 20.00 Sq.M./BU  BR=  2  LP= 5000 MU/Sq.M.  CPAR= 5000 MU/Sq.M.  K=
0.025
```

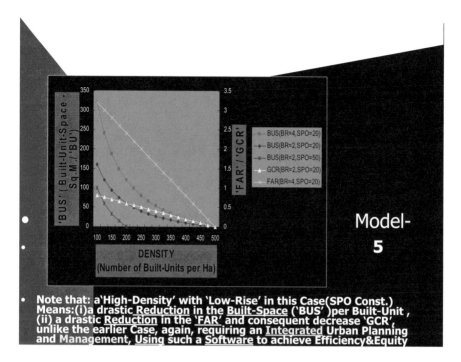

Fig. 6.6   Built-unit-density (BUD) and built-unit-space (also FAR, GCR) interaction

The above results (for BFE Set SPO = 20, BR = 2) indicate that for a given BR (building rise) a drastic increase in density (Built-Unit) means a drastic decrease in BUS and FAR but a gradual decrease in GCR (Ground Coverage Ratio). The above results along with such results for other BFE sets are shown graphically in Fig. 6.6 to give an insight to urban planners and urban managers about the interrelationships between different BFEs so that they can avoid giving any generalized planning policy prescriptions creating inefficiency in many cases. It is also desirable that all these interactions and sensitivities between BFEs should be considered in an integrated manner in a site-specific case, as part of integrated urban management instead of frequently giving a generalized urban planning policy prescriptions from top without considering the scientific relationships between different BFEs (built-form-elements) which may create inefficiency and inequity. The software CADUB-CAUB could be very useful tool in this effort.

In the above context, it is desirable that such urban planning and development problems are seen as part of integrated urban management and urban planners, civil engineers and other urban professionals start using management techniques like OR/CAD as above to attain a resource-efficient and equitable urban problem solutions keeping in view the resource-constraints and affordability of multiple stakeholders at various levels.

Software '**CAUB**' gives the interaction between FAR and BUC and using it the value of FAR at which BUC is minimum could be indirectly derived. However, if exact value of *Optimum* FAR giving *Minimum* BUC is desired the optimization model is to be solved. Such a model for optimal urban-built-form in a 'planning area' linking Built-Space (BUS), Open-Space (SPO), Building Rise (BR), land price, and construction costs, is developed and solved using a *nonlinear programming* technique in Chap. 7.

6.6.5.4 FAR and Built-Unit-Cost Interaction: Urban Management, Housing, and Urban Development Planning and Design Policy Implications

The above descriptive models could also incorporate easily the unit land price and building construction costs, to derive the total cost per built-unit (i.e. Building-Unit-Cost/Built-Unit-Cost, or BUC) considering the prevailing market land price and construction costs, and thus the effect of a prescribed planning regulation, like FAR, on the BUC could be determined instantly to confirm feasibility and acceptability of such planning regulations with people's (user citizens) participation (i.e. *bottom-up* approach), rather that imposing such regulations from top (usual *top-down* approach) which may not work. Accordingly, a number of models incorporating land price and construction costs have been developed and included in Application C++ software '**CAUB**'. Thus, a summary of ''CAUB'' results using **Model-4**, which incorporates the unit land price and building construction costs to derive the total cost per built-unit and the *scientific relationships* between built-form-elements, land price and building construction costs, is shown above. Similarly, a series of iterations using Model-4 giving the sensitivity of density/BU-Cost to the Floor Area Ratio is also shown in a Table above and also shown graphically in Fig. 6.5 (adopting land price of 10,000 Monetary Units (MU) per m^2, BUS = 110 m^2, SPO = 20 m^2). Figure 6.5 shows the relationship (*nonlinear*) between the '**FAR**' and '**Building-Unit Cost**', indicating that the respective *BU-Costs* at FAR of **0.846** (minimum feasible 'FAR' keeping above values of BUS and SPO, unchanged), and **0.5** (frequently specified by a Planning Authority or 'PLA', requiring an increased SPO = 110 m^2, even for a single-storey development), are **2.14 times** and **3.2 times the *Minimum-BU-Cost* occurring** *at the Optimum FAR* = **4.018** (indirectly derived by 'CAUB'). Such cost-variations only due to *Planning Regulations* may be even higher at higher land prices. Urban Planning Authorities (PLAs) in some countries continue to keep the *Planning Regulations* unchanged, say since the first Master Plan prepared in 1960s (thus insisting and trying to enforce a low FAR prescribed 50–60 years back) irrespective of the land price increase to astronomical figures since then (even by 1000 times in some cities, and consequent astronomical increase in *BU-Cost*), creating huge *inefficiency* and *inequity* and the consequent housing and urban problems leading to the public interest litigations in many cases (Fig. 6.7).

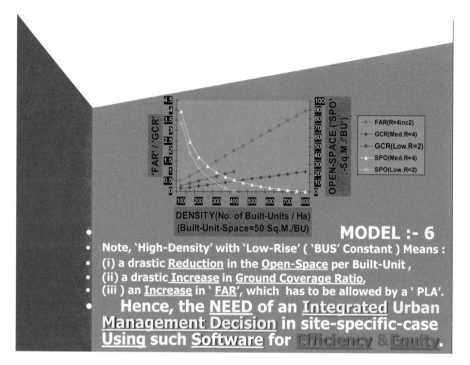

DENSITY(No. of Built-Units / Ha)
(Built-Unit-Space=50 Sq.M./BU)

MODEL :- 6

- Note, 'High-Density' with 'Low-Rise' ('BUS' Constant) Means :
- (i) a drastic Reduction in the Open-Space per Built-Unit ,
- (ii) a drastic Increase in Ground Coverage Ratio,
- (iii) an Increase in ' FAR', which has to be allowed by a ' PLA'.
- **Hence, the NEED of an Integrated Urban Management Decision in site-specific-case Using such Software for Efficiency & Equity.**

Fig. 6.7 Built-unit-density (BUD) and open-space (also FAR, GCR) interaction

6.6.5.5 Application of 'CAUB'/'CADUB' in Participatory-Planning to Decide Sustainable Planning Regulations Responsive to People's Need Using the Principles of Flexibility: A Case Study

Urban planners prescribe 'Planning Regulations' to achieve the goal of 'Public Welfare', to take care of deficiencies in the market and to give allowances for externalities, including adequate provision for open-space, and so on. Towards this goal, urban planners and Planning Authority (PLA) limit 'FAR' and other built-form-elements, often without establishing scientific relationships between these variables.

We may take, as example, a PLA specified limiting FAR = 0.65, land price of 3500 MU/m², and a BUC = 100,000 MU affordable by User Citizens (UC) of low-income community. Using Model-8 and selecting BR = 1 will show that available Built-Unit-Space (BUS) = 14.91 m², Semi-Public-Open-space (SPO) = 4.96 m² per Built-Unit (BU) and Ground Coverage Ratio (GCR) = 0.65—an infeasible solution as shown in graphic-output instantly given by software (Fig. 6.8-4). If the planning authority and the user citizen in a participatory-decision reduce FAR to 0.5, Model-8 will show that SPO increases to 11.06 m²/BU with BUS = 11.06 m²/BU, again, making it an unacceptable

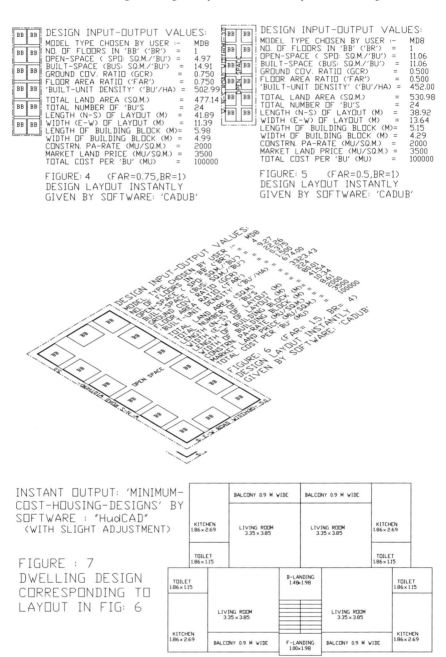

Fig. 6.8 (4, 5, 6, 7) Graphic-outputs instantly given by software 'CAUB'/'CADUB'—A case study

solution (Fig. 6.8-5). On the other hand, if PLA allows a higher FAR = 1.5 and, user citizens choose a BR = 4, Model-8 will show that BUS increases to 22.26 m^2/BU with SPO = 9.26 m^2/BU. These numerical input-output results of Model-8 could be instantly converted to a corresponding graphic output, i.e. layout (Fig. 6.8-6) instantly given by software 'CADUB' and CBDG (Graphic Component of CADUB), and the corresponding dwelling design can be obtained using the companion software 'HudCAD' is shown in Fig. 6.8-7. All the above numeric and graphic planning and design analysis could be carried out in an informed *Participatory Planning and Design Session* where planning authority and the user citizens could participate in a *transparent* manner discouraging illegal constructions and frequent corruptions.

Equipped with such information (using above software), rarely possible in conventional practice, as part of participatory-planning with flexibility, both PLA and UC can fulfill welfare objective of adequate open-space, less GCR of 0.365 and less depletion of land, with likely positive impact on ground-water-recharge and environment; a citizen gets 1.5 times higher BUS within his same affordable BUC. But, a rigid application of planning regulations (e.g. FAR = 0.65), irrespective of land price, construction cost, and affordability may price-out the urban poor(a large majority in many cities) adversely affecting the very 'Public Welfare' principle, on which basis such limiting values of FAR and other BFEs are specified, making the built-environment so planned UNSUSTAINABLE. In fact, illegal constructions and proliferation of slums can be reduced by such *participatory-planning* with *flexibility* to create public awareness, facilitating urban problem solution by *informed decisions* by *all stakeholders*.

This case study highlights the *perils* of imposing from top the various planning regulations restricting the various BFEs without a detailed planning and design analysis incorporating *informed-participation* of multiple stakeholders, if achieving *efficiency*, *equity*, and *public welfare* are the planning objectives.

6.6.5.6 Planning and Design Analysis and Optimization of Urban-Built-Form: Application of Model-9 and 10 to Derive Optimal FAR and Building Rise for Least-Cost and Maximum Benefit Built-Unit

The Model-4 results above showing that the FAR-BUC interaction indicates that for a given BUS, SPO, and Land Price (MU/sq.m.) as the FAR is reduced BUC increases, and as the FAR is increased, the BUC is reduced coming to a <u>*Minimum*</u> and thereafter any further increase of FAR increases the BUC. It may also be seen from Fig. 6.5 that the relationship between FAR and BUC is nonlinear and it is a *nonlinear programming optimization* problem which has been solved in Chap. 8 on optimization.

However, the solution of the above *nonlinear programming optimisation* problem leads to a simple algebraic expression for '*Optimum FAR*' (derived using *Nonlinear Programming* Technique), giving the *Least-Cost* per Building-Unit

(BU) for a given minimum acceptable *Built-Space* and *Open-Space* per BU, or, for 'Optimum FAR' for 'Maximizing-Built-Space per Unit' within a given maximum affordable-cost per built-unit. These expressions can be used by urban professionals and even by user citizens not familiar with *Nonlinear Programming* Techniques. Optimal FAR for Least-Cost per Built-Unit for a given input-set option is given instantly by Model-9 using such expressions.

The expression for Optimum Floor Area Ratio for Least-Cost Built-Unit Design used in Model-9 is shown below:

Optimum Floor Area Ratio (Chakrabarty, 2001) $=$

$$(c_3/c_1) * \left[\frac{(2^*k + \sqrt{k} - 1) * (c_1{}^*a_1 + c_3{}^*a_2)}{c_3{}^*a_2{}^*(4^*k - 1) + (c_1{}^*a_1 - c_3{}^*a_2) * (2^*k - \sqrt{k})} - 1 \right]$$

where

$c_1 =$ a cost coefficient, related to the building construction cost per unit of plinth area in the Urban Development (UD), in MU (monetary units) per sq.m.,

$c_2 =$ a cost coefficient (or a *High-Rise-Penalty-Factor* in a given context) related to the building rise, (effective building rise), in cost per unit of plinth area in the urban development, in MU per sq.m.,

$c_3 =$ a cost coefficient related to the market land price per unit of land area (monetary units/sq.m.),

$a_1 =$ the desired *Built-Unit-Space* in sq.m. per built-unit in the urban development,

$a_2 =$ the desired semi-public *Open-Space* in sq.m. per built-unit in the urban development, and

$k = \frac{c_2{}^*c_3{}^*a_1^2}{(c_1*a_1 + c_3*a_2)^2}$

The above expression for **Optimum FAR** (FAR) and other parameters has been further simplified which are presented below. The output results given by both sets of formulas (i.e. old and new simplified) are same, and hence, AEC professionals can use either set of formulas to get same results. However, simplified formulas given below are more convenient to use, and hence recommended.

Simplified Formula of Optimum FAR (FAR) $= \dfrac{c_3 a_1}{\left[c_3 a_2 + \sqrt{c_3 c_2} 3 a_1 \right]}$

Optimum Building Rise (BR) in Effective Number of Floors $= \sqrt{c_3/c_2}$

Simplified Formula for Least-Cost (LC) per Built-Unit in Monetary Units (MU)

$$LC = c_3 a_2 + 23 \sqrt{c_3 c_2} \, 3 a_1 + c_1 a_1$$

Similarly, Model-10 incorporates expression for Optimum Floor Area Ratio for 'Maximizing-Built-Space per Unit' within a given maximum affordable-cost per built-unit. However, these are also simplified, and the **Simplified Formula of**

Optimum FAR and other parameters for **Most-Benefit Optimal Design Model** are given in Sect. 6.5.2 above, and are reproduced below for ready reference.

Simplified Formula of Optimum FAR (FAR) is

$$= \frac{a_5 \times (1 - a_2 a_5)}{\sqrt{a_4 a_5} \times (1 + a_2 a_5) + a_2 a_3 a_5} \tag{6.9}$$

Formula for Maximum Built-Unit-Space (BUS) $= \dfrac{1 - a_2 a_5}{2 \times \sqrt{a_4 a_5} + a_3}$ (6.10)

Optimum Building Rise (BR) in Effective Number of Floors $= \sqrt{\dfrac{a_5}{a_4}}$ (6.11)

An Illustrative INPUT-OUTPUT result (Optimal FAR in *italics*) of MODEL-9 is shown below:

```
SUMMARY OF ''CAUB'' RESULTS IN THE INSTANT CASE ARE:
ITERATION MODEL CHOSEN        :        OPTIMISING-MODEL (LEAST-COST-
DESIGN)
INDEPENDENT DESIGN VARIABLES ('IDV'): BUILT-SPACE (BUS),OPEN-SPACE
(SPO),K,
                          LAND-PRICE (LP),BASE-PLNTH.AREA
RATE(CPAR)
DEPENDENT DESIGN VARIABLES ('DDV'):   LEAST BU-COST ('LBUC'), FAR, BR,
GCR, DENSITY
INPUT PARAMETERS WITH VALUES CHOSEN BY USER (LCO-MODEL):
BUILT-UNIT-SPACE ('BUS') =110.00 SQ.M./BU  OPEN-SPACE ('SPO')= 20.00
SQ.M./BU
LAND-PRICE ('LP')= 10000.0 MU/SQ.M.   CPAR= 4500.0 MU/SQ.M. K=0.0100
ITMODEL= 9 INCHOICE= 20  MU= MONETARY UNITS (i.e.Rupees, Dollars etc.)
OUTPUT PARAMETERS WITH OPTIMUM VALUES OF 'FAR' & 'BR':
OPTIMUM 'FAR' = 4.018  OPTIMUM 'BR'(BUILDING-RISE) = 15   br = 14.91
LEAST-BUILT-UNIT-COST('LBUC')=  842580 MU  DENSITY('BUD')= 365.24
BU/HECTARE
GROSS-PLINTH-AREA-RATE (GPAR)= 6659.8 MU/SQ.M.  GR.CV.RATIO ('GCR') =
0.260
```

Illustrative INPUT-OUTPUT results (Optimal FAR in *italics*) of MODEL-10 given below:

```
SUMMARY OF ''CAUB'' RESULTS IN THE INSTANT CASE ARE:
ITERATION MODEL CHOSEN       : OPTIMISING MODEL (MOST-BENEFIT-DESIGN)
INDEPENDENT DESIGN VARIABLES ('IDV'): B.UNIT-COST(ABUC),OPEN-SPACE
(SPO),K,
                          LAND-PRICE(LP),BASE-PLNTH.AREA
RATE(CPAR)
DEPENDENT DESIGN VARIABLES ('DDV') : BUILT-SPACE(MBUS), FAR, BR,GCR,
DENSITY
INPUT PARAMETERS WITH VALUES CHOSEN BY USER(MBO-MODEL):
```

```
AFFORD.BUILT-UNIT-COST(ABUC)=  842580 MU   OPEN-SPACE(SPO)=20.000
SQ.M./BU
LAND-PRICE ('LP')=10000.0 MU/SQ.M.      CPAR= 4500.0 MU/SQ.M.
K=0.0100
ITMODEL=10 INCHOICE=21 MU=MONETARY UNITS(i.e.Rupees,Dollars etc.)
OUTPUT PARAMETERS WITH OPTIMUM VALUES OF 'FAR' & 'BR':
OPTIMUM 'FAR' = 4.018 OPTIMUM 'BR'(BUILDING-RISE) =15 GC.RATIO ('GCR')
=0.260
GROSS-PLINTH-AREA-RATE(GPAR) =6659.8 MU/SQ.M. DENSITY (BUD) =365.24
BU/HECTARE
MOST-BENEFIT ('MBUS 'i.e.MAX.POSSIBLE BUILT-SPACE) =110.00 SQ.M./BU
br= 14.91
```

It is interesting to note from above that if input values are same except that BUILT-UNIT-SPACE ('BUS') is given as input in Model-9, and its Output LEAST-BUILT-UNIT-COST ('LBUC') is given as input in Model-10, the output *OPTIMUM 'FAR'* will be same in both cases as shown in the above illustrative INPUT-OUTPUT results (Optimal FAR in *italics*) of MODEL-9 and MODEL-10.

As already stated, the C++ computer program **OUBF.CPP** incorporates models which uses the above **Simplified** algebraic equations for **Least-Cost Optimal Design** as well as the **Most-Benefit Optimal Design**. However, using the above **Simplified** algebraic equations and formulas, AEC professionals in smallest planning, design, and construction office can derive **Optimal Design** parameters in a site-specific case even using the small handheld calculators, which are available everywhere.

6.6.5.7   Conventional Planning Practice and the Land Price-Optimum-FAR-Building-Rise Interaction Giving the Built-Unit-Cost

In the conventional planning practice, a number of limiting *Planning Regulations* (PRs) such as FAR (Floor Area Ratio), BR (Building Rise, i.e. limiting number of floors) are prescribed by the local Planning Authorities (PLAs) to ensure public welfare. Using the Model-9 and Model-6, the interaction between the land price and the optimal FAR and BR to give the Least-Cost per Built-Unit is generated and shown in Fig. 6.9. It may be seen from the Figure that higher the land price higher is the optimal FAR and BR which are unique values for each land price to give the Least-Cost per Built-Unit of given BUS and SPO. It may also be seen that optimal FAR may be as high as 6, and the optimal building rise may be as high 25 (Number of Floors) depending on the land price range chosen in this case. However, this scientific fact is not taken into account in conventional planning regulations in many countries where low rise and also low FAR of say 0.65–1.0 is prescribed by the local planning authorities even in high land price situations which may be even 40,000 MU/sq.m.

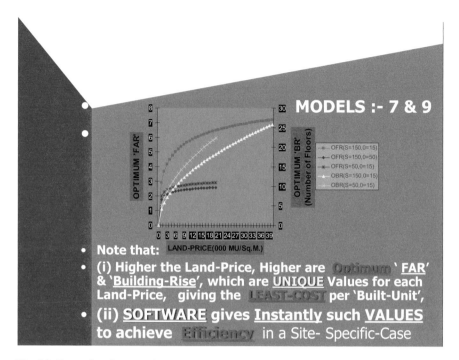

Fig. 6.9 Interactions between the land price, and the optimal 'FAR' and 'BR' to give the least-cost per built-unit

A result of Model-3 giving the BU-Cost, if low rise of BR = 1 and low FAR = 0.65 are adopted, where market land price = 40,000 MU/sq.m. is shown below:

```
SUMMARY OF ''CAUB'' RESULTS IN THE PRESENT CASE :
ITERATION MODEL CHOSEN         : INTERACTION/SENSITIVITY MODEL(COST
+PHY.)
INDEPENDENT DESIGN VARIABLES('IDV'): BUILDING-RISE(BR),BUILT-UNIT-
SPACE(BUS)
                                    & FLOOR-AREA-RATIO(FAR)
DEPENDENT DESIGN VARIABLES ('DDV') :  OPEN-SPACE(SPO),GR.CV.RATIO
(GCR),DENSITY
INPUT PARAMETERS WITH VALUES CHOSEN BY USER :
CONSTN.PLTH.-AREA-RATE(CPAR)= 5000 MU/SQ.M. LAND PRICE=40000 MU/SQ.
M. K=0.0250
BUILDING-RISE ('BR')  = 1  BUILT-UNIT-SPACE ('BUS') =150.00 SQ.M./BU
FLOOR-AREA-RATIO ('FAR') = 0.65  ITMODEL= 3  INCHOICE= 18
OUTPUT PARAMETERS WITH VALUES GIVEN BY COMPUTER :
TOTAL BU-COST(BUC)= 8668650 MU GROSS-PLINTH-AREA-RATE(GPAR)=58458.3
MU/SQ.M.
OPEN-SPACE (SPO)= 50.0000 SQ.M./BU  GROUND-COVERAGE-RATIO (GCR)= 0.650
DENSITY i.e. BUILT-UNIT-DENSITY (BUD)= 50.00 BU/HECTARE
```

In the same land price situation as above, i.e. land price of 40,000 MU/sq.m., if Model-9 is applied to give the optimal FAR and BR and the Least BU-Cost, output result of Model-9 is as follows:

```
SUMMARY OF ''CAUB'' RESULTS IN THE INSTANT CASE ARE:
ITERATION MODEL CHOSEN            : OPTIMISING-MODEL (LEAST-COST-
DESIGN)
INDEPENDENT DESIGN VARIABLES('IDV'): BUILT-SPACE(BUS),OPEN-SPACE
(SPO),K,
                        LAND-PRICE(LP),BASE-PLNTH.AREA RATE
(CPAR)
DEPENDENT DESIGN VARIABLES('DDV') : LEAST BU-COST('LBUC'),FAR,BR,
GCR,DENSITY
INPUT PARAMETERS WITH VALUES CHOSEN BY USER(LCO-MODEL):
BUILT-UNIT-SPACE ('BUS')=150.00 SQ.M./BU  OPEN-SPACE ('SPO')=15.00 SQ.
M./BU
LAND-PRICE ('LP')= 40000.0 MU/SQ.M.    CPAR= 5000.0 MU/SQ.M.
K=0.0250
ITMODEL= 9 INCHOICE= 20  MU= MONETARY UNITS(i.e.Rupees,Dollars etc.)
OUTPUT PARAMETERS WITH OPTIMUM VALUES OF 'FAR' & 'BR':
OPTIMUM 'FAR' = 6.414  OPTIMUM 'BR'(BUILDING-RISE)= 18    br= 16.89
LEAST-BUILT-UNIT-COST('LBUC')= 2020820 MU  DENSITY('BUD')= 426.62
BU/HECTARE
GROSS-PLINTH-AREA-RATE(GPAR)= 13462.1 MU/SQ.M.  GR.CV.RATIO ('GCR')
= 0.359
```

It may be seen that the Optimal FAR to give the Least-Cost per Built-Unit is as high as **6.414** for a land price of 40,000 MU/sq.m. if SPO is 15 sq.m. per BU and BUS is 150 sq.m. per BU giving the Least BU-Cost as **2,020,820 MU** and Optimal BR = 18 (Model-9 results as above), as against increasing the BU-Cost to **8,668,650 MU** (Model-3 results above), i.e. **4.34 times** that of **Least BU-Cost**, if a fixed and often unrealistic FAR = 0.65 is imposed by the local planning authority as per the conventional practice not recognizing the astronomical increase in the market land price in many countries. Similarly, Optimal Building Rise (BR), i.e. number of floors to give the Least-Cost per Built-Unit may be as high as 25 depending on the land price.

Solution of optimization models involves complicated mathematics of nonlinear programming with which urban professionals such as architects, engineers, and urban planners may not be conversant to enable them to apply such techniques to achieve efficiency and equity.

To obviate this difficulty such professionals can use Model-9 and Model-10 which are developed in such a way that a user can obtain optimal planning solutions even if not accustomed to such complicated mathematical optimization techniques.

Thus, a designer/planner instead of prescribing or adopting a fixed FAR in any land-price situation as usually the case in the conventional practice, and thus creating inefficiency and inequity in many cases adversely affecting the very objective of such planning regulations, i.e. '*public welfare*', could respond professionally, scientifically, and transparently and prescribe or adopt optimal BFEs including FAR after examining land price and other parameters in a site-specific case using above models in a *participatory* manner involving multiple stakeholders. Thus, if objective is to achieve a Least-Cost design of Optimal Built-Form (BF) with given desired minimum Built-Unit-Space (BUS), the Model-9 could be used prescribing the desired BUS and SPO as input. On the other hand, if the objective is to attain a most-benefit design of Optimal Built-Form (BF) giving maximum BUS (Benefit) within a given desired maximum Built-Unit-Cost (BUC), the Model-10 could be used prescribing the desired BUC and SPO as input, as shown in the above illustrative INPUT-OUTPUT results.

6.6.6 *Planning and Design Analysis and Integration of Numeric and Graphic Design*

Application of the different models is shown above in numeric forms each of which represents a graphic design which is more meaningful to the users and multiple stakeholders for interaction. To facilitate such meaningful interactions, the software CADUB-CAUB incorporates suitable formulations to generate in each of ten Models the input-output data in LISP form which can be used to convert instantly each numerical design into graphical design, i.e. integration of numeric and graphic design.

Procedure and computer prompt appearing in video screen while running the software: 'CADUB' – 'CAUB' are shown in Sect. 6.6.2 above where the computer prompts for the program 'CBDG', which is the GRAPHIC Component of Software CADUB (just as CAUB is the NUMERIC Component of CADUB), is not shown. These are presented below along with the numerical results for Model-1 obtained by running CAUB, and thereafter, converting it into graphic design.

While a schematic urban-built-form is displayed in the graphic part of computer screen to give an understanding to the user about some BFEs, and in the text part of screen, the following message appears for user response:

```
You want to RUN 'Numerical Iteration SESSION' (i.e. 'CAUB')? (Y/N)
```

If choice is 'Y', the following message appears in computer screen for response:

```
Type 'CAUB' for 'Num. Itn. Session' & After Results Load 'CBDG' for
'Drawing Session'
```

On typing 'CAUB', the C++ software: 'CAUB' is invoked and the 'Numerical Iteration Session' starts giving numerical planning solutions for each design condition, i.e. INPUT SET/Values selected including the MODEL Chosen. Accordingly, typing 'CAUB' against command line, choosing Model-1 and entering INPUT VALUES in response to the computer prompt, a numerical planning solution obtained by 'CADUB' is shown below:

```
SUMMARY OF ''CAUB'' RESULTS IN THE PRESENT CASE : (Model-1)
Iteration MODEL Chosen              : Sensitivity MODEL (Physical + Cost)
Independent Design Variables('IDV') : Building-Rise(BR), Open-Space
(SPO)
                                 & Built-Unit-Space(BUS)
Dependent Design Variables('DDV')  : FAR, Ground-Coverage-Ratio(GCR)&
DENSITY
```

INPUT Parameter-Values Chosen in the present Case, are:

```
CONSTN.PLTH.-AREA-RATE (CPAR)=5000 MU/SQ.M. LAND-PRICE=5000MU/SQ.M.
K=0.0250
Building-Rise ('BR')  = 4.00    Open-Space('SPO') = 25.00 Sq.m./BU
Built-Unit-Space('BUS') = 100.00 Sq.m./BU  ITMODEL= 1  INCHOICE= 9
```

OUTPUT Parameter-Values given instantly by the COMPUTER, are:

```
TOTAL BU-COST (BUC)= 800000 MU GROSS-PLINTH-AREA-RATE(GPAR)=8000.0
MU/SQ.M.
Floor-Area-Ratio (FAR)  = 2.00    Ground-Coverage-Ratio (GCR) = 0.500
DENSITY i.e. Built-Unit-Density ('BUD') = 200.0 BU/Hectare
You can RE-RUN the MODEL, keeping Constant the Values of any 2 IDV'S
(say-'BUS' & 'SPO') and, changing Value of the 3rd. 'IDV'( say- 'BR'), to
Determine the Sensitivity between the DDV'S (i.e. 'DENSITY', FAR & GCR
for above Choice) and, the 3rd. 'IDV'(i.e. 'BR' here) Chosen by you.
DO YOU WANT TO CHANGE 'INPUT VALUE(S)/INPUT SET/ITERATION MODEL', FOR
BETTER RESULTS/PARTICIPATORY-DECISION? Type y(if yes),or, n(if no) &
press <CR>
```

On typing 'N', the following message appears in computer screen for user response:

As the current solution is acceptable, Close the 'Numerical Iteration Session' and if desired start the 'Drawing Session' by Loading 'CBDG'

Accordingly, 'Num.Itn.Session' is closed returning computer command to screen of AutoCAD, graphic software: 'CBDG' loaded, 'CBDG' typed against command line, when the following message appears in computer screen for user response:

```
WELCOME TO: 'DRAWING SESSION' OF SOFTWARE: 'CADUB'. Does this 'DRAWING
SESSION' pertain to a Design Change? ( Y/N):
```

If user put 'Y', the following message is displayed for response:

```
Do you want the Changed Design(s) in Multiple Layers? (Y/N):
```

If above response is 'Y', multiple layers are continued serially without erasing DESIGNS generated by graphic software: 'CBDG'. If it is 'N', the computer prompts for 'erasing' option for earlier design(s), with the following message for user response (typing MD1/MD2 which are Model-1, Model-2, and so on in 'CAUB'):

```
Put the MODEL TYPE (MD1/MD2, so on) Chosen by you.
```

On typing MODEL type, software prompts USER to indicate change of preset text-height/location, preset road-widths, drawing starting point, and so on. It also asks for chosen number of DUs per floor, chosen number of Building Blocks (BBs) in different directions and displays the calculated number of DUs for acceptance or repeat the iteration cycle, till the calculated number of DUs is acceptable. Accordingly, here four BBs in N-S direction is chosen. It prompts USER to indicate change of preset value of set-back and separation-distance between BBs and displays crucial dimensions such as open-space-dimensions for acceptance or change. On giving response to above prompts, the graphic software: 'CBDG' starts displaying drawing and the user is also prompted for textual parameter display option by the following message:

```
Do you want MODEL-wise Textual Parameter Display in Drawing? (Y/N):
```

On response, the full GRAPHICAL PLANNING SOLUTION is DISPLAYED INSTANTLY, with or without textual parameter, depending on 'Y' or 'N' option (here, 'Y' response), as shown below as: **Graphical Planning Solution-I** in Fig. 6.10 for the above numerical planning solution:

Application of Software: 'CADUB'/'CBDG': Graphical Planning Solution-I

In the text part of computer screen, the following CHOICES appear for repeating the CYCLE OF ITERATION, until an acceptable DESIGN solution is derived:

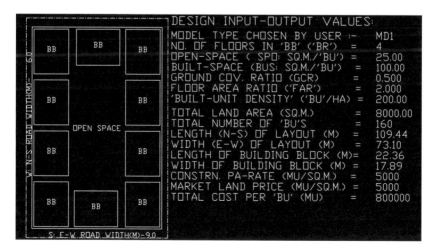

Fig. 6.10 Design layout instantly given by software 'CADUB-CBDG' corresponding to the numerical results of Model-1 shown above

(a) **Application of Software: 'CADUB'/'CBDG': Graphical Planning Solution-II**

```
Do you want to ERASE the DESIGN & START NEW 'Iteration' ? (Y/N):
Do you want DESIGN CHANGE with NEW 'Graphical Iteration'? (Y/N):
Do you want DESIGN CHANGE with NEW 'Numerical Iteration'? (Y/N):
Do you want to have HARD COPY of the DESIGN? (Y/N):
```

On selecting the OPTION, other steps of iteration follow as above to derive a DESIGN alternative. Accordingly, choosing the option of 'DESIGN CHANGE with NEW Graphical Iteration', and selecting three Building Blocks (BB) in N-S direction, two open courts, and the isometric drawing option, and giving interactive response to computer prompts, another graphic solution (graphical planning solution-II shown above) of the above numerical planning solution is INSTANTLY drawn by graphic software: 'CBDG' as shown in Fig. 6.11 as **Graphical Planning Solution-II**.

(b) **Textual Parameters of Graphical Planning Solution-II**

It may be seen that in both graphical planning solution-I and II, the Quantitative Design Efficiency Indicators (i.e. QDEIs such as FAR, BUD, GCR, BUC, and so on) are the same. Thus, using 'CADUB' and invoking graphic software: 'CBDG', a creative planner/designer could carry out a number of 'Graphical Iterations' to generate a 'Variety of Design Alternatives', fulfilling the same quantitative design efficiency indicators, given by a numerical planning solution obtained by invoking C++ software: 'CAUB' as above. Similarly, using 'CADUB' and invoking C++ software: 'CAUB', a planner/designer could carry out a number of 'Numeric Iterations' to generate a 'Variety of Numerical Planning Solutions', each of which can be converted to a 'Variety of Graphical Planning Solutions', giving in no time 'Numerous Design Alternatives' to multiple stakeholders for selection. This unique capability of the software for instant *integration* of numeric and graphic planning solutions in one platform indicating instantly the *Quantitative Design Efficiency Indicators* (QDEIs) of the solution and permitting *informed-participation* of multiple stakeholder in the planning and design process is rarely available in the conventional planning and design practice Thus, the software: 'CADUB',-'CBDG'-'CAUB' help enhance, creativity of the planners and designers while achieving *resource-efficient* planning and design solutions affordable and acceptable to the multiple stakeholders as per their priority and thereby improving both *efficiency* and *equity* in the urban development process.

As mentioned above, the software: 'CAUB', numeric version of CADUB, is also an independent and stand alone C++ software, to carry out any numeric iteration to generate a numerical planning and design solution for a given set of design conditions, using its various MODELS. Therefore, Software: 'CAUB' can be used by a user to generate numerous numerical planning solutions and carry out the above planning and design analysis tasks (except graphical planning solution), even if AutoCAD is not available.

a

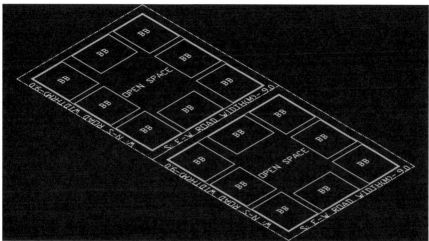

b

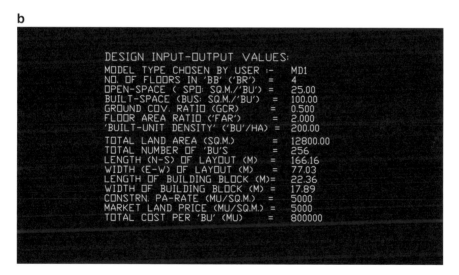

Fig. 6.11 Design layout instantly given by software 'CADUB-CBDG' corresponding to the numerical results of Model-1 shown above (Graphical Planning Solution-II)

6.6.7 Transfer of Development Rights and Improving Social Responsiveness in Urban Operations: Sale of FAR to Promote Social Housing

Operations of private urban organizations, including the private developers, will be primarily concerned with productivity and profitability. But, their operations also need to be evaluated in terms of their social responsiveness, particularly because of the interdependencies in respect of urban organizations and many groups in our society, as urban development is a joint public-private enterprise where decisions about the use of resources are made by the market place and by the governments (at various levels) in an interactive way, as discussed in Sect. 6.2 above. This is more so since private development is facilitated by the publicly provided infrastructure and services and by the public policy on land-use decisions, tax, and so on. It is desirable that efficiency and drive of private sector should be utilized to promote social housing by suitable mechanisms of public–private partnership in the urban development process taking into account the above factors.

We may take, as example; a private developer has a land admeasuring 10,000 sq. m. (with dimension 80 m × 125 m) where market land price is 10,000 MU/sq.m. and as per current planning regulations, the developer can carry out a commercial-cum-residential development within specified limiting FAR = 1.25 and a low-building-rise up to BR = 2. The prevailing building construction plinth area rate (CPAR) for such development is 5000 MU/sq.m. If the developer decides to build remunerative residential built-units of 120 sq.m./BU within the above planning regulations and uses Model-6, the INPUT-OUTPUT results in this case will be as follows:

```
YOU WANT TO CHANGE INPUT VALUE(S)/INPUT SET/NUM.ITERATION MODEL, FOR
BETTER
RESULT/PARTICIPATORY-DECISION (ALL STAKEHOLDERS)? Put 'y'(if yes)
or'n' (if no) y
ansch= y
Do you want to change (Initialize)'NUM.ITERATION MODEL 'earlier CHOSEN?
Type y (if 'yes') or n (if 'no') and press <CR>: n
YOU CAN CHOOSE ANY OF FOLLOWING 22 'INPUT SET' OPTIONS:
0. When you select the pre-set computer data input
1. When you want to change values of BUS, SPO, and BR
2. When you want to change values of SPO, FAR, BR
3. When you want to change values of FAR, BUS, BR
4. When you want to change values of BR, BUS
5. When you want to change values of BR, SPO
6. When you want to change values of BUS, SPO
6. When you want to change value of BR only
8. When you want to change value of BUS only
9. When you want to change value of SPO only
10. When you want to change value of FAR only
11. When you want to change value of BUD only
12. When you want to change value of CPAR only
13. When you want to change value of K only
14. When you want to change value of LP only
15. When you want to change value of BUC only
16. When you want to change values of BUD, SPO, and BR
16. When you want to change values of BUD, BUS, and BR
18. When you want to change values of LP, CPAR, and K
19. When you want to change values of FAR, BUS, SPO, LP, CPAR, K
20. When you want to change values of BUS, SPO, LP, CPAR, K (LCO-MODEL)
21. When you want to change values of BUC, SPO, LP, CPAR, K (MBO-MODEL)
PUT YOUR CURRENT'INPUT SET'OPTN. (0/21 ETC)-TDRP: 0/13/14:18
18
```

```
Type in same order (giving only space- and no ',' - between the input
values)
New input values for LP, CPAR, K:
10000 5000 0.025
You want OBR-INPUT OPTION (If NOT, it will be OSPO-INPUT OPTION)?
Type y (if 'yes') or n (if 'no') and press <CR>: y
Do you want to RUN 'TDRP' (or, SALE of 'FAR' Package) i.e.
Transfer of Development Rights Package to Analyze/Find
the most acceptable 'TDRP' for a Composite Development?
Type y (if 'yes') or n (if 'no') and press <CR>: n


SUMMARY OF ''CAUB'' RESULTS IN THE INSTANT CASE ARE:
ITERATION MODEL CHOSEN (find BUC)      :   INTERACTION/SENSITIVITY MODEL
(COST+PHY.)
INDEPENDENT DESIGN VARIABLES ('IDV'): FAR, BUILT-SPACE (BUS), OPEN-
SPACE (SPO), K,
                                LAND-PRICE (LP), BASE-PLNTH.AREA
RATE (CPAR)
DEPENDENT DESIGN VARIABLES ('DDV'): B.UNIT-COST (BUC), BR, GR.CV.
RATIO, DENSITY
INPUT PARAMETERS WITH VALUES CHOSEN BY USER:   ITMODEL=6 INCHOICE = 18
FAR= 1.25 BUILT-SPACE (BUS) = 120 SQ.M. /BU CPAR= 5000 MU/SQ.M. K=0.025
LAND-PRICE (LP) = 10000.0 MU/SQ.M. ITMODEL=  6 (NO 'TDR') INCHOICE = 18
BLDNG.RISE (BR) = 2
OUTPUT PARAMETERS WITH VALUES GIVEN BY THE COMPUTER:
GROSS-PLINTH-AREA-RATE (GPAR) =13250.0 MU/SQ.M.
B.UNIT-COST (BUC) =  1590000 MU   GR.CV.RATIO (GCR) =0.6250
DENSITY (i.e. 'BUD') = 104.16 BU/HECTARE
OPEN-SPACE (SPO) = 36 SQ.M. /BU
CAN RE-RUN MODEL WITH 5 IDV (say-BUS, SPO, LP, CPAR, K) VALUES CONSTANT &
6TH.IDV
(say-FAR) CHANGED TO FIND INTERACTION BETWEEN DDV'S (BUC, DENSITY, GCR,
BR here) & 6TH.IDV (FAR here)
```

It is clear from above that the developer can achieve a BUD (i.e. BU Density) of 104.16 BU/HECTARE with the limiting FAR of 1.25 prescribed by the PLA (i.e. Planning Authority) or he can build only 104 BUs of gross BUS (i.e. Built-Unit-Space) of 120 sq.m. each in his 10,000 sq.m. plot and the total cost per BU, i.e. BUC = 1,590,000 MU/BU as shown in the above OUTPUT results. The above planning and design given by the Model-6 in numerical terms can be converted instantly into a graphic design layout possible as per the current planning regulation, i.e. a Low FAR = 1.25 and a Low BR = 2.

If the local Planning Authority (PLA) decides to sell additional FAR to developers to promote *social housing* for the poor through the developers without any extra expenditure to the PLA but permitting some additional profit to the developer and he agrees to purchase additional FAR for this purpose on a mutually acceptable Transfer of Development Rights Package (TDRP), both PLA and developer can negotiate for such a package of TDRP by running a number of iterations both in numeric and graphic terms for a meaningful and *transparent* (and thus avoid

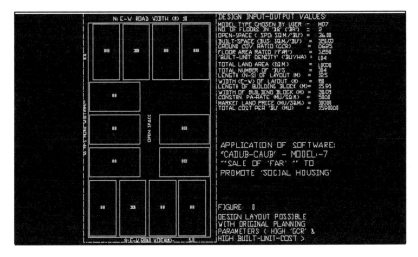

Fig. 6.12 Design layout instantly given by software 'CADUB-CBDG' corresponding to the first numerical results of Model-6 shown above

corruption) dialogue using the Software: CADUB-CAUB-CBDG before arriving at a suitable TDRP acceptable to both of them.

One such TDRP obtained using the software as above is shown below along with the computer prompts in this dialogue process (Fig. 6.12):

```
Do you want to RUN 'TDRP' (or, SALE of 'FAR' Package) i.e. Transfer of
Development
Rights Package to Analyze/Find the most acceptable 'TDRP' for a
Composite Development?
Type y (if 'yes') or n (if 'no') and press <CR>: y
Do you want to see Definitions of various TERMS in 'TDRP'?
Type y (if 'yes') or n (if 'no') and press <CR>: y
```

DEFINITIONS of TERMS used in Transfer of Development Rights Package (TDRP) are:

**LA = Project Land Area (square meter) proposed to be covered by the 'TDRP'

**OFAR = Original Floor Area Ratio permissible under the 'PLANNING REGULATIONS'

**EFAR = Enhanced Floor Area Ratio i.e. additional 'FAR' to be allowed (i.e. SALE) to a Developer to achieve Social Objective, i.e. 'Low-Income-Housing'

**EBUS = Built-Unit-Space (sq.m.) for remunerative BUs to be sold by Developer

**EBR = Building-Rise for remunerative BUs to be sold by the Developer

**ESPO = Semi-Public-Open-Space for remunerative BUs to be sold by the Developer

**EBUD = Built-Unit-Density for remunerative BU's to be sold by the Developer

**SBUS = Built-Unit-Space of Supported BUs to be handed over to the PLAN-NING AUTHORITY by the Developer in lieu of Enhanced 'FAR'

**SBUD = Built-Unit-Density for Supported BU's to be provided to PLANNING AUTHORITY

**CLA = Land Area Covered by the Commercial/Remunerative Built-Units

**SLA = Land Area Covered by Supported BUs provided to PLANNING AUTHORITY

**CSU = Cost saving per Built-Unit (Commercial) with Enhanced 'FAR' allowed

**CSR = Cost-Saving Ratio utilized for Supporting BUs for Low-Income-Households

**OBR = Original Building-Rise for the Commercial/Remunerative Built-Units

**OBUC = Original Built-Unit-Cost for Remunerative/Commercial BUs (MU)

**EBUC = Built-Unit-Cost for Commercial BUs with Enhanced Development Rights

**SBUC = Built-Unit-Cost for Supported BU's to be handed over by Developer

**ECN = Number of Commercial Built-Units to be sold by the Developer under the 'TDRP' i.e Enhanced Development Rights or Sale of 'FAR'

**SN = Number of Supported Built-Units for Poor, made available by Developer to the PLANNING AUTHORITY in lieu of Enhanced Development Rights

OPAR and EPAR are Construction Costs per unit of Plinth Area (Commercial BUs) for original and the development with enhanced 'FAR'. OBUS and OSPO are original number of BU, Built-Space, and Open-Space per BU(Commercial). SSPO, SPAR, and SBR are Open-Space per BU, Construction Cost per unit of Plinth Area, and the Building Rise, respectively, for the supported built-units.

One such result is shown below:

```
SUMMARY OF ''CAUB'' RESULTS IN THE INSTANT CASE ARE:
Iteration Model Chosen      :              Interaction/Sensitivity Model
(To find
                                 Cost & Physical Parameters of UBF)
Independent Design Variables ('IDV'):       FAR, Built-Unit-Space
(BUS), Open-Space (SPO),
                     Land-Price (LP), Base-Plinth-Area-Rate (CPAR)
Dependent Design Variables ('DDV'):       Built-Unit-Cost (BUC),
Building-Rise (BR)
                 Ground Coverage Ratio (GCR), Density (BUD)
```

INPUT PARAMETER-VALUES CHOSEN ARE:

```
LA= 10000.0 Sq.M. EBUS=120.0 Sq.M. /BU OBUS=120.0 Sq.M. /BU  ESPO=36.0
Sq.M. /BU
OSPO=36.0 Sq.M. /BU EPAR= 5500.0 MU/Sq.M. OPAR= 5000.0 MU/Sq.M.
Land-Price (LP) = 10000.0 MU/Sq.M. OFAR=1.25 EFAR=2.50
(i.e. additional FAR=1.25 for Sale)
SPAR= 2500.0 MU/Sq.M SBUS= 25.0 Sq.M. SSPO= 8.0 Sq.M. SBR= 5
CSR=0.6900  K=0.0250
```

ITMODEL= 6(TDR) INCHOICE= 0 MU= Monetary Units (i.e.Rupees, Dollars etc.)
OUTPUT PARAMETER-VALUES INSTANTLY GIVEN BY THE COMPUTER ARE:
EBUC= 1305000 MU OBUC= 1590000 MU EBR= 10 OBR= 2
EGCR=0.250 EBUD=208.3 BU/Hectare EBRF=10.0
OGCR= 0.625 OBUD= 104.2 BU/Hectare SGCR=0.385 OBRF= 2.0
ECN= 160 SN= 180 CSU= 285000 MU SBUC= 200312 MU SFAR= 1.92
ETN= 340 OTN= 104 ERN=3.256 ECNF= 159.61 SNF= 169.52 ONF= 104.16
CLA= 6666.26 Sq.M. SLA= 2333.63 Sq.M. SBUD= 669.2 BU/Hectare

The above result in numerical terms can be converted instantly into a graphic design layout possible as shown in Fig. 6.13.

The above numerical and graphical results show the following important *efficiency*, *equity,* and environmental indicators:

1. That the PLA can increase the housing supply by 3.26 times on the same land area of 10,000 sq.m. just by allowing additional FAR of 1.25 substantially improving the *efficiency* in land utilization
2. Additional FAR involves no reduction in open-space per built-unit and the same built-space per built-unit is maintained achieving the given environmental standard
3. As against a high Ground Coverage Ratio of 0.625 it is reduced to only 0.25 which will facilitate ground water recharging and help augment ground water table improving green coverage and environment

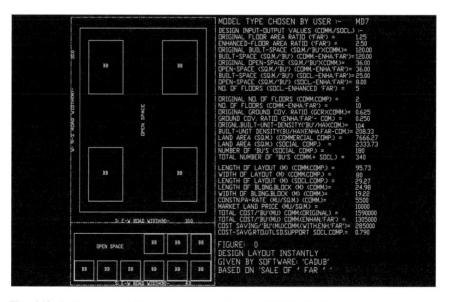

Fig. 6.13 Design layout instantly given by software 'CADUB-CBDG' corresponding to the second numerical results of Model-6 shown above

4. The PLA achieves a *social housing supply* of 180 built-units just only by the above planning decision involving no investment on its part, but improving *equity* substantially in the development process
5. The developer by supply of *social housing* of 180 built-units free of cost to the PLA improves and discharge its *social responsiveness* in the urban development process
6. The above *efficiency* and *equity* indicators could be further improved by changing the values of parameters such as CSR, SBR, SBUS, SSPO, EBUS, ESPO, and so on, through participatory and informed decision-making by the multiple stakeholders using the Software: CADUB-CAUB-CBDG, having both numeric and graphic components for a meaningful dialogue, to help decide in a *transparent* manner an *efficient* and *equitable* development package acceptable to all.

The above application results demonstrate the utility of the Software: CADUB-CAUB-CBDG in the planning and design analysis of a transfer of development rights packages for improving *Social Responsiveness* of various actors in the urban operations in a *transparent* manner, and thus reduce any scope of corruption and also avoid any controversy in this regard. Adopting such a procedure will help popularize the 'Sale of FAR' as above to promote *social housing*, and thus help solve the mounting housing problems in various countries through an *efficient* and effective public–private partnership mechanism.

6.7 Closure

Planning and design in the AEC (Architecture Engineering Construction) sector, involving building, housing, and urban services consuming huge resources and affecting the *quality of life* of all people, needs to be viewed as a problem-solving process involving search through numerous potential solutions to select a particular solution that meets a goal such as providing adequate shelter and urban services for all, within the *affordability* and *resource-constraints* at the individual, local, and national level. With increasing population and dwindling per capita resources, this view is equally applicable in all countries as we live in a *Finite Earth with Finite Resources*. Computers permit development of symbolic generative systems or models representing design by symbols, variables, and computer programs, producing potential solutions or goal set, help store, and retrieve data and graphics describing design, and integrate quantitative analysis, optimization, and graphics into the design process, and thus help achieve the design objectives and goals.

In this chapter, a generative system in the form of models representing designs by symbols, variables, and a Computer Software named as: 'CADUB'-'CAUB'-'CBDG' incorporating ten models to carry out planning and design analysis and different tasks to achieve problem solutions is presented. 'CAUB' which is the numeric component of the software developed in user-friendly C++ language, and the other two are graphic components of the software developed in Lisp language, and all the three components can be used in one graphic platform for planning and design analysis and for instant conversion of numeric solution to graphic design solutions as demonstrated in the example problems presented above. To facilitate problem solution, the function of planning and design in AEC sector needs to be part of holistic urban management, i.e. as *planning function* (one of the five traditional management functions as detailed in Chap. 2) to make cities viable and desirable places for people to live and work. Similarly, the principles of *efficiency* and *effectiveness* which are the essence of management, as well as the principles of *flexibility* and *navigational change* and the *principles of social responsiveness* should be equally applicable in urban development and management to achieve *resource-efficient* and *equitable* urban problem solutions as earlier explained in Chap. 2. The role of planning and design analysis using above software, in the application of the above important management principles in urban operations to achieve *resource-efficient* urban problem solution with *social equity* is outlined above using example problems and case studies.

As cited above, there can be *large variations* in the *efficiency indicators* of urban-built-form (depending on the rules such as *planning regulations* and the *dynamics* of input-prices), which are hard to be ignored by *User Citizens*, *Builders*, and other *Stakeholders*. In this context, a *Theory for Three-Dimensional Optimal Urban Built-Form* is presented, which incorporates many of the above factors to enable AEC professionals including planners, to develop effective response in a *transparent* manner, to the *urban dynamics*, in terms of the planning regulations, and planning and design solutions achieving *Resource-Efficiency*, *Social Equity*, and the desired and affordable environmental standards in the Urban Development process. *Algebraic formulas*, *which give results exactly same as the results obtained using geometric programming optimization technique*, are presented to enable AEC professionals, not having any knowledge of optimization techniques, to obtain optimal design of urban-built-form and to apply the *optimization principles* in their planning, design, and urban management practices to help achieve *efficiency* and *social equity* in the urban development process being managed by them.

As outlined above, there are ten 'Numerical Iteration Models' of '**CAUB**', depending on the set of design variables chosen as independent (i.e. to permit any desired input value), or, as dependent (i.e. dependent on values assigned). Illustrating applications of all the models of C++ software: 'CAUB' and the graphic software: 'CBDG' are shown demonstrating instant integration of numeric and graphic design in one platform. Illustrative application of 'CAUB' to determine *Optimum FAR* (unlike a <u>*Static*</u> FAR in conventional practice) to *Minimize* the Built-Unit-Cost for a given land price (which is *Dynamic* determined by the market beyond control of the *stakeholders*) in a site-specific-case, fulfilling the environmental standards like open-space, applying the traditional *Management Principle of Flexibility*. The expression for the 'Optimum FAR', (derived using the *Nonlinear Programming* Technique in Chap. 8), giving the *Least-Cost* per Building-Unit (BU) for a given minimum acceptable *Built-Space* and *Open-Space* per BU, is also presented.

It is shown above with illustrative examples that the **Conventional** *Planning Regulations* (PR) such as **FAR = 0.65**, (frequently adopted irrespective of the Market Land Price of 100 or 40,000 MU/sq.m.), may increase the Cost per BU even by **4.34 times** (with same *Built-Space* and *Open-Space* per *Built-Unit*) the **Least-Cost** given by the **Optimal FAR** of **6.414**, creating *Inefficiency* and *Inequity*. This may be one of the reasons for people to attempt *Optimization* by constructing more *Built-Space* and higher *Building Rise* in *violation* of such PRs. This highlights the need to include *Optimization Principles* in the conventional *Planning Regulations*, using this IT tool, to achieve *Efficiency* and *Equity* with people's (important *Stakeholders*) *Participation*, which may reduce such *violations*.

Application of Software: 'CADUB'-'CAUB'-'CBDG' to determine a site-specific Transfer of Development Rights Package (**TDRP**, i.e. Sale of FAR), acceptable to the *multiple stakeholders*, to promote social housing even in a high-land-price-case is also demonstrated above with an illustrative example problem. Hopefully, using the example problem as guide urban professionals (e.g. *Planners/Designers*) would be able to apply this concept in their own situation in a *transparent* manner to promote *social housing* without any investment on the part of their local planning authority. Use of the information technology tools (IT tools), i.e. Software: 'CADUB'-'CAUB'-'CBDG'(and 'HudCAD') for instant integration of numeric and graphic design is demonstrated above with illustrative examples. Such *instant visual display* of alternatives by the software, help both professionals (e.g. *Planners/Designers*) to reiterate more *transparently* the *merits* of their suggested planning and design solutions, and the *user citizens* (vital *stakeholders*) to interact (even through Internet, if such infrastructure and facility exist) more meaningfully with *Planners* and *Designers*, keeping in view their *Affordability* and *Priority* leading to *Efficient* and *Equitable* problem solution satisfactory to all.

In the conventional process, the architect/urban planner/engineer can obtain the value of *Quantitative Design Efficiency Indicators* (i.e. QDEIs such as cost, built-space, open-space, density, and so on) only after preparing the drawing. It is also shown above that there can be very large variations in the values of QDEIs depending on the planning and design decisions taken by the Architect/Planner/Engineer and it is difficult to change the drawing-design prepared by conventional time consuming manual process to get a better QDEI even if it is considered *inefficient*. Using such IT tools/software, the *design process is reversed* where architect/planner/engineer is the *Full-Controller* of *Design Efficiency Indicators* of the urban-development-system he/she is planning/designing and the drawing (or design in numeric terms) fulfilling the *design-efficiency-value* (e.g. cost, built-space, open-space, density, and so on) specified by them is given as output by the software. *Thus, using the IT tools above Urban Professionals can achieve* any feasible *Design-Efficiency-Value* (say any built-space with minimum cost, using optimizing models). *Urban Professionals* (architects, civil engineers, planners, and so on) *commit* huge *Resources* (money, land, materials, and so on) while taking *planning* and *design decisions* at each system level (project, 'spatial unit', so on), making them *Accountable* to the *Society* to achieve *Efficiency* and *Equity* in use of above *Resources* so *committed* by them. The above software are helpful IT tools to enable them to discharge their above *Ethical Responsibility* and *ACCOUNTABILITY* to the *Society* to achieve *Efficiency* and *Social Equity* in urban development.

As shown above, using this IT tool, a creative planner and designer could carry out a number of 'numeric iterations' (invoking any iteration model of 'CAUB') to generate a '*Variety of Numerical Planning Solutions*', each of which can be converted to a '*Variety of* Graphical Planning Solutions' (using 'CADUB'-'CBDG', and also 'HudCAD'), giving '*Numerous Design Alternatives*', with instant visual display to *multiple stakeholders*, for selection by *Informed-Participation* with *Transparency*. Thus, using these Information Technology Tools (IT tools), i.e. software: 'CADUB'-'CAUB'-'CBDG' (and 'HudCAD'), creativity of the Planners and Designers can be enhanced, while achieving *Optimal* and *Resource-Efficient* planning and design solutions *Acceptable* to the *multiple stakeholders* including the *user citizens*, and consequently, improving both *Efficiency* and *Equity* in the urban operations. The complete listings, i.e. *source code* of the software: 'CADUB'-'CAUB'-'CBDG' (and 'HudCAD') along with the Application Software: CAUB (i.e. C++ program: CAUB.CPP, and HUDCM.CPP and the Lisp programs: CADUB.LSP, CBDG.LSP) are placed in an online space connected to the book to help readers to use such techniques to achieve their own problem solutions.

Chapter 7
Optimization Techniques and Design Optimization

7.1 Introduction

The human mind confronting a task or problem can recognize more than one course of action (alternatives) followed by the selection (*decision*) of what is considered the best action, using a measure of effectiveness, often involving a value judgement. Operations Research (OR), which is often called 'quantitative common sense', is the application of scientific methods to the study of alternatives in a problem situation, with a view to obtaining a quantitative basis for arriving at a best solution (Koontz & Weihrich, 1990). This is quite relevant in AEC operations including architecture/ civil-structural engineering covering housing/urban development (an important area of AEC operations), where AEC professionals such as engineers, architects, and other *professionals,* and *user citizens* have to interact to select an alternative from among many alternatives, frequently with different *efficiency indicators.* Essential characteristics of OR (Operations Research) and *Optimization* are its emphasis on *models* (indicating physical depiction of a problem), *goals* (i.e. measure of effectiveness in terms of an accepted value scale), *variables, constraints,* and putting them in *mathematical terms,* i.e. a *generative system* discussed earlier. *Models* could be 'descriptive', if they are designed only to describe the relationships between *variables* in a problem, or '*optimizing'* designed to lead to selection of best course of action among available alternatives.

Optimization is a *three-step rational decision-making process,* i.e. knowledge of the system, finding a measure of system effectiveness and choosing those values of the system variables that yield *optimum* effectiveness. Developing the descriptive model could be the first step of the *Optimization* process, i.e. acquiring knowledge of the system. A problem can be *Maximized* for benefit, profit, and so on or *Minimized* for cost, loss, and so on. Which objective is desired as a matter of statement. Urban development is both a physical and a socio-economic system and may have conflicting goals and viewpoints, which need to be resolved using *Optimization* both in qualitative (judged by human preference) and quantitative (using exact mathematical means) terms. Economic *optimization* has to consider human values expressible in quantitative form even if indirectly. For example, the cost difference of given built-space between chosen development types, say between low-rise and medium-rise urban development, could be an indirect measure for qualitative evaluation of these two types of development, while selecting the preferred alternative, with the *participation* of *user citizen* using collaborative **IT** (Information Technology) **tools**.

The word optimum is a technical term connoting quantitative measurement and mathematical analysis, and the *Optimization* theory encompasses quantitative study of optima and methods for finding them (Beightler et al., 1979). Optimization theory is more than a set of numerical recipes for finding optima. By studying various optimization techniques, each suitable for different quantitative situations (even if idealized), one often discerns general *decision rules* appropriate to problems, and can develop an *insight* in recognizing the proper form of an optimal solution even when a problem is not completely formulated in mathematical terms. One advantage of such optimization study is that it can foster recognition of information valuable enough to be gathered to depict a system well enough for it to be optimized (say, bringing agreement on the measure of effectiveness and expressing its quantitative dependence on system variables), thus, help apply *Optimization* in a problem solution. Such an approach is *crucial* to achieve a *resource-efficient* and *equitable* solution of problems in the AEC fields such as housing and urban development problems.

7.2 Optimization: Its Philosophical Link and Ancient Origin

Earliest optimum principles concerned with the behaviour of light postulated around 100 B.C by Heron who asserted that light travels between two points by the shortest path (Beightler et al., 1979). Optimization has also philosophical link as the philosophers like Leibniz, who coined the word 'optimum' in 1710, sought philosophical truth through mathematics and speculated on the nature of the world and on its creation, i.e. its synthesis (Beightler et al., 1979). Leibniz's theological speculation indicates the rudiments of decision theory, i.e. description of phenomena in terms of an optimum principle and an attempt to prove the existence of an optimum leading to an optimization theory (Beightler et al., 1979). Leibniz's point of view influenced many scientists reflected by elegant formulations of a number of natural laws and economic principles such as single minimum principle linking light and mechanics together conceived by Hamilton (1735), Gauss's (1729) 'principle of least restraint' related to statics, and explaining the complex economic phenomena by Adam Smith (1776) in terms of the 'economic man' who always acts to maximize his personal profit (Beightler et al., 1979).

Even the *ancients*, who had little knowledge of *technology* and the *resource constraints* and *population pressure* were much less severe, applied *Optimization* and perceived that a circle has the greatest ratio of area to perimeter, which is the reason that many ancient cities are circular to minimize the length of city walls to enclose the city's fixed area (Beightler et al., 1979; Jelen & Black, 1983).

Thus, *Optimization* is not a new idea but a human trait that is as old as antiquity, and since it involves finding the best way to do things it has application in the practical world incorporating many fields including production, trade, urban services operations, and management, where sometime a small change in efficiency may imply the difference between success and failure of the operation. The unparalleled development in information technology and computer applications in our times should facilitate its development and popularize its application in all relevant fields including housing and urban development and management improving efficiency in various operation.

As mentioned above, ancients intuitively applied optimization principle making cities circular. On the same analogy, modern urban-built-form needs to be optimized considering the cost and availability of land as well as the structural feasibility and cost, which is elaborated further later. It is unfortunate that in spite of availability of modern highly powerful computers and graphic systems, such *Optimization* is rarely applied in our modern urban operations, causing a mismatch between the land price and the built-form, and thus creating many *inequities* and *inefficiencies* and the consequent urban problems (e.g. proliferation of slums, inadequate housing/urban services, environmental degradation, and so on), which needs urgent correction.

This is quite relevant in Architecture, Engineering and Construction (AEC) sector and particularly in housing and urban development at each system level, where building and urban professionals as well as the user citizens have to select an alternative from among many alternatives, frequently with different efficiency indicators and costs. Tedious and voluminous computations involved, in utilizing optimization and operations research techniques and to solve and use optimizing models, have prevented their applications in the AEC sector especially in traditional building design and urban development planning and design in the past. The continued development of cheap and powerful computers with interactive graphics systems has made it possible to perform many planning and design functions integrating graphics with complex computations, design analysis, cost engineering, and *Optimization* techniques on a computer making it feasible to achieve integrated computer-aided design by Optimization. This book is an attempt to promote application of such integrated CAD by *Optimization* in housing, urban development, and management.

7.3 Optimization in Planning and Design Problems

Nowadays many decisions are made by choosing a quantitative measure of system effectiveness in terms of some value scale, e.g. cost, profit, benefits, and so on, and choosing those values of system variables that yield optimum system effectiveness. Designing a facility system involves three steps. In the first step, one should know in quantitative terms how the system variables interact with each other. In the second step, the designer should decide a single measure of system effectiveness such as cost expressed in terms of the system variables. In the third step, the designer should choose those values of the system variables that yield optimum system effectiveness. A facility designer confronting a task or problem can recognize more than one course of action or alternative designs which have to be synthesized involving selection of an alternative out of a number of options some of which may perform better than the others, based on certain concepts. Thus, the designer tries to optimize the design followed by the selection (decision) of what is considered the desirable action, often involving a value judgement or qualitative and quantitative analysis of the facility design.

Moreover, engineering, architecture, and urban professionals commit huge resources (land, materials, money, and so on) in their planning and design decisions-making them ethically responsible and accountable to the society for efficiency and equity in the use of resources so committed (which are scarce in many cases). Design optimization is essential to attain such efficiency and equity to produce various products and services. In fact, no rational decision is really complete without optimization. Since all such operations involve numerous variables and many solutions, with large variations in efficiency indicators, search of optimal solution becomes quite complex, making it difficult to attain by conventional manual methods. Computer-Aided Design (CAD) is of particular importance for application of optimization involving hundreds of design variables and constraints although difficult to achieve by conventional manual design methods. Characteristics of optimization are discussed below.

7.4   Characteristics and Some Basic Concepts of Optimization

As discussed earlier, essential characteristics of OR (Operations Research) and *Optimization* are its emphasis on *models* (indicating physical depiction of a problem), *goals* (measure of effectiveness), or objective function, *variables*, and constraints, and putting them in *mathematical terms,* i.e. a *generative system.* The objective function represents the value criteria in terms of a measure of effectiveness or performance chosen in a problem situation. The effectiveness of an optimization study directly depends on our ability to measure and tailor our response according to the result of measurement adopting an acceptable scale of measurement which may be in terms of costs (monetary units), benefits, time, weight, and so on, depending on the type of a problem.

Optimization is a *three-step rational decision-making process,* i.e. knowledge of the system, finding a measure of system effectiveness, and choosing those values of the system variables that yield *optimum* effectiveness. Developing the descriptive model could be the first step of the *Optimization* process, i.e. acquiring knowledge of the system. A problem can be *Maximized* for benefit/profit or *Minimized* for cost/ loss, and which is desired as a matter of statement. Urban development is both a physical and a socio-economic system and may have conflicting goals/viewpoints, which need to be resolved using *Optimization* both in qualitative (judged by human preference) and quantitative (using exact mathematical means) terms. Economic *optimization* has to consider human values expressible in quantitative form even if indirectly. For example, the cost difference of given built-space between chosen development types, say between low-rise and medium-rise urban development, could be an indirect measure for qualitative evaluation of these two types of development, while selecting the preferred alternative, with the *participation* of *user citizen* using collaborative **IT** (Information Technology) **tools**.

Optimization theory is more than a set of numerical recipes for finding optima. By studying various optimization techniques, each suitable for different quantitative situations (even if idealized), one often discerns general *decision rules* appropriate to problems, and can develop an *insight* in recognizing the proper form of an optimal solution even when a problem is not completely formulated in mathematical terms. One advantage of such optimization study is that it can foster recognition of information valuable enough to be gathered to depict a system well enough for it to be optimized (say, bringing agreement on the measure of effectiveness and expressing its quantitative dependence on system variables), thus, help apply *Optimization* in a problem solution. Such an approach is *crucial* to achieve a *resource-efficient & equitable* solution of AEC professionals engineering (e.g. housing/urban development) problems. All these characteristics are further outlined below.

7.4.1 Models

Models are idealized mathematical representation of a problem such as urban services system of interest which is to be optimized. A modeller attempts to develop a simplest model which will predict and explain the problem with accuracy adequate for his purpose. Commonly models are classified as iconic, analog, and symbolic models.

Iconic models are scaled version of the real thing where relevant properties of the system are also represented in the model. Laboratory models of bridges and buildings are examples of iconic models where, for instance, model material is so chosen that modulus of elasticity of the model material is same as that of bridge material, for accurate study of complex structural behaviour of the bridge. In analogue models, one set of properties is represented by another set, such as, electric current can represent heat flow, fluid flow can represent traffic flow and so on.

Symbolic models use symbols to represent system variables and the relationships between them. Computer-aided design by optimization uses symbolic models to represent real buildings and structures which are to be optimized. Generality and refinement of symbolic models make them easy to manipulate, and therefore, adopted in the design optimization incorporated in this book.

7.4.2   Variables and Constraints

Variables in a design problem are those quantities that are not predetermined or predefined. Any facility design is a function of a number of variables many of which are within the control of the designer and some of which may be beyond control of the designer. Design variables play a crucial role in a design optimization problem and should be correctly identified as by changing the values of design variables, different design options can be generated to seek the optimal design amongst them. In the most design problems, there are constraints which have to be fulfilled to serve the design objective. Thus, the designer may specify the permissible range of values of some variables which are within his control, as constraints to the design problem. Accordingly, minimum acceptable width or area of a room may be specified as constraints and called as side constraints in a building design problem. Similarly, minimum acceptable width of a reinforced concrete beam may be named as a side constraint while maximum acceptable compressive stress of concrete may be named as a behaviour constraint in a reinforced concrete beam optimal design problem.

7.4.3   Objective Function and Optimization

Any optimization problem has an objective to be achieved. In fact, a problem exists where (1) some objective(s) is to be achieved in terms of some accepted 'value scale', (2) there are alternative ways to attain it, and (3) the optimal way(s) to achieve it is not apparent. In this context, there is need to arrive at the accepted 'value scale' in terms of cost, benefit, profit, loss, and so on, by which a facility design is to be evaluated. For numerical optimization, it is necessary that the design objective is expressed as a function of the design variables which is called objective function. The objective function of a problem expresses the value in terms of the accepted criteria of measure of performance for the facility being designed. Thus, the objective function for a building design may be to minimize the cost for a given building space or to maximize building space for a given amount of investment.

A general optimization model can be expressed as objective function:

Optimize: $z = f(x_1, x_2, x_3, \ldots x_n)$

Subject to the constraints: $g_i(x_1, x_2, x_3, \ldots x_n) \leq 0$ or ≥ 0 $i = 1,2,3,\ldots m$ (7.1)

$x_j \geq 0$ $j = 1,2,3,\ldots n$

When particular numbers or values are assigned to the components of $(x_1, x_2, x_3, \ldots x_n)$, the resulting vector is generally called a *policy* or *design*. In many cases, policies or designs incorporated in the objective function may be impossible, uneconomical, or for some given reasons cannot be included in the *feasible region*: F of solution. This region is generally defined mathematically by a set of simultaneous inequalities: $g(x) \geq 0$ and or equations: $g(x) = 0$. However, for the purpose of optimal design, a *closed region* will be adopted, that is, all inequalities will include the possibility of strict equalities, i.e. the region F contains all its boundary points. The objective function may be the minimum of z, denoted as $z*$ is identical with the greatest lower bound of z (Phillips & Beightler, 1976).

The constraints delineate the feasible region, and the objective function z is evaluated for values of x_i within the feasible region to yield the optimal solution.

7.5   Optimization Techniques

There are number optimization techniques in literature to solve optimization problems. Optimization technique can be separated broadly into two categories, namely: (1) direct methods and (2) indirect methods. Direct methods start at an arbitrary point and advance stepwise towards the optimum by successive improvement. A classical indirect optimization technique ultimately tries to find optimal solutions by solving systems of equations, frequently nonlinear, resulting from the algebraic expressions for the first partial derivatives which are equated to zero, and thus, reduced to root finding, rather than directly searching for an optimum.

A study of optimization theory gives a useful relationship between the minimization and a maximization problem which is that minimization techniques can be used in maximization problems (and vice versa) simply by changing the sign of the objective function.

7.6   Linear Programming Techniques

Linear programming is a special case of general optimization problems in which there are no terms of second degree or higher in the objective function or constraints. Thus, linear programming technique of optimization deals with the minimization or maximization of a linear function subject to the linear constraints. In this case, the linear function whose least or greatest value is sought is the objective function, and the collection of values of the variables at which the least or the greatest value is attained define optimal plan or solution. Any other collection of values of variables complying with the restrictions or constraints defines the feasible plan(s) or solution(s).

The standard form of linear programming can be denoted as:

$$\text{Minimize/maximize } y = c_1 x_1 + c_2 x_2 + \cdots + c_n x_n \tag{7.2}$$

Subject to the constraints:

$$a_{11}x_1 + a_{12}x_2 + \cdots + a_{1n}x_n = b_1$$
$$a_{21}x_1 + a_{22}x_2 + \cdots + a_{2n}x_n = b_2$$
$$a_{31}x_1 + a_{32}x_2 + \cdots + a_{3n}x_n = b_3$$
$$\cdots$$
$$a_{m1}x_1 + a_{m2}x_2 + \cdots + a_{mn}x_n = b_m$$

and

$$x_j \geq 0 \qquad j = 1,2,3,\ldots n$$

The above standard linear programming form can be expressed in an abbreviated form as follows:

$$\text{Minimize/maximize } y = \sum_{j=1}^{n} c_j x_j \tag{7.3}$$

Subject to the constraints:

$$\sum_{j=1}^{n} a_{kj}x_j = b_k \qquad k = 1,2,3 \ldots M \tag{7.4}$$

$$x_j \geq 0 \qquad j = 1,2,3 \ldots N \tag{7.5}$$

$$b_k \geq 0 \qquad j = 1,2,3 \ldots N \tag{7.6}$$

The quantities c_j, a_{kj}, and b_k are assumed as known constants, and m and n are given positive integers. As shown above, the constants b_k are assumed nonnegative, but, the quantities c_j and a_{kj} are unrestricted in sign.

In the above standard form of linear programming, the set of equations for constraints are assumed to be equality constraints. In reality, there may be both 'less than' and 'greater than' constraints. In case of 'less than' constraints adding a nonnegative 'slack variable' x_{n+1} converts the inequality constraint such as $a_{11}x_1 + a_{12}x_2 + \cdots + a_{1n}x_n \leq b_1$, to an equality constraint as follows:

$$a_{11}x_1 + a_{12}x_2 + \cdots + a_{1n}x_n + x_{n+1} = b_1$$

Similarly, in case of a 'greater than' constraint, one can subtract a nonnegative excess variable: x_{n+1} to attain an equality constraint as follows:

$$a_{11}x_1 + a_{12}x_2 + \cdots + a_{1n}x_n - x_{n+1} = b_1$$

It may be noted that if an inequality constraint is tight or binding at optimality, the value of the linked slack variable will be zero, while, in case the inequality constraint is loose or nonbinding at optimality, the value of the related slack variable will be positive. In the above abbreviated standard form of linear programming, the variable

x_j is defined to be a *basic* or solution variable if in a solution to the M equations in N unknowns that variable has a positive value. The set of all basic or solution variables in a particular solution set is defined as a *basis*. Those variables which are not basic (hence zero) are called nonbasic. A *feasible solution* or a *feasible vector* to the linear programming problem is defined as the set of nonnegative values of $x_1 \ldots x_n$ (consisting of the entire set of basic and nonbasic variables) that satisfies the original constraint set. The problem is to find a set of nonnegative values of $x_1 \ldots x_n$ or *optimal feasible vector* so that the linear objective function z is minimized (maximized) while satisfying the constraints. To solve linear programming problems with a large number of variables, the simplex method, which is an analytical method for solution of the main problem of linear programming, can be used.

7.6.1 Simplex Method in Linear Programming

A simplex is a N-dimensional geometrical figure consisting of $N + 1$ points or vertices, and all their interconnecting line segments, polygonal faces, and so on. Thus, in two dimensions, a simplex is a triangle, while in three dimensions, it is a tetrahedron. The simplex method of linear programming uses this geometrical concept. In such problems we start, say, with a full N-dimensional space of candidate vectors and slice away the regions that are eliminated in turn by each imposed constraint. As the constraints in such problems are linear each boundary imposed by this process is a plane or hyper-plane. When all the constraints in a problem are imposed, we are left with some feasible region or there may be none. Since a feasible region is bounded by hyper-planes, geometrically, it is a kind of simplex, and as the objective function is linear the *optimal feasible vector* cannot be in its interior away from the boundaries, and also, it means that we can always increase the objective function by running up the gradient until we hit a boundary, and finally arrive at a vertex of the original simplex. As a result, the optimization problem is reduced to a combinatorial problem to keep trying different combinations, and computing the objective function for each trial, until the best solution is found. Thus, the simplex method seeks a basic feasible solution that maximizes (or minimizes) the given objective function. Such sets of solution variables are finite and are represented by the vertices of the solution space.

Thus, a problem with M constraints and N variables, the maximum number of basic solutions (i.e. MS) is given by the following expression (Beightler et al., 1979):

$$MS = \frac{N!}{M!(N-M)!} \tag{7.7}$$

The simplex technique permits us to find out the optimal solution (or confirm its nonexistence) with a fraction of the above effort. This is because the simplex method organizes a procedure so that: (1) a series of combinations is tried to increase the objective function at each step, and (2) optimal physical vector is reached after a number of iterations which is not larger than of order of M or N, whichever is larger.

An optimal feasible vector or optimal solution may fail to exist because of two reasons, namely (1) there are no feasible vectors, i.e. the specified constraints are incompatible, or (2) there is no maximum which can happen when there is a direction in N space where one or more of the variables can be taken to infinity even while satisfying the constraints, giving an unbounded value for the objective function. Linear programming is very useful and relevant in many real life cases because (1) non-negativity constraint for any variable x_j as prescribed above (expression (7.6)) represents a real amount of some resources such as money, materials, and so on, (2) one is often interested in additive (or linear) limitations or bounds forced by man or nature such as maximum affordable cost, maximum labour, maximum capital, and so on as prescribed by expression (7.5), and (3) the objective function one wants to optimize may be linear or it can be approximated by a linear function as indicated by expression (7.4) above.

In the simplex method, the above system of constraints given by expression (7.5) are reduced to the unit basis by expressing $x_1, x_2, \ldots x_r (r \leq M)$ in terms of the rest of the variables as follows:

$$
\begin{aligned}
x_1 &= a'_{1,r+1} x_{r+1} + \cdots + a'_{1n} x_n + b'_1 \\
x_2 &= a'_{2,r+1} x_{r+1} + \cdots + a'_{2n} x_n + b'_2 \\
&\cdots \\
x_i &= a'_{i,r+1} x_{r+1} + \cdots + a_{in} x_n + b'_i \\
&\cdots \\
x_r &= a'_{r,r+1} x_{r+1} + \cdots + a'_{rn} x_n + b'_r
\end{aligned}
\tag{7.8}
$$

where $b'_1 \geq 0, b'_2 \geq 0, \geq 0, b'_3 \geq 0, \ldots b'_r \geq 0$.

If the given constraints are inequalities, these can be converted to equalities by introducing new nonnegative variables, e.g. slack variables as explained above. In the first step, assuming all the nonbasic variables as equal to zero and using the expression (7.8), we get the values of the basic variables or left-hand variables as: $x_1 = b'_1, x_1 = b'_2, \cdots x_1 = b'_r$. Thus, the solution $(b'_1, b'_2, \ldots b'_r, 0, \ldots 0)$ is a basic feasible solution. As per expression (7.5), there are exactly M numbers of left-hand variables or basic variables. The remaining $N - M$ variables are right-hand variables or nonbasic variables. A linear programming problem with only equality constraints as given by expression (7.4) above along with the non-negativity constraints is said to be of restricted normal form. Obviously, the above restricted normal form can be achieved only if $M \leq N$ in a particular case. The simplex method proceeds by a series of exchanges, and in each exchange, a right-hand variable and a left-hand variable change places. At each exchange, the problem is maintained in restricted normal form that is equivalent to the original problem. As explained above in restricted normal form, we can instantly find a feasible solution by setting all right variables equal to zero giving the values of left-hand variables for which the constraints are satisfied.

7.6.1.1   Application of Simplex Method in Linear Programming: Illustrative Example I

Application of the simplex method by reducing the system of constraints to the unit basis, as indicated by expression (7.5), is illustrated below using the following example problem of linear programming.

A ready-mix concrete company manufactures two types of concrete mixes, namely type A and B. To deliver one truckload of ready-mix concrete, the company consumes 2truck hours for type A and 4truck hours for type B mixes due to difference in delivery distance and logistics. The loading facility of the company cannot handle not more than 100 truck hours per day. The company earns a profit of 3 MU per truckload delivery of type A and 2MU per truckload delivery of type B under the condition that it cannot manufacture more than 40 truckloads of type A and not more than 20 truckloads of type B per day. What number of truckloads of each type of mixes should the company produce per day to maximize its profit?

Let us assume that the company produces x_1 truckloads of type A and x_2 truckloads of type B concrete mixes. Then we can express the objective function and constraints as follows:

$$\text{Objective function-maximize } y = 3x_1 + 2x_2$$

Subject to the constraints:

$$x_1 \leq 40$$
$$x_2 \leq 20$$
$$2\,x_1 + 4\,x_2 = 100$$

We can convert the '$\leq$' constraints to a system of equations by introducing the slack variables x_3 and x_4 as follows:

$$x_1 + x_3 = 40$$
$$x_2 + x_4 = 20$$
$$x_1 + 2\,x_2 = 50$$

In this problem, there are $M = 3$ constraints and $N = 4$ variables, i.e. $N - M = 1$, and therefore, any three variables can be expressed in terms of the remaining one variable.

Taking x_1, x_2, and x_3 as the basic variables, the unit basis will be in terms of the remaining nonbasic variable x_4 as follows:

$$x_2 = 20 - x_4$$
$$x_1 = 50-2^*(20 - x_4) = 10 + 2\ x_4$$
$$x_3 = 40 - x_1 = 40-10 - 2\ x_4 = 30 - 2\ x_4$$
$$y = 3(10 + 2x_4) + 2(20 - x_4) = 70 + 4\ x_4$$

Therefore, the first feasible solution will be $x_4 = 0$, $x_1 = 10$, $x_2 = 20$, and $x_3 = 30$ and for these values, the assessment of objective function $y = 3*10 + 2*20 = 70$.

The new expression for objective function given by equation (iv) above indicates that maximum increase of objective function is possible if the value of x_4 is maximized without making any of the variables x_1, x_2, and x_3 as negative (which is not allowed). It may be seen that the expression (iii) above gives such a maximum value when $x_3 = 0$ giving $2x_4 = 30$, i.e. $x_4 = 15$.

Therefore, taking x_3 (which is $= 0$) as the new nonbasic variable, the new unit basis will be in terms of x_3 as follows:

$$x_1 = 40 - x_3$$
$$x_2 = 25-\tfrac{1}{2}(40 - x_3) = 5 + \tfrac{1}{2}x_3$$
$$x_4 = 20 - x_2 = 20-5-\tfrac{1}{2}x_3 = 15-\tfrac{1}{2}x_3$$
$$y = 3(40 - x_3) + 2\ (5 + \tfrac{1}{2}x_3) = 120 + 10 - 2x_3 = 130 - 2x_3$$

Therefore, the second feasible solution is as follows:

$x_3 = 0$, $x_1 = 40$, $x_2 = 5$, and $x_4 = 15$. The value of objective function for these values of variables is $y = 130$.

Expression (iv) for objective function z above indicates that any increase in value of x_3 beyond 0 will decrease the value of objective function z below 130 given above. Thus, here the maximum possible value of z is 130.

Hence, the optimum production schedule for the company is 40 (i.e. x_1) truck-loads of type A and 5 (i.e. x_2) truckloads of type B concrete mixes, giving the maximum profit (i.e. objective function z) of 130 MU (Monetary Units).

7.6.1.2 Application of Simplex Method in Linear Programming: Illustrative Example II

As explained above, linear programming involves a number of iterations although not larger than of order of M or N, whichever is larger. Even the above small sized linear programming example, problem concerning only two variables indicates that the solution effort needed is quite involved. In real life problems, the number of variables N and the number of constraints M are generally quite large when adoption of the above solution procedure becomes unmanageable. The simplex procedure becomes easier even for a large value of N and M if it is represented through the use of a tableau (which is also easily amenable to computer programming for solution).

Use of such tableau to find out the optimal solution is illustrated taking the following linear programming problem as an example:

$$\text{Objective function-maximize } z = -x_4 + x_5$$

Subject to the constraints:

$$\left.\begin{array}{l} x_1 + x_4 - 2x_5 = 1 \\ x_2 - 2x_4 + x_5 = 2 \\ x_3 + 3x_4 + x_5 = 3 \\ x_1, x_2, x_3, x_4, x_5 \geq 0 \end{array}\right\} \tag{7.9}$$

Here, $N = 5$, $M = 3$, and $N - M = 2$. Therefore, any three variables can be expressed in terms of the remaining two variables. In this procedure, generally, the objective function should be written so as to depend only on right-hand variables. Hence, here right-hand variables are x_4 and x_5, and the left-hand variables are x_1, x_2, and x_3.

Therefore, from expression (7.9) above, we can express x_1, x_2, and x_3 in terms of x_4 and x_5 and thus reduce the system to the unit basis as follows:

$$\left.\begin{array}{l} x_1 = 1 - x_4 + 2x_5 \\ x_2 = 2 + 2x_4 - x_5 \\ x_3 = 3 - 3x_4 - x_5 \end{array}\right\} \tag{7.10}$$

The objective function and the information content in expression (7.10) above can be recorded in the form of a tableau as follows:

		x_4	x_5
Z	0	-1	1
x_1	1	-1	2
x_2	2	2	-1
x_3	3	-3	-1

The first step in the simplex method is to inspect the top row or z-row of the tableau and also examine the entries in columns labelled by right-hand variables or right columns. We are to examine in turn the effect of increasing each right-hand variable from its present value of zero, while keeping the value of all the other right-hand variables at zero. At each stage of examination, we are to assess whether the objective function is increasing or decreasing indicated by the sign of the entry in the z-row. Since, in this case, we want to only increase the value of objective function, only the right-hand columns with positive z-row entries are of interest. In the above tableau, there is only one such column with positive sign z-row entry which is 1.

The second step is to examine the column entries below each z-row entry selected in step one, for the purpose of finding how much we can increase the value of the right-hand variable before one of the left-hand variables attains a negative value which violates the specified non-negativity constraint, and therefore, cannot be allowed. There is no bound on the objective function if all the entries in any right-hand column are positive. If a tableau element at the intersection of a right-hand column and the row of a left-hand variable is positive, then the corresponding left-hand variable will be driven more and more positive. If one or more entries below a positive z-row entry are negative, then we have to select the element in the right-hand column that first limits the increase of that column's right-hand variable fulfilling the above criteria, and this element is called the pivot element. This is done by evaluating repeatedly the increase in the objective function for all possible right-hand columns to find the pivot element which gives the largest such increase fulfilling the non-negativity criteria, as illustrated in the example below. Thus, in the second step, we complete the selection of pivot element.

In the third step, the selected right-hand variable is increased making it a left-hand variable, and simultaneously, alter all the left-hand variables. The corresponding pivot-row element is converted to be a right-hand variable. Thus, at each such exchange, a right-hand variable and a left-hand variable change places, while the problem is maintained in normal form that is equivalent to the original problem, as shown in example below.

In the above example, there is only one right-hand column with positive sign z-row entry which is 1, and below it there are two negative elements in the column: -1 and -1. We divide the corresponding constant column entry (i.e. left most entry) by these numbers, giving the quotients: -2 and -3 and the maximum value of x_5 as: 2 and 3 respectively. If the value of x_5 is taken as 3, it will give the value of x_2 as -1, i.e. negative which is not allowed. Hence, we choose the value of x_5 as 2 and the pivot element will be at the intersection of x_5-column and x_2-row. Thus, x_5 and x_2 will change places making x_5 a left-hand variable and x_2 a right-hand variable.

At first, we solve the pivot row equation for the new left-hand variable x_5 in favour of the old one x_2 as follows:

$$x_2 = 2 + 2x_4 - x_5 \quad \text{or,} \quad x_5 = 2 + 2x_4 - x_2$$

Substituting this into the old z-row, we get:

$$\begin{aligned} z &= -x_4 + x_5 \\ &= -x_4 + 2 + 2x_4 - x_2 \\ &= 2 + x_4 - x_2 \end{aligned}$$

Similarly, substituting into all other left-hand variables, we get the modified left-hand variables as:

$$x_1 = 1 - x_4 + 2x_5 = 1 - x_4 + 2[2 + 2x_4 - x_2]$$
$$= 1 - x_4 + 4 + 4x_4 - 2x_2 = 5 + 3x_4 - 2x_2$$
$$x_3 = 3 - 3x_4 - x_5 = 3 - 3x_4 - [2 + 2x_4 - x_2]$$
$$= 1 - 5x_4 + x_2$$

Using the above equations, a new tableau is formed as follows:

		x_4	x_2
Z	2	1	-1
x_1	5	3	-2
x_3	1	-5	1
x_5	2	2	-1

Here, again since we want to increase the objective function, only the right columns having positive z-row entries are of interest. Here, there is only one such right column with entry of 1. The column entries below this z-row entry of 1 are 3, -5, and 2. Here, the pivot element -5 will give the largest increase in the objective function, and hence chosen as the pivot element. As before, solving this new pivot row equation for the new left-hand variable x_4 in favour of the old one x_3, we get:

$$x_3 = 1 - 5x_4 + x_2 \quad \text{or} \quad x_4 = 1/5 - 1/5x_3 + 1/5x_2$$

Substituting this expression for x_4 into all other left-hand variables, we get the modified left-hand variables as:

$$x_5 = 2 - x_2 + 2[1/5 - 1/5x_3 + 1/5x_2]$$
$$= 2 - x_2 + 2/5 + 2/5x_3 - 2/5x_2$$
$$= 12/5 - 2/5x_3 - 3/5x_2$$
$$x_1 = 5 + 3[1/5 - 1/5x_3 + 1/5x_2] - 2x_2$$
$$= 5 + 3/5 - 3/5\,x_3 + 3/5x_2 - 2x_2$$
$$= 27/5 - 3/5x_3 - 7/5x_2$$

Substituting the above values into the old z-row, we get:

$$z = -1/5 + 1/5x_3 - 1/5x_2 + 12/5 - 2/5x_3 - 3/5x_2$$
$$= 11/5 - 1/5x_3 - 4/5x_2$$

Using the above equations, we get a new tableau as follows, where the variables x_3 and x_4 change places:

		x_3	x_2
Z	11/5	−1/5	−4/5
x_4	1/5	−1/5	1/5
x_5	12/5	−2/5	−3/5
x_1	27/3	−3/5	−7/5

In the fourth step, we go back to the first step and repeat the process searching for another possible increase in the objective function. This process is repeated as many times as required until all the right-hand entries in the z-row are negative, indicating that no further increase in the objective function is possible. In the current example, it may be seen from the above tableau that all the right-hand entries in the z-row are negative indicating that the maximum value of z is reached.

Thus, the solution of the problem can be read from the constant column of the final tableau as above. Here, we can see that the objective function z is maximized to a value of 11/5, with values of other variables as: $x_1 = 27/5$, $x_2 = 0$, $x_3 = 0$, $x_4 = 1/5$, $x_5 = 12/5$.

7.6.2 Application of Linear Programming in Building Floor Plan Synthesis Problems

Use of linear programming has been attempted in building floor plan synthesis problems (Mitchell, 1977). This is applied in floor planning of a mobile trailer unit. In the above problem, it is recognized that the floor plans involve areas, i.e. products of lengths and widths, which are nonlinear quantities and therefore, linear programming technique cannot be applied. This limitation is overcome by assuming known values of some dimensions, as given for the overall width of the trailer. The objective function of the linear programming problem is assumed as minimizing the length of the trailer, subject to the given dimensional constraints of upper and/or lower limits of values of the length, width, and so on, of various rooms. The optimum dimensions of the rooms and the minimum length of the trailer are found using standard linear programming system.

Optimization of real life building floor plans, physical layout plans, and dwelling layout systems, and so on will involve area constraints, cost constraints, and so on, which are important objectives but most of them are nonlinear functions and therefore, linear programming cannot be applied. It is necessary to use nonlinear programming techniques to solve such optimization problems which are attempted in subsequent chapters of the book. The concept of nonlinear programming is discussed and elaborated below.

7.7 Nonlinear Programming Techniques and Some Basic Concepts

In the preceding section, optimization problems with linear objective function and constraints are dealt with using linear programming techniques. However, when the objective function and constraint equations are nonlinear above, linear programming technique cannot be applied, and to solve such problems, nonlinear programming techniques have to be applied.

The general optimization model shown earlier in expression (7.2) can be represented graphically, say, for a two-dimensional nonlinear optimization problem as shown in Fig. 7.1 below. Here, the objective function $y = f(x_1, x_2)$ is evaluated for different values of x_1 and x_2 to yield the contours of g_3, g_2, g_1 for the two-dimensional nonlinear optimization problem as shown in Fig. 7.1 below, which displays a minimum point g_0 also. By changing the sign of the objective function 'y', this point becomes a maximum.

Although the subject area of nonlinear programming is under constant development, it is useful to categorize nonlinear programming problems and the techniques to solve them into two broad sections, namely: (1) classical optimization techniques which seeks optimal solutions by solving systems of equations, (2) search techniques. Again, classical optimization techniques may cover the following broad topics (Stark & Nichols, 1972):

1. Differential Calculus or Differential Approach
2. Lagrange Multiplier
3. Geometric Programming

The differential calculus can be applied in classical optimization problems as an indirect method which reduces the original optimization problem to the solution of simultaneous equations usually nonlinear. If such equations are too difficult to solve, other methods need to be used. Lagrange multiplier technique is useful in problems

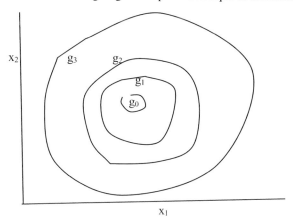

Fig. 7.1 Minimum and the contours of two-dimensional nonlinear optimization problem

with several equality constraints which are nonlinear. Geometric programming is relatively new technique, and differs from other optimization techniques in that instead of searching first for the optimal solution vector and then determining the resultant relative magnitudes of the terms in the objective function of the optimization problem, it searches first for the optimal cost distribution and then determines the resulting values of the solution variables. Before describing these optimization techniques in more details, the significance of some useful concepts in optimization problems is outlined below.

7.7.1 Significance of Concave and Convex Functions in Optimization

In certain cases of optimization, any solution identified as locally optimal is also a global optimal solution. Hence, it is very important to examine such cases to find that this property is true, and thus, enhance the utility of the nonlinear programming algorithmic procedures. A special type of function which is extremely useful in optimization is a *concave* or *convex* function. A *convex* function is defined mathematically as follows (Phillips & Beightler, 1976):

Given any two points in a N-dimensional space $(\hat{x}, \ \hat{\underline{x}})$, if the following inequality holds at all pairs of points, the function f is said to be *convex* (Phillips & Beightler, 1976):

$$f((1-\varnothing)\,\hat{x} + \varnothing\underline{\hat{x}}) \le (1-\varnothing)\,f(\hat{x}) + \varnothing f(\hat{x})$$
$$0 \le \varnothing \le 1$$

Similarly, a concave function is defined mathematically with the above inequality sign reversed.

Considering the calculus point of view, a single variable function $z = f(x)$ is said to be:

1. Convex when the second derivative of the function is positive, i.e. when $\frac{\delta^2 z}{\delta x^2} \ge 0$,
2. Concave when the second derivative of the function is negative, i.e. when $\frac{\delta^2 z}{\delta x^2} \le 0$, and.
3. Linear when it is zero in the interval of interest.

Geometrically, the definition of convex and concave function means that if a function is convex, then a line, drawn between any two points on the surface of the function, will lie entirely above that function. Similarly, if a function is concave, then a line drawn between any two points on the surface of the function will lie entirely below that function. The above points mean that (Stark & Nichols, 1972):

1. The convex or concave function definition is not dependent upon the function being continuous.

2. A function can be convex over one region and concave over another region.
3. A linear function is both convex and concave.

Thus, if a stationary point (defined below) is found for a convex (concave) function, it is guaranteed that the point is a global minimum (maximum), which is a very useful concept to simplify solution of real life nonlinear optimization problems.

If we take a two variable function $f(x_1, x_2)$, it is *convex* if the determinant of the following matrix of partial derivatives (which is called Hessian Matrix 'H') and the pure derivatives of the function as shown are nonnegative, i.e. *positive* (Stark & Nichols, 1972):

$$\begin{pmatrix} \dfrac{\partial^2 f(x_1,\ x_2)}{\partial x_1^2} & \dfrac{\partial^2 f(x_1,\ x_2)}{\partial x_1 \partial x_2} \\ \dfrac{\partial^2 f(x_1,\ x_2)}{\partial x_2 \partial x_1} & \dfrac{\partial^2 f\ (x_1,\ x_2)}{\partial x_2^2} \end{pmatrix} / 0 \tag{7.11}$$

and $\dfrac{\partial^2 f(x_1,\ x_2)}{\partial x_1^2} / 0$ and also $\dfrac{\partial^2 f(x_1,\ x_2)}{\partial x_2^2} / 0$.

If the pure derivatives are **nonpositive**, i.e. if $\dfrac{\partial^2 f(x_1,\ x_2)}{\partial x_1^2} \leq 0$ and $\dfrac{\partial^2 f(x_1,\ x_2)}{\partial x_2^2} \leq 0$, the function $f(x_1, x_2)$ is concave.

If we take a function of several variables, i.e. $f(x_1, x_2, x_3 \ldots x_n)$, it is *convex* if the determinant of the '$n \times n$' Hessian Matrix 'H', (which consists of the second partial derivative—here subscripts of f: $x_1 x_1$, $x_1 x_2$, and so on, denote the respective partial derivatives—of the function) is *nonnegative*, as indicated below (Stark & Nichols, 1972):

$$H = \begin{pmatrix} f_{X_1 X_1} & f_{X_1 X_2} & f_{X_1 X_3} \cdots f_{X_1 X_n} \\ f_{X_2 X_1} & f_{X_2 X_2} & f_{X_2 X_3} \cdots f_{X_2 X_n} \\ & \cdots & \\ f_{X_n X_1} & f_{X_n X_2} & f_{X_n X_3} \cdots f_{X_n X_n} \end{pmatrix} / 0 \tag{7.12}$$

More description of this subject is available in literature, specifically in Beightler et al. (1979) and Stark and Nichols (1972). The very brief discussion presented above and below is similar to theirs which is offered only with the purpose of acquainting readers with this topic.

7.7.2 Necessary and Sufficient Conditions for Optimality

Vanishing of the first derivative of a function at a point is a necessary condition for the point to be a local minimum, but not sufficient condition for local minimum, as the tangent at a point could also be horizontal at a maximum or at an inflection point. A gradient is defined by the vector of first partial derivatives evaluated at a particular point in the solution space. Any point where the gradient vanishes is called a *stationary point* which is denoted by $\underline{x}^0$ and the *optimum point* is denoted by $\underline{x}^*$. Figure 7.2 shows the minimum of a single variable (one dimension) function where $x^* = \underline{x}^0$. By changing the sign of the objective function 'y', the point becomes a maximum. A valley (ridge) is formed by a set of adjacent local minima (maxima) of a single variable function as shown in Fig. 7.3. This situation arises when minimum of 'y' is satisfied for some $x \neq \underline{x}^*$ in the neighbourhood of $\underline{x}^*$.

Figure 7.4 above shows saddle points: x^0 of a single variable function, where $\frac{\partial y}{\partial x_n} = 0$ for some points $\underline{x}$ in every feasible neighbourhood of the stationary point $\underline{\boldsymbol{x}}^{\boldsymbol{0}}$. Here, any stationary point is denoted by $\underline{\boldsymbol{x}}^{\boldsymbol{0}}$ until it is definitely proved to be an optimum, in which case it is denoted as $\underline{x}^*$.

Taking a single variable hypothetical function $y = f(x)$, some of its critical points (i.e. *stationary points*) such as global maximum, valley, local maximum, local minimum, and global minimum can be represented as shown in Fig. 7.5 below:

Fig. 7.2 Minima of one variable function

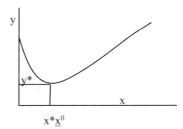

Fig. 7.3 Valley of one variable function

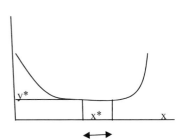

Fig. 7.4 Saddle points of one variable function

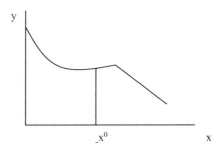

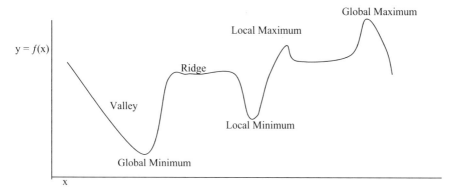

Fig. 7.5   Some stationary points of a single variable hypothetical function $y = f(x)$

As already stated above, a stationary point denoted by x^0 is any point where the gradient vanishes, i.e. $dy/dx = 0$ in the case of single variable function. However, to prove its optimality, higher derivatives have to be analysed. If the second derivative at x^0 is positive, the stationary point is at least a local minimum, and if it is negative, the point is at least a local maximum. If the second derivative vanishes and changes sign about the stationary point, it is a point of inflection. If the second derivative vanishes but does not change sign, it corresponds to valleys and ridges. A global minimum is the smallest of all local minimums, and a global maximum is the largest of all local maximums.

An important property of the function is that if $y = f(x)$ is known to be convex, any point x^0, for which the first derivative is zero, is a global minimum, and if it is concave and the first derivative vanishes, it is a global maximum. Thus, in this case the proviso that the first derivative vanishes is both *necessary and sufficient conditions* for a *global minimum* when the function is *convex*, and when the function is known to be *concave* the above proviso is both *necessary and sufficient conditions* for a *global maximum*.

7.8   Use of Lagrange Multiplier Approach to the Optimization of Functions of Several Variables

In case of a general optimization problems given by expression (7.1) above, if there are several nonlinear equality constraints $g_j(x_1, x_2, x_3, \ldots x_n)$, it becomes difficult to solve for the variables explicitly. In such cases, the **Lagrange Multiplier Technique** is very useful. Here, an attempt is made only to explain this concept in a simplified manner without going into the details of intricate theoretical presentation which is available in literature. In this procedure, each of the constraints $i(1, 2, 3 \ldots M)$ are multiplied by $\lambda_1, \lambda_2, \lambda_3 \ldots \lambda_M$, respectively, and added to the objective function. Thus, the problem is reduced to solving M+ N simultaneous equations in the $M + N$ unknowns (i.e. M unknown's $\lambda_k + N$ unknowns' x_n). Symmetric Lagrange formulation sometimes exposes an advantageous order of solution unlike other methods and in many cases, it is easier to solve such $M + N$ simultaneous equations incorporating λ_k which are originally conceived as: 'undetermined multipliers', but can be conceived as *sensitivity coefficients* into the solution (Beightler et al., 1979). Thus, in this case the objective function, which is often called the Lagrangian and denoted by L, can be formed as follows (Stark & Nichols, 1972):

$$L = f(x_1, x_2, x_3, \ldots x_n) + \sum_{I=1}^{M} \lambda_i g_i (x_1, x_2, x_3, \ldots x_n) \qquad (7.13)$$

Here, L is considered as an objective function in $M + N$ variables and to solve it we set the derivatives equal to zero and get the following $M + N$ equations:

$$\frac{\partial L}{\partial X_J} = \frac{\partial Z}{\partial X_J} + \sum_{I=1}^{M} \lambda_I \frac{\partial g_I}{\partial X_J} = 0 \quad j = 1,2,3 \ldots N$$

and

$$\frac{\partial L}{\partial \lambda_i} = 0 \qquad i = 1,2,3 \ldots M$$

Solving these simultaneous equations in $(N + M)$ unknown variables, we can get the solution.

7.8.1 Application of Lagrange Multiplier Technique: An Example Problem for Optimal Design of a Box-Container to Carry Materials in a Construction Site

To illustrate the application of Lagrange multiplier technique in optimization, we can take the following example:

A construction engineer desires to design a box-container which is to be used to carry construction materials with a capacity of 20 cu.m. which is to be square shape with suitable height so as to minimize its cost. Manufacturing costs of the container will be 25 Monetary Units (MU) per sq.m. of its vertical surface area as well as the top and bottom surface of the container. Objective is to determine the length, width, and height of the container so as to minimize the cost while fulfilling the capacity requirement of 20 cu.m. as cited above.

Here, let us assume x_1 is the length of the container which will also be the width of the container which is to be of square in shape. Let us also assume the height of the container as x_2 and since the volume of the container is specified as 20 cu.m, we get $x_1^{2*}x_2 = 20$.

Here, the objective function y is the total cost, and the expression for the total cost of the container will be given by,

$$y = 2*25* \, x_1^2 + 25*4*x_1*x_2$$

or $y = 50*x_1^2 + 100*x_1*x_2$.

Thus, the optimization problem can be formulated as:

$$\text{Minimize } y = 50*x_1^2 + 100*x_1*x_2$$

Subject to

$$x_1^{2*}x_2 = 20$$

Here, the Lagrangian can be expressed as:

$$L = 50*x_1^2 + 100*x_1*x_2 + \lambda^* \left[x_1^{2*}x_2 - 20 \right]$$

By differentiating, we get:
From (ii) above, we get $x_1 = -100/\lambda$

$$\frac{\partial L}{\partial x_1} = 100x_1 + 100x_2 + 2\lambda x_1 x_2 = 0 \quad (i)$$

$$\frac{\partial L}{\partial x_2} = 100x_1 + \lambda x_1^2 = 0 \quad (ii)$$

$$\frac{\partial L}{\partial \lambda} = \left[x_1^{2*}x_2 - 20 \right] = 0 \quad (iii)$$

From (iii) above, we get $x_2 = \lambda^2 \, 20/10{,}000 = \lambda^2/500$
From (i) above, we get

$$100*(-100/\lambda) + 100*\lambda^2/500 + 2\lambda \, (-100/\lambda)\lambda^2/500 = 0$$

Or $100 - \lambda^3/500 + 2\lambda^3/500 = 0$
Or $50000 - \lambda^3 + 2\lambda^3 = 0$
Therefore, $\lambda^3 = -50{,}000$
Or, $\lambda = -36.74$
Therefore, $x_1 = -100/-36.74 = 2.714$ m

$$x_2 = \lambda^2/500 = 1357.21/500 = 2.714 \text{ m}$$

The value of the objective function giving the minimum cost $= 50*$ $(2.714)^2 + 100*(2.714)^2 = 1105.21$ MU.

It may be seen that the result is same as given by application of geometric programming technique in the same problem presented later. Additional factor λ that is the **Lagrange Multiplier** derived here can be interpreted as a *sensitivity coefficient* (Beightler et al., 1979) for cost of the container, that is, the additional cost which would be incurred for the unit increase in the volume of the container.

The above simple example of only two variables and one constraint is taken merely to demonstrate the application of Lagrange Multiplier Technique in nonlinear optimization, and does not have much advantage compared to other methods of solution. However, Lagrange's method is more useful in other complicated situations (Beightler et al., 1979), say, involving more variables and constraints.

7.9 Kuhn–Tucker Conditions for Optimality

Although Lagrangian method can be used to solve constrained nonlinear programming problem, the presence of complimentary slackness conditions, which state that for each constraint, either the slack variable or the Lagrange multiplier (or both) must be zero, leads to a trial and error solution involving search over a total of 2^M possible policies where M is the number of constraints (Phillips & Beightler, 1976). Kuhn–Tucker conditions provide clues to the nature of the optimal solutions and enable an analyst to check the optimality of a proposed solution.

Description of this subject is available in literature, specifically in Phillips and Beightler (1976) and the very brief discussion presented here is similar to theirs which is offered only with the purpose of acquainting readers with the Kuhn–Tucker theory in relation to optimization problems.

$$\text{Minimize } y = f(x)$$

If we consider the following general nonlinear programming problem:
Subject to:

$$\left. \begin{array}{ll} h_i(\underline{x}) = 0; & i = 1, 2, 3 \ldots M \\ g_i(\underline{x})/0; & i = M + 1, \ldots P \end{array} \right\}$$

if x^* is a solution to the problem, and the functions $f(\underline{x})$, $h_i(\underline{x})$, and $g_i(\underline{x})$ are once differentiable, then there exists a set of vectors λ^* and ν^* such that $\underline{x}^*$, ν^*, and λ^* satisfy the following relations:

$$h_i(\underline{x}) = 0; \quad i = 1, 2, 3 \ldots M \tag{7.14}$$

$$g_i(\underline{x})/0 \tag{7.15}$$

$$\nu_i[g_i(\underline{x})] = 0 \quad i = M + 1, \ldots P \tag{7.16}$$

$$\nu_i/0 \tag{7.17}$$

$$\frac{\partial f(x)}{\partial x_k} + \sum_{i=1}^{M} \lambda_i \left(\frac{\partial h_i(x)}{\partial x_k} \right) - \sum_{i=M+1}^{P} \nu_i \left(\frac{\partial g_i(x)}{\partial x_k} \right) = 0; \quad k = 1,2,3 \ldots N$$

$$(7.18)$$

The relationships represented by the Eqs. (7.14)–(7.18) are known as the **Kuhn–Tucker Conditions**. When these conditions are satisfied, these are necessary conditions for an optimal solution of the above nonlinear programming problem. It may be noted that the Eqs. (7.14) and (7.15) specify the primal feasibility, Eq. (7.16) is complementary slackness conditions, Eq. (7.17) is non-negativity conditions on the dual variables ν_i which corresponds to the Lagrangian multipliers discussed earlier.

7.9.1   Kuhn–Tucker sufficiency conditions for optimality

Let the objective function $f(x)$ be convex, $g_i(x)$ be concave for all $i = M + 1, \ldots P$, and $h_i(x)$ be linear for all $i = 1, 2, 3 \ldots M$. if there exists a solution ($\underline{x}^*$, $\underline{v}^*$, and $\underline{\lambda}^*$) satisfying the **Kuhn–Tucker Conditions,** then $\underline{x}^*$ is an optimal solution to the nonlinear programming problem presented above.

7.10   Differential Approach and Classical Optimization

The general optimization model presented in expression (7.1) above becomes a classical optimization problem when the objective function and all constraints are continuous and possess partial derivatives at least up to second order, and the constraints are equations whose number does not exceed that of the number of independent variables, and also the non-negativity constraints on the variables is removed (Stark & Nichols, 1972). Differential approach to optimization of function of single independent variable and to optimization of function of several independent variables is outlined below with application examples.

7.10.1   Functions of Single Variable: Application Example of Differential Approach to Architectural Design Optimization

If we take a function of single variable $y = f(x)$ having a stationary point x^0, then a necessary condition for x^0 to be optimum is that dy/dx at x^0 must vanish, i.e. $dy/dx = 0$. This corresponds to a horizontal tangent at this point. Figure 7.6 above shows a number of such critical or stationary points for a hypothetical function. It is necessary to derive higher derivatives to know the nature of optimization of these

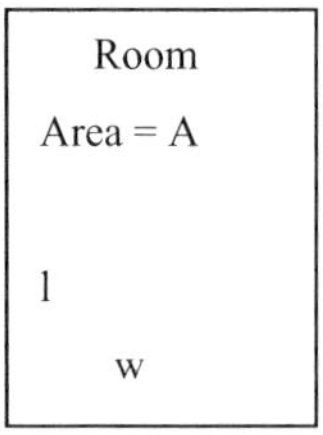

Fig. 7.6 A typical
detached room

```
┌─────────────┐
│   Room      │
│   Area = A  │
│             │
│ l           │
│      w      │
└─────────────┘
```

points. Thus, if the second derivative is positive, i.e. $\frac{\delta^2 f}{\delta x^2} \geq 0$, the function is *convex* giving a *global minimum*, and on the other hand, if the second derivative is negative i.e. $\frac{\delta^2 f}{\delta x^2} \leq 0$, the function is *concave* giving a *global maximum*.

We can take a simple example of a room in a building design optimization where the designer may like to find the optimum proportion of a rectilinear shape room of fixed area so as to minimize the length of the wall enclosing the room as shown in Fig. 7.6 below:

We can assume y or L as the total length of walls enclosing the room, A as the fixed area of the room, w as the width of the room, l as the length of the room, and R as the room proportion, i.e. length-width ratio of the room ($= l/w$). We can also assume sx and sy as the value of wall sharing coefficients of the room in the X and Y direction, respectively. The relationship between the above variables can be expressed as:

$$R = l/w,\ l = R^*w,\ A = l^*w = w^*R^2$$

Therefore, $w = (A/R)^{1/2}$, and length of the room $l = R^*w = (A^*R)^{1/2}$.

The value of wall sharing coefficients $sx = sy = 1$ in case of a detached room (HUDCO, 1982) adopted above. Obviously, the total length of walls L enclosing the room can be expressed in terms of A (fixed) and R (variable) as follows, where L is a function of a single variable R while A is fixed:

$$L = 2^* \left[(A/R)^{1/2} + (A * R)^{1/2} \right], \quad \text{or} \quad L = 2^* A^{1/2*} \left(R^{-1/2} + R^{1/2} \right)$$

Differentiating with respect to R and putting it to zero, we get:

$$\frac{dL}{dR} = 2^* A^{1/2*} \left(-1/2^* R^{-3/2} + 1/2^* R^{-1/2} \right) = 0$$

Therefore, $1/2 * R^{-1/2} = 1/2 * R^{-3/2}$.

Dividing both sides by $1/2 * R^{-3/2}$, we get $R = 1$.

As stated above, a necessary condition for a stationary point x^0 to be optimum is that dy/dx at x^0 must vanish, i.e. $dy/dx = 0$. It is necessary to derive higher derivatives, i.e. $\frac{\delta^2 f}{\delta x^2}$ to know the nature of optimization of the stationary point. Here, the stationary point $x^0 = R_0 = (1)$. To know the nature of optimality of this

stationary point, we have to find the value of second derivative $\frac{\delta^2 L}{\delta R^2}$ at the above
stationary point which is $= 2^* A^{1/2*} \left(3/4^* R_0^{-5/2} - 1/4^* R_0^{-3/2} \right) > 0$. Since, here the
second derivative is positive, i.e. $\frac{\delta^2 L}{\delta R^2} \geq 0$, the function is *convex* giving a *global
minimum*.

Thus, the above results show that optimum room ratio to give minimum length of
wall enclosing a fully detached room is '1', i.e. in this case, a square room gives the
minimum length of wall for a given room area. The value of this optimum room ratio
R will vary depending on the wall sharing options and the consequent value of wall
sharing coefficients in each case, which are '1' in both X and Y direction in the
above case. By adopting similar procedure, it can be shown that the value of
optimum room ratio R can be expressed as: $R = sx/sy$, where sx and sy are the
values of wall sharing coefficients in the X and Y direction of the room. Thus, in case
of semi-detached room, the value of $sx = 2$ and $sy = 1.5$ and, therefore, optimum
room ratio $R = 2/1.5 = 1.33$. Similarly, in case of row-housing room the value of
$sx = 2$ and $sy = 1$ and therefore, optimum room ratio $R = 2/1 = 2$. Adopting this
concept a building consisting of a number of rooms could be optimized which is
elaborated in Chap. 9 of this book.

7.10.2   Functions of Several Variables: Application Example of Differential Approach to Optimization of Function of Several Variables

In case of functions of several variables, i.e. $f(x_1, x_2, x_3, \ldots x_n)$, a necessary
condition for the stationary points $x_{10}, x_{20}, x_{30}, \ldots x_{n0}$ to be an optimum is that:

$$\frac{\partial f}{\partial x_j} = 0 \quad \text{for} \quad j = 1,2,3 \ldots n \text{ at } [x_{10}, x_{20}, x_{30} \ldots x_{n0}]$$

Solutions of these N equations in N variables give the *stationary points* as in case
of a single variable function. Similarly, using the expression: (7.12) above, we can
find the determinant of the Hessian Matrix of partial second derivatives, and if it is
positive definite, i.e. $H > 0$, and if the pure derivatives are positive, the optimum is at
least a local minimum. Similarly, in case the determinant of the Hessian Matrix of
partial second derivatives is negative definite, i.e. $H < 0$, and if the pure derivatives
are negative, the optimum is at least a local maximum. Such analytic considerations
are available in literature (Hadley, 1962). The local optima are to be examined to
select the global optimum from such local optima. As discussed above if the function
is known to be strictly convex or concave, a sufficient condition for global optimum
is established.

To illustrate the application of differential approach to optimization of functions
of several variables, we can take an example function of two independent variables
$z = f(x_1, x_2)$ as shown below:

$$z = x_1^3 + x_2^3 - 15\,x_1\,x_2$$

As stated earlier, the points where the gradient vanishes or the partial derivatives are equal to zero are called *stationary points*. Assuming $S_0(x_{10}, x_{20})$ to be a stationary point of the above function $f(x_1, x_2)$, we can find the stationary point by making use of the necessary conditions for optimization, that is, by setting the partial derivatives of the above function equal to zero as follows and get two equations in two variables:

$$\frac{\partial z}{\partial x_1} = 3x_1^2 - 15x_2 = 0 \quad \text{(i)}$$

$$\frac{\partial z}{\partial x_2} = 3x_2^2 - 15x_1 = 0 \quad \text{(ii)}$$

From (i), we get $x_2 = x_1^2/5$.

Substituting the above value of x_2 in (ii), we get $x_1^3 = 125$ or, $x_1 = 5$ and $x_2 = x_1^2/5 = 25/5 = 5$.

Thus, we get a stationary point: $S_0(x_{10}, x_{20})$ of the above function as: $S_0(5, 5)$.

Now to examine the optimality of the above stationary point, we can simplify the expression (7.11) by designating A, B, C, and D as follows:

$$A = \frac{\partial^2 f(x_{10}, x_{20})}{\partial x_{10}^2}, \quad B = \frac{\partial^2 f(x_{10}, x_{20})}{\partial x_{10}\partial x_{20}}, \quad C = \frac{\partial^2 f(x_{10}, x_{20})}{\partial x_{20}\partial x_{10}}, \quad D = \frac{\partial^2 f(x_{10}, x_{20})}{\partial x_{20}^2}$$

Thus, the Hessian Matrix can be expressed as:

$$H = \begin{pmatrix} A & B \\ C & D \end{pmatrix} \quad \text{and its determinant } \mathrm{DET} = A^*D - B^*C$$

With these designations, the conditions for optimality become:

(a) If $\mathrm{DET} > 0$, then the function is optimum at the point S_0, which is:

   (i) *minimum* for $A > 0$ or $D > 0$
   (ii) *maximum* for $A < 0$ or $D < 0$

(b) If $\mathrm{DET} < 0$, then there is no optimum at the point S_0.

Now we find the second order partial derivatives at the point S_0 as follows:

$$A = \frac{\partial^2 f(x_{10}, x_{20})}{\partial x_{10}^2} = 6x_{10} = 6^*5 = 30$$

$$B = \frac{\partial^2 f(x_{10}, x_{20})}{\partial x_{10} \partial x_{20}} = -15$$

$$C = \frac{\partial^2 f(x_{10}, x_{20})}{\partial x_{20} \partial x_{10}} = -15$$

$$D = \frac{\partial^2 f(x_{10}, x_{20})}{\partial x_{20}^2} = 6^*5 = 30$$

Using the above values, we derive the determinant of the Hessian Matrix: H, i.e. DET as follows:

$$\text{DET} = A^*D - B^*C = 30^*30 - (-15) * (-15) = 900 - 225 = 675$$

Here, the determinant (= DET) of the Hessian Matrix: $H = 675 > 0$, and both the pure derivatives $A = 30 > 0$ and also $D = 30 > 0$, and consequently, the given function is *convex* and is *minimum* at the stationary point $S(5, 5)$.

The value of the function at this point is $z_{min} = x_1^3 + x_2^3 - 15 x_1 x_2 = 125 + 125 - 15^*5^*5 = -125$.

7.11   Unconstrained and Constrained Nonlinear Optimization

Earlier the variables and constraints were broadly explained. Here, the significance of constraints in optimization problems is outlined with example problems. Unconstrained optimization signifies an optimization problem where there are no constraints and refers to the technique of finding optimum solution in this condition. Unconstrained optimization is very rare in practical design optimization problems. However, many constrained design optimization problems can be converted to unconstrained optimization problem by suitable manipulation and optimum solution found by simpler optimization techniques, as unconstrained optimization is far simpler compared to constrained optimization. This is outlined below with an example problem of optimization of FAR (Floor Area Ratio) in a dwelling-layout system, where a constrained problem converted to a single variable problem and solved graphically.

7.11.1   *Illustrative Examples of Constrained and Unconstrained Optimization in Urban-Built-Form Design*

A model for *Optimal* Design of Urban-Built-Form in a '*planning area*' linking built-space (BUS), open-space (SPO), Building Rise (BR), land price, and construction

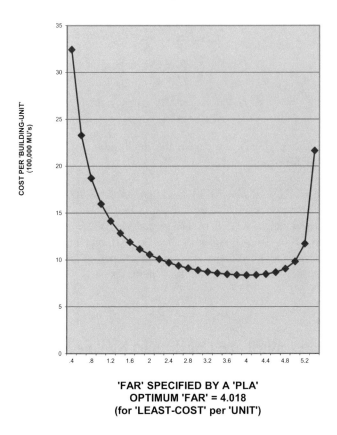

'FAR' SPECIFIED BY A 'PLA'
OPTIMUM 'FAR' = 4.018
(for 'LEAST-COST' per 'UNIT')

◆— Land Price=10,000 MU per Sq.M.

Fig. 7.7 Variation in cost per 'building-unit' with 'far' (same built and open space per 'unit')

costs, can be developed as follows (its geometric explanation presented later below) (Fig. 7.7):

$$\text{Minimize the built-unit-cost} = y(x) = c_1\,x_1 + c_2\,x_1\,x_4 + c_3\,x_1\,x_3^{-1} \qquad (7.19)$$

Subject to:

$$x_1^{-1}\,x_2\,x_3 + x_3\,x_4^{-1} \le 1 \qquad (7.20)$$

(FAR, built-unit-space, open space, and building rise compatibility *constraint*)

$$a_1 \, x_1^{-1} \leq 1 \qquad\qquad (7.21)$$

(built-unit-space *constraint*)

$$a_2 \, x_2^{-1} \leq 1 \qquad\qquad (7.22)$$

(open space *constraint*)
  Here,

$x_1 = optimum$ space in sq.m. per built-unit (BU, i.e. dwelling unit, commercial unit, so on) in *planning area*,

$x_2 = optimum$ semi-public open space (SPO), i.e. open-to-sky space in SQ.M/BU in the *planning area*,

$x_3 =$ a variable giving the *optimum* FAR, i.e. the ratio between the total built-space on all floors of an urban development and its land area,

$x_4 =$ variable giving *optimum* building rise (effective), i.e. number of floors in the *planning area*,

$c_1 =$ cost coefficient relating building construction cost per unit of plinth area (Monetary Units or MU/sq.m.),

$c_2 =$ a cost coefficient related to building rise, in terms of the increased cost per unit of plinth area (monetary units/m^2) in the urban development. This factor can also be used as a *social-cost-coefficient* (say, a *Penalty Factor* for High Building Rise determined by the **Society** *uniformly*, instead of *arbitrary imposition from top*, for determining the **Optimal Building-Rise** (effective) based on the ***social-cost-benefit-analysis*** *principles*, in a *site-specific-case*,

$c_3 =$ a cost coefficient related to land price per unit of land area (Monetary Units or MU/sq.m.),

$a_1 =$ the desired built-unit space in sq.m. per built-unit in the urban development,

$a_2 =$ the desired semi-public open space in sq.m. per built-unit in the urban development.

  Input values in the above Table CASE are as follows:

$$x_1 = 110.0 \text{ Sq.M./BU} \quad x_2 = 20.0 \text{ Sq.M./BU} \quad K = 0.01$$
$$c_3 = 10{,}000 \text{ MU/Sq.M.} \quad c_1 = 4500 \text{ MU/Sq.M.} \quad c_2 = 4500^*0.01 = 45.$$

  The above tabular results are also shown **graphically** in Fig. 7.8 above, *clearly depicting that it is a case of optimization*, from where the **Optimal** FAR (i.e. x_3), giving the **Least-Value** of **BUC** (Built-Unit-Cost), can be read.

  The model for **Optimal** Design of Urban-Built-Form defined by expressions (7.19)–(7.22) above is a typical constrained *Nonlinear Programming* (NLP) problem and requires sophisticated NLP techniques such as geometric programming, which is very relevant and has a number of other advantages as discussed later, to solve and find the optimum FAR as shown later. However, this problem can also be converted to a simple one variable unconstrained optimization problem by removing

Fig. 7.8 A schematic urban-built-form (single-building-block)

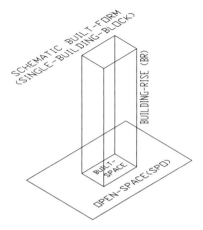

the constraints and replacing the constrained values of other variables into the one variable unconstrained objective function as shown below.

Now, if we assume constant values for variables x_1 and x_2 along with suitable assignment of values for model cost coefficients, the above constrained optimization problem is converted to a simple one variable (i.e. x_3) unconstrained optimization problem and can be solved by computer graphics to find optimum FAR, i.e. x_3 as outlined below.

Here, we assume the values of c_1 = 4500 MU/SQ.M, c_2 = $K * c_1$ (here $K = 0.01$) = 45, and c_3 = 10,000 MU/SQ.M. and also the values x_1 = 110 SQ.M/BU, x_2 = 20 SQ.M/BU. Thus, we get the expression for the built-unit-cost = $y(x)$ in terms of only one variable x_3, i.e. FAR making it a one variable unconstrained optimization problem as follows:

$$y(x) = 4500^*110 + 45^*110^*(x_3^*110^*/(110 - x_3^*20)) + 10000^*110^*x_3^{-1}$$

Now, using the above equation and by changing the value of FAR, i.e. x_3 we can get the value of the corresponding Built-Unit-Cost (BUC) = $y(x)$ and thus, find graphically the value of FAR giving the Least BUC which is the optimum FAR. The results are shown in Table 7.1 above.

It may be seen that BUC goes on reducing till FAR = 4.0, and thereafter, BUC increases to 746,377 MU at FAR = 4.25 and it further increases to 977,250 MU at FAR = 5.0. Thus, although the Built-Unit-Cost (BUC) varies from 1,601,050 MU to 977,250 MU for different FARs, a graph plotted with the above results will indicate that the BUC reaches a *Minimum* 742,600 MU at an *Optimum* FAR of around 4.0 which can be found approximately using such *computer graphics* without resorting to complicated mathematical programming techniques. However if an exact *Optimum* FAR value (which is 4.017 in this case) for a general case is desired it can be found using *Geometric Programming Techniques* as shown later.

Table 7.1 Sensitivity of built-unit-cost to the floor area ratio (FAR, i.e. x_3)

FAR (x_3)	Built-Unit-Cost (BUC) (MU)	GPAR (MU/Sq.M.)
1.000	1,601,050	14,555
1.250	1,373,007	12,573
1.500	1,237,543	11,259
1.750	1,136,276	10,330
2.00	1,060,557	96,413
2.25	1,002,737	9116
2.50	957,677	7706
2.75	22,225	7374
3.00	794,337	7130
3.25	772,777	7934
3.50	756,929	7790
3.75	746,673	7697
4.00	**742,600**	**7660**
4.25	746,377	7694
4.50	761,957	7736
4.75	799,004	7173
5.00	977,250	7975

7.12   Geometric Programming

Geometric programming is developed as a powerful tool for solving nonlinear programming problems in general and engineering design problems in particular which sometimes consists of a sum of component costs and could be minimized by inspection under suitable conditions as explained later. Geometric programming first conceived by Zener in 1961 and its mathematical base was provided by Duffin by application of arithmetic–geometric mean inequality relationship between sums and products of positive numbers, which prompted Duffin to call this technique 'geometric programming'. Many problems of optimization in AEC sector also consists of a sum of component costs and are amenable to easy solution by this technique of 'geometric programming', and therefore, this technique has been adopted extensively in this book for solving various nonlinear programming optimization problems. In view of this application orientation, this section of the book outlines only main ideas and some concepts of geometric programming and also giving some compact formulations; patterned after Beightler et al. (1979), Phillips and Beightler (1976) and Stark and Nichols (1972) without proof and with minor symbolic details. Readers interested in more details of the development of these formulations may find them in Beightler et al. (1979) and Phillips and Beightler (1976) and other publications available in literature.

7.12.1   An Illustrative Example of Optimization to Explain Some Concepts in Geometric Programming

In order to explain the basic underlying principles of geometric programming, we can take the simplified example of optimal design of a container to carry materials in a construction site as presented in Sect. 7.7.1 above and reproduced below for ready reference.

A construction engineer desires to design a container which is to be used to carry construction materials with a capacity of 20 cu.m. which is to be square shape with suitable height so as to minimize its cost. Manufacturing costs of the container will be 25 Monetary Units (MU) per sq.m. of its vertical surface area as well as the top and bottom surface of the container. Objective is to determine the length, width, and height of the container so as to minimize the cost while fulfilling the capacity requirement of 20 cu.m. as cited above.

As before, let us assume x_1 is the length of the container which will also be the width of the container which is to be of square in shape. Let us also assume the height of the container as x_2, and since the volume of the container is specified as 20 cu.m, we get the relation: $x_1^{2^*} x_2 = 20$ or, $x_2 = 20/x_1^2$.

Therefore, the expression for the total cost of the container will be given by,

$$y = 2^* 25^* x_1^2 + 25^* 4^* x_1^* 20/x_1^2$$

or $y = 50^* x_1^2 + 2000^* x_1^{-1}$.

If we designate the cost coefficients 50 and 2000 in the above expression as c_1 and c_2, respectively, we get the expression for total cost to be minimized as follows:

$$\text{Minimize } y = \underbrace{c_1 x_1^2}_{A} + \underbrace{c_2 x_1^{-1}}_{B}$$

It may be noted from above that the total cost of the container is composed of the sum of two cost components, namely (1) component A which is the cost of bottom and top surface of the container and (2) component B which is the cost of side surfaces of the container.

Using differential calculus, the solution of the problem could be easily found as follows:

$$\frac{\partial y}{\partial x} = 2c_1 x_1 - c_2 x_1^{-2} = 0 \quad \text{or, } 2c_1 - c_2 x_1^{-3} = 0 \quad \text{or, } x_1^{-3} = \frac{2c_1}{c_2}$$

From above, we can easily find the optimum values of x_1 and x_2 as follows:

$$x_1^* = \left(\frac{c_2}{2c_1}\right)^{1/3} = 2.714\,\text{m} \quad x_2^* = 20\left(\frac{2c_1}{c_2}\right)^{2/3} = 2.714\,\text{m}$$

Therefore, the optimum or minimum cost of the container is given by:

$$y^* = c_1\left(\frac{c_2}{2c_1}\right)^{2/3} + c_2\left(\frac{c_2}{2c_1}\right)^{-1/3}$$

or

$$y^* = 2^{-2/3}\underbrace{\left(c_2^{2/3}\,c_1^{1/3}\right)}_{A} + 2^{1/3}\underbrace{\left(c_2^{2/3}\,c_1^{1/3}\right)}_{B}$$

It may be noted from above that at the optimum, the minimum cost is again composed of two components, i.e. cost A and cost B as shown above. It may also be noted that the cost term $c_2^{2/3}c_1^{1/3}$ is common to both cost components A and B and, as a result, the ratio between these cost components A and B is independent of the economic cost coefficients c_1 and c_2 derived from the construction material costs. Consequently, the *optimal cost distribution* between the components is invariant under the change in the values of economic cost coefficients c_1 and c_2. Similarly, algebraic analysis of the component costs B and A shows that B is always twice the value of A. In other words, the *relative contribution* to the total optimal cost by the components A and B is invariant to the changes in the values of economic cost coefficients c_1 and c_2.

Thus, at the optimality, one third of the total cost will be contributed by the cost component A, and two third of the total cost will be contributed by the cost component B. It may be noted that any other distribution between these components will yield a total cost higher than the optimum, that is, minimum cost. Using prior knowledge of this fact, we can get the following relationships:

$$1/3y^* = c_1x_1^2 \text{ and } 2/3y^* = c_2x_1^{-1}$$

$$\text{Or } y^* = 3c_1x_1^2 = 3/2\left(c_2x_1^{-1}\right)$$

$$\text{Or, } 3^*2/3\left(x_1^3\right) = c_2/c_1$$

$$\text{Or, } x_1^3 = c_2/2c_1$$

$$x_1^* = \left(\frac{c_2}{2c_1}\right)^{1/3}$$

This result is same as optimal solution previously obtained above using differential calculus, and leads us to an entirely different way of solving mathematical programming problems. Here, instead of searching first for the optimal solution vector and then determining the resultant optimal cost, one searches first for the *optimal cost distribution* and then determines the values of the resulting optimal solution variables. This method can be very useful particularly in the Architecture, Engineering and Construction (AEC) planning and design field, and therefore, adopted extensively in this book as elaborated in subsequent chapters.

7.12.2 Mathematical Formulation of Geometric Programming Problem

A general mathematical formulation of geometric programming problem can be written as follows (Phillips & Beightler, 1976):

$$\text{Minimize } y_o(x) = \sum_{t=1}^{T_o} \eta_{ot}\, c_{ot} \prod_{n=1}^{N} x_n^{a_{otm}}$$

Subject to:

$$y_m(x) = \sum_{t=1}^{T_m} \eta_{mt}\ c_{mt} \prod_{n=1}^{N} x_n^{a_{mtn}} \leq \eta_m; \quad m = 1,2,3 \ldots M$$
$$x_n > 0; \qquad\qquad n = 1,2,3 \ldots N$$

where
$\eta_{ot} = \pm 1, \quad \eta_{mt} = \pm 1, \quad \eta_m = \pm 1, \text{and} c_{ot} > 0,\ c_{mt} > 0;$
here, the terms a_{otm} and a_{mtn} are unrestricted in sign.

To = total number of terms in the objective function,
T_m = total number of terms in the *m*th constraint, where $m = 1, 2 \ldots M$.

In terms of engineering and architectural designs the terms c_{ot} and c_{mt} represent the economic coefficients for engineering and architectural design, x_n are engineering and architectural design decision variables, the a_{mtn} represent the technological exponents of the decision variables, and η vector has elements binary variables ($\pm$), whose signs represent the sign of each term and inequality in the geometric programming problem statement. If all the η are positive, then the program is called a *posynomial* geometric programming problem. If one or more of η is negative, then the program is called a *signomial* geometric programming problem.

It will be clear from the examples presented later that when certain favourable properties are available, problems with complex nonlinear objective functions and constraints can be solved through the solution of a set of well-defined linear equations. Such results provide a remarkably simple solution technique. Thus, geometric programming ich generally provides a vehicle through which simpler formulations of the original problem can be derived.

7.12.3   *Posynomial Geometric Programming-Unconstrained Case*

A posynomial function is defined as:

$$y \equiv y(\underline{x}) = \sum_{t=1}^{T} c_t p_t(\underline{x}) \tag{7.23}$$

where the coefficients c_t are positive numbers, $\underline{x} = (x_1, x_2 \ldots x_N)$, and the functions: $p_t(\underline{x})$ are defined as:

$$p_t(\underline{x}) = \prod_{n=1}^{N} x_n^{a_{tn}} \tag{7.24}$$

where the a_{tn} being any real numbers.

Thus, the function to be optimized is as follows:

$$y = \sum_{t=1}^{T} c_t \prod_{n=1}^{N} x_n^{a_{tn}} \tag{7.25}$$

It may be noted here that y is <u>not</u> a polynomial, as the coefficients of the terms must be positive, while the exponents a_{tn} are not subjected to any such restrictions to be positive integers.

If y is the objective function to be minimized, it is useful to think it as a sum of a number of component costs.

Thus, in an unconstrained optimization problem, it is desired to find the positive values of $x_1, x_2, \ldots x_N$, which minimize y.

At the minimum, the first derivative of the function y must vanish, i.e.

$$\frac{\partial y}{\partial x_k} = \sum_{t=1}^{T} c_t \, a_{tk} x_k^{a_{t,k}-1} \prod_{\substack{n=1 \\ n \neq k}}^{N} x_n^{a_{tn}} = 0; \quad k = 1,2 \ldots N$$

Since all $x_n > 0$, these equations become:

$$\sum_{t=1}^{T} c_t a_{tk} \prod_{n=1}^{N} x_n^{a_{tn}} = 0; \quad k = 1,2 \ldots N \tag{7.26}$$

Equation (7.26) is a system of N nonlinear simultaneous equations in N unknowns: x_n. These equations can be solved using the Newton–Raphson method discussed and presented in Chap. 3, along with C++ program Listing for solving one, two, and three variable nonlinear simultaneous equations. However, in some cases, such solutions may not converge particularly if the initial guess of values of variables given as input to the C++ program is not correctly chosen. Since, the solution of the simultaneous nonlinear equations is necessary in the classical method to find the optimal solution of a multivariable function, it is essential that these initial guess values should be carefully selected, and if required changed and the program re-run, so that the solution converges and correct roots are obtained.

An alternative method is to find the optimal way to distribute the total cost between the various terms of the objective function, instead of finding first the optimal values of the independent variables in the objective function. Using these optimal allocations, the minimum cost y^* can be found without knowing the optimal values of the independent variables x_n^*.

To find the optimal way to distribute the total cost, we can define the optimal weights, w_t, as:

$$w_t = \frac{c_t\, p_t(x^*)}{y^*}; \quad t = 1,2 \dots T \tag{7.27}$$

Obviously, the sum of all the weights is unity, i.e.

$$\sum_{t=1}^{T} w_t = 1 \tag{7.28}$$

The Eq. (7.28) is called the *normality condition*.
From Eqs. (7.26) and (7.27), we get:

$$\sum_{t=1}^{T} a_{tn} w_t\, y^* = 0; \quad n = 1,2 \dots N$$

Or, as y^* must be positive,

$$\sum_{t=1}^{T} a_{tn}\, w_t = 0; \quad n = 1,2 \dots N \tag{7.29}$$

The Eq. (7.29) is called the *orthogonality condition*.
The Eqs. (7.28) and (7.29) are also referred to as *dual constraints*.
From the knowledge of the values of w_t, the value of y^* can be found as shown below:

$$y^* = \prod_{t=1}^{T} (y*)^{w_t} = \prod_{t=1}^{T} \left[\frac{c_t p_t(y^*)}{w_t} \right]^{w_t}$$

$$= \prod_{t=1}^{T} \left[\frac{c_t}{w_t} \right]^{w_t} \prod_{t=1}^{T} [p_t(x^*)]^{w_t} \tag{7.30}$$

But from Eq. (7.19) and Eq. (7.29), i.e. $\sum_{t=1}^{T} a_{tn}\, w_t = 0$, we get:

$$\prod_{t=1}^{T} [p_t(x*)]^{w_t} = \prod_{n=1}^{N} \prod_{t=1}^{T} (x_n^*)^{a_{tn}.w_t} = \prod_{N=1}^{N} (x_n^*)^{\sum_{t=1}^{T} a_{tn}.w_t} = 1$$

Therefore, Eq. (7.30) becomes

$$y^* = \prod_{t=1}^{T} \left[\frac{c_t}{w_t} \right]^{w_t} \tag{7.31}$$

Thus, the value of y^* can be found from the values of cost coefficients: c_t, and the optimal weights: w_t. Now, the problem of optimizing objective function 'y' becomes a problem to find the values of w_t, as the cost coefficients: c_t, can be calculated from the given design conditions for a computer-aided design problem. It may be noted that Eqs. (7.28) and (7.29) consist of $(N + 1)$ simultaneous linear equations in T unknowns. In case, $T = N + 1$, the optimal solution, i.e. the values of optimal weights w_t, will be uniquely determined by the equations set (7.28) and (7.29), as long as the coefficient matrix of these equations has full rank.

It may be noted that the only nonlinear equations that are to be solved in geometric programming algorithm are those in equation: (7.26), each containing only one term, and therefore, is linear if the logarithm of x_n is taken. Thus, in this solution method, it is not necessary to solve any nonlinear equation particularly when degree of difficulty (discussed later) in the problem is less. The crucial scheme in the development of this method is the concept of weights introduced into the equation, without which we would have to solve a set of N nonlinear simultaneous equations presenting serious difficulties. It may also be noted that the dimensional analysis used in developing Eqs. (7.28) and (7.29) did not include c_t, and therefore the values of w_t represent invariance properties of a system being optimized, and must hold for any values of the term coefficients (Phillips & Beightler, 1976).

7.12.4   Degrees of Difficulty

Conventionally, the difference between the number of variables and the number of independent linear equations is called the number of degrees of freedom. In a posynomial geometric programming problem, there are N orthogonality conditions, one for each variable x_n, one normality condition, and T weights, one for each term. Hence, following the above convention, these equations have $T - (N + 1)$ degrees of freedom. However, in geometric programming technique, this is called the number of *degrees of difficulty*, since larger this number, more difficult is the geometric programming problem to solve.

As discussed earlier, the function to be optimized is as follows:

$$y = \sum_{t=1}^{T} c_t \prod_{n=1}^{N} x_n^{a_{tm}}$$

To illustrate clearly the technique to solve a geometric programming problem, one example with zero degrees of difficulty is presented below, and also later a problem with the presence of the degrees of difficulty is presented.

7.12.4.1   An Example Problem with Zero Degrees of Difficulty

In this problem, it is desired to find the values of the positive numbers x_1, x_2, x_3, and x_4 that minimize the value of the following objective function:

$$\text{Minimize } y = x_3^{0.7} \, x_4^{1.4}$$

Subject to the following constraints:

$$x_1^{-2} \, x_3^{-1} + x_2 \, x_3^{-1} \leq 1 \quad \text{(i)}$$
$$x_1 \, x_4^{-1} + x_2^{-1} \, x_4^{-1} \leq 1 \quad \text{(ii)}$$

We can designate the weights in the objective function as: w_{0i}, where '0' indicates the objective function weight, and 'i' indicates the location of the objective function term, i.e. first ($i = 1$), second ($i = 2$), and so on. Similarly, we can designate the weights in the constraints as: w_{ji}, where 'j' indicates the constraint number and 'i' indicates the location of the constraint term, i.e. first ($i = 1$), second ($i = 2$), and so on.

In this problem, there are $T = 5$ terms, and $N = 4$ variables.

Hence, the degree of difficulty in this problem is $= T - (N + 1) = 5 - (4 + 1) = 0$, and the orthogonal and normality conditions in this case are as follows:

$$
\begin{aligned}
w_{01} &&&= 1 \\
-2w_{11} && + w_{21} &= 0 \\
& w_{12} & - w_{22} &= 0 \\
0.7w_{01} - w_{11} - w_{12} &&&= 0 \\
1.4w_{01} && -w_{21} - w_{22} &= 0
\end{aligned}
\qquad (7.32)
$$

In the equation set (7.32), there are five independent linear equations in the five unknown variables. Hence, there are exactly enough equations to determine the unique optimal values of the variables without recourse to further analysis.

From the equation set (7.32), we get the optimal values of dual weights as follows:

$$
\begin{aligned}
w_{01} &= 1 \\
w_{11} &= 0.6 & w_{12} &= 0.2 \\
w_{21} &= 1.2 & w_{22} &= 0.2
\end{aligned}
\qquad (7.33)
$$

Now from the first and second constraints, we get the equations relating the optimal values of the variables x_1, x_2, x_3, and x_4 as follows:

$$x_1^{-2} x_3^{-1} = \frac{w_{11}}{w_{11.} + w_{12}} \quad (i)$$

$$x_2 x_3^{-1} = \frac{w_{12}}{w_{11.} + w_{12}} \quad (ii)$$

$$x_1 x_4^{-1} = \frac{w_{21}}{w_{21.} + w_{22}} \quad (iii) \qquad\qquad (7.34)$$

$$x_2^{-1} x_4^{-1} = \frac{w_{22}}{w_{21.} + w_{22}} \quad (iv)$$

Dividing equation (ii) by equation (i) above, we get:

$$x_2 x_1^2 = \frac{w_{12}}{w_{11}} \quad (v)$$

Similarly dividing equation (iii) by equation (iv) above, we get:

$$x_1 x_2 = \frac{w_{21}}{w_{22}} \quad (vi)$$

Again dividing equation (v) by equation (vi) above, we get optimal value of x_1 as:

$$x_1^* = \frac{w_{12}}{w_{11}} \times \frac{w_{22}}{w_{21}} = (0.2 \times 0.2)/(0.6 \times 1.2) = 0.0555$$

Using this value of x_1 and the above equations, we get the optimal values of other variables as follows:

$$x_2^* = \frac{w_{21}}{w_{22}} \times \frac{1}{x_1} = 107, \ x_3^* = (w_{11} + w_{12}) \times x_2/w_{12} = 432,$$

$$x_4^* = (w_{21} + w_{22}) \times x_1/w_{21} = 0.0647$$

Substituting the above optimal values of x_3^* and x_4^* in the above objective function, we get the minimum value of the objective function y_o^* as:

$$y_o^* = 432^{0.7} \times 0.0647^{1.4} = 2.7735.$$

'It may be noted that the optimal distribution represented by w_t is totally unaffected by the cost coefficients c_1, c_2 and so on. This separation of technological effects, as reflected by the exponents a_{tn}, from the economic effects, as measured by the coefficients c_t, is one of the valuable features of the geometric programming approach.' Put this in suitable example later.

7.12.5 *Duality Concept and the Primal Dual Relations in Geometric Programming*

Duality is a basic concept in mathematics performing an important role in optimization theory and in both linear and nonlinear programming problem and forms the basis for the geometric programming algorithms (Beightler et al., 1979).

Linked with every primal problem, there is another dual problem and the solution to either is sufficient for obtaining the solution to the other. A crucial result of these formulations is that the *maximum (minimum) of the dual function is equal to the minimum (maximum) of the primal function*. This theorem enables us to achieve optimization either by minimizing the primal function or maximizing the dual function whichever is easier to perform.

As the generalized arithmetic-geometric inequality is fundamental to the geometric programming technique, it is outlined here very briefly, patterned after Beightler et al. (1979), Phillips and Beightler (1976), and Stark and Nichols (1972) without proof. Readers interested in more details of the development of these formulations may find them in Beightler et al. (1979) and Phillips and Beightler (1976) and other publications available in literature.

The general geometric inequality can be stated as (Phillips & Beightler, 1976; Stark & Nichols, 1972):

$$w_1 v_1 + w_2 v_2 + w_3 v_3 + \cdots + w_t v_t \geq v_1^{W_1} \, v_2^{W_2} \, v_3^{W_3} \cdots v_t^{W_t} \tag{7.35}$$

with equality only when $v_1 = v_2 = \ldots = v_t$.

To use the above inequality, the posynomial objective function 'y' given by Eq. (7.17) can be expressed as (Phillips & Beightler, 1976):

$$y = \sum_{t=1}^{T} w_t \left[\frac{c_t p_t}{w_t} \right]$$

Therefore, from the expression (7.35), we get (Phillips & Beightler, 1976):

$$\sum_{t=1}^{T} c_t \, p_t \Big/ \prod_{t=1}^{T} \left[\frac{c_t p_t}{w_t} \right]^{.w_t} = \prod_{t=1}^{T} \left[\frac{c_t}{w_t} \right]^{.w_t} \tag{7.36}$$

The third expression above resulting from the fact that the weights are chosen so that the exponents vanish making the function p_t dimensionless with respect to the original variables x_n as indicated in expression (7.30).

The expression:

$$\prod_{t=1}^{T} \left[\frac{c_t p_t}{w_t} \right]^{.w_t} \tag{7.37}$$

is referred to as the *predual function,* and the variables w_t are referred to as the *dual variables.*

Similarly, the expression:

$$d(w) = \prod_{t=1}^{T} \left[\frac{c_t}{w_t}\right]^{.w_t}$$ (7.38)

is referred to as the *dual function.*

We can state the dual function $d(w)$ to be optimized as:

$$\text{Optimize } d(w) = \prod_{t=1}^{T} \left[\frac{c_t}{w_t}\right]^{.w_t}$$ (7.39)

Subject to:

$$\sum_{t=1}^{T} w_t = 1$$
$$\sum_{t=1}^{T} a_{tn} \, w_t = 0 \quad ; n = 1,2 \ldots N$$
$$w_t > 0$$

Since all the constraints in the dual problem are linear, the dual constraint set forms a convex region. To study the nature of the dual objective function $d(w)$, we may make a substitution in terms of $z(w) = \ln[d(w)]$ to facilitate a better understanding of the problem, particularly when it is known that all values of w_t should lie between zero and one, and the $\ln[d(w)]$ is a monotonic function in the dual variables w_t.

Therefore,

$$Z(w) = \ln \prod_{t=1}^{T} \left[\frac{c_t}{w_t}\right]^{.w_t} = -\sum_{t=1}^{T} w_t \, \ln\left(\frac{w_t}{C_t}\right)$$ (7.40)

The function $z(w)$ is concave with respect to weights w_t, since it is the negative of a sum of functions that are themselves convex (Phillips & Beightler, 1976). Thus, the dual problem is to find a stationary point for a concave objective function subject to a set of convex constraints. Therefore, the function has a unique stationary point, that is, a *global maximum.* Thus, the dual problem becomes a problem of maximizing $z(w)$, subject to the normality and orthogonality constraints, while the weights w_t are constrained to be positive. Thus, in this case, the *global minimum* for the primal problem equals the *global maximum* for the dual problem, and also, the maximization of the dual function subject to the normality and orthogonality conditions is a sufficient condition for the primal objective function to be a *global minimum.* Since it is generally easier to solve dual problems, this is a very important result, particularly because it gives many other advantages such as invariance properties of the weights as indicated earlier.

The arithmetic–geometric mean inequality relationships presented above can also be written as:

$$y(x)/y^* = d^*/d(w) \qquad (7.41)$$

Here, the asterisk mark denotes a global optima—a minimum for the primal function 'y' and a maximum for the dual function 'd'. Sometimes the Eq. (7.41) can be used to get a quick estimate of the optimal value of an objective function when the degree of difficulty in the problem is less. Thus, for example, a quick estimate can be obtained for a problem by neglecting one of its terms making it a zero degree of difficulty problem and solving for the weights using the normality and orthogonality conditions omitting the corresponding weight. These calculated weights are then substituted into the dual function for this reduced degree of difficulty problem to obtain a lower bound on the true optimal cost. The corresponding x_i values can be used to construct an upper bound by substituting them into the objective function for the original problem.

7.12.5.1 Duality Concept and the Primal Dual Relations: Container Example Problem

The above concepts will be clearer if we explain these using the container example problem at Sect. 7.12.1 presented earlier. Thus, the *predual function,* i.e. Eq. (7.37) in this example is given by:

$$\left(\frac{c_1 x_1^2}{w_1}\right)^{w_1} \left(\frac{c_2 x_1^{-1}}{w_2}\right)^{w_2}$$

In this example, there is only one x_i variable, i.e. x_1, so that the exponent of this only one variable x_1 is $2w_1 - w_2$ which are the expressions (7.29): $\sum_{t=1}^{T} a_{it}\, w_t$, which must be set to zero, that is $2w_1 - w_2 = 0$ to fulfill the orthogonality condition.

Similarly, to fulfill the normality condition: $\sum_{t=1}^{T} w_t = 1$, that is $w_1 + w_2 = 1$.

Thus, we have the following two equations for two variables:

$$w_1 + w_2 = 1$$
$$2w_1 - w_2 = 0$$

From which we get $3w_1 = 1$ giving the value of $w_1 = 1/3$ and $w_2 = 2/3$.

Thus, without knowing the optimal values of decision variables, the optimal cost can be determined as shown below.

In this example problem, the *dual function* is given by:

$$d(w) = \left(\frac{c_1}{w_1}\right)^{w_1} \left(\frac{c_2}{w_2}\right)^{w_2} = \{50/(1/3)\}^{1/3} \times \{2000/(2/3)\}^{2/3} = 1105.32 = y_o$$

This is the expression:

$\prod_{t=1}^{T} \left[\frac{c_t}{w_t}\right]^{.w_t}$ i.e. Eq. (7.38) dual function in this case.

Thus, for this problem:

$$y_1 = x_1^2, y_2 = x_1^{-1}$$
$$c_1 y_1 / w_1 = c_2 y_2 / w_2 = y_o$$

Here, $c_1 = 50$ $c_2 = 2000$.

Thus, as required by the arithmetic-geometric mean equality, for this problem: $c_1 x_1^2 / w_1 = c_2 x_1^{-1} / w_2 = y_o = 1105.32$, giving the optimum value of $x_1 = 2000 \times 3/(1105.32 \times 2) = 2.714$.

This is the same value derived earlier in Sect. 7.12.1 of illustrative example.

Here, optimal component costs are $1105.32/3 = 367.44$ and $1105.32*2/3 = 736.77$.

Thus, using geometric programming and the *dual function* with *normality* and *orthogonality* conditions, the optimal value of the objective function can be derived even before solving fully the optimization problem.

It may be noted that the dimensional analysis used in developing the Eqs. (7.28) and (7.29) did not include c_t, and therefore, the values found for w_1 and w_2 in the above example signify the invariance properties of the system being optimized, and must sustain for any values of cost or other parameter coefficients of a term. Thus, for example, if we change the value of c_1 from 50 to 300, then from Eq. (7.38), we get:

$$d(w) = \left(\frac{c_1}{w_1}\right)^{w_1} \left(\frac{c_2}{w_2}\right)^{w_2} = \{300/(1/3)\}^{1/3} \times \{2000/(2/3)\}^{2/3} = 2007.76$$

The new optimal policy would be found by keeping the optimal weights unchanged as before:

$300 \times x_1^2 = 2007.76 \times (1/3) = 669.62$ giving value of $x_1 = 1.49$

$2000 \times x_1^{-1} = 2007.76 \times (2/3) = 1339.24$ again giving value of $x_1 = 1.49$

As earlier mentioned, the total cost of the container is composed of the sum of two cost components, namely (1) component A which is the cost of bottom and top surface of the container and (2) component B which is the cost of side surfaces of the container. But, without knowing the optimal cost of component A and B, using the geometric programming technique, it is possible to say definitely that these optimal values must always be such that the cost of component B (1339.24 and 736.77 as above) would be always two times that of component A (669.62 and 367.44) as indicated above. Additionally, this optimal distribution is totally unaffected by changes in the cost coefficients c_1 and c_2, i.e. c_t representing the economic effects. This separation of technological effects, represented by the exponent's $\underline{a}_{it}$ from the economic effects as measured by the coefficients c_t, is one of the most attractive features of the geometric programming technique.

7.12.6   Posynomial Geometric Programming: Inequality Constrained Case

In the above discussion, we have dealt with unconstrained posynomial geometric programming problem. In real world problems, in many cases there are constraints that must be satisfied to fulfill the design objectives while optimizing a design. Optimization problems with inequality constraints can be dealt with by methods similar to those adopted for unconstrained problems as outlined above. Posynomial geometric programming problems with inequality constraints are outlined here entirely patterned after Phillips and Beightler (1976) for ready reference to readers.

This constraint posynomial geometric programming problem can be formulated as follows:

$$\text{Minimize } y_{\text{o}}(x) = \sum_{t=1}^{T_o} c_{ot} \prod_{i=1}^{N} x_i^{a_{oti}} \tag{7.42}$$

Subject to:

$$y_m(x) = \sum_{t=1}^{T_m} c_{mt} \prod_{i=1}^{N} x_i^{a_{mti}} \leq 1; \quad m = 1,2,3 \ldots M \tag{7.43}$$
$$x_i > 0; \quad i = 1,2,3 \ldots N$$

It may be noted that there are $M + 1$ posynomial expressions, namely one for objective function and M for the inequality constraints. c_t's are coefficients which are single subscripted for each term in objective function. Here, T_{o} is the total number of terms in the objective function, T_m is the number of terms in each posynomial constraint which can vary for each $m = 1, 2 \ldots M$. The coefficient of each term in constraints is double subscripted one subscript designating the constraint equation and the other for term. Similarly, each exponent is triple subscripted identifying the equation, variable and the posynomial term to which each belongs. Here, the optimization problem is to minimize $y_{\text{o}}(x)$ subject to M linear or nonlinear inequality constraints, $y_m(x) \leq 1$; $m = 1, 2, 3 \ldots M$. An equivalent dual problem having all constraints linear could be constructed exploiting special structure of this formulation.

Thus, using the generalized arithmetic-geometric mean inequality as presented in expression (7.34) above, this inequality can be given by the following expression:

$$\delta_1 v_1 + \delta_2 v_2 + \delta_3 v_3 + \cdots + \delta_t \, v_t \geq v_1^{\delta_1} \, v_2^{\delta_2} \, v_3^{\delta_3} \cdots v_t^{\delta_t} \tag{7.44}$$

Here, the variables δ_1, δ_2, δ_3 ... δ_t are nonnegative weights chosen such that: $\delta_1 + \delta_2 + \delta_3 + \cdots + \delta_t = 1$, and the variables: v_1, v_2, v_3, ... v_t are arbitrary positive numbers.

Using an equivalent expression: $u_i = \delta_i \, v_i$, $i = 1, 2, 3 \ldots t$, we get:

$$u_1 + u_2 + u_3 \ldots + u_t \geq \left[\frac{u_1}{\delta_1}\right]^{\delta_1} \left[\frac{u_2}{\delta_2}\right]^{\delta_2} \left[\frac{u_3}{\delta_3}\right]^{\delta_3} \cdots \left[\frac{u_t}{\delta_t}\right]^{\delta_t} \tag{7.45}$$

$$\delta_1 + \delta_2 + \delta_3 + \cdots + \delta_t = 1 \tag{7.46}$$

The above inequality can be written as the generalized arithmetic-geometric mean inequality with weights that are not normalized. If we define such weights as: w_1, w_2, w_3 ...w_t, and represent their sum as:

$$\lambda = w_1 + w_2 + w_3 \ldots + w_t \tag{7.47}$$

Thus, δ_1, δ_2, δ_3, ..., δ_t could be defined as:

$$\delta_i = \frac{w_i}{\lambda}; \quad i = 1,2,3,\ldots,t \tag{7.48}$$

Substituting Eq. (7.48) into Eq. (7.45), we get:

$$u_1 + u_2 + u_3 \ldots + u_t \geq \left[\frac{u_1}{w_1}\right]^{w_1/\lambda} \left[\frac{u_2}{w_2}\right]^{w_2/\lambda} \left[\frac{u_3}{w_3}\right]^{w_3/\lambda} \cdots \left[\frac{u_t}{w_t}\right]^{w_t/\lambda} . \lambda \tag{7.49}$$

$$(u_1 + u_2 + u_3 \ldots + u_t)^\lambda \geq \left[\frac{u_1}{w_1}\right]^{w_1} \left[\frac{u_2}{w_2}\right]^{w_2} \left[\frac{u_3}{w_3}\right]^{w_3} \cdots \left[\frac{u_t}{w_t}\right]^{w_t} . \lambda^\lambda \tag{7.50}$$

The problem represented by Eqs. (7.42) and (7.43), and using only a single constraint for illustration purpose, we get:

$$\text{Minimize } y_o(x) = \sum_{t=1}^{T_o} c_{ot} p_t(x) \tag{7.51}$$

Subject to:

$$y_1(x) = \sum_{t=1}^{T_1} c_{1t} q_t(x) \leq 1 \tag{7.52}$$
$$x > 0$$

where

$$p_t(x) = \prod_{n=1}^{N} x_n^{a_{otn}} \quad t = 1,2,3 \ldots T_o \tag{7.53}$$

$$q_t(x) = \prod_{n=1}^{N} x_n^{a_{1tn}} \quad t = 1,2,3 \ldots T_1 \tag{7.54}$$

The inequality relationships for $y_o(x)$ and $y_1(x)$ can be written using the generalized arithmetic-geometric mean inequality (Phillips & Beightler, 1976).

Thus, the objective function inequality will be given by:

$$[y_0(x)]^{\lambda_0} \geq \left[\frac{c_{01}\, p_1(x)}{w_{01}}\right]^{w_{01}} \left[\frac{c_{02}\, p_2(x)}{w_{02}}\right]^{w_{02}} \cdots \left[\frac{c_{0_{t_0}}\, p_{t_0}(x)}{w_{0_{t_0}}}\right]^{w_{0_{t_0}}} . \lambda_0^{\lambda_0} \tag{7.55}$$

The constraint inequality will be given by:

$$1 \geq [y_1(x)]^{\lambda_1} \geq \left[\frac{c_{11}\, q_1(x)}{w_{11}}\right]^{w_{11}} \left[\frac{c_{12}\, q_2(x)}{w_{12}}\right]^{w_{12}} \cdots \left[\frac{c_{1_{t_1}}\, q_{t_1}(x)}{w_{1_{t_1}}}\right]^{w_{1_{t_1}}} . \lambda_1^{\lambda_1} \tag{7.56}$$

By relating algebraically the above two inequalities and by suitable manipulations, Phillips and Beightler (1976) derived the following results:

$$y_0(x) \geq \left[\frac{c_{01}}{w_{01}}\right]^{w_{01}} \left[\frac{c_{02}}{w_{02}}\right]^{w_{12}} \cdots \left[\frac{c_{0_{t_0}}}{w_{0_{t_0}}}\right]^{w_{0_{t_0}}} . \left[\frac{c_{11}}{w_{11}}\right]^{w_{11}} \left[\frac{c_{12}}{w_{12}}\right]^{w_{12}} \cdots \left[\frac{c_{1_{t_1}}}{w_{1_{t_1}}}\right]^{w_{1_{t_1}}} . \lambda_1^{\lambda_1}$$
$$\tag{7.57}$$

where

$$\lambda_1 = w_{11} + w_{12} + \cdots + w_{1_{t_1}}$$

Or, equivalently,

$$y_0(x) \geq \left[\frac{c_{01}}{w_{01}}\right]^{w_{01}} \left[\frac{c_{02}}{w_{02}}\right]^{w_{02}} \cdots \left[\frac{c_{0_{t_0}}}{w_{0_{t_0}}}\right]^{w_{0_{t_0}}}$$
$$. \left[\frac{c_{11}\, \lambda_1}{w_{11}}\right]^{w_{11}} \left[\frac{c_{12}\, \lambda_1}{w_{12}}\right]^{w_{12}} \cdots \left[\frac{c_{1_{t_1}}\, \lambda_1}{w_{1_{t_1}}}\right]^{w_{1_{t_1}}} \tag{7.58}$$

where

$$w_{01} + w_{02} + \cdots + w_{0_{t_0}} = 1 \tag{7.59}$$

$$\sum_{m=0}^{1} . \sum_{t=1}^{t_m} a_{mtn} w_{mt} = 0; \quad n = 1,2,3, \ldots, N \tag{7.60}$$

$$\lambda_1 = \sum_{t=1}^{t_1} w_{1t} \tag{7.61}$$

The right-hand side of the Eq. (7.58), which is an inequality, provides a lower bound for the objective function $y_0(x)$ for any choice of $\underline{w}$ that satisfy Eqs. (7.59) through (7.61). It may also be noted that proper choice for $\underline{w}$ would be that set of variables which maximizes the right-hand side. It is to be noted that the constraint $y_1(x)$ contributes exactly T_1 product terms to the inequality, one for each term in the constraint plus the factor $\lambda_1^{\lambda_1}$.

Thus, the dual geometric program for the general posynomial case would be as follows:

$$\text{Maximize } d(w) \prod_{m=0}^{M} \cdot \prod_{t=1}^{T_m} \cdot \left[\frac{C_{mt} w_{mo}}{w_{mt}} \right]^{w_{mt}} \tag{7.62}$$

Subject to:

$$\sum_{t=1}^{T_0} w_{ot} = 1 \qquad \text{i.e. normality condition} \tag{7.63}$$

$$\sum_{m=1}^{M} \cdot \sum_{t=1}^{T_m} a_{mtn} w_{mt} = 0; \\ n = 1,2,3, \ldots, N \qquad \text{i.e. orthogonality condition} \tag{7.64}$$

$$w_{mo} = \sum_{t=1}^{T_m} w_{mt} \quad m = 1,2,3, \ldots M \tag{7.65}$$

Provided we define $w_{00} = 1$ and $w_{mo} = \lambda_m$.

It may be noted there are exactly $(N + 1)$ independent dual constraint equalities, and exactly T independent dual variables, one for each term of primal problem. When the number of degrees of difficulty, i.e. $T - (N + 1)$ is zero, then only these equalities are to be solved giving unique solution for the unknown dual variables. Even in case the value of $T - (N + 1) > 0$, a true optimization problem exists. But, by adopting the geometric programming technique, a nonlinear programming problem with nonlinear constraints is transformed to a concave programming problem with linear constraints, making it much easier to solve by establishing relationship between T dual variables and N primal decision variables as shown below.

However, it is necessary to identify the exact relationship that exists between T dual variables in expression (7.62) and the N primal decision variables. It is clear from the Eq. (7.27) that the optimal values, $w_{ot}^*, t = 1,2,3, \ldots T_o$, are the true weights for the optimal primal objective function, $y_0(x^*)$. Again, through an extension of Eq. (7.58) to the general case, $d(w^*) \equiv y_0(x^*)$, that is $d(\underline{w}^*)$ and $y_0(x^*)$ are identical. Therefore, for the terms of the objective function, the following relationships hold at optimality:

$$w_{ot}^*(y_0(x^*)) = c_{ot} \prod_{n=1}^{N} x_n^{a_{otn}}; \quad t = 1,2,3 \ldots T_o \tag{7.66}$$

as regard the constraint terms, it may be noted that the Eq. (7.65) is a direct generalization of the Eq. (7.61) for multiple constraints, where $\lambda_m \equiv w_{mo}$; $m = 1$, 2, 3, ..., M. Applying Eq. (7.48), to the general case, the generalized weight for the tth term in the mth constraint, we get:

$$w_{mt} = \frac{w_{mt}}{\sum_{t=1}^{T_m} w_{mt}}$$

However, it may be noted that the weights: w_{mt} are simply the values of the terms themselves for active constraints, since if a constraint is tight, the terms must sum to unity. Hence, for the tth term in the mth constraint, the following relationships hold:

$$w_{mt} = c_{mt} \prod_{n=1}^{N} x_n^{a_{mtn}}; \quad \begin{array}{l} m = 1,2,3,\ldots M \\ t = 1,2,3,\ldots T_m \end{array} \tag{7.67}$$

The above completes the development of the formulas necessary to solve the constrained posynomial geometric programming problem. Proofs for the above formulations are available in Phillips and Beightler (1976).

It is to be noted that if a primal constraint is tight at optimality, then all dual variables associated with that constraint will be strictly positive at optimality. On the other hand, if a primal constraint is loose at optimality, then all dual variables associated with that constraint will be zero at optimality. In such a case of loose constraints, the dual objective is undefined for those dual variables that assume a zero value. In fact, if it is known in advance of problem solution that a constraint will be loose at optimality it could be dropped from the problem formulation without affecting the optimal solution. This means that those terms associated with the loose constraints should not appear in the dual program. To avoid such a problem during problem formulation, one can carry out some test for any dual variable that is driven to zero (Beightler et al., 1979).

To sum up, the above discussion we can conclude about some basic results which will be useful while dealing with many optimization problems. One very important basic result is that the maximum of the dual function equals the minimum of the primal function. This enables us to attain optimization either by minimizing the primal objective function or maximizing the dual objective function whichever is easier. It is also to be noted that the maximization of the dual objective function subject to the normality and orthogonality conditions is a *sufficient condition* for the primal objective function to be a *global minimum* (Mark et al., 1970).This is also applicable in case of constrained geometric programming problem with 'less than inequalities' constraints with all positive terms in the constraints. The optimum solution is unique in problems with zero degrees of difficulty (Mark et al., 1970). Even in case of problems with degrees of difficulty, the dual objective function can be maximized using differential calculus to derive the *global minimum* of the primal objective function, particularly when the degree of difficulty is comparatively less. An example problem with presence of degrees of difficulty is presented below.

7.12.7 Urban-Built-Form Design Optimization: An Example of Posynomial Problem with Inequality Constraints and with Presence of Degrees of Difficulty

A descriptive model of urban-built-form giving scientific relationship between different built-form-elements is presented in Chap. 7, which is reproduced below for ready reference.

Here, the Built-Form-Elements (BFEs), such as built-space per built-unit, i.e. built-unit-space (BUS), building rise (BR—which is assumed as effective number of floors having continuous value as explained later), Semi-Public-Open space (SPO—which is the open-to-sky space per built-unit and may include recreation space, road space, and so on in a layout), and the Floor Area Ratio (FAR—which is the ratio between the total built-space covering all floors and the total land area in a layout), are related mathematically as follows:

$$FAR = \frac{BUS \cdot 3BR}{BUS + BR \cdot 3SPO}; \quad BUD = \frac{FAR}{BUS}; \quad GC = \frac{FAR}{BR}; \quad (7.68)$$

Designating BUS, SPO, FAR, and BR as: x_1, x_2, x_3, and x_4, respectively, the above expression for FAR become:

$$x_3 = x_1 x_4 / (x_1 + x_4 x_2)$$

OR,

$$x_3 x_4 x_2 + x_3 x_1 = x_1 x_4$$

Thus, to ensure compatibility between FAR (Floor Area Ratio), BUS (Built-Unit-Space), SPO (Semi-Public-Open space), and the BR (Building Rise) in a layout, the above expression can be taken as a constraint as follows:

$$x_3 x_4 x_2 + x_3 x_1 \leq x_1 x_4$$

Dividing both sides by $x_1 x_4$, the above constraint expression becomes as follows:

$$x_1^{-1} x_2 x_3 + x_3 x_4^{-1} \leq 1$$

In the model for optimal design of Urban-Built-Form presented below, the above constraint is included as constraint-1, as shown below. Since the objective function is to minimize the cost per built-unit, obviously the value of BUS or x_1 tends to be reduced to even zero to reduce cost, which is not a feasible or practicable solution. Hence, minimum acceptable value of x_1 is included as constraint-2. Similarly, a constraint defining the minimum acceptable value of open-space per built-unit (SPO), i.e. x_2 is added as constraint-3.

Using FAR, the area of land consumed per built-unit of built-area BUS can be calculated as $= \mathrm{BUS/FAR} = x_1 \, x_3^{-1}$.

As explained below, the first term of the objective function represents the building construction cost per built-unit, the second term represents the increased building construction cost per built-unit due to increase in building rise, and the third term represents the land cost per built-unit based on consumption of land as calculated above.

Using these relationships a model for optimal design of urban-built-form in a 'planning area' linking Built-Space (BUS), Open-Space (SPO), Building Rise (BR), land price, construction costs, can be developed as follows:

$$\text{Minimize built-unit-cost} = y(x) = c_{01}x_1 + c_{02}x_1x_4 + c_{03}x_1x_3^{-1} \qquad (7.69)$$

Subject to:

$$x_1^{-1}x_2x_3 + x_3x_4^{-1} \le 1 \quad (1)$$

(FAR, built-unit-space, open space, and building rise compatibility)

$$c_{21}\,x_1^{-1} \le 1 \quad (2)$$

(built-unit-space constraint)

$$c_{31}\,x_2^{-1} \le 1 \quad (3)$$

(open space constraint)

$$x_1, x_2, x_3, x_4 > 0$$

(non-negativity constraint)

Here,

In terms of the use of the equations presented above, and as done in case of above example, we can designate the decision variables, constant coefficients, and weights as follows:

Decision variables are designated as: x_i's. Here, there are x_1, x_2, x_3, and x_4, i.e. four decision variables as stated above.

The objective function constant coefficients are designated as: c_t's where 't' indicates the location of the constant coefficient term, i.e. first ($t = 1$), second ($t = 2$), and so on. Thus, here c_{01}, c_{02}, and c_{03} are the three constant coefficients for the three terms in the objective function. Alternatively, the objective function constant coefficients can also be designated as: c_{0t}'s where '0' indicates the objective function and 't' indicates the location of the constant coefficient term as mentioned above. However, there is no need for such double subscripting of objective function constant coefficients as constraint function constant coefficients are already double subscripted to distinguish them from the objective function constant coefficients.

There are three constraint functions in this problem, i.e. $m = 3$.

Constraint function constant coefficients are designated as c_{mt}'s where 'm' indicates the constraint number and 't' indicates the location of the constraint term, i.e. first ($t = 1$), second ($t = 2$), and so on. Thus, here c_{11}, c_{12}, c_{21}, and c_{31} are the four constraint function constant coefficients for the four terms in three constraints.

Objective function weights are designated as: w_{0t}'s where '0' indicates the objective function and 't' indicates the location of the objective function term, i.e. first ($t = 1$), second ($t = 2$), and so on. Thus, here w_{01}, w_{02}, and w_{03} are the three weights for the three terms of the objective function.

Constraint function weights are designated as: w_{mt}'s, where 'm' indicates the constraint number and 't' indicates the location of the constraint term, i.e. first ($t = 1$), second ($t = 2$), and so on. Thus, here w_{11}, w_{12}, w_{21}, and w_{31} are the four weights for the four terms in the three constraint functions.

Here,

$x_1 = $ optimum space per Built Unit (BU, i.e. dwelling unit, commercial unit, so on) in planning area,

$x_2 = $ optimum Semi-Public Open space (SPO), i.e. open-to-sky space per BU in the planning area,

$x_3 = $ a variable giving the optimum FAR, i.e. the ratio between the total built-space on all floors of an urban development and its land area,

$x_4 = $ variable giving optimum building rise (effective), i.e. number of effective floors in the planning area. Ideally this should be an integer value, and consequently, nonlinear integer programming which is quite complicated without commensurate benefit, should be applied (interested readers can try this). But, literature indicates that efficient solution methods are not available even for linear integer programming (Mark et al., 1970). However, the problem can be overcome by adopting the concept of 'effective building rise', and thus, permitting application of nonlinear continuous geometric programming as adopted here. This concept is outlined in more details in Chap. 9.

$c_1 = $ cost coefficient relating building construction cost per unit of plinth area (monetary units/m^2),

$c_2 = $ a cost coefficient related to building rise, in terms of the increased cost per unit of plinth area (monetary units/m^2) in the urban development. This factor can also be used as a social-cost-coefficient (say, a Penalty Factor for High Building Rise determined by the Society uniformly, instead of arbitrary imposition from top) for determining the optimal building-rise (effective) based on the social-cost-benefit-analysis principles, in a site-specific-case,

$c_3 = $ a cost coefficient related to land price per unit of land area (monetary units/sq.m.),

$c_{21} = $ the desired built-unit space in sq.m. per built-unit in the urban development,

$c_{31} = $ the desired semi-public open space in sq.m. per built-unit in the urban development.

Here, $c_{11} = c_{12} = 1$.

The normality and orthogonality conditions in this case are as follows:

$$
\begin{aligned}
w_{01} + w_{02} + w_{03} &= 1 \quad \text{normality condition} \\
w_{01} + w_{02} + w_{03} - w_{11} - w_{21} &= 0 \quad \text{orthogonality condition for } x_1 \\
w_{11} \qquad\quad - w_{31} &= 0 \quad \text{orthogonality condition for } x_2 \\
-w_{03} + w_{11} + w_{12} &= 0 \quad \text{orthogonality condition for } x_3 \\
w_{02} - w_{12} &= 0 \quad \text{orthogonality condition for } x_4
\end{aligned}
\tag{7.70}
$$

In the equation set (7.69), there are 5 independent linear equations in the 7 unknown variables. Hence, there are not enough equations to determine the unique optimal values of the variables and recourse to further analysis is required. Here, there are 7 terms, i.e. $T = 7$, 4 variables, i.e. $N = 4$ giving the *degree of difficulty* $= T - (N + 1) = 7 - 5 = 2$.

The dual problem is to find the values of nonnegative $\underline{w}$, i.e. w_{01}, w_{02}, w_{03}, w_{11}, w_{12}, w_{21}, and w_{31} that maximizes the dual function $d(w)$ given by:

$$
\begin{aligned}
d(w) = &\left(\frac{c_{01}}{w_{01}}\right)^{w_{01}} \left(\frac{c_{02}}{w_{02}}\right)^{w_{02}} \left(\frac{c_{03}}{w_{03}}\right)^{w_{03}} \left(\frac{C_{11}(W_{11} + W_{12})}{w_{11}}\right)^{w_{11}} \\
&\left(\frac{C_{12}(W_{11} + W_{12})}{w_{12}}\right)^{w_{12}} (c_{21})^{w_{21}} (c_{31})^{w_{31}}
\end{aligned}
\tag{7.71}
$$

As the equation set (7.71) has seven variables and five equations, the dual constraint set has no unique solution. To solve such problems, the following two approaches can be adopted (Phillips & Beightler, 1976):

1. If it is known that dropping of some terms will have little effect in the solution of the problem until the problem is converted to a zero degree of difficulty problem. The restricted problem can be solved for the optimal value of the restricted dual objective function. The computed values of the primal variables are used in the original objective function to derive its value. The two solutions provide the bounds for the true optimal solution. If the bounds are within acceptable limit, the solution can be accepted.

2. In the second approach, we can solve for $(N + 1)$ dual weights in terms of the remaining $T - (N + 1)$ weights. Substituting these new values into the dual objective function, we can iterate on these $T - (N + 1)$ weights until the dual objective function is maximized. The substituted dual objective function can also be maximized by setting the first partial derivative to zero in respect of the remaining weights referred above, and thus, get additional equations to get the unique solution. This method is very useful when the degree of difficulty is less, say up to three.

Thus, adopting the second approach, we can express 5 dual weights in terms of the remaining $T - (N + 1) = 7 - (4 + 1) = 2$ dual weights. Here, we have selected w_{11} and w_{12} as the two dual weights, and, based on the dual constraint equation set

(7.69), the values of the 5 dual weights in terms of these two dual weights are as follows:

$$w_{02} = w_{12}; \ w_{03} = (w_{11} + w_{12}); \ w_{01} = (1 - w_{11} - 2w_{12}); \ w_{21} = (1 - w_{11}); \ w_{31} = w_{11};$$

$$(7.72)$$

Thus, the dual objective function $d(w)$ is now a function of dual weights w_{11} and w_{12} as follows:

$$d(w) = \left(\frac{c_{01}}{(1 - w_{11} - 2w_{12})}\right)^{(1 - w_{11} - 2w_{12})} \left(\frac{c_{02}}{w_{12}}\right)^{w_{12}} \left(\frac{c_{03}}{(w_{11} + w_{12})}\right)^{(w_{11} + w_{12})}$$
$$\left(\frac{C_{11}(W_{11} + W_{12})}{w_{11}}\right)^{w_{11}} \left(\frac{C_{12}(W_{11} + W_{12})}{w_{12}}\right)^{w_{12}} (c_{21})^{(1 - w_{11})} (c_{31})^{w_{11}}$$

$$(7.73)$$

To maximize the dual objective function $d(w)$, we can set the partial derivatives of the substituted dual objective function $d(w_{11}, w_{12})$ with respect to w_{11} and w_{12}, respectively, to zero, and thus, get two equations to determine the unique values of dual weights w_{11} and w_{12} to maximize $d(w)$. Since it is easier to work with logarithm, we can substitute $z(w_{11}, w_{12}) = \ln d(w_{11}, w_{12})$ and then set the first derivatives $dz./w_{11} = 0$ and $dz./w_{12} = 0$ as follows:

Here, $c_{11} = c_{12} = 1$, hence omitted in subsequent expressions. Thus, taking logarithm of expression (7.73), we get:

$$
\begin{aligned}
z(w_{11}, w_{12}) &= \ln d(w_{11}, w_{12}) \\
&= (1 - w_{11} - 2w_{12})[\ln c_{o1} - \ln(1 - w_{11} - 2w_{12})] + w_{12}[\ln c_{o2} - \ln w_{12}] \\
&\quad + (w_{11} + w_{12})[\ln c_{o3} - \ln(w_{11} + w_{12})] \\
&\quad + w_{11}[\ln(w_{11} + w_{12}) - \ln w_{11}] + w_{12}[\ln(w_{11} + w_{12}) - \ln w_{12}] \\
&\quad + (1 - w_{11}) \ln c_{21} + w_{11} \ln c_{31}
\end{aligned}
$$

Setting the first partial derivative $dz./dw_{11}$ to zero, we get:

$$
\begin{aligned}
dz/dw_{11} &= -\ln c_{o1} + \ln(1 - w_{11} - 2w_{12}) \\
&\quad - (1 - w_{11} - 2w_{12})/(1 - w_{11} - 2w_{12})3(-1) \\
&\quad + \ln c_{o3} - \ln(w_{11} + w_{12}) - (w_{11} + w_{12})/(w_{11} + w_{12}) \\
&\quad + \ln(w_{11} + w_{12}) - \ln w_{11} + w_{11}/(w_{11} + w_{12}) + w_{12}/(w_{11} + w_{12}) \\
&= -\ln c_{o1} + \ln(1 - w_{11} - 2w_{12}) + \ln c_{o3} - \ln w_{11} - \ln c_{21} + \ln c_{31} = 0
\end{aligned}
$$

It may be seen that the non-logarithmic terms cancel each other; this is always true because of the normality and orthogonality conditions. Removing the logarithms, we finally get the following equation:

$$c_{03}c_{31}(1 - w_{11} - 2w_{12}) - c_{01}c_{21}w_{11} = 0 \qquad (7.74)$$

Similarly, setting the first partial derivative $dz./d\, w_{12}$ to zero, we get:

$$
\begin{aligned}
dz/d\, w_{12} = &-2\, \ln c_{01} + 2\ln(1 - w_{11} - 2w_{12}) \\
&- (1 - w_{11} - 2w_{12})/(1 - w_{11} - 2w_{12})3(-2) \\
&+ \ln c_{02} - \ln w_{12} - w_{12}/w_{12} + \ln c_{03} - \ln(w_{11} + w_{12}) \\
&- (w_{11} + w_{12})/(w_{11} + w_{12}) + w_{11}/(w_{11} + w_{12}) \\
&+ w_{12}/(w_{11} + w_{12}) - w_{12}/w_{12} + \ln(w_{11} + w_{12}) - w_{12} \\
= &\ 0
\end{aligned}
$$

After the non-logarithmic terms cancel each other, we get the following equation:

$$
\begin{aligned}
&-2\ln c_{01} + 2\ln(1 - w_{11} - 2w_{12}) + \ln c_{02} - \ln w_{12} + \ln c_{03} - \ln(w_{11} + w_{12}) \\
&+ \ln(w_{11} + w_{12}) - w_{12} = 0
\end{aligned}
$$

Removing the logarithms, we finally get the following equation:

$$c_{02}\, c_{03}(1 - w_{11} - 2w_{12})^2 - c_{01}^2 w_{12}^2 = 0 \qquad (7.75)$$

Thus, we have got two Eqs. (7.75) and (7.74) with two unknown dual variables, namely w_{11} and w_{12}. These equations could be easily solved to find the values of w_{11} and w_{12} using the Newton–Raphson method for which the computer program NR123.CPP is developed and presented in Chap. 3 for easy application. However, a computer program OUBF.CPP which is a modified version of the computer program NR123.CPP is also developed specifically for optimal design of urban build form. Knowing the values of w_{11} and w_{12}, the values of other dual variables could be determined, and the problem could be completely solved.

Thus, assigning specific values to the objective function constant coefficients c_{01}, c_{02}, and c_{03}; as well to the constraint function constant coefficients c_{21} and c_{31}, the optimization model can be calibrated and problem solved completely by solving the Eqs. (7.74) and (7.75) giving a *stationary point* and a *local minimum*. Since the model for optimal design of urban built form represented by the equation set (7.69) is a posynomial problem with only 'less than inequality' constraints, this *stationary point* and *local minimum* are also a *Global Minimum* requiring no further tests (Beightler et al. 1976).

In Sect. 6.6.5.2 of Chap. 6, the *Minimum* cost per built-unit was found approximately as: 842600 MU at an *Optimum* FAR of around 4.0 which is derived approximately using such *computer graphics* without resorting to complicated mathematical programming techniques, where the values of model constant coefficients adopted were as follows:

$c_{01} = 4500$, $P = 0.01$, $c_{02} = 450,030.01 = 450$, and $c_{03} = 10,000$, $c_{21} = 110$, and $c_{31} = 20$.

Adopting the same values of model constant coefficients, the Eqs. (7.75) and (7.74) are solved using the computer program OUBF.CPP which is a modified version of the computer program NR123.CPP presented in Chap. 3, incorporating the Eqs. (7.75) and (7.74) derived above.

Complete listing of the C++ program OUBF.CPP is placed in an online space connected to the book to help readers to use such techniques to achieve their own problem solutions. The program has two models—one model designated as Model-1 uses compact algebraic formulas developed below solving completely the optimal design of 'Urban-Built-Form' (UBF), and the Model-2 uses the Newton–Raphson Method to solve *Nonlinear Equations* for *Optimal Design* of UBF, which can be applied by any user for his/her specific problem situation/equation(s) by inserting the same in the above Model-2 Program.

To show the interaction between the Urban 'Built-Form-Elements' (BFE) and the *sensitivity* of the *objective function* to the values of some of these factors, three case studies are presented below using the computer program OUBF.CPP. Typical Interactive Input-Output dialogue of the computer program when run, giving the output results solving the above complicated equations in fraction of a second-in just 7 iterations as in Case 7.1 presented below. While running the program, the user should choose only feasible and realistic input values to get realistic results. For example, if input value of land price is given as zero, the program will give all model weights linked with land price including the FAR as zero, which are obviously not realistic and feasible; and in such a case optimization is also not applicable.

Case 7.1

```
          WELCOME TO PROGRAM: ''OUBF''
   Using ''OUBF'', you can CARRY OUT the following TASKS:
** (1) DETERMINE 'Optimal Design' of URBAN-BUILT-FORM in SITE-
SPECIFIC-CASE.
** (2) DETERMINE Interaction between the URBAN 'BUILT-FORM-
ELEMENTS' (BFE),
**    SUCH AS FLOOR AREA RATIO(FAR),LAND PRICE,BUILT-UNIT-COST(BUC),
BUILDING
**   RISE(BR),GROUND COVERAGE RATIO(GCR),AND BUILT-UNIT-DENSITY(BUD).
** (3) FIND Instantly,'LEAST-COST' per BUILT-UNIT of GIVEN PLINTH-AREA
('BUS'.)
** (4) DETERMINE Optimal 'PLANNING REGULATIONS' including 'FAR',
RESPONSIVE to
   PEOPLE'S NEED,based on 'CITIZEN-Planning-Authority-PARTICIPATORY-
   -INTERACTIVE' Session,in the CONTEXT of 'URBAN-DYNAMICS'.

PRESET VALUE OF LIMIT OF ITERATION NUMBER 'ITRL' = 50
PRESET VALUE OF ACCEPTABLE MARGIN OF ERROR 'ITOL' = 0.0001
DO YOU WANT TO CHANGE THE ABOVE PRESET VALUES (Y/N) ?
n
YOU HAVE NOT CHANGED THE VALUES OF 'ITRL' & 'ITOL'
PRESET VALUES OF CPAR= 4500 LP=10000 P=0.01 BUS=110 SPO=20
DO YOU WANT TO CHANGE THE ABOVE PRESET VALUES (Y/N) ?
n
YOU HAVE NOT CHANGED THE PRESET VALUES OF 'CPAR, LP, P, BUS& 'SPO'
```

'ITERATION MODELS' AVAILABLR HERE ARE SHOWN BELOW :
MODEL USING ALGEBRAIC FORMULAS SOLVING COMPLETELY OPTIMAL DESIGN OF
'UBF'
(ITMODEL- 1)
MODEL USING NR-METHOD TO SOLVE NONLINEAR EQUATIONS FOR OPTIMAL DESIGN OF
'UBF'
(ITMODEL- 2)

PUT YOUR CURRENT CHOICE OF ITMODEL (1/2)
2
PUT THE CURRENT VALUE OF X .5
PUT THE CURRENT VALUE OF Y .5
CURRENT VALUE OF ITERATION NUMBER IS = 1
CURRENT VALUE OF fx IS = -695000
CURRENT VALUE OF fy IS = -400000
CURRENT VALUE OF gx IS = 450000
CURRENT VALUE OF gy IS = -1.935e+07
CURRENT VALUE OF f IS = -347500
CURRENT VALUE OF g IS = -4.95e+06
CURRENT VALUE OF x IS = 0.5
CURRENT VALUE OF y IS = 0.5
PUT ANY CHARACTER TO CONTINUE n
NEW VALUE OF ITERATION NUMBER IS = 2
NEW VALUE OF f IS = 1.09637e-11
NEW VALUE OF g IS = -1.06511e+06
NEW VALUE OF x IS = 0.15179
NEW VALUE OF y IS = 0.23609
NEW VALUE OF ITERATION NUMBER IS = 3
NEW VALUE OF f IS = 6.10356e-12
NEW VALUE OF g IS = -217470
NEW VALUE OF x IS = 0.212926
NEW VALUE OF y IS = 0.130042
NEW VALUE OF ITERATION NUMBER IS = 4
NEW VALUE OF f IS = 1.05445e-11
NEW VALUE OF g IS = -25797.7
NEW VALUE OF x IS = 0.233979
NEW VALUE OF y IS = 0.0934446
NEW VALUE OF ITERATION NUMBER IS = 5
NEW VALUE OF f IS = 2.77112e-12
NEW VALUE OF g IS = -632.795
NEW VALUE OF x IS = 0.237271
NEW VALUE OF y IS = 0.0777236
NEW VALUE OF ITERATION NUMBER IS = 6
NEW VALUE OF f IS = -5.00222e-12
NEW VALUE OF g IS = -0.417795
NEW VALUE OF x IS = 0.237366
NEW VALUE OF y IS = 0.0775766
NEW VALUE OF ITERATION NUMBER IS = 7
NEW VALUE OF f IS = -2.91323e-12
NEW VALUE OF g IS = -1.72542e-07
NEW VALUE OF x IS = 0.237366
NEW VALUE OF y IS = 0.0775765
CURRENT VALUE OF ITERATION NUMBER IS = 7

```
CURRENT VALUE OF f IS = -2.91323e-12
CURRENT VALUE OF g IS = -1.72542e-07
CURRENT VALUE OF x IS = 0.237366
CURRENT VALUE OF y IS = 0.0775765


INPUT MODEL CONST.VALUES CHOSEN ARE:
CONSTRCTN.PLINTH AREA RATE:CPAR=4500 MU/SQ.M. LAND PRICE=10000 MU/SQ.
M. P=0.01
BUILT-UNIT-SPACE ('BUS')=110 SQ.M./BU OPEN-SPACE ('SPO')=20 SQ.M./BU
OUTPUT OPTIMAL VALUES ARE:
MODEL WEIGHTS: w11= 0.2374 w12= 0.0776 w01= 0.5775 w02= 0.0776 w03 =
0.3249
 w21= 0.7626 w31=0.2374
OPTIMAL FAR= 4.017   LEAST B-UNIT-COST= 742570 MU  OPTIMAL BLDG.RISE=
14.91
GROUND COVERAGE RATIO (GCR) = 0.270
BUILT-UNIT-DENSITY (BUD)=365.243 BUILT UNITS/HECTARE
THE VALUE OF ITOL CHOSEN IS = 0.0001
 SINCE THE CURRENT SOLUTION MEETS THE VALUE OF CHOSEN
 MARGIN OF ERROR, THE SOLUTION ACCEPTED & PROGRAM TERMINATED
'ITERATION CLOSED'
```

In the above computer output given in C++ language, the optimal model weights w_{01}, w_{02}, w_{03}, w_{11}, w_{12}, w_{21}, and w_{31} are shown without subscript.

It may be seen that the exact solution presented above gives the optimal FAR as 4.017 and the least built-unit-cost as 742,570 MU as against approximate FAR value of about 4 and minimum BU-Cost of 742,600 MU given by graphic solution presented in Table 7.1 and Fig. 7.8 in Sect. 7.11.1 above. Thus, there is only a marginal difference between the exact solution and the approximate solution given by the computer graphics.

It may be observed that the exact solution gives the value of optimum Building Rise (BR) as 14.91 number of floors which should be in whole number. As earlier explained, in geometric programming values of all variables are assumed continuous giving fractional values of number of floors which ideally should be in integer values requiring application of nonlinear integer programming, which is quite complicated to apply without commensurate benefit (interested readers can try this). But, litera-ture indicates that efficient solution methods are not available even for linear integer programming (Mark et al., 1970). However, the problem can be overcome by adopting the concept of 'effective building rise' where the fractional value of Building Rise (BR) may be interpreted as: 'whole number part' has the whole 'Built-Unit-Space' (BUS) while the fractional part has corresponding fractional 'Built-Unit-Space' (BUS), which is also in line with the usual urban planning regulations permitting a fractional coverage at the top floor of a building. However, the exact integer value of BR can be obtained using an appropriate model of application software sub-module 'CAUB' presented in Chap. 6, barely affecting the optimization results.

Now, if we study the output optimal weights w_{01}, w_{02}, and w_{03}, these give some interesting insight to the interactions between the cost related to building construction cost per unit of plinth area, cost related to building rise in terms of the increased cost per unit of plinth area, and the cost related to land price per unit of land area, which are rarely considered in conventional urban development planning, design and management while taking management decisions. For example, the Model for Optimal Design of Urban-Built-Form presented above representing such interactions clearly indicates that land price has a profound influence on the optimal Urban-Built-Form to *Minimize* the Built-Unit-Cost (BUC), which tends to be more vertical (i.e. higher building rise) with the increase in land price creating a higher Floor Area Ratio (FAR), lower Ground Coverage Ratio (GCR), and higher Built-Unit-Density (BUD) while the Built-Unit-Space (BUS) and Semi-Public Open Space (SPO) per Built-Unit (BU) are kept unchanged. On the other hand, for lower land price the optimal Urban-Built-Form tends to be less vertical (i.e. lower building rise) with lower FAR, higher GCR, and lower BUD as shown below. But, in the conventional urban planning and design practice rarely such interactions are considered while prescribing such planning regulations which are frequently imposed on arbitrary basis creating *inefficiency* and *inequity* in the urban development process, promoting growth of slums in many cases. In view of acute pressure on land in many countries where per capita land is reducing fast due to increase in population while land mass remains same, a *Planning Policy* to increase built-unit-density without decreasing per built-unit semi-public open space is imperative, if all *stakeholders* including the urban low-income population are to be accommodated in the urban development process.

In the computer results presented in Case 7.1 above solving the *Optimization Model*, the land price is assumed as 10,000 MU/SQ.M., and thus, the building construction cost, the increased cost related to building rise, and the cost related to land price constitutes $0.5775(w_{01})$, $0.0776(w_{02})$ and $0.3249(w_{03})$, i.e. 57.75%, 7.76%, and 32.49% of the total cost per built-unit. On the other hand, if the land price is reduced to only 1000MU/SQ.M., keeping values of other input parameters unchanged, the computer results of this Case 7.2 solving above equations in just 7 iterations are given below.

Case 7.2

```
SUMMARY OF ''oubfm2.cpp'' RESULTS IN THE INSTANT CASE ARE:
INPUT MODEL CONSTANT VALUES CHOSEN BY USER ARE:
CONSTRUCTION PLINTH AREA RATE: CPAR= 4500 MU/SQ.M. LAND PRICE= 1000
MU/SQ.M. P=0.010
BUILT-UNIT-SPACE ('BUS')=110.00 SQ.M./BU  OPEN-SPACE ('SPO')= 20.00
SQ.M./BU
CURRENT VALUE OF ITERATION NUMBER IS = 7
CURRENT VALUE OF f IS =    0.000000000
CURRENT VALUE OF g IS =   -0.000000033
CURRENT VALUE OF x IS = 0.035607
CURRENT VALUE OF y IS = 0.041545
OUTPUT OPTIMAL MODEL WEIGHTS: w11= 0.0356 w12= 0.0415 w01= 0.7713 w02=
0.0415
w03 = 0.0772 w21= 0.9644 w31=0.0356
OPTIMAL FAR= 2.537  LEAST B-UNIT-COST= 561669 MU  OPTIMAL BLDG.RISE=
4.71
```

```
GROUND COVERAGE RATIO (GCR) = 0.537
BUILT-UNIT-DENSITY (BUD) =230.763 BUILT UNITS/HECTARE
THE VALUE OF ITOL CHOSEN IS =0.000100
 SINCE THE CURRENT SOLUTION MEETS THE VALUE OF CHOSEN
 MARGIN OF ERROR, THE SOLUTION ACCEPTED & PROGRAM TERMINATED
```

It may be seen that as a result of reducing land price from 10,000 MU/SQ.M in the former case to 1000 MU/SQ.M in the latter case, the component of the building construction cost has increased to 77.13% (w_{01} in latter case as shown above) from 57.75% in the former case. Interestingly, the component of the cost related to land price in the latter case has come down to only 7.72% (w_{03} in latter case as shown above) from 32.49% in the former case. Again, as a result of reduction in land price, the value of optimal FAR has reduced to 2.537(as shown above) from 4.017 in the former case, and also, the value of optimal effective Building Rise (BR) has reduced to 4.71 only (as shown above) from 14.91 in the former case. Similarly, there are large variations in values of other planning and design parameters such as least built-unit-cost, ground coverage ratio, and Built-Unit-Density (BUD).

Again, if the land price is further reduced to only100 MU/SQ.M and the openspace SPO is increased to 30 SQ.M. per built-unit, keeping values of other input parameters unchanged, the computer results solving above equations in just 10 iterations is given below as Case 7.3(a), and in Case 7.3(b) presented below SPO value is retained at 20 SQ.M. per built-unit, keeping values of other input parameters unchanged.

Case 7.3(a)

```
SUMMARY OF ''oubfm2.cpp'' RESULTS IN THE INSTANT CASE ARE:
INPUT MODEL CONSTANT VALUES CHOSEN BY USER ARE:
CONSTRUCTION PLINTH AREA RATE: CPAR= 4500 MU/SQ.M. LAND PRICE=
100 MU/SQ.M. P=0.010
BUILT-UNIT-SPACE ('BUS')=110.00 SQ.M./BU  OPEN-SPACE ('SPO')= 30.00
SQ.M./BU
CURRENT VALUE OF ITERATION NUMBER IS = 10
CURRENT VALUE OF f IS =    0.000000000
CURRENT VALUE OF g IS =    0.000000000
CURRENT VALUE OF x IS = 0.005751
CURRENT VALUE OF y IS = 0.014391

OUTPUT OPTIMAL MODEL WEIGHTS: w11= 0.0059 w12= 0.0144 w01= 0.9654 w02=
0.0144
w03 = 0.0202 w21= 0.9941 w31=0.0059
OPTIMAL FAR= 1.060  LEAST B-UNIT-COST= 512757 MU  OPTIMAL BLDG.RISE=
1.49
GROUND COVERAGE RATIO (GCR) = 0.711
BUILT-UNIT-DENSITY (BUD) =96.347 BUILT UNITS/HECTARE
THE VALUE OF ITOL CHOSEN IS =0.000100

 SINCE THE CURRENT SOLUTION MEETS THE VALUE OF CHOSEN
 MARGIN OF ERROR, THE SOLUTION ACCEPTED & PROGRAM TERMINATED
```

Case 7.3(b)

```
SUMMARY OF ''oubfm2.cpp'' RESULTS IN THE INSTANT CASE ARE:
INPUT MODEL CONSTANT VALUES CHOSEN BY USER ARE:
CONSTRUCTION PLINTH AREA RATE: CPAR= 4500 MU/SQ.M. LAND PRICE=
100 MU/SQ.M. P=0.010
BUILT-UNIT-SPACE ('BUS')=110.00 SQ.M./BU  OPEN-SPACE ('SPO')= 20.00
SQ.M./BU
CURRENT VALUE OF ITERATION NUMBER IS = 9
CURRENT VALUE OF f IS =    0.000000000
CURRENT VALUE OF g IS =   -0.000097999
CURRENT VALUE OF x IS = 0.003907
CURRENT VALUE OF y IS = 0.014419


OUTPUT OPTIMAL MODEL WEIGHTS: w11= 0.0039 w12= 0.0144 w01= 0.9673 w02=
0.0144
w03 = 0.0173 w21= 0.9961 w31=0.0039
OPTIMAL FAR= 1.173  LEAST B-UNIT-COST= 511757 MU  OPTIMAL BLDG.RISE=
1.49
GROUND COVERAGE RATIO (GCR) = 0.777
BUILT-UNIT-DENSITY (BUD) =106.621 BUILT UNITS/HECTARE
THE VALUE OF ITOL CHOSEN IS =0.000100

 SINCE THE CURRENT SOLUTION MEETS THE VALUE OF CHOSEN
MARGIN OF ERROR, THE SOLUTION ACCEPTED & PROGRAM TERMINATED
```

It may be seen that in Case 7.3(a) where land price assumed is only 100 MU/SQ. M, the component of the building construction cost has increased to 96.54%, and the component of the cost related to land price has come down to only 2.02%, even though the open space SPO is increased to 30 SQ.M. per built-unit. Similarly, the value of optimal FAR has reduced to only 1.060 (as shown in Case 7.3(a) above) from 4.017 in Case 7.1, and also, the value of optimal effective Building Rise (BR) has reduced to 1.49 only (as shown in Case 7.3(a) above). In Case 7.3 (b) where land price assumed is 100 MU/SQ.M and open space SPO is retained at 20 SQ.M. per built-unit, the value of optimal FAR is 1.173 and Ground Coverage Ratio (GCR) is increased to 0.777 while optimal effective Building Rise (BR) remains same at 1.49 as in Case 7.3(a). Normally, in conventional practice, an FAR of around 1 is prescribed, with the implied assumption that the land price will be very low which the case is rarely. In Case 7.3(a), the Ground Coverage Ratio (GCR) = 0.711, which in Case 7.3(b) is further increased to 0.777 (as against 0.270 in Case 7.1) which is quite high affecting the openness of a layout even after increasing the SPO value by 50%. Again, in this case, the Built-Unit-Density (BUD) is only 96.347 Built Units/Hectare, i.e. only about 26% of 365.243 Built Units/Hectare achieved in Case 7.1, seriously affecting the efficiency of land utilization which is becoming very scarce and costly with rise in population in many countries while land mass is constant.

Such large variations in the values of optimal parameters only due to change of value of only one parameter (here land price) emphasize the need for such scientific planning and design analysis and design optimization to achieve *efficiency* and *equity* in urban development process, and thus, attain a *resource-efficient* solution of our mounting urban problems.

7.12.7.1 Compact Formulas Developed for Urban-Built-Form Design Optimization

It may be seen that the optimization model incorporates a number of important urban development planning and design parameters such as optimum FAR, optimum building rise which are generally adopted on *arbitrary basis* without any *scientific analysis*, affecting seriously the *efficiency* and *equity* in an urban development which should provide for all stakeholders in the urban sector. This is mainly because *optimization techniques* and *computer-aided design* rarely form a part of educational *curricula* of AEC professional courses. As a result, AEC professionals are generally not familiar with such techniques, and naturally, are reluctant to apply such techniques to achieve *efficiency* and *equity* in their planning and design solutions. This needs correction by suitable modification of educational *curricula* of AEC professional courses.

Fortunately, in the instant case, while solving the *nonlinear programming optimization model*, it is possible to develop compact formulas for optimum FAR, optimum building rise, and other planning parameters in terms of site-specific data. These can be applied in site-specific *planning* and *design* by AEC professionals without developing an optimization model or without any knowledge of optimization techniques. To help this process, a number of compact formulas for optimal FAR, optimal BR, least-built-unit-cost, and so on, as developed are presented below.

From (7.74) we get $(1 - w_{11} - 2w_{12}) = (c_{01}c_{21}/c_{03}c_{31})w_{11}$.

From (7.75) we get $w_{12}^2 = (c_{02}c_{03}/c_{01}^2)(1 - w_{11} - 2w_{12})^2 = (c_{02}c_{03}/c_{01}^2)$ $(c_{01}c_{21}/c_{03}c_{31})^2 w_{11}^2$

$$\text{Therefore, from above we get } w_{12}\sqrt{\frac{c_{02}}{c_{03}}}3\left(\frac{c_{21}}{c_{31}}\right) = 3w_{11} \qquad (7.76)$$

Again, from (7.74) using the above value of w_{12}, we get:

$$c_{03}c_{31}\left(1 - w_{11} - 2\sqrt{\frac{c_{02}}{c_{03}}}\left(\frac{c_{21}}{c_{31}}\right)w_{11}\right) - c_{01}c_{21}w_{11} = 0$$

or,

$$c_{03}c_{31} - c_{03}c_{31}w_{11} - 2\,c_{03}c_{31}\sqrt{\frac{c_{02}}{c_{03}}}\left(\frac{c_{21}}{c_{31}}\right)w_{11} - c_{01}c_{21}w_{11} = 0$$

or,

$$w_{11} = [c_{03}c_{31}/(c_{03}c_{31} + 2\sqrt{c_{03}c_{02}}3c_{21} + c_{01}c_{21})] \qquad (7.77)$$

Therefore,

$$\sqrt{\frac{c_{02}}{c_{03}}}3\left(\frac{c_{21}}{c_{31}}\right)3 \quad c_{03}\,c_{31} = \sqrt{c_{03}c_{02}}3\,c_{21}$$

$$w_{12} = \sqrt{\frac{c_{02}}{c_{03}}}3\left(\frac{c_{21}}{c_{31}}\right)3\left[c_{03}\,c_{31}/(c_{03}\,c_{31} + 2\sqrt{c_{03}c_{02}}3\,c_{21} + c_{01}c_{21})\right]$$

$$w_{12} = \sqrt{c_{03}c_{02}}3c_{21}/(c_{03}\,c_{31} + 2\sqrt{c_{03}c_{02}}3c_{21} + c_{01}c_{21}) \qquad (7.78)$$

Therefore, using the above values of w_{11} and w_{12} and the equation set (7.72) we can determine the values of all the dual weights or variables: w_{01}, w_{02}, w_{03}, w_{21}, and w_{31}.

Hence, using the above values of dual weights w_{11} and w_{12} in terms of the constant coefficients and the equation set (7.72) we can determine the values of all the dual variables in terms of the constant coefficients, and thus, solving completely the dual problem.

Again, instead of complete solution of each problem with specific values of constant coefficients for each case, we can express the optimal design decision variables x_1, x_2, x_3, and x_4, i.e. four decision variables in terms of the constant coefficients as illustrated below, and thus, help derive optimal values of design decision variables in a site-specific case even without any knowledge of complicated optimization techniques such as geometric programming.

Thus, referring to the objective function $y(x)$ given by expression (7.72) its optimum value, i.e. $y_0(x)$ is given by:

$$y_0(x) = c_{01}\,x_1^* + c_{02}\,x_1^*\,x_4^* + c_{03}\,x_1^*\,x_1^{*-1},$$

where a design decision variable with asterisk mark indicates their optimum values.

Again, each term of the objective function $= y_0(x)$ 3 weight in respect of the term.

Thus, the first term $c_{01}\,x_1^* = y_0(x)\,3w_{01}$.

Similarly, third term $c_{03}\,x_1^*\,x_3^{*-1} = y_0(x)\,3w_{03}$.

Dividing first term by third term, we get $c_{01}\,x_1^*x_3^*/c_{03}\,x_1^* = y_0(x)3w_{01}/y_0(x)3w_{03}$.

Hence, optimal value of FAR is given by:

$$x_3^* = (c_{03}/c_{01}) * (w_{01}/w_{03})$$

From Eqs. (7.77), (7.78), and (7.72), we get:

$$w_{03} = w_{11} + w_{12}$$
$$= \left[c_{03}\,c_{31} + \sqrt{c_{03}c_{02}}3c_{21}\right]/\left[(c_{03}\,c_{31} + 23\sqrt{c_{03}c_{02}}3\,c_{21} + c_{01}c_{21})\right] \qquad (7.79)$$

$$w_{01} = (1 - w_{11} - 2w_{12}) = c_{01}c_{21}/[(c_{03}\,c_{31} + 23\sqrt{c_{03}c_{02}}3c_{21} + c_{01}c_{21})] \qquad (7.80)$$

Therefore, $w_{01}/w_{03} = c_{01}c_{21}/\left[c_{03}\,c_{31} + \sqrt{c_{03}c_{02}}3c_{21}\right]$

$$w_{02} = w_{12} = \sqrt{\frac{c_{02}}{c_{03}}}3\left(\frac{c_{21}}{c_{31}}\right)3\left[c_{03}\,c_{31}/(c_{03}\,c_{31} + 2\sqrt{c_{03}c_{02}}3c_{21} + c_{01}c_{21})\right]$$

$$= \left[\sqrt{c_{03}c_{02}}3c_{21}\right]/[c_{03}\,c_{31} + 2\sqrt{c_{03}c_{02}}3c_{21} + c_{01}c_{21})] \tag{7.81}$$

Therefore,

$$(w_{02}/w_{01}) = \left[\sqrt{c_{03}c_{02}}3c_{21}\right]/(c_{01}\,c_{21}) = \sqrt{c_{03}c_{02}}/c_{01}$$

Thus, by simplifying the above expressions, the **Optimum Floor Area Ratio** (FAR) to give the **Least-Cost** per Built-Unit is given by the following simplified expression linked only with *constant coefficients* of the *Optimization Model*:

Simplified Formula of __Optimum FAR__$(FAR) = (c_{03}/c_{01}) * (w_{01}/w_{03})$

$$= \frac{c_{03}c_{21}}{\left[c_{03}c_{31} + \sqrt{c_{03}c_{02}}3c_{21}\right]} \tag{7.82}$$

Similarly, the optimal value of the effective building rise is given by:

$$y_0(x)3w_{02}/y_0(x)3w_{01} = c_{02}\,x_1{}^* \,x_4{}^*/c_{01}\,x_1{}^*$$

$$x_4{}^* = (c_{01}/c_{02}) * (w_{02}/w_{01}) = (c_{01}/c_{02}) * \sqrt{c_{03}c_{02}}/c_{01} = \sqrt{c_{03}/c_{02}}$$

__OptimumBuilding Rise__ (BR) in Effective Number of Floors

$$= \sqrt{c_{03}/c_{02}} \tag{7.83}$$

$$x_2{}^* = c_{31}$$

$$x_1{}^* = c_{21}$$

$$y_0(x) = c_{01}\,x_1{}^*/w_{01} = c_{01} * c_{21}/w_{01}$$

$$= (c_{01} * c_{21}) * [(c_{03}\,c_{31} + 23\sqrt{c_{03}c_{02}}3c_{21} + c_{01}c_{21})]/(c_{01}c_{21})$$

Simplified Formula for Least-Cost (LC) per Built-Unit in Monetary Units (MU)

$$\mathbf{LC} = c_{03}\,c_{31} + 23\sqrt{c_{03}c_{02}}3c_{21} + c_{01}c_{21} \tag{7.84}$$

$$\textbf{Ground Coverage Ratio (GCR)} = FAR/BR \tag{7.85}$$

Built-Unit-Density (BUD) $= 100003FAR/BUS$ Built-Units/Hectare   (7.86)

It may be seen that the expressions (7.76)–(7.86) give the complete solution of the Model for Optimal Design of Urban-Built-Form presented in expression (7.69) above, only in terms of the objective function constant coefficients and the constraint function constant coefficients. Thus, the AEC professionals such as engineers, architects, urban planners, and urban managers can solve the model for optimal design of urban-built-form in their site-specific situation by applying these formulas, and even using only a hand-held calculator.

However, as already stated the C++ Computer Program OUBF.CPP incorporates one Model (designated as Model-1) which uses the above algebraic equations giving instantly the complete solution of the Model for Optimal Design of Urban-Built-Form as offered above. AEC professionals can use this program to find instantly the complete solution of the Model for Optimal Design of Urban-Built-Form. A typical result of the C++ Computer Program OUBF.CPP solving the above algebraic formulas (7.76)–(7.86) giving the complete solution of the Model for Optimal Design of Urban-Built-Form is also presented below, taking Input Data as in Case 7.1 above:

```
SUMMARY OF ''oubfm1.cpp'' RESULTS IN THE INSTANT CASE ARE:
INPUT MODEL CONSTANT VALUES CHOSEN BY USER ARE:
CONSTRUCTION PLINTH AREA RATE: CPAR= 4500 MU/SQ.M. LAND PRICE= 10000
MU/SQ.M. P=0.010
BUILT-UNIT-SPACE ('BUS')=110.00 SQ.M./BU  OPEN-SPACE ('SPO')= 20.00
SQ.M./BU

OUTPUT OPTIMAL MODEL WEIGHTS: w11= 0.2374 w12= 0.0776 w01= 0.5775 w02=
0.0776
w03 = 0.3249 w21=  0.7626 w31=0.2374
OPTIMAL FAR= 4.017  LEAST B-UNIT-COST= 742570 MU  OPTIMAL BLDG.RISE=
14.91
GROUND COVERAGE RATIO (GCR) = 0.270
BUILT-UNIT-DENSITY (BUD) =365.243 BUILT UNITS/HECTARE
```

It may be seen that the above output computer results using the algebraic formulas (7.76)–(7.86) are exactly same as the output computer results given by solving the nonlinear Eqs. (7.74) and (7.75) presented as: Case 7.1 above, as in both cases input model constant values chosen are same. *Thus, results of algebraic formulas are exactly same as the optimal solution represented by the solution of the system of nonlinear equations, and a user can choose any of these methods for his/her problem solution.*

Therefore, AEC professionals not having any knowledge of *Optimization and Computer-Aided Design Techniques* should have no inhibition in applying this model for optimal design of Urban-Built-Form and the *optimization principles* in their planning, design, and urban management practices to help achieve *efficiency* and *equity* in the urban development process although acquiring such expert knowledge is desirable for AEC professionals to be more effective in solving our mounting urban problems. To help this process, complete listing of the C++ program OUBF.CPP is presented in an online space attached to this book.

In Chaps. 8–11, similar *optimization models* for integrated computer-aided structural design optimization, building design optimization, layout design optimization, and the dwelling-layout system design optimization are presented along with C++ Computer Programs (presented in an online space attached to this book) to solve such optimization models both in numeric forms and also in graphic forms using AutoLISP language.

7.12.8 *Signomial Geometric Programming*

The posynomial **geometric** programming presented above was such that every term in both objective function and in constraints had only positive coefficients. Additionally, each constraint was of 'less than inequalities' type and written as: $y_m(x) \leq 1$, $m = 1, 2, 3 \ldots M$. In many real world problems, there may be terms with negative coefficients or there may be constraints of 'more than inequalities' type written as: $y_m(x)/1$, $m = 1, 2, 3 \ldots M$. If a geometric programming problem has terms with negative coefficients or if any constraint is of the form $y_m(x)/1$, such problems are called signomial geometric programming problems. Such occurrence of terms with both positive and negative coefficients creates the need to expand the Eqs. (7.62) through (7.67) as discussed below.

The constraint function in Eq. (7.43) presented earlier can also be expressed as:

$$y_m(x) = \sum\nolimits_{t=1}^{T_m} \sigma_{mt} c_{mt} \prod\nolimits_{i=1}^{N} x_i^{a_{mti}} \overline{1}; \qquad m = 1,2,3 \ldots M \qquad (7.87)$$
$$x_i > 0; \qquad i = 1,2,3 \ldots N$$

We can now define a signum function σ_m for the mth constraint such that it takes the value of +1 when $y_m(x) \leq 1$ and the value -1 when $y_m(x)/1$, respectively. Similarly, signum function can be used to represent negative sign of terms in objective function.

Using such concepts of signum function, we can modify the posynomial geometric programming problem given by expressions (7.42) and (7.43) presented previously, and define a nonlinear objective function with N variables and T terms in the following form converting it to a signomial geometric programming problem:

$$\text{Minimize } y_o(x) = \sum\nolimits_{t=1}^{T_o} \sigma_{ot} c_{ot} \prod\nolimits_{i=1}^{N} x_i^{a_{oti}} \qquad (7.88)$$

Subject to M inequality constraints given by:

$$y_m(x) = \sum\nolimits_{t=1}^{T_m} \sigma_{mt} c_{mt} \prod\nolimits_{i=1}^{N} x_i^{a_{mti}} \leq \sigma_m; \qquad m = 1,2,3 \ldots M \qquad (7.89)$$
$$x_i > 0; \qquad i = 1,2,3 \ldots N$$

where

$$\sigma_{mt} = 61; \qquad t = 1,2,3 \ldots T_m; \quad m = 0.1,2, \ldots M$$
$$\sigma_m = 61; \qquad m = 1,2,3 \ldots M;$$
$$c_{mt} > 0; \qquad t = 1,2,3 \ldots T_m; \quad m = 0,1,2 \ldots M$$
$$x_i > 0; \qquad i = 1,2,3 \ldots N$$

Here, the T signum functions σ_{mt} for $t = 1, 2, 3 \ldots T_m$; $m = 0.1, 2 \ldots M$, are added to absorb the sign of each term, and the signum functions σ_m are added to generalize the constraint right-hand side to 61. These variables in the problem formulation are referred as signum functions.

As done in posynomial case, it is possible to exploit the special form of this problem and to obtain an equivalent problem with linear constraints which is much easier to solve. This has been done by Beightler et al. (1979) who have provided the following compact formulas for signomial geometric programming, which have been applied for solution of optimization problems in relevant chapters of this book. The derivation and proof of these compact formulas are also given by Beightler et al. in their above books and interested readers can refer to them.

To summarize, a signomial geometric programming problem can be stated as:

I. When the objective function is posynomial in form (all signum functions are positive)

$$\text{Minimize } y_o(x) = \sum_{t=1}^{T_o} c_{ot} \prod_{i=1}^{N} x_i^{a_{oti}}$$

II. When all the constraints are signomial in form (one or more of the signum functions are negative)

$$y_m(x) = \sum_{t=1}^{T_m} \sigma_{mt} \ c_{mt} \prod_{i=1}^{N} x_i^{a_{mti}} \leq \sigma_m; \quad m = 1,2,3 \ldots M$$

The solution to this geometric programming problem can be derived by solving equivalent dual geometric program as presented by Phillips and Beightler (1976).

A general signomial problem can be written as:

$$y_m(x) = \sum_{t=1}^{T_m} \sigma_{mt} \ c_{mt} \prod_{i=1}^{N} x_i^{a_{mti}}; \quad m = 1,2,3 \ldots M \tag{7.90}$$

with

$$\sigma_{mt} = 61$$

and

$$c_{mt} > 0$$

The primal problem is to minimize

$$y_o(x) \tag{7.91}$$

Subject to M inequality constraints given by:

$$y_m(x) \leq \sigma_m(=61); \quad m = 1,2,3 \ldots M \tag{7.92}$$

and $x_i > 0 \quad i = 1, 2, 3 \ldots N$.

Now, considering a set of signum functions $\sigma_m (m = 0, 1, 2 \ldots M)$, and a set of $T = \sum_{m=0}^{M} T_m$ dual variables designated as 'w' that satisfy the generalized normality conditions given by:

$$\sum_{t=1}^{T_0} \sigma_{ot} w_{ot} = \sigma_o \tag{7.93}$$

N orthogonality conditions given by:

$$\sum_{m=0}^{M} \sum_{t=1}^{T_m} \sigma_{mt} a_{mti} w_{mt} = 0; \quad i = 1,2,3 \ldots N \tag{7.94}$$

T non-negativity conditions given by:

$$w_{mt}/0; \quad t = 1,2,3 \ldots T_m; \quad m = 0,1,2 \ldots M$$

as well as 'M' linear inequality constraints given by:

$$w_{mo} = \sigma_m \sum_{t=1}^{T_m} \sigma_{mt} w_{mt} \geq 0; \quad m = 0,1,2 \ldots M \tag{7.95}$$

The definitions of the signum functions in this case are as follows:

$$\begin{aligned}
\sigma_m &\equiv 61; \quad m = 0,1,2 \ldots M \\
\sigma_m &\equiv 61; \quad t = 1,2,3 \ldots T_m; \quad m = 0,1,2 \ldots M
\end{aligned} \tag{7.96}$$

Using these variables, the coefficients c_{mt}, and the signum functions, the equivalent dual function is given by:

$$d(w) = \sigma_o \left(\prod_{m=0}^{M} \cdot \prod_{t=1}^{T_m} \cdot \left(\frac{c_{mt} w_{mo}}{w_{mt}} \right)^{\sigma_{mt} w_{mt}} \right)^{\sigma_o} \tag{7.97}$$

It may be noted that as a result of Eqs. (7.93) and (7.95):

$$w_{oo} = 1 \tag{7.98}$$

Thus, for every point x^0, where $y_o(x)$ is locally minimum there exists a set of dual variables σ^0 and w^0 satisfying Eqs. (7.93)–(7.96) so that:

$$d(w^0) = y_o(x^0) \tag{7.99}$$

Therefore, as the dual variables w and σ are known, the corresponding primal variables x are found from the following relationships:

$$c_{0t} \prod_{i=1}^{N} x_i^{a_{0ti}} = w_{0t} \sigma_o y_o(x^0); \quad t = 1,2,3 \ldots T_0 \tag{7.100}$$

$$c_{mt} \prod_{i=1}^{N} x_i^{a_{mti}} = \frac{w_{mt}}{w_{m0}} \quad t = 1,2,3 \ldots T_m$$

$$m = 0,1,2 \ldots M \tag{7.101}$$

Thus, we can find N equations solvable for the N primal variables provided enough primal constraints are tight at optimality, and also because there are always more primal terms than primal variables x. Although the Eqs. (7.100) and (7.101) are nonlinear, each has only one term, and therefore, we can take logarithms giving linear simultaneous equation in log x_i and can easily be solved.

7.12.8.1   An illustrative Example of Signomial Geometric Programming Problem

To illustrate the application of the above equations in solving a signomial geometric programming problem we can use the following problem (taken from Beightler et al., 1979):

$$\text{Minimize } y_0 = -5x_1^2 + x_2^2 x_3^4$$

Subject to

$$(5/2) \, x_1^2 x_2^{-2} - (3/2) \, x_2^{-1} x_3 \leq -1$$

$$x_1, x_2, x_3 > 0$$

The problem is already in standard geometric programming format.

The problem has the number of terms $T = 4$, and the number of variables $N = 3$. Hence, the degree of difficulty in the problem is $= T - (N + 1) = 4-4 = 0$.

Thus, for this problem, the value of signum functions is as follows:

$$\sigma_{01} = -1.0; \quad \sigma_{02} = +1.0; \quad \sigma_{11} = 1.0; \quad \sigma_{12} = -1.0;$$
$$c_{01} = 5.0; \quad\quad c_{02} = 1.0; \quad\quad c_{11} = 5/2; \quad c_{12} = 3/2;$$

We can guess the value of signum function for objective function $\sigma_0 = -1.0$.

In this case, using the generalized normality condition given by the Eq. (7.93) above, we get:

$$w_{01} + w_{02} = \sigma_0 = -1.0; \text{or } w_{02} = w_{01} - 1.0; \quad (1)$$

Similarly, in this case, using the N orthogonality conditions given by Eq. (7.94), we get:

$$2w_{02} - 2w_{11} + w_{12} = 0; \quad \text{orthogonality condition for } x_1 \quad (2)$$

$$-2w_{01} + 2w_{11} = 0; \quad \text{orthogonality condition for } x_2 \quad (3)$$

$$4w_{02} - w_{12} = 0; \quad \text{orthogonality condition for } x_3 \quad (4)$$

In this case, there are four equations for four dual variables, i.e. w_{01}, w_{02}, w_{11}, and w_{12}, giving unique value for each variable which are derived as follows:

From (3) $w_{11} = w_{01}$;
Therefore from (1) and (2) $w_{12} = 2w_{11} - 2w_{02} = 2w_{01} - 2w_{01} + 2 = 2$;
From (4) $4w_{02} = w_{12}$ or $w_{02} = 2/4 = 0.5$;
From (1) $w_{01} = w_{02} + 1.0 = 1.50$; and $w_{11} = w_{01} = 1.50$.

Since the values of all the dual variables are positive, the initial guess of value for signum function for the objective function $\sigma_0 = -1.0$ was correct.

The above dual variables 'w' should also satisfy the linear inequality constraints, i.e. the additional non-negativity constraint given by Eq. (7.95) above which in this case is as follows:

$$w_{10} = \sigma_1 [\sigma_{11} w_{11} + \sigma_{12} w_{12}] = (-1.0)[(1)(1.5) + (-1.0)(2)] = +0.50/0;$$

Thus, additional non-negativity constraint is satisfied (7.97).
The dual objective function in this case using the Eq. (7.80) is given by:

$$d(w^*) = (\sigma_0)\left(\left(\left(\frac{c_{01}}{w_{01}}\right)^{\sigma_{01}w_{01}}\left(\frac{c_{02}}{w_{02}}\right)^{\sigma_{02}w_{02}}\left(\frac{(c_{11})(w_{10})}{w_{11}}\right)^{\sigma_{11}w_{11}}\left(\frac{(c_{12})(w_{10})}{w_{12}}\right)^{\sigma_{12}w_{12}}\right)^{(\sigma_0)}\right)$$

Putting the values of coefficients, dual variables, and the signum functions, the dual objective function in this case using the Eq. (7.97) is given by:

$$d(w^*) = (-1.0)\left(\left(\left(\frac{5}{1.5}\right)^{-1.5}\left(\frac{1.0}{0.50}\right)^{0.50}\left(\frac{(0.50)(5/2)}{1.5}\right)^{1.5}\left(\frac{(0.50)(3/2)}{2.0}\right)^{-2.0}\right)^{-1}\right)$$
$$= -0.7954951 = y_0(x^*)$$

The values of primal solution variables can now be obtained by simultaneous solution of the Eqs. (7.100) and (7.101) presented above and as shown below.
Thus, using Eq. (7.100), we get:

$$-5x_1^2 = (\sigma_0)(w_{01})(\sigma_{01})y_0(x^*) = (-1.0)(1.5)(-1.0)(-0.7954951)$$

Or $x_1 = 0.4775167$.
Again, using Eq. (7.101) and the first term of constraint -1, we get:

$$(5/2)x_1^2 x_2^{-2} = (w_{11})(\sigma_{11})/w_{10} = 1.5/0.5 = 3;$$

Or $x_2^{-2} = 3/(2.530.477516730.4775167)$

$$x_2 = 0.4459527$$

Now, using Eq. (7.100) and the second term of objective function, we get:

$$x_2^2 x_3^4 = (\sigma_0)(w_{02})(\sigma_{02})y_0(x^*) = (-1.0)(0.5)(+1.0)(-0.7954951)$$
$$x_3^4 = [(0.5)(0.7954951)]/x_2^2$$

or, $x_3 = 1.1792071$.

Thus, the optimal primal variable values are as follows: $x_1^* = 0.4775167$, $x_2^* = 0.4459527$, $x_3^* = 1.1792071$.

7.12.8.2 Affordable Dwelling-Layout System Design Optimization: An Example of Signomial Geometric Programming Problem

As stated above in many real world problems, there may be terms with negative coefficients or there may be mixed constraints requiring the application of signomial geometric programming to achieve optimal design. As for example, in case of *Design Optimization* of **Dwelling-Layout System**, the AEC professionals may have to achieve the optimal design within the constraint of investment or cost which intended beneficiaries and stakeholders can afford, and at the same time maximize the *dwelling space* which is of high priority particularly for low-income households. In this context, optimal design model for affordable dwelling-layout-system is developed as follows.

The *dwelling space* can be expressed as $= x_1 x_2$, where these variables x_1 and x_2 represent the width and depth, respectively, of a residential lot. Since, the objective is to maximize the *dwelling space*, the objective function can be written as:

$$\text{Maximize } y_o(x) = x_1 x_2$$

This objective function can be expressed in standard geometric form as:

$$\text{Minimize } y_o(x) = -x_1 x_2$$

Since the objective function has negative coefficient, it is a signomial geometric programming problem and the entire design optimization model can be developed as follows:

$$\text{Minimize } y_o(x) = \sigma_0 x_1 x_2$$

Subject to:

$$
\left.
\begin{array}{l}
\sigma_{11}\, c_{11}\, x_2\, x_3^{-1} + \sigma_{12}\, c_{12}\, x_6\, x_3^{-1} \le 1 \\[4pt]
\sigma_{21}\, c_{21}\, x_5\, x_4^{-1} + \sigma_{22}\, c_{22}\, x_6\, x_4^{-1} \le 1 \\[4pt]
\sigma_{31}\, x_1\, x_7^{-1} + \sigma_{32}\, x_7\, x_7^{-1} \le 1
\end{array}
\right\}
\text{Geometric Compatibility Constraints}
$$

$$\sigma_{41}\, c_{41}\, x_3\, x_4\, x_5^{-1}\, x_7 \le 1 \qquad\qquad \text{Density Constraint}$$

$$\sigma_{51}\, c_{51}\, x_2^{-1}\, x_7^{-1} \le 1 \qquad\qquad\quad \text{Open Space Constraint}$$

$$\sigma_{61}\, c_{61}\, x_4\, x_6^{-1} \le 1 \qquad\qquad\quad \text{Circulation Interval Ratio Constraint}$$

$$\sigma_{71}\, c_{71}\, x_4\, x_5^{-1}\, x_7 + \sigma_{72}\, c_{72}\, x_3\, x_5^{-1}\, x_7 + \sigma_{73}\, c_{73}\, x_3\, x_4\, x_5^{-1}\, x_7 + \sigma_{74}\, c_{74}\, x_2 + \sigma_{75}\, c_{75}\, x_1$$
$$\le 1 \quad \text{Affordable Cost Constraint}$$

$$x_1, x_2, x_3, x_4, x_5, x_6, x_7, x_7 > 0 \quad \text{Nonnegativity Constraint}$$

Here,

$x_1, x_2, x_3, x_4, x_5, x_6, x_7$, and x_7 are primal variables related to lot width, lot depth, layout module width, layout module length, number of lots in layout module, width of subsidiary main street, allocated lot width for open-space, and gross width of lot, respectively.

Coefficients c_{11}, c_{12}, and so on are model constants related to various design parameters regarding number of rows of lots, width of streets and their hierarchy, dwelling or population density, unit costs of streets, unit cost of land, unit costs of building walls, and service connection lines.

The problem has $T = 15$ terms and $N = 7$ variables giving degrees of difficulty $= T - (N + 1) = 15 - 9 = 6$.

This dwelling-layout system design model for optimization is presented here only to show applicability and the relevance of signomial geometric programming in the real world problems solution such as housing, urban development, and management. Complete details and solution of the model are available in a paper of the author (Chakrabarty, 1977a, 1977b) and also incorporated in Chap. 11 of the book.

7.13   Search Techniques

Search techniques are generally employed when in an optimization problem, an analytic expression for the objective function is either unavailable or it is too complicated to manipulate. In some cases from the specific set of values of independent variables, one can compute the corresponding value of objective function and the computation device may be a graph, table of numbers, and so on, permitting evaluation of objective function at many points, giving an indication of optimum value of objective. One such case is shown in Sect. 7.11.1 above where a search technique using computer graphics to find optimum point is shown. Many search techniques try to accumulate information and move into regions where optimum may lie and eliminates the areas where it is not likely. In this process, there can be two approaches to optimization. In the first approach, one may evaluate objective at many points and try to approximate it by an expression which are called approximation techniques. In the second approach, one may drive directly towards peak using incomplete information generated along the way, which is called direct method. The direct methods have two major categories, namely (1) elimination techniques which try to shrink the region where the peak may lie and (2) climbing procedure, which try to move in the directions where the objective seems improving based on local measurements. In literature, there are a number of search techniques available involving single variables or multi-variables (Beightler et al., 1979) which can be referred by interested readers. Here, only one search technique along with an illustrative example is presented just to give an idea of such techniques.

7.13.1 Steepest Gradient Search

The steepest gradient search is one of the efficient techniques developed. It is a climbing technique where the vector of first partial derivatives $\left(\frac{\partial y}{\partial x_1}, \frac{\partial y}{\partial x_2} \cdots \frac{\partial y}{\partial x_N}\right) = \nabla y$, are called gradients, signifying that it points in the direction in which the response surface has the steepest slope (Beightler et al., 1979). If we take an objective function $f(x, y)$ and the coordinate (x_1, y_1) as a trial point in a two-dimensional unconstrained steepest gradient search problem, the direction of the steepest slope (which may be ascending or descending) in terms of partial derivatives $\frac{\partial f}{\partial x_1}$ and $\frac{\partial f}{\partial y_1}$ of the function at this point is given by (Mark et al., 1970):

$$\frac{\nabla x1}{\nabla y1} = \frac{\frac{\partial f}{\partial x_1}}{\frac{\partial f}{\partial y_1}}$$

The new coordinate values of x and y giving the best distance for the next trial point can be defined as:

$$(x)_2 = (x)_1 - p\left(\frac{\partial f}{\partial x_1}\right)$$
$$(y)_2 = (y)_1 - p\left(\frac{\partial f}{\partial y_1}\right)$$

(7.102)

Here, p is a distance proportionality factor.
The objective function at the ith trial can be expressed as:

$$f_{i+1} = f\left(\left\{(x)_i - p\left(\frac{\partial f}{\partial x_i}\right)\right\}, \left\{(y)_i - p\left(\frac{\partial f}{\partial y_i}\right)\right\}\right)$$

(7.103)

to get the value of p where f is extremum (i.e. minimum or maximum), we can differentiate the objective function f with respect to p and setting the derivative to zero, and solving for the value of p^*. On getting the value of p, we can determine the new value of x and y using the Eq. (7.102) and begin the next iteration.

As an example we can consider a two-dimensional unconstrained optimization problem given by the following function which is to be minimized.

$$f(x, y) = 2x^2 + 2xy + 5y^2$$

We can choose the starting point for iteration (x_1, y_1) as: $(2, -2)$ giving the value of $f(x, y)_1$ as:

$$f(x, y)_1 = 234 + 232(-2) + 534 = 20$$

The partial derivatives are as follows:

$$\frac{\partial f}{\partial x} = 4x + 2y$$

$$\frac{\partial f}{\partial y} = 2x + 10y$$

Using the Eq. (7.102), we get:

$$(x)_2 = 2 - p(432 + 23(-2)) = 2{-}4p$$
$$(y)_2 = -2{-}p(4 - 20) = -2 + 16p$$

Therefore,

$$\begin{aligned}
f(x, y)_2 &= 2(2{-}4p)^2 + 2(2{-}4p)(-2 + 16p) + 5(-2 + 16p)^2\\
&= 2(4 - 16p + 16p^2) + 2(-4 + 32p + 8p - 64\,p^2) + 5(4 + 256\,p^2{-}64p)\\
&= 8 - 32p + 32\,p^2 - 8 + 64p + 16p - 128\,p^2 + 20 + 1280\,p^2{-}320p\\
&= -32p + 32\,p^2 + 64p + 16p{-}128\,p^2 + 20 + 1280\,p^2{-}320p\\
&= 20 - 272\,p + 1184\,p^2
\end{aligned}$$

Thus, the objective function $f(x, y)_2$ is expressed in terms of a single variable p, and the value of p giving an extreme or extremum value of $f(x, y)$ can be obtained by differentiating this equation with respect to p and setting it to zero as follows:
$\frac{\partial f}{\partial p} = -272 + 2368\,p = 0$; giving the value of $p = 0.115$.

Using this value of p, the value of $(x)_2$ and $(y)_2$ to be used for next iteration is given by:

$$(x)_2 = 2{-}4p = 2 - 430.115 = 1.54$$
$$(y)_2 = -2 + 16p = -2 + 1630.115 = -0.16$$

and the value of objective function $f(x, y)_2$ after second iteration using the above values of $(x)_2$ and $(y)_2$ is given by:

$$f(x, y)_2 = 4.377$$

Here, the value of $\frac{\partial^2 f}{\partial p^2} = +2368$ (i.e. positive), and therefore, p tends to minimize the objective function $f(x, y)$.

Again, using Eq. (7.102) in the next iteration, we get:

$$(x)_3 = (x)_2 - p \left(\frac{\partial f}{\partial x_2} \right) = 1.54 - p \, (431.54 - 230.16) = 1.54 - 5.74p$$

$$(y)_3 = (y)_2 - p \left(\frac{\partial f}{\partial y_2} \right) = -0.16 - p \, (231.54 - 1030.16) = -0.16 - 1.47p$$

Therefore,

$$f(x, y)_3 = 2(1.54 - 5.74p)^2 + 2 \, (1.54 - 5.74p) \, (-0.16 - 1.47p) + 5 \, (-0.16 - 1.47p)^2$$
$$= 96.456p^2 - 36.294p + 4.372$$

$\frac{\partial f}{\partial p} = 192.912 \, p - 36.274 = 0$; giving the value of $p = 0.177$.

Using this value of p, we get the value of $(x)_3$ and $(y)_3$ as follows which can be used for next iteration:

$$(x)_3 = 1.54 - 5.7430.177 = 0.442$$
$$(y)_3 = -0.16 - 1.4730.177 = -0.437$$

Putting these values of $(x)_3$ and $(y)_3$, we get the value of the objective function $f(x, y)_3$ as follows:

$$f(x, y)_3 = 0.967$$

Thus, the value of objective function is reduced considerably as compared to earlier iterations, indicating that the iterations are moving in the direction of 'steepest descent' and progressing towards the minimum. However, more iteration could be carried out to observe the convergence.

7.14   Closure

Optimization is decisive and often generates information about sensitivity of optimum conditions to various factors, and no rational decision-making process is really complete without optimization (Beightler et al., 1979). Optimization has an important role both in synthesis, i.e. design and decision-making; and in analysis, i.e. understanding how a system behaves, as also elaborated in Chap. 6. A large variation in resource-efficiency-indicators, depending on planning-design-decisions (as shown in many examples above in Chap.6), creates a large scope for *resource-optimization* in the AEC sector; making it imperative to apply optimization, particularly because *we live in a finite earth with finite resources with* many claimants. But, *Optimization* is rarely applied in our urban operations, creating many *inequities* and *inefficiencies* and the consequent urban problems (e.g. proliferation of slums, inadequate housing/urban services, environmental degradation, and so on), which needs correction.

In this context, planning and design in the AEC sector need to be viewed as a *problem-solving process*, forming part (i.e. *planning* function) of *integrated urban management* as elaborated in Chap. 2, to achieve the goal of *resource-efficient* and *equitable* solution of our mounting housing and urban problem. This process involves search through numerous potential solutions, with very large variations in *resource-efficiency indicators* including costs (even 2–4 times as shown in example problems, and also briefly outlined below), to select a particular solution that meets the goal as above. Moreover, operations research, *optimization,* and Computer-Aided-Design are basic techniques to improve *efficiency* and *effectiveness* (Koontz & O'Donnel, 1976; Koontz & Weihrich, 1990), and therefore, it is basic that the AEC professionals use such techniques to discharge their *ethical responsibility* to the society to achieve *efficiency*, *effectiveness,* and *equity* in the use of large resources committed by them in their plans and designs.

Tedious and voluminous computations involved, in utilizing optimization and operations research techniques and to solve and use optimizing models, have prevented their applications in the AEC sector especially in traditional building design and urban development planning and design in the past. The continued development of cheap and powerful computers with interactive graphics systems has made it possible to perform many planning and design functions integrating graphics with complex computations, design analysis, cost engineering, and *Optimization* techniques on a computer, making it feasible to achieve integrated computer-aided design by optimization. This book is an attempt to promote application of such integrated CAD by *Optimization* in housing, urban development, and management. In this chapter, some characteristics and basic concepts of optimization are outlined, and a few optimization techniques for design optimization are presented, to give an understanding, and help AEC professionals to apply such techniques in their planning and design function, and thus facilitate to achieve a *resource-efficient* AEC operations including housing and urban problem solution with *equity*.

Linear programming technique of optimization which deals with the minimization or maximization of a linear function subject to the linear constraints is presented with some example problems. Application of linear programming in building floor plan synthesis problems as available in literature (Mitchell, 1977) is discussed. It is pointed that optimization of real life building floor plans, physical layout plans, dwelling-layout systems, and so on involves area constraints, cost constraints, and so on, which are important objectives but most of them are nonlinear functions, and therefore, linear programming cannot be applied. It is necessary to use nonlinear programming techniques to solve such optimization problems.

Hence, some basic concepts of nonlinear programming such as concave and convex functions, *stationary points*, *local minimum/maximum*, *global minimum/ maximum*, and the *necessary and sufficient conditions* for *optimality* are outlined, with specific reference to differential calculus or differential approach, Lagrange multiplier, and geometric programming. Application of Lagrange multiplier technique for optimal design of a container to carry materials in a construction site is presented. Differential approach to optimization of functions of single and multiple variables is elaborated with examples along with an application example of differential approach to architectural design optimization. Concepts and illustrative examples of constrained and unconstrained *Design Optimization* in Urban-Built-Form are presented with optimal solution using computer graphics, and relevance of geometric programming in such problems is discussed.

Geometric programming is a relatively new nonlinear programming technique which evolved in 1960s (Beightler et al. 1976). It differs from other optimization techniques in that, instead of searching first for the optimal solution vector and then determining the resultant value of objective function, it emphasize on the relative magnitudes of the terms in the objective function, and thereafter, derives the values of solution variables. Thus, even before complete solution of the problem, geometric programming gives indications about the optimal way to distribute the resource or total cost among the various terms or components in the objective function contributing to the design objective. These concepts are explained with several example optimization problems. Such features of geometric programming make it a very important technique in planning and engineering design in general, and extremely useful in attaining *resource-efficient* problem solution in the AEC sector.

Therefore, geometric programming is presented in detail (mainly based on Beightler et al., 1976, 1979) including general mathematical formulation covering both posynomial and signomial types, and both unconstrained and inequality constrained cases for posynomial geometric programming. Duality concept and the primal dual relations in geometric programming are illustrated with a container example problem for easy understanding. Concept of degree of difficulty is explained by means of examples of posynomial problems with both zero degree of difficulty, and with problems having presence of degree of difficulty. A real world problem of *Urban-Built-Form Design Optimization*, which is an example of posynomial problem with inequality constraints and with presence of *degrees of difficulty* of two, giving two nonlinear simultaneous equations, is solved. Similarly, dwelling-layout system design optimization, which is an example of signomial geometric programming problem with inequality constraints and with presence of *degrees of difficulty* of six, is presented which is solved in Chap. 11.

Search techniques are generally employed when an analytic expression for the objective function is either unavailable or it is too complicated to manipulate. A number of search techniques are available in literature (Beightler et al., 1979). In Sect. 7.11.1, a search technique using computer graphics to find optimum point is shown. The steepest gradient search is one of the efficient search techniques developed, which is presented with an example problem solution.

In Sect. 7.12.7, the mathematical relationships between the Built-Form-Elements (BFEs), such as Built-Unit-Space (BUS), Building Rise (BR—effective number of floors), Semi-Public-Open space (SPO—open-to-sky space per built-unit), Floor Area Ratio (FAR), are shown in equation set (7.68), and a *Nonlinear Programming Model* for urban-built-form design optimization represented by the equation set (7.69), with *degrees of difficulty* of two, giving two nonlinear simultaneous equations, is presented. The model is solved in fraction of a second by the C++ computer program OUBF.CPP developed and source code included in the chapter. Typical interactive input-output dialogue giving output results instantly are shown in three case studies.

In Sect. 7.11.1, Table 7.1 showing the *Sensitivity* of Built-Unit-Cost (BUC) to the Floor Area Ratio (FAR), it is already revealed that a *variation* of nearly *two times in BUC* can happen only due to change of FAR, all other factors such as Built-Unit-Space (BUS), Open Space (SPO), remaining unchanged. To study this matter in more details three case studies are presented in Sect. 7.12.7 which are discussed below.

In Case 7.1, where land price assumed as 10,000 MU/SQ.M., the building construction cost, the increased cost related to building rise, and the cost related to land price constitutes $0.5775(w_{01})$, $0.0776(w_{02})$, and $0.3249(w_{03})$, i.e. 57.75%, 7.76%, and 32.49% of the total cost per Built-Uni. In Case 7.2, where land price is reduced to 1000MU/SQ.M., keeping values of other input parameters unchanged, the component of the building construction cost increases to 77.13% from 57.75%; the component of the cost related to land price comes down to only 7.72% from 32.49%. Again, as a result of reduction in land price in Case 7.2, the value of Optimal FAR has reduced to 2.537 from 4.017 in Case 7.1, and also, the optimal effective building rise (BR) has reduced to 4.71 from 14.91 in the former case. Similarly, there are large variations in values of other planning and design parameters such as least built-unit-cost, ground coverage ratio, and Built-Unit-Density (BUD). Case 7.3(a) where the land price is further reduced to 100 MU/SQ.M, reducing the value of Optimal FAR to 1.060 from 4.017 in Case 7.1, and also, the value of optimal effective building rise (BR) reduced to 1.49 only as against14.91 in Case 7.1. Similarly, the component of the building construction cost increases to 96.54%, and the component of the cost related to land price comes down to only 2.02%, even though the open space SPO is increased to 30 SQ.M. per built-unit. In Case 7.3(b) if open space SPO is retained at 20 SQ.M. per built-unit, there is increase in value of Optimal FAR to 1.173 and Ground Coverage Ratio (GCR) is increased to 0.777 while optimal effective Building Rise (BR) remains same at 1.49 as in Case 7.3(a). Normally, in conventional practice, an FAR of around 1 is prescribed, with the implied assumption that the land price will be very low which the case is rarely. In Case 7.3, the Ground Coverage Ratio (GCR) = 0.711 (as against 0.270 in Case 7.1) which is quite high affecting the openness of a layout even after increasing the SPO value by 50%. Again, in this case the Built-Unit-Density (BUD) is only 96.347 Built Units/Hectare, i.e. only about 26% of 365.243

Built Units/Hectare achieved in Case 7.1, seriously affecting the *efficiency* of land utilization which is becoming very scarce and costly with rise in population in many countries while land mass remains constant.

Such large variations in the values of optimal parameters due to change of value of only one parameter (here land price) in one *Design Optimization Model* only emphasize the need for such scientific planning and design analysis and design optimization covering all types of planning and design problems in the AEC sector, to achieve *efficiency* and *equity* in the urban development process, and thus, attain a *resource-efficient* solution of our mounting housing and urban problems. Therefore, in Chaps. 8–11, a number of models for *Design Optimization* covering integrated computer-aided structural design, building design, layout analysis and design, and dwelling-layout system design, are presented along with listing of C++ and Lisp programs developed to achieve integrated optimal design in these areas. To help readers of this book to use such techniques to achieve their own problem solutions, complete listings of the respective C++ and AutoLISP programs are included in an online space connected to this book.

It may be seen that the optimization model incorporates a number of important urban development planning and design parameters such as optimum FAR, optimum building rise which are generally adopted on *arbitrary basis* without any *scientific analysis*, affecting seriously the *efficiency* and *equity* in an urban development, which should provide for all stakeholders in the urban sector. This is mainly because *optimization techniques* and *computer-aided design* rarely form a part of educational *curricula* of AEC professional courses. As a result, AEC professionals are generally not familiar with such techniques, and naturally, are reluctant to apply such techniques to achieve *efficiency* and *equity* in their planning and design solutions. This needs correction by suitable modification of educational *curricula* of AEC professional courses, so that optimization theories reinforce these professions rather than supplant them by optimization specialists. Fortunately, while solving the *Nonlinear Programming Model* for Urban-Built-Form Design Optimization represented by the equation set (7.69) above, it is possible to develop compact formulas for optimum FAR, optimum building rise, and other planning parameters in terms of site-specific data which can be applied by AEC professionals in site-specific *planning and design*, and achieve *optimal solutions* even without any knowledge of optimization techniques. To help this process, a number of compact formulas for optimal FAR, optimal BR, built-unit-cost, and so on, are developed and presented in Sect. 7.12.7 above.

As mentioned above, in spite of availability of modern highly powerful computers and graphic systems, *Optimization* is rarely applied in our modern urban operations, creating many *inequities* and *inefficiencies*, although many centuries' back even ancients intuitively applied optimization principle making many ancient cities circular. Moreover, AEC professionals such as engineers, architects, and planners commit large resources (e.g. money, land, materials) in their *planning-design-decisions*, i.e. while discharging *planning function* forming part of integrated urban management, for urban problem solution as outlined in Chap. 2 of the book. This enjoins on them an *ethical responsibility* to the society to achieve *efficiency* and *effectiveness* in resource-use so committed by them, for a *resource-efficient* problem solution with *equity*. Use of *Optimization Techniques* and *Design Optimization* is essential to achieve this objective, and these are presented in this chapter to help AEC professionals to discharge their above *ethical responsibility* to the society, to achieve the goal of *resource-efficient* and *equitable* solution of our mounting housing and urban problem.

Chapter 8
Integrated Computer-Aided Structural Design by Optimization

8.1 Introduction

Optimization is a crucial requirement in AEC sector, as already stated in Chap. 1; since resources are always inadequate compared to the need, and AEC professionals commit large resources in their planning and design-decisions making it obligatory on them to attain efficiency in resource-use in such decisions. Various structures of reinforced concrete, steel, and so on are important components in AEC operations, and therefore structural design optimization should be of special interest to AEC professionals and structural engineers to achieve efficiency in resource-use in such components. This is particularly because no rational decision is really complete without optimization and structural design involves many variables giving numerous solutions with large variations in resource-efficiency indicators making it imperative to apply optimization techniques to search optimal structural design solution using CAD.

Structures are important cost component of buildings, and the structural optimization plays a significant role in today's highly competitive AEC operations including real-estate, housing, and urban development, where there is continuous increase in customer demand for improved quality, better safety, and affordable cost. Conventional ways of design development in structural engineering frequently lead to excess material usage, high design margins, again, ending up consuming more material into the structures and buildings. As earlier stated because of Information Technology Revolution in recent times, computational power is becoming more and more efficient and affordable to everyone. This availability of high capacity computational power and graphics facility gives designers the opportunity for evaluating multiple options during the design development phase itself, and also, applies innovative and complicated optimization techniques and algorithms for optimizing the multiple design variables considering given constraints, scenarios at the same time. The combination of this high power computation, graphics, and algorithms (like finite element method) is giving the designers limitless opportunities to use optimization to achieve resource-efficient and cost-effective structural design with better safety. Integration of computer-aided structural design with graphics makes it more efficient and effective. Therefore, it is imperative that AEC professionals exploit these potential benefits by more widespread implementation of optimization within the architecture-civil-structural engineering industry.

Reinforced concrete (RC) is widely used in AEC facilities, and reinforced concrete beams are important structural concrete members, and basic elements of such facilities involving considerable costs and consumption of materials. But, conventional design methods of RCC and steel structures, generally, do not incorporate any cost function for such structural members. To promote integrated CAD by optimization in this area, the author has developed a number of such *optimizing models* published elsewhere.

In this chapter, structural engineering and the concept of structural design optimization are discussed, optimization types are outlined along with their relevance to optimum design of reinforced concrete structures. Size and shape optimization of reinforced concrete beams are discussed, and two optimization models for optimal design of reinforced concrete beams are presented. It is possible to apply the concept of integrated computer-aided design by optimization in structural engineering by integration of numerical optimal design and graphics to attain more productivity, i.e. efficiency and effectiveness in AEC operations. This is demonstrated by taking the optimal design of reinforced concrete beams as examples, where optimal design in numerical terms is derived by a C++ program using geometric programming, which is instantly converted to a graphical design drawing by means of LISP program.

8.2 Structural Engineering and Structural Design Optimization

Structural engineering is concerned with understanding, predicting, and calculating the stability, strength, and rigidity of built structures for buildings and non-building structures. Structural engineers are also responsible to develop structural designs and to integrate their design with that of other designers, and to supervise construction of projects on site. The structural design developed for a building must ensure that the building is able to stand up safely, able to function without excessive deflections or movements which may cause fatigue of structural elements, cracking, or failure of fixtures, fittings or partitions, or discomfort for occupants. Structural engineering is based upon theories of strength of materials and applied knowledge of the structural performance of different materials and geometries. Structural engineering design utilizes a number of relatively simple structural elements such as beams, columns, and so on, to build complex structural systems. Structural engineers are responsible for making creative and efficient use of funds and other resources, structural elements, and materials to achieve these goals.

8.2.1   *Structural Design Optimization*

Structural design optimization can be described as the process of selecting the 'best' structural design defined by the optimization statement (i.e. objective function explained in Chap. 8), within the available means, i.e. constraints or design requirements. Different designs are described by the design variables. Structural optimization is concerned with maximizing the utility of a fixed quantity of resources, i.e. materials, labour, and so on to fulfill a given objective which may be the minimization of stresses, weight, or compliance (a global measure of the displacements, i.e. strain energy of the structure under the prescribed boundary conditions) for a given amount of material and the boundary conditions. Any optimization problem has an objective to be achieved. In fact, a problem exists where (1) some objective(s) is to be achieved in terms of some accepted 'value scale', (2) there are alternative ways to attain it, and (3) the optimal way(s) to achieve it is not apparent. In this context, there is need to arrive at the accepted 'value scale' in terms of cost, benefit, profit, loss, and so on, by which a facility design is to be evaluated. This is equally in case of structural design optimization, where this 'value scale' may be in terms of cost, displacement, mass, natural frequency, and so on. In civil and structural optimization, the objective function can be set as minimizing the volume of materials, thus the problem can also be stated as weight minimization or minimum weight design fulfilling the given constraints in a given situation. In this way, reducing the cost as well as ensuring optimum utilization of materials. Constraints in structural optimization can be set as restriction on deflection, yielding stress, buckling stress, and so on, which must be satisfied while minimizing the set objective function such as cost, volume, and so on. Fully stressed design is the most commonly used for structures in which strength considerations govern over stiffness, such as small and medium size frames. Structural optimization has over the past decades qualified as an important tool in the design process, and it is desirable that this should be applied in all AEC operations involving structures, and thus improving efficiency. The structural design optimization method can be grouped into topology, size, and shape optimization as explained later.

As presented in Chap. 8, computer-aided design by optimization involves symbolic models, incorporating symbols to signify system variables and the relationships between them are used to represent real buildings and structures which are to be optimized. Generally, the objective function designated as: y or z related to optimization, depends on N real scalar independent variables: $x_1, x_2, x_3 \ldots x_N$, which may be abbreviated and assembled into N-component vector: $\underline{x}$ as follows, where superscript 'denotes transposition.

$$\underline{x} = \begin{pmatrix} x_1 \\ x_2 \\ x_N \end{pmatrix} = (x_1, x_2, x_3 \ldots x_N), \tag{8.1}$$

When particular numbers or values are assigned to the components of $\underline{x}$, the resulting vector is generally called a *policy* or *design*. The *feasible region*: F of solution, incorporating acceptable policies or designs, is generally defined mathematically by a set of simultaneous inequalities: $y(\underline{x}) \geq 0$ and or equations: $y(\underline{x}) = 0$. However, for the purpose of optimal design, a *closed region* will be adopted, that is, all inequalities will include the possibility of strict equalities, i.e. the region F contains all its boundary points. Since the objective function is continuous, the minimum of y, denoted as y^* or $y(x^*)$, is identical with the greatest lower bound of y (Phillips & Beightler, 1976). If we denote $\underline{x}^*$ as the optimizing policy, then for a minimization problem, $y^* \equiv y(x^*) \leq y(\underline{x})$, for all $\underline{x}$ in a feasible neighbourhood of $\underline{x}^*$. In certain cases, any solution identified as locally optimal is also a global optimal solution for which some tests are to be carried out.

As already presented in Chap. 8, the general optimization model can be expressed as:

$$\text{Optimize: } y = f(x_1, x_2, x_3, \ldots x_n) \tag{8.2}$$

$$\text{Subject to the constraints: } g_i(x_1, x_2, x_3, \ldots x_n) < 0 \text{ or } > 0 \quad i = 1,2,3 \ldots m$$

$$x_j \geq 0 \qquad\qquad\qquad\qquad j = 1,2,3 \ldots n$$

The constraints delineate the feasible region, and the objective function z is evaluated for values of x_i within the feasible region to yield the optimal solution.

The above general optimization model is equally applicable in structural design optimization which will entail

1. selecting a set of design variables to describe the design alternatives.
2. selecting an objective function articulated by some accepted 'value scale', expressed in terms of the design variables, which is to be minimized or maximized.
3. choosing a set of constraints, expressed in terms of the design variables, which must be satisfied by any acceptable design.
4. finding a set of values for the design variables, which minimize (or maximize) the objective function, while satisfying all the constraints.

Three categories of structural optimization exist as stated below:

1. Size optimization
2. Shape optimization
3. Topology optimization

Detailed presentation of the above topics is beyond the scope of this book, and interested readers may refer to literature for more details. However, only the salient features and a broad idea of the main differences among size, shape, and topology optimization are presented below.

8.2.1.1 Size Optimization

Early structural optimization was concerned with the dimensional proportion of structures referred to as 'sizing optimization'. The objective of sizing optimization is to minimize the objective as weight or displacement, by changing the dimensions of the members of structure, treated as design variables. Usable material or allowable stress is usually taken as constraints in sizing optimization. Thus, a size optimization problem, which is also called sizing optimization, deals with finding the optimum values of size variables, i.e. the required cross-sectional area or thickness of a structural member to minimize the objective. Hence, it may be thickness of a plate or membrane, depth, width, or radius of the cross section of a beam. Size optimization problems can easily be expressed mathematically and frequently solved by deterministic methods. Using sizing variables such as cross sections and thicknesses of finite elements (in shell structures for example), many mathematical programming approaches have been tested and implemented into finite element programs like MSC/NASTRAN, PERMAS, and so on. Minimum weight design is the one in which every member is subjected to the maximum permissible stress in at least one load case, i.e. fully stressed. Calculation of the sensitivities for sizing optimization purposes is comparatively easy permitting handling of many realistic problems.

A diagrammatical representation of size optimization process is shown in Fig. 8.1 above (). Size optimization problems involving assignment of member section-sizes are more easily solvable in structural optimization problems, due to its mathematically well-defined nature. This is revealed by the large amount of academic research in this area and by many examples of structural optimization available in literature. However, even in this class of problems, there are issues like local minima, discrete variables on account of section catalogues, multiple constraints, multiple variables for a given cross section, and possible non-linearity (e.g. in the case of concrete structures). Despite examples in high-profile projects optimization are not commonplace in structural design practice. Moreover, the efficient integration of size optimization into topology optimization routines presents a challenge, where more research is needed. However, increased use of computer-aided design methods and CAD software (including the purpose written types) with appropriate capabilities

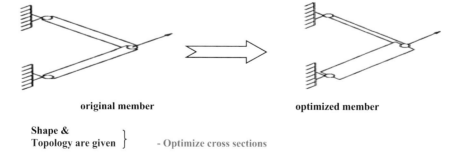

Fig. 8.1 Diagrammatical representation of size optimization process

will help electronic data-document interchange between architects and engineers, and thus raising the potential for more parametric design investigation, and ultimately leading to more efficient computational structural design optimization in AEC operations.

8.2.1.2 Shape Optimization

A shape optimization problem deals with finding the optimal shape (outer or inner shape) of a structure to minimize the objective function. The shape of a building could be curved, rectangular, circular, triangular, and trapezoidal and so on. Compared with the sizing optimization, the shape optimization is more complex. Shape optimization in structural optimization attempts to alter the shape of the boundary of a two- or three-dimensional structure. The objective of shape optimization is to minimize the mass of the structure subject to geometry or response constraints, such as displacement, stress, and natural frequencies. Shape optimization adds flexibility to sizing optimization as both dimension and shape of the structure may be changed, although the configuration or topology of the structure remains unchanged in shape optimization.

In shape optimization, two approaches are used, namely:

 (i) shape optimization based on Finite Element (FE) models, and
(ii) shape optimization based on geometry models. In this method, the coordinates of the surface nodes are regarded as design variables which will be modified during the optimization. This usually leads to a large number of design variables which might cause considerable mathematical difficulties.

In shape optimization method based on geometry models, the linkage of an FE model and a geometry model is maintained, and in this case, the parameters of the geometry model are the design variables. Complex geometry changes can be described using suitable software for shape optimization. The results of such optimization really depend on the number and selection of the design variables. In case of free form surfaces, the selection of the design variables is very difficult. If the mathematical programming methods are used, and if the number of design variables is too large, the numerical efforts needed is increased drastically.

8.2.1.3 Topology Optimization

The term 'topology' is derived from the Greek word 'topos', meaning position/place. The application of topology optimization extends to the number of holes, their location, their shape, and the connectivity of the structural domain. Shape optimization and sizing optimization are more limited than topology optimization in the respect that the designer must specify the topology of the proposed structure which is then fixed throughout the optimization process. The general form of the topology optimization problem is to determine the optimum distribution of material within a designable domain to fulfill a given objective.

Topology optimization is the technique that finds the optimal layout of the structure within a specified design domain, with the possible constraints such as the usable volume or material, design restrictions (or geometry constraints, such as holes or non-design regions) and subjected to certain applied loads and boundary conditions. Topology optimization is the most general optimization problem of a structure such as a truss, where the optimal configuration of a truss is found in order to minimize the objective function. Some well-known topologies are cross-bracings, diagonal bracings, and k-bracings. Topology optimization techniques can be applied to generalized problems through the use of the Finite Element (FE) methods.

A diagram explaining basic terms used in topology optimization is shown in Fig. 8.2 above. Suitably designating the void (no material) and solid material in design space or design domain, one can develop '0-1' design, where the final distribution of material within the design space is comprised entirely of either solid material or voids. The solution of the '0-1' problem has been attempted; but, in general the application of such techniques is computationally prohibitive due to the number of finite elements necessary to model the design space. This problem is addressed by SIMP (solid isotropic material with penalization) technique, where the material within each of the finite elements is defined as a continuous design variable. Conversion of the design variable from discrete to continuous makes it possible to use more computationally efficient mathematical programming methods for the solution of the original problem. Out the three categories of optimization, the structural topology optimization is the most general yielding information on the configuration including the number, location, size, and shape of 'openings' within a continuum, subject to a series of loads and boundary conditions. In this process, areas may arise with no more mass at all (openings and holes) and areas which contain high-density mass (bars and struts).

Compared with the sizing and shape optimization, the numerical efforts in topology optimization increase enormously. In such optimization problems, the

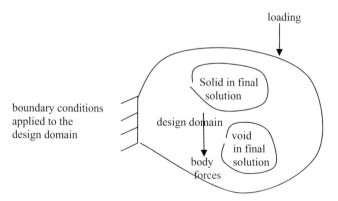

Fig. 8.2 Schematic diagram explaining basic terms used in topology optimization

number of design variables is typically between 5.000 and 100.000 requiring large efforts in sensitivity analysis. Due to the above difficulties regarding mathematical approaches, no method can be considered as a standard for calculating the optimum topology. Topology optimization achieves improved weight-to-stiffness ratio and the perceived aesthetic appeal of specific structural forms. Aerospace and automotive engineers routinely employ topology optimization and have reported significant structural performance gains such as weight savings in structures. However, application of topology optimization in the field of civil/structural engineering, where weight savings are seen as less critical due to the one-off nature of building structures, is generally in low key. In this era of sustainable infrastructures, attempts should be made to optimize every single structure to the best of its efficiency. Moreover, in many structural design problems, minimizing material volume or mass may be of primary concern in these days of resource constraint. In view of above, even the one-off nature of civil-structural engineering projects demands the use of rigorous optimization techniques including topology optimization, for improved efficiencies in increasingly complex projects of today, in the context of reducing resources. Topology optimization offers significant opportunities in civil/structural design and architecture to attain specified objectives, and it can be used as a tool to help greater collaboration between engineers and architects during the conceptual design process. However, only a limited number of examples of topology optimization being used in structural engineering and architecture can be found in the literature. This may be because research is driven by limited resources in the research area of civil engineering although technological competition and sustainable development demand lightweight, low cost, and high-performance structures.

Recently, designers of buildings and structures have also started investigating the use of topology optimization, for design of efficient and aesthetically pleasing developments. Bruggi (2016) presented an analysis and design of reinforced concrete structures as a topology optimization problem. It is pointed out that technical code for buildings deals with cracked reinforced concrete structures assuming concrete as a compression–only material, whereas rebar (the representation of discrete reinforcement bars) provides the structural component with the required tensile strength. The topology optimization problem is formulated as an alternative approach to cope with the analysis of cracked reinforced concrete structures, assuming the hyper-elastic no-tension model for concrete. The analysis of a 2D no-tension continuum is re-formulated as a topology optimization problem for minimum compliance (a global measure of the displacements is the strain energy, also called *compliance*, of the structure under the prescribed boundary conditions). The equilibrium of any compression-only structure is solved seeking for the distribution of an 'equivalent' orthotropic material that minimizes the potential energy of the hyper-elastic solid. An energy-based formulation has been extended to handle compression-only composite structures embedding bars of steel reinforcement, and a problem of size optimization has been outlined to cope with the optimal rebar of RC sections. Preliminary numerical simulations are shown to assess the proposed procedure outlining the ongoing research for assessment of the analysis problem.

It is desirable that such investigation and research in this area should be promoted in various AEC operations. Hopefully, with the development of powerful iterative solvers and increased computer and graphics capacities, topology optimization may be the state-of-the-art very soon even in AEC operations including buildings and structures.

8.2.1.4 Integration of Size, Shape, and Topology Optimization

The 'optimum structure' from sizing and shape optimization emanates only from the topology of the initial design model, and only change in shape & size may not lead to the design objective such as reduction of structural weight. Therefore, if the initial design model does not have optimal topology, the final structure from sizing and shape optimization will not be the true optimum structure. Hence, it is imperative to find the optimum topology of the structure before undertaking sizing and shape optimization, and also integrate topology optimization with the shape and sizing optimization. Ideally, the structural optimization should be an integrated process permitting selection of the best structural design pertaining to each of the following categories:

(a) Size optimization
(b) Shape optimization
(c) Topology optimization

Although Baldock cites some difficulties in achieving an integrated process, he has carried out simultaneous optimization of size and topology by performing a single iteration of the optimality criteria method at each topological step, offering substantial volume reductions when compared to sequential topology and size optimization. Even though in most structural design problems, minimizing the mass or volume of materials consumed may be of primary concern; it is desirable to find the trade-off between minimizing mass and minimizing piece count in the design of a bracing system for example. Integration of topology with sizing and shape optimization has been done using 'Approximation Concepts Approach'. In this approach, an approximate analysis model is created and optimized at each design cycle. The design solution of the approximate optimization is then used to update the finite element model, and a full system analysis is performed to create the next approximate analysis model. The implementation is in the GENESIS program. Graphics facilities have been used to integrate the topology results with boundary shape optimization. A design process for optimization of various aircraft components has been used, incorporating topology model building, topology optimization, size and shape model building, size and shape optimization, assembly of optimized size and shape model, and solid geometry extraction. In this process, topology optimization is used to suggest good initial designs for aircraft components followed by a detailed sizing and shape optimization to provide efficient aircraft component designs satisfying manufacturing, stability, and stress constraints. The full process, starting from finite element model generation through to the generation of a final design and import this design back into a CAD system, is included in the optimization process.

Most of the literature on the above three sets of structural design optimization problems; the objective function is selected in terms of displacement (a global measure of the displacements is the strain energy, also called *compliance*, of the structure under the prescribed boundary conditions), mass or volume of materials, natural frequency (vibration design), but, rarely in terms of cost. This is because such optimization is routinely employed by aerospace (with aggressive weight targets) and automotive engineers in the design of aircrafts and automobiles, where the above factors including weight minimization are of prime concerns for efficiency of the products designed.

It is to be noted that if the initial design model does not have optimal topology, the final structure from sizing and shape optimization will not be the true optimum structure. However, finding the optimum topology of the structure before sizing and shape optimization, and integration of topology optimization with shape and sizing optimization are still important research issues for two decades. Similarly, very few studies are available to exploit concrete strength by conceiving shapes other than rectangular whereby ineffective concrete in tension zone is reduced and incorporated in compression zone where it is effective (although some study gives indication that a trapezoidal beam is less costlier than rectangular shape, more research is needed to be conclusive).

It is desirable that similar research and application in AEC operations are encouraged so that such optimization could be routinely employed in structural engineering and architecture, as is the present case in aerospace and automotive engineering. Moreover, most AEC operations involve extensive use of cement-based components promoting higher cement production which is a major source of CO_2 emissions. There is a need to reduce cement consumption (thus discouraging higher cement production) by facilitating production of lighter concrete structures while maintaining the required load carrying capacity (consequently improve load-bearing capacity per unit weight of concrete), using the above integrated design procedure, including topology optimization, which can be a crucial step towards such sustainable structural design.

Again, optimization in AEC operations should ideally be driven by cost models. The holistic objective in a AEC project should be to minimize the total cost incurred in design, materials, labour, construction, and so on; and where applicable, maximizing the revenue from letting or selling of floor space. This approach has been adopted in various optimization models related to AEC operations presented in this book, incorporating an integrated process for computer-aided design by optimization including graphics and cost optimization, i.e. least-cost design or most-benefit design (e.g. maximizing built-space in a building within a given cost). Similarly, in view of the fact that finding the optimum topology of concrete structures, and integration of topology optimization with shape and sizing optimization are still in the research in the research stage; various optimization models presented in this book attempt to optimize cross sections (i.e. size optimization) of reinforced concrete sections , assuming a given topology and shape (i.e. rectangular shape).

8.2.1.5 Need for Expanding Applications of Optimization in Structural Engineering and Architecture

Structural optimization methodologies are now considered an integral part of the design process of structural components in the aerospace and automotive industries. Despite their rapid development in other areas, application of optimal design techniques is comparatively rare in traditional structural engineering as practiced in the building industry. Several studies reported in literature reveal the potential of incorporating optimization in the disciplines of structural engineering and architectural design which are driven by two main factors:

i. Growing interest within the architectural community in structural optimization, predominantly topology optimization, as a means of generating aesthetic and efficient structural forms; and
ii. Rising awareness for sustainable development, calling for reducing the consumption of materials and resources used by the building industry as well as for achieving higher energy efficiency of buildings.

There is need to bring together researchers and practicing professionals from the disciplines of structural engineering, architectural design, and engineering optimization, and develop an international consensus for promoting the integration of advanced optimization techniques into design disciplines related to the building industry and AEC operations, and also for encouraging research in the relevant areas. The curricula of the professional courses for the above disciplines can be modified accordingly, where the following topics, supplemented by advanced research and investigations as may be required, could be covered as applicable:

- Structural optimization of civil engineering structures and practical applications in the building industry.
- Multidisciplinary design optimization in the building industry and AEC operations.
- Optimization for conceptual architectural design.
- Size, shape, and topology optimization, and form-finding, integrating all three components.
- Optimization of earthquake-resisting structures.
- Optimization for sustainable building design.

Depending on the stage of research and finality of investigation in each of the above topics related to building industry and AEC operations, and the status of an international consensus for promoting the integration of such advanced optimization techniques into design disciplines related to the building industry and AEC operations, various national building codes should also be modified incorporating such techniques. This will go a long way to make application of such advanced optimization techniques an integral part of the design process of structural components in traditional structural engineering as practiced in the building industry and AEC operations, and derive the above-mentioned benefits.

As reinforced concrete structures are widely used in AEC operations including in housing and building; efficient optimal design of these components is essential to achieve cost reduction and the cost-effective use of materials which are becoming scarce. Present state-of-art in optimum design of reinforced concrete structures is discussed below.

8.3   Optimum Design of Reinforced Concrete Structures

As earlier stated in this era of sustainable infrastructures, diminishing resources, and rising cost of construction, attempts should be made to optimize every single structure to the best of its efficiency. In view of above, even the one-off nature of civil-structural engineering projects demands the use of rigorous optimization techniques, for improved efficiencies in increasingly complex projects of today. Since reinforced concrete structures form an important component of modern construction projects such optimization techniques should also be applied in the design of above structures.

Because of its significance in the industry, optimization of reinforced concrete structures has been the subject of multiple earlier studies. Moharrami and Grierson (1883) presented a computer-based method for the optimal design of reinforced concrete building frameworks, taking width, depth, and longitudinal reinforcement of member sections as the design variables, and applied an optimality criteria method to minimize the cost of concrete, steel, and formwork subject to constraints on strength and stiffness. Balling and Yao (1887) obtained optimum results from a multilevel method that simultaneously optimizes concrete-section dimensions and the number, diameter, and topology of the reinforcing bars The design method provides an iterative optimization strategy that converges in relatively few cycles to a least-cost design of reinforced concrete frameworks satisfying relevant requirements of the governing design code. Sarma and Adeli (1888) mentioned that as concrete structures involve at least three different materials: concrete, steel, and formwork, design of these structures should be based on cost rather than weight minimization. They presented a review of structural concrete optimization including beams, slabs, columns, frame structures, bridges, water tanks, folded plates, shear walls, pipes, and tensile members and concluded that there is a need to perform research on cost optimization of realistic three-dimensional structures, especially large structures with hundreds of members where optimization can result in substantial savings.

Lee and Ahn (2003) used a genetic algorithm (GA) to perform the discrete optimization of reinforced concrete plane frames subject to combinations of gravity and lateral loads. By constructing data sets, which contain a finite number of sectional properties of beams and columns, an attempt is made to alleviate difficulties in finding optimum sections from a semi-infinite set of member sizes and reinforcement arrangements. Camp et al. (2003) also implemented GA for discrete optimization of reinforced concrete frames, and examples are presented to demonstrate the efficiency of the procedure for the flexural design of simply supported beams, uniaxial columns, and multi-story frames. Guerra and Kiousis (2006) presented a study using MATLAB`s sequential quadratic programming algorithm for the design optimization of RC structures, and obtained optimal sizing and reinforcing of multi-bay and multi-story RC structures incorporating optimal stiffness correlation among structural members, and showing costs savings of up to 23% over a typical design method. Ahab et al. (2005) presented a model using the equivalent frame method for cost optimization of reinforced concrete flat slab buildings, selecting objective function as the total cost of the building including the cost of floors, columns, and foundations and handling the process in three different levels. In the first level, the optimum column layout is attained by an exhaustive search. In the second level, the optimum dimensions of columns and slab thickness for each column layout are found using a hybrid algorithm including genetic algorithm and discretized form of the Hook and Jeeves method. In the third level, the optimum number and size of reinforcing bars of reinforced concrete members are found employing an exhaustive search. In this study cost of materials, forming, and labour based on member dimensions are incorporated, and a structural model with distinct design variables for each member is implemented.

Vidosa et al. applied Simulated Annealing (SA) procedure, a search algorithm for the four case studies of reinforced concrete retaining walls, portal, and box frames used in road construction and building frames and concluded that optimization of the building frame indicates that instability in columns and flexure, shear and deflections in beams are the main restrictions that condition its design. Xiong et al. (2015) presented an optimization method for high-rise residential reinforced concrete building, based on decomposition of the main structure into substructures: floor system, vertical load resisting system, lateral load resisting system, and foundation system; then each of the subsystems using the design criteria established at the building codes is improved.

Ferreira et al. (2003) developed expressions to obtain the analytical optimal design of the reinforcement of a reinforced concrete T-section considering nonlinear behaviour of concrete and steel, and applied to examples. Tliouine et al. developed a cost optimization process for reinforced concrete T-beams under ultimate design loads, using the generalized reduced gradient technique with a reduced number of design variables, and presented typical design examples showing significant cost and materials savings. Bekdaş et al. (2015) used a random search technique to find optimum design of uniaxial RC columns with minimum cost. Narayan (2007) attempted shape optimization of beams using an analytical investigation employing Sequential Unconstrained Minimization Technique (SUMT) and found results showing that trapezoidal beams happen to be less costly than their rectangular counterparts.

The methods of structural optimization may be classified into two broad groups, namely: exact methods and heuristic methods. The exact methods are the traditional approach such as the differential approach and classical optimization techniques, including the Kuhn–Tucker conditions, linear and nonlinear programming. Some of these are presented in this chapter and available in literature. As explained in this chapter, an extremum may be local minimum (maximum) or global minimum (maximum) and Kuhn–Tucker conditions provide clues to the nature of the optimal solutions and enable an analyst to check the optimality or non-optimality of a proposed solution in exact methods. A literature review indicates that a very limited number of research deals with the exact methods which are based on the nonlinear and linear programming approach. It is desirable that such research studies and application for optimization of reinforced structures at various levels including structural element (e.g. beams, columns) level should be encouraged, so that the benefits of optimization, which are acceptable to all (since it is based on exact mathematics), are derived at the project level with less dependence on costly black-box type commercial software.

The second group of structural optimization mainly comprised of the heuristic methods, whose recent development is linked to the evolution of artificial intelligence procedures and FE methods. This group includes a number of heuristic search algorithms such as genetic algorithms (GA), simulated annealing, and artificial neural network and so on. The search for discrete valued solutions in GA is difficult because of the large number of combinations of possible member dimensions in the design of RC structures. Similarly, the difficulties in NLP techniques also arise from the need to round continuous-valued solutions to constructible solutions. Also, NLP techniques can be computationally expensive for large models. In general, most studies on optimization of RC structures, whether based on discrete- or continuous-valued searches, have found success with small RC structures using reduced structural models.

A literature review suggests that a number of investigators have applied different optimization methods to obtain the optimal design of RC beams, whether as isolated structural components or as part of a structural frame. It is also pointed out that the objective of optimization (e.g. minimum cost or weight), the design variables, and the constraints considered by different studies vary widely. Ima et al. (2014) also presented an implementation of cost optimization of RC beams in the MS Excel environment to illustrate the utility of the exhaustive enumeration method for such small discrete search spaces. The review of literature also suggests that nonlinear deterministic approaches can be efficiently employed to provide optimal design of RC beams, given the small number of variables (Ima et al., 2014).

As mentioned above, application of advanced optimization techniques, covering all structural components in an integrated manner including all categories of optimization, is still an issue of research studies and investigation in the building industry and AEC operations. However, many application results of optimization of standard, repeated members such as beams, columns, and slabs, showing considerable saving in cost and material consumption, are available in literature as outlined above. Development and applying structural optimization in the design of such elements may also be tackled by more simple approaches by an individual structural designer with complete understanding of the process, instead of depending on costly black-box type commercial software, which often does not give such insight.

Structural designers can make a significant contribution to improve efficiency and achieve material consumption reduction in engineering practices. Reinforced concrete beams form an important component of structural optimization, and are widely used in housing and building. Therefore, elements of optimum design of reinforced concrete beams, taking cost as the criteria for optimization and adopting geometric programming technique, an exact method of optimization as outlined in this chapter, are presented below.

8.4 Optimum Design of Reinforced Concrete Beams

The cost of a reinforced concrete element such as beam is dependent not only on the structural design of the beam cross section, but also on the unit costs of concrete, steel reinforcement, and shuttering. Therefore, the optimal design model for a reinforced concrete beam should consider all the above factors in a combined manner, taking into account the design economics of the beam as discussed below. Optimization in structural engineering is of special interest to AEC professionals/ structural engineers. It is possible to apply the concept of integrated computer-aided design by optimization in structural engineering to attain more productivity in AEC operations. Reinforced concrete (RC) is widely used in AEC facilities, and reinforced concrete beams are important structural elements of such facilities involving considerable costs and consumption of materials. To promote integrated CAD by optimization in this area, the author has developed a number of such *optimizing models* published elsewhere.

8.4.1 Design Economics of a Reinforced Concrete Beam

The proportion of tensile reinforcement in a RC beam which gives a 'balanced design' for a given ratio of permissible stresses is usually called 'economic percentage' which may be a misnomer. This is because the relative amounts of steel and concrete in the most economical design of a beam are dependent not only on the permissible stresses, but also on the costs of materials and shuttering. Therefore, in any economical design of beams, the cost of concrete, steel, and shuttering should also be considered, along with the structural design requirements for different design conditions. However, this is rarely done in conventional structural design practice.

In conventional design practice, a trial section of beam is chosen and checked for strength, serviceability, and other requirements as per applicable design code where the self-weight of the beam is assumed. Thus, for flexural design, the moment of resistance of the chosen section is determined to check its suitability against the applied bending moment. The process is repeated choosing new trial sections spending considerable time, until a trial section is found suitable. This procedure often creates difficulty in exactly matching the moment of resistance of the section with the total applied bending moment, including that due to the self-weight, which may be quite substantial in many cases. Moreover, this process does not take into account the cost of different elements of the beam. Structural optimization is an effective technique which takes into account costs of different elements, yielding economical and rational design and leading to effective cost reductions.

Adopting the strength design method, the expression for the nominal moment capacity: M_n, of a rectangular RC beam, with tension reinforcement only, is given by:

$$M_n = bd^2 f_c{}^1 w \left(1 - 0.58w\right) \tag{8.3}$$

where
  b = width of the beam
  d = the distance from the extreme compressive fibre to the centroid of tension reinforcement

$$w = A_s f_y / bd f_c{}^1 \tag{8.3a}$$

  A_s = area of tension reinforcement
  $f_c{}^1$ = compressive strength of the concrete
  f_y = yield strength of reinforcement
There are an infinite number of solutions to the equation set (8.3) that yield the same value of M_n (Everard et al., 1887). Hence, a designer has to assume the value of one or more of the variables b, d, and A_s in a trial section, and start an iterative design process to evaluate its suitability; and if not satisfied, a new section is adopted and the process is repeated. But, only a few trial sections can be evaluated in this manner. Moreover, the Eq. (8.3) does not incorporate any cost parameter, and therefore, even the use of usual design tables does not help in achieving the *unique least-cost* design of a beam. It is necessary to develop models for optimal design, as also outlined in this chapter, of a reinforced concrete beam considering all the above factors in a combined manner, to achieve the *least-cost* design for different design conditions.

As mentioned earlier, the majority of the structural optimization applied on RC structures involves heuristic methods (e.g. genetic algorithms) which are not exact, but very useful in dealing with structural optimization involving numerous variables (typically between 5.000 and 100.000 cited above). But, optimization of standard, repeated members such as beams, columns, and slabs involves comparatively less number of variables with relatively less complexity. Therefore, it is desirable that more emphasis should be given to apply exact methods of optimization in such problems, which will also enhance the understanding of structural designers to the interaction between various design variables towards optimization, and thus help them to contribute more to improve efficiency in building and AEC operations. Accordingly, models for optimal design of a reinforced concrete beam, using geometric programming technique (as outlined in this chapter), an exact method of optimization, are presented below.

Here,

k_u = a factor defining the neutral axis as shown in above figure

ϵ_y = ultimate strain in tensile steel reinforcement

ϵ_u = ultimate strain in concrete in compression

a = depth of equivalent rectangular stress block = $\beta_1\, c$;

β_1 to be taken as 0.85 for concrete strengths of 30 MPA or less

c = distance from extreme compression fibre to neutral axis at ultimate strength

8.5   Optimization Models for Reinforced Concrete Beams

As already mentioned, a number of **Optimization Models** for optimal design of reinforced concrete beams have been developed to achieve directly the optimal, i.e. the *least-cost* design of a reinforced concrete beam, for different design conditions and some of which are published elsewhere. Two of these models are also presented below as part of an integrated computer-aided optimal design software for reinforced concrete beams, incorporating both numerical optimal design and graphics, enabling a structural designer to obtain **instantly** the *least-cost* design drawing in response to the input of loading and other design conditions provided by the designer. Application of the Software for **Integrated Computer-Aided Optimal Design of Structures (RC beams)** is also illustrated in Sects. 8.7–8.11 below.

8.5.1   Least-Cost Design of Reinforced Concrete Beam-Width of Beam as Constraint: Model-I

Schematic section of a reinforced concrete beam is shown in Fig. 8.1 below. The cost per unit length of the beam will be given by the following expression, which will be the objective function for the optimization model:

$$y(x) = c_1\, x_1 + c_2\, x_2\, x_3 + c_3\, x_2 + c_4\, x_3 \qquad (8.4)$$

Here $y(x)$ is the cost (in terms of Monetary Units or MU) per unit length of beam (say in MU/cm), and c_1, c_2, c_3, and c_4 are cost coefficients due to volume of tensile steel reinforcement (say MU/cm³), volume of concrete (say MU/cm³), shuttering along the vertical surfaces of bean (say MU/cm²), and the shuttering along the horizontal surface of bean (say MU/cm²), respectively. Similarly, x_1, x_2, and x_3 are the variables giving the area of tensile steel reinforcement (say cm²), the effective depth of the beam (say cm), and the width of the beam (say cm), respectively, as shown in Fig. 8.3. The variables x_1, x_2, and x_3 not only affect the cost of a beam but will also determine its moment of resistance. An optimizing model should give the

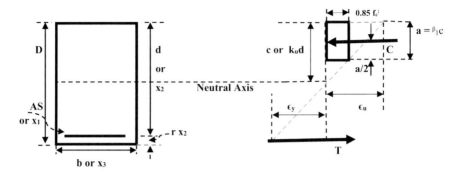

Fig. 8.3 Schematic cross section of a RC beam with equivalent rectangular stress distribution (Singly reinforced concrete beam showing strain and resistive forces)

optimum value of the variables x_1, x_2, and x_3 so as to minimize the cost of the beam, i.e. the value of $y(x)$, the objective function in the optimization model, and simultaneously, achieve the required moment of resistance to ensure safety.

Accordingly, the model for the optimal design of the beam could be developed as follows:

Minimize the Cost per Unit Length of Beam $= y(x) = c_1 x_1 + c_2 x_2 x_3 + c_3 x_2 + c_4 x_3$

Subject to:

$$a_1 x_1^{-1} x_3 x_5 \leq 1 \tag{8.5}$$
(*Constraint* for Equilibrium)

$$a_2 x_4^{-1} + a_3 x_2 x_3 x_4^{-1} \leq 1 \tag{8.6}$$
(*Constraint* for bending moment compatibility)

$$a_4 x_2^{-1} x_3^{-1} x_4 x_5^{-1} + a_5 x_2^{-1} x_5 \leq 1 \tag{8.7}$$
(*Constraint* for resisting bending moment)

$$a_6 x_3^{-1} \leq 1 \tag{8.8}$$
(*Constraint* for Beam width)

$$x_1, x_2, x_3, x_4, x_5 > 0 \tag{8.9}$$
(Non-negativity *constraint*)

The above optimization problem is a posynomial geometric programming problem with inequality constraints as explained in Sect. 7.12.6. As per terminology used in this section, the subscript o for c should be used to designate the objective function coefficients, and the coefficient of each term in constraints should be double subscripted one subscript designating the constraint equation and the other for specific term of constraint. Similarly, each exponent should be triple subscripted identifying the equation, variable and the posynomial term to which each belong. However, since the above problem is comparatively simpler with less number of variables and constraints; coefficients of all terms in objective function and constraint are single subscripted as indicated above, where 'c' with appropriate subscript indicates objective function coefficient and 'a' with appropriate subscript indicates constraint coefficient.

Here, $c_1, c_2, c_3, c_4, a_1, a_2, a_3, a_4, a_5,$ and a_6; are the model constants related to the cost parameters and the prescribed design conditions. Here, $c_1, c_2, c_3,$ and c_4 are cost coefficients related to the volume of tensile steel reinforcement, volume of concrete, area of shuttering along the vertical surfaces of beam, and the area of shuttering along the horizontal surface of beam, respectively. Here, a_1 is related to yield strength of tensile steel reinforcement and compressive strength of concrete; a_2 relates to applied bending moment; a_3 relates to beam span and density of concrete for self-weight bending moment; a_4 relates to the compressive strength of concrete and the capacity reduction factor; a_5 defines the centroid of concrete compression; and a_6 defines the minimum acceptable width of beam.

Here, the model *variables* are:

$x_1 =$ Optimum area of tensile steel reinforcement
$x_2 =$ A variable giving the optimum effective depth of the beam
$x_3 =$ A variable giving the optimum width of the beam
$x_4 =$ A variable defining the total applied bending moment including that due to self-weight of beam
$x_5 =$ A variable defining the depth of equivalent rectangular stress block

As mentioned above, the value of the variables $x_1, x_2,$ and x_3 not only affect the cost of the beam but also determines its moment of resistance. Therefore, the solution of an optimization model should give the optimum value of each of the variables: $x_1, x_2,$ and x_3 so as to minimize the cost of the beam, i.e. the value of the objective function $y(x)$, and simultaneously achieve the required moment of resistance to ensure safety. Accordingly, expressions for the constraints 1, 2, 3, and 4 of the model are developed and shown above as expressions (8.5), (8.6), (8.7), and (8.8), respectively, indicating within bracket of each expression, the specific design condition to be fulfilled by the model. Since the model has to give only the realistic values of the variables, none of the model variables can be negative which design condition is specified by the nonnegativity constraint-5, i.e. expression (8.9) above.

8.5.1.1 Model Solution

Many design problems in AEC sector involve nonlinear terms needing nonlinear programming techniques for solution. In such design problems, the objective function is usually a summation of product functions and the constraints also contain the product functions, as is the case in the above model. The well-known geometric programming technique is an efficient tool in solving such problems as explained in this chapter. In this technique, emphasis is placed on the relative magnitudes of the terms of the objective function rather than on the variables, and instead of finding the optimal values of the design variables directly, the optimal way to distribute the total cost, among the various terms or components in the objective function, is found first. This indicates how resources should be apportioned among the components contributing to the design objective, which is very useful in engineering and architectural design. Even in the presence of great degree of difficulty, the geometric programming formulation uncovers important invariant properties of the design problem which can be used as design rules for optimality (soc86). In this technique, if the primal geometric programming problem has N variables and T terms, in the dual geometric programming problem, there will be N orthogonality conditions-one for each variable x_n, one normality condition and T weights, one for each term. Thus, there will be $(N + 1)$ independent linear constraint equations as against T independent dual variables which are unknown. In case $T - (N + 1)$, which is the degree of difficulty in the problem, is equal to zero, the $(N + 1)$ linear equations uniquely solve for the dual variables. In case $T - (N + 1)$ is greater than zero, but small, a set of $(N +1)$ dual weights could be expressed in terms of the remaining $T - (N + 1)$ dual weights, and substituted into the dual objective function, which is then maximized by setting the first derivative to zero with respect to each of the dual weights coming into the substituted dual objective function which is being maximized. Thus, a sufficient number of equations are available for solving the problem. Application of the geometric programming technique in various AEC design problems has been shown in a number of papers of the author published elsewhere () and also presented in this chapter with a number of illustrative examples.

Using this technique, and taking the terminology used in literature (Beightler et al., 1878) on geometric programming as explained in Sect. 8.12.5, the dual objective function and the associated normality and orthogonality constraints for the above primal problem are given by:

$$
\begin{aligned}
d(w) = {} & \left(\frac{c_{01}}{w_{01}}\right)^{w_{01}} \left(\frac{c_{02}}{w_{02}}\right)^{w_{02}} \left(\frac{c_{03}}{w_{03}}\right)^{w_{03}} \left(\frac{c_{04}}{w_{04}}\right)^{w_{04}} (c_{11})^{w_{11}} \left(\frac{C_{21}(W_{21} + W_{22})}{w_{21}}\right)^{w_{21}} \\
& \times \left(\frac{C_{22}(W_{21} + W_{22})}{w_{22}}\right)^{w_{22}} \left(\frac{C_{31}(W_{31} + W_{32})}{w_{31}}\right)^{w_{31}} \left(\frac{C_{32}(W_{31} + W_{32})}{w_{32}}\right)^{w_{32}} (c_{41})^{w_{41}}
\end{aligned}
$$

$$(8.10)$$

As discussed above, the coefficients for objective function and constraints are simplified as: $c_{01} = c_1$, $c_{02} = c_2$, and so on for objective function coefficients; and $c_{11} = a_1$, $c_{21} = a_2$, and so on for constraint coefficients. Accordingly, the dual objective function to be maximized can be expressed in a simplified form as follows:

Maximise $d(w) = [c_1/w_{01}]^{w_{01}} [c_2/w_{02}]^{w_{02}} [c_3/w_{03}]^{w_{03}} [c_4/w_{04}]^{w_{04}}$

$$\times [a_1]^{w_{11}} [a_2(w_{21}+w_{22})/w_{21}]^{w_{21}} [a_3(w_{21}+w_{22})/w_{22}]^{w_{22}} \quad (8.11)$$

$$\times [a_4(w_{31}+w_{32})/w_{31}]^{w_{31}} [a_5(w_{31}+w_{32})/w_{32}]^{w_{32}} [a_6]^{w_{41}}$$

Subject to:

$$w_{01} + w_{02} + w_{03} + w_{04} = 1 \quad \text{(normality condition)} \quad (8.12)$$

$$w_{01} - w_{11} = 0 \quad \text{(orthogonality condition for } x_1) \quad (8.13)$$

$$w_{02} + w_{03} + w_{22} - w_{31} - w_{32} = 0 \quad \text{(orthogonality condition for } x_2) \quad (8.14)$$

$$w_{02} + w_{04} + w_{11} + w_{22} - w_{31} - w_{41} = 0 \quad \text{(orthogonality condition for } x_3) \quad (8.15)$$

$$- w_{21} - w_{22} + w_{31} = 0 \quad \text{(orthogonality condition for } x_4) \quad (8.16)$$

$$w_{11} - w_{31} + w_{32} = 0 \quad \text{(orthogonality condition for } x_5) \quad (8.17)$$

Here, w_{01}, w_{02}, $w_{03,}$ and w_{04} represent the four cost weights related to the four terms in the objective function $y(x)$, i.e. expression (8.4) above. Similarly, w_{11}, w_{21}, w_{22}, w_{31}, w_{32}, and w_{41} represent the dual variables related to the single term in constraint-1, i.e. expression (8.5), to the first term in constraint-2, i.e. expression (8.6), to the second term in constraint-2, i.e. expression (8.6), to the first term in constraint-3, i.e. expression (8.7), to the second term in constraint-3, i.e. expression (8.7), and to the single term in constraint-4, i.e. expression (8.8) above, respectively.

There are ten terms (T) and five primal variables (N) in the primal objective function and constraints [expressions (8.4)–(8.8)], and therefore, the degree of difficulty in the problem is $= T - (N + 1) = 4$. Thus, any six of the dual variables can be expressed in terms of the remaining four dual variables as shown below, choosing w_{01}, w_{03}, w_{04}, and w_{41} as the four dual variables:

$$w_{02} = 1 - w_{01} - w_{03} - w_{04}; \quad w_{11} = w_{01}; \quad w_{21} = 1 - w_{03} - w_{41};$$

$$w_{22} = 2\,w_{03} - w_{04} - 1 + 2\,w_{41}; \quad w_{31} = w_{03} - w_{04} + w_{41}; \quad (8.18)$$

$$w_{32} = w_{03} - w_{04} - w_{01} + w_{41}.$$

The cost weights w_{01}, w_{02}, w_{03}, and w_{04} represent the optimum proportion of total cost per unit length of beam, distributed between the tensile steel reinforcement, concrete, shuttering along the vertical surfaces of the beam, and the shuttering along the bottom horizontal surface of the beam, respectively. The set of Eq. (8.18) represent the relationship between these cost distributions among the above components, which are independent of the values of the cost coefficients c_1, c_2, c_3, and c_4. This invariant property of geometric programming could be used for development of suitable *design rules* for *optimality* even in the presence of a great degree of difficulty, having general application in any problem of this nature.

If we substitute the expressions in equations set (8.18) into the dual objective function [Eq. (8.11)] and maximize this substituted dual objective function over the four variables w_{03}, w_{04}, w_{01}, and w_{41} (i.e. by setting the first derivative of the function with respect to each variable to zero), we will get four additional equations, thus providing sufficient conditions for optimality. The process is simplified by maximizing the natural logarithm of the dual function; that is, by substituting

$$\ln d(w_{01}, w_{03}, w_{04}, w_{41}) = z(w_{01}, w_{03}, w_{04}, w_{41}).$$

Thus, by above substitution and taking logarithm on both sides of Eq. (8.10), we get the following equation:

$$
\begin{aligned}
z(w) = {} & w_{01} \left(\ln c_1 - \ln w_{01} \right) + \left(1 - w_{01} - w_{03} - w_{04} \right) \\
& \left(\ln c_2 - \ln \left(1 - w_{01} - w_{03} - w_{04} \right) \right) + w_{03} \left(\ln c_3 - \ln w_{03} \right) \\
& + w_{04} \left(\ln c_4 - \ln w_{04} \right) + w_{01} \ln a_1 + \left(1 - w_{03} - w_{41} \right) \\
& \left(\ln a_2 + \ln \left(w_{03} - w_{04} + w_{41} \right) - \ln \left(1 - w_{03} - w_{41} \right) \right) + \left(2w_{03} - w_{04} - 1 + 2w_{41} \right) \\
& \left(\ln a_3 + \ln \left(w_{03} - w_{04} + w_{41} \right) - \ln \left(2 w_{03} - w_{04} - 1 + 2 w_{41} \right) \right) \\
& + \left(w_{03} - w_{04} + w_{41} \right) \left(\ln a_4 + \ln \left(2 w_{03} - 2w_{04} - w_{01} + 2 w_{41} \right) \right) \\
& - \ln \left(w_{03} - w_{04} + w_{41} \right) + \left(w_{03} - w_{04} - w_{01} + w_{41} \right) \\
& \left(\ln a_5 + \ln \left(2 w_{03} - 2w_{04} - w_{01} + 2 w_{41} \right) - \left(w_{03} - w_{04} - w_{01} + w_{41} \right) \right) \\
& + w_{41} \ln a_6.
\end{aligned}
\tag{8.19}
$$

By setting the first derivative $dz/dw_{01} = 0$, we get:

$$
\begin{aligned}
& (c_1 a_1 / c_2 a_5)(1 - w_{01} - w_{03} - w_{04}) (w_{03} - w_{04} - w_{01} + w_{41}) \\
& - (2w_{03} - 2w_{04} - w_{01} + 2w_{41}) w_{01} = 0
\end{aligned}
\tag{8.20}
$$

Similarly, by setting the first derivative $dz/dw_{03} = 0$, we get:

$$
\begin{aligned}
& (c_3 a_3^2 a_4 a_5 / c_2 a_2)(1 - w_{01} - w_{03} - w_{04})(1 - w_{03} - w_{41}) (2w_{03} - 2w_{04} - w_{01} + 2w_{41})^2 \\
& - (2w_{03} - w_{04} - 1 + 2w_{41})^2 (w_{03} - w_{04} - w_{01} + w_{41}) w_{03} = 0
\end{aligned}
\tag{8.21}
$$

Again, by setting the first derivative $dz/dw_{41} = 0$, we get:

$$
\begin{aligned}
& (a_3^2 a_4 a_5 a_6 / a_2)(1 - w_{03} - w_{41}) (2w_{03} - 2w_{04} - w_{01} + 2w_{41})^2 \\
& - (2w_{03} - w_{04} - 1 + 2w_{41})^2 (w_{03} - w_{04} - w_{01} + w_{41}) = 0
\end{aligned}
\tag{8.22}
$$

Similarly, setting the first derivative $dz/dw_{04} = 0$, we get:

$$(c_4/c_2 a_3 a_4 a_5)(1 - w_{01} - w_{03} - w_{04})(2w_{03} - w_{04} - 1 + 2w_{41})(w_{03} - w_{04} - w_{01} + w_{41})$$
$$- (2w_{03} - 2w_{04} - w_{01} + 2w_{41})^2 w_{04} = 0$$

$$(8.23)$$

By dividing expression (8.21) by expression (8.22), we get $(c_3/c_2 a_6)\,(1 - w_{01} - w_{03} - w_{04}) - w_{03} = 0$, giving the value of w_{03} in terms of w_{01} and w_{04} as:

$$w_{03} = \{c_3/(c_3 + c_2 a_6)\}\,(1 - w_{01} - w_{04}) \tag{8.24}$$

Substituting this expression for w_{03} in expression (8.20), we get:

$$w_{02} = \{c_2 a_6/(c_3 + c_2 a_6)\}\,(1 - w_{01} - w_{04}). \tag{8.25}$$

From Eqs. (8.20), (8.22), (8.23), and (8.24), by substitution of w_{02} in terms of w_{01} and w_{04} given by Eq. (8.25), and by suitable manipulation, we get the following three equations in terms of three variables w_{01}, w_{04} and w_{41}:

$$[c_1 a_1 a_6/a_5 (c_3 + c_2\, a_6)](1 - w_{01} - w_{04})$$
$$[\{c_3/(c_3 + c_2 a_6)\}(1 - w_{01} - w_{04}) - w_{04} - w_{01} + w_{41}] \tag{8.26}$$
$$- [\{2c_3/(c_3 + c_2 a_6)\}\,(1 - w_{01} - w_{04}) - 2w_{04} - w_{01} + 2w_{41}]\,w_{01} = 0$$

$$(c_4/c_1 a_1 a_3 a_4)\,[\{2c_3/(c_3 + c_2 a_6)\}\,(1 - w_{01} - w_{04}) - w_{04} - 1 + 2w_{41}]\,w_{01}$$
$$- [\{2c_3/(c_3 + c_2 a_6)\}\,(1 - w_{01} - w_{04}) - 2w_{04} - w_{01} + 2w_{41})^2 w_{04} = 0 \tag{8.27}$$

$$(a_3^2 a_4 a_5 a_6/a_2)[1 - w_{41} - \{c_2 a_6/(c_3 + c_2 a_6)\}\,(1 - w_{01} - w_{04})]$$
$$\times [\{2c_3/(c_3 + c_2 a_6)\}\,(1 - w_{01} - w_{04}) - 2w_{04} - w_{01} + 2w_{41}]^2$$
$$- [\{2c_3/(c_3 + c_2 a_6)\}\,(1 - w_{01} - w_{04}) - w_{04} - 1 + 2w_{41}]^2 \tag{8.28}$$
$$\times [\{c_3/(c_3 + c_2 a_6)\}(1 - w_{01} - w_{04}) - w_{04} - w_{01} + w_{41}] = 0$$

Equations (8.26), (8.27), and (8.28) are three nonlinear simultaneous equations in terms of three unknown variables w_{01}, w_{04}, and w_{41} which can be solved by the Newton–Raphson method as explained and illustrated in Chap. 3 and this chapter. The author has developed C++ computer programmes for solving such nonlinear simultaneous equations covering many variables, and the computer code of the programme is presented in appendix. Application of the geometric programming technique gives the following simple expressions for determining the optimum values of the model variables, thus, fully solving the model.

$$x_2 = (c_4 w_{02})/(c_2 w_{04}),$$
$$x_3 = (c_3 w_{02})/(c_2 w_{03}),$$
$$y(x) = (c_2 x_2 x_3)/w_{02},$$
$$x_1 = y(x)w_{01}/c_1, \tag{8.29}$$
$$x_4 = a_2 + a_3 x_2 x_3,$$
$$x_5 = x_1/(a_1 x_3).$$

8.5.1.2   Model Calibration

The model needs to be calibrated relating it to the site-specific cost parameters and desired design conditions to obtain the optimal design fulfilling these conditions. Thus, the model constants, i.e. cost coefficients: c_1, c_2, c_3, and c_4 are related to the cost of labour, materials, and shuttering in the given site condition and could be determined from relevant schedule of rates or market rates of materials and labour as illustrated in Chap. 4.

Accordingly, the expressions for the values of c_1, c_2, c_3, and c_4, with reference to the objective function Eq. (8.4) are developed below, where cs, cc, and csh are unit costs of reinforcing steel, concrete, and shuttering, and r is the cover ratio (see Fig. 8.3).

The expression for the value of model constants a_1 as shown below is derived from the constraint expression (8.5) defining the equilibrium condition between compressive strength of concrete and yield strength of steel. Similarly, the expression for the value of model constants a_2 and a_3 is determined from the bending moment compatibility condition between the applied bending moment, beam span, and the density of concrete giving the self-weight bending moment, and the total applied bending moment represented by the variable x_4 in the constraint expression (8.6). Again, at the equilibrium the total resisting moment capacity of the beam section, applying appropriate capacity reduction factor as prescribed in the relevant national code of practice, must match with the total applied bending moment, which condition is represented by the constraint expression (8.7). Accordingly, from this constraint, the expression for the value of a_4 is developed as shown below. The value of model constant a_5 is 0.5 assuming that centroid of concrete compression is at half the depth of rectangular stress block. The value of model constant a_6 is the minimum acceptable width of the beam.

To sum up, the expressions for all the above model constants are developed below in equation set (8.30), without specifying the units of measurement, which may be chosen by a designer as applicable in his/her situation, and determine the numerical values of the model constants which can be given as input to find the optimal solution in a specific case.

$$c_1 = ws \times cs, \; c_2 = cc(1 + r), \; c_3 = 2(1 + r)csh, \; c_4 = csh$$
$$a_1 = 0.85 \, f'_c / f_y$$
$$a_2 = \text{UABM}$$
$$a_3 = \text{LFDL}(1 + r)wc \times L^2 \times \text{BMC} \tag{8.30}$$
$$a_4 = 1.0 / \left(0.85 \, f'_c \times \text{CRF} \right)$$
$$a_5 = 0.5$$
$$a_6 = \text{Minimum acceptable beam width (MBW)}.$$

Here, 'ws' is the unit weight of steel reinforcement, cs is the unit cost of steel reinforcement, cc is the unit cost of concrete, csh is unit cost of shuttering, wc is unit weight of concrete, f'_c is compressive strength of concrete, f_y is yield strength of steel reinforcement, UABM is ultimate applied bending moment, LFDL is load factor for dead load as prescribed in relevant code of practice, r is cover ratio, L is beam span, BMC is bending moment coefficient, and CRF is the capacity reduction factor as prescribed in relevant code of practice.

If we adopt the metric units of measurement with ws in kg/cm^3, wc in kg/cm^3, cs in Monetary Units (MU) per kg, cc in MU per m^3, csh in MU/m^2, wc in kg/m^3, f'_c in N/cm^2, f_y in N/cm^2, UABM in N-cm, L in cm, the expressions for the calibration of the above model constants can be developed as below in equation set (8.32):

$$c_1 = (7.85/1000)cs \text{ in MU/cm}^3; \text{ assuming ws as 7.85 g/cm}^3$$
$$c_2 = cc \times (1 + r)/(1,000,000) \text{ in MU/cm}^3$$
$$c_3 = (2 \times (1 + r) \times csh)/(10,000) \text{ in MU/cm}^2 \tag{8.31}$$
$$c_4 = csh/10,000 \text{ in MU/cm}^2$$

$$a_1 = 0.85 f'_c / f_y$$
$$a_2 = \text{UABM in N.cm}$$
$$a_3 = \text{LFDL}(1 + r)wc \times L^2 \times \text{BMC}$$
$$\quad = \text{LFDL} \times (1 + r) \times 0.002323 \times 8.8 \times L^2 \times \text{BMC in N.cm}, \tag{8.32}$$
$$a_4 = 1.0 / \left(0.85 f'_c \times \text{CRF} \right)$$
$$a_5 = 0.5$$
$$a_6 = \text{MBW in cm}.$$

$$y(x) = c_1 \, x_1 + c_2 \, x_2 \, x_3 + c_3 \, x_2 + c_4 \, x_3 \text{ in MU per 'cm' length of the beam}$$

The value of objective function $y(x)$, as given by Eq. (8.4) and reproduced above, will be in MU per 'cm' length of the beam, where the units of measurements of variables: x_1, x_2, and x_3 will be in cm^2, cm, and cm, respectively.

As explained in Sect. 8.12.5 of this chapter, it is to be noted that the maximization of the dual objective function subject to the normality and orthogonality conditions is a *sufficient condition* for the primal objective function to be a *global minimum* (Mark et al., 1970). This is also applicable in case of constrained geometric programming problem with 'less than inequalities' constraints with all positive terms in the constraints. The optimization Model-I presented above is a constrained geometric programming problem with 'less than inequalities' constraints with all positive terms in the constraints. Hence, the optimum solution derived above is a *global minimum*. In case of problems with zero degrees of difficulty, unique optimum solution is directly determined. Even in case of problems with degrees of difficulty, the dual objective function can be maximized using differential calculus to derive the *global minimum* of the primal objective function, particularly when the degree of difficulty is comparatively less, as is the case in Model-I above.

8.5.2 Least-Cost Design of Reinforced Concrete Beam-Width: Depth Ratio as Constraint: Model-II

Adopting the same schematic section of a reinforced concrete beam as shown in Fig. 8.1, the cost per unit length of the beam will be given by the expression (8.4) as in Model-I, which will be the objective function for this optimization model also:

$$y(x) = c_1 x_1 + c_2 x_3 + c_3 x_2 + c_4 x_3 \tag{8.4}$$

Minimize the Cost per Unit Length of Beam $= y(x) = c_1 x_1 + c_2 x_2 x_3 + c_3 x_2 + c_4 x_3$

Subject to:

$$a_1 x_1 x_3^{-1} x_5^{-1} \qquad\qquad \leq 1 \tag{8.33}$$
(*Constraint* for Equilibrium)

$$a_2 x_4^{-1} + a_3 x_2 x_3 x_4^{-1} \qquad\qquad \leq 1 \tag{8.34}$$
(*Constraint* for bending moment compatibility)

$$a_4 x_1^{-1} x_2^{-1} x_4 + a_5 x_2^{-1} x_5 \qquad \leq 1 \tag{8.35}$$
(*Constraint* for bending moment)

$$a_6 x_2 x_3^{-1} \qquad\qquad\qquad \leq 1 \tag{8.36}$$
(*Constraint* for Beam width : depth ratio)

$$x_1, x_2, x_3, x_4, x_5 \qquad\qquad\qquad > 0 \tag{8.37}$$
(*Non-negativity constraint*)

Here, c_1, c_2, c_3, and c_4 are cost coefficients due to volume of tensile steel reinforcement (MU/cm^3), volume of concrete (MU/cm^3), shuttering along the vertical surfaces of bean (MU/cm^2), and the shuttering along the horizontal surface of bean (MU/cm^2), respectively. Here, a_1 is related to yield strength of tensile steel reinforcement and compressive strength of concrete; a_2 relates to applied bending moment; a_3 relates to beam span and density of concrete for self-weight bending moment; a_4 relates to the compressive strength of concrete and the capacity reduction factor; a_5 defines the centroid of concrete compression; and a_6 defines the desired beam width:depth ratio.

Here, the model *variables* are:

$x_1 =$ Optimum area of tensile steel reinforcement (cm^2)
$x_2 =$ A variable giving the optimum effective depth of the beam (cm)
$x_3 =$ A variable giving the optimum width of the beam (cm)
$x_4 =$ A variable defining the total applied bending moment including that due to self-weight of beam
$x_5 =$ A variable defining the depth of equivalent rectangular stress block

8.5.2.1   Model Solution

As in case of Model-I, the dual objective function and the associated normality and orthogonality constraints for the above primal problem can be expressed taking the terminology used in literature on geometric programming as explained in Sect. 8.12.5. However, as discussed earlier in the case of Model-I, the coefficients for objective function and constraints can be simplified as: $c_{01} = c_1$, $c_{02} = c_2$ and so on, for objective function coefficients; and $c_{11} = a_1$, $c_{21} = a_2$ and so on, for constraint coefficients. Accordingly, the dual objective function to be maximized in this case also can be expressed in a simplified form as follows:

$$\text{Maximize } d(w) = [c_1/w_{01}]^{w_{01}} [c_2/w_{02}]^{w_{02}} [c_3/w_{03}]^{w_{03}} [c_4/w_{04}]^{w_{04}}$$
$$\times [a_1]^{w_{11}} [a_2(w_{21} + w_{22})/w_{21}]^{w_{21}} [a_3(w_{21} + w_{22})/w_{22}]^{w_{22}} \quad (8.38)$$
$$\times [a_4(w_{31} + w_{32})/w_{31}]^{w_{31}} [a_5(w_{31} + w_{32})/w_{32}]^{w_{32}} [a_6]^{w_{41}}$$

Subject to:

$$w_{01} + w_{02} + w_{03} + w_{04} = 1 \quad \text{(normality condition)} \quad (8.39)$$

$$w_{01} + w_{11} - w_{31} = 0 \quad \text{(orthogonality condition for } x_1) \quad (8.40)$$

$$w_{02} + w_{03} + w_{22} - w_{31} - w_{32} + w_{41}$$
$$= 0 \quad \text{(orthogonality condition for } x_2) \quad (8.41)$$

$$w_{02} + w_{04} - w_{11} + w_{22} - w_{41} = 0 \quad \text{(orthogonality condition for } x_3) \quad (8.42)$$

$$- w_{21} - w_{22} + w_{31} = 0 \quad \text{(orthogonality condition for } x_4) \quad (8.43)$$

$$- w_{11} + w_{32} = 0 \quad \text{(orthogonality condition for } x_5) \quad (8.44)$$

Here, w_{01}, w_{02}, w_{03}, and w_{04} represent the four cost weights related to the four terms in the objective function $y(x)$, i.e. expression (8.4) above. Similarly, w_{11}, w_{21}, w_{22}, w_{31}, w_{32}, and w_{41} represent the dual variables related to the single term in constraint-1, i.e. expression (8.33), to the first term in constraint-2, i.e. expression

(8.34), to the second term in constraint-2, i.e. expression (8.34), to the first term in constraint-3, i.e. expression (8.35), to the second term in constraint-3, i.e. expression (8.35), and to the single term in constraint-4, i.e. expression (8.36) above, respectively. There are ten terms (T) and five primal variables (N) in the primal objective function and constraints [expressions (8.4) and (8.33)–(8.36)], and therefore, the degree of difficulty in the problem is $= T - (N + 1) = 4$. Thus, any six of the dual variables can be expressed in terms of the remaining four dual variables as shown below, choosing w_{01}, w_{03}, w_{04}, and w_{41} as the four dual variables:

$$w_{02} = 1 - w_{01} - w_{03} - w_{04}; \quad w_{11} = w_{03} - w_{04} - w_{01} + 2w_{41}; \quad w_{21} = 1 - w_{03} - w_{41};$$
$$w_{22} = 2w_{03} - w_{04} - 1 + 3w_{41}; \quad w_{31} = w_{03} - w_{04} + 2w_{41}; \quad w_{32} = w_{11}.$$

$$(8.45)$$

as in case of Model-I, the cost weights w_{01}, w_{02}, w_{03}, and w_{04} represent the optimum proportion of total cost per unit length of beam, distributed between the tensile steel reinforcement, concrete, shuttering along the vertical surfaces of the beam, and the shuttering along the bottom horizontal surface of the beam, respectively. The set of Eq. (8.45) represent the relationship between these cost distributions among the above components, which are independent of the values of the cost coefficients c_1, c_2, c_3, and c_4. This invariant property of geometric programming could be used for development of suitable *design rules* for *optimality* even in the presence of a great degree of difficulty, having general application in any problem of this nature.

If we substitute the expressions in equations set (8.45) into the dual objective function [Eq. (8.38)] and maximize this substituted dual objective function over the four variables w_{03}, w_{04}, w_{01}, and w_{41} (i.e. by setting the first derivative of the function with respect to each variable to zero), we will get four additional equations, thus providing sufficient conditions for optimality. The process is simplified by maximizing the natural logarithm of the dual function; that is, by substituting:

$$\ln d(w_{01}, w_{03}, w_{04}, w_{41}) = z(w_{01}, w_{03}, w_{04}, w_{41}).$$

Thus, by above substitution and taking logarithm on both sides of Eq. (8.38), we get the following equation:

$$\begin{aligned}
z(w) = {}& w_{01} \left(\ln c_1 - \ln w_{01} \right) + \left(1 - w_{01} - w_{03} - w_{04} \right) \\
& \left(\ln c_2 - \ln \left(1 - w_{01} - w_{03} - w_{04} \right) \right) + w_{03} \left(\ln c_3 - \ln w_{03} \right) \\
& + w_{04} \left(\ln c_4 - \ln w_{04} \right) + \left(w_{03} - w_{04} - w_{01} + 2w_{41} \right) \ln a_1 + \left(1 - w_{03} - w_{41} \right) \\
& \left(\ln a_2 + \ln \left(w_{03} - w_{04} + 2w_{41} \right) - \ln \left(1 - w_{03} - w_{41} \right) \right) + \left(2w_{03} - w_{04} - 1 + 3w_{41} \right) \\
& \left(\ln a_3 + \ln \left(w_{03} - w_{04} + w_{41} \right) - \ln \left(2w_{03} - w_{04} - 1 + 3w_{41} \right) \right) \\
& + \left(w_{03} - w_{04} + 2w_{41} \right) \left(\ln a_4 + \ln \left(2w_{03} - 2w_{04} - w_{01} + 4w_{41} \right) \right. \\
& \left. - \ln \left(w_{03} - w_{04} + 2w_{41} \right) \right) + \left(w_{03} - w_{04} - w_{01} + 2w_{41} \right) \\
& \left(\ln a_5 + \ln \left(2w_{03} - 2w_{04} - w_{01} + 4w_{41} \right) - \left(w_{03} - w_{04} - w_{01} + 2w_{41} \right) \right) \\
& + w_{41} \ln a_6.
\end{aligned}$$

$$(8.46)$$

As done in case of Model-I, by setting the first derivatives $dz/dw_{01} = 0$, $dz/dw_{03} = 0$, $dz/dw_{04} = 0$, and $dz/dw_{41} = 0$; we will get four nonlinear simultaneous equations in terms of four unknown variables w_{01}, w_{03}, w_{04}, and w_{41} as shown below:

By setting the first derivatives $dz/dw_{01} = 0$, we get:

$$
\begin{aligned}
&(c_1/c_2a_1a_5)(1 - w_{01} - w_{03} - w_{04})\,(w_{03} - w_{04} - w_{01} + 2w_{41}) \\
&- (2\,w_{03} - 2w_{04} - w_{01} + 4w_{41})\,w_{01} = 0
\end{aligned}
\tag{8.47}
$$

By setting the first derivatives $dz/dw_{03} = 0$, we get:

$$
\begin{aligned}
&(c_3a_1a_3^2a_4a_5/c_2a_2)(1 - w_{01} - w_{03} - w_{04})\,(1 - w_{03} - w_{41}) \\
&(2w_{03} - 2w_{04} - w_{01} + 4w_{41})^2 - (2w_{03} - w_{04} - 1 + 3w_{41})^2 \\
&(w_{03} - w_{04} - w_{01} - 2w_{41})\,w_{03} = 0
\end{aligned}
\tag{8.48}
$$

By setting the first derivatives $dz/dw_{04} = 0$, we get:

$$
\begin{aligned}
&(c_4/c_2a_1a_3a_4a_5)\,(1 - w_{01} - w_{03} - w_{04})\,(2w_{03} - w_{04} - 1 + 3w_{41}) \\
&(w_{03} - w_{04} - w_{01} + 2w_{41}) - (2w_{03} - 2w_{04} - w_{01} + 4w_{41})^2 w_{04} = 0
\end{aligned}
\tag{8.49}
$$

By setting the first derivatives $dz/dw_{41} = 0$, we get:

$$
\begin{aligned}
&(a_1^2a_3^3a_4^2a_5^2a_6/a_2)\,(1 - w_{03} - w_{41})\,(2w_{03} - 2w_{04} - w_{01} + 4w_{41})^4 \\
&- (2w_{03} - w_{04} - 1 + 3w_{41})^3\,(w_{03} - w_{04} - w_{01} + 2w_{41})^2 = 0
\end{aligned}
\tag{8.50}
$$

From Eqs. (8.48) and (8.50), we get:

$$
\begin{aligned}
&(c_3/c_2a_1a_3a_4a_5a_6)(2w_{03} - w_{04} - 1 + 3w_{41})(w_{03} - w_{04} - w_{01} + 2w_{41}) \\
&(1 - w_{01} - w_{03} - w_{04}) - (2w_{03} - 2w_{04} - w_{01} + 4w_{41})^2 w_{03} = 0
\end{aligned}
\tag{8.51}
$$

From (8.49) and (8.51), we get:

$$
w_{04} = (c_4a_6/c_3)w_{03}
\tag{8.51}
$$

By expressing $k1 = (c_4a_6)/c_3$ in equation (8.51), we get $w_{04} = k1\,w_{03}$

Substituting $w_{04} = k1\,w_{03}$ in Eqs. (8.47), (8.48) and (8.49), these equations become:

$$
\begin{aligned}
&(c_1/c_2a_1a_5)(1 - w_{01} - w_{03} - k1\,w_{03})\,(w_{03} - k1\,w_{03} - w_{01} + 2w_{41}) \\
&- (2w_{03} - 2k1\,w_{03} - w_{01} + 4w_{41})\,w_{01} = 0
\end{aligned}
\tag{8.53}
$$

$$\left(c_3 a_1 a_3^2 a_4 a_5 / c_2 a_2\right)\left(1 - w_{01} - w_{03} - k1\, w_{03}\right)\left(1 - w_{03} - w_{41}\right)$$
$$\left(2w_{03} - 2k1\, w_{03} - w_{01} + 4w_{41}\right)^2 - \left(2w_{03} - k1\, w_{03} - 1 + 3w_{41}\right)^2 \qquad (8.54)$$
$$\left(w_{03} - k1\, w_{03} - w_{01} + 2w_{41}\right) w_{03} = 0$$

$$\left(c_4 / c_2 a_1 a_3 a_4 a_5\right)\left(1 - w_{01} - w_{03} - k1\, w_{03}\right)\left(2w_{03} - k1\, w_{03} - 1 + 3w_{41}\right)$$
$$\left(w_{03} - k1\, w_{03} - w_{01} + 2w_{41}\right) - \left(2w_{03} - 2k1\, w_{03} - w_{01} + 4w_{41}\right)^2 k1\, w_{03} = 0 \qquad$$
$$(8.55)$$

We can simplify the above three equations by expressing $k2$, $k3$, and $k4$ in terms of the model constants as shown below:

$$k2 = c_1 / (c_2 a_1 a_5)$$
$$k3 = \left(c_3 a_1 a_3^2 a_4 a_5\right) / (c_2 a_2)$$
$$k4 = c_3 / (c_2 a_1 a_3 a_4 a_5 a_6)$$

To simplify further we can substitute $x = w_{01}$, $y = w_{03}$ and $z = w_{41}$, and using the above constants, we can express the common components of the above three equations as r, p, m, w, and u as follows:

$$r = 1 - x - (k1 + 1) * y;$$
$$p = 2^* z - x + (1 - k1) * y;$$
$$m = 4^* z - x + 2^* (1 - k1) * y;$$
$$w = 3^* z + (2 - k1) * y - 1;$$
$$u = 1 - y - z;$$
$$w_{04} = k1\, y$$

Putting the above expressions for w_{04} and for r, p, m, w, and u in the Eqs. (8.47), (8.48), and (8.49) we can express these equations in a simplified form as $f(x, y, z)$— replacing Eq. (8.47), $g(x, y, z)$—replacing Eq. (8.48), and $h(x, y, z)$—replacing Eq. (8.49) as follows:

$$f(x, y, z) = k2rp - mx = 0 \qquad (8.56)$$

$$g(x, y, z) = k3rum^2 - w^2 py = 0 \qquad (8.57)$$

$$h(x, y, z) = k4rwp - m^2 y = 0 \qquad (8.58)$$

Equations (8.56), (8.57), and (8.58) are three nonlinear simultaneous equations in terms of three unknown variables w_{01}, w_{03}, and w_{41} which can be solved by the Newton–Raphson method as explained and illustrated in and Chap. 3 and in this chapter. The author has developed C++ computer programmes for solving such nonlinear simultaneous equations covering many variables and the computer code of the C++ programme '**cbm.cpp**' presented in appendix incorporates the solution of this Model-II also, and using this programme a designer can **instantly** solve this model using his/her site-specific design condition. Even without solving this model, as in case of Model-I, application of the geometric programming technique gives the following simple expressions for determining the optimum values of the model variables, also giving an insight to the designer regarding the interrelationship between these variables for an optimal solution.

$$
\begin{aligned}
x_2 &= (c_4 w_{02})/(c_2 w_{04}), \\
x_3 &= (c_3 w_{02})/(c_2 w_{03}), \\
y(x) &= (c_2 x_2 x_3)/w_{02}, \\
x_1 &= y(x)\, w_{01}/c_1, \\
x_4 &= a_2 + a_3 x_2 x_3, \\
x_5 &= a_1 x_1/x_3.
\end{aligned}
\tag{8.59}
$$

8.5.2.2   Model Calibration

As in case of Model-I, the Model-II also needs to be calibrated relating it to the site-specific cost parameters and desired design conditions to obtain the optimal design in these conditions. As before, the model constants i.e. cost coefficients: c_1, c_2, c_3, and c_4 are related to the cost of materials and shuttering in the given site condition and could be determined from relevant schedule of rates or market rates of materials and labour as illustrated in Chap. 4. The expressions giving the values of c_1, c_2, c_3, and c_4 with reference to the Eq. (8.4) and the Fig. 8.3 will be same as given for Model-I which are also included in the **C++ Computer Programme** (cbm.**cpp**) presented in appendix.

The expression for the value of model constants a_1 as shown below is derived from the constraint expression (8.33) defining the equilibrium condition between compressive strength of concrete and yield strength of steel. Similarly, as in case of Model-I, the expression for the value of model constants a_2 and a_3 is determined from the bending moment compatibility condition represented by the constraint expression (8.34). Again, at the equilibrium the total resisting moment capacity of the beam section, applying appropriate capacity reduction factor as prescribed in the relevant national code of practice, must match with the total applied bending moment, which condition is represented by the constraint expression (8.35). Accordingly, from this constraint, the expression for the value of a_4 is developed as shown below. The value of model constant a_5 is 0.5 as in case of Model-I. The value of model constant a_6 is the beam width:depth ratio desired by the designer, represented by the constraint expression (8.36). Parameters ws, r, cs, cc, and csh are same as explained in Model-I.

To sum up, the expressions for all the above model constants are developed below in equation set (8.57), without specifying the units of measurement, which may be chosen by a designer as applicable in his/her situation, and determine the numerical values of the model constants which can be given as input to find the optimal solution in a specific case.

$$
\begin{aligned}
a1 &= f_y/0.85 f_c' \\
a_2 &= \text{UABM} \\
\text{UUL} &= \text{LFDL} \times \text{DDL} + \text{LFLL} \times \text{DLL} \\
\text{UABM} &= \text{UUL} \times \text{BMC} \times \text{L}^2
\end{aligned} \tag{8.60}
$$

$a3 = \text{LFDL}(1 + \text{r})\text{wc} \times \text{L}^2 \times \text{BMC}$

$a4 = 1/\text{CRF} \times f_y$

$a5 = 0.5$

$a6 = $ desired ratio between width (b) and effective depth of beam (d) $= b/d$

If we adopt the metric units of measurement with 'ws' in kg/cm^3, 'cs' in Monetary Units (MU) per kg, 'cc' in MU per m^3, 'csh' in MU/m^2, 'wc' in kg/m^3, f'_c in N/cm^2, f_y in N/cm^2, and UABM in N-cm the value of the model constants will be given by:

$$
\begin{aligned}
c_1 &= (7.85/1000)cs; \text{ assuming ws as } 7.85 \text{ g/cm}^3 \\
c_2 &= (cc \times cr)/(1,000,000) \\
c_3 &= (2 \times cr \times csh)/(10,000) \\
c_4 &= csh/10,000.
\end{aligned}
$$

$a_1 = f_y/0.85\,f_c'$

$a_2 = \text{UABM in N.cm}$

$a_3 = \text{LFDL}(1 + r)wc \times \text{L}^2 \times \text{BMC}$

$\quad = \text{LFDL} \times (1 + r) \times 2323.0 \times 8.8 \times \text{L}^2 \times \text{BMC} \times 10^{-2} \text{ in N.cm};$

$\quad \quad \text{if } wc \text{ is in kg/m}^3 \text{ and } L \text{ is in meter}$

$a_4 = 1.0/\left(f_y \times \text{CRF}\right); \text{if } f_y \text{ is in N/cm}^2$

$a_5 = 0.5$

$a_6 = $ desired ratio between width (b) and effective depth of beam (d) $= b/d$

8.6 Conventional Design Method of Reinforced Concrete Beam and Sensitivity Analysis: Model-III

Using the schematic section of a reinforced concrete beam shown in Fig. 8.1, the cost per unit length of the beam will be same as the objective function Eq. (8.4) for the optimization models presented earlier, i.e. $y(x)$ can be expressed as:

$$
Y(x) = c_1 x_1 + c_2 x_2 x_3 + c_3 x_2 + c_4 x_3 \tag{8.4}
$$

Similarly, using the Eqs. (8.3) and (8.3a), the nominal moment capacity M_n and the ultimate moment capacity M_u of a reinforced concrete beam section are given by:

$$M_n = bd^2 f_c{}' \, w \, (1 - 0.58w) = ptbd^2 f_y \left(1 - 0.58 \, pt \, \left(f_y/f_c{}'\right)\right) \qquad (8.61)$$

$$M_u = \phi bd^2 f_c{}' w \, (1 - 0.58w) = \phi ptbd^2 f_y \left(1 - 0.58 \, pt \, \left(f_y/f_c{}'\right)\right) \qquad (8.62)$$

Thus, assuming specified values of M_u, f_y, $f_c{}'$, and b, the sensitivity of $y(x)$ to the variation of pt can be determined using the Eq. (8.62). Similarly, assuming specified values of M_u, f_y, $f_c{}'$, and pt, the sensitivity of $y(x)$ to the variation of b can be determined using the Eq. (8.62). All of these methods and the Eq. (8.62) can be used for conventional design of reinforced concrete beams.

It is difficult to carry out manually the sensitivity analysis of reinforced concrete beam cost to various design parameters such as steel ratio, beam width, and so on which is necessary so as to develop an insight into the interactions of these parameter, and this help achieve efficiency in design. Using computers, it is very easy to carry out such sensitivity analysis and derive benefits. Therefore, the C++ computer program cbm.cpp includes the Model-III to perform such tasks, which incorporates the following definition of different beam elements.

```
DLL  = DESIGN Uniform LIVE LOAD (in kN/m)
DDL  = DESIGN Uniform DEAD LOAD (in kN/m)
BMC  = DESIGN BENDING MOMENT CO-EFFICIENT (0.125 for simply sup. so on)
BSPN = SPAN of the BEAM (in m)
BWT  = ASSUMED WEIGHT of the BEAM (in kN/m)
MBW  = MINIMUM ACCEPTABLE BEAM WIDTH (in cm)
WB   = ANY CHOSEN WIDTH OF BEAM (in cm)
BMWDR= DESIRED VALUE OF BEAM WIDTH:DEPTH (W/D) RATIO
LFDL = LOAD FACTOR for DEAD LOAD (normally 1.4)
LFLL = LOAD FACTOR for LIVE LOAD (normally 1.7)
CRF  = CAPACITY REDUCTION FACTOR (normally 0.8)
SYS  = DESIGN STEEL YIELD STRENGTH (in MPA: 300 etc.)
pt   = DESIGN STEEL RATIO
AS   = AREA OF TENSION STEEL REINFORCEMENT (in Sq.Cm.)
CCS  = DESIGN CONCRETE COMPRESSIVE STRENGTH (in MPA: 30 etc.)
UWC  = UNIT WEIGHT of CONCRETE (2323 kg/cu.m.)
cr   = DESIRED CONCRETE COVER RATIO (1.08/1.10 etc)
cc   = UNIT COST OF CONCRETE (IN MONETARY UNITS i.e.MU/Cubic Meter)
cs   = UNIT COST OF STEEL RE-INFORCEMENT (MU/Kilogram)
csh  = UNIT COST OF SHUTTERING (MU/Sq.M.)
```

8.6.1   Sensitivity Analysis of Reinforced Concrete Beam Cost to the Variation of Steel Ratio

To carry out sensitivity analysis of reinforced concrete beam cost to the variation of steel ratio specified values of M_u, f_y, $f_c{}'$, and b are assumed, and the sensitivity of $y(x)$ to the variation of pt can be determined using the Eq. (8.62). Thus, the beam width x_3 in expression (8.4) is designated as: BW and given a specified value, to carry out sensitivity analysis to the variation of steel ratio. The relationship between the other elements is as follows.

```
UUL=LFDL*(DDL+ BWT)+ LFLL*DLL; //in kN/m
UUL=LFDL*(DDL+ ABWT)+ LFLL*DLL; //in kN/m
TABM=UUL×BMC×BSPN×BSPN×100= ; //in kN.cm
For each value of pt the following relationships hold:
k= (1- (0.58× pt ×SYS) /(CCS))
x₃ = BW
K= (TABM×1000)/(k × CRF × pt × SYS×100× x₃);
Thus, from the equation (8.57) the value of  x₂ = √K
Therefore,  x₁= pt × x₂×x₃;
AS=x₁;
D=cr×x₂ ; D is the overall depth of RC beam as shown in Figure 8.3
```

Since all values of x_1, x_2, and x_3 are known the value of $y(x)$ as per Eq. (8.4) can be calculated for each interval value of pt within the range of low and high value of pt specified by the designer, and thus the sensitivity of beam cost to the variation of steel ratio, i.e. pt can be determined. This is shown in Sect. 8.8.3 of example application results using Model-III of C++ programm cbm.cpp.

8.6.2 Sensitivity Analysis of Reinforced Concrete Beam Cost to the Variation of Beam Width

To carry out sensitivity analysis of reinforced concrete beam cost to the variation of beam width:BW, specified values of M_u, f_y, f_c^1, and steel ratio are assumed. Thereafter, the sensitivity of $y(x)$ to the variation of BW can be determined using the Eq. (8.62). The relationship between the other elements is same as above.

In this case, the range of low, high, and the interval value of beam width, i.e. BW is specified by the designer. Since all values of x_1, x_2, and x_3 are known the value of $y(x)$ as per Eq. (8.4) can be calculated for each interval value of BW within the range of low and high value of BW specified by the designer. Thus, the sensitivity of beam cost to the variation of beam width, i.e. BW can be determined. This is shown in Sect. 8.8.3 of example application results using Model-III of C++ programm cbm.cpp.

8.6.3 Conventional Design Method of Reinforced Concrete Beam Including Cost

Earlier adopting the strength design method for flexural computation, the Eq. (8.3) for the nominal moment capacity: M_n, of a rectangular RC beam, with tension reinforcement only, was presented, which is reproduced below.

$$M_n = bd^2 f_c^1 w (1 - 0.58w) \tag{8.3}$$

Equation (8.61) above is developed by substituting the basic variables in the above equation. Similarly, Eq. (8.62) above is developed by substituting the basic variables which is reproduced below:

$$M_u = \phi bd^2 f_c{}^1 w \, (1 - 0.58w) = \phi ptbd^2 f_y \left(1 - 0.58 \, pt \, \left(f_y/f_c{}^1\right)\right) \qquad (8.62)$$

where

  b = width of the beam
  d = the distance from the extreme compressive fibre to the centroid of tension
     reinforcement
  $w = A_s f_y/bd f_c{}^1$
  A_s = area of tension reinforcement
  $f_c{}^1$ = compressive strength of the concrete
  f_y = yield strength of reinforcement

  M_n, the nominal (or design) resisting moment of a cross section is related to M_u, the ultimate resisting moment of a cross section, by the equation $M_n = M_u/\phi$, where ϕ is the capacity reduction factor, which is 0.8 for flexure as per 1883 ACI Code. This is incorporated in Eq. (8.62) above and can be used for design of a beam cross section. As already mentioned, there are an infinite number of solutions to the equation set (8.3) that yield the same value of M_n. It may be seen from Eqs. (8.61) and (8.62) that once type and strength of concrete and steel reinforcement is decided, M_n is a function of three variables, i.e. b, d, and A_s. Hence, a designer has to assume the value of two of these variables in a trial section and find the value of third variable, and start an iterative design process to evaluate its suitability; and if not satisfied, a new section is adopted and the process is repeated.

  For solution of the Eq. (8.62) for singly reinforced concrete beams, the following three methods are suggested:

  (i)   direct substitution into formulas,
 (ii)   use of tables for formulas and constants, and
(iii)   use of curves or charts for the formulas and constants.

  However, in these days of Information Technology Revolution and easy availability of computers at affordable cost, such tedious and time-consuming manual methods can be avoided using computers. To help this process, the C++ computer program cbm.cpp is developed, which incorporates the relationship between the beam elements as shown above, and thus, start the conventional general iterative design process for solution of the above equation and to evaluate various design alternatives using the Model-III. In Sect. 8.8.3, example application results of conventional design method, for singly reinforced concrete beam using the Model-III, are shown.

8.7   Computer-Aided Optimal Design of Structures

The optimization models presented in Sect. 8.5 clearly show that it is nearly impossible to solve these nonlinear programming models by manual methods. No doubt, because of this reason both development and application of such optimization methods have been seriously impeded in the past. In fact, tedious and voluminous computations involved, in utilizing *Optimization* and operations research techniques and to solve and use *Optimizing Models* like above, have prevented their applications in building/urban *planning* and *design* and even in structural engineering in the past. As mentioned in Chap. 1, the continued development of cheap and powerful computers with interactive graphics systems has made it possible to perform many planning and design functions integrating graphics with complex computations, design analysis, cost engineering, and *Optimization* techniques (such as the optimization models presented above) on a computer. As earlier discussed, application of computers has revolutionized the fields of design and engineering particularly in the area of aerospace and automotive engineering, and AEC sector can derive great benefits by adopting such an approach keeping in view many factors as listed in Chap. 1, which make 'PC-CADD' a more productive approach. Many advantages of this approach, including instant linkage/conversion of optimal design in alphanumeric terms into a graphic output, facilitating application of optimization techniques for resource-efficiency, is demonstrated in later sections.

CAD primarily depends on (1) system software—supplied with the hardware and (2) application software— available commercially or to be developed in-house. The area of 'Architectural design and drafting-engineering-construction (AEC)' ranks relatively low in PC-CADD systems by application, as integrated CAD rarely forms part of research/education curriculum in many countries, and commercial CAD systems incorporating all components of integrated CAD are also rare. An integrated CAD system should have six components: (1) geometric modeling, (2) cost engineering, (3) planning and design analysis, (4) design optimization, (5) drafting and drawing, and (6) data management, storage and transfer. But, most microcomputer-based commercial CAD systems are constructed primarily to perform geometric modeling and drafting, and rarely include cost engineering and optimization. Use of such modern planning, design, and management techniques can increase in AEC only with the development of more and more sophisticated software for optimal design, decision support systems, and database management, using information technology.

Towards this effort, the author is carrying out research and developed/published a number of descriptive and nonlinear programming optimizing models for building/urban development, and structural engineering and now devising software, linking these optimizing models, as an integrated computer-aided planning and design system, to help better *efficiency* and *social responsiveness* in urban and building operations, and also to improve *efficiency* in structural engineering. This book is an attempt to promote this approach in AEC operations and to empower AEC professionals to apply such techniques and to contribute to their further development so that *Computer-Aided Design (CAD) by Optimization* becomes routine in AEC operations with consequent benefits, as in the other fields mentioned above.

Moreover, building/urban operations (e.g. housing development) consume huge *Resources* involving many variables giving numerous solutions with large variations in *resource-efficiency* indicators, making *search* of optimal solution very complex to be solved manually, but not so with computers. Same is the case with structural engineering problems. Drawing is the language of architecture, engineering, and physical planning, where ideas are created, evaluated, and developed. Therefore, it is desirable to develop application software not only for numerical solution of optimizing models presented above, but also, to convert instantly the optimal designs in alphanumeric terms (given as output by the optimizing models) into 'graphic form', i.e. dimensioned drawing including the working drawings to be meaningful to the users. The software should also incorporate cost engineering to derive the cost coefficients for the optimizing models, using any time-specific, site-specific, and country-specific basic unit cost data on labour, materials, and so on. Accordingly, an application software:'**RCCAD'** is being developed as an *Integrated Computer-Aided Design and Drafting* (CADD) *System* to achieve resource-efficient and optimal structural engineering designs. Integration of numerical optimal design and graphics is discussed in Sect. 8.8, and integrated computer-aided optimal design of structures taking reinforced concrete beam as example is demonstrated in Sect. 8.10 below.

Application of computer-aided optimal design, in example problems both for Model-I and Model-II, is demonstrated below. Similarly, application of computer-aided design and sensitivity analysis is illustrated in example problems for Model-III below.

8.8   Application of Computer-Aided Optimal and Conventional Design in Example Problems

Computers can be used as a tool to automatically generate design solutions *integrating* sophisticated analysis and *optimization* techniques into the *design process* making it possible to maintain much more *effective cost control* fulfilling specified design constraints and exploring more alternative solutions to derive benefits. As stated above, it is nearly impossible to solve nonlinear programming models by manual methods; and tedious, voluminous, and time consuming computations involved, in utilizing *optimization* and operations research techniques and to solve and use *Optimizing Models* like above, have prevented their applications in building/urban *planning* and *design* and even in structural engineering in the past. To overcome such problems in case of optimal design of beams, the author has developed a C++ program named as cbm.cpp which can solve the above nonlinear optimization models in flip of a few seconds. Example application results of computer-aided optimal design are illustrated below for each of the optimization models.

8.8.1   *Example Application Results of LEAST-COST R.C. Beam Design: Model-I*

As earlier stated, it is rather impossible to solve manually the nonlinear programming optimization models presented in Sects. 8.5.1 and 8.5.2. To solve such nonlinear programming models for optimal design of reinforced concrete beams a C++ program named as: **cbm.cpp** is developed, which is the *numerical part* of the

Integrated Computer-Aided Design Drafting (CADD) Software: **RCCAD**, to carry out numerical analysis including *optimization*. The CADD software has also a *graphic part*, named as: bmdg.lsp, to carry out graphic and drafting tasks with developing working drawings of the RC beam as numerically analysed and optimized.

Application of the optimization model is illustrated using the following case study problem taken from Everard and Tanner, without assuming beam weight beforehand, showing that the optimal design leads to considerable saving in cost and material consumption as compared to the conventional design.

Example problem is to design a **Least-Cost** reinforced concrete beam simply supported over a span of 10 m supporting a uniform dead load of 15 kN/m and uniform live load of 20 kN/m. The concrete strength $fc' = 30$ MPa and the steel yield strength $fy = 300$ MPa. The unit cost of steel (CS), concrete (CC), and shuttering (CSH) adopted here are 0.72 MU (Monetary Unit)/kg, 64.5 MU/m^3 and 2.155 MU/m^2, respectively. A cover ratio (r) of 0.10, unit weight of 2323 kg/m^3 for concrete and a capacity reduction factor of 0.80 are assumed here. Assume the value of load factor for dead load, i.e. LFDL= 1.4, and the value of load factor for live load, i.e. LFLL $= 1.7$.

The ultimate uniform load $= 1.4 \times 15 + 1.7 \times 20 = 55$ kN/m

The ultimate applied bending moment $= 55 \times 10^2/8 = 687.5$ kNm $= 687.5 \times 10^5$ Ncm

Using the above input data, the values of cost coefficients and other model constants and model outputs for this case can be derived as follows, instantly given as output by the **C++ Computer Programme (cbm.cpp)** developed by the author:

$c_1 = 0.00565200$,　$c_2 = 0.000070850$,　$c_3 = 0.000474100$,　$c_4 = 0.000215500$,

$a_1 = 0.0850$,　$a_2 = 68750000.00$,　$a_3 = 4382.3385$,　$a_4 = 0.00043573$,　$a_5 = 0.50$,

$a_6 = 30.00$.

Putting these values of model constants in Eqs. (8.4)–(8.8), (8.11), and in Eqs. (8.18)–(8.29) and also solving the above three nonlinear simultaneous equations, using the C++ software sub-module '**cbm.cpp**' of Software: '**RCCAD**', which solves, in no time, this *Optimization Model* for optimal design of reinforced concrete beam. Thus, we get the optimum cost weights, optimum beam dimensions, optimum area of tensile steel reinforcement, least-cost of the beam, and other details which appear in the computer screen as the program '**cbm.cpp**' of Software: '**RCCAD**' is run in graphic mode.

On loading and typing '**RCCAD**' in the text part of the computer screen (24,25), Application Software: '**RCCAD**', developed for Integrated Computer-Aided Optimal Design and Drafting (CAODD) of RCC sections, prompts the designer/user to select a specific module, with the following message appearing in the graphic part of the computer screen (text in *italics* are comments of the author):

WELCOME TO ::

'' RCCAD ''

A SOFTWARE FOR ''REINFORCED CONCRETE (RC)
COMPUTER-AIDED-DESIGN''TO ACHIEVE LEAST-COST.
''RCCAD'',HAS 2 PARTS: (I)''NUMERIC OPTIMIZATION'',
AND (II)''GRAPHIC COMPONENT'', WHICH CONVERTS
INSTANTLY THE ''NUMERIC DESIGN'' TO ''DRAWING''.
SOFTWARE ''RCCAD'' HAS NUMBER OF MODELS.
SELECT 'NUMERIC' ,OR, 'GRAPHIC' OPTION AND,
A MODEL TO RUN AS PER YOUR REQUIREMENT.

Do you want to RUN 'Numeric SESSION' for RCC BEAM DESIGN ? (Y/N):
On putting 'y' response, the following message appears in text part of computer screen:
Type 'CBM' for 'Numeric SESSION'. On Results Load 'BMDG' to START'DRAWING SESSION'
On putting 'n' response, the following message appears in text part of computer screen:
Load 'BMDG' to START BEAM 'DRAWING SESSION', if Numeric Results already Exist.
On exercise of 'y' response, for "NUMERIC" session, and on Typing CBM, the C ++ Program, for Reinforced Concrete Beam Optimizing Model and Conventional Design Solution, is opened with the following messages for response:

        WELCOME TO SOFTWARE: ''RCCAD''
(Reinforced-Concrete-Computer-Aided-analysis-Design)

Using ''RCCAD''(This Numeric Module Version),you can CARRY OUT
following TASKS:

(1) FIND Least-Cost-Design of Reinforced-Concrete-Beam(Given Beam
    Width)
(2) FIND Least-Cost-Design of Reinforced-Concrete-Beam
    (Given Beam Width:Depth Ratio)
(3) FIND Conventional Design and CARRY OUT 'Sensitivity-Analysis'(SA),
    choosing an Appropriate SET of Independent/Dependent DESIGN
    VARIABLES and the 'ITERATION MODEL'.

Do you want to see Definitions of Beam Elements?

Type y(if 'yes') or n(if 'no')and press <CR>:y

On exercise of 'y' response, definitions of beam elements are displayed (which are same as presented in Sect. 8.6) as below:

```
DEFINITION OF DIFFERENT BEAM ELEMENTS ARE :
DLL  = DESIGN Uniform LIVE LOAD(in kN/m)
DDL  = DESIGN Uniform DEAD LOAD(in kN/m)
BMC  = DESIGN BENDING MOMENT CO-EFFICIENT(0.125 for simply sup. so on)
BSPN = SPAN of the BEAM(in m)
BWT  = ASSUMED WEIGHT of the BEAM(in kN/m)
MBW  = MINIMUM ACCEPTABLE BEAM WIDTH(in cm)
WB   = ANY CHOSEN WIDTH OF BEAM(in cm)
BMWDR= DESIRED VALUE OF BEAM WIDTH:DEPTH(W/D) RATIO
LFDL = LOAD FACTOR for DEAD LOAD(normally 1.4)
LFLL = LOAD FACTOR for LIVE LOAD(normally 1.7)
CRF  = CAPACITY REDUCTION FACTOR(normally 0.8)
SYS  = DESIGN STEEL YIELD STRENGTH(in MPA: 300 etc.)
pt   = DESIGN STEEL RATIO
AS   = AREA OF TENSION STEEL REINFORCEMENT(in Sq.Cm.)
CCS  = DESIGN CONCRETE COMPRESSIVE STRENGTH(in MPA: 30 etc.)
UWC  = UNIT WEIGHT of CONCRETE(2323 kg/cu.m.)
cr   = DESIRED CONCRETE COVER RATIO(1.08/1.10 etc)
cc   = UNIT COST OF CONCRETE(IN MONETARY UNITS i.e.MU/Cubic Meter)
cs   = UNIT COST OF STEEL RE-INFORCEMENT(MU/Kilogram)
csh  = UNIT COST OF SHUTTERING(MU/Sq.M.)
```

Do you want to START the 'NUM.ITERATION SESSION' ?

```
Type y(if 'yes') or n(if 'no')and press <CR>:y
```

On exercise of 'y' response as above, the following message appears in computer screen for user response:

```
AS SHOWN BELOW,THERE ARE 3 'ITERATION MODELS',
DEPENDING ON THE SET OF DESIGN CONDITION YOU CHOOSE:
LEAST-COST FLEXURAL STRENGTH DESIGN-GIVEN MIN.ACCEPT.BEAM-WIDTH
(ITMODEL-1)
LEAST-COST FLEXURAL STRENGTH DESIGN-GIVEN B.WIDTH:DEPTH RATIO
(ITMODEL-2)
ITERATIVE FLEXURAL STRENGTH DESIGN i.e. CONVENTIONAL METHOD,INCLUDING
SENSITIVITY ANALYSIS:-(i)TO STL.RATIO (ii)TO B.WIDTH (iii)GEN.
ITMODEL-3
PUT YOUR CURRENT CHOICE OF 'NUM.ITERATION MODEL'(1,2..)
1
```

On putting value of '1' as above, the Model-1 is selected and the following message appear in computer screen for response by the user/designer:

```
YOU CAN CHOOSE ANY OF FOLLOWING 15 'INPUT SET' OPTIONS:
0.When you select the pre-set computer data input
1.When you want to change value of DLL(Design Live Load)only
2.When you want to change value of DDL(Design Dead Load)only
```

```
3.When you want to change value of MBW(MIN.BEAM WIDTH)only
4.When you want to change value of BSPN(BEAM SPAN)only
5.When you want to change values of DLL,DDL
6.When you want to change values of DLL,DDL & MBW
7.When you want to change values of DLL,DDL & BSPN
8.When you want to change values of DLL,DDL,BSPN & BMC
8.When you want to change values of ALL Physical Design Data
10.When you want CALIBRATE MODEL with INPUT Element-Unit-COSTS
11.When you want MODEL CALIBRATION with INPUT Values for ALL Elements
12.When you want to change ALL MODEL CONSTANTS(i.e.c1,c2..,a1,a2..)
13.When you want to change Initial Trial-Values of Variables
14.When you want change Initial VALUES of Variables: BWT & pt(ITMODEL-3)
PUT YOUR CURRENT CHOICE OF 'INPUT SET' CASE(0/1 ETC)
0
```

It may be seen that there 15 Input Set Cases giving numerous choice to the designer to fulfill his/her need.To facilitate an easy application demonstration of each of 3 Models, the case 0, i.e. preset computer data input for a specific design is provided needing no new INPUT to produce design drawing for this case. However, during the program run these can be changed as per requirement choosing appropriate specific input case.

On putting the value of '0' as above, following message and results appear in computer screen:

```
Since case 0 i.e. Pre-Set Computer Data Input is selected, no new INPUT.
a1=0.085000 a2=68750000.00000000 a3=4382.33850000 a4= 0.00043573
a5= 0.50
a6=30.00 c1= 0.00565200 c2= 0.00007085 c3= 0.00047410 c4= 0.00021550
CALCULATED MODEL CONSTs.ABOVE AS PER USER-INPUT-CHECK/PUT ANY CHAR.TO
CONTINUE:
c
 1    0.5248    0.0857    0.6585    0.85513   -0.22231   -0.05405
 2    0.5260    0.0500    0.5688    0.06184   -0.02636   -0.01357
 3    0.5024    0.0287    0.5055   -0.01768   -0.00326   -0.00385
 4    0.4876    0.0184    0.4688   -0.00718   -0.00047   -0.00106
 5    0.4815    0.0153    0.4554   -0.00210   -0.00010   -0.00024
 6    0.4804    0.0146    0.4528   -0.00033   -0.00001   -0.00003
 7    0.4804    0.0146    0.4527   -0.00001    0.00000    0.00000
 8    0.4804    0.0146    0.4527    0.00000    0.00000    0.00000
 8    0.4804    0.0146    0.4527    0.00000    0.00000    0.00000
10    0.4804    0.0146    0.4527    0.00000    0.00000    0.00000
11    0.4804    0.0146    0.4527    0.00000    0.00000    0.00000
12    0.4804    0.0146    0.4527    0.00000    0.00000    0.00000
13    0.4804    0.0146    0.4527    0.00000    0.00000    0.00000
14    0.4804    0.0146    0.4527    0.00000    0.00000    0.00000
15    0.4804    0.0146    0.4527    0.00000    0.00000    0.00000
16    0.4804    0.0146    0.4527    0.00000    0.00000    0.00000
17    0.4804    0.0146    0.4527    0.00000    0.00000    0.00000
18    0.4804    0.0146    0.4527    0.00000    0.00000    0.00000
18    0.4804    0.0146    0.4527    0.00000    0.00000    0.00000
```

```
20   0.4804   0.0146   0.4527   0.00000   0.00000   0.00000
21   0.4804   0.0146   0.4527   0.00000   0.00000   0.00000
22   0.4804   0.0146   0.4527   0.00000   0.00000   0.00000
23   0.4804   0.0146   0.4527   0.00000   0.00000   0.00000
24   0.4804   0.0146   0.4527   0.00000   0.00000   0.00000
25   0.4804   0.0146   0.4527   0.00000   0.00000   0.00000
26   0.4804   0.0146   0.4527   0.00000   0.00000   0.00000
27   0.4804   0.0146   0.4527   0.00000   0.00000   0.00000
28   0.4804   0.0146   0.4527   0.00000   0.00000   0.00000
28   0.4804   0.0146   0.4527   0.00000   0.00000   0.00000
30   0.4804   0.0146   0.4527   0.00000   0.00000   0.00000
31   0.4804   0.0146   0.4527   0.00000   0.00000   0.00000
32   0.4804   0.0146   0.4527   0.00000   0.00000   0.00000
33   0.4804   0.0146   0.4527   0.00000   0.00000   0.00000
34   0.4804   0.0146   0.4527   0.00000   0.00000   0.00000
35   0.4804   0.0146   0.4527   0.00000   0.00000   0.00000
36   0.4804   0.0146   0.4527   0.00000   0.00000   0.00000
37   0.4804   0.0146   0.4527   0.00000   0.00000   0.00000
38   0.4804   0.0146   0.4527   0.00000   0.00000   0.00000
38   0.4804   0.0146   0.4527   0.00000   0.00000   0.00000
40   0.4804   0.0146   0.4527   0.00000   0.00000   0.00000
41   0.4804   0.0146   0.4527   0.00000   0.00000   0.00000
42   0.4804   0.0146   0.4527   0.00000   0.00000   0.00000
43   0.4804   0.0146   0.4527   0.00000   0.00000   0.00000
44   0.4804   0.0146   0.4527   0.00000   0.00000   0.00000
45   0.4804   0.0146   0.4527   0.00000   0.00000   0.00000
46   0.4804   0.0146   0.4527   0.00000   0.00000   0.00000
47   0.4804   0.0146   0.4527   0.00000   0.00000   0.00000
48   0.4804   0.0146   0.4527   0.00000   0.00000   0.00000
48   0.4804   0.0146   0.4527   0.00000   0.00000   0.00000
Number of Iterations is: 50
```

In the above iteration, the first column gives the iteration number, next 3 columns give the current values of the unknown variables w_{01}, w_{04}, and w_{41} in three nonlinear simultaneous Equations (8.26), (8.27), and (8.28) presented earlier. The last 3 columns give the current value of Equations (8.26), (8.27), and (8.28), respectively, using the current value of three unknown variables w_{01}, w_{04}, and w_{41} in respective iteration number, determined by using the Newton–Raphson method for 3 variables as explained and illustrated in Chap. 3 and this chapter. It may be seen that the iterations converge rapidly to '0' values, even at the early stage of iteration number 7 and 8. If there is no such convergence, the user/designer is given the following option to change: (1) limit of number of iterations and (2) error tolerance value, and re-run the program until convergence is obtained. As shown below, the user/designer is also given the option to change the initial trial values of unknown variables of w_{01}, w_{04}, and w_{41}, designated here as: x, y, and z, respectively, to facilitate convergence solution within prescribed margin of error.

```
Want change No.of ITERNS.LIMIT/ITERTN.Tolerance(if no Convergance)?
Type y(if 'yes') or n(if 'no')and press <CR>: n
```

You have chosen not to change ITRL,ITOL

(Initial-Trial-Values of NLP Indpdnt.Variables are:x=0.3,y=0.3,z=0.8)
Want Continue Iteration with NEW Values(if no Convergance)?
Type y(if 'yes') or n(if 'no')and press <CR>: n

Since, the Optimization Model-I is solved in just 7/8 Iterations as shown above, there is no need to change: (1) limit of number of iterations, (2) error tolerance value or (3) initial trial values of unknown variables, and accordingly, response is given as 'n' in above dialogue process. Thus, the following final results are obtained after solving the Model-I.

SWBM= 11314.68 kNcm,uabm= 68750.00 kNcm,TABM= 80064.68 kNcm,
x4= 80064.68 kNcm

CALIBRATED MODEL CONSTANTS FOR CHOSEN DESIGN CONDITIONS (ITMODEL= 1)
are:
a1=0.0850 a2= 68750000.0000 a3=4382.3385000 a4= 0.0004357 a5=0.50
a6=30.00
c1= 0.00565200 , c2= 0.00007085 , c3= 0.00047410 , c4= 0.00021550
DESIGN Values of OPTIMAL SECTION:
LA= 78.6721 cm, C= 1130.78 kN, T= 1130.78 kN, a3=4382.33850000
MNC= 88860.50 kN.cm, MNS= 88860.50 kN.cm, MR= 80064.45 kN.cm
COM= 1.0000
SOME KEY INPUTS: DLL=20.0 kN/m, DDL=15.0 kN/m, BSPN=1000.0 cm

LEAST-COST OPTIMAL RESULTS AFTER SOLVING THE 'NLP' MODEL:
The values of Optimal Cost Weights are:
w01= 0.4804, w02=0.4131, w03=0.0820, w04=0.0146, t=1.0000,
w11=0.4804
w21= 0.4552, w22=0.0748, w31=0.5302, w32=0.0488, w41=0.4527
Optimal Area of Tensile Steel Re-inforcement= 37.6825 Sq.Cm.
Optimal Effective Depth of the RC Beam = 86.0628 Cm
Optimal Width of the RC Beam   = 30.0000 Cm
Total Applied Bending Moment(inc.Self.wght)= 80064.6820 kN.Cm
Calculated Moment of Resistance of OPTIMAL SECTION= 80064.4468 kN.Cm
Optimal Depth of Eq.Rect.Stress Block = 14.7814 Cm
The LEAST-COST per Unit-Length of BEAM = 0.4435 MU/Cm
The Calculated WEIGHT per Unit-Length of BEAM = 6.5875 kg/Cm

 adif= 0.0218
 rlay= 2 dia1= 3.000 bn1= 4 dia2= 2.000 bn2= 3 ccv1= 6.500
 AS= 37.6858 ta= 37.7143 adif= 0.0218

PRE-INPUT:Layer No,Dia.of bars,No.of Re-inf.bars of each category,
clear Cover,design reinf.area,actual reinf.area & diff.(adif)between
design reinf.area & actual reinf.area are shown above

You want to CHOOSE NEW Dia./No.of Re-inf.bars & clear Cover?
(IF '0'INPUT OPTION CHOSEN,CAN USE COMPUTER DATA PUTTING 'N0'BELOW)
Type y(if 'yes') or n(if 'no')and press <CR>: n

```
CHOSEN NO NEW REINF.BAR DIA/NO./LAYER,OR,RESULTS ARE WITHIN TOL.
```

The C++ program cbm.cpp calculates the total area of reinforcement bars as per the Pre-Input (or New Input) layer number,diameter and number of bars of each category; and compares with design steel reinforcement area i.e. 'AS' as well as the area difference (adif), and continue to calculate 'adif' when response to CHOOSE NEW Dia etc. is put as 'y', and until 'adif' is within tolerable limit fixed by designer, with Pre-Input or New Input of dia and number of bars. In the current case with '0' INPUT OPTION CHOSEN 'adif' is within tolerable limit, hence, response to CHOOSE NEW Dia etc. is put as 'n', as above. It desirable that steel manufacturers provide a wide variety of bar diameters, so that it becomes easy to provide reinforcement area matching with the optimal area, thus, promoting optimal design with consequent benefit of cost and material consumption reduction.

```
YOU WANT TO CHANGE 'INPUT VALUE(S)/INPUT SET/
NUM.ITERATION MODEL',FOR BETTER RESULTS ?
Type y(if 'yes') or n(if 'no')and press <CR>:y
```

On putting response 'y' as above, the user/designer is given the following option to re-run the program with new input values specifying the desired/site-specific design conditions, or select a new Model i.e. Optimal Design Model-II or Conventional Model-III and re-run the program to solve these models in no time, as done in case of Optimal Design Model-I as above.

```
AS SHOWN BELOW,THERE ARE 3 'ITERATION MODELS',
DEPENDING ON THE SET OF DESIGN CONDITION YOU CHOOSE:

LEAST-COST FLEXURAL STRENGTH DESIGN-GIVEN MIN.ACCEPT.BEAM-WIDTH
(ITMODEL-1)

LEAST-COST FLEXURAL STRENGTH DESIGN-GIVEN B.WIDTH:DEPTH RATIO
(ITMODEL-2)

ITERATIVE FLEXURAL STRENGTH DESIGN i.e.CONVENTIONAL METHOD,INCLUDING
SENSITIVITY ANALYSIS:-(i)TO STL.RATIO (ii)TO B.WIDTH (iii)GEN.
ITMODEL-3

PUT YOUR CURRENT CHOICE OF 'NUM.ITERATION MODEL'(1,2..)
```

Alternatively, on putting a response of 'n' to the above dialogue, the following message appears in the computer screen,and the program/iteration is terminated.

```
AS CURRENT SOLUTION ACCEPTABLE,CLOSE 'NUMERICAL ITERATION
SESSION',LOAD 'BMDG'& START 'DRAWING SESSION'IF DESIRED
'ITERATION CLOSED'
```

On closing the iteration as above, the user/designer can obtain the output file bm1out.cpp of PROGRAM: cbm.cpp giving full results of solution of the Optimization Model-I as shown below:

```
INPUT-Design-Decisions/OUTPUTS(SOME)in Current-Case are:
ITERATION MODEL FOR LEAST-COST RC-BEAM DESIGN CHOSEN IS: 'ITMODEL- 1'
SOME RC-BEAM DESIGN CONDITIONS CHOSEN ARE:
DLL=20.000 DDL=15.000 BMC= 0.125 BSPN=1000.000 LFLL= 1.700
LFDL= 1.400 SYS=300.000 CCS=30.000 cs= 0.7200 cc= 64.5000
cr= 1.100 D= 84.668 cg1= 2.106 csh= 2.1550

Some Design Values Chosen/Calculated:
covc= 8.606 covl= 5.000 ccv1= 6.500 cov2=13.825
ccv2=14.825 cc2v=14.825 layc= 8.425 laycv= 5.825
rlay= 2 dia1= 3.000 bn1= 4 dia2= 2.000 bn2= 3
AS = 37.6858   ta= 37.7143   adif= 0.0218

OUTPUT-OPTIMAL-VALUES(some)given by RCCAD(Numerical Part):
SOME OPTIMAL RESULTS AFTER SOLVING THE LEAST-COST 'NLP' MODEL:
The values of Optimal Geometric Programming Cost Weights are:
w01= 0.4804, w02=0.4131, w03=0.0820, w04=0.0146
t=1.0000, w11=0.4804, w21= 0.4552, w22=0.0748,
w31=0.5302, w32=0.0488, w41=0.4527, pt=0.0146,

Area of Tensile Steel Re-inforcement= 37.6825 Sq.Cm.
Optimal Effective Depth of the RC Beam = 86.0628 Cm
Optimal Width of the RC Beam       = 30.0000 Cm
Total Applied Bending Moment (inc.Self.wght)= 80064.6820 kN.Cm
Calculated Moment of Resistance of OPTIMAL SECTION= 80064.4468 kN.Cm
Optimal Depth of Eq.Rect.Stress Block = 14.7814 Cm
The LEAST-COST per Unit-Length of BEAM = 0.4435 MU/Cm
The Calculated WEIGHT per Unit-Length of BEAM = 6.5875 kg/Cm

OUTPUT FILE: bm1out.cpp of PROGRAM: cbm.cpp
```

8.8.2 Example Application Results of LEAST-COST R.C. Beam Design: Model-II

Application of the Optimization Model-II is illustrated using the same case study problem taken in case of Optimization Model-I as shown above. Using the same input data the ultimate load, applied bending moment, cost coefficients, and other model constants are derived as follows, which are instantly given as output by the **C ++ Computer Programme (cbm.cpp)**.

The ultimate uniform load $= 1.4 \times 15 + 1.7 \times 20 = 55$ kN/m

The ultimate applied bending moment $= 55 \times 10^2/8 = 687.5$ kNm $= 687.5 \times 10^5$ Ncm

$a_1 = 11.76470588, \quad a_2 = 68750000.0000, \quad a_3 = 4382.33850000,$

$a_4 = 0.00003704, \quad a_5 = 0.50, \quad a_6 = 0.40$

$c_1 = 0.00565200, \quad c_2 = 0.00007085, \quad c_3 = 0.00047410, \quad c_4 = 0.00021550$

Putting these values of model constants in Eqs. (8.4), (8.33)–(8.36), (8.38), and in Eqs. (8.46)–(8.69); and also solving the above three nonlinear simultaneous Eqs. (8.56), (8.57), and (8.58) in terms of three unknown variables w_{01}, $w_{03,}$ and w_{41} using the Newton–Raphson method as explained and illustrated in Chap. 3 and in this chapter. As in case of Model-I, the C++ software sub-module '**cbm.cpp**' of software: '**RCCAD**' solves, in no time, this *Optimization Model* for optimal design of reinforced concrete beam. Thus, we get the optimum cost weights, optimum beam dimensions, optimum area of tensile steel reinforcement, **Least-Cost** of the beam, and other details which appear in the computer screen as partly shown below, as the program '**cbm.cpp**' of Software: '**RCCAD**' is run, as demonstrated in case of Model-I, but choosing the Model-II in this case.

```
Number of Iterations is: 50

Want change No.of ITERNS.LIMIT/ITERTN.Tolerance(if no Convergance)?

Type y(if 'yes') or n(if 'no')and press <CR>: n
You have chosen not to change ITRL,ITOL

(Initial-Trial-Values of NLP Indpdnt.Variables are:x=0.3,y=0.3,
z=0.8)
Want Continue Iteration with NEW Values(if no Convergance)?
Type y(if 'yes') or n(if 'no')and press <CR>: n
SWBM= 8008.85 kNcm,uabm= 68750.00 kNcm,TABM= 77758.85 kNcm,x4=
77758.85 kNcm

CALIBRATED MODEL CONSTANTS FOR CHOSEN DESIGN CONDITIONS(ITMODEL= 2)
are:
a1= 11.764706 a2= 68750000.0000 a3= 4382.3385 a4= 0.00003704 a5=0.50
a6= 0.40
c1= 0.00565200 , c2= 0.00007085 , c3= 0.00047410 , c4= 0.00021550
DESIGN Values of OPTIMAL SECTION:
LA= 62.2106 cm, C= 1388.81 kN, T= 1388.81 kN
MNC= 86404.75 kN.cm, MNS= 86404.75 kN.cm, MR= 77764.27 kN.cm
COM= 1.000
SOME KEY INPUTS: DLL=20.0 kN/m, DDL=15.0 kN/m, BSPN=10.0 m,
BMWDR=0.400,CRF=0.80

LEAST-COST OPTIMAL RESULTS AFTER SOLVING THE 'NLP' MODEL:
The values of Optimal Geometric Programming Cost Weights are:
w01= 0.5846, w02=0.3257, w03=0.0758, w04=0.0138, t=1.0000,
w11=0.1054
w21= 0.6101, w22=0.0788, w31=0.6801, w32=0.1054, w41=0.3140

Optimal Area of Tensile Steel Re-inforcement= 46.2868 Sq.Cm.
Optimal Effective Depth of the RC Beam = 71.7104 Cm
```

Optimal Width of the RC Beam   = 28.6672 Cm
Total Applied Bending Moment(inc.Self.wght)= 77758.8482 kN.Cm
Calculated Moment of Resistance of OPTIMAL SECTION= 77764.2735 kN.Cm
Optimal Depth of Eq.Rect.Stress Block = 18.8887 Cm
The LEAST-COST per Unit-Length of BEAM = 0.4476 MU/Cm
The Calculated WEIGHT per Unit-Length of BEAM = 5.2530 kg/Cm

 adif= 0.4688
 rlay= 2 dia1= 3.000 bn1= 4 dia2= 2.800 bn2= 3 ccv1= 6.500
 AS= 46.2868 ta= 46.7657 adif= 0.4688

PRE-INPUT:Layer No,Dia.of bars,No.of Re-inf.bars of each category,
clear Cover,design reinf.area,actual reinf.area & diff.(adif)between
design reinf.area & actual reinf.area are shown above
You want to CHOOSE NEW Dia./No.of Re-inf.bars & clear Cover?
(IF '0'INPUT OPTION CHOSEN,CAN USE COMPUTER DATA PUTTING 'NO'BELOW)
Type y(if 'yes') or n(if 'no')and press <CR>: n

CHOSEN NO NEW REINF.BAR DIA/NO./LAYER,OR,RESULTS ARE WITHIN TOL.

YOU WANT TO CHANGE 'INPUT VALUE(S)/INPUT SET/
NUM.ITERATION MODEL',FOR BETTER RESULTS ?
Type y(if 'yes') or n(if 'no')and press <CR>:n

*On putting a response of 'n' to the above dialogue, the following message
appears in the computer screen,and the program/iteration is terminated.*

AS CURRENT SOLUTION ACCEPTABLE,CLOSE 'NUMERICAL ITERATION
SESSION',LOAD 'BMDG'& START 'DRAWING SESSION'IF DESIRED

'ITERATION CLOSED'

On closing this iteration as done in case of Model-I above, the user/
designer can obtain the output file of Model-II i.e. bm2out.cpp file of
PROGRAM: cbm.cpp, giving full results of solution of the Optimization
Model-II as shown below:

INPUT-Design-Decisions/OUTPUTS(SOME)in Current-Case are:
ITERATION MODEL FOR LEAST-COST RC-BEAM DESIGN CHOSEN IS: 'ITMODEL- 2'

SOME RC-BEAM DESIGN CONDITIONS CHOSEN ARE:
DLL=20.000 DDL=15.000 BMC= 0.125 BSPN=10.000 LFLL= 1.700
LFDL= 1.400 SYS=300.000 CCS=30.000 cs= 0.7200 cc= 64.5000
cr= 1.100 D= 1.400 cg1= -76.810 csh= 2.1550

Some Design Values Chosen/Calculated:
covc=-70.310 covl= 5.000 ccv1= 6.500 cov2=-188.277
ccv2=-187.877 cc2v=-187.877 layc=-184.377 laycv=-187.277
rlay= 2 dia1= 3.000 bn1= 4 dia2= 2.800 bn2= 3
AS = 46.2868 ta= 46.7657 adif= 0.4688

OUTPUT-OPTIMAL-VALUES(some)given by RCCAD(Numerical Part):
SOME OPTIMAL RESULTS AFTER SOLVING THE LEAST-COST 'NLP' MODEL:

```
The values of Optimal Geometrical Programming Cost Weights are:
w01= 0.5846, w02=0.3257, w03=0.0758, w04=0.0138
t=1.0000, w11=0.1054, w21= 0.6101, w22=0.0788,
w31=0.6801, w32=0.1054, w41=0.3140, pt=0.0225,

Area of Tensile Steel Re-inforcement= 46.2868 Sq.Cm.
Optimal Effective Depth of the RC Beam = 71.7104 Cm
Optimal Width of the RC Beam       = 28.6672 Cm
Total Applied Bending Moment(inc.Self.wght)= 77758.8482 kN.Cm
Calculated Moment of Resistance of OPTIMAL SECTION= 77764.2735 kN.Cm
Optimal Depth of Eq.Rect.Stress Block = 18.8887 Cm
The LEAST-COST per Unit-Length of BEAM =  0.4476 MU/Cm
The Calculated WEIGHT per Unit-Length of BEAM =  5.2530 kg/Cm

OUTPUT FILE: bm2out.cpp of PROGRAM: cbm.cpp
```

8.8.3   Example Application Results of Conventional R.C. Beam Design: Model-III

Application of Model-III, for conventional R.C. beam design, is illustrated using the same case study problem as in case of Model-I shown above, with same input data of ultimate load, applied bending moment, and cost coefficients, but choosing the Model-III, i.e. '3' in this case. We can carry out three tasks using the Model-III, namely: (1) sensitivity analysis of beam cost to the steel ratio, (2) sensitivity analysis of beam cost to the width of beam, and (3) iterative flexural strength design of singly reinforced concrete beam using conventional method. These are illustrated below.

8.8.3.1   Example Application Results of Sensitivity Analysis of Beam Cost to the Steel Ratio

In this case, we get the beam depth, area of tensile steel reinforcement, and cost of the beam as the value of 'pt' (steel ratio) is changed, which appear in the computer screen as shown below, as the program '**cbm.cpp**' of Software: '**RCCAD**' is run, as demonstrated in case of Model-I, but choosing the Model-III, i.e. '3' in this case, as shown below.

```
AS SHOWN BELOW,THERE ARE 3 'ITERATION MODELS',
DEPENDING ON THE SET OF DESIGN CONDITION YOU CHOOSE:

LEAST-COST FLEXURAL STRENGTH DESIGN-GIVEN MIN.ACCEPT.BEAM-WIDTH
(ITMODEL-1),
LEAST-COST FLEXURAL STRENGTH DESIGN-GIVEN B.WIDTH:DEPTH RATIO
(ITMODEL-2),
ITERATIVE FLEXURAL STRENGTH DESIGN i.e.CONVENTIONAL METHOD,INCLUDING
SENSITIVITY ANALYSIS:-(i)TO STL.RATIO (ii)TO B.WIDTH (iii)GEN.
```

ITMODEL-3

PUT YOUR CURRENT CHOICE OF 'NUM.ITERATION MODEL'(1,2..)
3

On putting value of '3' as above, the Model-III is selected and the following message appears in computer screen for response by the user/designer(0 option means preset computer data input):

PUT YOUR CURRENT CHOICE OF 'INPUT SET' CASE(0/1 ETC)
0

Since case 0 i.e. Pre-Set Computer Data Input is selected, no new INPUT.
cr= 1.10 c1=0.00565200 c2=0.00007085 c3=0.00047410 c4=0.00021550
TABM=83625000.0
WB=40.000 pt=0.018600 AS= 50.8761 CCS= 3000.0 SYS= 30000.0 CRF=
0.8000

CALCULATED MODEL CONSTs.ABOVE AS PER USER-INPUT-CHECK/PUT ANY CHAR.TO
CONTINUE:
c
Want Run Series of ITERATIONS for STEEL RATIO SENSITIVITY ANALYSIS?
Type y(if 'yes') or n(if 'no')and press <CR>: y

On choosing 'y', series of iterations for steel ratio sensitivity analysis is run, giving the following output:

CURRENT ASSUMED VALUE OF DESIGN STEEL RATIO: 'pt =0.0186'
PUT DESIRED 'pt'RANGE-LOW,HIGH,ITRN.INTERVAL IN SAME ORDER
0.001   0.04   0.002

SENSITIVITY OF BEAM-DEPTH COST TO THE DESIGN STEEL-RATIO(pt):'ITMODEL- 3'

pt (Design Steel-Ratio)	d (Eff.Depth -cm)	D (Gross Depth-cm)	AS (TENSION STEEL-sq.cm)	COST (BEAM-COST) (MU/cm)
0.0010	278.08	307.00	11.1635	0.8861
0.0030	162.10	178.31	18.4516	0.6554
0.0050	126.32	138.85	25.2641	0.5688
0.0070	107.42	118.16	30.0763	0.5344
0.0080	85.32	104.85	34.3151	0.5183
0.0110	86.76	85.44	38.1754	0.5118
0.0130	80.32	88.35	41.7654	0.5107
0.0150	75.25	82.78	45.1527	0.5131
0.0170	71.15	78.27	48.3828	0.5177
0.0180	67.75	74.52	51.4886	0.5240
0.0210	64.87	71.36	54.4840	0.5315
0.0230	62.41	68.65	57.4180	0.5388
0.0250	60.28	66.30	60.2752	0.5488
0.0270	58.41	64.25	63.0778	0.5586
0.0280	56.76	62.43	65.8361	0.5687
0.0310	55.28	60.82	68.5582	0.5782
0.0330	53.88	58.38	71.2515	0.5801
0.0350	52.80	58.08	73.8225	0.6013

```
    0.0370            51.74         56.82         76.5768              0.6128
    0.0380            50.78         55.86         78.2184              0.6246
```

```
INPUT:WB= 40.00 CM. cs= 0.7200 MU/KG cc=64.5000 MU/CU.M. csh= 2.1550
MU/Sq.M.
ANY NEGATIVE OUTPUT-VALUE MEANS INFEASIBLE INPUT-VALUE-CHOICE & NEEDS
CHANGE
You want to change above Input-Values of WB,cs,cc,csh ?
Type y(if 'yes') or n(if 'no')and press <CR>: n

YOU WANT TO CHANGE 'INPUT VALUE(S)/INPUT SET/
NUM.ITERATION MODEL', FOR BETTER RESULTS ?
Type y(if 'yes') or n(if 'no')and press <CR>:n

AS CURRENT SOLUTION ACCEPTABLE,CLOSE 'NUMERICAL ITERATION
SESSION',LOAD 'BMDG'& START 'DRAWING SESSION'IF DESIRED

'ITERATION CLOSED'
```

It may be seen from the above output results that at 'pt' value of 0.001, the beam cost is 0.8861 which reduces to 0.5107 at the 'pt' value of 0.013, and again, the beam cost increases to 0.5131 at a 'pt' value of 0.015 and finally beam cost increases to 0.6246 MU/cm at the 'pt' value of 0.0380. Thus, the above results clearly show that the *reinforced concrete beam design is an optimization problem*; and that there is an *optimum* value of 'pt' which reduces the beam cost to the *minimum* for given design conditions. Model-I and Model-II presented earlier solve this optimization problem of reinforced concrete beam design.

8.8.3.2   Example Application Results of Sensitivity Analysis of Beam Cost to the Width of Beam

In this case, we get the beam depth, area of tensile steel reinforcement, and cost of the beam as the value of 'WB' (beam width) is changed, which appear in the computer screen as shown below, as the program '**cbm.cpp**' of Software: '**RCCAD**' is run.

```
Want Run Series of ITERATIONS for STEEL RATIO SENSITIVITY ANALYSIS?
Type y(if 'yes') or n(if 'no')and press <CR>: n
Want Run Series of ITERATIONS for BEAM WIDTH SENSITIVITY ANALYSIS?
Type y(if 'yes') or n(if 'no')and press <CR>: y
```

On choosing 'y', series of iterations for beam width sensitivity analysis is run, giving the following output:

```
CURRENT ASSUMED VALUE OF BEAM WIDTH : 'WB =40.000'

PUT DESIRED 'WB'RANGE-LOW,HIGH,ITRN.INTERVAL IN SAME ORDER
20  100  5
```

```
SENSITIVITY OF BEAM-DEPTH COST TO THE BEAM-WIDTH(WB):'ITMODEL- 3'
WB          d           D           AS          COST
(Beam-      (Eff.Depth  (Gross      (Tension    (Beam-Cost)
Width-cm)   -cm)        Depth-cm)   Steel-sq.cm) (MU/cm)
 20.00      68.38        75.22       50.8761      0.4213
 25.00      86.71       106.38       35.8748      0.4261
 30.00      86.50        85.15       40.2211      0.4588
 35.00      78.86        86.86       44.0600      0.4801
 40.00      73.10        80.41       47.5803      0.5187
 45.00      68.38        75.22       50.8761      0.5480
 50.00      64.47        70.82       53.8623      0.5750
 55.00      61.16        67.28       56.8813      0.6010
 60.00      58.32        64.15       58.6576      0.6260
 65.00      55.83        61.42       62.3103      0.6501
 70.00      53.64        58.01       64.8546      0.6735
 75.00      51.68        56.86       67.3028      0.6861
 80.00      48.84        54.83       68.6650      0.7181
 85.00      48.35        53.18       71.8487      0.7385
 80.00      46.81        51.60       74.1641      0.7604
 85.00      45.58        50.15       76.3142      0.7807
100.00      44.37        48.81       78.4054      0.8006
```

INPUT:pt= 0.0186 CM. cs= 0.7200 MU/KG cc=64.5000 MU/CU.M. csh= 2.1550
MU/Sq.M.
ANY NEGATIVE OUTPUT-VALUE MEANS INFEASIBLE INPUT-VALUE-CHOICE & NEEDS
CHANGE
You want to change above Input-Values of pt,cs,cc,csh ?
Type y(if 'yes') or n(if 'no')and press <CR>: n

YOU WANT TO CHANGE 'INPUT VALUE(S)/INPUT SET/
NUM.ITERATION MODEL', FOR BETTER RESULTS ?
Type y(if 'yes') or n(if 'no')and press <CR>:n

AS CURRENT SOLUTION ACCEPTABLE, CLOSE 'NUMERICAL ITERATION
SESSION', LOAD 'BMDG'& START 'DRAWING SESSION'IF DESIRED

'ITERATION CLOSED'

It may be seen from the above output results that at 'WB' value of 20, the beam cost is 0.4213 MU/cm which increases with increase of 'WB' value l; and that, at 'WB' value of 100 the beam cost is increased to 0.8006 MU/cm. Thus, the above results show that if the cost reduction is the objective, the width of beam should be chosen as low as possible and as practicable. It may be noted that some of the output results may not be practicable, but still shown just for the purpose of sensitivity analysis to give an understanding of interaction between different beam elements including cost. Based on such sensitivity analysis and optimization where possible, a designer can build up an insight to the interaction of various beam elements to the beam cost, and develop simple design rules for cost and material consumption reduction. Adopting such an approach, particularly for all repeated members such as beams, columns, and slabs which can be tackled by simple approaches as above by an individual structural designer with complete understanding of the process, will lead to considerable saving in cost and materials consumption in structural engineering with attendant benefits.

8.8.3.3 Example Application Results of Conventional Iterative Flexural Strength Design of Beam

Model-III, i.e. choosing '3' when the program '**cbm.cpp**' of Software: '**RCCAD**' is run, can be used for conventional iterative flexural strength design of singly reinforced concrete beam. This is illustrated using the same case study problem taken from Everard and Tanner as in Model-I, assuming beam weight as 8.5 kN/m as in case study problem, with same input data of ultimate load, applied bending moment, and cost coefficients, but choosing the Model-III, i.e. '3'. In this case, we get the beam depth, area of tensile steel reinforcement, and cost of the beam, choosing different values of 'WB' (beam width) and pt (reinforcement ratio) as Input (if 0 case selected no new Input), which appear in the computer screen as shown below, as the program '**cbm.cpp**' of Software: '**RCCAD**' is run.

```
PUT YOUR CURRENT CHOICE OF 'INPUT SET' CASE(0/1 ETC)
0
Since case 0 i.e. Pre-Set Computer Data Input is selected, no new INPUT.
cr= 1.10 c1=0.00565200 c2=0.00007085 c3=0.00047410 c4=0.00021550
TABM=83625000.0
WB=40.000 pt=0.018600 AS= 50.8761 CCS=   30.0 MPA SYS=   300.0 MPA CRF=
0.8000
CALCULATED MODEL CONST.ABOVE AS PER USER-INPUT-CHECK PUT ANY CHAR.TO
CONTUE:
c
Want Run Series of ITERATIONS for STEEL RATIO SENSITIVITY ANALYSIS?
Type y(if 'yes') or n(if 'no')and press <CR>: n
Want Run Series of ITERATIONS for BEAM WIDTH SENSITIVITY ANALYSIS?
Type y(if 'yes') or n(if 'no')and press <CR>: n
```

On choosing 'n' for both series of iterations for sensitivity analysis as above; general conventional design model is run, giving the following output:

```
RESULTS OF GENERAL CONVENTIONAL DESIGN MODEL FOR RC BEAM:'ITMODEL- 3'
SOME KEY INPUTS: DLL=20.0 kN/m, DDL=15.0 kN/m, BSPN=10.0 m
Chosen Width of the RC Beam(i.e. WB) = 40.0000 Cm
Chosen Steel Ratio          = 0.01860

SOME MODEL OUTPUTS FOR CHOSEN DESIGN CONDITIONS ARE:
D= 75.22,UUL= 66.80,k= 0.88026,K=4676.08687,TABM=83625000.00
c1=0.0056520000,c2=0.0000708500,c3=0.0004741000,c4=0.0002155000

SOME RESULTS AFTER SOLVING CONVENTIONAL DESIGN MODEL:
Area of Tensile Steel Re-inforcement= 50.8761 Sq.Cm.
Effective Depth of the RC Beam = 68.3818 Cm
Total Applied Bending Moment(inc.Self.wght)= 83625.0000 kN.Cm
Calculated Moment of Resistance of DESIGN SECTION= 83655.8322 kN.Cm
```

The COST per Unit-Length of BEAM = 0.5227 MU/Cm
The Calculated WEIGHT per Unit-Length of BEAM = 6.8885 kg/Cm
The Assumed WEIGHT per Unit-Length of BEAM = 8.6735 kg/Cm
 adif= 0.0381
 rlay= 1 dia1= 3.600 bn1= 5 dia2= 0.000 bn2= 0 ccv1= 6.838
 AS= 50.8761 ta= 50.8143 adif= 0.0381

PRE-INPUT:Layer No,Dia.of bars,No.of Re-inf.bars of each category,
clear Cover,design reinf.area,actual reinf.area & diff.(adif)between
design reinf.area & actual reinf.area are shown above

You want to CHOOSE NEW Dia./No.of Re-inf.bars & clear Cover?
(IF '0'INPUT OPTION CHOSEN,CAN USE COMPUTER DATA PUTTING 'N0'BELOW)
Type y(if 'yes') or n(if 'no')and press <CR>: n
CHOSEN NO NEW REINF.BAR DIA/NO./LAYER,OR,RESULTS ARE WITHIN TOL.

YOU WANT TO CHANGE 'INPUT VALUE(S)/INPUT SET/
NUM.ITERATION MODEL',FOR BETTER RESULTS ?
Type y(if 'yes') or n(if 'no')and press <CR>:n

AS CURRENT SOLUTION ACCEPTABLE,CLOSE 'NUMERICAL ITERATION
SESSION',LOAD 'BMDG'& START 'DRAWING SESSION'IF DESIRED
'ITERATION CLOSED'

On closing this iteration as done in case of Model-I and II above, the user/
designer can obtain output files such as: snstypt.cpp and snstybw.cpp incorporating
the results of sensitivity analysis as shown above. Similarly, the general output file of
Model-III, i.e. bm3out.cpp file of PROGRAM: cbm.cpp, giving full results of
general conventional design model solution for RC beam can be obtained as
shown below:

INPUT-Design-Decisions/OUTPUTS(SOME)in Current-Case are:

ITERATION MODEL(NO OPTIMISATION)FOR RC-BEAM DESIGN CHOSEN IS:
'ITMODEL- 3'
SOME RC-BEAM DESIGN CONDITIONS CHOSEN ARE:
DLL=20.000 DDL=15.000 BMC= 0.125 BSPN=10.000 LFLL= 1.700
LFDL= 1.400 SYS=300.000 CCS=30.000 cs= 0.7200 cc= 64.5000
cr= 1.100 D= 75.220 cg1= 0.000 csh= 2.1550
Chosen Steel Ratio = 0.01860 UUL= 66.8000 kN/m

Some Design Values Chosen/Calculated:
covc= 6.838 covl= 5.038 ccv1= 6.838 cov2= 0.000
ccv2= 0.000 cc2v= 6.838 layc= 0.000 laycv= 0.000
rlay= 1 dia1= 3.600 bn1= 5 dia2= 0.000 bn2= 0
AS = 50.8761 ta= 50.8143 adif= 0.0381

OUTPUT-Values(some)by RCCAD(CONVENTIONAL-NO OPTIMISATION)-'ITMODEL-
3'
Area of Tensile Steel Re-inforcement= 50.8761 Sq.Cm.
Effective Depth of the RC Beam = 68.3818 Cm
Width of the RC Beam   = 40.0000 Cm

```
Total Applied Bending Moment(inc.Self.wght)= 83625.0000 kN.Cm
Calculated Moment of Resistance of DESIGN SECTION= 83655.83 kN.Cm
Depth of Eq.Rect.Stress Block = 14.8636 Cm
The COST per Unit-Length of BEAM =  0.5227 MU/Cm
The Calculated WEIGHT per Unit-Length of BEAM =  6.8885 kg/Cm
The Assumed  WEIGHT per Unit-Length of BEAM =  8.6735 kg/Cm

OUTPUT FILE: bm3out.cpp of PROGRAM: cbm.cpp
```

8.9 Integration of Computer-Aided Numerical Optimal Design and Graphics

Drawing is the language of architecture, engineering, and physical planning, where ideas are created, evaluated, and developed. This is equally applicable in structural engineering. Therefore, it is desirable to develop application software not only for numerical solution of optimizing models for optimal design of reinforced concrete members such as beams, but also, to convert instantly such optimal designs in alphanumeric terms (given as output by the numerical optimizing models) into 'graphic form', i.e. dimensioned drawing including the working drawings to be meaningful to the users. The software should also incorporate cost engineering to derive the cost coefficients for the optimizing models, using any time-specific, site-specific, and country-specific basic unit cost data on labour, materials, and so on. Accordingly, **Application Software: 'RCCAD'** is being developed as an *Integrated Computer-Aided Design and Drafting* (CADD) *System*, for design of reinforced concrete sections.

The software is broadly divided into two parts: (1) **Numeric** and (2) **Graphic**. The numeric part is developed as an independent **Software** in user-friendly **C++** language which solves the nonlinear programming optimizing models, i.e. least-cost optimizing models as well as the conventional design models presented above. Designer can also use cost engineering models presented in Chap. 3 to derive the cost coefficients for the above models, using any time-specific, site-specific, and country-specific basic unit cost data on labour, materials, and so on. Numeric part of the software generates optimal solution in *alphanumeric* form, and also generates a LISP file for each model incorporating numerical data output given by each of the optimal or conventional design models. Thus, the output LISP files 'bm1.lsp', 'bm2.lsp', and 'bm3.lsp' incorporated in the C++ cbm.cpp program are the LISP output files generated by the Model-I, Model-II, and Model-III, respectively, described in Sect. 8.8 above. As an example, the output LISP file 'bm1.lsp' is shown below, which is generated by the Model-I for the same example application results of LEAST-COST R.C. Beam Design: Model-I, presented in Sect. 8.8.1.

```
(setq dlis
(list 20.00 15.00 0.125 1000.00 1.7 1.4  300   30
    37.683  86.063 30.000 1.100  84.668 2.186
```

```
8.606 5.000 6.420 11.558 12.878 6.558 3.718
4 1.42 2 1.42    80064.68 14.7814  0.4435 2 ) )
```

The numerical values shown in each row of the above LISP file indicates the value of each design parameter (which are already defined in the C++ program) located in the same order in the respective row shown below:

```
DLL, DDL, BMC, BSPN, LFLL, LFDL, SYS, CCS
x₁, x₂, x₃, cr, D, cg1
covc, cov1, ccv1, cov2, ccv2, layc, laycv
bn1, rrad1, bn2, rrad2, x₄, x₅, y(x), rlay
```

Thus, the value of DLL (i.e. DESIGN Uniform LIVE LOAD in kN/m defined in Sect. 8.8.1 above) is 20 as shown in the first row and first column above. Similarly, the value of CCS is 30 MPA (as defined in Sect. 8.8.1 above) as shown in the first row and last column above. Again, the values of x_1, x_2, x_3, cr, and D are 37.683 sq. cm, 86.063 cm, 30 cm, 1.1 and 84.668 cm, respectively, as shown in column 1 to 5, respectively, of row 2 of the list above. Similarly, column 1-4 of last row gives the number and radius of reinforcement bars of type-1 and then type-2 respectively. Thus, the user/designer can read all the design parameter values by relating each as above.

The graphic part developed in LISP language for graphic operations uses these numerical design data output incorporated in the above LISP file generated by the numeric part of the **Application Software: 'RCCAD'**, to generate the corresponding drawings. As drawing is the most important medium of a design, any *numeric design solution* needs to be converted ***Instantly*** into drawings to be meaningful to all concerned. The graphic program carries out this task, and can call the numeric C++ program when required to generate an acceptable numeric solution before converting it into a graphic solution. As shown in Sect. 8.8.1 above, the C++ program of this software module has 15 input set cases (extendable as per requirements) giving numerous choices to the designer to fulfill his/her need. Thus, using this *Technique designers/users* could generate/analyse/evaluate *numerous potential solutions* both in numeric and graphic forms (as each *numeric solution can be converted to graphic solution, thus, permitting both numeric and graphic iterations*) to select a particular solution that meets a specified criteria, which is rarely possible in the conventional practices.

8.10   Integrated Computer-Aided Optimal Design and Drafting in Structural Engineering-RC Beam

As presented in Sect. 8.8.1, on loading and typing '**RCCAD**' in the text part of the computer screen (24,25), Application Software: '**RCCAD**', developed for Integrated Computer-Aided Optimal Design and Drafting (CAODD) of RCC sections, prompts the designer/user to select a specific module, with the following message appearing in the graphic part of the computer screen (text in *italics* are comments of the author):

```
WELCOME TO ::
              '' RCCAD ''
A SOFTWARE FOR ''REINFORCED CONCRETE (RC)
COMPUTER-AIDED-DESIGN''TO ACHIEVE LEAST-COST.
''RCCAD'',HAS 2 PARTS: (I)''NUMERIC OPTIMIZATION'',
AND (II)''GRAPHIC COMPONENT'', WHICH CONVERTS
INSTANTLY THE ''NUMERIC DESIGN'' TO ''DRAWING''.
       SOFTWARE ''RCCAD'' HAS NUMBER OF MODELS.
       SELECT 'NUMERIC' ,OR, 'GRAPHIC' OPTION AND,
       A MODEL TO RUN AS PER YOUR REQUIREMENT.
```

Do you want to RUN 'Numeric SESSION' for RCC BEAM DESIGN ? (Y/N):

On putting 'y' response to above, the following message appears in text part of computer screen:

Type 'CBM' for 'Numeric SESSION'. On Results Load 'BMDG' to START 'DRAWING SESSION'

On putting 'n' response to above, the following message appears in text part of computer screen:

Load 'BMDG' to START BEAM 'DRAWING SESSION', if Numeric Results already Exist.

On loading and typing '**BMDG**' in the text part of the computer screen the following message appear in computer screen for user response:

```
WELCOME TO:'DRAWING SESSION' (MODEL-1/2 etc.) OF SOFTWARE: ''RCCAD''
   You want Earlier Design(s)/Drawing in Screen to be ERASED ?(Y/N):
```

On putting 'y' response earlier drawings are erased making available a clear screen
```
Enter MODEL TYPE (MD1/MD2) Chosen by you:
```

On choosing a specific model say MD1 i.e. Model-I in C++ cbm.cpp program, the drawing process is started, first by loading the bm1.lsp file, generated by the C++ program and presented above, and then asking for drawing starting point as follows:

Put the drawing starting point (list):

On supply of drawing starting point by the user, the drawing is started and completed as shown in Fig. 8.4(a) and (b) below.

It may be seen that the values of x_2 (i.e. effective depth d), x_3(i.e. width of beam), and D (overall depth of beam) adopted in drawing Fig. 8.4 are same, i.e. 86.06,

a

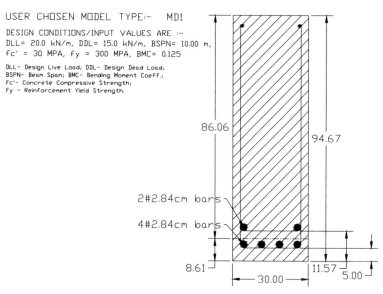

b

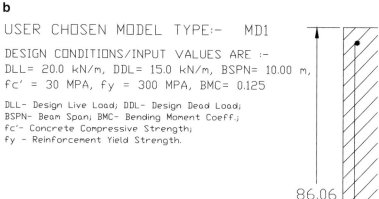

Fig. 8.4 (**a**) Least-cost optimal design drawing of RCC beam given by 'RCCAD'. (**b**) Least-cost optimal design drawing of RCC beam given by 'RCCAD'—zoom view indicating input design conditions

30, and 84.67 cm (correct to 2 decimal places) as in the LISP output file bm1.lsp produced by the C++ cbm.cpp program. Thus, the drawing Fig. 8.4 are completely integrated with the numerical results of Optimizing Model-I given by the C++ cbm. cpp program.

As in case of Model-I above, if we choose MD3, i.e. Model-III in C++ cbm.cpp program, the drawing process is started, first by loading the bm3.lsp file, generated by the C++ program and presented below.

```
(setq dlis
 (list  20.00 15.00 0.125 10.00  1.7 1.4  300   30
        50.876 68.382 40.000 1.100  75.220 0.000
        6.838 5.038 6.838 0.000 0.000 0.000 0.000
        5 1.80 0 1.40   77758.85 14.8636  0.5227 0.01860 1 ) )
```

As in case of Model-I, the numerical values shown in each row of the above LISP
file indicate the value of each design parameter (which are already defined in the C+
+ Program) located in the same order in the respective row shown below:

```
DLL, DDL, BMC, BSPN, LFLL, LFDL, SYS, CCS
x₁, x₂, x₃, cr, D, cg1
covc, cov1, ccv1, cov2, ccv2, layc, laycv
bn1, rrad1, bn2, rrad2, x₄, x₅, y(x), rlay
```

In this case, reinforcements are placed only in one layer as the width of the beam
chosen is higher. Therefore, no values are assigned (i.e. 0 values) to cg1, cov2, ccv2,
layc, and laycv.

On choosing Model-III, i.e. MD3 for conventional design, and supply of drawing
starting point by the user, as shown above in case of Model-I, the drawing for
conventional design is started and completed as shown in Fig. 8.5(1) below. In this
integrated computer-aided optimal as well conventional design and drafting process,
it is possible to change the design parameter values such as the number and radius of
reinforcement bars of type-1 and type-2 so as to provide reinforcement area minimizing the difference with reinforcement area as per design calculation. The choice

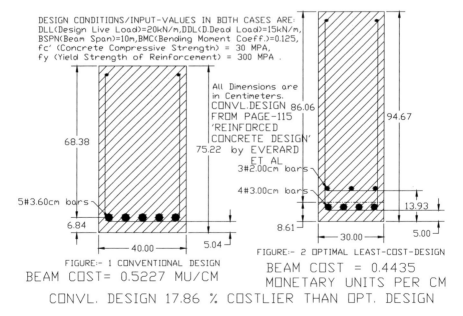

Fig. 8.5 Comparative design drawings of RCC beams given by software 'RCCAD' using:
(1) Conventional design method—Model-III, (2) optimal least-cost design method—Model-I

of values of 4 1.42 2 1.42 for bn1, rrad1, bn2, rrad2, respectively, as adopted for Model-I above gives an area difference (adif) of + 0.3310 sq.cm with the optimum reinforcement area of 37.683 sq.cm. To reduce adif further we can choose the values of 4 1.50 3 1.00 for bn1, rrad1, bn2, rrad2, respectively, reducing the value of adif to + 0.0218 only. Accordingly, the new output LISP file 'bm1.lsp' is shown below, which is generated by the Model-I.

```
(setq dlis
(list 20.00 15.00 0.125 1000.00 1.7 1.4  300   30
    37.683  86.063 30.000 1.100  84.668 2.106
    8.606 5.000 6.500 13.825 14.825 8.425 5.825
    4 1.50 3 1.00   80064.68 14.7814  0.4435 2 ) )
```

As before, numerical values shown in each row of the above LISP file indicate the value of each design parameter located in the same order in the respective row shown below:

```
DLL, DDL, BMC, BSPN, LFLL, LFDL, SYS, CCS
x₁, x₂, x₃, cr, D, cg1
covc, cov1, ccv1, cov2, ccv2, layc, laycv
bn1, rrad1, bn2, rrad2, x₄, x₅, y(x), rlay
```

Choosing these new values of design parameters for Model-I or MD1, and adopting the same drawing procedure as above, a new drawing is started and completed for Model-I or MD1, as shown in Fig. 8.5(2) below.

It may be seen that the optimal design gives a deep beam section reducing deflection problems besides achieving economy. It is shown elsewhere that in some cases the balanced design can be costlier by 34% compared to the optimal design with same width of beam and unit costs of materials and shuttering. This is quite contrary to the usual notion of economy in balanced design, which tends to reduce the lever arm compared to the optimal design, creating dis-economy. Similarly, optimal design gives comparatively less reinforcement ratio, i.e. a more ductile section, which is recommended in strength design method. Here the optimal reinforcement ratio is 0.0146, which is the output given by the optimal design Model-I; and not pre-assumed as in case of the conventional design method. Thus, in Model-III—the conventional design method—a reinforcement ratio is arbitrarily pre-assumed, which is 0.0186 here. This is about 27% higher than the optimal steel ratio of 0.0146 given as output by the optimal design Model-I.

8.11   Optimization of Other Repeated Reinforced Concrete Members

As earlier stated, many application results of optimization of standard, repeated members such as beams, columns, and slabs; showing considerable saving in cost and material consumption, are available in literature. Development and application of structural optimization in the design of such repeated elements can be undertaken by more simple approaches by an individual structural designer with complete understanding of the process, instead of depending on costly black-box type commercial software, which often does not give such an insight. Adopting such an approach structural designers can make a significant contribution to improve efficiency and achieve material consumption reduction in engineering practices. To help this process, simple optimization models for singly reinforced concrete beams are presented above and complete computer codes of C++ and LISP program for solving such optimal design and drafting/drawing problems are also provided in appendix. These are expected to give a complete understanding of the process to the user structural designer, enabling him/her to develop and solve structural optimization models of such repeated reinforced concrete elements in his/her own specific situation leading to application of optimization in the design of all standard, repeated members, deriving considerable saving in the costs and material consumptions. Typical optimization models for some other frequently repeated reinforced concrete sections are discussed below.

8.11.1   Optimization of Rectangular Concrete Beams with Compression Reinforcement: Model-IV

Schematic diagrams of some other frequently repeated reinforced concrete sections are shown in Fig. 8.6 below.

Typical optimization model with objective function and constraints, for doubly reinforced concrete beams with compression reinforcement as shown in Fig. 8.6(a) above, could be developed as follows:

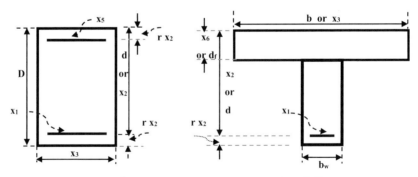

(a)   **Rectangular Beam With Compression Reinforcement**   (b)   **T- Beam**
       Or Rectangular Column With Reinforcement at Two Edges

Fig. 8.6 Schematic diagrams of some other frequently repeated reinforced concrete sections. (**a**) Rectangular beam with compression reinforcement. (**b**) T-beam or rectangular column with reinforcement at two edges

Minimize the Cost per Unit Length of Beam $= y(x) = c_1 x_1 + c_2 x_2 x_3$
$+ c_3 x_2 + c_4 x_3 + c_5 x_5$ (8.63)

Subject to:

$$a_1 x_3^{-1} \leq 1$$
(*Constraint* for Beam width)   (8.64)

$$a_2 x_2^{-1} x_3^{-1} x_4 x_6^{-1} + a_3 x_2^{-1} x_6 \leq 1$$
(*Constraint* for concrete bending moment)   (8.65)

$$a_4 x_2^{-1} x_4 x_5^{-1} \leq 1$$
(*Constraint for compression steel bending moment*)   (8.66)

$$a_5 x_1^{-1} x_3 x_6 + a_6 x_1^{-1} x_5 \leq 1$$
(*Constraint* for equilibrium)   (8.67)

$$a_7 x_4^{-1} + a_8 x_2 x_3 x_4^{-1} \leq 1$$
(*Constraint* for bending moment compatibility)   (8.68)

$$x_1, x_2, x_3, x_4, x_5, x_6 > 1$$
(Non-negativity *constraint*)   (8.69)

Here, $c_1 (= c_5)$, c_2, c_3, and c_4 are cost coefficients due to volume of tensile/compression steel reinforcement (MU/cm^3), volume of concrete (MU/cm^3), shuttering along the vertical surfaces of bean (MU/cm^2), and the shuttering along the horizontal surface of bean (MU/cm^2), respectively. Here, a_1 is a model constant defining the minimum acceptable width of beam; a_2 is a model constant related to the compressive strength of concrete, capacity reduction factor and the design proportion of applied bending moment resisted by the concrete; a_3 is a model constant defining the centroid of concrete compression; a_4 is a model constant related to the yield strength of compression steel, capacity reduction factor, and the design proportion of applied bending moment resisted by the compression steel; a_5 is a model constant related to yield strength of tensile steel reinforcement and the compressive strength of concrete; a_6 is a model constant related to the yield strength of compression and tensile steel reinforcement; a_7 is a model constant related to the applied bending moment as given; a_8 a model constant related to beam span and density of concrete for self-weight bending moment.

Here, the model *variables* are:

x_1 = a variable giving the optimum area of tensile steel reinforcement (cm^2)
x_2 = A variable giving the optimum effective depth of the beam (cm)
x_3 = A variable giving the optimum width of the beam (cm)
x_4 = A variable defining the total applied bending moment including that due to self-weight of beam

$x_5 = $ A variable giving the optimum area of compression steel reinforcement (cm^2)
$x_6 = $ A variable related to the depth of equivalent rectangular stress block (neglecting reduction in area due to compression steel which is comparatively small)

Using the geometric programming technique as explained in Sect. 8.12.5 and as illustrated in case of solution of Model-I in Sect. 8.5.1.1 above, the dual objective function and the associated normality and orthogonality constraints for the above primal problem can be developed in terms of the dual variables.

It may be seen that there are 13 terms (T) and 6 primal variables (N) in the primal objective function and constraints [expressions (8.63)–(8.68)], and therefore, the degree of difficulty in the problem is $= T - (N+1) = 6$. Thus, any 7 of the dual variables can be expressed in terms of the remaining 6 dual variables. Choosing w_{01}, w_{02}, w_{03}, w_{05}, w_{21}, and w_{41} as the 6 dual variables (nomenclature of such dual variables related to the respective term in objective function and constraints are explained in Sect. 8.5.1.1 above), and substituting these dual variables into the dual objective function, we can maximize this substituted dual objective function over the 6 dual variables w_{01}, w_{02}, w_{03}, w_{05}, w_{21}, and w_{41} (i.e. by setting the first derivative of the function with respect to each dual variable to zero), similar to Model-I solution. Thus, we will get 6 providing sufficient conditions for optimality. To simplify, we can also make the following substitutions:

$$
\begin{aligned}
k &= a_1 a_2 a_3 a_8{}^2/a_7 \\
k1 &= c_2 a_1/c_3 \\
k2 &= a_4 a_6/a_2 a_5 \\
k3 &= c_5 a_4/c_1 a_2 a_5 \\
k4 &= k2 + k3 \\
k5 &= c_1 a_2 a_5 a_8/c_4 \\
k6 &= c_3 a_7/c_4 a_1^2 a_8 \\
k7 &= 1 + k1
\end{aligned}
\tag{8.70}
$$

From the 6 additional equations referred above, we can derive the following expressions for the value of the 3 dual variables w_{01}, w_{02}, and w_{05} in terms of the remaining 3 dual variables w_{03}, w_{21}, and w_{41} as follows:

$$
\begin{aligned}
w_{01} &= w_{41} + k2\, w_{21} w_{41} (2w_{21} - w_{41})^{-1} \\
w_{02} &= k1\, w_{03} \\
w_{05} &= k3\, w_{21} w_{41} (2w_{21} - w_{41})^{-1}
\end{aligned}
\tag{8.71}
$$

Thus, by using the above stated 6 additional equations along with the above expression for the dual variables w_{01}, w_{02}, and w_{05}; and by suitable substitution and manipulation, we get the following three simultaneous equations in terms of the three dual variables w_{21}, w_{03}, and w_{41}:

$$k(2w_{21} - w_{41})^2 \left(k4w_{41}(2w_{21} - w_{41})^{-1} + 1\right) (k7w_{03} + w_{41} - w_{21})$$
$$- \left(k4w_{21}\,w_{41}(2w_{21} - w_{41})^{-1} + 2w_{21} - k7w_{03} - w_{41}\right)^2 (w_{21} - w_{41}) = 0 \tag{8.72}$$

$$k6\left(1 - k7w_{03} - w_{41} - k4w_{21}w_{41}(2w_{21} - w_{41})^{-1}\right)$$
$$\left(k4w_{21}\,w_{41}(2w_{21} - w_{41})^{-1} + 2w_{21} - k7w_{03} - w_{41}\right) \tag{8.73}$$
$$- (k7w_{03} + w_{41} - w_{21})\,w_{03} = 0$$

$$k5\left(1 - k7w_{03} - w_{41} - k4w_{21}w_{41}(2w_{21} - w_{41})^{-1}\right)\left(k4w_{41}(2w_{21} - w_{41})^{-1} + 1\right)$$
$$- \left(k4w_{21}w_{41}(2w_{21} - w_{41})^{-1} + 2\,w_{21} - k7w_{03} - w_{41}\right)(2w_{21} - w_{41})^{-1}w_{41} = 0$$
$$\tag{8.74}$$

Equations (8.72), (8.73), and (8.74) are three nonlinear simultaneous equations in terms of three unknown variables w_{03}, w_{21}, and w_{41} which can be solved by the Newton–Raphson method as explained and illustrated in Chap. 3 and this chapter, and also in case of Model-I. The author has developed C++ computer programmes for solving such nonlinear simultaneous equations covering many variables and the computer code of the programme is presented in appendix. Application of the geometric programming technique gives the following simple expressions for determining the optimum values of the model variables, thus fully solving the model.

$$w_{04} = 1 - w_{01} - w_{02} - w_{03} - w_{05}$$
$$x_2 = (c_4 w_{02})/(c_2 w_{04}),$$
$$x_3 = (c_3 w_{02})/(c_2 w_{03}),$$
$$y(x) = (c_2 x_2 x_3)/w_{02},$$
$$x_1 = y(x)\,w_{01}/c_1,$$
$$x_4 = a_7 + a_8 x_2 x_3,$$
$$x_5 = y(x)\,w_{05}/c_5,$$
$$x_6 = (w_{22} \times x_2)/(a_3 \times (w_{22} + w_{21}))$$

8.11.1.1 Model Calibration and Some Example Application Results

Similar to models presented above, this model also needs to be calibrated relating it to the site-specific cost parameters and desired design conditions to obtain the optimal design fulfilling these conditions. Thus, the model constants, i.e. cost coefficients: c_1, c_2, c_3, and c_4 are related to the cost of labour, materials, and shuttering in the given site condition and could be determined from relevant schedule of rates or market rates of materials and labour as illustrated in Chap. 4. Here, the value of additional model cost coefficient c_5 relates to the compression steel reinforcement, and will be equal to c_1 adopted for tensile steel reinforcement. Taking the same unit costs for cc, cs, and csh as in Model-I, the cost coefficients for this model will be as follows:

$$c_1 = c_5 = 0.00565200, \quad c_2 = 0.000070850, \quad c_3 = 0.000474100, \quad c_4 = 0.000215500.$$

The expressions for other model constants will be as follows (where numerical values taken for each parameter are same as adopted in example problem of Model-I presented earlier):

$a_1 = $ minimum acceptable beam width $= 30$ cm
$a_1 = rc/(0.85 f'_c \; \phi \beta_1)$, where rc is the design proportion of compressive moment resisted by concrete, $\beta_1 = 0.85$ as per test results (everard), other parameters as explained in earlier models.
$\quad = rc/(0.85 \times 3000 \times 0.8 \times 0.85) = rc/1850.75$
$a_3 = \frac{1}{2} \beta_1 = \frac{1}{2} \times 0.85 = 0.425$
$a_4 = rcs/(\phi \, (1 - r) \, f_y)$; where rcs is the design proportion of compressive moment resisted by compression steel
$\quad = rcs/(0.8 \times 0.8 \times 30,000) = rcs/24,300$
$a_5 = 0.85 f'_c \beta_1/f_y = 0.85 \times 30 \times 0.85/300 = 0.07225$
$a_6 = 1$
$a_7 = $ applied bending moment $= 68,750,000.00$ N-cm
$\quad a_8 = $ LFDL $\times (1 + r) \times 2323.0 \times 8.8 \times L^2 \times$ BMC $\times 10^{-2}$ in N.cm; if 'wc' is in kg/m^3 and 'L' is in metre
$\quad = 4382.3385$
$y(x) = c_1 x_1 + c_2 x_2 x_3 + c_3 x_2 + c_4 x_3 + c_5 x_5$ in MU per 'cm' length of the beam.

Putting the above values of model constants in the Model-IV represented by the Eqs. (8.53)–(8.64), we can solve this model by solving the nonlinear simultaneous Eqs. (8.72), (8.73), and (8.74) presented above.

If we take the same case study as in Model-I, and put the value of rc, i.e. the design proportion of applied bending moment resisted by concrete in compression as $= 0.888888 \approx 1$, and the value of rcs, i.e. the design proportion of applied bending moment resisted by steel in compression as $= 0.000001 \approx 0$, and solve the Model-IV putting values of other parameters as above the optimal results for the Model-IV will be as follows:

$w_{01} = 0.4804, \quad w_{02} = 0.4131, \quad w_{03} = 0.0820, \quad w_{04} = 0.0146,$
$w_{05} = 4.8781 \times 10^{-11} = w_{31} = w_{42}$
$w_{11} = 0.4527, \quad w_{21} = 0.5302, \quad w_{22} = 0.0488, \quad w_{41} = 0.4804, \quad w_{51} = 0.4552,$
$w_{52} = 0.0748$
$x_1 = 37.6826$ sq.cm, $x_2 = 86.0628$ cm, $x_3 = 30.0$ cm, $x_4 = 80,064.71$ kN cm,
$x_5 \approx 0, \quad x_6 = 17.3871$ cm
$y(x) = 0.4435.$

Depth of equivalent rectangular stress block $= \beta_1 \, x_6 = 0.85 \times 17.3871 =$ 14.78 cm

It may be seen that the constraint-1,2, 3, and 4 in Model-I corresponds to constraint-3, 5, 2, and 1 in Model-IV, and therefore, geometric programming weights w_{11}, w_{21}, w_{22}, w_{31}, w_{32}, and w_{41} in Model-I corresponds to geometric programming weights of w_{31}, w_{51}, w_{52}, w_{21}, w_{22}, and w_{11} in Model-IV. Similarly, there is some change in variable names. Thus, x_5 in Model-I designates the depth of equivalent rectangular stress block, while it designates area of compression reinforcement in Model-IV.

Noting the above changes in nomenclature of geometric programming weights and variables, it may be seen that in a special case of Model-IV, when $rc \approx 1$, and $rcs \approx 0$; its results are exactly same (with some minor difference due to choice of decimal place) as the results given by the Model-I presented earlier and reproduced below:

$w_{01} = 0.4804$, $w_{02} = 0.4131$, $w_{03} = 0.0820$, $w_{04} = 0.0146$ $w_{21} = 0.4552$,

$w_{22} = 0.0748$ $w_{41} = 0.4527$ $w_{31} = 0.5302$,

$w_{32} = 0.0488$, $x_1 = 37.6825$ Sq.cm. $x_2 = 86.0628$ cm $x_3 = 30.00$ cm

$x_4 = 80064.68$ kN cm $x_5 = 14.7814$ cm

$y(x) = 0.4435$ MU/cm.

Thus, the Optimization Model-IV for doubly reinforced concrete beams can be considered as a *general optimization model* both for singly reinforced concrete beams and doubly reinforced concrete beams; and only the values of rc and rcs can be changed to use the Model-IV either for optimization of singly reinforced concrete beam or for doubly reinforced concrete beam. However, it is necessary to check whether the additional compression steel reinforcement reach its yield strength at the ultimate strength of the reinforced concrete member, using appropriate formulations prescribed in codes.

8.11.2 Optimization of Reinforced Concrete T-Beams: Model-V

In the strength design for a flanged reinforced concrete section such as I or T-sections, it can be assumed that its resisting moment is composed of two components, namely: (1) moment capacity of the rectangular portion of the concrete, i.e. b_w times d in Fig. 8.6(b) above, and the corresponding amount of tensile reinforcement, and (2) moment capacity of the overhang portion of the concrete, i.e. $(x_3 - b_w) \times x_5$ as shown in Fig. 8.6(b) above, with corresponding additional tensile reinforcement for equilibrium. Thus, it is assumed that the effect of the overhanging portion of the flange is the same as the effect of compression reinforcement in a doubly reinforced concrete beam, which is already presented in Sect. 8.11.1 above, and can be designed accordingly for optimality. It is necessary to check whether (1) the neutral axis falls within the flange in which case the T-beam can be designed as a rectangular beam with tensile reinforcement as presented in Sect. 8.5 above; or (2) the neutral axis falls outside the flanged section, in which case the procedure presented in Sect. 8.11.1 above can be followed to obtain the optimal design.

The neutral axis falls within the flange if flange thickness of a T or I–section d_f/1.18 wd/β. It is often suggested that it is more practical to check whether the flange of a T-beam is capable of resisting the applied bending moment. If not the neutral axis falls outside of the flange and the section does not act as a rectangular beam, but as a T-beam. Thus, if the nominal moment capacity of the flange M_{nf} given by:

$M_{nf} = 0.85\ fc'\ b\ d_f\ (d - 0.5\ d_f)$ is greater than the applied bending moment, then the member is designed/proportioned as a rectangular beam.

However, a simpler model with typical objective function and constraints, for a reinforced concrete T-beam as shown in Fig. 8.6(b) above, could be developed as follows on the lines of Model-I presented earlier in the Sect. 8.5.1, with some modifications as shown below:

Minimize the Cost per Unit Length of T-Beam $= y(x) = c_1 x_1 + c_2 x_3 x_5$
$\quad + c_3 x_2 + c_4 x_3$ (8.75)

Subject to:

$$a_1 x_1^{-1} x_3 x_5 \leq 1$$
(*Constraint* for Equilibrium); and   (8.76)

$$a_2 x_2^{-1} x_5 + a_3 x_2^{-1} x_3^{-1} x_4 x_5^{-1} \leq 1$$
(*Constraint* for resisting bending moment); and   (8.77)

$$a_4 x_4^{-1} + a_5 x_2 x_4^{-1} + a_6 x_3 x_5 x_4^{-1} \leq 1$$
(*Constraint* for total bending moment compatibility); and   (8.78)

$$a_7 x_2 \leq 1$$
(*Constraint* for T-beam web depth); and   (8.79)

$$x_1, x_2, x_3, x_4, x_5 > 0$$
(Non-negativity *constraint*)   (8.80)

Here, c_1, c_2, c_3, and c_4 are cost coefficients due to volume of tensile steel reinforcement (MU/cm³), volume of concrete (MU/cm³), shuttering along the vertical surfaces of beam with some adjustment (MU/cm²), and the shuttering along the horizontal surface of beam (MU/cm²), respectively. Here, a_1 is related to yield strength of tensile steel reinforcement and compressive strength of concrete; a_2 defines the centroid of concrete compression; a_3 relates to the compressive strength of concrete and the capacity reduction factor prescribed in codes; a_4 relates to the applied bending moment; a_5 relates to the beam span and density of concrete for self-weight bending moment-first component; a_6 also relates to beam span and density of concrete for self-weight bending moment-second component; and a_7 defines the desired maximum effective depth of the stem of T-beam.

Here, the model *variables* are:

x_1 = optimum area of tensile steel reinforcement (cm^2)

x_2 = a variable giving the optimum effective depth of stem of the T-beam (cm)

x_3 = a variable giving the optimum flange width of the T-beam (cm)

x_4 = a variable defining the total applied bending moment including that due to self-weight of T-beam

x_5 = a variable defining the depth of equivalent rectangular stress block. Here, a simplified assumption is made that the depth of flange of the T-beam is equal to the depth of equivalent rectangular stress block.

Adopting the same procedure and nomenclatures as in Model-I presented earlier in Sect. 8.5.1, the normality and orthogonality conditions of dual problem for this model would be as follows:

$$w_{01} + w_{02} + w_{03} + w_{04} = 1 \qquad \text{(normality condition)}$$

$$w_{01} - w_{11} = 0 \qquad \text{(orthogonality condition for } x_1\text{)}$$

$$w_{03} - w_{21} - w_{22} + w_{32} + w_{41} = 0 \quad \text{(orthogonality condition for } x_2\text{)} \quad (8.81)$$

$$w_{02} + w_{04} + w_{11} - w_{22} + w_{33} = 0 \quad \text{(orthogonality condition for } x_3\text{)}$$

$$w_{22} - w_{31} - w_{32} - w_{33} = 0 \qquad \text{(orthogonality condition for } x_4\text{)}$$

$$w_{02} + w_{11} + w_{21} - w_{22} + w_{33} = 0 \quad \text{(orthogonality condition for } x_5\text{)}$$

This is a posynomial geometric programming problem with a degree of difficulty of $T - (N + 1)$, i.e. $11 - 6 = 5$. It may be seen that there are 11 dual variables for 11 terms of the problem. Thus, as done in Sect. 8.5.1.1, a set of $(N + 1)$ dual weights, i.e. 6 here, could be expressed in terms of the remaining $T - (N + 1)$ dual weights, i.e. 5 here. Using the above equations; and choosing $w_{01}, w_{03}, w_{04}, w_{33}$, and w_{41} as the 5 dual variables, the remaining 6 dual variables can be expressed in terms of these 5 chosen dual variables as shown below:

$$w_{02} = 1 - w_{01} - w_{03} - w_{04}; \quad w_{11} = w_{01}; \quad w_{21} = w_{04}; \quad w_{22} = 1 - w_{03} + w_{33};$$

$$w_{31} = w_{03} - w_{04} - w_{33} + w_{41}; \quad w_{32} = 1 - 2w_{03} + w_{04} + w_{33} - w_{41};$$

$$(8.82)$$

Substituting these chosen dual variables into the dual objective function, and to maximize it, we can set the first derivative of this dual objective function to zero with respect to each of these dual weights coming into the substituted dual objective function, as done in Sect. 8.5.1.1. Thus, a sufficient number of equations are available for solving the problem. Using this procedure and by suitable manipulations and substitutions, we will get the following simple equations set:

$$k_1 = \frac{c_3(1 - a_3a_6a_7)}{a_3a_7(a_5 + a_4a_7)(c_2 + c_1a_1)} \qquad k_2 = \frac{c_4a_2a_7}{(c_2 + c_1a_1)(1 - a_3a_6a_7)}$$

$$w_{01} = \frac{c_1a_1}{(c_2 + c_1a_1)}(1 - w_{03} - w_{04}); \qquad w_{33} = \frac{a_3a_6a_7}{(1 - a_3a_6a_7)}(1 - w_{03} + w_{04});$$

$$(8.83)$$

$$w_{41} = \frac{a_4a_7}{(a_5 + a_4a_7)}(1 - w_{03}) - w_{03} + w_{04} + w_{33};$$

$$k_1(1 - w_{03})(1 - w_{03} - w_{04}) - (1 - w_{03} + w_{04})w_{03} = 0 \qquad (8.84)$$

$$k_2(1 - w_{03} - w_{04})(1 - w_{03} + w_{04}) - w_{04}^2 = 0 \qquad (8.85)$$

Equations (8.84) and (8.85) are two nonlinear simultaneous equations in two unknown variables, i.e. w_{03} and w_{04} and can be easily solved using the C++ computer programs nrndf.cpp or nr123.cpp presented in Chap. 3, giving the values of w_{03} and w_{04}. Knowing the values of w_{03} and w_{04} and using the equation set (8.82), we can first find the values of w_{01}, w_{33}, and w_{41}, and using these values and the equation set (8.82), the values of the remaining 6 dual variables w_{02}, w_{11}, w_{21}, w_{22}, w_{31}, and w_{32} can be determined. Using the values of these dual variables, the values of model *variables* x_1, x_2, x_3, x_4, x_5, and $y(x)$ could be determined using the procedure adopted in Sect. 8.5.1.1, and thus, the complete Model-V is solved.

In T- and I-beam design, it is necessary to limit the area of tension steel, available to develop the concrete of the web in compression, to $0.75\ p_b$ fulfilling the following relationship:

$$As/b_wd - Asf/b_wd \leq 0.75p_b$$

Compliance of this relationship also guard against possible brittle or compressive failure of the concrete. In the above relationship, '*As*' defines the total area of tensile steel, '*A_sf*' is the area of reinforcement to develop compressive strength of the overhanging flanges in the I- and T-sections, '*bw*' is the width of web in the I- and T-sections as shown in Fig. 8.6(b) above, '*d*' is the distance from extreme compression fibre to centroid of tension reinforcement as shown in Fig. 8.6, p_b is the reinforcement ratio producing balanced conditions at the ultimate strength of a section.

8.11.2.1   Model Calibration and Example Application Results of Least-Cost RC T-Beam

Calibration of the above model is illustrated with an example application problem below.

Design a **Least-Cost** reinforced concrete T-beam to resist an ultimate moment (Mu) of 600 ft-kips or 81,348.0 kN.Cm, i.e. 81,348,000.0 Ncm. The concrete strength $f'_c = 30$ MPa and the steel yield strength $f_y = 300$ MPa. The unit cost of steel (cs), concrete (cc), and shuttering (csh) adopted here are 0.72 MU (Monetary Unit)/kg, 64.5 MU/m^3 and 2.155 MU/m^2, respectively. A web width b_w of 30 cm, cover ratio (r) of 0.0873, ws the unit weight of steel reinforcement is 0.00785 kg/cm^3, and a capacity reduction factor of 0.80(ϕ) is assumed here.

Using the above data, the model can be calibrated expressing the model constants (in 'cm' units as the objective function, i.e. $y(x)$ is in MU/cm length of beam) as follows:

$$c_1 = 0.72 \times 0.00785 = 0.005652; \quad c_2 = 0.0000645; \quad c_4 = 0.0002155;$$

Referring to Fig. 8.6(b), the concrete volume per unit length of the T-beam is given by:

$$x_3 x_6 + (1 + r)\, x_2 b_w {-} b_w x_6 = x_3 x_6 + x_2 b_w \text{ (assuming } rx_2 \approx x_6 \approx x_5 \text{ to simplify)}$$

Thus, the cost of concrete per unit length of T-beam $= c_2(x_3 x_6 + x_2 b_w) = c_2 x_3 x_6 + c_2 b_w x_2$

Therefore, the coefficient $c_2 b_w$ in second term in the above expression, which is associated with x_2, should be added to c_3, which is also associated with x_2 in the third term of objective function $y(x)$. Similarly, the shuttering cost for the vertical surfaces of web portion of T-beam $= 2c_4 (x_2 + r\, x_2 - x_6) \approx 2c_4 x_2$, considering the above simplified assumption. Thus, the value of the cost coefficient c_3 in the third term of objective function $y(x)$ which is associated with the variable x_2 will be summation of $c_2 b_w$ and $2c_4$, i.e. the expression for c_3 will be as follows (here b_w or BW is assumed as $= 30$ cm):

$$c_3 = c_2 b_w + 2c_4 = 0.0000645 \times 30 + 2 \times 0.0002155 = 0.002366.$$

As already mentioned, the above model is simplified by making a simplified assumption that the depth of flange of the T-beam is equal to the depth of equivalent rectangular stress block. This can be easily achieved by selection of suitable value of cover ratio 'r' and the desired maximum depth of web of the T-beam (determining the value model constant $a_7 = 1/d$), which are assumed here as: 0.0873 and 75 cm, respectively. With this assumption and the modification of the value of c_3 as above, the values of all cost coefficients and other model constants in this case (adopting relevant units of measurements) are:

$$c_1 = ws \times cs = 0.005652, \quad c_2 = cc = 0.0000645, \quad c_4 = csh = 0.0002155,$$
$$c_3 = c_2\, b_w + 2c_4 = 0.002366,$$

$$a_1 = 0.85 f'_c / f_y = 0.085, \quad a_2 = 0.50, \quad a_3 = 1/(0.85\, f'_c\, \phi) = 0.00043573$$
$$a_4 = Mu = 81348000.0, a_5 = k\, wc\, b_w\, L^2$$

(k is bending moment coefficient for beam self-weight bending moment calculation, 'wc' weight of concrete, L span of T-beam) $= 0.0$ (as no span specified), $a_6 = k\, wc$ $L^2 = 0.0$ (as no span specified), $a_7 = 1/d = 0.0133333$.

Putting these input values, the following final output results are obtained after solving the two nonlinear simultaneous Eqs. (8.84) and (8.85) of Model-V presented above, using the Newton–Raphson approximation technique incorporated in the C++ computer programs developed and discussed earlier.

```
LEAST-COST FLEXURAL STRENGTH DESIGN OF T-BEAM-MAX.ACCEPT WEB DEPTH-
MODEL-V

SOME KEY INPUT VALUES CHOSEN(INDEPENDENT VARIABLES):
BW= 30.00 CM. BMC= 0.1250 CRF=0.80 CCS= 30.00 MPA SYS=300.00 MPA
cr=1.0873 cs= 0.7200 MU/KG cc= 64.5000 MU/CU.M. csh= 2.1550 MU/SQ.M.

SOME KEY OUTPUT VALUES (DEPENDENT VARIABLES):
LA=71.3520, C=1267158.07, MR= 81372.81, CRF=0.800,MNC=80414344.88
SWBM=   0.00, UABM=81348000.00, TABM= 81348.00, x4=81348000.00

CALIBRATED MODEL CONSTANTS(for the Chosen DESIGN CONDITION)are:
a1=0.0850 a2=0.5000 a3=0.00043573 a4=81348000.00 a5= 0.00 a6= 0.00
a7= 0.0133
c1=0.00565200 , c2= 0.00006450 , c3=0.00236600 , c4=0.00021550
DESIGN Values of OPTIMAL SECTION:
LA= 71.3520 cm, C=1267158.07 N, T=1267158.07 N, a3= 0.0004, x6= 7.30
MNC=80414344.88 N.cm, MNS=80414344.88 N.cm, MR= 81372.81 kN.cm
UABM=81348000.00 N.cm

OPTIMAL RESULTS AFTER SOLVING THE 'NLP' MODEL:
The values of Optimal Cost Weights are:
w01= 0.51572, w02=0.06824, w03=0.38333, w04=0.03171, t=1.0000,
w11=0.51572
w21= 0.03171, w22=0.61667, w31=0.61667, w32=0.00000, w33=0.00000,
w41=0.26504

Optimal Area of Tensile Steel Re-inforcement=  42.24 sq.cm.
Optimal Effective Depth of Web of the T=Beam  =  75.00 cm
Optimal Width of Flange of the RC T-Beam    = 68.11 cm
Total Applied Bending Moment(inc.Self.wght)  =  81348.00 kN.cm
Calculated Moment of Resistance(ap.CRF)of Optimal Section=  81372.81
kN.cm

Optimal Depth of Eq.Rect.Stress Block    =   7.30 cm
The LEAST-COST per Unit-Length of T-BEAM  =  0.4628 MU/cm
The Calculated WEIGHT/Unit-Length of T-BEAM  =  6.6780 kg/cm
```

The above results show the optimal values of some basic model variables as follows:

$x_1 =$ a variable giving the optimum area of tensile steel reinforcement $(cm^2) = 42.24$ sq.cm.

$x_2 =$ a variable giving the optimum effective depth of stem of the T-beam (cm) $= 75.00$ cm

$x_3 =$ a variable giving the optimum flange width of the T-beam (cm) $= 68.11$ cm

$x_5 =$ a variable defining the depth of equivalent rectangular stress block $= 7.30$ cm

The above results also show that the optimal value of T-beam flange thickness d_f or x_6 (see Fig. 8.6(b)) = 7.30 cm, while the value of optimal depth of equivalent rectangular stress block $x_5 = 7.30$ cm given by the model, i.e. both are equal (correct to one decimal place) matching with the assumption earlier made. Thus, the results given by the model are consistent and the initial assumption made is correct. It is necessary to check whether the neutral axis of the T-beam falls within the flange fulfilling the following condition:

$$d_f/1.18 \; wd/\beta$$

In the above case, the member acts as a rectangular beam and can be proportioned/optimally designed according to requirements of rectangular beams with tension reinforcement as in Optimization Model-I and Model-II.

In the present case as shown above, As, i.e. $x_1 = 42.24$, d i.e. $x_2 = 75$, and $b_w = 30$. Therefore, $w = As\,f_y/b_w df'_c = (42.24 \times 300)/(30 \times 75.0 \times 30) = 0.188$. Hence, in this case, the neutral axis of the T-beam will fall within the flange if $d_f/1.18 \times 0.188 \times 75.0/0.85 = 18.57$.

But, here as shown above $d_f = 7.30$ which is less than the above value of 18.57. Hence, in this case neutral axis of the T-beam falls outside the flange, and therefore, it acts as a T-beam and should be designed accordingly.

In the conventional design of T-beams, the overhanging portion of the flange of T-beam is considered the same as an equivalent amount of compression steel A_{sf} which is given by the expression:

$$A_{sf} = 0.85(x_3 - b_w) \, d_f f'_c/f_y$$

Therefore, in this case $A_{sf} = 0.85(68.11 - 30) \, 7.3 \times 30/300 = 23.647$ sq.cm., and the nominal moment capacity of the overhanging portion is $M_{nf} = A_{sf} \times f_y(d - 0.5\,d_f)$ = $23.647 \times 30,000 \, (75 - 0.5 \times 7.3)/1000 = 50,616.404$ kN-cm. Similarly, we can calculate the nominal moment capacity of the rectangular portion b_w times d, i.e. web portion of the T-beam as follows using the Eq. (8.3):

$$M_{nw} = b_w d^2 f'_c \; w \, (1 - 0.58w)$$

Here, the total area of tension reinforcement is = 42.24 sq.cm. as shown above. Therefore, the reinforcement area available to develop the web $A_{sw} = 42.24 - A_{sf} = 42.24 - 23.65 = 18.58$. Hence, the values of w and M_{nw} are given by the following expressions:

$$w = A_{sw} f_y/b_w df'_c = 18.58 \times 300)/(30 \times 75.0 \times 30) = 0.0826 \approx 0.083 \text{ and}$$

$$M_{nw} = 30 \times 75^2 \times 3000 \times 0.083(1 - 0.58 \times 0.083)/1000 = 38,861.082 \text{ kN-cm}$$

Therefore, the total nominal moment capacity of the T-beam = 38,861.082 + 50,616.404 = 80,577.486 kN-cm. Applying the capacity reduction factor of 0.8, the moment of resistance of the section is 81,518.75 kN.cm. which is 171.75 kN.cm.

higher than the given ultimate moment of 81,348.00 kN.cm. to be resisted by the T-beam. Thus, the conventional design method as adopted above slightly overestimate the total nominal moment capacity of the T-beam. However, it may be seen from output results of the T-beam optimal design Model-V presented above that the total nominal moment capacity of the T-beam = 80,414,344.88 N.cm,. i.e. 80,414.34 kN.cm. and applying the capacity reduction factor of 0.8 the moment of resistance of the section is 81,372.81 kN.cm. which is only 24.81 kN.cm. higher than the given ultimate moment of 81,348.00 kN.cm. to be resisted by the T-beam, and considering decimal error the Optimal Design Model-V presented above can be said to give more correct results.

If the concrete and the reinforcing steel are stressed to the allowable ultimate values at the same time; then the section is said to have balanced reinforcement. Assuming the modulus of elasticity of steel reinforcement as Es = 200,000 MPa, the ultimate concrete strain as = 0.003, and the ultimate steel strain = fy/Es; the balanced steel ratio p_b is given by the following expression:

$$\mathcal{P}_b = \frac{0.85f'_c\beta1}{f_y}\frac{600}{(600+f_y)}$$

Therefore, if $f_c' = 30$ MPa, $f_y = 300$ MPa, and $\beta_1 = 0.85$ the value of $p_b = 0.0482$ and 0.75 $p_b = 0.0361$ in this example.

Codes require that in I- and T-sections, the area of the tension reinforcement available to develop the web portion of the beam should be limited, to avoid the possibility of brittle or compression failure of the concrete. Therefore, in this example $p_w = A_{sw}/b_wd$ must be ≤ 0.75 p_b. Here, the value of $p_w = 18.58/(30 \times 75) = 0.0083$ which value is < 0.75 $p_b = 0.0361$ and fulfills the above code requirement. In case any code requirement is violated the designer can change the input value related to relevant model constant(s) such as a_7.

As in case of other Optimization Models-I, II, and IV for rectangular RC beams presented earlier in Sects. 8.5.1, 8.5.2, and 8.11.1, the Optimization Model-V for RC T-beams also finds the unique least-cost section instantaneously, out of an infinite number of possible solutions, obviating the need of any design chart or trial and error solution in conventional design practice, which does not consider any cost factor. This is demonstrated by the example application results of **Least-Cost** T-Beam design using the Model-V presented above.

8.11.3   Optimization of Reinforced Concrete Continuous Beams: Model-VI

In the optimization models for flexural design of reinforced concrete beams presented above, we adopted only single span to find the optimal section on flexural consideration. In multiple spans with different support conditions, it is desirable to carry out mathematical analysis such as slope deflection and moment distribution methods to find moments and shears at different sections. However, instead of exact mathematical analysis of statically indeterminate structures, approximate coefficients of shear and bending moment can be utilized for continuous beams and slabs if the following conditions are satisfied:

1. Adjacent clear spans may not differ in length by more than 20% of the shorter span
2. Ratio between the live load and dead load should not exceed 3
3. Loads must be uniformly distributed

When the above conditions are satisfied, a list of approximate formulas in the form of coefficients are given in ACI Code to determine shear forces and bending moments in continuous beams and in one way slabs. These coefficients can be used in optimization models for continuous beams. In order to demonstrate the applicability of optimization in continuous beams, we can take a simple example of a two-span continuous beam as shown in Fig. 8.7 below:

To simplify the model development, in the above schematic diagram of the continuous beam, the total length of reinforcing bars are assumed as some fraction (0.3,0.6,0.8) of the span L of the two-span continuous beam with equal span of length L. However, for more accurate results, these lengths should be modified if necessary considering the development length of bars in tension and compression, and the need to extend reinforcement beyond the point at which it is not required to resist flexure, as per provisions in various national codes. Similarly, instead of determining the ultimate positive and negative moments and shears based on critical loading conditions for maximum bending moment and shears, only bending moments based on approximate coefficients as prescribed above are considered in the model, assuming that the above 3 conditions are satisfied. It is possible to develop optimization model for reinforced concrete beam for *Least-Cost* design, considering both bending moment and shear as presented earlier (Chakrabarty 1982). Interested readers can develop more accurate models considering all the above factors. The simplified model for optimization of reinforced concrete continuous beams based on the above assumptions would be same as Model-I or Model-II with respective design condition of minimum acceptable beam width or desirable beam width-depth ratio, but model constants for the continuous beam will be different, and the objective function will be to minimize the total cost of the continuous beam considering all spans. Thus, taking the Model-II as the basic model, the optimization model for the continuous beam as shown in Fig. 8.7, would be as follows:

Minimize the Total Cost of the Continuous Beam $= y(x) = c_1 x_1 + c_2 x_2 x_3 + c_3 x_2 + c_4 x_3$

Subject to:

$$a_1 x_1 x_3^{-1} x_5^{-1} \qquad\qquad \le 1 \qquad\qquad (8.33)$$
(*Constraint* for Equilibrium)

$$a_2 x_4^{-1} + a_3 x_2 x_3 x_4^{-1} \qquad\qquad \le 1 \qquad\qquad (8.34)$$
(*Constraint* for bending moment compatibility)

$$a_4 x_1^{-1} x_2^{-1} x_4 + a_5 x_2^{-1} x_5 \qquad\qquad \le 1 \qquad\qquad (8.35)$$
(*Constraint* for bending moment)

$$a_6 x_2 x_3^{-1} \qquad\qquad \le 1 \qquad\qquad (8.36)$$
(*Constraint* for Beam width : depth ratio)

Here, the expressions for the objective function and constraints are same as the basic Model-II presented earlier in Sect. 8.5.2, but the value of some model constants is different as shown below. The approximate bending moment coefficients in different sections of the continuous beam shown in Fig. 8.7 above are taken as follows:

$M_D = wL^2/8$ negative moment; $M_E = M_F = wL^2/11$ positive moment; $M_A = M_C = wL^2/24$ negative moment;

Here, w is the uniformly distributed load in kN/m, and L is the span length in metre. It would be seen that the critical section is D, giving the maximum moment coefficient. Hence, Model-II is applied in this section to get the optimum values of the design variables x_1, x_2, and x_3 at this section. Thereafter, the same values of x_2 and x_3 are assumed in other sections of the continuous beam, but the value of x_1 is changed in these sections proportionate to the bending moments as per the bending moment coefficients in these sections as adopted above. Since, in these cases b (i.e. x_3) and d (i.e. x_2) are given, the value of x_1, i.e. A_s = area of tension reinforcement is the only unknown in the Eq. (8.3) reproduced below:

$$M_n = bd^2 f_c^{-1} w (1 - 0.58w) \qquad\qquad (8.3)$$

The above equation could be further modified, in terms of the ultimate resisting moment of a cross section, or, $M_u = M_n/\phi$, where ϕ is the capacity reduction factor, as follows:

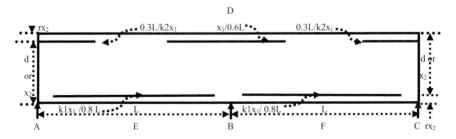

Fig. 8.7 Schematic diagram of a two-span reinforced concrete continuous beam

$$\left(\phi^* 0.58^* f_y^2/b^* f'_c\right) As^2 - \left(\phi d f_y\right) As + M_u = 0; \left(\phi^* 0.58^* f_y^2/b^* f'_c\right) = a$$

This is a quadratic equation giving the value of As as:

$$As = \frac{\left(\phi d f_y\right) - \sqrt{\left(\phi d f_y\right)^2 - 4 \times a \times Mu}}{2 \times a} \tag{8.86}$$

Solving the Eq. (8.86) adopting the values of $b = 30$ cm, $d = 78.86$ cm. and varying the value of M_u from 83,625.00 Kn.cm to 8362.50 Kn.cm, gives the following variation of tension steel area As to the value of M_u or Total Applied Bending Moment (TABM):

```
VARIATION OF TENSION STEEL AREA(AS) TO 'TABM'    :
TABMR    TABM        AS            ASR    x2      x3
(TABM-   (Kn.cm)     (TENSION             (cm)    (cm)
RATIO)               STEEL-sq.cm)
0.10     8362.50     3.8616        0.1    78.86   30.00
0.20     16725.00    8.0046        0.2    78.86   30.00
0.30     25087.50    12.1342       0.3    78.86   30.00
0.40     33450.00    16.3563       0.4    78.86   30.00
0.50     41812.50    20.6773       0.5    78.86   30.00
0.60     50175.00    25.1047       0.6    78.86   30.00
0.70     58537.50    28.6465       0.7    78.86   30.00
0.80     66800.00    34.3123       0.8    78.86   30.00
0.80     75262.50    38.1127       0.8    78.86   30.00
1.00     83625.00    44.0600       1.0    78.86   30.00
```

It would be seen from above that the area of tension reinforcement As (i.e. x_1), is directly proportionate to the value of M_u; and the values of TABM ratio and the As ratio are more or less same in all cases. Hence, changing the value of x_1 in different sections, proportionate to the bending moments as per the bending moment coefficients in these sections, is in order. This is applied in solving the above optimization model for the two-span continuous beam. However, if more accurate results are desired, the value of x_1 in different sections can be determined using the Eq. (8.86). Accordingly, the proportionate steel area coefficients $k1$ for sections E and F would be $k1 = M_E/M_D = 8/11$, and area coefficients $k2$ for sections A and C would be $k2 = M_A/M_D = 8/24$.

8.11.3.1 Optimization of Reinforced Concrete Continuous Beam: An Example Problem

To illustrate the application of the above model, we can take an example problem to design a **Least-Cost** reinforced concrete two-span continuous beam (as shown in Fig. 8.7) with each span L of length of 10 m supporting a total ultimate uniform load of 55 kN/m. The concrete strength $f'_c = 30$ MPa and the steel yield strength $f_y = 300$ MPa. The unit cost of steel (CS), concrete (CC), and shuttering (CSH) adopted here are 0.72 MU (Monetary Unit)/kg, 64.5 MU/m^3 and 2.155 MU/m^2, respectively.

A cover ratio (r) of 0.10, unit weight of 2323 kg/m^3 for concrete, and a capacity reduction factor (CRF) of 0.80 are assumed here.

Based on the above assumptions, the value of model constants in this example problem could be determined as follows:

$$c_1 = (1.8L \times 8/11 + 0.6\ L \times 8/24 + 0.6L) \times CS \times WS$$
$$= (1.8 \times 1000 \times 8/11 + 0.6 \times 1000 \times 8/24 + 0.6 \times 1000) \times 0.72 \times 0.00785$$
$$= 12.887$$
$$c_2 = 2L \times (1+r) \times CC = 2000 \times 1.1 \times 64.5/1{,}000{,}000 = 0.1418$$
$$c_3 = 4L \times (1+r) \times CSH = 4000 \times 1.1 \times 2.155/10{,}000 \quad = 0.8482$$
$$c_4 = 2L \times CSH = 2000 \times 2.155/10{,}000 \quad = 0.431$$
$$M_D = wL^2/8 = 55 \times 1000 \times 1000 \times 1000/8 = 61{,}111{,}111.11 \ \text{N.cm} = \text{uabm}$$
$$a_1 = f_y/0.85f'_c = 300/0.85 \times 30 = 11.76471$$
$$a_2 = \text{uabm}$$
$$a_3 = \text{LFDL}(1+r)wc \times L^2 \times \text{BMC} = 1.4 \times 1.1 \times 2323 \times 8.8/8 = 3885.4128$$
$$a_4 = 1/\text{CRF} \times f_y = 1/0.8 \times 30{,}000 = 0.00003704$$

$a_5 = 0.50$; assuming the centroid compressive force at half the depth of equivalent rectangular stress block

$a_6 = 0.40$; continuous beam width : depth (effective depth) ratio, chosen by the designer.

Using the above model constants, the Model-II (presented in Sect. 8.5.2 above) as the basic model, and the C++ program cbm.cpp discussed earlier; the output optimal results instantly given by the optimization model for the continuous beam (as shown in Fig. 8.7) are as follows:

```
LEAST-COST FLEXURAL STRENGTH DESIGN-GIVEN BEAM-WIDTH:DEPTH
RATIO:'ITMODEL- 2'
REINFORCED CONCRETE CONTINUOUS BEAM

SOME KEY INPUT VALUES CHOSEN(INDEPENDENT VARIABLES):
DLL=20.0 kN/m, DDL=15.0 kN/m, LFLL= 1.7 LFDL= 1.4 BSPN=10.0 m
BMWDR= 0.40 BMC= 0.1111 CRF=0.80 CCS=3000.00 MPA SYS=30000.00 MPA
cr= 1.10 cs= 0.7200 MU/KG cc= 64.5000 MU/CU.M. csh= 2.1550 MU/SQ.M.
SOME KEY OUTPUT VALUES (DEPENDENT VARIABLES):
LA=62.2583, C=1230.0588, MR=68823.1882, CRF=0.800,MNC=76581.3482
SWBM= 7788.51 kNcm, UABM= 61111.11 kNcm, TABM= 68808.62 kNcm, x4=
68808.62
CALIBRATED MODEL CONSTANTS(for the Chosen DESIGN CONDITION)are:
a1=11.7647 a2= 61111111.1000 a3= 3885.4128 a4= 0.0000370 a5=0.50
a6= 0.40
```

```
c1= 12.88700000 , c2=  0.14180000 , c3= 0.84820000 , c4= 0.43100000
DESIGN Values of OPTIMAL SECTION:
LA= 62.2583 cm, C=  1230.06 kN, T=  1230.06 kN, a3=3885.4128
MNC= 76581.35 kN.cm, MNS= 76581.32 kN.cm, MR= 68823.18 kN.cm
COM=1.000, UABM=61111.1111 kN.cm
OPTIMAL RESULTS AFTER SOLVING THE 'NLP' MODEL:
The values of Optimal Cost Weights are:
w01= 0.5846, w02=0.3168, w03=0.0748, w04=0.0136, t=1.0000,
w11=0.0845
w21= 0.6112, w22=0.0780, w31=0.6882, w32=0.0845, w41=0.3140
Optimal Area of Tensile Steel Re-inforcement= 41.0018 Sq.Cm.
Optimal Effective Depth of the RC Beam = 70.7863 Cm
Optimal Width of the RC Beam       = 28.2818 Cm
Total Applied Bending Moment(inc.Self.wght)= 68808.62 kN.Cm
Calculated Moment of Resistance of OPTIMAL SECTION= 68823.18 kN.Cm
Optimal Depth of Eq.Rect.Stress Block = 17.0560 Cm
LEAST-COST for the Total-Length(all Spans) of Continuous BEAM =
885.4768 MU/Cm
The Calculated WEIGHT per Unit-Length of BEAM =  5.1156 kg/Cm
```

The optimal area of tensile steel reinforcement= 41.0018 sq.cm. (x_1) shown above is for the section D (top of length = $0.6 \times 1000 = 600$ cm), and the area of tensile steel reinforcement at other sections are as follows:

```
Section E (bottom) = k1× x1 = 8/11×41.0018 = 33.547 sq.cm for length of =
0.8×1000= 800 cm
Section F (bottom) = k1× x1 = 8/11×41.0018 = 33.547 sq.cm for length of =
0.8×1000= 800 cm
Section A ( top) = k2× x1 = 8/24×41.0018 = 15.376 sq.cm for length of =
0.3×1000   = 300 cm
Section B ( top) = k2× x1  = 8/24×41.0018 = 15.376 sq.cm for length of =
0.3×1000   = 300 cm
```

As already mentioned in Sect. 8.8.1, the C++ program cbm.cpp calculates the total area of reinforcement bars as per the pre-input (or new input) layer number, diameter, and number of bars of each category and compares with design steel reinforcement area, i.e. 'AS' as well as the area difference (adif), and continue to calculate 'adif' when response to CHOOSE NEW Dia, etc. is put as 'y', and until 'adif' is within tolerable limit fixed by designer, with pre-input or new input of dia and number of bars. Following this procedure, the designer can choose layer number, diameter, and number of bars of each category matching with the area of tensile steel reinforcement as above at each section of the continuous beam. in this optimal design solution as shown above the optimal effective depth of the beam = 70.7863 cm, and optimal width of the beam = 28.2818 cm, which are adopted in all sections of the continuous beam. Accordingly, the beam width:depth ratio of $28.2818/70.7863 = 0.4$ is achieved, as specified by the designer by choosing the value of $a_6 = 0.40$. Model gives the **Least-Cost** for the total-length of the *Optimum Continuous Beam* considering all spans as: 885.4768 MU as shown. It may be seen that cost of reinforcement is highest component, i.e. 58.46% (w_{01}), horizontal shuttering is lowest (1.36%), while concrete cost is 31.68% of total cost; requiring careful selection of span fractional values cited above.

As mentioned above, optimization of the above continuous beam is done on flexural strength design considerations. However, shearing force at different sections of the beam could be calculated, and if needed shearing reinforcement in the form of stirrups and so on, could be provided. It is also possible to develop optimization model for reinforced concrete beams for *Least-Cost* design, considering both bending moment and shear as presented earlier (Chakrabarty, 1982).

8.11.4   Optimization of Reinforced Concrete Columns: Model-VII

There are a number of general requirements stipulated in ACI code and in national codes with regard to reinforced concrete columns, which are to be followed in column design. Similarly, there are a number of special conditions and safety provisions in the strength design of short columns, which are defined by the length criteria as presented in Everard et al., 1887. The basic problem in column design is to find the values of *Ast* (total area of longitudinal reinforcement) and *pt* (i.e. *Ast/Ag*, where *Ag* is the gross area of column section) corresponding to the given loading conditions and satisfying all provisions of the ACI code, which is facilitated by use of interaction diagrams. Again, compatibility between the strain in steel and the compression concrete have to be ensured, limiting the reinforcing steel stress below the yield strength. Thus, there are several interacting factors involved in the design of columns as mentioned above, making the development of optimization model for optimal design of reinforced concrete columns a fairly complicated task. However, to demonstrate the applicability of optimization even in such complex problems, a model for optimal design of reinforced concrete columns with some simplified assumptions is presented below, taking an example of a short column with ties subjected to uniaxial bending and axial load as shown in Fig. 8.8 below:

Adopting the schematic section of a reinforced concrete column shown in Fig. 8.8 above, a model for **Least-Cost Design of Reinforced Concrete Column** could be developed as outlined below. Here, the cost per unit length of the column will be given by the expression (8.4) as in Model-I, which will be the objective function for this column optimization model also as shown below:

Minimize the Cost per Unit Length of Column $= y(x) = c_1 x_1 + c_2 x_2 x_3 + c_3 x_2$
$+ c_4 x_3$

Subject to:

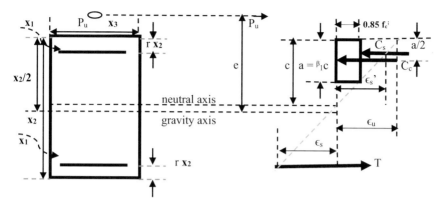

Fig. 8.8 Schematic diagram of a reinforced concrete column subjected to uniaxial bending

$$a_1 x_1 x_3^{-1} x_5^{-1} + a_2 x_3^{-1} x_5^{-1} \qquad \leq 1 \qquad (8.87)$$
(*Constraint* for ultimate axial force Pu)

$$a_3\, x_3^{-1} x_5^{-1} \qquad \leq 1 \qquad (8.88)$$
(*Constraint* for bending moment resisted by concrete)

$$a_4 x_1^{-1} x_2^{-1} x_4^{-1} \qquad \leq 1 \qquad (8.89)$$
(*Constraint* for bending moment resisted by reinforcing steel)

$$a_5 x_2^{-1} x_5 + a_6 x_2^{-1} x_4 x_5 \qquad \leq 1 \qquad (8.90)$$
(*Constraint* for strain compatibility)

$$x_1, x_2, x_3, x_4, x_5 > \quad 0$$
(Non-negativity *constraint*)

Here, c_1, c_2, c_3, and c_4 are cost coefficients due to volume of tensile steel reinforcement (MU/cm^3), volume of concrete (MU/cm^3), shuttering along the vertical surfaces of x_2 side of column (MU/cm^2), and the shuttering along the vertical surfaces of x_3 side of column (MU/cm^2), respectively. Here, a_1 is related to the ratio between the depth of equivalent rectangular stress block and the depth of neutral axis, a_2 relates to applied axial force and capacity reduction factor; a_3 relates to applied bending moment resisted by concrete and the capacity reduction factor, a_4 relates to applied bending moment resisted by reinforcing steel and the capacity reduction factor; a_5 related to concrete cover ratio; and a_6 is related to the concrete ultimate strain and the modulus of elasticity of reinforcing steel.

Here, the model *variables* are:

$x_1 = $ Optimum area of steel reinforcement at each edge of column (cm^2)
$x_2 = $ A variable giving the optimum overall depth of the column (cm)

x_3 = A variable giving the optimum overall width of the column (cm)
x_4 = A variable defining the stress in reinforcing steel (N/cm^2)
x_5 = A variable defining the depth of equivalent rectangular stress block (cm).

8.11.4.1   Model Solution

As in case of Model-I, the dual objective function and the associated normality and orthogonality constraints for the above primal problem can be expressed taking the terminology used in literature on geometric programming as explained in Sect. 8.12.5. However, as discussed earlier in the case of Model-I, the coefficients for objective function and constraints can be simplified as: $c_{01} = c_1$, $c_{02} = c_2$, and so on, for objective function coefficients; and $c_{11} = a_1$, $c_{21} = a_2$, and so on, for constraint coefficients. Accordingly, the dual objective function to be maximized in this case also can be expressed in a simplified form as follows:

$$\text{Maximize } d(w) = [c_1/w_{01}]^{w_{01}} [c_2/w_{02}]^{w_{02}} [c_3/w_{03}]^{w_{03}} [c_4/w_{04}]^{w_{04}}$$
$$\times [a_1]^{w_{11}} [a_2(w_{21} + w_{22})/w_{21}]^{w_{21}} [a_3(w_{21} + w_{22})/w_{22}]^{w_{22}} \quad (8.38)$$
$$\times [a_4(w_{31} + w_{32})/w_{31}]^{w_{31}} [a_5(w_{31} + w_{32})/w_{32}]^{w_{31}} [a_6]^{w_{41}}$$

Subject to:

$w_{01} + w_{02} + w_{03} + w_{04} = 1$	(normality condition)	(8.39)
$w_{01} + w_{11} - w_{31} = 0$	(orthogonality condition for x_1)	(8.40)
$w_{02} + w_{03} + w_{22} - w_{31} - w_{32} + w_{41} = 0$	(orthogonality condition for x_2)	(8.41)
$w_{02} + w_{04} - w_{11} + w_{22} - w_{41} = 0$	(orthogonality condition for x_3)	(8.42)
$- w_{21} - w_{22} + w_{31} = 0$	(orthogonality condition for x_4)	(8.43)
$- w_{11} + w_{32} = 0$	(orthogonality condition for x_5)	(8.44)

Here, w_{01}, w_{02}, w_{03}, and w_{04} represent the four cost weights related to the four terms in the objective function $y(x)$, i.e. expression (8.4) above. Similarly, w_{11}, w_{21}, w_{22}, w_{31}, w_{32}, and w_{41} represent the dual variables related to the single term in constraint-1, i.e. expression (8.33), to the first term in constraint-2, i.e. expression (8.34), to the second term in constraint-2, i.e. expression (8.34), to the first term in constraint-3, i.e. expression (8.35), to the second term in constraint-3, i.e. expression (8.35), and to the single term in constraint-4, i.e. expression (8.36) above, respectively. There are ten terms (T) and five primal variables (N) in the primal objective function and constraints [expressions (8.4) and (8.33)–(8.36)], and therefore, the degree of difficulty in the problem is = $T - (N + 1) = 4$. Thus, any six of the dual variables can be expressed in terms of the remaining four dual variables as shown below, choosing w_{01}, w_{03}, w_{04}, and w_{41} as the four dual variables:

$$w_{02} = 1 - w_{01} - w_{03} - w_{04}; \; w_{11} = w_{03} - w_{04} - w_{01} + 2w_{41}; \; w_{21} = 1 - w_{03} - w_{41};$$
$$w_{22} = 2w_{03} - w_{04} - 1 + 3w_{41}; \; w_{31} = w_{03} - w_{04} + 2w_{41}; \; w_{32} = w_{11}.$$

$$(8.45)$$

as in case of Model-I, the cost weights w_{01}, w_{02}, w_{03}, and w_{04} represent the optimum proportion of total cost per unit length of beam, distributed between the tensile steel reinforcement, concrete, shuttering along the vertical surfaces of the beam, and the shuttering along the bottom horizontal surface of the beam, respectively. The set of Eq. (8.45) represent the relationship between these cost distributions among the above components, which are independent of the values of the cost coefficients c_1, c_2, c_3, and c_4. This invariant property of geometric programming could be used for development of suitable *design rules* for *optimality* even in the presence of a great degree of difficulty, having general application in any problem of this nature.

If we substitute the expressions in equations set (8.45) into the dual objective function [Eq. (8.38)] and maximize this substituted dual objective function over the four variables w_{03}, w_{04}, w_{01}, and w_{41} (i.e. by setting the first derivative of the function with respect to each variable to zero), we will get four additional equations, thus providing sufficient conditions for optimality. The process is simplified by maximizing the natural logarithm of the dual function; that is, by substituting:

$$\ln \, d(w_{01}, w_{03}, w_{04}, w_{41}) = z(w_{01}, w_{03}, w_{04}, w_{41}).$$

Thus, by above substitution and taking logarithm on both sides of Eq. (8.38), we get the following equation:

$$
\begin{aligned}
z(w) = & \; w_{01}(\ln c_1 - \ln w_{01}) + (1 - w_{01} - w_{03} - w_{04}) \\
& (\ln c_2 - \ln(1 - w_{01} - w_{03} - w_{04})) + w_{03}(\ln c_3 - \ln w_{03}) \\
& + w_{04}(\ln c_4 - \ln w_{04}) + (w_{03} - w_{04} - w_{01} + 2w_{41}) \ln a_1 \\
& + (1 - w_{03} - w_{41})(\ln a_2 + \ln(w_{03} - w_{04} + 2w_{41}) - \ln(1 - w_{03} - w_{41})) \\
& + (2w_{03} - w_{04} - 1 + 3w_{41})(\ln a_3 + \ln(w_{03} - w_{04} + w_{41}) - \ln(2w_{03} - w_{04} - 1 + 3w_{41})) \\
& + (w_{03} - w_{04} + 2w_{41})(\ln a_4 + \ln(2w_{03} - 2w_{04} - w_{01} + 4w_{41}) - \ln(w_{03} - w_{04} + 2w_{41})) \\
& + (w_{03} - w_{04} - w_{01} + 2w_{41})(\ln a_5 + \ln(2w_{03} - 2w_{04} - w_{01} + 4w_{41}) \\
& - (w_{03} - w_{04} - w_{01} + 2w_{41})) + w_{41} \, \ln a_6.
\end{aligned}
$$

$$(8.46)$$

As done in case of Model-I, by setting the first derivatives $dz/dw_{01} = 0$, $dz/dw_{03} = 0$, $dz/dw_{04} = 0$, and $dz/dw_{41} = 0$, we will get four nonlinear simultaneous equations in terms of four unknown variables w_{01}, w_{03}, w_{04}, and w_{41} and can be solved adopting the procedure earlier presented above.

8.12   Optimal Design Method and Conventional Design Method: A Comparison

In Sect. 8.6 a model (Model-III) for conventional design of reinforced concrete beam and sensitivity analysis is presented which also form part of C++ software sub-module 'cbm.cpp' of software: 'RCCAD'. Application results of the Model-III, i.e. conventional design method for the same example problem to design a reinforced concrete beam simply supported over a span of 10 m supporting a uniform dead load of 15 kN/m and uniform live load of 20 kN/m are shown in Sect. 8.6. Here also the same concrete strength $f'_c = 30$ MPa, steel yield strength $f_y = 300$ MPa, same unit cost of steel (CS), concrete (CC), and shuttering (CSH) of 0.72 MU/kg, 64.5 MU/m^3 and 2.155 MU/m^2, respectively, same cover ratio (r) of 0.10, unit weight of 2323 kg/m^3 for concrete and a capacity reduction factor of 0.80, are assumed. It may be seen that the conventional design method gives beam cost as 0.5227 MU/cm, calculated beam weight as 6.8885 kg/cm, while the optimum design method gives beam cost as 0.4435 MU/cm and calculated beam weight as 6.5876 kg/cm. Thus, in this case by optimization, a designer can achieve a saving of 15.15% (i.e. conventional design is 17.86% costlier than optimal design and beam weight is also higher by 6% increasing cement consumption and the consequent CO_2 emission problem discussed earlier), which may be even higher depending on the site-specific unit costs.

In this optimization method, a user is only to give input data as per his/her requirement and the software: 'RCCAD' gives instantly the least-cost design drawing as output, and the user can review it and put new input to get better results until a satisfactory solution is found. This is rarely possible in conventional design practice.

It may also be seen that the total applied bending moment (**80072.3416 kN.cm**) including that due to the self-weight of the beam (which is not known before hand and assumed in conventional practice) exactly matches the calculated moment of resistance of optimal section using the prescribed capacity reduction factor.

Sometimes, because of unfamiliarity, conventional designers may have some wrong impression that the optimization method is focused only on optimizing the cross-sectional dimensions and total quantity of reinforcement, and does not deal with the selection of bar diameters and the number of bars, positioning of bars, etc. creating applicability problem. This notion would appear to be completely misplaced if one studies carefully the models and the computer codes of the C++ program cbm.cpp (presented in appendix) which incorporates a 'goto' statement with a *label* 'dn' within it, where the designer is alerted about the given PRE-INPUT values of area of steel (AS), Layer No, Diameter of bars, No.of Re-inf.bars of each size, clearCover, reinforcement area diff.(adif) between design and actual area; and is given the option to CHOOSE NEW Dia./No.of Re-inf.bars and clear Cover if the value of adif is higher; and the designer can repeat the process giving new set of values, until the area difference is below the tolerable limit. Such a facility is not available in conventional method of design creating *inefficiency*, cost increase, and other problems. This is demonstrated in the example problems presented above as part of **Integrated Computer-Aided Optimal Design Procedure** with graphics.

Again, it may be noted that the optimization models presented above incorporate exactly the standard expressions for the nominal moment capacity of a rectangular beam for flexural strength design, as used in the conventional design practice, without any simplification. Therefore, the so-called applicability problem if any should be same both in conventional design practice and in optimal design method. But, the optimal design gives large savings in costs and in material consumption as shown in example problems presented above. Yet again in this new technique of design, the designer has to give only his/her design conditions as input and in response, the unique least-cost beam section is found by the model instantaneously, out of an infinite number of possible solutions, which is hardly possible in conventional design practice. This obviates the need for any chart or trial-and-error solution, as adopted in the conventional engineering design practice.

Moreover, it should be noted that both in conventional method and in optimization method, the design steel area obtained by a designer is a *continuous decimal value* and not integer value, whereas the number of reinforcement bars (with associated reinforcement area) and reinforcement layers to be chosen by a designer (in both cases) are *integer values*. Thus, the problem of matching these two values is equally applicable both in conventional method and in optimal design method. However, this problem can be easily resolved in optimization method using the C ++ program cbm.cpp as explained above.

Again, in the computer-aided optimal design method proposed above, the cost of various items of works such as concrete, steel reinforcement can be derived instantly by a designer using the C++ program costeng.cpp presented in Chap. 4, and adopting the site-specific cost of labour and materials, making it more realistic cost assessment. This is rarely done in conventional design method.

In view of above, it is obvious that the optimization method proposed is not merely a theoretical exercise but is sufficiently applicable in structural engineering practice. Therefore, a designer interested in achieving *efficiency* should have no inhibition in applying optimization methods in engineering practice, taking advantage of the tremendous development in modern information technology revolution, and computer applications in all sectors of society discussed in Chap. 1, and the AEC sector cannot remain isolated from this process. Moreover, it is to be noted that, if such optimization procedure is adopted in design of all structural elements in various projects, the total savings and other benefits will be huge. These factors obviate the urgent need to apply such ***optimization*** techniques in all construction projects in the AEC sector.

8.13   Closure

As stated above, the combination of this high power computation, graphics, and algorithms is giving the designers limitless opportunities to use optimization to achieve resource-efficient and cost-effective structural design with better safety. Integration of computer-aided structural design with graphics makes it more efficient and effective. Therefore, it is imperative that AEC professionals exploit these potential benefits by more widespread implementation of optimization within the architecture-civil-structural engineering industry.

Repeatable Reinforced concrete (RC) sections like beams, columns are widely used in AEC facilities involving considerable costs and consumption of materials. But, conventional design methods of RCC and steel structures, generally, do not incorporate any cost function for such structural members. To promote integrated CAD by optimization in this area, the author has developed a number of such *optimizing models* published elsewhere.

In this chapter, structural engineering and the concept of structural design optimization are discussed, optimization types are outlined along with their relevance to optimum design of reinforced concrete structures. Size and shape optimization of reinforced concrete beams are discussed, and seven optimization models for optimal design of reinforced concrete rectangular beams, continuous beams, T-beams, and columns are presented. It is possible to apply the concept of integrated computer-aided design by optimization in structural engineering by integration of numerical optimal design and graphics to attain more productivity, i.e. efficiency and effectiveness in AEC operations. It is rather impossible to solve manually the nonlinear programming optimization models presented in Sects. 8.5.1 and 8.5.2. To solve such models for optimal design of reinforced concrete beams, a C++ program named as: **cbm.cpp** is developed, which is the *numerical part* of the Integrated Computer-Aided Design Drafting (CADD) Software: **RCCAD**, to carry out numerical analysis including *optimization*, and also drafting-drawing. This is demonstrated by taking optimal design of reinforced concrete beams as examples, where optimal design in numerical terms is derived by a C++ program using geometric programming, which is instantly converted to a graphical design drawing by AutoLISP program.

To help readers of this book to use such techniques to achieve their own problem solutions, complete listings of the respective C++ programs and AutoLISP programs are included in an online space connected to this book.

Chapter 9
Integrated Computer-Aided Buildings and Housing Design by Optimization

9.1 Introduction

As stressed in earlier chapters, planning and design is a *problem-solving process* involving *search* through *numerous* potential solutions to select a particular solution that meets a specified criteria or *goal*, making computer use indispensable. Moreover, building and housing operations consume large *Resources* involving many variables giving numerous solutions with large variations in *resource-efficiency* indicators, making *search* of optimal solution very complex. Again, AEC professionals (e.g. architects, civil, and building engineers) commit huge *Resources* (e.g. money, materials, land, and so on) while taking various design-decisions in AEC operations, enjoining on them an *ethical responsibility* to the *society* to achieve *efficiency*, *effectiveness,* and *social equity* in the use of *resources* (which are *scarce-quantities* with many claimants). *Operations research* (and *optimization*) and *Computer-Aided Design (CAD)* are essential *integrative techniques* to improve *efficiency* and *effectiveness* (Koontz & Weihrich, 1990). This is equally applicable in urban planning and management as discussed in Chap. 2. Hence, it is vital that designers apply such techniques and achieve *optimization* to discharge their above *responsibility* to the society and people, for a *resource-efficient* and *equitable* solution of our mounting housing and urban problems. As cited earlier, even the ancient's applied *optimization*, which is rarely applied in our building, housing, and urban operations creating many *inefficiencies* and *inequities*. With increase in population and dwindling *resources* applying such techniques in all operations has become more *crucial* (facilitated by modern computers), as we *live in a finite earth with finite resources*.

To promote application of CAD by *optimization* in housing and urban development, a number of models are developed many of which are published elsewhere (Chakrabarty 1986a, 1986b, 1987a, 1987b, 1991, 1996, 1998, 2007). Because of limitations of space, the above models do not incorporate the detailed methodologies for developing cost functions, cost coefficients, constraint functions, constraint-coefficients, and the procedure for appropriate model calibration with reference to

B. K. Chakrabarty, *Integrated CAD by Optimization*,
https://doi.org/10.1007/978-3-030-99306-1_9

the design-specific and the site-specific data. Because of the same reason, the supporting computer programs to solve the *nonlinear programming optimization* models were also omitted. However, such details are essential if AEC professionals are to be fully conversant with such techniques, enabling them to derive maximum benefit. Hence, in this chapter, the above details are presented elaborating the concept of integrated computer-aided design by optimization for buildings and housing (taking small multi-family dwelling development as examples), the module concept, cost function, typology of optimization models, and the need for application software to solve optimization models. To illustrate the detailed model development process for the benefit AEC professionals and other users, two representative *universal optimization models* (which can be related to site-specific conditions by appropriate calibration), namely: (1) Least-Cost Design Model and (2) Most-Benefit Design Model are presented below.

It is possible to apply the concept of integrated computer-aided design by optimization in building and housing development system by integration of numerical optimal design and graphics to attain more productivity, i.e. *efficiency* and *effectiveness* in AEC operations. This is demonstrated by taking optimal design of building and housing development system as examples, where optimal design in numerical terms is derived by a C++ program using geometric programming (a nonlinear programming technique), which is instantly converted to a graphical design drawing by means of LISP program.

To enable any AEC professional to develop such skills in his/her own situation, the application software: '**HUDCAD**', integrating building design optimization and drafting along with complete computer code of **C++** and **Lisp programmes** (unlike the usual black-box software), is presented. The application results show that there can be very large variations in *efficiency indicators* depending on the planning and design-decisions by AEC professionals, which are rarely considered in an integrated manner in conventional planning and design practice in building and housing, creating many *inefficiencies* and social *inequities* and consequent urban problems which needs to be corrected by applying such planning, design, and management techniques. These concepts and methodologies can also be applied in all types of buildings (e.g. schools, offices, hospitals, and so on), if *optimization* and *resource-efficiency* are the goals.

9.2   Building and Housing Design as a Problem-Solving Process: Useful Role of Computers

As discussed in earlier chapters, building and housing design can also be viewed as a special kind of *problem-solving process* involving *search* through alternative solutions to discover a solution that meets certain specified criteria or goals. Although this view has its limitations, it is quite relevant in many countries to achieve the national *goals* of providing adequate shelter and urban services for all, within the affordability and resource-constraints at the individual, local, and national level. Accordingly, planning and design of building, housing, and urban services needs to be viewed as a *problem-solving process* involving *search* through *numerous* potential solutions (because of numerous design variables) to select a particular solution that meets a specified criteria or *goal* (say, affordable housing design). This makes computer use indispensable for problem solution in building and housing and in fact, in architecture, planning, and in all branches of civil engineering, i.e. AEC sector as a whole, particularly in the context of *search* through *numerous* potential solutions with large variations in *resource-efficiency* indicators. This view for planning and design in civil engineering/urban development is quite relevant especially in developing countries to achieve the national *goals* of providing adequate shelter and urban services for all, within the affordability and *Resource-Constraints* at the individual, local, and national level. With increasing population and dwindling per capita *Resources*, this view is equally applicable even in the developed countries, as *we live in a Finite Earth with Finite Resources*.

If the goal is to obtain something that is as yet non-existent, such as the design for a building then the question arise how potential solutions may be generated, which is answered by constructing a *generative system* (traced to Aristotle) to produce a variety of potential solutions (Mitchell 1977). There can be different categories of generative systems to represent potential solutions to a design problem. Symbolic generative systems represent potential solutions by means of symbols such as mathematical equations and design variables which can take different values to represent a particular design proposal, and a computer program (as presented later) to model a design can constitute a generative system. Data structures describing a practical building design might consist of many hundreds and thousands of variables, making computer use indispensable to store, retrieve and operate on such data with great speed.

Thus, a design problem can be characterized in terms of *data structures, constraints, objectives,* and a *solution generation procedure* to produce potential solutions for consideration, and a computer can be used in the design process to *store and retrieve data* describing a design, to *test potential solutions* for membership of the goal set, and to *automatically generate solutions* to well defined problems (Mitchell 1977). By storing data describing a design, a computer performs as a data processing machine to retrieve information *efficiently* and to generate alphanumeric or graphic displays with great speed replacing and augmenting conventional manual drawings/ drafting, specifications writing, and so on, improving **productivity**. Since, building and urban operations (e.g. housing development) consume large *Resources* involving many variables giving numerous solutions with large variations in *resource-efficiency* indicators, making *search* of optimal solution very complex, it is desirable to use computers to automatically generate optimal solutions adopting the above procedure. Mitchell pointed out that the ability to *integrate* sophisticated analysis and *optimization* techniques into the *design process* would make it possible to maintain much more *effective cost control* and response to the complex design constraints exploring more alternatives, making people to demand them, and agencies not able to provide them would find it difficult to compete (Mitchell 1977).

Towards developing such capabilities, the author has developed a number of *optimizing models* some of which are published elsewhere as cited above and a few of these are also integrated into a computer-aided design system software presented later. The concept of operations research and *Optimization* and the utility of Computer-Aided Design and Drafting (CADD) to apply such techniques to achieve *efficiency* and *social equity* in the building, housing, and urban operations are discussed below.

9.3   Optimization and CAD in AEC Operations: Building, Housing, and Urban Development Systems

Optimization is a crucial requirement in AEC sector, as already stated in Chap. 1; since resources are always inadequate compared to the need, and AEC professional commits large resources in their planning and design-decisions making it obligatory on them to attain *efficiency* in resource-use in such decisions. Building and housing are important components in AEC operations, and therefore, design *optimization* of such components should be of special interest to AEC professionals to achieve *efficiency* in resource-use and cost reduction in such components. This is particularly because, no rational decision is really complete without optimization as the designs of such components involve many variables giving numerous solutions with large variations in resource-efficiency indicators making it imperative to apply *optimization techniques* to search *optimal* building and housing design solution, and Computer-Aided Design (CAD) can facilitate this process.

Essential characteristics of Operations Research (OR) and *Optimization* are their emphasis on *models* indicating physical depiction of a problem and describing the relationships between *goals*, *variables*, and *constraints* in *mathematical terms*. Development of *small-scale* models to address a particular policy problem is often suggested. This is very relevant in urban management, involving building and urban development, requiring both 'descriptive' and 'optimizing' models. The author is working in the field of urban development modeling for planning analysis and design optimization and developed earlier a number of descriptive models for shelter design and analysis mostly for single-family dwelling, where analysis of such single-family dwelling unit design module structure with different wall-sharing options is also presented and not repeated here(HUDCO 1982). Using knowledge given by descriptive models cited above, a number of design analysis and *Optimizing* models for housing and urban development systems are also developed, covering a wide domain of problems. The relationships between many design variables connected to cost are *nonlinear*, time dependent, and physical design related. Hence, it is desirable to develop *small-scale nonlinear programming* urban development *models*, linked with physical design and adopting a shorter planning horizon, to permit *optimal design* with more realistic analysis of costs. Accordingly, a number of such optimizing models for housing and urban development systems are developed and published elsewhere, covering different areas of problems in built-environment. These models give stress on efficient and effective use of resources at urban development project level, to enhance both *Efficiency* and *Effectiveness*, and also social *Responsiveness* to help solve urban problems *Resource-Efficiently*, improving the performance of urban organizations in this regard including profitability.

9.4   Optimization of Building and Housing Development Systems

Optimization of building-housing including dwelling development systems can be at different system levels. In view of the desirability to develop *small-scale nonlinear programming* urban development *models* as stated above, this approach is adopted here for development of models for optimization of building-housing including dwelling development systems. There are many components and steps in the development of such optimization models as discussed below. The first step in this approach is the module concept in building and housing development system design as outlined below.

9.4.1   Module Concept in Building and Housing Development System Design

The module concept in layout planning is outlined in Chap. 10. Such concepts are equally applicable in building and housing development system design. For example, in a single-family dwelling development for low-income housing, a designer generally adopts some repeatable basic units of designs with different wall sharing options for cost reduction. These can be called unit design modules for dwelling designs forming the dwelling development system. Dimensionless schematic diagrams of a set of 'unit design modules' for a single-family dwelling development are shown in Fig. 9.1. As in case of single-family dwelling development, a designer generally adopts some repeatable basic units of designs or dwelling 'unit design modules' with different combination or typology of grouping patterns in a multi-family dwelling development system also. Dimensionless schematic diagrams of a set of unit design modules for a multi-family dwelling development are shown in Fig. 9.2 below. Such module concept can also be applied in planning and design of public buildings like schools, hospitals, and also the commercial buildings like shopping centre and so on. The above schematic diagrams of unit design modules are presented only for the purpose of illustration. A creative designer generally develops and adopts a variety of such unit design modules in building and housing projects depending on the site-specific situation.

In many cases, building and housing development takes place in phases in accordance with the basic 'structure plan', in response incremental effective demand from time to time, as also discussed in Chap. 10. Design and development of building and housing at the level of 'unit design modules' is generally within the immediate sphere of influence of the planner-designer, in response to the immediate or existing effective demand. Depending on the phase of development of building and housing, the adjacency conditions of such 'unit design modules' may be completely flexible or somewhat constrained. Both conditions permit optimization of such 'unit design modules', but, the benefit of optimization is generally more if the adjacency conditions of such 'unit design modules' are more flexible. Hence, it is desirable to strive for such flexibility to derive maximum benefit from optimization. We can consider the dimensionless schematic building or dwelling unit design modules as shown in Figs. 9.1 and 9.2 below, along with their layout and the corresponding streets, utility networks and other services, as the building or dwelling-layout development systems. Thereafter, the building or dwelling-layout development systems can be optimized to achieve the desired objective such as minimizing the costs fulfilling the given constraints such as the desired built-space, in a site-specific situation as presented in Chap. 12. Sometimes, building or dwelling unit design modules, for single-family or multi-family may be located in a parcel of developed land or in a serviced lot where streets, utility networks, and other services are already available and included in the cost of this developed land or lot, which is part of the building or dwelling development system to be optimized. Thus, in this case, only the unit cost of prepared land can be included in the optimization model and the building or housing development system can be optimized accordingly.

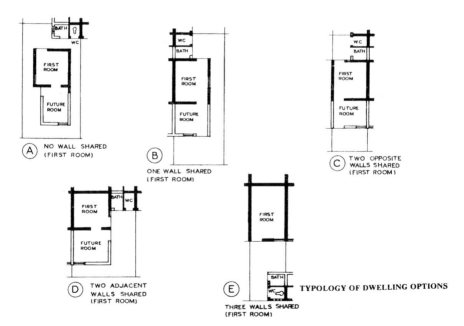

Fig. 9.1 Schematic diagrams of a set of single-family dwelling unit design modules

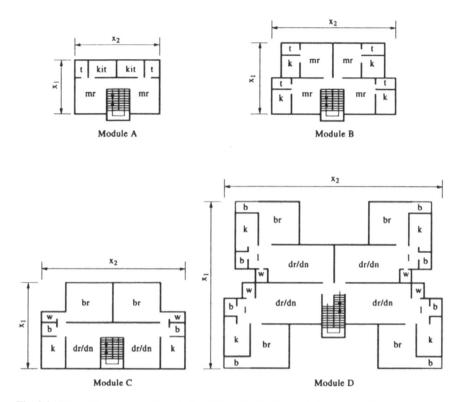

Fig. 9.2 Schematic diagrams of a set of multi-family dwelling unit design modules

To help application of integrated CAD by optimization in housing and urban development system, the Application Software: 'HUDCAD' is developed, and one of its modules dedicated to optimization of building or housing development located in a parcel of developed land where streets, utility networks, and other services are already available and included in the cost of this developed land. Application of this 'HudCAD' software module in the optimization of housing development designs as above is demonstrated later. Essential characteristics of Operations Research (OR) and *Optimization* are their emphasis on *models* indicating physical depiction of a problem and describing the relationships between *goals*, *variables*, and *constraints* in *mathematical terms*. Development of *small-scale* models to address a particular policy problem is often suggested which is adopted in this chapter.

9.4.2   Optimization of a Single-Family Dwelling Shell Without Adopting Module Concept

Even in case of a single-family dwelling located in a service lot of fixed size, optimization can be applied without adopting the module concept. Such an optimized single-family dwelling shell is shown in Fig. 9.3, below. Expressions for wall

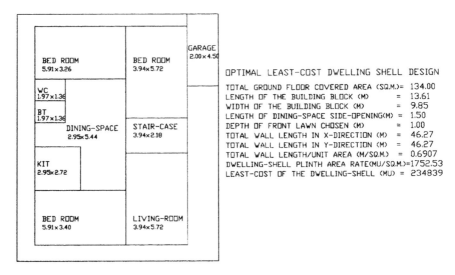

Fig. 9.3 Optimal design of a single-family dwelling shell without adopting module concept

lengths can be easily developed using the width and depth of a lot as specified, along with the prescribed lot coverage and other parameters as prescribed in building regulations. Using such parameters, we will get an equation for wall length in terms of only one variable and the minimum wall length can be derived adopting differential calculus in terms of this single variable.

It is interesting to note that the wall lengths in x-direction is = 46.27 m and in y-direction = 46.27 m as shown in Fig. 3, and are equal in both direction to give the minimum length of walls and hence, the **Least-Cost** of the built-space. Thus, a planner-designer can adopt this simple thump rule to attain optimization in such cases.

9.4.3 Relevance of Cost Engineering in Optimization

Cost engineering is an important constituent in the development of optimization models. There are many components of cost engineering such as cost indices, elemental estimating, building specification; elemental unit cost, and economic analyses of projects, cost control, optimization, management aspects, and a wide range of other topics. Computerized information systems and cost estimation are very important elements in Computer-Aided Design (CAD) by optimization.

Cost is a major consideration in all human activities and is of paramount importance in all AEC activities including building and housing development where many disciplines have to interact together in a project, making cost as the common language for such interactions specifically in the computer-aided design process. Cost engineering involves using engineering judgement and experience in the application of scientific principles and techniques to problems of cost estimating, cost control, planning, and scheduling including productivity analysis, optimization, and operations research. As stated earlier, there are unparalleled transformations in computers in terms of their capacity to capture data, to analyse, manipulate, and store them, to disseminate and communicate information, and to apply these capabilities in different areas and fields of specialization including CAD. Cost engineering using such techniques can greatly contribute in solving various societal problems such as housing and urban development in an efficient and effective manner.

9.4.3.1 Cost Estimating

Cost estimating is a process to predict the final outcome of a future expenditure, taking into account the parameters and conditions concerning a project which are known when the cost estimate is prepared. In computer-aided design in the AEC sector cost is a vital parameter and in fact the optimal design and analysis presented in this book is based mainly on cost considerations. Therefore, it is imperative to incorporate suitable methods and techniques of cost estimating so that the costs predicted are realistic as per the site-specific conditions. The unit rate technique is one of the most widely used estimating techniques both in preliminary and detailed estimate (Jelen and Black 1983). Preliminary estimate of building cost can be done by multiplying the area (say square metre) or volume (cubic metre) of a building by the unit cost data based on average unit costs for major cities in a country. However, variations in labour and material costs and other local factors will cause variations in such unit costs and consequently in preliminary estimates. Using the unit costs of building components such as walls, floors, and foundations may give more accurate cost estimates. This approach may incorporate both bare unit cost and the complete unit cost. Labour wage costs per day, material cost per unit of weight or volume are bare unit costs. Installed costs of specific diameter pipeline per unit of length, completed cost of specified quality of concrete per unit volume are the complete unit costs.

In the unit rate technique, two basic factors which are to be considered are labour and material consumption and their unit prices leading to the estimation of the total cost in a project. The input data required for such cost estimates are:

1. Units of measurement for various items which may be in terms of centimetre, metre, square centimetre, square metre, cubic centimetre, cubic metre, hour, day, kilogram, tonne, and so on.
2. Labour unit quantities of various skills which are consumed to produce a unit output of various items of work such as earthwork, masonry, plain cement concrete, reinforced cement concrete, and so on.
3. Rates per unit of labour of various un-skilled and skilled categories such as mason, plumber, and so on.
4. Material unit quantities of various categories such as cement, steel, bricks, sand, and so on which are consumed to produce a unit output of various items of work

such as concrete, masonry, and so on of various specifications as per standard practice.

5. Rates per unit of material of various categories such as cement, steel, bricks, sand, and so on.
6. Miscellaneous and sundries such as hire and running charges of plants and so on.
7. Specifications of various building components such as masonry, concrete, and so on.

The above input data can be assembled in the form of a matrix to generate a schedule of rates for various items of work. Generally, Public Work Department of various countries collects such data and incorporates them in the form of a Schedule of Rates (SOR) for various items of works in a book form for a particular year. These rates are updated periodically or a percentage is added to the Schedule of Rates (SOR) for a particular year to make it at par with the current market rates. All such activities are generally carried out manually making it very difficult to respond correctly to the market conditions from time to time. In these days of Information Technology Revolution and cheap availability of computers, it is desirable that all such Schedule of Rates (SOR) should be computerized so that these can be constantly updated without any loss of time leading to more efficient and accurate estimating of costs. To help this process, a C++ program on cost engineering, giving stress on the Schedule of Rates, is developed, and included in Chaps. 4 and 6 presented earlier. Salient features of the computer-aided cost estimating included in the software module of 'HudCAD' discussed above are outlined below.

9.4.4 Computer-Aided Cost Estimating

Computer-Aided Estimating is a technique to reduce time and cost of preparing cost estimates and also to increase the accuracy of such estimates, and should be an integral part of a computer-aided design system integrating planning and design by optimization as attempted in this book. There are many examples of computer-aided estimating applications (Jelen and Black 1983). To illustrate such applications and also to help achieve optimal design considering costs, the unit cost estimating method (both for various items of work and for various components of a building) are taken as examples here. In this example, 15 items of work and 16 items of labour and material coefficients are considered in the form of a *Matrix* just for illustration. This can be extended to cover hundreds of items of works as in a standard Schedule of Rates, adopting the same technique presented in this example. Similarly, only building components such as foundation, wall, roof, and floor types are considered to derive component unit costs, which can be extended to cover other components such as pipelines for water supply of various sizes and specification, sewerage networks of various sizes and specifications, roads and pavements of various dimensions and specifications and so on, adopting the same principles and technique presented in this example.

In this software-model example, the following labour and material items are considered:

```
b[0] = Labour:Un-Skilled-Category-1(unskl1) (rate-each MU/day)
b[1] = labour:un-skilled-category-2(unskl2) such as: coolie(rate-each
MU/day)
b[2] = labour:skilled-category-1(skl1) such as:mason-class-1(rate-
each MU/day)
b[6] = material:-portland cement (rate MU per metric tonne)
........................
b[15] = Other3:-Miscellaneous/Sundries/Hire & Running Charges of
Plants etc
```

The input basic-cost engineering (quantities) coefficients (per unit) both for labour and material for each item are usually generated by works and method study and its design specifications. These are generally constants for each item of work irrespective of the time and location of the work, and therefore, can be constant input to the cost engineering model. On the other hand, the rate or price per unit of labour (in terms of hour, day, and so on) and per unit of material (in terms of cubic metre, kilogram, quintal, and so on) may vary with time and location of the work, and hence, may have to be given as fresh input every time a program is run in a site-specific case, if there is variation. Once the inputs are given correctly to the C++ program, named as: COSTENG.CPP available in pen drive, it will instantly generate the Site-Specific Schedule of Rate for various items of work incorporated, and also the building component unit costs which are accurate and realistic with reference to the *site-specific data input*, and consequently, the optimal design generated using such costs will also be realistic as illustrated in later chapters of the book.

The application result of the program along with its *dialogue structure* when it is run is shown below:

```
 WELCOME to SOFTWARE 'HudCAD-Cost Engineering Model'. Using it one can
find UNIT-COSTS of Building Items and Components, given INSTANTLY on
selection
of SITE-SPECIFIC-INPUT-DATA in terms of Labour and Material Costs.Use
of such
Realistic UNIT-COSTS in the Design Process makes the Designs more
Realistic
Solutions in a SITE-SPECIFIC CONTEXT.
```

The above cost engineering program also forms a sub-model, in the optimization models CLC1.CPP (LCO Model) and CMB1.CPP (MBO Model), incorporated in the Module-I of HudCAD discussed and presented later, to derive **instantly** the unit elemental costs to be input to the above optimization models .

Typical dialogue structure of the above optimization models to run the cost engineering sub-model is outlined below.

On opening of the above models (i.e. CLC1.CPP or CMB1.CPP), the software prompts the designer/user with the following option to run cost engineering model to derive site-specific unit costs for (Y/N) response.

```
"Do you want to run COST ENGINEERING MODEL to Derive Unit Costs ? (put 'y'
if yes/ 'n' if no)"
```

On 'y' response Cost Engineering Model is RUN to derive unit elemental costs based on the *site-specific* unit material/labour costs and construction specification, with the following message:

```
"Do you want the individual Basic Results displayed? (put 'y' if yes/'n'
if no)"
```

On 'y' response the individual Basic Results displayed with the following message:

```
"Do you want the individual Item Quantities displayed? (put 'y' if yes/'n'
if no)"
```

On 'y' response the individual Item Quantities displayed with the following message:

```
"You want to see Definitions of Labour & Material Items? (put 'y' if yes/'n'
if no)"
```

On 'y' response the labour and material items considered in the module are displayed followed by a message:

```
"IN THIS SOFTWARE-MODULE, the FOLLOWING LABOUR & MATERIAL ITEMS ARE
CONSIDERED:
  "b[0] = LABOUR:Un-Skilled-Category-1(UNSKL1) such as:BELDAR(RATE-
EACH MU/PER DAY)
  "b[1]=LABOUR:Un-Skilled-Category-2(UNSKL2) such as:COOLIE(RATE-
EACH MU/PER DAY)
  "b[2] = LABOUR:Skilled-Category-1(SKL1) such as:MASON-CLASS-1(RATE-
EACH MU/DAY)\n"
  "b[3] = LABOUR:Skilled-Category-2(SKL2) such as:MASON-CLASS-2(RATE-
EACH MU/DAY)
  "b[4] = LABOUR:Skilled-Category-3(SKL3) such as:MATE(RATE-EACH
MU/PER DAY)
  "b[5] = LABOUR:Skilled-Category-4(SKL4) like BHISTI/OTHER-1(RATE-
EACH MU/DAY)
  "b[6] = MATERIAL:-Portland Cement (RATE MU PER METRIC TONNE)
  "b[7] = MATERIAL:-Fine Sand(F.SAND) (RATE MU PER CUBIC METER)
  "b[8] = MATERIAL:-Course Sand(C.SAND) (RATE MU PER CUBIC METER)
  "b[9] = MATERIAL:-Stone Aggregate 10 mm Nominal Size(RATE MU PER CUBIC
METER)
```

"b[10] = MATERIAL:-Stone Aggregate 20 mm Nominal Size (RATE MU PER CUBIC METER)
"b[9] = MATERIAL:-Stone Aggregate 40 mm Nominal Size (RATE MU PER CUBIC METER)
"b[12] = MATERIAL:-1st Class Bricks (RATE MU PER 1000 Nos)
"b[13] = MATERIAL:-Mild Steel/Torsteel Round Bars (RATE MU PER Kilo Grams (Kg))
"b[14] = OTHER2:-White Lime/OTHER MATERIALS etc. (RATE MU PER QUINTAL)
"b[15] = OTHER3:-MISCELLANEOUS/SUNDRIES/Hire & Running Charges of Plants etc
"MU = 'Monetary-Unit' in any Currency (Rupees, Dollars and so on)

"Want to change ALL Input Cost-Engineering (Qties) Co-Efficients? (put 'y' if yes/'n' if no)"

On 'y' response the engineer/designer is given option to change each of the 240 values (Qties) 16 per row preset in the Model (here, engineer/designer can change and give his/her *country-specific* and *site-specific* 240 values as INPUT to get such results using the same Model), afterwards following message appears for response:

"Want to change ALL Input Labour-Material-Unit-Cost (MKT.RATES) Co-Efficients? (put 'y' if yes/'n' if no)"

On 'y' response the engineer/designer is given option to change each of the 16 values i.e. Labour-Material-Unit-Costs (Market Rates)1 per column replacing the preset in the Model (here, engineer/designer can change and give his/her *country-specific* and *site-specific* 16 values as INPUT to get such results using the same Model), then the following message appears for response:

"Want to change Some Input Cost-Engineering (Qties) Co-Efficients? (put 'y' if yes/'n' if no)"

On 'y' response the engineer/designer is given option to change specific value (s) (Qties), and then the following message appears for response:

"Want to change Some Input Labour-Material-Unit-Cost (MKT.RATES) Co-Efficients? (put 'y' if yes/'n' if no)"

On 'y' response the engineer/designer is given option to change specific value (s) (i.e. specific Labour-Material-Unit-Costs (Market Rates), and then the following message appears for response:

"Do you want to see the Basic Input Data? (put 'y' if yes/'n' if no)"

On 'y' response the Basic Input Data considered in the Module are displayed followed by a message:

"Are the Input Data as above acceptable? (put 'y' if yes/'n' if no)"

On 'n' response the engineer/designer is given, again as above, the option to change specific value(s) (Qties/Labour-Material-Unit-Cost), and when the Input Data acceptable the following messages are displayed:

```
"Computed Basic Item Rates given INSTANTLY by COST ENGG.MODEL are:"

"Want to see Foundtn./Wall/Roof/Floor Qty.Input Data?(put'y'if yes/'n'
if no)"
```

On 'y' response the Foundtn./Wall/Roof/Floor Qty.Input Data considered in the Module are displayed followed by a message:

```
"Are the Input Data as above acceptable? (put 'y' if yes/'n' if no)"
```

On 'n' response the engineer/designer is given the option to change specific value (s), and when the Input Data acceptable the following messages are displayed:

```
"Choose Foundation/Wall/Roof/Floor Types i.e. choose any of: 1/2/3
(i.e. specification name for Each of 4 Components) and Put 4 values in same
order"
```

On response the Unit Costs of the Chosen Foundation/Wall/Roof/Floor Types are displayed and the Cost Engineering Model is ended.

It would be seen from the *dialogue structure* that the program provides for change of values of all or some 'Input Cost-Engineering (Qties) Co-Efficients' and 'Input Labour-Material-Unit-Cost (MKT. RATES) Co-Efficients' as per *country-specific* and *site-specific* requirements in a particular location. For example, 230 mm long brick sizes are adopted here adopting the corresponding number of bricks of 494 per cubic metre of brickwork under item 6 and 7 in the 15×16 *Matrix* of item-labour-material coefficients. If the brick sizes are different, corresponding number of bricks per cubic metre should be calculated and given as input in response to the *dialogue* prompt, replacing the above number of bricks and accordingly changing the *Matrix* while running the program giving correct results. Thus, the *Software* is a *Universal Model* to generate item wise and building component wise unit costs permitting generation of optimal designs in any situation in any country on cost considerations.

A Schedule of Rates (SOR) and also the building component unit costs *instantly* given by the cost engineering model is given below:

Cost is of paramount importance in all AEC activities including housing development, where many disciplines have to interact together in a project, making cost as the common language for such interactions specifically in the computer-aided design process. In this chapter, the imperative need to incorporate suitable methods and techniques of cost estimating, so that the costs predicted are realistic as per the site-specific conditions, is stressed. Computer-aided estimating is a technique to reduce time and cost of preparing cost estimates and also to increase the accuracy of such estimates. Need to make it part of a computer-aided design system integrating planning and design by optimization is stressed and also attempted in this book.

To help achieve optimal design considering costs, as also presented later, the Unit Cost Estimating both for various items of work and for various components of a building is taken as examples here. The Unit Rate Technique (URT) is one of the most widely used estimating techniques both in preliminary and detailed estimate (Jelen & Black, 1983). Therefore, in this book, use of URT is stressed. To facilitate this process, the C++ listing of SOFTWARE 'HudCAD-Cost Engineering Model' is presented in this chapter. Using it, one can find unit-costs of building items and components, given INSTANTLY on selection of site-specific-input-data in terms of labour and material costs. Use of such realistic unit-costs in the design process makes the designs more realistic solutions in a site-specific context.

9.5   Creating Models for Building and Housing Development System Design Optimization

As earlier stated, a design problem can be characterized in terms of *data structures, constraints, objectives,* and a *solution generation procedure* to produce potential solutions for consideration, and selecting the best solution. This is best achieved by developing models and computers can be used in the design process to solve such optimization models and also *store and retrieve data* describing a design. These models need to incorporate various components and steps as outlined above, and also, some basic parameters such as design module coefficients, cost coefficients, cost functions, and constraints depending on the specific optimal design problem to be solved, as discussed below.

After the unit design modules for dwelling designs forming the dwelling development system are decided, the first step in developing models to obtain the optimal design solution, i.e. least-cost or most-benefit design of building or housing development system, is the derivation of unit design module coefficients for different unit dwelling design modules, conceived by a planner/designer keeping in view the different contexts, one of which can be the intended target income group of beneficiaries. Again, it is necessary to formulate the building or dwelling cost coefficients based on the respective unit design module coefficients derived as above, which can be used in the development of cost function for the optimization model for least-cost or most-benefit design of building or housing development system, as presented later. Development of cost coefficients and cost functions for an optimization model is presented later. As an example, we can take the case of building or dwelling unit design modules located in a parcel of developed land. In this case streets, utility networks and other services are already available, and their costs are included in the unit cost of this developed land.

Thus, the optimization models for least-cost or most-benefit design of building or housing development system presented below incorporate the following main cost components:

i. Cost of developed land which is part of the building or dwelling development system to be optimized,

ii. Cost of walls and their foundations related to the dwelling unit design module, and

iii. Cost of floors and roof related to the dwelling unit design module.

Once the dwelling unit design modules are selected keeping in view the different contexts, unit design module coefficients for different unit dwelling design modules need to be derived as outlined below.

9.6 Deriving Unit Design Module Coefficients for Building and Housing Development System Design Optimization

It is well known that resources (money, materials, land, and so on) are always inadequate compared to the need whether in low- or high-income countries. No doubt low-income countries are usually subjected to severe resource-constraints necessitating scaling-down of needs in the context of available resources. But, even in affluent countries such resource-constraints are applicable because of scaling-up of needs to have higher standard of living consuming more resources. Derivation of unit design module coefficients is illustrated below using different dwelling design modules usually applicable in low/middle countries and the high-income countries, which are designated as Context A and B, respectively.

However, in view of the acute resource-constraints and struggle for affordable housing, for example 'dwelling unit design modules' are provided below mostly for the context of low-income countries and for the middle-income countries indicating the relevance of the results with reference to the national, local, and individual contexts in these countries. Nevertheless, one example 'dwelling unit design module', which may be pertinent to the context of high-income countries, is presented below just to show the applicability of the *technique* of *quantitative analysis* and *optimization* in building and housing in these countries to achieve more *resource-efficient* planning and design solutions. Thus, it is obvious that this *technique* is applicable in all countries covering various income or development statuses, including high-income countries.

9.6.1 Context A: Low- and Medium-Income Countries

In case of low-income countries, one of the prime considerations is the affordability of the prospective beneficiary, and therefore, the built-unit-area and its cost have to be low to make them affordable to the respective income category. Therefore, if necessary, planning standards may have to be adjusted depending on the priority given on particular aspects of design by the intended beneficiary group, and also keeping in view the safety considerations which cannot be compromised. We can take the multi-family dwelling unit design Module-D shown in Fig. 9.2 above as the example, to illustrate the process of formulating dwelling coefficients as part of developing the cost function for such building or housing development system as outlined below. We can take the variables x_1 in the direction of staircase and x_2 in the direction perpendicular to the staircase, as the basic variables and express all other variables in terms of these basic variables as their proportions. Thus, width and length of kitchen can be expressed as $k_2 x_2$ and $k_1 x_1$, and so on. A schematic diagram of the dwelling unit design Module-D indicating the proportion coefficients with respect to x_1 and x_2 for various rooms and other components of the unit design module is shown in Fig. 9.4 above.

Figure 9.4(a) and (b) above show two alternative schematic diagrams of dwelling unit design Module-D presenting proportion coefficients related to the x_1 and x_2 dimensions of the design module as shown in the above figures. Using these proportion coefficients, we can express the total length of walls (including their foundation) of the dwelling unit design module in the direction of x_1 in terms x_1, and in the direction of x_2 in terms x_2. Similarly, using these proportion coefficients, we can express the total built-up area of the dwelling unit design module in terms of the two design variables x_1 and x_2 dimensions of the design module. We can also designate cx_1 and cx_2 as the wall-length coefficients of the dwelling unit design module in the direction of x_1 (i.e. along the direction of staircase) and in the direction of x_2 (i.e. perpendicular to the direction of staircase), respectively. Similarly, the built-area coefficients of the dwelling unit design module can be designated as: *cba*.

Thus, there will be three dwelling unit design module coefficients designated as cx_1, cx_2, and *cba* giving the total length of walls (including their foundation) of the dwelling unit design module in the direction of x_1, the total length of walls (including their foundation) of the dwelling unit design module in the direction of x_2, and the total built-up area of the dwelling unit design module, respectively. Obviously, all these dwelling coefficients will be expressed in terms of only these two design variables x_1 and x_2 dimensions of the design module. Thus, the total cost of the dwelling structure including foundations could also be expressed only in terms of the above two variables only, using the unit costs of foundations, walls, floors, and roof derived on the basis of local market rates of labour and materials as shown in Chap. 3. Example development of dwelling coefficients, cost coefficients, and the expression for cost function for the optimization model, using the above derived dwelling unit design module coefficients is illustrated below.

The derivation of the unit dwelling design module coefficients cx_1, cx_2, and *cba* for different dwelling design modules shown in Fig. 9.2, and Fig. 9.4 is illustrated below using the various proportion coefficients, shown in above figures and in the diagrams below for respective unit dwelling design module. Proportion coefficients can be assigned values by the planners and designers as per experience, preferences, and the feasibility. The planners and designers can change the input chosen values if some values chosen do not give feasible and desirable output results in a cycle of iterations as per the *general schematic* optimum housing and urban development design procedure together with calibration of different optimization models, including those for optimal layout planning and design, optimal building and housing design and so on, as shown in Fig. 10.7 in Chap. 10.

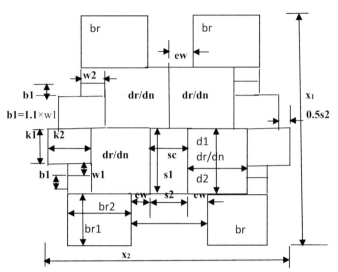

(a) Dwelling Unit Design Module Showing Various Proportion Coefficients: Alternative-I

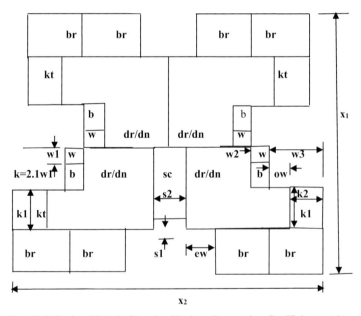

(b) Dwelling Unit Design Module Showing Various Proportion Coefficients: Alternative-II

Fig. 9.4 Schematic diagrams of dwelling unit design module-d showing proportion coefficients. (**a**) Dwelling unit design module showing various proportion coefficients: alternative-I. (**b**) Dwelling ing unit design module showing various proportion coefficients: alternative-II

9.6.1.1   Deriving Unit Design Module Coefficients for Dwelling Module-D

(a) **Alternative-II—Figure 9.4(b)**

Referring to the Fig. 9.4(b), we can easily derive the expressions for cx_1, cx_2, and cba as follows:

 i. Expression for cx_1
    if $(s1 \geq 0)$
    $cx_1 = 5.5 + 2 \times s1 + k1$;
    if $(s1 < 0)$
    $cx_1 = 5.5 + k1$
 ii. Expression for cx_2
    if (n i.e. number of rooms $=3$)
    $cx_2 = 5 + 8 \times w2 + 2 \times w3 - 3 \times s2 - 4 \times ew$;
    Otherwise
    $cx_2 = 5 + 4 \times w2 + 2 \times w3 - 3 \times s2 - 4 \times ew$;
 iii. Expression for cba
    $cba = 1 - s2 \times (0.5 + k1 - s1) - 4 \times ew \times k1 - 2 \times w3 \times k + 4 \times w3 \times k1 \times k$

(b) **Alternative-I—Figure 9.4(a)**

Referring to the Fig. 9.4(a), the expressions for cx_1, cx_2, and *cba* can be derived as follows:

$cx_1 = 4 + 6.3 \times w1 + 3 \times k1$ or $cx1 = 4 + 6.3 \times w1 + 2.5 \times k1$ (if one $k1$ wall is shared)

$cx_2 = 5 + 12 \times w2 - 3 \times s2 - 4 \times k2 - 4 \times ew$

$cba = 1 - (s2 + 4ew)(0.5 - k1 - 2.1w1) - 4(k2 - w2)(0.5 - k1) - 0.5s2$

Similarly, the diagrams of other dwelling unit design modules showing various proportion coefficients and the expressions for the dwelling unit design module coefficients: cx_1, cx_2, and cba for these dwelling unit design modules are presented below.

9.6.1.2   Deriving Design Module Coefficients for Dwelling Module: C—Figure 9.2

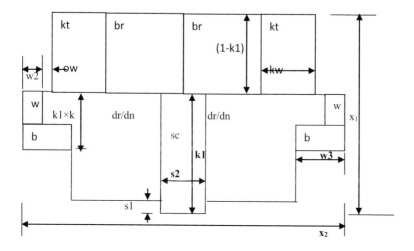

Schematic diagram of dwelling unit design **module: C** showing various proportion coefficients is shown above with slight variations in arrangement of rooms shown in Fig. 9.2, but the basic character of the module, i.e. two dwelling unit design modules served by one staircase remains unchanged from dwelling module: C shown in Fig. 9.2. The expressions for the dwelling unit design module coefficients: cx_1, cx_2, and cba for the above dwelling unit design module could be developed as follows:

$$cba = 1 - 2w3(k1 - k \times k1) - 2(ow + w2)(1 - k1) + s1 \times s2$$

$$cx_2 = 3 + 2 \times (w3 - w2 - ow)$$

$$cx_1 = 5 + 2 \times s1 + k1 \times (2 \times k - 1) \text{ (if } s1 > 0); \text{ otherwise } cx_1 = 5 + k1 \times (2 \times k - 1);$$

9.6.1.3 Deriving Design Module Coefficients for Dwelling Module: B—Figure 9.2

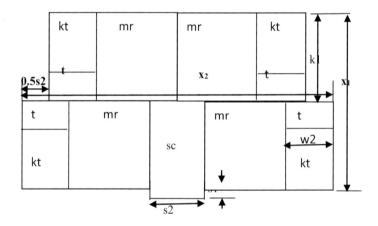

Schematic diagram of dwelling unit design **Module: B** showing various proportion coefficients is shown above with slight variations in arrangement of rooms but the basic character of the module, i.e. four dwelling unit design modules served by one staircase remains unchanged from dwelling module: B shown in Fig. 9.2. The expressions for the dwelling unit design module coefficients: cx_1, cx_2, and cba for the above dwelling unit design module could be developed as follows:

$$cx_1 = 6 + 2 \times s1 - k1 \text{ (if } s1 > 0), \text{ otherwise } cx_1 = 6 - k1; ew = 0; ow = 0; k = 0;$$

$$cx_2 = 3 + 4 \times w2 - s2; \quad cba = 1 - s2 \times k1 + s1 \times s2 \text{ (if } s1 > 0);$$

9.6.1.4   Deriving Design Module Coefficients for Dwelling Module: A—Figure 9.2

Schematic diagram of dwelling unit design **Module: A** showing various proportion coefficients is shown below with slight variations in arrangement of rooms; but the basic character of the module, i.e. two dwelling unit design modules served by one staircase remains unchanged from dwelling Module: A shown in Fig. 9.2.

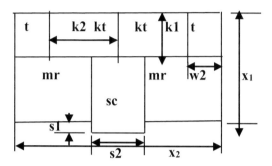

The expressions for the dwelling unit design module coefficients: cx_1, cx_2, and cba for the above dwelling unit design module could be developed as follows:

$$cba = 1 + s1 \times s2;$$
$$cx_2 = 3 + 2 \times w2;$$
$$cx_1 = 3 + 2 \times k1 \ \ (\text{if} \ \ s2 = 0);$$
$$cx_1 = 4 + k1 + 2 \times s1 \ (\text{if} \ \ s1 \geq 0);$$

9.6.2   Context-High-Income Countries

Schematic diagrams of unit design modules presented above are mainly to facilitate *optimal resource-efficient* design solutions in the context of severe *resource-constraints* and *affordability* struggle specifically in the low-income countries.

However, such *optimal resource-efficient* design solutions are equally applicable in high-incomes countries as no rational decision is really complete without *optimization* and no *stakeholder* would like to have an *inefficient* design solution, although the context in basic *value scale* may be different in these countries. Moreover, there is a complex interaction between the land price, utility network cost, service connection cost, and layout physical parameters; and the building and housing cost value is determined by the combined effect of all these factors, indicating the *efficiency* achieved in the use of scarce resources committed by AEC professionals in their building and housing design in all countries.

It is necessary to examine the complex interaction between the land price, utility network cost, service connection cost, layout and building physical parameters (say building rise, planning regulations, and so on), at a building and housing system level, considering all involved disciplines (architecture, engineering, planning, and so on) to derive the *optimal* building and housing design. As mentioned above in low-income countries, the built-unit-area and other planning standards are generally low to fit with the very low affordability of prospective beneficiaries of housing and urban development projects in these countries. But in case of high-income countries, the built-unit-area and other planning standards can be high to fit with the high affordability of prospective beneficiaries in these countries. But, even in this case there can be large variation in costs, land utilization ratios, optimal lot/built-unit densities, and so on, although, the scale in terms of absolute value may be different as shown in examples below. Here, Monetary Units (MU) such as USDs are assumed with representative unit costs, but a planner-designer can provide as input the unit costs as per the local situation and get the output results accordingly. Development of dwelling unit design module for high-income housing and deriving design module coefficients for such dwelling module is outlined below.

9.6.2.1 A Schematic Dwelling Unit Design Module: E for High-Income Housing

A schematic dwelling unit design module for high-income housing, i.e. dwelling unit design module: E is presented in Fig. 9.5 only for the purpose of illustration, to show how the concept of *quantitative analysis* and *optimization* can be applied even in high-income housing to derive **efficiency** and other benefits. A creative designer generally develops and adopts a variety of unit design modules in building and housing projects depending on the site-specific situation, both for single-family and multi-family dwelling development system. The above figure shows the designated proportion coefficients and using these values the design module coefficients for the dwelling Module: E can be developed as shown below.

9.6.2.2 Deriving Design Module Coefficients for the Dwelling Module: E

The expressions for the dwelling unit design module coefficients: cx_1, cx_2, and cba, for the multi-family dwelling unit design module for high-income housing shown in Fig. 9.5 above could be developed as follows, using the proportion coefficients indicated in the diagram:

$$k1 = (1 - 2(ew + b1 + t1))/2, \quad k2 = (k - 2 \times t2)/2;$$
$$w2 = 0.5 \times (k - ow), \quad s2 = (1 - 2 \times k);$$

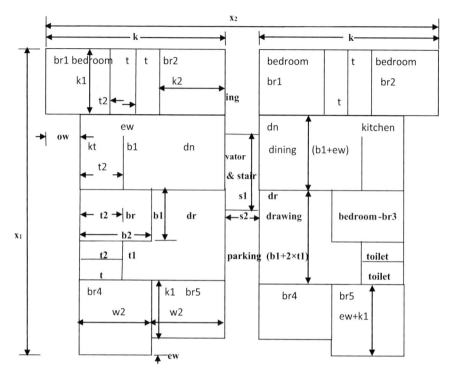

Fig. 9.5 Schematic unit design module: e for high-income housing showing proportion coefficients

$$cba = 1 - 2 \times ow \times (1 - k1) - s2 \times 1 - ew \times w2 \times 2 + s1 \times s2;$$

$$cx1 = 4 + 8 \times k1 + 4 \times b1 + 4 \times t1;$$

$$cx2 = 2 + 8 \times (k - ow) + 2 \times (b2 + t2).$$

In the above expressions, the proportion coefficients k, $b1$, $b2$, $t1$, $t2$, ow, ew, and $s1$ are independent design variables, and a designer can assign a value to each of them based on his/her experience, desire, and feasibility. In fact, the suitability and feasibility of such values chosen can be verified in each cycle of iteration as per the optimum housing and urban development design procedure shown in Fig. 10.7 in Chap. 10. Representative input and output results of *optimization* given by the **C++ program: clc1.cpp** and **cmb1.cpp** using the above expressions for the dwelling design module coefficients cx_1, cx_2, and cba for dwelling design Module: E are shown and discussed later in Sect. 9.10.1.2 giving some applications example results.

9.7   Developing Cost Coefficients and Cost Functions in Building and Housing Development Systems Optimization

A number of techniques are available for efficient structural designs of walls and floors of dwellings and also optimal design of reinforced concrete elements outlined in Chap. 9, thereby improving *efficiency* and reducing costs (Chakrabarty, 1977a, 1977b, 1987a, 1987b, 1992a, 1992b; Sahlin, 1971). Cost coefficients and cost functions in building and housing development systems need to consider all these aspects. In residential structures, the thickness of walls and floors will be usually constant, although, for larger buildings with larger spans these may be more (and accordingly, unit costs of such elements may be more) compared to those for small and medium size dwellings, where the span of floors and laterally unsupported lengths of load bearing walls may be smaller. Similarly, the width of street pavements and the size of utility networks (e.g. water supply, sewerage lines) are generally constant for each branch of networks, at the level of modules mentioned above. The unit costs of the aforesaid elements could be derived considering the above factors (HUDCO, 1982). Planning and design analysis models linking the physical parameters of the layout unit design modules, and dwelling unit design modules, with the quantitative efficiency indicators including costs have been developed (HUDCO, 1982). These models incorporate most basic types of land use and physical parameters generally encountered in a residential layout such as private plottable land or saleable land, circulation space covered by streets and walkways, public or semi-public open space, and density of development. The above models can be used only for planning and design analysis of numerous possible design solutions to get an insight about the interaction between various design variables, and cannot be used to obtain optimal solution, i.e. least-cost or most-benefit design solution, which require mathematical programming approach as discussed earlier.

Cost coefficients and cost functions in building and housing development systems can be derived using the building component unit costs, which again, depend on the site-specific-input-data in terms of labour and material costs. A cost engineering model is presented in Chap. 4 which can generate item wise and building component wise unit costs. The model provides for change of values of all or some 'Input Cost-Engineering Co-Efficient' and also 'Input Labour-Material-Unit-Cost (MKT. RATES) Co-Efficient' as per the site-specific requirement in a particular location in a particular country. Thus, it permits use of realistic unit costs in the design process generating the optimal designs more realistic in any country. An example of Schedule of Rates (SOR) and also the building component unit costs instantly given by the cost engineering model is also presented in the above chapter. To illustrate the process of developing **Cost Coefficients**, four building components are considered, namely: foundation, wall, roof, and flooring. The unit costs of these building components are designated as: '*cf*' in MU (Monetary Units) per metre, '*cw*' in

MU per metre, '*cr*' in MU per square metre, and '*cfl*' in MU per square metre; for the components: foundation, wall, roof, and floor, respectively.

It may be noted that the parameter '*cw*' includes joinery items such as doors and windows which are not shown separately in various **Built-Unit-Design Patterns and Schematic Diagrams of different Dwelling Unit Design Modules** shown in this chapter. To simplify, it is assumed that the unit cost of joinery items and wall material is same, and accordingly, the value of '*cw*' is derived adopting the procedure presented in Chap. 4, and incorporated in cost functions and *optimization models*. However, considering the desired ratio of joinery area and room area and the unit cost of wall and joinery separately, as presented in reference HUDCO, 1982, one can develop more refined cost function and optimization models. Interested readers can consider developing such cost function and optimization models. A factor designated as: '*fic*' is included to allow for the increase in unit costs due to building rise. Suitable value can also be assigned to this factor to incorporate the cost of elevators in a building depending on its rise and size. The number of floors in a building is designated as: '*nf*'. Thus, using the above designated terms and the dwelling unit design module coefficients: cx_1, cx_2, and cba developed earlier for various dwelling unit design modules, expressions for the **Cost Coefficients:** c_1 and c_2, representing the total unit costs of walls in the direction of x_1 and x_2, respectively, can be developed as follows:

$$c_1 = cx1 \times ((nf \times (1 + (nf - 1) \times fic)) \times cw + (1 + (nf - 1) \times fic) \times cf); \quad (9.1)$$

$$c_2 = cx2 \times c_1/cx1. \quad (9.2)$$

Similarly, using the dwelling unit design module coefficients: '*cba*' developed earlier for various dwelling unit design modules, the expression for the cost coefficients representing the total unit cost of floors and roof designated as: c_3, can be developed as follows:

$$c_3 = (cba \times cr \times cif \times nf + cba \times cfl \times cif \times 1)/nf; \quad (9.3)$$

Thus, the total cost of walls (including foundation) in the direction of x_1 will be $= c_1 x_1$

Similarly, the total cost of walls (including foundation) in the direction of x_2 will be $= c_2 x_2$

Again, using the above coefficients, the total cost of roof and floors will be $= c_3 x_1 x_2$

Thus, three terms of the **Cost Function** in building and housing development systems will incorporate $c_1 x_1$, $c_2 x_2$, and $c_3 x_1 x_2$ representing the total cost of walls in the direction of x_1, the total cost of walls in the direction of x_2, and the total cost of floors and roof, respectively. The relationship between Built-Unit-Space (BUS), Building Rise (BR), Semi-Public Open-Space (SPO), and Floor Area Ratio (FAR) is shown in Chap. 7. Using this relationship, the allocated land area per Built-Unit

(BU) can be calculated, wherefrom the total cost of developed land per BU can be found as follows:

$$c_4 = cba \times cl, \text{ where } cl \text{ is the unit cost of developed land} \qquad (9.4)$$

This will be the fourth term in the **Cost Functions** in building and housing development systems.

Thus, the complete **Cost Function** of the building and housing development system incorporating all the four terms mentioned above, giving the total cost per BU, will be as follows:

$$\text{The total Cost per Built-Unit} = y(x) = c_1 x_1 x_3^{-1} x_5^{-1} + c_2 x_2 x_3^{-1} x_5^{-1} + c_3 x_1 x_2 x_5^{-1}$$
$$+ c_4 x_1 x_2 x_4^{-1} x_5^{-1}$$

The terms x_3, x_4, and x_5 represent the number of floors, FAR, and the number of BUs per floor, respectively, as also explained later.

9.8   Typology of Optimizing Models for Optimal Building and Housing Design

As mentioned above, essential characteristics of OR/*Optimization* are its emphasis on *models* indicating physical depiction of a problem and describing the relationships between *goals*, *variables*, and *constraints* in *mathematical terms*. Many authors have attempted modeling the urban system and a number of *large-scale* urban models are available in the literature (McLoughlin, 1969; Foot, 1981). Development of *small-scale* models to address a particular policy problem is often suggested. This is very relevant in urban management, involving building and urban development, requiring both 'descriptive' and 'optimizing' models. The author developed earlier a number of descriptive models for shelter design and analysis (HUDCO, 1982). Application of the concept of OR in **Urban Development Planning/Design Policy Analysis** is illustrated earlier in Chap. 7 (see Sect. 7.6), using a 'descriptive' model developed for urban-built-form.

The descriptive model presented earlier in terms of mathematical equations can be used by the decision maker both in *analysis*—understanding how the system behaves—and in *synthesis*—design and *decision-making*. Thus, it helps in acquiring knowledge of the system (i.e. urban-built-form)—the *first-step* of the three-step decision process mentioned above. To make *Optimization* possible, in the *second-step*, it is necessary to develop a *measure of effectiveness*, which may be related to cost, space, density, and so on, which is to be *Minimized or Maximized*. Here, we can choose it to be the *Minimization* of the *Cost* per Built-Unit fulfilling the chosen *minimum acceptable* built-space per built-unit, i.e. Built-Unit-Space (i.e. BUS expressed as a function of the system variables such as x_1, x_2, and x_5 explained below and, represented by the *constraint* given in expression (9.9) below), and the Open-To-Sky-Space per Unit (i.e. SPO mentioned above represented by the system variable such as x_6 explained below), and accordingly, the *objective function*, $y(x)$, to be *Minimized* could be expressed in terms of the system *variables* as also shown in expression (9.4) below.

The relationships between many design variables connected to cost are *nonlinear*, time dependent, and physical design related. Hence, it is desirable to develop *small-scale nonlinear programming* urban development *models*, linked with physical design and adopting a shorter planning horizon, to permit *optimal design* with more realistic analysis of costs. Accordingly, a number of such optimizing models for **Housing and Urban Development Systems** are developed and published elsewhere, covering a wide domain of problems in built-environment. Some example *optimizing models* are presented below. These models give stress on efficient and effective use of resources at the urban development project level, to enhance both *Productivity* (i.e. *Efficiency & Effectiveness*) and Social *Responsiveness* to help solve urban problems *Resource-Efficiently*, improving performance of urban organizations in this regard including profitability.

Thus, *Optimizing Models* are of following two types:

1. Least-Cost Optimizing Models (**LCO Models**)—this can be used to obtain **Least-Cost Design** of urban development system, for a given minimum acceptable built-unit-space, i.e. built-space per built-unit.
2. Most-Benefit Optimizing Models (**MBO Models**)—this can be used to obtain **Maximum-Built-Unit-Space (i.e. Most-Benefit) Design** of urban development system, for a given investment/affordable cost limit. As sale—proceeds of a real estate are generally proportionate to built-space, which is maximized by a MBO model, it can also be called '***Maximum Profit Model***' for developers/builders in building/urban sector.

To illustrate the concept and its applicability in urban development and management to achieve a *resource-efficient* urban problem solution attaining *social equity*, and thus enhance *social responsiveness* of urban organizations; two *Optimizing Models*: one to obtain the **Least-Cost Design**, and the other to find the **Most-Benefit Design** of building and housing in the urban development system, are presented below:

9.8.1 An Example Least-Cost Optimizing Model (LCO Model) for Optimal Building-Housing Design

In this case, we may choose the *measure of effectiveness* as the *Minimization* of the *Cost* per Built-Unit fulfilling the chosen *minimum acceptable* built-space per built-unit, i.e. Built-Unit-Space (i.e. BUS expressed as a function of the system variables x_1, x_2, and x_5 explained below and, represented by the *constraint* given in expression

(9.9) below), and the Open-To-Sky-Space per Unit (i.e. SPO mentioned above represented by the system variable x_6 explained below), and accordingly, the *objective function*, $y(x)$, to be *Minimized* could be expressed in terms of the system *variables* as shown in expression (9.4) below. When particular values are assigned to the system *variables* $x\times$ (i.e. x_1, x_2, and so on), fulfilling the given *constraints*, it completes the *third-step* and gives the *optimal design*. Accordingly, the **Least-Cost** *Optimizing Model* (LCO Model) to obtain **Optimal Building Design**, giving the Least-Cost per Built-Unit fulfilling the chosen *minimum acceptable* Built-Space per Built-Unit and, linking the desired Open-Space, Building Rise, and the given land price and the elemental construction costs, can be developed as follows.

$$\textit{Minimize the Cost per Built-Unit} = y(x) = c_1 x_1 x_3^{-1} x_5^{-1} + c_2 x_2 x_3^{-1} x_5^{-1}$$

$$+ c_3 x_1 x_2 x_5^{-1} + c_4 x_1 x_2 x_4^{-1} x_5^{-1} \tag{9.5}$$

Subject to:

$$a_1 x_1^{-1} x_2^{-1} x_4 x_5 x_6 + a_2 x_3^{-1} x_4 \leq 1 \tag{9.6}$$

(FAR, built-unit-space, open-space and building rise compatibility *constraint*)

$$a_3 x_3 \leq 1 \tag{9.7}$$

(Building rise, i.e. number of floors *constraint*)

$$a_4 x_5 \leq 1 \tag{9.8}$$

(Number of built-units per floor *constraint*)

$$a_5 x_6^{-1} \leq 1 \tag{9.9}$$

(Minimum acceptable open-space per built-unit *constraint*)

$$a_{10} x_1^{-1} x_2^{-1} x_5 \leq 1 \tag{9.10}$$

(Minimum acceptable built-unit-space *constraint*)

$$x_1, x_2, x_3, x_5, x_6 > 0$$

(Non-negativity *constraint*)

Here, c_1, c_2, c_3, c_4, a_1, a_2, a_3, a_4, a_5, and a_{10} are the model constants, related to the cost parameters (giving unit cost coefficients), *dimensionless* physical design patterns (giving built-unit design coefficients), and the design standards, mostly dependent on the input design conditions, decided by an architect, engineer, urban planner/

urban manager as per requirements, and on the *constraints* beyond control (say market price of land) as may be applicable. Using the dwelling unit design module coefficients: cx_1, cx_2, and cba developed earlier in Eqs. (9.1), (9.2), and (9.3) for various dwelling unit design modules, and the site-specific cost parameters (giving the unit cost coefficients using the cost engineering model as incorporated in the C++ program discussed above), values of Model Constants, for calibrating the above optimal built-unit design model, can be calculated as follows:

$$c_1 = cx1((nf(1 + (nf - 1)fic))cw + (1 + (nf - 1)fic)cf), \quad c_2 = cx2 \times c_1/cx1,$$
$$c_3 = (cba \times cr \times cif \times nf + cba \times cfl \times cif \times 1)/nf, \quad c_4 = cba \times cl;$$
$$cif = 1 + (nf - 1)fic;$$

$$(9.11)$$

$$a_1 = a0 = 1/cba; \quad a_2 = 1.0; \quad a_3 = 1.0/nf; \quad a_4 = 1.0/nd; \quad a_5 = spo; \quad a_{10}$$
$$= gpa/cba;$$

where

nf = number of floors specified by the designer,
fic = coefficient of increase in construction cost per unit of building rise,
cw = cost of load bearing brick wall (with all finishing and joinery) per unit length,
cf = cost of foundation per unit length,
cr = cost of concrete floor/roof (with all finishing and reinforcements) per unit area,
cfl = cost of flooring (with all finishing) per unit area,
cl = cost of developed land (with all services) per unit area,
nd = number of dwelling (built-unit) per floor specified by the designer,
spo = semi-public open-to-sky space (in sq.m) per built-unit, specified by the designer,
gpa = gross plinth area (in sq.m) per built-unit, specified by the designer.

Here, the model *variables* are:

x_1 = Optimum depth of the built-unit block
x_2 = Optimum width of the built-unit block
x_3 = A variable giving the optimum building rise, i.e. number of floors in the built-unit block
x_4 = A variable giving the Optimum Floor Area Ratio (FAR) in any layout of the built-unit blocks, for the design conditions adopted
x_5 = A variable giving the optimum number of built-units per floor in the block
x_6 = A variable giving the optimum open space per built-unit in any layout of the built-unit blocks, for the design conditions adopted

Choosing those values of the system *variables* (i.e. x_1, x_2, x_3, x_4, x_5, and x_6) that yield *optimum* effectiveness, is the *third-step* of the *three-step rational decision-making process* (i.e. mathematical *Optimization*), and to complete this *third-step,* it is necessary to solve the above *Nonlinear Programming Problem*, which is a typical geometric programming problem and can be solved accordingly (Beightler et al., 1979).

As mentioned earlier, *Optimizing* models are of two types. Similar to **Least-Cost** *Optimizing Model* (LCO Model) shown above, we can develop a **Most-Benefit** *Optimizing Model* (MBO Model) presented below to obtain *Optimal Building Design*. This is quite relevant when a building design like housing development has to comply with the limit of Investment *Affordable* by a *stakeholder*, say the low-income/medium-income households, while *Maximizing* the built-space——a high *priority* with such important *stakeholders* in the urbanization and development process.

9.8.1.1   Model Solution

The above optimal design model is a typical geometric programming problem. Thus, it may be seen that the objective function in the above problem is a summation of product functions and the constraints also contain the product functions. The geometric programming technique is an efficient tool in solving such problems as explained in Chap. 8. In this technique, emphasis is placed on the relative magnitudes of the terms of the objective function rather than on the variables, and instead of finding the optimal values of the design variables directly, the optimal way to distribute the total cost, among the various terms or components in the objective function, is found first. This indicates how resources should be apportioned among the components contributing to the design objective, which is very useful in engineering and architectural design. Even in the presence of great degree of difficulty, the geometric programming formulation uncovers important invariant properties of the design problem which can be used as design rules for optimality as discussed above (soc86).

In this technique, if the primal geometric programming problem has N variables and T terms; in the dual geometric programming problem, there will be N orthogonality conditions-one for each variable x_n, one normality condition and T weights, one for each term. Thus, there will be $(N + 1)$ independent linear constraint equations as against T independent dual variables which are unknown. In case $T - (N + 1)$, which is the degree of difficulty in the problem. Accordingly, using the geometric programming technique, and taking the terminology used in literature (Beightler et al., 1979) on geometric programming as explained in Sect. 8.12.5 of Chap. 8, the dual objective function and the associated normality and orthogonality constraints for the above primal problem are given by:

$$d(w) = \left(\frac{c_{01}}{w_{01}}\right)^{w_{01}} \left(\frac{c_{02}}{w_{02}}\right)^{w_{02}} \left(\frac{c_{03}}{w_{03}}\right)^{w_{03}} \left(\frac{c_{04}}{w_{04}}\right)^{w_{04}} \left(\frac{C_9(W_9 + W_{12})}{w_9}\right)^{w_9} \left(\frac{C_{12}(W_9 + W_{12})}{w_{12}}\right)^{w_{12}}$$
$$\times (c_{21})^{w_{21}} (c_{31})^{w_{31}} (c_{41})^{w_{41}} (c_{51})^{w_{51}}$$

$$(9.12)$$

As discussed above, the coefficients for the objective function are simplified as:

$$c_{01} = c_1, \; c_{02} = c_2, \; c_{03} = c_3, \; c_{04} = c_4.$$

Similarly, the coefficients for the constraint functions are simplified as:

$$c_9 = a_1, \; c_{12} = a_2, \; c_{21} = a_3, \; c_{31} = a_4, \; c_{41} = a_5, \; c_{51} = a_{10}.$$

Accordingly, the dual objective function to be maximized can be expressed in a simplified form as follows:

$$
\begin{aligned}
\text{Maximize } d(w) \; = \; & [c_1/w_{01}]^{w}{}_{01} [c_2/w_{02}]^{w}{}_{02} [c_3/w_{03}]^{w}{}_{03} [c_4/w_{04}]^{w}{}_{04} \\
& \times [a_1(w_9 + w_{12})/w_9]^{w}{}_9 [a_2(w_9 + w_{12})/w_{12}]^{w}{}_{12} \\
& \times [a_3]^{w}{}_{21} [a_4]^{w}{}_{31} [a_5]^{w}{}_{41} [a_{10}]^{w}{}_{51}
\end{aligned}
\tag{9.13}
$$

Subject to:

$$w_{01} + w_{02} + w_{03} + w_{04} = 1 \quad \text{(normality condition)} \tag{9.14}$$

$$w_{01} + w_{03} + w_{04} - w_9 - w_{51} = 0 \quad \text{(orthogonality condition for } x_1) \tag{9.15}$$

$$w_{02} + w_{03} + w_{04} - w_9 - w_{51} = 0 \quad \text{(orthogonality condition for } x_2) \tag{9.16}$$

$$- w_{01} - w_{02} - w_{12} + w_{21} = 0 \quad \text{(orthogonality condition for } x_3) \tag{9.17}$$

$$- w_{04} + w_9 + w_{12} = 0 \quad \text{(orthogonality condition for } x_4) \tag{9.18}$$

$$- w_{01} - w_{02} - w_{03} - w_{04} + w_9 + w_{31} + w_{51} = 0 \quad \text{(orthogonality condition for } x_5) \tag{9.19}$$

$$w_9 - w_{41} = 0 \quad \text{(orthogonality condition for } x_6) \tag{9.20}$$

Here, $w_{01}, w_{02}, w_{03},$ and w_{04} represent the four cost weights related to the four terms in the objective function $y(x)$, i.e. expression (9.5) above. Similarly, $w_9, w_{12}, w_{21}, w_{31}, w_{41},$ and w_{51} represent the dual variables related to the first term in constraint-1, i.e. expression (9.6), to the second term in constraint-1, i.e. expression (9.6), to the single term in constraint-2, i.e. expression (9.7), to the single term in constraint-3, i.e. expression (9.8), to the single term in constraint-4, i.e. expression (9.9), and to the single term in constraint-5, i.e. expression (9.10) above, respectively.

In the present optimal design problem, there are ten terms (T) and six variables (N) in the primal objective function and constraints [expressions (9.5)–(9.10)]. Thus, the degree of difficulty in the problem is $= T - (N + 1) = 10 - (6 + 1) = 3$, which is comparatively small, and therefore, a set of $(N + 1)$ dual weights could be expressed in terms of the remaining $T - (N + 1)$ dual weights, and substituted into the dual objective function, which is then maximized by setting the first derivative to zero with respect to each of the dual weights coming into the substituted dual objective function, which is being maximized. Thus, a sufficient number of equations are available for solving the problem. Therefore, any seven of the dual variables can be expressed in terms of the remaining three dual variables as shown below, choosing w_{01}, w_{04}, and w_9 as the three dual variables:

$$w_{02} = w_{01}; w_{03} = 1 - w_{01} - w_{04}; w_{12} = w_{04} - w_9; w_{21} = 2w_{01} + w_{04} - w_9;$$
$$w_{31} = w_{01}; w_{41} = w_9; w_{51} = 1 - w_{01} - w_9. \tag{9.21}$$

As earlier mentioned, using *optimization concepts*, one can discern general *decision rules* appropriate to problems, and develop an *insight* regarding proper form of an optimal solution even when a problem is not completely formulated mathematically. This is illustrated below, noting that the cost function in the above expression (9.5) literally means that: the total Cost per Built-Unit = Cost contributed by all walls (including foundation and finishes) in x_1 direction (say, along the staircase) + Cost contributed by all walls (including foundation and finishes) in x_2 direction (say, perpendicular to the staircase) + Cost contributed by floors/roof (including finishes) + Cost contributed by developed land (i.e. land and services).

Thus, the cost weights w_{01}, w_{02}, w_{03}, and w_{04} represent the optimum proportion of total cost per built-unit, distributed between the cost contributed by all walls (including foundation and finishes) in x_1 direction (say, along the staircase), cost contributed by all walls(including foundation and finishes) in x_2 direction (say, perpendicular to the staircase), cost contributed by floors/roof (including finishes), and the cost contributed by developed land (i.e. land and services), respectively.

In this case by developing only the exponent matrix, without solving the model completely, one can discern the following *decision rules* if one has to operate *optimally*:

  i. no term of the cost function can contribute more than 100% of the total cost nor less than 0% at the *optimality*,
 ii. the cost contributed by all walls in the x_1 direction must be equal to the cost contributed by all walls in the x_2 direction, at the *optimality*,
iii. sum of the twice the cost contributed by all walls in the x_1 direction and the cost contributed by the developed land (raw land and services) must be less than the total cost, at the *optimality*.

The set of Eq. (9.21) represent the relationship between the cost distributions among the above components, which are independent of the values of the cost coefficients c_1, c_2, c_3, and c_4. This invariant property of geometric programming could be used for development of suitable *design rules* for *optimality* even in the presence of a great degree of difficulty, having general application in any problem of this nature. If we substitute the expressions in equations set (9.21) into the dual objective function [Eq. (9.13)] and maximize this substituted dual objective function over the three variables w_{01}, w_{04}, and w_9 (i.e. by setting the first derivative of the function with respect to each variable to zero), we will get three additional equations, thus providing sufficient conditions for optimality. The process is simplified by maximizing the natural logarithm of the dual function; that is, by substituting

$$\ln d(w_{01}, w_{04}, w_9) = z(w_{01}, w_{04}, w_9).$$

Thus, by above substitution and taking logarithm on both sides of Eq. (9.13), we get the following equations:

$$
\begin{aligned}
z(w) = {} & w_{01}(\ln c_1 - \ln w_{01}) + w_{01}(\ln c_2 - \ln w_{01}) + (1 - 2w_{01} - w_{04}) \\
& (\ln c_3 - \ln(1 - 2w_{01} - w_{04})) \\
& + w_{04}(\ln c_4 - \ln w_{04}) + w_9(\ln a_1 + \ln w_{04} - \ln w_9) + (w_{04} - w_9) \\
& (\ln a_2 + \ln w_{04} - \ln(w_{04} - w_9)) \\
& + (2w_{01} + w_{04} - w_9)\ln a_3 + w_{01}\ln a_4 + w_9\ln a_5 + (1 - w_{01} - w_9)\ln a_{10}.
\end{aligned}
$$
$$(9.22)$$

By setting the first derivative $dz/dw_9 = 0$, we get:

$$
\begin{aligned}
& (\ln a_1 + \ln w_{04} - \ln w_9) - w_9/w_9 - \ln a_2 - \ln w_{04} + \ln(w_{04} - w_9) + 1 - \ln a_3 \\
& + \ln a_5 - \ln a_{10} = 0
\end{aligned}
$$

Or, $a_1 a_5(w_{04} - w_9) - a_2 a_3 a_{10} w_9 = 0$ or, $w_9 = a_1 a_5 w_{04}/(a_1 a_5 + a_2 a_3 a_{10})$
$$(9.23)$$

By setting the first derivative $dz/dw_{01} = 0$, we get:

$$
\begin{aligned}
& \ln c_1 - \ln w_{01} - 1 + \ln c_2 - \ln w_{01} - 1 - 2\ \ln c_3 + 2\ln(1 - 2w_{01} - w_{04}) + 2 \\
& + 2\ \ln a_3 + \ln a_4 - \ln a_{10} = 0
\end{aligned}
$$

or, $c_1 c_2 a_3^2 a_4 (1 - 2w_{01} - w_{04})^2 - c_3^2 a_{10} w_{01}^2 = 0$
$$(9.24)$$

Again, by setting the first derivative $dz/dw_{04} = 0$, we get:

$$
\begin{aligned}
& -\ln c_3 + \ln(1 - 2w_{01} - w_{04}) + 1 + \ln c_4 - \ln w_{04} - 1 + \ln a_2 + \ln w_{04} + 1 \\
& -\ln(w_{04} - w_9) - 1 + \ln a_3 = 0
\end{aligned}
$$

Or, $c_4 a_2 a_3(1 - 2w_{01} - w_{04}) - c_3(w_{04} - w_9) = 0$
$$(9.25)$$

From (9.23), we get:

$$(w_{04} - w_9) = w_{04} - a_1 a_5 w_{04}/(a_1 a_5 + a_2 a_3 a_{10}) = w_{04}(a_2 a_3 a_{10})/(a_1 a_5 + a_2 a_3 a_{10})$$
$$(9.26)$$

From (9.25) and (9.26), we get:

$$
\begin{aligned}
& c_4 a_2 a_3(1 - 2w_{01}) - w_{04} c_4 a_2 a_3 - w_{04} c_3(a_2 a_3 a_{10})/(a_1 a_5 + a_2 a_3 a_{10}) = 0 \\
& (1 - 2w_{01})c_4 a_2 a_3/c_3 - w_{04}(c_4 a_2 a_3(a_1 a_5 + a_2 a_3 a_{10}) + c_3(a_2 a_3 a_{10})) \\
& /c_3(a_1 a_5 + a_2 a_3 a_{10}) = 0
\end{aligned}
$$

$$\text{Or,} \quad w_{04} = \frac{c_4(a_1 \, a_5 + a_2 \, a_3 \, a_{10})(1 - 2w_{01})}{c_3 \, a_{10} + c_4(a_1 \, a_5 + a_2 \, a_3 \, a_{10})} \tag{9.27}$$

Therefore, $1 - 2w_{01} - w_{04} = (1 - 2w_{01})\left(1 - \frac{c_4(a_1 \, a_5 + a_2 \, a_3 \, a_{10})}{c_3 \, a_{10} + c_4(a_1 \, a_5 + a_2 \, a_3 \, a_{10})}\right)$

$$= (1 - 2w_{01})\left(\frac{c_3 a_{10}}{c_3 a_{10} + c_4(a_1 a_5 + a_2 a_3 a_{10})}\right) \tag{9.28}$$

We may make the following substitutions: $a_9 = c_1 c_2 a_3^2 a_4 a_{10}$, $a_{12} = (c_3 a_{10} + c_4(a_1 a_5 + a_2 a_3 a_{10}))^2$ and $a_{13} = a_9/a_{12}$: and then, again, substituting the value of a_{13} in Eq. (9.24), we get

$$a_{13}(1 - 2w_{01})^2 - w_{01}^2 = 0$$

Thus, the above becomes a quadratic equation of the following form which can be easily solved:

$$(4a_{13}-1)w_{01}^2 - 4a_{13}w_{01} + a_{13} = 0 \tag{9.29}$$

Expressions for the optimal values of other model variables for optimal built-unit design are:

$w_{04} = c_4 a_9 (1 - 2w_{01})/a_{12};$

$w_9 = a_1 a_5 w_{04}/a_9; w_{02} = w_{01}; w_{03} = 1 - 2w_{01} - w_{04}; w_{12} = w_{04} - w_9;$

$w_{21} = 2w_{01} + w_{04} - w_9; w_{31} = w_{01}; w_{41} = w_9; w_{51} = 1 - w_{01} - w_9; \tag{9.29a}$

$x_3 = 1/a_3; x_5 = 1/a_4; x_6 = a_5; x_1 = c_2 w_{03}/(w_{02} c_3 x_3);$

$x_2 = x_1 c_1/c_2; ac = c_1 x_1/(x_3 x_5 w_{01}); x_4 = c_4 x_1 x_2/(a c w_{04} x_5)$

Putting the value of w_{01} determined by Eq. (9.29) in the equation set (9.29a), we can determine the optimal values of other model variables for optimal built-unit design, and thus solving the optimization model.

9.8.2   An Example Most-Benefit Design Model (MBO Model) for Optimal Building-Housing Design

In this case, we may choose the *measure of effectiveness* as *Maximization* of the Built-Space per Built-Unit, i.e. Built-Unit-Space (i.e. BUS) fulfilling the *constraints* such as the chosen limit of *Affordable Cost* (or Investment) per Built-Unit (represented by the *constraint* given in expression (9.24) below), the desired *minimum acceptable* Open-To-Sky-Space per Unit, and accordingly, the *objective function*, $y(x)$, to be *Maximized* could be expressed in terms of the system *variables* as shown in expression (9.19) below. Here also when particular values are assigned to the system *variables* x× (i.e. x_1, x_2, and so on), fulfilling the given *constraints*, it completes the *third-step* of optimization process and gives the *optimal design*. Accordingly, the **Most-Benefit** *Optimizing Model* (MBO Model) for ***Optimal Building Design*** giving the *Maximum* possible built-unit-space within the chosen limit of *Affordable Cost* per built-unit, linking the desired open-space, building rise, and the given land price and the elemental construction costs, is developed as follows.

$$\text{\textit{Maximize} Built-Unit-Space} = y(x) = a_0 x_1^{-1} x_2^{-1} x_5 \qquad (9.30)$$

Subject to:

$$a_1 x_1^{-1} x_2^{-1} x_4 x_5 \, x_6 + a_2 x_3^{-1} x_4 \leq 1 \qquad (9.31)$$

(FAR, built-unit-space, open-space and building rise compatibility *constraint*)

$$a_3 x_3 \leq 1 \qquad (9.32)$$

(Building rise, i.e. number of floors *constraint*)

$$a_4 x_5 \leq 1 \qquad (9.33)$$

(Number of built-units per floor *constraint*)

$$a_5 x_6^{-1} \leq 1 \qquad (9.34)$$

(Minimum acceptable open-space per built-unit *constraint*)

$$a_6 x_1 \, x_3^{-1} x_5^{-1} + a_7 x_2 x_3 x_5^{-1} + a_8 x_1 x_2 x_5^{-1} + a_9 x_1 x_2 x_4^{-1} x_5^{-1} \leq 1 \qquad (9.35)$$

(Maximum investment or affordable cost per built-unit *constraint*)

$$x_1, x_2, x_3, x_4, x_5, x_6 > 0$$

(Non-negativity *constraint*)

Here, a_0, a_1, a_2, a_3, a_4, a_5, a_6, a_7, a_8, and a_9 are the model constants, related to *dimensionless* physical design patterns (giving built-unit design coefficients), cost parameters (giving unit cost coefficients), and the design standards, mostly dependent on the input design conditions, decided by civil engineer, architect, urban planner/manager as per need, and on the *constraints* beyond control as may be applicable. Here, the expressions for model constants: a_1, a_2, a_3, a_4, and a_5 will be same as presented in equation set (9.9) above and using the expressions for model constants: c_1, c_2, c_3, and c_4 given above, the model constants: a_6, a_7, a_8, and a_9 can be defined as follows:

$a_6 = c_1/ac$, $a_7 = c_2/ac$, $a_8 = c_3/ac$, and $a_9 = c_4/ac$; where $ac =$ affordable or limit of investment per built-unit in monetary units (MU).

Here, the model *variables* (i.e. x_1, x_2, x_3, x_4, x_5, and x_6) are same as defined in Sect. 9.6.1 above. Thus, using these values of **Model Constants**, the above optimal built-unit design model for **Most-Benefit** can be easily calibrated for model solution in a site-specific case as outlined below.

9.8.2.1   Model Solution

The *Objective: Maximize* Built-Unit-Space $= y(x) = a_0 x_1^{-1} x_2^{-1} x_5$ can also be written in standard geometric programming form as:

$$\text{Minimize } y(x) = - a_0 x_1^{-1} x_2^{-1} x_5$$

Thus, this optimization model is a typical signomial geometric programming problem as outlined in Chap. 8, and can be solved accordingly using the procedure suggested in the above Chap. 8. To solve the problem, let us guess the signum function $\sigma_0 = -1.0$ for $y(x)$ the negative value of $y(x)$, since the original objective function is to *Maximize. Similarly, the value of the signum function σ_{01} is guessed as $= -1.0$. The values of other signum functions in the problem are guessed as: $\sigma_m = +1.0$ and $\sigma_{mt} = +1.0$. To simplify the notations, we can denote the sum of dual variables w_{51}, w_{52}, w_{53}, and w_{54} by w_5. With these simplified notations and other simplifications outlined in Chap. 8, the dual objective function, and the associated normality and orthogonality constraints for the above primal problem can be expressed as:

Optimise: $d(w) = (-1) \left(\left(\dfrac{a_1(w_9 + w_{12})}{w_9} \right)^{w_9} \left(\dfrac{a_2(w_9 + w_{12})}{w_{12}} \right)^{w_{12}} (a_3)^{w_{21}} (a_3)^{w_{31}} \right.$

$$\left. \times (a_5)^{w_{41}} \left(\dfrac{a_6(W_5)}{w_{51}} \right)^{w_{51}} \left(\dfrac{a_7(W_5)}{w_{52}} \right)^{w_{52}} \left(\dfrac{a_8(W_5)}{w_{53}} \right)^{w_{53}} \left(\dfrac{a_9(W_5)}{w_{54}} \right)^{w_{54}} \right)$$

$$(9.36)$$

Subject to:

$$w_{01} = (-1) \times (-1) = 1 \quad \text{(normality condition)}$$

$$-w_{01} - w_9 + w_{51} + w_{53} + w_{54} = 0 \quad \text{(orthogonality condition for } x_1) \quad (9.37)$$

$$-w_{01} - w_9 + w_{52} + w_{53} + w_{54} = 0 \quad \text{(orthogonality condition for } x_2) \quad (9.38)$$

$$-w_{12} + w_{21} - w_{51} - w_{52} = 0 \quad \text{(orthogonality condition for } x_3) \quad (9.39)$$

$$w_9 + w_{12} - w_{54} = 0 \quad \text{(orthogonality condition for } x_4) \quad (9.40)$$

$$w_{01} + w_9 + w_{31} - w_{51} - w_{52} - w_{53} - w_{54} = 0 \quad \text{(orthogonality condition for } x_5)$$

$$(9.41)$$

$$w_9 - w_{41} = 0 \quad \text{(orthogonality condition for } x_6) \quad (9.42)$$

Here, w_{01} represent the dual variable related to the single term in the objective function $y(x)$, i.e. expression (9.30) above. Similarly, w_9, w_{12}, w_{21}, w_{31}, w_{41}, w_{51}, w_{52}, w_{53}, and w_{54} represent the dual variables related to the first term in constraint-1, i.e. expression (9.31), to the second term in constraint-1, i.e. expression (9.31), to the single term in constraint-2, i.e. expression (9.32), to the single term in constraint-3, i.e. expression (9.33), to the single term in constraint-4, i.e. expression (9.34), to the first term in constraint-5, i.e. expression (9.35), to the second term in constraint-5, i.e. expression (9.35), to the third term in constraint-5, i.e. expression (9.35), and to the fourth term in constraint-5, i.e. expression (9.35) above, respectively.

There are ten terms (T) and six primal variables (N) in the primal objective function and constraints [expressions (9.30)–(9.35)], and therefore, the degree of difficulty in the problem is $= T - (N + 1) = 3$. As shown above, the value of $w_{01} = 1$ is already determined leaving remaining nine variable values to be determined. Therefore, any six of these nine dual variables can be expressed in terms of the remaining three dual variables as shown below, choosing w_{41}, w_{42}, w_{43}, and w_{44} as the four dual variables.

From Eqs.(9.37) and (9.38) we get : $w_{52} = w_{51}$

From Eqs.(9.37) and (9.41) we get : $w_{31} = w_{52} = w_{51}$

From Eq.(9.41) we get : $w_9 = w_{51} + w_{53} + w_{54} - 1$

From Eq.(9.40) we get : $w_{12} = w_{54} - w_9 = 1 - w_{51} - w_{53}$ (9.43)

From Eq.(9.39) we get : $w_{21} = w_{12} + w_{51} + w_{52} = 1 + w_{51} - w_{53}$

From Eq.(9.42) we get : $w_{41} = w_9 = w_{51} + w_{53} + w_{54} - 1$

$w_9 + w_{12} = w_{54}; w_{51} + w_{52} + w_{53} + w_{54} = 2\, w_{51} + w_{53} + w_{54};$

If we substitute the expressions in equations set (9.44) into the dual objective function [expression (9.37)] and maximize this substituted dual objective function over the four variables w_{51}, w_{53}, and w_{54} (i.e. by setting the first derivative of the function with respect to each variable to zero), we will get three additional equations, thus providing sufficient conditions for optimality. The process is simplified by maximizing the natural logarithm of the dual function; that is, by substituting

$$\ln d(w_{51}, w_{53}, w_{54}) = z(w_{51}, w_{53}, w_{54}).$$

Thus, by the above substitutions and taking logarithm on both sides of Eq. (9.36), we get the following equation:

$$z(w) = (w_{51} + w_{53} + w_{54} - 1)(\ln a_1 + \ln w_{54} - \ln (w_{51} + w_{53} + w_{54} - 1))$$
$$+ (1 - w_{51} - w_{53})(\ln a_2 + \ln w_{54} - \ln (1 - w_{51} - w_{53}))$$
$$+ (1 + w_{51} - w_{53}) \ln a_3 + w_{51} \ln a_4 + (w_{51} + w_{53} + w_{54} - 1) \ln a_5$$
$$+ w_{51}(\ln a_6 + \ln (2 w_{51} + w_{53} + w_{54}) - \ln w_{51}))$$
$$+ w_{51}(\ln a_7 + \ln (2 w_{51} + w_{53} + w_{54}) - \ln w_{51}))$$
$$+ w_{53}(\ln a_8 + \ln (2 w_{51} + w_{53} + w_{54}) - \ln w_{53}))$$
$$+ w_{54}(\ln a_9 + \ln (2 w_{51} + w_{53} + w_{54}) - \ln w_{54}))$$

$$(9.44)$$

By setting the first derivative $dz/dw_{51} = 0$, we get:

$$\ln a_1 + \ln w_{54} - \ln (w_{51} + w_{53} + w_{54} - 1) - 1 - \ln a_2 - \ln w_{54}$$
$$+ \ln (1 - w_{51} - w_{53}) + 1 + \ln + \ln + \ln + \ln a_6 + \ln (2 w_{51} + w_{53} + w_{54})$$
$$- \ln w_{51} - 1 + \ln a_7 + \ln (2w_{51} + w_{53} + w_{54}) - \ln w_{51} - 1$$
$$+ (w_{51} + w_{51} + w_{53} + w_{54})2/(2 w_{51} + w_{53} + w_{54}) = 0$$
$$\text{Or, } (a_1 a_3 a_4 a_5 a_6 a_7/a_2)(2w_{51} + w_{53} + w_{54})^2(1 - w_{51} - w_{53})$$
$$- (w_{51} + w_{53} + w_{54} - 1)w_{51}^2 = 0$$

$$(9.45)$$

By setting the first derivative $dz/dw_{53} = 0$, we get:

$$\ln a_1 + \ln w_{54} - \ln (w_{51} + w_{53} + w_{54} - 1) - 1 - \ln a_2 - \ln w_{54}$$
$$+ \ln (1 - w_{51} - w_{53}) + 1 - \ln a_3 + \ln a_5 + \ln a_8 + \ln (2w_{51} + w_{53} + w_{54})$$
$$- \ln w_{53} - 1 + (w_{51} + w_{51} + w_{53} + w_{54})/(2w_{51} + w_{53} + w_{54}) = 0$$
$$\text{Or, } (a_1 a_5 a_8/a_2 a_3)(2w_{51} + w_{53} + w_{54})(1 - w_{51} - w_{53})$$
$$- (w_{51} + w_{53} + w_{54} - 1)w_{53} = 0$$

$$(9.46)$$

By setting the first derivative $dz/dw_{54} = 0$, we get:

$$\ln a_1 + \ln w_{54} + (w_{51} + w_{53} + w_{54} - 1 + 1 - w_{51} - w_{53})/w_{54}$$
$$- \ln (w_{51} + w_{53} + w_{54} - 1) - 1 + \ln a_5 + \ln a_9$$
$$+ \ln (2 w_{51} + w_{53} + w_{54}) - \ln w_{54} - 1 + (w_{51} + w_{51} + w_{53} + w_{54}) \quad (9.47)$$
$$/(2 w_{51} + w_{53} + w_{54}) = 0$$
$$\text{Or,} (a_1 a_5 a_9)(2w_{51} + w_{53} + w_{54}) - (w_{51} + w_{53} + w_{54} - 1) = 0$$

Dividing (9.46) by (9.47) we get: $(a_1a_5a_8/a_2a_3\ a_1a_5a_9\,)\,(1-w_{51}-w_{53})-w_{53}=0$

Or, $(a_8/a_2a_3a_9)-(a_8/a_2a_3a_9)w_{51}-(a_8/a_2a_3a_9)w_{53}-w_{53}=0$

Or, $w_{53}(a_8+a_2a_3a_9)/(a_2a_3a_9)=(a_8+a_2a_3a_9)/(1-w_{51})$

Or, $\boldsymbol{w_{53}=(a_8/(a_8+a_2a_3a_9))\ (1-w_{51})}$

To simplify we can substitute: $a_3^2a_4a_6a_7=a_{14}$, $(a_8\ +\ a_2a_3a_9)\ =\ a_{15}$, $(1-a_1a_5a_9)=a_{16}$, $a_{15}a_{16}+a_{14}=a_{17}$, and $a_{14}/a_{17}=a_{18}$.

$$\text{Then } \boldsymbol{w_{53}=(a_8/a_{15})(1-w_{51})} \tag{9.48}$$

Dividing (9.45) by (9.46), we get:

$$(a_1a_3a_4a_5a_6a_7a_2a_3/a_2a_1a_5a_8)\,(2w_{51}+w_{53}+w_{54})-w_{51}{}^2/w_{53}=0$$

Or, $(a_3^2a_4a_6a_7/a_8)\,(2w_{51}+w_{53})+(a_3^2a_4a_6a_7/a_8)w_{54}-w_{51}^2/w_{53}=0$

Or, $w_{54}=(w_{51}^2/w_{53})a_8/(a_3^2a_4a_6a_7)-(2w_{51}+w_{53})$

$$\text{Or, } \boldsymbol{w_{54}=(w_{51}^2/w_{53})(a_8/a_{14})-(2w_{51}+w_{53})} \tag{9.49}$$

Therefore, putting the above value of w_{54} in $(w_{51}+w_{53}+w_{54}-1)$, we get:
$(w_{51}+w_{53}+w_{54}-1)=a_8/a_{14}(w_{51}{}^2/w_{53})-(2w_{51}+w_{53})+w_{51}+w_{53}-1=a_8/a_{14}(w_{51}{}^2/w_{53})-w_{51}-1$, and
Similarly, $(2w_{51}+w_{53}+w_{54})=2w_{51}+w_{53}+a_8/a_{14}(w_{51}^2/w_{53})-(2w_{51}+w_{53})=a_8/a_{14}(w_{51}^2/w_{53})$
Thus, putting the above values of two expressions, the Eq. (9.47) becomes:

$$(a_1a_5a_9)/(a_8/a_{14}(w_{51}^2/w_{53}))-a_8/a_{14}(w_{51}^2/w_{53})+w_{51}+1=0,$$

Putting the value of w_{53} given by (9.48) in above equation and with substitution of a_{14}, a_{15} and a_{16}, we get:

$$(a_1a_5a_9a_8)\,(w_{51}^2)\,a_{15}/(a_{14}a_8(1-w_{51}))-(w_{51}^2)\,a_{15}/(a_{14}a_8(1-w_{51}))+w_{51}+1=0$$

Or, $a_{14}=a_{14}w_{51}^2+a_{15}w_{51}^2(1-a_1a_5a_9)=a_{14}w_{51}^2+a_{15}w_{51}^2a_{16}$

$=(a_{14}+a_{15}\times a_{16})w_{51}^2$

Thus, finally we get:

$$w_{51} = \sqrt{\frac{a_{14}}{a_{14} + a_{15} \times a_{16}}} = \sqrt{\frac{a_{14}}{a_{17}}} = \sqrt{a_{18}} \tag{9.50}$$

Expressions for the optimal values of other model variables for optimal built-unit design are:

$$w_{53} = (a_8/a_{15})(1 - w_{51}); \quad w_{52} = w_{51}; \quad w_{31} = w_{51};$$
$$w_{54} = (w_{51}^2/w_{53})(a_8/a_{14}) - (2w_{51} + w_{53});$$
$$w_9 = (w_{51} + w_{53} + w_{54}) - 1; \quad w_{12} = 1 - w_{51} - w_{53}; \quad w_{21} = 1 + w_{51} - w_{53}; \quad w_{41} = w_9;$$
$$x_4 = (a_9 \times w_{53})/(a_8 \times w_{54}); \quad x_3 = 1/a_3;$$
$$x_2 = (w_{54} \times x_4 \times a_6)/(w_{51} \times x_3 \times a_9);$$
$$x_1 = (w_{54} \times x_4 \times a_7)/(w_{51} \times x_3 \times a_9); \quad x_5 = 1/a_4; \quad x_6 = a_5;$$
$$gpa = (x_1 \times x_2)/(a_0 \times x_5);$$

$$\tag{9.51}$$

Putting the value of w_{51}, determined by Eq. (9.50), in the equation set (9.51) we can determine the optimal values of other model variables for optimal built-unit design, and thus, solving the optimization model.

9.9 Calibration of the Least-Cost and Most-Benefit Optimal Design Models

The above models need to be calibrated before being used for optimal built-unit design. As the first step in this process dwelling unit design module coefficients: cx_1, cx_2, and cba need to be calculated as per the expressions developed earlier for various dwelling unit design modules in Sect. 9.5.4. A creative designer can also build such dwelling unit design modules and develop expressions for unit design module coefficients to calculate their values accordingly. Using these unit design module coefficients, the calibration of the least-cost optimal design models is done by calculating the cost coefficients: c_1, c_2, c_3, and c_4 using the expression set (9.9) given above. These cost coefficients are related to the site-specific cost parameters such as cost of labour, materials, and so on; and also to the geometric properties of the dwelling unit design modules as per the expressions (9.1), (9.2), (9.3), and (9.4) presented above. Similarly, the design parameters a_1 **and** a_2 (which can also be designated as: d_1, d_2 etc.) are related to the theory of optimal urban built-form for resource-efficiency presented earlier in Sect. 7.5 of Chap. 7 defining the relationships between BUS, BR, SPO, and FAR by the expression (7.1). Again, the design parameters: a_3, a_4, and a_5 are related to the building rise, number of built-units per floor, minimum acceptable open-space per built-unit, respectively, which needs to be calculated using the expression set (9.9) given above, so that the optimal built-unit design obtained fulfills these design conditions. the expressions for model constants:

a_1, a_2, a_3, a_4, and a_5 are same for both the least-cost and most-benefit design models as presented in equation set (9.9) above. using the expressions for model constants: c_1, c_2, c_3, and c_4 given above, the model constants: a_6, a_7, a_8, and a_9 are also defined earlier and reproduced below:

$a_6 = c_1/ac$, $a_7 = c_2/ac$, $a_8 = c_3/ac$, and $a_9 = c_4/ac$; where ac = affordable or limit of investment per built-unit in monetary units (MU).

A planner-designer can choose or develop any other dwelling unit design module, and accordingly, calculate the cost coefficients, design parameter as also the module parameters which are dependent on the specific geometric design of the module, and also on the site conditions for the relevant cost coefficients. It may be noted that there are two sets of model constants. One set relates to the design parameter values, externally decided by the planner or designer such as built-unit-space, number of built-units, semi-public open space, and so on (if desired for better clarity, a planner-designer may prefix design parameter name by the letter 'd' indicating that the design parameter value to be externally decided by the planner or designer). The other set relates to the specific geometric design of the dwelling unit design module and also depends on the site conditions for the relevant cost coefficients. These parameter values are to be calculated by the planner or designer by developing suitable expression for each dwelling unit design module pattern based on its geometry (for better clarity a planner-designer may prefix these module parameter names by the letter 'm' indicating that the parameter value is to be internally calculated).

All the design parameter and module parameter values are to be given as input to the model, by the planner or designer. It may be noted that these optimization models are *Universal Models* independent of the geometry of the layout module chosen, provided it is of rectilinear shape, and the same model can be used for different geometrical patterns of dwelling unit design module, by developing expressions and calculating values of model constants accordingly as discussed above, which are given as input to the model. The **C++** programs named as: **CLC1.CPP** (for **Least-Cost** design) and **CMB1.CPP** (for **Most-Benefit** design) developed by the author, automatically calculate the model constants and calibrate the model *instantly* if the required design-specific and site-specific inputs as discussed above are provided by the planner or designer as input, while running the C++ program as illustrated below. Even the width of streets, built-unit/lot density, and so on can be optimized if desired, by developing suitable models for single-family and multi-family dwelling development systems, which are available elsewhere (Chakrabarty, 1987a, 1987b, 1988a, 1988b, 1990, 1996, 1998, 2007), where a variety of *optimization models* in two sets (one set for **Least-Cost** design and the other for **Most-Benefit** design) adopting an array of constraint sets decided by the planner or designer, representing different design conditions such as built-unit/lot area, built-unit/lot density, built-unit-cost, number of built-units/building blocks, open-space standard, circulation interval ratio, and so on are presented. Adopting the same principles and methods, a creative planner or designer can develop similar *optimization models* according to his or her requirements, and thus help promote a ***resource-efficient*** urban development process achieving ***social equity***, in all countries.

9.10   Application of the Least-Cost and Most-Benefit Optimal Design Models: Illustrative Numerical Examples of Computer-Aided Building and Housing Design by Optimization

Numerical application examples of computer-aided building and housing design are presented below, using the above **Least-Cost** and **Most-Benefit** Optimal Design Models, to demonstrate the *resource-efficiency* achieved by such techniques. Resources are always inadequate compared to the need, whether in low- or high-income countries. No doubt low-income countries are usually subjected to severe resource-constraints necessitating scaling-down needs in the context of available resources. But, even in affluent countries such resource-constraints are applicable because of scaling-up of needs to have higher standard of living consuming more resources. It is desirable that large variations in QDEIs (discussed earlier) are presented by AEC professionals in all countries in a transparent manner to all stakeholders including user citizens belonging to both the low- income and the high-income categories, for appropriate planning and design-decisions. This will help achieve a *resource-efficient* problem solution with *social equity* with *stakeholder participation* in the building and housing development process. A general schematic optimum housing and urban development design procedure, which includes optimal building and housing design, is shown in Fig. 10.7 of Chap. 10.

Two representative optimization model examples: (1) one Least-Cost Optimizing Model (**LCO Model)** and the other (2) a Most-Benefit Optimizing Model (**MBO Model)** for optimal building/housing design are offered above, using the cost coefficients and cost functions developed previously. These models are also solved by means of **C++ program: clc1.cpp** and **cmb1.cpp** for **LCO Model** and **MBO Model,** respectively, developed and presented in Appendix/CD/Pen Drive. Illustrative numerical application results of the **C++ programs: clc1.cpp** for **Least-Cost Design** and **cmb1.cpp** for **Most-Benefit Design** are presented below separately for each program. As discussed above, low-income countries are usually subjected to severe resource-constraints necessitating scaling-down needs in the context of available resources, and even in affluent countries such resource-constraints are applicable because of scaling-up of needs to have higher standard of living consuming more resource. Therefore, results are presented below both for the middle/low-income countries designated as Context A, and for the high-income countries designated as the Context B. This also highlights the general applicability of the computer-aided building and housing planning and design by optimization technique in all countries for all income categories, to help achieve *resource-efficiency* and *social equity* in the building and housing development process.

9.10.1 Least-Cost Design

Representative input and output results given by the **C++ program: clc1.cpp**, using the above expressions for the **Cost Coefficients** and the dwelling design **Module Coefficients:** cx_1, cx_2, and cba for various dwelling design modules, are shown below, separately for each Context A and B.

9.10.1.1   Context: A—Middle/Low-Income Countries

Here, three examples are taken—one for **Dwelling Design Module: C—Figure 9.2** and two for **Dwelling Design Module-D—Figure 9.4(b)** shown above. The results instantly given by the **C++ Program: clc1.cpp** is reproduced below, showing both the input parameter values chosen by the planner-designer and the output optimal results given by the model. The program used the above expressions for cost coefficients, and the dwelling design module coefficients cx_1, cx_2, and cba for the specific dwelling design module adopted.

Example: LCL1 Dwelling Design Module: D with Two Bedrooms and a Staircase serving Four DUs

```
INPUT-Design-Decisions(some)in Current-Case are:
BUILT-UNIT DESIGN PATTERN CHOSEN IS: 'BU-PATTERN- 1'
Some Input Built-Unit Building Co-efficients adopted are:
s1= 0.010 s2= 0.100 k1= 0.230 w2= 0.050 flw= 1.100
w3= 0.160 ow= 0.060 k = 0.418 ew= 0.120 kw= 0.180 gpa= 72.530 npa= 69.805
bt1= 0.550 cx1= 5.750 cx2= 4.940 cba= 0.74537 n= 3 ns= 8
Some Input Cost Co-efficients adopted are:
cl=5000.00 cf=1076.00 cw= 837.00 cr= 737.00 cfl= 737.00
c1=27345.85 c2=23493.65 c3= 738.17 c4=3726.85
SOME OUTPUT OPTIMAL GEOMETRIC PROGRAMMING WEIGHTS ARE:
w9= 0.25000 w12= 0.30221 w21= 0.51057 w31= 0.10418 w03= 0.23943
w41= 0.25000 w51= 0.64582 w04= 0.55221 w01/02= 0.10418
OPTIMAL DIMENSIONS OF THE BUILT-UNIT COMPONENTS ARE(M):
lbr= 4.206 wbr= 3.193 llr= 4.937 wlr= 5.108
lk = 4.206 wk = 3.831 lbt= 2.874 wbt= 2.128
lwc= 2.064 wwc= 1.064 lsc= 5.120 wsc= 2.128
ssf= 0.183 owl= 2.554 wo = 1.277 blw= 2.420
OTHER OPTIMAL OUTPUT VALUES GIVEN BY THE MODEL ARE:
LEAST-POSSIBLE COST PER BUILT-UNIT: 299999.844 Monetary Unit
Length of Built-Unit Block along the staircase: 18.287 M
Width of Built-Unit Block across the staircase: 21.285 M
Optimum Plinth Area Rate given by the MODEL is= 4136.219 Monetary Units/
SQ.M.
Optimum Number of Floors given by the MODEL(same as the
Maximum Number of Floors specified)is : 4
Optimum Floor Area Ratio given by the MODEL is= 2.189
Optimum Open Space Standard(same as the Minimum Acceptable
Standard Specified) given by the MODEL is : 15.000 SQ.M./Built Unit
Optimum Ground Coverage Ratio given by the MODEL is= 0.547
Optimum Built-Unit Density achieved(in BU/HA)is ==301.819
```

Example: LCL2 Dwelling Design Module: D with Two Bedrooms and a Staircase serving Four DUs

```
INPUT-Design-Decisions (some) in Current-Case are:
BUILT-UNIT DESIGN PATTERN CHOSEN IS: 'BU-PATTERN- 1'
Some Input Built-Unit Building Co-efficients adopted are:
s1= 0.080 s2= 0.150 k1= 0.230 w2= 0.070 flw= 1.100
w3= 0.120 ow= 0.050 k = 0.400 ew= 0.100 kw= 0.150 gpa= 28.760 npa= 26.770
bt1= 0.550 cx1= 5.890 cx2= 4.670 cba= 0.75866 n=  2 ns=  8
Some Input Cost Co-efficients adopted are:
cl=5000.00 cf=1076.00 cw= 837.00 cr= 737.00 cfl= 737.00
c1=16602.44 c2=13163.56 c3= 859.67 c4=3793.30
SOME OUTPUT OPTIMAL GEOMETRIC PROGRAMMING WEIGHTS ARE:
w9= 0.50000 w12= 0.23967 w21= 0.39137 w31= 0.07585 w03= 0.10863
w41= 0.50000 w51= 0.42415 w04= 0.73967 w01/02= 0.07585
OPTIMAL DIMENSIONS OF THE BUILT-UNIT COMPONENTS ARE (M):
lbr= 2.522 wbr= 2.420 llr= 2.961 wlr= 3.250
lk = 2.522 wk = 2.074 lbt= 1.776 wbt= 0.968
lwc= 1.184 wwc= 0.968 lsc= 3.838 wsc= 2.074
ssf= 0.877 owl= 1.383 wo = 0.691 blw= 1.138
OTHER OPTIMAL OUTPUT VALUES GIVEN BY THE MODEL ARE:
Least-Possible Cost per Built-Unit: 299999.844 Monetary Unit
Length of Built-Unit Block along the staircase: 10.965 M
Width of Built-Unit Block across the staircase: 13.829 M
Optimum Plinth Area Rate given by the MODEL is= 10431.141 Monetary
Units/SQ.M.
Optimum Number of Floors given by the MODEL (same as the
Maximum Number of Floors specified) is :  2
Optimum Floor Area Ratio given by the MODEL is= 0.648
Optimum Open Space Standard (same as the Minimum Acceptable
Standard Specified) given by the MODEL is : 30.000 SQ.M./Built Unit
Optimum Ground Coverage Ratio given by the MODEL is= 0.324
Optimum Built-Unit Density achieved (in BU/HA) is ==225.327
```

Example: LCL3 Dwelling Design Module: C with Two Bed Rooms and a Staircase Serving Two DU's

```
INPUT-Design-Decisions (some) in Current-Case are:
BUILT-UNIT DESIGN PATTERN CHOSEN IS: 'BU-PATTERN- 2'
Some Input Built-Unit Building Co-efficients adopted are:
s1=-0.140 s2= 0.120 k1= 0.600 w2= 0.080 flw= 1.100
w3= 0.160 ow= 0.060 k = 0.450 ew= 0.000 kw=0.150 gpa= 50.000 npa=46.395
bt1= 0.550 c1= 4.940 c2= 3.040 c3= 0.76560 ns=  8
Some Input Cost Co-efficients adopted are:
cl=5000.00 cf=1076.00 cw= 837.00 cr= 737.00 cfl= 737.00
SOME OUTPUT OPTIMAL GEOMETRIC PROGRAMMING WEIGHTS ARE:
w9= 0.31292 w12= 0.26077 w21= 0.48047 w31= 0.10985 w03= 0.20660
w41= 0.31292 w51= 0.57722 w04= 0.57369 w01/02= 0.10985
OPTIMAL DIMENSIONS OF THE BUILT-UNIT COMPONENTS ARE (M):
lbr= 3.586 wbr= 3.059 llr= 5.379 wlr= 4.079
lk = 3.586 wk = 2.185 lbt= 1.331 wbt= 2.331
lwc= 1.089 wwc= 1.166 lsc= 4.124 wsc= 1.748
ssf=-1.255 owl= 0.874 wo = 2.959 blw= 1.424
OTHER OPTIMAL OUTPUT VALUES GIVEN BY THE MODEL ARE:
```

```
LEAST-POSSIBLE COST PER BUILT-UNIT: 239675.078 Monetary Unit
Length of Built-Unit Block along the staircase:  8.965 M
Width of Built-Unit Block across the staircase: 14.569 M
Optimum Plinth Area Rate given by MODEL is=4793.500Monetary Units/SQ.M.
Optimum Number of Floors given by the MODEL (same as the
Maximum Number of Floors specified) is :  4
Optimum Floor Area Ratio given by the MODEL is= 1.818
Optimum Open Space Standard (same as the Minimum Acceptable
Standard Specified) given by the MODEL is: 15.000 SQ.M./Built Unit
Optimum Ground Coverage Ratio given by the MODEL is= 0.455
Optimum Built-Unit Density achieved (in BU/HA) is ==363.636
```

Example: LCL4 Dwelling Design Module: D with Two Bedrooms and a Staircase Serving Four DUs

```
INPUT-Design-Decisions (some) in Current-Case are:
BUILT-UNIT DESIGN PATTERN CHOSEN IS: 'BU-PATTERN- 1'
Some Input Built-Unit Building Co-efficients adopted are:
s1= 0.010 s2= 0.100 k1= 0.230 w2= 0.050 flw= 1.100
w3= 0.160 ow= 0.060 k = 0.418 ew= 0.120 kw= 0.180 gpa= 72.530 npa= 69.805
bt1= 0.550 cx1= 5.750 cx2= 4.940 cba= 0.74537 n= 3 ns= 8
Some Input Cost Co-efficients adopted are:
cl=5000.00 cf=1076.00 cw= 837.00 cr= 737.00 cfl= 737.00
c1=16207.81 c2=13924.62 c3= 844.61 c4=3726.85
SOME OUTPUT OPTIMAL GEOMETRIC PROGRAMMING WEIGHTS ARE:
w9= 0.18177 w12= 0.43946 w21= 0.61904 w31= 0.08979 w03= 0.19919
w41= 0.18177 w51= 0.72844 w04= 0.62123 w01/02= 0.08979
OPTIMAL DIMENSIONS OF THE BUILT-UNIT COMPONENTS ARE (M):
lbr= 4.206 wbr= 3.193 llr= 4.937 wlr= 5.108
lk = 4.206 wk = 3.831 lbt= 2.874 wbt= 2.128
lwc= 2.064 wwc= 1.064 lsc= 5.120 wsc= 2.128
ssf= 0.183 owl= 2.554 wo = 1.277 blw= 2.420
OTHER OPTIMAL OUTPUT VALUES GIVEN BY THE MODEL ARE:
LEAST-POSSIBLE COST PER BUILT-UNIT: 412607.688 Monetary Unit
Length of Built-Unit Block along the staircase: 18.287 M
Width of Built-Unit Block across the staircase: 21.285 M
Optimum Plinth Area Rate given by the MODEL is= 5688.790 Monetary Units/
SQ.M.
Optimum Number of Floors given by the MODEL (same as the
Maximum Number of Floors specified) is :  2
Optimum Floor Area Ratio given by the MODEL is= 1.415
Optimum Open Space Standard (same as the Minimum Acceptable
Standard Specified) given by the MODEL is: 15.000 SQ.M./Built Unit
Optimum Ground Coverage Ratio given by the MODEL is = 0.707
Optimum Built-Unit Density achieved (in BU/HA) is =195.065
```

9.10.1.2 Context: B—High-Income Countries

Here, two example results are presented for the **Dwelling Design Module-E** shown in Fig. 9.5 above, where the planner/designer specifies the maximum number of floors as 2 and 4. The results instantly given by the **C++ Program: clc1.cpp** are reproduced below, showing both the input parameter values chosen by the planner-designer and the output optimal results given by the model. The program used the general expressions for cost coefficients as presented above, and the specific 'dwelling design Module Coefficients' cx_1, cx_2, and cba as developed in Sect. 9.6.2.2 for the dwelling design Module-E.

Example: LCH1 Dwelling Design Module-E: Five Bedrooms One Staircase/ Elevator Serving Two DUs

```
INPUT-Design-Decisions (some) in Current-Case are:
BUILT-UNIT DESIGN PATTERN CHOSEN IS: 'BU-PATTERN- 5'
Some Input Built-Unit Building Co-efficients adopted are:
s1= 0.300 s2= 0.200 k1= 0.150 t2= 0.090 t1 = 0.90
b1= 0.180 ow= 0.090 k = 0.400 ew= 0.060 k2=0.90 gpa=300.000 npa= 273.852
b2= 0.200 cx1= 6.360 cx2= 5.060 cba= 0.68840 n= 5
Some Input Cost Co-efficient adopted are:
cl=5000.00 cf=1076.00 cw= 837.00 cr= 737.00 cfl= 737.00
SOME OUTPUT OPTIMAL GEOMETRIC PROGRAMMING WEIGHTS ARE:
w9= 0.18450 w12= 0.46126 w21= 0.60643 w31= 0.07258 w03= 0.20907
w41= 0.18450 w51= 0.74291 w04= 0.64576 w01=w02= 0.07258
OPTIMAL DIMENSIONS OF THE BUILT-UNIT COMPONENTS ARE (M):
ldn= 6.320 wdn= 7.282 ldr=10.533 wdr= 3.641
lk = 4.74 wk = 2.98 lbt= 2.90 wbt= 2.98 tlbt= 3.95 twbt= 2.98
ewl=1.580 wwc= 0.000 lsc= 7.900 wsc= 6.620
lbr1= 3.950 owl= 0.000 wo = 2.979 wbr1= 3.641
lbr2= 3.950 lbr3= 4.740 lbr4 = 3.950 lbr5= 5.530
wbr2= 3.641 wbr3= 6.620 wbr4= 5.130 wbr5= 5.130
OTHER OPTIMAL OUTPUT VALUES GIVEN BY THE MODEL ARE:
```
Least-Possible Cost per Built-Unit: 1625981.250 Monetary Unit
Length of Built-Unit Block along the staircase: 26.333 M
Width of Built-Unit Block across the staircase: 33.099 M
Optimum Plinth Area Rate given by MODEL is= 5419.94 Monetary Units/SQ.M.
Optimum Number of Floors given by the MODEL (same as the
Maximum Number of Floors specified) is : 2
Optimum Floor Area Ratio given by the MODEL is= 1.429
Optimum Open Space Standard (same as the Minimum Acceptable
Standard Specified) given by the MODEL is : 60.000 SQ.M./Built Unit
Optimum Ground Coverage Ratio given by the MODEL is= 0.714
Optimum Built-Unit Density achieved (in BU/HA) is ==47.619

Example: LCH2 Dwelling Design Module-E: Five Bedrooms One Staircase/ Elevator Serving Two DUs

```
INPUT-Design-Decisions (some) in Current-Case are:
BUILT-UNIT DESIGN PATTERN CHOSEN IS: 'BU-PATTERN- 5'
Some Input Built-Unit Building Co-efficients adopted are:
s1= 0.300 s2= 0.150 k1= 0.220 t2= 0.060 t1 = 0.100
b1= 0.200 ow= 0.100 k= .400 ew= 0.050 kw=0.000 gpa=300.000 npa= 281.224
cx1= 6.960 cx2= 4.620 cba= 0.7190 n= 5
Some Input Cost Co-efficients adopted are:
cl=5000.00 cf=1076.00 cw= 837.00 cr= 737.00 cfl= 737.00
SOME OUTPUT OPTIMAL GEOMETRIC PROGRAMMING WEIGHTS ARE:
w9= 0.25710 w12= 0.32137 w21= 0.48828 w31= 0.08345 w03= 0.25462
w41= 0.25710 w51= 0.65945 w04= 0.57847 w01=w02= 0.08345
OPTIMAL DIMENSIONS OF THE BUILT-UNIT COMPONENTS ARE (M):
```

ldn= 5.884 wdn= 8.510 ldr= 9.414 wdr= 3.546
lk = 4.707 wk = 2.127 lbt= 2.354 wbt= 2.127
lwc= 0.000 wwc= 0.000 lsc= 7.061 wsc= 5.318
lbr1= 5.178 owl= 0.000 wo = 3.546 wbr1= 7.091
lbr2= 5.178 lbr3= 4.707 lbr4 = 5.178 lbr5= 6.355
wbr2= 7.091 wbr3= 7.091 wbr4 = 7.091 wbr5= 7.091
OTHER OPTIMAL OUTPUT VALUES GIVEN BY THE MODEL ARE:
Least-Possible Cost per Built-Unit: 966863.625 Monetary Unit
Length of Built-Unit Block along the staircase: 23.536 M
Width of Built-Unit Block across the staircase: 35.456 M
Optimum Plinth Area Rate given by MODEL is=3889.54 Monetary Units/SQ.M.
Optimum Number of Floors given by the MODEL (same as the
Maximum Number of Floors specified) is : 4
Optimum Floor Area Ratio given by the MODEL is= 2.222
Optimum Open Space Standard (same as the Minimum Acceptable
Standard Specified) given by the MODEL is: 60.000 Sq.M./Built Unit
Optimum Ground Coverage Ratio given by the MODEL is= 0.556
Optimum Built-Unit Density achieved (in BU/HA) is ==74.074

A diagram of the dwelling unit design module: E for high-income housing, using the output **Optimum Block** and **Room Dimensions** given by the Model **instantly** for the **Least-Cost** as in **Example: LCH1** above, is presented in Fig. 9.6. *Instant production of such drawings can be automated for each input planning and design-*

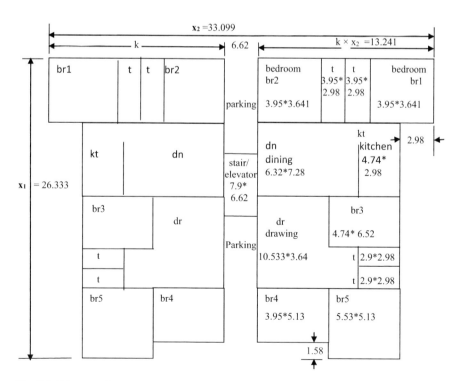

Fig. 9.6 Diagram of dwelling unit design module: E for high-income housing, showing optimum block/room dimensions (not to scale-Example: LCH1) given by the model

decision set given by the planner-designer while running the C++ programs discussed above. This is demonstrated later in Sect. 9.9, presenting a technique for integrated computer-aided design by optimization including graphics.

9.10.2   Most-Benefit Design

Representative input and output results given by the **C++ program: cmb1.cpp**, using the above expressions for the **Cost Coefficients** and the dwelling design **Module Coefficients: cx$_1$, cx$_2$,** and cba for various Dwelling Design Modules, are shown below, separately for each Context A and B.

9.10.2.1   Context: A—Middle/Low-Income Countries

Here, four examples are taken—two for **Dwelling Design Module-D—Figure 9.4 (b),** one for **Dwelling Module: B** and one for **Dwelling Module: A** shown above. The results instantly given by the **C++ Program: cmb1.cpp** is reproduced below, showing both the input parameter values chosen by the planner-designer and the output optimal results given by the model. The program used the above expressions for cost coefficients, and the dwelling design module coefficients cx$_1$, cx$_2$, and cba for the specific dwelling design module adopted.

Example:MBL1 Dwelling Module-D: Alternative: I—Four Double Bedroom Dwelling Served by One Staircase

```
INPUT-Design-Decisions(some)in Current-Case are:
BUILT-UNIT DESIGN PATTERN CHOSEN IS: 'BU-PATTERN- 1'
Built-Unit Physical-Design-Co-efficients Chosen:
s1= 0.010 s2= 0.100 k1= 0.230 w2= 0.050 flw= 1.100
w3= 0.160 ow= 0.060 k = 0.418 ew= 0.120 kw= 0.180
bt1= 0.550 cx1= 5.750 cx2= 4.940 cba= 0.74537 ns= 8
Some Built-Unit Cost-Co-efficients Chosen:
ac= 300000.00 cl=5000.00 cf=1076.00 cw= 837.00 cr= 737.00 cfl= 737.00
OUTPUT-Optimal-Parameter-Values(some)given by HudCAD:
Optimal Geometrical Programming Weights:
w9= 0.38710 w12= 0.46795 w21= 0.79057 w31= 0.16131
w41= 0.38710 w53= 0.37074 w54= 0.85505 w51/52= 0.16131
Optimal Dimensions of the Built-Unit Components(M):
lbr= 4.206 wbr= 3.193 llr= 4.937 wlr= 5.108
lk = 4.206 wk = 3.831 lbt= 2.874 wbt= 2.128
lwc= 2.064 wwc= 1.064 lsc= 5.120 wsc= 2.128
ssf= 0.183 owl= 2.554 wo = 1.277 blw= 2.420
Other Optimal OUTPUT Values given by the SOFTWARE:
Maximum Possible Built-Unit-Space is:  72.530 SQ.M.
Length of Built-Unit Block along the staircase: 18.287 M
Width of Built-Unit Block across the staircase: 21.285 M
Optimum Plinth Area Rate given by MODEL= 4136.217 Monetary Units/SQ.M.
Optimum Number of Floors given by the MODEL(same as the
Maximum Number of Floors specified)is=  4
```

Optimum Floor Area Ratio given by the MODEL is= 2.189
Optimum Open Space Standard(same as the Minimum Acceptable
Standard Specified) given by the MODEL is= 15.000 SQ.M./Built-Unit
Optimum Ground Coverage Ratio given by the MODEL is= 0.547
Optimum Built-Unit DENSITY achieved(in BU/HA)= 301.818

Example: MBL2 Dwelling Module-D: Alternative: II—Four Double Bed Room Dwelling Served by one Stair-case

INPUT-Design-Decisions (some) in Current-Case are:
BUILT-UNIT DESIGN PATTERN CHOSEN IS: **'BU-PATTERN- 1'**
Built-Unit Physical-Design-Co-efficients Chosen:
s1= 0.080 s2= 0.150 k1= 0.230 w2= 0.070 flw= 1.100
w3= 0.120 ow= 0.050 k = 0.400 ew= 0.100 kw= 0.150
bt1= 0.550 n=2 ns= 8
Some Built-Unit Cost-Co-efficients Chosen:
ac= 300000.00 cl=5000.00 cf=1076.00 cw= 837.00 cr= 737.00 cfl= 737.00
OUTPUT-Optimal-Parameter-Values(some)given by HudCAD:
cx1= 5.890 cx2= 4.670 cba= 0.75866
Optimal Geometrical Programming Weights:
w9= 1.17883 w12= 0.56505 w21= 0.92272 w31= 0.17883
w41= 1.17883 w53= 0.2569 w54= 1.74389 w51/52= 0.17883
Optimal Dimensions of the Built-Unit Components (M):
lbr= 2.522 wbr= 2.420 llr= 2.961 wlr= 3.250
lk = 2.522 wk = 2.074 lbt= 1.776 wbt= 0.968
lwc= 1.184 wwc= 0.968 lsc= 3.838 wsc= 2.074
ssf= 0.877 owl= 1.383 wo = 0.691 blw= 1.138
Other Optimal OUTPUT Values given by the SOFTWARE:
Maximum Possible Built-Unit-Space is: **28.760** SQ.M.
Length of Built-Unit Block along the staircase: 10.965 M
Width of Built-Unit Block across the staircase: 13.829 M
Optimum Plinth Area Rate given by MODEL=10431.132 Monetary Units/SQ.M.
Optimum Number of Floors given by the MODEL(same as the
Maximum Number of Floors specified)is= **2**
Optimum Floor Area Ratio given by the MODEL is= 0.648
Optimum Open Space Standard(same as the Minimum Acceptable
Standard Specified) given by the MODEL is= 30.000 SQ.M./Built-Unit
Optimum Built-Unit DENSITY achieved(in BU/HA)= 225.327

Example: MBL3 Dwelling Module-D: Alternative: II—Four Double Bedroom Dwelling Served by One Staircase with SPO = 30 as in MBL2, but changed BR = 4

INPUT-Design-Decisions(some)in Current-Case are:
BUILT-UNIT DESIGN PATTERN CHOSEN IS: 'BU-PATTERN- 1'
Built-Unit Physical-Design-Co-efficients Chosen:
s1= 0.080 s2= 0.150 k1= 0.230 w2= 0.070 flw= 1.100
w3= 0.120 ow= 0.050 k = 0.400 ew= 0.100 kw= 0.150
bt1= 0.550 cx1= 5.890 cx2= 4.670 cba= 0.75866 n=2 ns= 8
Some Built-Unit Cost-Co-efficients Chosen:
ac= 300000.00 cl=5000.00 cf=1076.00 cw= 837.00 cr= 737.00 cfl= 737.00
OUTPUT-Optimal-Parameter-Values(some)given by HudCAD:

```
Optimal Geometrical Programming Weights:
w9= 1.19165 w12= 0.45102 w21= 0.83431 w31= 0.19165
w41= 1.19165 w53= 0.35733 w54= 1.64267 w51/52= 0.19165
Optimal Dimensions of the Built-Unit Components(M):
lbr= 3.169 wbr= 3.041 llr= 3.720 wlr= 4.084
lk = 3.169 wk = 2.607 lbt= 2.232 wbt= 1.217
lwc= 1.488 wwc= 1.217 lsc= 4.823 wsc= 2.607
ssf= 1.102 owl= 1.738 wo = 0.869 blw= 2.123
Other Optimal OUTPUT Values given by the SOFTWARE:
Maximum Possible Built-Unit-Space is:  45.418 SQ.M.
Length of Built-Unit Block along the staircase: 13.779 M
Width of Built-Unit Block across the staircase: 17.379 M
Optimum Plinth Area Rate given by MODEL= 6605.252 Monetary Units/SQ.M.
Optimum Number of Floors given by the MODEL(same as the
Maximum Number of Floors specified)is=  4
Optimum Floor Area Ratio given by the MODEL is= 1.098
Optimum Open Space Standard(same as the Minimum Acceptable
Standard Specified) given by the MODEL is=  30.000 SQ.M./Built-Unit
Optimum Ground Coverage Ratio given by the MODEL is= 0.275
Optimum Built-Unit DENSITY achieved(in BU/HA)= 241.89
```

Example: MBL4 Dwelling Module: B—Four Single Bedroom Dwelling Served by one Staircase

One representative input and output results given by the **C++ program: cmb1. cpp** using the above expressions for the dwelling design module coefficients cx_1, cx_2, and cba for Dwelling Design Module: B is shown below.

```
INPUT-Design-Decisions(some)in Current-Case are:
BUILT-UNIT DESIGN PATTERN CHOSEN IS: 'BU-PATTERN- 3'
Built-Unit Physical-Design-Co-efficients Chosen:
s1= 0.050 s2= 0.180 k1= 0.500 w2= 0.150 flw= 1.000
w3= 0.150 ow= 0.000 k = 0.000 ew= 0.000 kw= 0.000
bt1= 0.300 cx1= 5.600 cx2= 3.420 cba = 0.91900 ns= 8
Some Built-Unit Cost-Co-efficients Chosen:
ac= 80000.00 cl=2000.00 cf=1076.00 cw= 837.00 cr= 737.00 cfl= 737.00
OUTPUT-Optimal-Parameter-Values (some)given by HudCAD:
Optimal Geometrical Programming Weights:
w9= 0.42528 w12= 0.24295 w21= 0.79465 w31= 0.27585
w41= 0.42528 w53= 0.48120 w54= 0.66823 w51/52= 0.27585
Optimal Dimensions of the Built-Unit Components(M):
lbr= 3.897 wbr= 3.318 llr= 0.000 wlr= 0.000
lk = 2.728 wk = 1.914 lbt= 1.169 wbt= 1.914
lwc= 0.000 wwc= 0.000 lsc= 4.286 wsc= 2.297
ssf= 0.390 owl= 0.000 wo = 0.000 blw= 1.766
Other Optimal OUTPUT Values given by the SOFTWARE:
Maximum Possible Built-Unit-Space is:  22.850 SQ.M.
Length of Built-Unit Block along the staircase:  7.794 M
Width of Built-Unit Block across the staircase: 12.761 M
Optimum Plinth Area Rate given by MODEL= 3501.056 Monetary Units/SQ.M.
Optimum Number of Floors given by the MODEL (same as the
Maximum Number of Floors specified)is=  4
Optimum Floor Area Ratio given by the MODEL is= 1.454
```

```
Optimum Open Space Standard (same as the Minimum Acceptable
Standard Specified) given by the MODEL is=  10.000 SQ.M./Built-Unit
Optimum Ground Coverage Ratio given by the MODEL is= 0.364
Optimum Built-Unit DENSITY achieved (in BU/HA)= 636.433
```

Using input values of $s1,s2,w2$, and $k1$ in above expressions give output values of $cx1 = 5.6$, $cx2 = 3.42$, $cba = 0.919$, as shown.

```
INPUT-Design-Decisions (some) in Current-Case are:
BUILT-UNIT DESIGN PATTERN CHOSEN IS: 'BU-PATTERN- 3'
Built-Unit Physical-Design-Co-efficients Chosen:
s1= 0.050 s2= 0.180 k1= 0.500 w2= 0.150 flw= 1.000
w3= 0.150 ow= 0.000 k = 0.000 ew= 0.000 kw= 0.000
bt1= 0.300 cx1= 5.600 cx2= 3.420 cba= 0.91900 n= 1 ns= 8
Some Built-Unit Cost-Co-efficients Chosen:
ac=  80000.00 cl=2000.00 cf=1076.00 cw= 837.00 cr= 737.00 cfl= 737.00
OUTPUT-Optimal-Parameter-Values (some) given by HudCAD:
Optimal Geometrical Programming Weights:
w9= 0.42528 w12= 0.24295 w21= 0.79465 w31= 0.27585
w41= 0.42528 w53= 0.48120 w54= 0.66823 w51/52= 0.27585
Optimal Dimensions of the Built-Unit Components (M):
lbr= 3.897 wbr= 3.318 llr= 0.000 wlr= 0.000
lk = 2.728 wk = 1.914 lbt= 1.169 wbt= 1.914
lwc= 0.000 wwc= 0.000 lsc= 4.286 wsc= 2.297
ssf= 0.390 owl= 0.000 wo = 0.000 blw= 1.766
Other Optimal OUTPUT Values given by the SOFTWARE:
Maximum Possible Built-Unit-Space is:  22.850 SQ.M.
Length of Built-Unit Block along the staircase:  7.794 M
Width of Built-Unit Block across the staircase: 12.761 M
Optimum Plinth Area Rate given by MODEL= 3501.055 Monetary Units/SQ.M.
Optimum Number of Floors given by the MODEL (same as the
Maximum Number of Floors specified) is=  4
Optimum Floor Area Ratio given by the MODEL is= 1.454
Optimum Open Space Standard (same as the Minimum Acceptable
Standard Specified) given by the MODEL is=  10.000 SQ.M./Built-Unit
Optimum Ground Coverage Ratio given by the MODEL is= 0.364
Optimum Built-Unit DENSITY achieved (in BU/HA)= 636.434
```

Example: MBL5 Dwelling Module: A—Two Single Bedroom Dwelling Served by one Staircase

One representative input and output results given by the **C++ program: cmb1. cpp** using the above expressions for the dwelling design module coefficients cx_1, cx_2, and cba for Dwelling Design Module: A is shown below.

```
INPUT-Design-Decisions (some) in Current-Case are:
BUILT-UNIT DESIGN PATTERN CHOSEN IS: 'BU-PATTERN- 4'
Built-Unit Physical-Design-Co-efficients Chosen:
s1= 0.120 s2= 0.000 k1= 0.320 w2= 0.150 flw= 1.000
w3= 0.000 ow= 0.000 k = 0.000 ew= 0.000 kw= 0.000
bt1= 0.000 ns= 8
Some Built-Unit Cost-Co-efficients Chosen:
ac=  80000.00 cl=2000.00 cf=1076.00 cw= 837.00 cr= 737.00 cfl= 737.00
OUTPUT-Optimal-Parameter-Values (some) given by HudCAD:
```

```
cx1= 3.640 cx2= 3.300 cba= 1.00000
Optimal Geometrical Programming Weights:
w9= 0.18399 w12= 0.40995 w21= 0.98577 w31= 0.28791
w41= 0.18399 w53= 0.30214 w54= 0.59394 w51/52= 0.28791
Optimal Dimensions of the Built-Unit Components(M):
lbr= 3.056 wbr= 2.479 llr= 0.000 wlr= 0.000
lk = 1.438 wk = 1.735 lbt= 0.000 wbt= 0.744
lwc= 1.438 wwc= 0.744 lsc= 3.596 wsc= 0.000
ssf= 0.539 owl= 0.000 wo = 0.000 blw= 1.076
Other Optimal OUTPUT Values given by the SOFTWARE:
```
Maximum Possible Built-Unit-Space is: 9.141 Sq.M.
```
Length of Built-Unit Block along the staircase:  4.494 M
Width of Built-Unit Block across the staircase:  4.958 M
Optimum Plinth Area Rate given by MODEL= 7180.768 Monetary Units/SQ.M.
Optimum Number of Floors given by the MODEL(same as the
Maximum Number of Floors specified)is :  1
Optimum Floor Area Ratio given by the MODEL is= 0.690
Optimum Open Space Standard (same as the Minimum Acceptable
Standard Specified) given by the MODEL is:  5.000 SQ.M./Built-Unit
Optimum Ground Coverage Ratio given by the MODEL is=0.690
Optimum Built-Unit DENSITY achieved (in BU/HA)= 619.545
```

9.10.2.2   Context: B—High-Income Countries

Here, two application examples are presented to show the sensitivity, i.e. increase of BUS (Built-Unit-Space) due to the increase of BR (Building Rise), keeping the BUC (Built-Unit-Cost) unchanged in both cases.

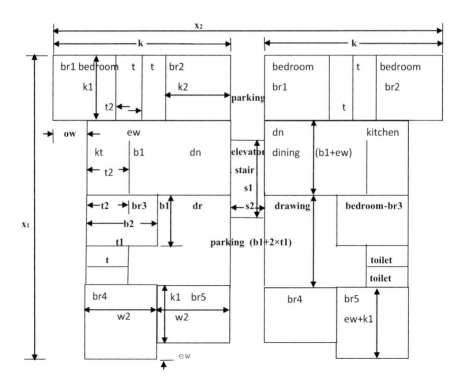

Example: MBH1 Dwelling Module: E-Two Five Bedroom Dwelling—One Staircase/Elevator (BR = 2)

```
INPUT-Design-Decisions(some)in Current-Case are:
BUILT-UNIT DESIGN PATTERN CHOSEN IS: 'BU-PATTERN- 5'
Built-Unit Physical-Design-Co-efficients Chosen:
s1= 0.300 s2= 0.200 k1= 0.150 w2= 0.155 flw= 0.000
w3= 0.000 ow= 0.090 k = 0.400 ew= 0.060 kw= 0.000
bt1= 0.000 cx1= 6.360 cx2= 5.060 cba= 0.68840 n= 5 ns= 10
Some Built-Unit Cost-Co-efficients Chosen:
ac=1625981.25 cl=5000.00 cf=1076.00 cw= 837.00 cr= 737.00 cfl= 737.00
OUTPUT-Optimal-Parameter-Values(some)given by HudCAD:
Optimal Geometrical Programming Weights:
w9= 0.24835 w12= 0.62088 w21= 0.81628 w31= 0.09770
w41= 0.24835 w53= 0.28142 w54= 0.86923 w51/52= 0.09770
Optimal Dimensions of the Built-Unit Components(M):
lbr1= 3.950 wbr1= 3.641 ldr=10.533 wdr= 3.641 ldn= 6.320
lbr2= 3.950 wbr2= 3.641 lbr3= 4.740 wbr3= 6.620 wdn= 7.282
lbr4= 3.950 wbr4= 5.130 lbr5= 5.530 wbr5= 5.130
lk = 4.740 wk = 2.979 lbt= 2.897 wbt= 2.979
lbr= 0.000 wbr= 0.000 llr= 0.000 wlr= 0.000
lk = 4.740 wk = 2.979 lbt= 2.897 wbt= 2.979
lwc= 0.000 wwc= 0.000 lsc= 7.900 wsc= 6.620
ssf= 0.000 owl= 0.000 wo = 2.979 blw= 0.000
Other Optimal OUTPUT Values given by the SOFTWARE:
Maximum Possible Built-Unit-Space is: 300.000 SQ.M.
Length of Built-Unit Block along the staircase: 26.333 M
Width of Built-Unit Block across the staircase: 33.099 M
Optimum Plinth Area Rate given by MODEL= 5419.937 Monetary Units/SQ.M.
Optimum Number of Floors given by the MODEL(same as the
Maximum Number of Floors specified)is= 2
Optimum Floor Area Ratio given by the MODEL is= 1.429
Optimum Open Space Standard(same as the Minimum Acceptable
Standard Specified) given by the MODEL is= 60.000 SQ.M./Built-Unit
Optimum Ground Coverage Ratio given by the MODEL is= 0.714
Optimum Built-Unit DENSITY achieved(in BU/HA)= 47.619
```

Example: MBH2 Dwelling Module: E—Two Five Bedroom Dwelling—One Staircase/Elevator (BR=4)

```
INPUT-Design-Decisions(some)in Current-Case are:
BUILT-UNIT DESIGN PATTERN CHOSEN IS: 'BU-PATTERN- 5'
Built-Unit Physical-Design-Co-efficients Chosen:
s1= 0.300 s2= 0.200 k1= 0.150 w2= 0.155 flw= 0.000
w3= 0.000 ow= 0.090 k = 0.400 ew= 0.060 kw= 0.000
bt1= 0.000 cx1= 6.360 cx2= 5.060 cba= 0.68840 ns= 10
Some Built-Unit Cost-Co-efficients Chosen:
ac=1625981.25 cl=5000.00 cf=1076.00 cw= 837.00 cr= 737.00 cfl= 737.00
OUTPUT-Optimal-Parameter-Values(some)given by HudCAD:
Optimal Geometrical Programming Weights:
w9= 0.24998 w12= 0.49943 w21= 0.70920 w31= 0.10488
w41= 0.24998 w53= 0.39569 w54= 0.74941 w51/52= 0.10488
Optimal Dimensions of the Built-Unit Components(M):
```

```
lbr1= 4.994 wbr1= 4.603 ldr=13.317 wdr= 4.603 ldn= 7.990
lbr2= 4.994 wbr2= 4.603 lbr3= 5.992 wbr3= 8.369 wdn= 9.206
lbr4= 4.994 wbr4= 6.486 lbr5= 6.991 wbr5= 6.486
lk = 5.992 wk = 3.766 lbt= 3.662 wbt= 3.766
lbr= 0.000 wbr= 0.000 llr= 0.000 wlr= 0.000
lk = 5.992 wk = 3.766 lbt= 3.662 wbt= 3.766
lwc= 0.000 wwc= 0.000 lsc= 9.987 wsc= 8.369
ssf= 0.000 owl= 0.000 wo = 3.766 blw= 0.000
Other Optimal OUTPUT Values given by the SOFTWARE:
Maximum Possible Built-Unit-Space is: 479.498 SQ.M.
Length of Built-Unit Block along the staircase: 33.292 M
Width of Built-Unit Block across the staircase: 41.845 M
Optimum Plinth Area Rate given by MODEL= 3391.010 Monetary Units/SQ.M.
Optimum Number of Floors given by the MODEL(same as the
Maximum Number of Floors specified)is=  4
Optimum Floor Area Ratio given by the MODEL is= 2.666
Optimum Open Space Standard(same as the Minimum Acceptable
Standard Specified) given by the MODEL is=  60.000 SQ.M./Built-Unit
Optimum Ground Coverage Ratio given by the MODEL is= 0.666
Optimum Built-Unit DENSITY achieved (in BU/HA)= 55.594
```

Increased Built-Unit-Area Ratio = 479.498/300 = 1.598 i.e. about 60% increase in Built-Unit-Area (BUA), keeping the same Built-Unit-Cost (BUC) = 1,625,981.25 MU, just by changing the number of floor from 2 to 4.

9.11   Integrated Numeric-Analysis-Graphics in Building-Housing Design by Optimization Including all Five Components–Need/Utility of Application Software for Resource-Efficient Problem Solution

Tedious and voluminous computations involved, in utilizing *Optimization* and Operations Research Techniques and to solve and use *Optimizing Models* like above, have prevented their applications in building, housing, and urban *planning* and *design* in the past. The continued development of cheap and powerful computers with interactive graphics systems has made it possible to perform many planning and design functions integrating graphics with complex computations, design analysis, cost engineering, and *Optimization* techniques on a computer. Development of MICROCADD/'PC-CADD' has revolutionized the fields of design, engineering, and drafting because of many advantages over the traditional manual methods, substantially improving **Productivity** (Goetch, 1990). 'PC-CADD' is a more productive approach to design and drafting than traditional manual techniques because of:

1. Faster production of design and data creation
2. Faster data manipulation including graphic data, taking little time in correcting and revising designs and documentation of those designs
3. Faster, more convenient data storage including graphic data

4. Faster data output including alphanumeric and graphic outputs
5. Instant linkage/conversion of optimal design in alphanumeric terms into a graphic output, facilitating application of optimization techniques for resource-efficiency

CAD primarily depends on (1) system software—supplied with the hardware, and (2) application software—available commercially or to be developed in-house. The area of '*Architectural design and drafting-engineering-construction* (AEC)' ranks relatively low in PC-CADD systems by application, as *Integrated CAD rarely forms part of education, research and full-length subject of study in AEC professional education curriculum in many countries, and commercial CAD systems incorporating all components of integrated CAD are also rare. An integrated CAD system for AEC operations should have five components: (1) Geometric Modeling, (2) Cost Engineering, (3) Design Analysis Including Design Optimization, (4) Drafting/Drawing, and (5) Data Management, Storage and Transfer* (Hsu & Sinha, 1992). But, most microcomputer-based commercial CAD systems such as AutoCAD are constructed primarily to perform only *geometric modeling* and *drafting* and their performance is effectively limited to what is included in them (Hsu & Sinha, 1992). Use of such *modern planning, design*, and *management techniques* can increase in AEC sector only with the development of more and more sophisticated software for *optimal design*, decision support systems, and database management, using information technology. Towards this effort the author is carrying out research and developed and published a number of descriptive and *Nonlinear Programming Optimizing Models* for building, housing and urban development, and now devising software, linking these <u>*Universal Optimizing Models*</u>, as an *Integrated Computer-Aided Planning & Design System*, to help better **Efficiency & Social Responsiveness** in building, housing, and urban development operations. This is outlined below presenting a software for *Integrated Computer-Aided Design by Optimization*.

9.11.1   *Application Software: 'HudCAD' for Integrated Computer-Aided Building-Housing Design by Optimization and Drafting-Drawing—Salient Features and Brief User Guide*

Mitchell pointed out that computers can be used as a design synthesis tool to automatically generate solutions and the ability to *integrate* sophisticated analysis and *optimization* techniques into the *design process* would make it possible to maintain much more *effective cost control* and response to the complex design constraints exploring more alternatives, making people to demand them (Mitchell, 1977). Moreover, building, housing, and urban operations (e.g. housing development) consume huge *Resources* involving many variables giving numerous solutions with large variations in *resource-efficiency* indicators, making *search* of optimal solution very complex to be solved manually, but not so with computers. Drawing is the language of architecture, engineering, and physical planning, where ideas are created, evaluated, and developed. Therefore, it is desirable to develop application software not only for numerical solution of optimizing models, but also, to convert instantly the optimal designs in alphanumeric terms (given as output by the optimizing models) into 'graphic form' i.e. dimensioned drawing including the working drawings to be meaningful to the users. The software should also incorporate cost engineering to derive the cost coefficients for the optimizing models, using any time-specific, site-specific, and country-specific basic unit cost data on labour, materials, and so on.

Accordingly, **Application Software: 'HudCAD'** is partially developed as an *Integrated Computer-Aided Design and Drafting* (CADD) *System,* i.e. a UNIVERSAL Building/Urban Planning, Design and Management PC-CADD TOOL to improve *Efficiency, Equity,* and ***Social Responsiveness*** in housing and urban development operations. As shown below with examples, '**HudCAD**' can be used as a '*generative system*' (Mitchell, 1977) or '*generative model*' to produce a variety of potential solutions and help select a solution fulfilling the specified *objectives* and *constraints*. Software '**HudCAD**' has different modules covering building design, area planning, built-form analysis (software modules '**CADUB**'/'**CAUB**' presented in Chap. 6). It has different components covering geometric modeling, cost engineering, design analysis, optimization, graphics, and so on, incorporating both descriptive and optimizing models for different types of building, housing, and urban development problems. Thus, each module for integrated computer-aided design covers all five components: (1) Geometric Modeling, (2) Cost Engineering, (3) Design Analysis Including Design Optimization, (4) Drafting-Drawing, and (5) Data Management, Storage and Transfer, with compatibility between all the five components. As stated above, the Software: '**HudCAD**' is in modular form and relates to the building, housing, layout planning, and design optimization. Geometric modeling, which defines an object in terms of geometric properties and arrangement of spaces, is done by including *dimensionless* building-unit or built-unit patterns in different groups in the graphic part of the software. The *design analysis* and *design optimization* component can include both descriptive and *optimizing* models. *Thus, the Software Module-I of '**HudCAD**' covers all the five components of an integrated computer-aided design system including optimization, rarely available in commercial CAD systems cited above.* Operation of the application software module for *computer-aided building design by optimization*, indicating its salient features and brief user guide is outlined below.

9.11.2   Operation of the Application Software

Operation of the software is broadly divided into two parts, namely: (1) **Numeric Part** (C++ programs, e.g. LCO and MBO models described above) and the (2) **Graphic Part** (Lisp Graphic programs, e.g. GMOD1.LSP described later). Software's **Numeric Part** incorporates the optimizing models to generate *optimal designs* in *alphanumeric* form, and the **Graphic Part** of the software converts the *alphanumeric* optimal design solution so generated into a *graphic* form or drawing, complying with the specified *objective* (such as least-cost or most-benefit design) and *constraints*, linking various building and urban planning parameters such as built-space, density, open space, floor area ratio, ground coverage ratio and incorporates cost engineering (including construction specifications, elemental costs) to permit *integrated computer-aided design by optimization.*

Here, the Module-I is presented where the nonlinear programming optimizing models, i.e. least-cost optimizing model:CLC1.CPP and the most-benefit optimizing model:CMB1.CPP outlined above, are included in the numeric part which also store and transfer data of interest creating data files. Drafting and drawing functions are carried out by a graphic program which also generates, stores, and transfers data of interest in the output drawing. In numeric part of the software, a cost engineering program also carries out data management, storage and transfer functions related to costs and efficiency indicators. The graphic program carries out the graphics task, and can call the numeric C++ program when required to generate an acceptable numeric solution before converting it into a graphic solution. As shown below, the C ++ program of this software module has 19 input set options (extendable). Thus, using this technique building designers, users, and multiple stakeholders could generate, analyse, and evaluate numerous potential solutions both in numeric and graphic forms (as each numeric solution can be converted to graphic solution, thus permitting both numeric and graphic iterations) to select a particular solution that meets a specified criteria or goal, which is rarely possible in the conventional practices.

On loading and typing 'HudCAD' in the text part of computer screen, the software prompts the designer/user to select a specific module, with the following message appearing in the graphic part of the computer screen:

9.11.3 Optimizing Iteration Dialogue Structure of 'HudCAD': Module-I, Displayed in Computer Screen During Run of the Application Software

Typical optimizing dialogue structure of '**HudCAD**': **Module-I** is outlined below.

On loading '**HudCAD**' in the text part of computer screen, the software prompts the designer/user with following message for (Y/N) response.

```
DO YOU WANT TO RUN THE MODULE-I OF 'HudCAD'?
        Type Y (if 'yes') or N (if 'no') and press <CR>:
 Ans. = N (Message "Select another Module" appears in computer screen
for response)
M Ans. = Y i.e. selection of Module-I: the dimensionless design patterns
(for building-units or built-units) included in it are displayed in
graphic form prompting the user to choose a specific pattern, with the
following message in the Text Part of the Computer Screen for response by
the Designer/User:

BUILT-UNIT PATTERNS INCLUDED IN GROUP - I, ARE DISPLAYED ABOVE. SEE/
ENTER YOUR CURRENT "BU-PATTERN" CHOICE (D4/D2 so on)
```

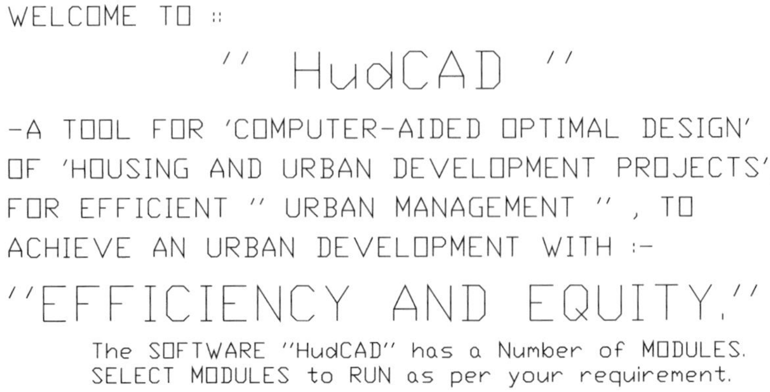

```
WELCOME TO ::
          '' HudCAD ''
-A TOOL FOR 'COMPUTER-AIDED OPTIMAL DESIGN'
OF 'HOUSING AND URBAN DEVELOPMENT PROJECTS',
FOR EFFICIENT '' URBAN MANAGEMENT '' , TO
ACHIEVE AN URBAN DEVELOPMENT WITH :-
''EFFICIENCY AND EQUITY.''
          The SOFTWARE ''HudCAD'' has a Number of MODULES.
          SELECT MODULES to RUN as per your requirement.
```

Fig. 9.7 Message appearing in graphic part of the computer screen on loading 'HudCAD'

Some typical residential built-unit design patterns included in the Group-I of the software module-I are shown in Fig. 9.7. However, it can include different design patterns conceived by a creative architect-designer, and derive the model constants accordingly, while using the same LCO or MBO models which are universal models. In fact, architect/building designers could create a **LIBRARY** of dimensionless built-unit-design-patterns for different building types, and include them in the software to improve its versatility.

On exercise of the 'Bu-Pattern' choice referred above, say, by typing D4 or any other pattern name, the following message appears in text part of the computer screen for response to start 'Optimization Session':

```
Type 'CMB1/CLC1 ' on command line & Start ' Optimizing Iteration Session'
```

On Typing CMB1, the C++ program for MBO model solution is opened with the following messages for response:

```
WELCOME to SOFTWARE 'HudCAD':'optimising Iteration Session'
(using'Geometric Programming', a NLP Optimisation\n Technique),to
achieve INSTANTLY the "MOST-BENEFIT'' or "LEAST-COST" Optimal Design of
an Urban Development. This 'HudCAD-MODULE' gives 'MAXIMUM SPACE
PHYSICAL DESIGN', within a given INVESTMENT per BUILT-UNIT
(Residential).OBSERVE that the OUTPUT OPTIMAL DESIGNS, given INSTANTLY
on selection of INPUT Values, shows a LARGE VARIATION in the BUILT-UNIT-
SPACE for a given COST per BUILT-UNIT, depending on Input Values(CHOSEN
BY DIFFERENT DISCIPLINES) Hence,carefully select INPUT Values of your
Specialisation, during the Cycle of Iteration, to achieve the OUTPUT
Optimal Design, which is both MOST COST-EFFECTIVE, EFFICIENT, and
ACCEPTABLE (EQUITY)Solution in a site-specific CONTEXT.
```

The C++ program gives the following option to run cost engineering model to derive site-specific unit costs.

```
"Do you want to run COST ENGINEERING MODEL to Derive Unit Costs ? (put 'y'
if yes/ 'n' if no)
```

On 'y' response Cost Engineering Model is RUN to derive unit elemental costs based on the *site-specific* unit material/labour costs and construction specification Input given by Engineer-Designer, who may also preset such values based on experience.

Similarly, planners can provide input related to environmental planning standards such as open-space, density, FAR, so on, and interact along with above disciplines in the building design/development process for realistic/desirable planning parameters adopting flexibility principles. Thus, the software integrates all the three elements, i.e. cost, design optimization, and drawing in a combined-platform covering architectural, physical planning, and engineering components, to maximize productivity, *resource-efficiency* and improve *social equity* in the building, housing, and urban development process, with interactive and *informed-participation* of multiple stakeholders in planning-design-decision-making (with *transparency*) in a site-specific-project, in context of urban dynamics and the individual, local, and national *resource-constraints*.

Earlier 'HudCAD' had included only four BU patterns as displayed in graphic part of 'HudCAD' shown in Fig. 9.8 above. These four BU patterns are mainly applicable to low- and middle-income housing, although, in very high land price and land scarcity situations some of these BU patterns may be applied even in high-income housing. However, to stress the applicability of the CAD technique

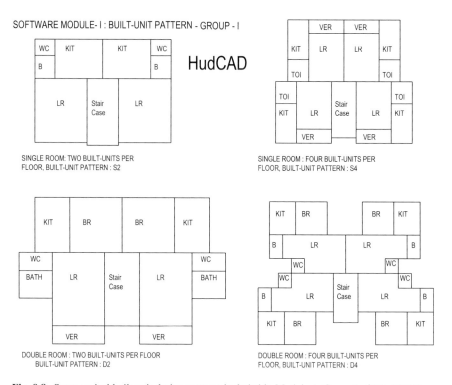

Fig. 9.8 Some typical built-unit-design patterns included in Module-1, Group-I of 'HudCAD'

presented in the book in housing for all income categories in all countries, a BU pattern with five bedrooms (as shown in Fig. 9.5 above giving the **Schematic Unit Design Module:E for High-Income Housing Showing**) is included in this Module-I of 'HudCAD'. Expressions for the built-unit design coefficients for the **Unit Design Module:E** are also presented above (see Sect. 9.6.2.2 above) and included in the current C++ programs of CMB1 and CLC1 for MBO and LCO Model Solution incorporated in Module-I of 'HudCAD'. It may be noted that the same C ++ programs (here CMB1 and CLC1) can be used for *optimal design* of any **Built-Unit Design Module** (preferably of rectilinear shape) developed by any creative designer, just by adding the expressions for the built-unit design coefficients for the respective **Built-Unit Design Module**, in the respective C++ program as clarified above. Thus, the optimization models presented are *Universal Optimization Models* which can be applied by any planner-designer in any **Built-Unit Design Module** conceived by a creative designer with the above additions, and obtain the *optimal design solutions*.

On finding the unit elemental costs either using cost engineering model referred above or on selecting of the preset unit elemental costs, the following message is displayed prompting the designer to enter the current choice of built-unit design pattern.

```
FOLLOWING FIVE BUILT-UNIT DESIGN PATTERNS AVAILABLE IN THIS MODULE/
GROUP
(FOUR DESIGNS SHOWN IN'BUILT-UNIT PATTERN: GROUP-I'DWG.-YELLOW PLATE)

1.DOUBLE ROOM DESIGN:4-BU'S PER FLOOR(PATTERN-'D4')
2.DOUBLE ROOM DESIGN:2-BU'S PER FLOOR(PATTERN-'D2')
3.SINGLE ROOM DESIGN:4-BU'S PER FLOOR(PATTERN-'S4')
4.SINGLE ROOM DESIGN:2-BU'S PER FLOOR(PATTERN-'S2')
5.FIVE BED ROOM DESIGN:2-BU'S PER FLOOR(PATTERN-'F2', presently not
included in Yellow Plate)

ENTER YOUR CURRENT CHOICE OF DESIGN PATTERN (1/2 so on)
```

On selection of design pattern, the following input choice is given:

```
Do you want to CONTINUE ITERATION SESSION WITH DIFFERENT DATA SETS?
    Type Y (if 'yes') or N (if 'no') and press <CR>:
  Ans. = N (Message "Close this Session" appears in computer screen for
response)
  Ans. = Y (The following message appears in computer screen for
appropriate response)
S.   YOU CAN CHOOSE ANY OF THE FOLLOWING 19 INPUT SET OPTIONS (FOR CHANGE
OF INPUT VALUES) WHICH ARE AVAILABLE ( EXTENDABLE) IN THIS MODULE:
 0. When you want to select the pre-set computer data input (for
    initializing the iteration)
 1. When you want to change values of all input variables
 2. When you want to change values of AC, BR, SPO, and all built-unit
    design coefficients
 3. When you want to change values of BR, SPO, and all built-unit design
    coefficients
```

```
4. When you want to change values of BR, and all built-unit design
   coefficients
5. When you want to change values of all built-unit design coefficients
6. When you want to change values of some built-unit design coefficients
7. When you want to change values of AC and all cost coefficients
8. When you want to change values of AC, BR, SPO and a few built-unit
   design coefficients
9. When you want to change values of BR, SPO, LP and a few built-unit
   design coefficients
10. When you want to change values of BR, LP
9. When you want to change values of BR, SPO
12. When you want to change value of NR (number of rooms) only
13. When you want to change value of BR only
14. When you want to change value of SPO only
15. When you want to change value of LP only
16. When you want to change value of AC only
17. When want change value of 's1, s2' (say no staircase) only
18. When you want to change value of 'fic' only
ENTER YOUR CURRENT CHOICE OF INPUT SET OPTION (0,1, 2 and so on):
(On selection of input set, computer prompts user to put the values of
corresponding input parameters; on putting these input values the C++
program gives instantly the output optimal results, including
generation of the output *cpp and *lsp files, followed by messages
appearing in screen prompting the user for apt response)
OUTPUT OPTIMAL GEOMETRIC PROGRAMMING WEIGHTS ARE:
OUTPUT OPTIMAL DIMENSIONS/VALUES OF BUILT-UNIT COMPONENTS ARE:
DO YOU WANT TO CHANGE INPUT OPTION/VALUES TO GET BETTER RESULTS?
          Type Y (if 'yes') or N (if 'no') and press <CR>:
 Ans. =   Y (Program goes back to S and the iteration restarted and
continued as above)
 Ans. =   N (Following message appears in computer screen for response
accordingly)
AS ABOVE OPTIMAL SOLUTION IS ACCEPTABLE, CLOSE "OPTIMIZING ITERATION
SESSION" /LOAD "GMOD1" AND START "DRAWING SESSION" TO GET THE SOFT / HARD
COPY OF OPTIMAL PHYSICAL DESIGN DRAWING OF BUILT-UNIT.
```

On closing '*OPTIMIZING ITERATION SESSION*' the '*DRAWING SESSION*' is started by loading the graphic program 'GMOD1' for the Module-I OF 'HudCAD'. While running the graphic program 'GMOD1', it loads the C++ output LISP file incorporating the numerical output *optimal* values of different planning and design parameters generated by the '*OPTIMIZING ITERATION SESSION*', and thus, integrate the *numerical optimal design* with the corresponding graphic *optimal design*.

9.11.4 Output Lisp Files Generated by the Numeric Part of the Software

Typical *cpp output files given by the C++ programs: CLC1.cpp (Least-Cost Optimizing Model), and CMB1.cpp (Most-Benefit Optimizing Model) are shown above, respectively, in Sect. 9.10.1 giving output optimal results of LCO model, and in Sect. 9.10.2 giving the output optimal results of MBO model. During running of these models, they also generate the 'schedule of rates' and 'detailed estimate' files, if the cost engineering option is chosen by the user, outlined later. Similarly, the LCO and MBO models generate the output LISP files named as: dg1.lsp and dg2.lsp, corresponding to the output *cpp file of these models, *generating the numerical data of optimal design* produced, for the purpose of *data transfer* to the graphic part of the software, which utilizes these data to produce the corresponding optimal design in graphic or drawing form, thus, achieving **instantly**, the generation and integration of both numeric and graphic optimal design in one platform. Typical output LISP files: dg1.lsp and dg2.lsp, corresponding to the output optimal results in Sect. 9.10.2 generated by the MBO Model: CMB1.cpp during its run for the **Example: MBL2** dwelling Module-D: Four Double Bedroom dwelling Served by one Staircase, is shown below just as illustrative examples.

```
File:dg1.lsp
(setq dlis1
(list 10.965 13.829 2.522 2.420 3.838
 2.074  28.760   300000 5000 2  4
30.000 5.890 4.670 0.75866 1076.000 837.000
 737.000   1.100  26.770   0.000   0.000 ) )
```

```
File:dg2.lsp
(setq dlis2
(list 2.522 2.074 0.968 1.776 0.968 1.184
 3.250 2.961 0.877 1.383 0.691 0.080
 0.150 0.230 0.070 0.120 0.400 0.100 0.550 0.150
 0.050  2 0.000 0.000 0.000  0 31025 30266 520
 0.000 0.000 0.000 0.000 0.000 0.000 2.700
 ) )
```

It may be seen that the output values of 10.965, 13.829, 2.522, 2.420, 3.838, 2.074 incorporated in the output LISP file dg1.lsp shown above are the numerical values (in metres) of lb (i.e. value of x_1), wb (i.e. value of x_2), lbr, wbr, lsc, and wk (i.e. width of bedroom), respectively (see Figs. 9.4(b) and 9.9). Again, the output values of 28.760, 300,000, 5000, 2, and 4 are the numerical values of gpa, ac, cl, nf (number of floors) and nd (number of dwelling units per floor), respectively. Similarly, the output values of 2.522, 2.074, 0.968, 1.776, 0.968, 1.184, 3.250, and 2.961,....incorporated in the output LISP file dg2.lsp shown above, are the numerical values (in metres) of lk (i.e. length of bed room), wsc, wbt, lbt, wwc, lwc, wlr, and llr, respectively (see Figs. 9.4(b) and 9.9). Thus, while running module-I of HudCAD its numerical *optimization* components CMB1.CPP and CLC1.CPP, generates **instantly** the *optimal numerical data* output results which are also *stored* instantly in a *cpp file as well as in a LISP output file for *data transfer* as input to the component graphic program 'GMOD1' in LISP language for drafting and drawing.

It may be seen that the module-I of HudCAD incorporates geometric modeling (see Sect. 9.6 and Fig. 9.8 above), cost engineering (see Sects. 9.4.2, 9.4.3, 9.4.4, and 9.7 above), design analysis, and *design optimization* (see Sects. 9.8, 9.9, and 9.10

above, and also the *design analysis* and *optimization* procedure outlined in Fig. 10.7 in Chap. 10). The drafting and drawing task is carried out by the LISP program 'GMOD1' forming graphic component of module-I of HudCAD. The task of data management, storage and transfer is carried out at various stages of running module-I of HudCAD as outlined above. Thus, all the five components of integrated CAD system discussed earlier are incorporated in the module-I of HudCAD, and both its numerical **optimization** part (i.e. CMB1.CPP and CLC1.CPP cited above) and the graphic part program (i.e. 'GMOD1'), are **instantly** and fully **integrated** producing the *integrated computer-aided optimal design by optimization* in no time, for the desired building and housing planning and design solution. These are presented in more details in Sect. 9.15 below with application example.

Typical iteration dialogue structure of the software outlined above shows that **Mostly YES/NO Options** are to be exercised by a user at various stages of its RUN, and hence, it is a very USER-FRIENDLY SOFTWARE. As shown at **M** above (of the Dialogue Structure of '**HudCAD**': module-I presented **above**) of the current C++ Programs of CMB1 and CLC1 (for MBO and LCO Model Solution included in module-I of 'HudCAD'), five built-unit design patterns are included (out of which first four BU patterns are displayed in graphic part of 'HudCAD' shown in Fig. 9.8 above). Interested architects and designers may consider enlarging and including additional BU patterns as per their requirements.

9.11.5 Some Application Results of Integrated CAD by Optimization in Housing and Urban Development Using Application Software: 'HUDCAD'

Application of the software '**HudCAD**' is illustrated using its module-I outlined above and taking optimal design of the residential built-unit-development as example. The design patterns shown in Fig. 9.7 and the MBO model presented above are used for optimal design of affordable housing development for low/medium income families (in the context of a developing country) even in sites subjected to high market land price, as is the case in many cities. The comparative optimal built-unit (BU) design drawings (Design-1 and Design-2), obtained as output by running '**HudCAD**' are shown in Figs. 9.9 and 9.10 (selecting the design pattern 'D4'), where some input *design-decisions* (such as SPO, i.e. Open-Space per BU, BR, i.e. Building Rise) are also shown. Here, the same affordable cost/investment per BU (AC = 300,000 monetary units/BU), market unit land price (LP = 5000 monetary units per m^2), and the construction specifications/elemental unit costs are chosen in both the designs, to facilitate comparison.

Figures 9.9 and 9.10 show that the design-1, adopting a higher SPO = 30 m^2/BU and a 'low-rise' (BR = 2), gives the *Maximum Possible Built-Space* (BUS) of *28.76* m^2 only (permitting only 1 bedroom), with optimal FAR = 0.648, GCR = 0.324, and the Built-Unit Density (BUD) of 225.327 BU/Hectare in any layout.

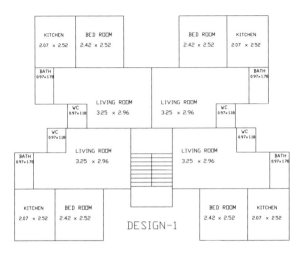

SOME INPUT PARAMETER VALUES GIVEN AS DESIGN CONSTRAINTS TO THE SOFTWARE :-
    NO. OF FLOORS ('BUILDING RISE' OR 'BR') = 2   OPEN SPACE ('SPO' : SQ.M./'BU') =   30.000
    CEILING COST/INVESTMENT (MONETARY UNITS PER 'BUILT-UNIT' OR 'BU')   =   300000.00
    MARKET LAND PRICE (MONETARY UNITS/SQ.M.) APPLICABLE IN THE ABOVE COST =   5000.00
SOME OUTPUT OPTIMAL PHYSICAL DESIGN VALUES GENERATED BY THE SOFTWARE :-
    MAXIMUM POSSIBLE BUILT-UNIT-SPACE (OR 'BUS') WITHIN CEILING INVESTMENT =   28.760   SQ.M.
    GROUND COVERAGE RATIO ACHIEVED IN ANY LAYOUT OF THE BUILT-UNITS   =   0.324
    'FAR' MATCHING WITH THE ABOVE 'SPO' WHICH CAN BE ALLOWED IN ANY LAYOUT =   0.648
    'BUILT-UNIT DENSITY'('BUD') ACHIEVED (NO OF BU'S/HECTARE) IN ANY LAYOUT =   225.327

Fig. 9.9 Optimal built-unit design drawing instantly given as output by 'HudCAD' depending on the input design-decisions (HOUSING: 'DESIGN-1', BR = 2)

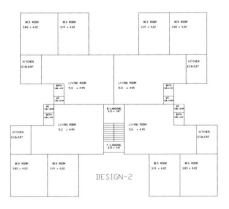

SOME INPUT PARAMETER VALUES GIVEN AS DESIGN DECISIONS TO SOFTWARE :-
    NO. OF FLOORS ('BUILDING RISE' OR 'BR')   = 4   OPEN SPACE ('SPO'SQ.M./'BU') = 15.000
    CEILING COST/INVESTMENT ('MONETARY UNITS' PER 'BUILT-UNIT' OR 'BU')   =   300000.00
    MARKET LAND PRICE (MONETARY UNITS/SQ.M.) APPLICABLE IN THE ABOVE COST =   5000.00
SOME OUTPUT OPTIMAL DESIGN VALUES GENERATED BY THE SOFTWARE :-
    MAXIMUM POSSIBLE BUILT-UNIT-SPACE('BUS') WITHIN CEILING INVESTMENT   =   72.534 SQ.M.
    GROUND COVERAGE RATIO ACHIEVED IN ANY LAYOUT OF BUILT-UNITS   =   0.547
    OPTIMUM 'FAR' ACHIEVING ABOVE 'SPO' IN ANY LAYOUT OF BUILT-UNITS   =   2.189
    'BUILT-UNIT DENSITY'('BUD') ACHIEVED (NO OF BU'S/HECTARE) IN ANY LAYOUT =   301.809

Fig. 9.10 Optimal built-unit design drawing instantly given as output by 'HudCAD' depending on the input design-decisions (HOUSING: 'DESIGN-2', BR = 4)

Design-2, instantly obtained by changing only the *Input Design-Decisions* related to SPO and BR, i.e. adopting a lower SPO = 15 m^2/BU and a 'medium-rise' (BR = 4) with corresponding change in relevant BU-coefficients, gives the *Maximum Possible Built-Space* (BUS) of *72.53* m^2 per BU (permitting 2 bedrooms as shown).

Thus, there is *increase of Built-Space per BU by* **2.52 times**, of optimal FAR by 3.38 times to 2.189, and of BUD by 1.34 times to 301.809 BU/Hectare in Design-2 compared to the Design-1, *even though* the input unit land price and the *investment/ cost per BU* (Built-Unit) are kept *unchanged* by the planner-designer in both the Designs-1 and 2. Such a large cost and design *efficiency* difference cannot be ignored by AEC professionals, multiple stakeholders and the *user citizens* who would like to maximize benefits in terms of space and other design parameters for each unit of money spent or invested in a project, fulfilling the desired environmental standards. This is possible only by design analysis and *optimization* in the form of a iteration process as proposed above.

In the iteration dialogue process (Sect. 9.13.1) showing **Mostly YES/NO Options**, all the stakeholders, such as the user citizens, architect, engineer, urban planner, urban manager, and so on, can ***participate*** and have a say in the planning and design process. Thus, they can contribute by giving appropriate input values as per their need and expertise, so as to obtain a most acceptable and optimal physical design of an urban development in a site-specific context, availing the *opportunities* and respecting the *constraints* (say the '*affordability*') beyond control at that decision level. *Participation* in such a dialogue process by the urban local bodies (including 'Building'/'Planning-Regulation' Authorities) and the real estate developers/builders, using such scientific tools, may lead to more realistic development control regulations, more *efficiency* and improved *social responsiveness* of various actors in the urban sector. Such a tool should have a wide selection of input set options, covering concerns of the above interest groups, so as to have a meaningful dialogue process. Accordingly, as shown in Sect. 9.13.1, in this software module 19 input set options are available, where AC, LP, BR, and SPO are affordable cost or investment per Built-Unit (BU), market land price, building rise, and semi-public open space per BU, respectively. The input set options can be enlarged/modified as per the priority of the concerned interest group(s) including the professional groups, i.e. architects, engineers, and so on.

9.12 Compatibility Between all Five Components of Integrated CAD by Optimization, and Its Utility in AEC Operations (Building and Housing) in all Countries with Diverse Income Statuses

As discussed in earlier chapters, an integrated CAD system should have five components: (1) geometric modeling, (2) cost engineering, (3) design analysis and *design optimization*, (4) drafting and drawing, and (5) data management, storage and transfer, and all the five components need to combine keeping compatibility with each other. It may be seen that the software module-I of '**HudCAD**' presented in earlier sections covers all the above five components including cost engineering. There is considerable difference in concepts and standards related to building and housing, particularly housing, in very low-income and high-income countries, often leading to the erroneous notion that *design analysis* and *optimization* are applicable only in the context of very low-income countries. To highlight the importance, relevance, general applicability, and the utility of integrated CAD by optimization in all countries, here, three application example cases, namely: (1) very low-income case, (2) low/middle-income case, and (3) high-income case, are presented, demonstrating that the software module-1 of '**HudCAD**' ensures compatibility between the *site-specific* cost engineering data, *optimal housing design* in numeric and graphic-drawing, and the short item wise detailed estimate; all generated **instantly** by '**HudCAD**', for all types of dwelling modules in the above cases. Application results of the software module-1 of '**HudCAD**' in the above three cases are presented below.

9.12.1 Context: Very Low-Income Countries Case

In mass urban housing in very low-income countries, struggling with acute affordability problem of prospective beneficiaries, stress is given to reduce Built-Unit-Area (BUA). Here, a least-cost design of dwelling module: B with a minimum acceptable Built-Unit-Space (BUS) of only 15.744 m^2 per Built-Unit (BU) is taken as an example. The expressions for deriving module coefficients for dwelling module: B presented in Sect. 9.6.1.3 above, are included in the C++ programmes: CLC1.CPP and CMB1.CPP which derive instantly the corresponding model constants. As outlined in Sect. 9.9.3 above, giving the optimizing iteration dialogue structure of '**HudCAD**': **module-I**, the above C++ programs incorporated in the **module-I**, offers the following option to run cost engineering model to derive *site-specific* unit costs.

```
"Do you want to run COST ENGINEERING MODEL to Derive Unit Costs ? (put 'y'
if yes/ 'n' if no)
```

On 'y' response cost engineering model is RUN to derive the Site-Specific Schedule of Rates (SOR) and the unit elemental costs, based on the *site-specific* unit material/labour costs and construction specification input given by the engineer-designer, who may also preset such values based on experience. Accordingly, opting to run cost engineering model, firstly, the site-specific Schedule of Rates (SOR) is generated instantly, by the cost engineering model, using the *site-specific* unit material/labour costs given as input by the engineer designer, as shown below.

9.12.1.1 Site-Specific Schedule of Rates (SOR) Instantly Given by the Cost Engineering Model

The Basic Rates (B.Rate) or SOR generated by Cost Engineering Model for the chosen Input parameters are:

Item	Description of Item	Unit	Basic Rate (Monetary Units)
1.	Earthwork (ordinary soil) in Wall Foundation	Cum	24.020
2.	Plain Cement Concrete 1:4:8 (1 Cement:4 Coarse Sand:8 Stone Agg.40 mm) in Foundation	Cum	1002.010
3.	Brick-work in Foundation in 1:8 (1 Cement:8 Fine Sand) cement mortar	Cum	809.300
4.	Brick-work in Foundation in 1:6 (1 Cement:6 Fine Sand) cement mortar	Cum	842.000
5.	Providing/Laying DPC with Cement Conc. 1:2:4 (1Cem:2C.Sand:4S.AG10)	Cum	1780.065
6.	Brick-work in Superstructure in 1:8 (1 Cement:8 Fine Sand) cement mortar	Cum	846.700
7.	Brick-work in Superstructure in 1:6 (1 Cement:6 Fine Sand) cement mortar	Cum	879.400
8.	Cement plaster 15 mm thick (rough side of wall) in 1:6 (1Cem.:6F.Sand) cement mortar	Sqm	36.067
9.	Cement plaster 12 mm thick in wall in 1:6 (1Cem.:6F.Sand) cement mortar	Sqm	30.462
10.	Cement plaster 6 mm thick (roof/ceiling) in 1:4 (1Cem.:4F.Sand) cement mortar	Sqm	24.127
9.	Re-infrcd.cem.conc.in roof/floor 1:3:5 (1Cem.:3C.Sand:5S.AG20) excldng.shutrng./re-infrcmnt.	Cum	1404.820
12.	Re-infrcd.cem.conc.in roof/floor 1:2:4 (1Cem.:2C.Sand:4S.AG20) excldng.shutrng./re-infrcmnt.	Cum	1682.725
13.	Mild Steel Reinforcement for RCC Work incl. bending,binding,placing in position	Kg	16.075
14.	Colour washing (green,blue,buff) on New Work (2 or more coats) including a base coat of white washing with lime to give even shade	Sqm	3.120
15.	Cement Concrete Flooring 1:2:4 (1 Cement:2 Coarse Sand:4 Stone Aggregate 20 mm Nom.Size) finished with a floating coat neat cement	Cum	2338.925

INPUT MARKET RATES (+CENTAGE) OF ABOVE BASIC COST ENGINEERING ITEMS ARE:
UNSKL1=70.000, UNSKL2=60.000, SKL1=120.000, SKL2=100.000, SKL3=90.000, SKL4=80.000
CEMENT=2700.000, SAND.F=200.000, SAND.C=357.000, A10=600.000, A20=518.000
A40=178.000, BRICK=900.000, STEEL=13.500, LIME=140.000

9.12.1.2   Site-Specific Building Component Unit Costs and the Output Optimal Results Instantly Given by the Cost Engineering Model

Here, three construction specifications, which are designated by three numerals, i.e. 1, 2, and 3, are chosen for each building component, i.e. foundation, wall, roof, and floor types. Based on this site-specific 'SOR' and choosing the foundation, wall, roof, and floor types, i.e. choosing 4 numeral values of any of 1/2/3 designating the type of each component (say: 2, 2, 2, and 2 in this case), the building component unit costs of the chosen foundation, wall, roof, and floor types instantly generated by the CE model are as follows:

$$cf = 287.600 \text{ MU per RM}, \quad cw = 722.277 \text{ MU per RM}, \quad cr = 326.330 \text{ MU per Sq.M}, \quad cfl = 136.026 \text{ MU per Sq.M}$$

On starting numerical 'Optimising Iteration Session' with different data sets and choosing a specific built-unit design pattern (here chosen pattern: B or 3, out of 5 BU Patterns included in the '**HudCAD': Module-I**), but using the same C++ programs for optimization) and a data set, the model constants c_1, c_2, and c_3 are derived using the above unit elemental costs given by the SOR and the expressions (9.1), (9.2), and (9.3) respectively, presented earlier, using a cost factor designated as: 'fic', and the number of floors in a building designated as: 'nf', to allow for the increase in unit costs due to building rise as explained earlier (see Sect. 9.7). The model constants c_4 is derived using the expression (9.4) given above. Accordingly, the output optimal results given by the C++ program CLC1.CPP in this case are presented below.

Output Optimal Results of C++ Program CLC1.CPP for Given Input Design-Decisions

```
INPUT-Design-Decisions (some) in Current-Case are:
BUILT-UNIT DESIGN PATTERN CHOSEN IS: 'BU-PATTERN- 3'
Some Input Built-Unit Building Co-efficients adopted are:
s1= 0.050 s2= 0.180 k1= 0.500 w2= 0.150 flw= 1.000
w3= 0.150 ow= 0.000 k = 0.000 ew= 0.000 kw= 0.000 gpa= 15.744 npa= 14.048
bt1= 0.300 cx1= 5.600 cx2= 3.420 cba= 0.91900 n= 1 ns= 8
Some Input Cost Co-efficients adopted are:
 cl=2000.00 cf= 287.60 cw= 722.28 cr= 326.33 cfl= 136.03
 c1=19123.79 c2=9679.17 c3= 355.99 c4=1838.00
SOME OUTPUT OPTIMAL GEOMETRIC PROGRAMMING WEIGHTS ARE:
w9= 0.40457 w12= 0.15924 w21= 0.47206 w31= 0.15641 w03= 0.12337
w41= 0.40457 w51= 0.43902 w04= 0.56381 w01/02= 0.15641
OPTIMAL DIMENSIONS OF THE BUILT-UNIT COMPONENTS ARE (M):
lbr= 3.235 wbr= 2.754 llr= 0.000 wlr= 0.000
lk = 2.264 wk = 1.589 lbt= 0.970 wbt= 1.589
lwc= 0.000 wwc= 0.000 lsc= 3.558 wsc= 1.907
ssf= 0.323 owl= 0.000 wo = 0.000 blw= 1.038
```

```
OTHER OPTIMAL OUTPUT VALUES GIVEN BY THE MODEL ARE:
LEAST-POSSIBLE COST PER BUILT-UNIT: 49434.836 Monetary Unit
Length of Built-Unit Block along the staircase: 6.469 M
Width of Built-Unit Block across the staircase: 10.593 M
Optimum Plinth Area Rate given by the MODEL is= 3139.940 Monetary Units/
SQ.M.
Optimum Number of Floors given by the MODEL(same as the
Maximum Number of Floors specified) is : 4
Optimum Floor Area Ratio given by the MODEL is= 1.130
Optimum Open Space Standard(same as the Minimum Acceptable
Standard Specified) given by the MODEL is : 10.000 SQ.M./Built Unit
Optimum Ground Coverage Ratio given by the MODEL is= 0.282
Optimum Built-Unit Density achieved(in BU/HA) is ==717.568
```

In this Cost Engineering Software-Module, the following labour and material items are considered (interested readers of the book can enlarge or modify it as per their requirements):

```
LABOUR:Un-Skilled-Category-1(UNSKL1) such as:BELDAR(RATE-EACH MU/PER
DAY)
LABOUR:Un-Skilled-Category-2(UNSKL2) such as:COOLIE(RATE-EACH MU/PER
DAY)
LABOUR:Skilled-Category-1(SKL1) such as:MASON-CLASS-1(RATE-EACH
MU/DAY)
LABOUR:Skilled-Category-2(SKL2) such as:MASON-CLASS-2(RATE-EACH
MU/DAY)
LABOUR:Skilled-Category-3(SKL3) such as:MATE(RATE-EACH MU/PER DAY)
LABOUR:Skilled-Category-4(SKL4) like BHISTI/OTHER-1(RATE-EACH
MU/DAY)
MATERIAL:-Portland Cement (RATE MU PER METRIC TONNE)
MATERIAL:-Fine Sand(SAND.F) (RATE MU PER CUBIC METER)
MATERIAL:-Course Sand (SAND.C) (RATE MU PER CUBIC METER)
MATERIAL:-Stone Aggregate 10 mm (A10) Nominal Size(RATE MU PER CUBIC
METER)
MATERIAL:-Stone Aggregate 20 mm (A20) Nominal Size(RATE MU PER CUBIC
METER)
MATERIAL:-Stone Aggregate 40 mm (A40) Nominal Size (RATE MU PER CUBIC
METER)
MATERIAL:-1st Class Bricks (RATE MU PER 1000 Nos)
MATERIAL:-Mild Steel/Torsteel Round Bars (RATE MU PER Kilo Grams (Kg)
OTHER2:-White Lime/OTHER MATERIALS etc. (RATE MU PER QUINTAL)
OTHER3:-MISCELLANEOUS/SUNDRIES/Hire & Running Charges of Plants etc.
MU = 'Monetary-Unit' in any Currency (Rupees, Dollars and so on)
```

To allow for increase in the unit costs due to building rise explained earlier, each basic rate (i.e. B.Rate) given by the SOR as above, is multiplied by a factor '$(nf(1- + (nf - 1)fic)$' to derive the effective rate (i.e. E.Rate), which is used in the detailed estimate as shown below.

9.12.1.3 Detailed Estimate per Dwelling Unit INSTANTLY Given by the Cost Engineering Model

While running the cost engineering model, it generates instantly the detailed estimate per dwelling unit, based on the input data chosen by the engineer designer, which is exactly compatible with the above output optimal results, as shown below:

Site-Specific Detailed Estimate per Dwelling Unit INSTANTLY Given by Cost Engineering Model

Costs given by Cost Engineering MODEL for the Chosen Foundation/Wall/Roof/ Floor Types are:

$cf = 287.600$ MU per RM, $cw = 722.277$ MU per RM $cr = 326.330$ MU per SQM, $cfl = 136.026$ MU per SQM

   Built-Unit Design Selected is:-SINGLE ROOM DESIGN:4-BU'S PER FLOOR (PATTERN-'S4' or 3)
   Dwelling Specifications Selected: Foundation Type = 2, Wall Type = 2, Roof Type = 2, Floor Type = 2
   Detailed Estimate per Dwelling Unit INSTANTLY given by Cost Engineering Model:

```
Item Description of Item     Quantity  B. Rate    E. Rate    Amount(MU)
                             /D.Unit   (MU/Unit)  (MU/Unit)
1. Earthwork(ordinary soil) 2.296 Cum  24.020 25.822 59.274
   in Wall Foundation
2. Plain Cement Concrete 1:4:8(1 Cement:4  0.612 Cum  1002.010   1077.161
659.423
   Coarse Sand:8 Stone Agg.40 mm)in Foundation
3. Brick-work in Foundation in 1:8  1.069 Cum  809.300  869.997  929.769
   (1 Cement:8 Fine Sand)cement mortar
4. Brick-work in Foundation in 1:6   0.000 Cum  842.000  905.150  0.000
   (1 Cement:6 Fine Sand)cement mortar
5. Providing/Laying DPC with Cement Conc.  0.042 Cum  1780.065 1913.570
79.722
   1:2:4(1Cem:2C.Sand:4S.AG10)
6. Brick-work in Superstructure in 1:8  9.249 Cum  846.700  910.203
10238.463
   (1 Cement:8 Fine Sand)cement mortar
7. Brick-work in Superstructure in 1:6  0.000 Cum  879.400 945.355  0.000
   (1 Cement:6 Fine Sand)cement mortar
8. Cement plaster 15 mm thick(rough side of 48.907 Sqm  36.067  38.772
1896.214
   wall)in 1:6(1Cem.:6F.Sand)cement mortar
9. Cement plaster 12 mm thick in wall  48.907 Sqm 30.462 32.747  1601.533
   in 1:6(1Cem.:6F.Sand)cement mortar
```

10. Cement plaster 6 mm thick (roof/ceiling) in   15.744 Sqm 24.127
25.937   408.341
  1:4 (1Cem.:4F.Sand) cement mortar
9. Re-infrd.cem.conc.in roof/floor 1:3:5 (1Cem.   0.000 Cum   1404.820
1510.182   0.000
  :3C.Sand:5S.AG20) excldng.shutrng./re-infrcmnt.
12. Re-infrd.cem.conc.in roof/floor 1:2:4 (1Cem.   1.574 Cum   1682.725
1808.929   2847.956
  :2C.Sand:4S.AG20) excldng.shutrng./re-infrcmnt.
13. Mild Steel Reinforcement for RCC Work incl. 70.847 Kg   16.075   17.281
1224.288
  bending,binding,placing in position
14. Colour washing (green,blue,buff) on New Work 93.557 Sqm   3.120   3.354
380.872
  (2 or more coats) including a base coat of
  white washing with lime to give even shade
15. Cement Concrete Flooring 1:2:4 (1 Cement:2 0.492 Cum 2338.925 2514.344
1237.048
  Coarse Sand:4 Stone Aggregate 20 mm Nom.Size)
  finished with a floating coat neat cement
16. Cost per Dwelling Unit (As Per Above Detailed Estimate For Building
Cost) =21562.902
17. Land Cost Per Dwelling Unit (As Per Far, Gpa & Land Price In Model)
=27871.932
18. Cost Per D.Unit (Difference Between Least Cost (Ac) & Land Cost/D.
Unit)   =21562.904
19. Total Cost Per D.Unit (Bldg.Cost Per Dwelling Unit + Land Cost/D.
Unit)   =49434.836
20. Total Least Cost Per Dwelling Unit (As Given By The Lco Model)
=49434.836
21. Minimum Acceptable Built-Unit-Space (Given As Input To The Lco
Model)   = 15.744 Sq.M.
INPUT MARKET RATES (+CENTAGE) OF BASIC COST ENGINEERING ITEMS ARE:
UNSKL1=70.000, UNSKL2=60.000, SKL1=120.000, SKL2=100.000,
SKL3=90.000, SKL4=80.000
CEMENT=2700.000, SAND.F=200.000, SAND.C=357.000, A10=600.000,
A20=518.000
A40=178.000, BRICK=900.000, STEEL=13.500, LIME=140.000, OTHER2=1.0

9.12.1.4   Site-Specific Building Design Drawing Compatible with the Above Output Optimal Results, Instantly Given by the Component Graphic Program of 'HudCAD': Module-I

As discussed in Sect. 9.13.1, the drafting and drawing task is carried out by the LISP program 'GMOD1' forming graphic component of MODULE-I of HudCAD. The procedure to convert numerical optimal design given by the C++ program (as presented in Sect. 9.13.1.2 above for this case) is also outlined in above section. Adopting this procedure, the least-cost optimal built-unit design drawing instantly given by '**HudCAD**', compatible with the above output optimal results, is shown in Fig. 9.11 above.

Fig. 9.11 Least-cost optimal built-unit design drawing (case-very low-income housing, BU pattern: B) instantly given by 'HudCAD', with all five components of integrated CAD

9.12.2   Context: Low- or Middle-Income Countries Case

In urban housing in low/middle-income countries, generally, moderate Built-Unit-Area (BUA) compatible with the affordability and priority of prospective beneficiaries is adopted. Here, a least-cost design of dwelling Module-D, shown in Fig. 9.8 above, with a minimum acceptable Built-Unit-Space (BUS) of 72.319 m^2 per Built-Unit (BU) (see also Sect. 9.6.1.1 and Fig. 9.4(b) above) is taken as an example. The expressions for deriving module coefficients for dwelling design module: D presented in Sect. 9.6.1.1 above, and is included in the C++ programmes: CLC1. CPP and CMB1.CPP which derives instantly the corresponding model constants.

9.12.2.1   Site-Specific Building Component Unit Costs and the Output Optimal Results Instantly Given by the Cost Engineering Model

As discussed earlier in Sect. 9.13.1 above, the optimizing iteration dialogue structure of '**HudCAD': Module-I**, the above C++ programs incorporated in the **Module-I**, offers the option to run cost engineering model to derive the *site-specific* Schedule of Rates (SOR) and the unit elemental costs, based on the *site-specific* unit material/labour costs and construction specification input given by the engineer-designer. Here, the same input materials and labour costs and the corresponding Schedule of Rates (SOR) instantly given by the cost engineering model as shown in Sect. 9.13.1.1 above are used. Similarly, based on this site-specific 'SOR' and choosing the foundation, wall, roof, and floor types as: 2, 2, 2, and 2, respectively, in this case also, the building component unit costs of the chosen foundation, wall, roof, and floor types instantly generated by the CE model are as follows:

$cf = 287.600$ MU per RM, $cw = 722.277$ MU per RM, $cr = 326.330$ MU per Sq.M, $cfl = 136.026$ MU per Sq.M

On starting numerical 'Optimising Iteration Session' with different data sets and choosing a specific built-unit design pattern (here chosen pattern: D or 1, out of 5 BU patterns included in the '**HudCAD**': **Module-I**), but using the same C++ programs for optimization) and a data set, the model constants c_1, c_2, and c_3 are derived using the above unit elemental costs given by the SOR and the expressions (9.1), (9.2), and (9.3) respectively, presented earlier, using a cost factor designated as: 'fic', and the number of floors in a building designated as: 'nf', to allow for the increase in unit costs due to building rise as explained earlier (see Sect. 9.7). The model constants c_4 is derived using the expression (9.4) given above. Accordingly, the output optimal results given by the C++ program CLC1.CPP in this case is presented below.

Output Optimal Results of C++ Program CLC1.CPP for Given Input Design-Decisions

```
INPUT-Design-Decisions (some) in Current-Case are:
BUILT-UNIT DESIGN PATTERN CHOSEN IS: 'BU-PATTERN- 1'
Some Input Built-Unit Building Co-efficients adopted are:
s1= 0.010 s2= 0.100 k1= 0.240 w2= 0.060 flw= 1.100
w3= 0.160 ow= 0.060 k = 0.400 ew= 0.120 kw= 0.180 gpa= 72.319 npa= 69.699
bt1= 0.550 cx1= 5.760 cx2= 5.020 cba= 0.74524 ns = 8
Some Input Cost Co-efficients adopted are:
cl=5000.00 cf= 287.60 cw= 722.28 cr= 326.33 cfl= 136.03
c1=19670.18 c2=17143.9 c3= 288.68 c4=3726.20
SOME OUTPUT OPTIMAL GEOMETRIC PROGRAMMING WEIGHTS ARE:
w9= 0.31429 w12= 0.37881 w21= 0.56832 w31= 0.09475 w03= 0.9739
w41= 0.31429 w51= 0.59096 w04= 0.69310 w01/02= 0.09475
OPTIMAL DIMENSIONS OF THE BUILT-UNIT COMPONENTS ARE (M):
lbr= 4.414 wbr= 3.166 llr= 4.782 wlr= 4.854
lk = 4.414 wk = 3.799 lbt= 2.869 wbt= 2.90
lwc= 1.913 wwc= 1.266 lsc= 4.966 wsc= 2.90
ssf= 0.184 owl= 2.532 wo = 1.266 blw= 2.266
OTHER OPTIMAL OUTPUT VALUES GIVEN BY THE MODEL ARE:
LEAST-POSSIBLE COST PER BUILT-UNIT: 238636.094 Monetary Units (MU)
Length of Built-Unit Block along the staircase: 18.393 M
Width of Built-Unit Block across the staircase: 21.104 M
Optimum Plinth Area Rate given by the MODEL is= 3299.772 Monetary Units/
SQ.M.
Optimum Number of Floors given by the MODEL (same as the
Maximum Number of Floors specified) is: 4
Optimum Floor Area Ratio given by the MODEL is= 2.186
Optimum Open Space Standard (same as the Minimum Acceptable
Standard Specified) given by the MODEL is : 15.000 SQ.M./Built Unit
Optimum Ground Coverage Ratio given by the MODEL is= 0.547
Optimum Built-Unit Density achieved (in BU/HA) is ==302.300
```

9.12.3   Detailed Estimate per Dwelling Unit INSTANTLY Given by the Cost Engineering Model

To allow for increase in the unit costs due to building rise explained earlier, each basic rate (i.e. B.Rate) given by the SOR as above is multiplied by a factor '(nf (1+(nf-1)fic)' to derive the effective rate (i.e. E.Rate), which is used in the detailed estimate as shown below.

9.12.3.1   Site-Specific Detailed Estimate per Dwelling Unit INSTANTLY Given by Cost Engineering Model:

As mentioned above, while running the cost engineering model, it generates instantly the detailed estimate per dwelling unit, based on the input data chosen by the engineer designer, which is exactly compatible with the above output optimal results, as shown below:

Unit costs given by CE model for chosen foundation/wall/roof/floor types are:

$cf = 287.600$ MU per RM, $cw = 722.277$ MU per RM $cr = 326.330$ MU per SQM, $cfl = 136.026$ MU per SQM

```
Built-Unit Design Selected is:-DOUBLE ROOM DESIGN: 4- BU'S PER FLOOR
(PATTERN- 'D4 or 1')
Dwelling Specifications Selected: Foundation Type= 2, Wall Type= 2, Roof
Type= 2, Floor Type= 2
Detailed Estimate per Dwelling Unit INSTANTLY given by Cost Engineering
Model:
Item  Description of Item     Quantity/    B.Rate    E.Rate     Amount
                             D.Unit      (MU/Unit) (MU/Unit)   (MU)
1. Earthwork (ordinary soil) 7.370 Cum    24.020    25.822    190.304
   in Wall Foundation
2. Plain Cement Concrete 1:4:8
   (1 Cement:4            2.283 Cum 1002.010  1077.161  2459.136
   Coarse Sand:8 Stone Agg.40 mm) in Foundation
3. Brick-work in Foundation in 1:8  3.125 Cum  809.300  869.997  2719.012
   (1 Cement:8 Fine Sand) cement mortar
4. Brick-work in Foundation in 1:6  0.000 Cum  842.000  905.150    0.000
   (1 Cement:6 Fine Sand) cement mortar
5. Providing/Laying DPC with Cement Conc.  0.122 Cum 1780.065  1913.570
233.138
   1:2:4(1Cem:2C.Sand:4S.AG10)
6. Brick-work in Superstructure in 1:8   32.895 Cum  846.700   910.203
29941.303
   (1 Cement:8 Fine Sand) cement mortar
7. Brick-work in Superstructure in 1:6 0.000 Cum  879.400  945.35    0.000
   (1 Cement:6 Fine Sand) cement mortar
8. Cement plaster 15 mm thick (rough side of  143.023 Sqm   36.067    38.772
5545.278
```

wall) in 1:6 (1Cem.:6F.Sand) cement mortar
9. Cement plaster 12 mm thick in wall   143.023 Sqm   30.462   32.747
4683.513
  in 1:6 (1Cem.:6F.Sand) cement mortar
10. Cement plaster 6 mm thick (roof/ceiling) in 72.319 Sqm   24.127
25.937   1875.703
  1:4 (1Cem.:4F.Sand) cement mortar
9. Re-infrd.cem.conc. in roof/floor 1:3:5 (1Cem.   0.000 Cum   1404.820
1510.182   0.000
  :3C.Sand:5S.AG20) excldng.shutrng./re-infrcmnt.
12. Re-infrd.cem.conc. in roof/floor 1:2:4 (1Cem.   7.232 Cum   1682.725
1808.929   13081.988
  :2C.Sand:4S.AG20) excldng.shutrng./re-infrcmnt.
13. Mild Steel Reinforcement for RCC Work incl. 325.435 Kg   16.075   17.281
5623.725
  bending,binding,placing in position
14. Colour washing (green,blue,buff) on New Work 358.364 Sqm   3.120
3.354   1201.954
  (2 or more coats) including a base coat of
  white washing with lime to give even shade
15. Cement Concrete Flooring 1:2:4 (1 Cement:2   2.260 Cum 2338.925
2514.344   5682.336
  Coarse Sand:4 Stone Aggregate 20 mm Nom.Size)
  finished with a floating coat neat cement
16. Cost per Dwelling Unit (As per Above Detailed Estimate for Building Cost)
= 73237.391
17. Land Cost per Dwelling Unit (As per FAR, GPA & Land Price in Model)   =
165398.703 MU
18. Cost per D.Unit (Difference between Least Cost (ac) & Land Cost/D.Unit)
= 73237.391 MU
19. Total Cost per D.Unit (Bldg.Cost per Dwelling Unit + Land Cost/D.Unit)
= 238636.094 MU
20. Total Least Cost per Dwelling Unit (As Given By the LCO Model)   =
238636.094 MU
21. Minimum Acceptable Built-Unit-Space (Given as Input to the LCO Model) =
72.319 Sq.M.

The corresponding unit item wise short detailed estimate is presented above, which is generated **instantly** and stored by the C++ program as above, along with the *site-specific* input unit market rates (placed below) of basic cost engineering items of labour and materials given as input by the engineer-designer, while running the C++ programs CLC1.CPP or CMB1.CPP cited earlier. The dwelling construction specifications input given by the engineer-designer for the four components: foundation, wall, roof, and floor are shown in Fig. 9.11 above, and also in the detailed estimate below by types designated by numerals 1, 2, 3 for each of the above 4 dwelling components. Thus, Foundation Type = 2 means: predesigned Foundation for 230 mm thick brick wall in 1:8 Cem.sand mortar (while running cost engineering model option to redesign foundation is given), Wall Type = 2 means: 230 mm thick Brick Wall (2.7 M High) in 1:8 Cem.sand mortar, Roof Type = 2 means: RCC Roof in MARK 200, i.e. in 1:2:4 Mix with 6mm cement sand plaster, Floor Type = 2 means: Flooring 25 mm thick CC(1:2:4) on 1:4:8 PCC finished with a floating coat cement.

The units applicable for the respective basic cost engineering items referred above are given in Sect. 9.4.4 **Computer-Aided Cost Estimating** outlined above. Here, input data regarding both the input cost engineering coefficients (quantities) and the input labour-material-unit-cost (MKT.RATES) coefficients are taken as per the Indian Public Work Department practice, just to demonstrate the feasibility of **Instant** computer-aided cost estimating using the *country-specific* and *site-specific* data to make the *optimal* output results given by the '**HudCAD**' most realistic in terms of costs. The dialogue structure in Sect. 9.4.4 above would show that the cost engineering model is so designed that the users have the option to put their own *country-specific* and *site-specific* data if so desired, and use the cost engineering model accordingly. Such cost engineering data can also be generated by cost engineering research and by *works and method studies* (as done in industrial engineering) in AEC operations in different countries, to make such computer-aided cost estimating more scientific. As mentioned above while running the C++ programs CLC1.CPP or CMB1.CPP cited earlier, the engineer/designer is given the option to run cost engineering model to derive the 'unit elemental costs'. Accordingly, in this case the 'unit elemental costs' of $cf = 287.60$, $cw = 722.28$, $cr = 326.33$, and $cfl = 136.03$ for the chosen foundation, wall, roof, and floor types, respectively, shown in the above 'Output Optimal Results of the C++ program CLC1.CPP', are generated **instantly** by the same C++ program. Similarly, the site-specific Schedule of Rates (SOR) **instantly** given by the cost engineering model in this case is also shown above. basic unit rates (i.e. B.Rate in MU/Unit) given by site-specific schedule of rates is multiplied by a cost factor (function of 'fic', 'nf') due to building rise to derive R.Rate in MU/Unit for each basic unit rate as shown above.

9.12.3.2   Site-Specific Building Design Drawing Compatible with the Above Output Optimal Results, Instantly Given by the Component Graphic Program of 'HudCAD': Module-I

As outlined in Sect. 9.11.4 above, the drafting and drawing task is carried out by the LISP program 'GMOD1' forming graphic component of Module-I of HudCAD. Typical *cpp output files given by the C++ programs CLC1.cpp and CMB1.cpp are shown, respectively, in Sects. 9.10.1 and 9.10.2 above. It is pointed out that while running the above C++ programs CLC1.cpp and CMB1.cpp, these programs also generate the output LISP files named as: dg1.lsp and dg2.lsp, incorporating the numerical data of the *optimal design solution*, corresponding to the output *cpp file generated. The procedure to convert these *numerical optimal design solution* data given by the C++ program is also outlined in the above sections with application examples. Adopting this procedure, the **Least-Cost** *optimal* built-unit design drawing instantly given by '**HudCAD**', compatible with the above output *optimal* numerical results, is shown in Fig. 9.12 above.

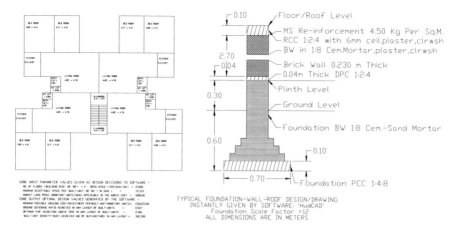

Fig. 9.12 Least-cost optimal built-unit design drawing (case-low/middle-income housing, BU pattern: D) given by 'HudCAD', incorporating all five components of integrated CAD

9.12.4 Context: High-Income Countries Case

In residential schemes in high-income countries, generally affordability problem is not as acute as in very low-, low-, and middle-income countries. Here, a least-cost design of dwelling Module-E, shown in Fig. 9.5 above, with a minimum acceptable Built-Unit-Space (BUS) of 300 m^2 per Built-Unit (BU) is taken as an example. The expressions for deriving design module coefficients for dwelling design module: E presented in Sect. 9.6.2.2 above, and is included in the C++ programmes: CLC1. CPP and CMB1.CPP referred above, which derives instantly the corresponding model constants, i.e. dwelling unit design module coefficients: cx_1, cx_2, and cba, for the multi-family dwelling unit design module for high-income housing shown in Fig. 9.5 above, using the proportion coefficients indicated in the diagram.

9.12.4.1 Site-Specific Building Component Unit Costs and the Output Optimal Results Instantly Given by the Cost Engineering Model

As discussed earlier in Sect. 9.11.3 above, the optimizing iteration dialogue structure of 'HudCAD': **Module-I**, the above C++ programs incorporated in the **Module-I**, offers the option to run cost engineering model to derive the *site-specific* Schedule of Rates (SOR) and the unit elemental costs, based on the *site-specific* unit material/labour costs and construction specification Input given by the engineer-designer. Here, the same input materials and labour costs and the corresponding Schedule of Rates (SOR) instantly given by the cost engineering model as shown in Sects. 9.12.1 and 9.12.2 above, are used. Similarly, based on this site-specific 'SOR' and choosing the foundation, wall, roof, and floor types as: 3, 3, 3, and 3 respectively, in this case also, the building component unit costs of the chosen foundation, wall, roof, and floor types instantly generated by the CE model are as follows:

$$cf = 549.05 \text{ MU per RM}, cw = 1016.08 \text{ MU per RM}, cr = 49.14 \text{ MU per Sq.M},$$
$$cfl = 197.12 \text{ MU per Sq.M}$$

On starting numerical 'Optimising Iteration Session' with different data sets and choosing a specific built-unit design pattern (here pattern: D or 1–5 BU patterns included in '**HudCAD**': **Module-I**), but using the same C++ programs for optimization) and a data set, model constants c_1, c_2, and c_3 are derived using above unit elemental costs given by SOR and the expressions (9.1), (9.2), and (9.3), respectively, given earlier, using cost factor(function of 'fic', building rise 'nf', to allow for increase in unit costs due to building rise as explained earlier (Sect. 9.7). The model constants c_4 is derived using expression (9.4) above. Accordingly, output optimal results **instantly** given by C++ program CLC1.CPP in this case are presented below.

Output Optimal Least-Cost Results of C++ Program CLC1.CPP for Given Input Design-Decisions

```
INPUT-Design-Decisions (some) in Current-Case are:
BUILT-UNIT DESIGN PATTERN CHOSEN IS: 'BU-PATTERN- 5 or E '
Some Input Built-Unit Building Co-efficients adopted are:
s1= 0.300 s2= 0.200 k1= 0.150 t2= 0.090 t1 = 0.90
b1= 0.180 ow= 0.090 k = 0.400 ew= 0.060 k2= 0.90 gpa= 300.000 npa=
273.852
b2= 0.200 c1= 6.360 c2= 5.060 c3= 0.68840 n= 5 ns= 10
Some Input Cost Co-efficients adopted are:
 cl=5000.00 cf= 549.05 cw=1016.08 cr= 49.14 cfl= 197.12
 C1=16826.89 C2=13387.43 C3= 359.65 C4=3442.00
SOME OUTPUT OPTIMAL GEOMETRIC PROGRAMMING WEIGHTS ARE:
w9= 0.21004 w12= 0.52510 w21= 0.68022 w31= 0.07756 w03= 0.10974
w41= 0.21004 w51= 0.71240 w04= 0.73515 w01=w02= 0.07756
OPTIMAL DIMENSIONS OF THE BUILT-UNIT COMPONENTS ARE (M):
ldn= 6.320 wdn= 7.282 ldr=10.533 wdr= 3.641
lk = 4.74 wk = 2.98 lbt= 2.90 wbt= 2.98 tlbt= 3.95 twbt= 2.98
ewl= 1.580 wwc= 0.000 lsc= 7.900 wsc= 6.620
lbr1= 3.950 owl= 0.000 wo = 2.979 wbr1= 3.641
lbr2= 3.950 lbr3= 4.740 lbr4 = 3.950 lbr5= 5.530
wbr2= 3.641 wbr3= 6.620 wbr4 = 5.130 wbr5= 5.130
OTHER OPTIMAL OUTPUT VALUES GIVEN BY THE MODEL ARE:
LEAST-POSSIBLE COST PER BUILT-UNIT: 1428284.875 Monetary Unit
Length of Built-Unit Block along the staircase: 26.333 M
Width of Built-Unit Block across the staircase: 33.098 M
Optimum Plinth Area Rate given by MODEL is= 4760.96 Monetary Units/SQ.M.
Optimum Number of Floors given by the MODEL (same as the
Maximum Number of Floors specified) is : 2
Optimum Floor Area Ratio given by the MODEL is= 1.429
```

```
Optimum Open Space Standard (same as the Minimum Acceptable
Standard Specified) given by the MODEL is : 60.000 SQ.M./Built Unit
Optimum Ground Coverage Ratio given by the MODEL is= 0.714
Optimum Built-Unit Density achieved (in BU/HA) is ==47.619
```

9.12.4.2   Detailed Estimate per Dwelling Unit INSTANTLY Given by the Cost Engineering Model

To allow for increase in the unit costs due to building rise explained earlier, each basic rate (i.e. B.Rate) given by the SOR as above is multiplied by a factor '(nf $(1+(nf-1)$fic)' to derive the effective rate (i.e. E.Rate), which is used in the detailed estimate as shown below.

Site-Specific Detailed Estimate per Dwelling Unit INSTANTLY Given by Cost Engineering Model

As mentioned above, while running the cost engineering model, it generates instantly the detailed estimate per dwelling unit, based on the input data chosen by the engineer designer, which is exactly compatible with the above output optimal results, as shown below:

Unit Costs given by CE MODEL for Chosen Foundation/Wall/Roof/Floor Types are:

```
cf= 549.053 MU per RM, cw=1016.077 MU per RM, cr= 49.143 MU per SQM, cfl=
197.121 MU per SQM
Built-Unit Design Selected is:-FIVE BED ROOM DESIGN: 2-BU'S PER FLOOR
(PATTERN-'F2' or 5)
Dwelling Specifications Selected: Foundation Type= 3, Wall Type= 3, Roof
Type= 3, Floor Type= 3
Detailed Estimate per Dwelling Unit INSTANTLY given by Cost Engineering
Model:
Item Description of Item     Quantity/    B.Rate      E.Rate      Amount
                             D.Unit     (MU/Unit)   (MU/Unit)     (MU)
1. Earthwork (ordinary soil) 91.341 Cum   24.020      24.621    2248.858
   in Wall Foundation
2. Plain Cement Concrete 1:4:8 (1 Cement:4  25.551 Cum 1002.010  1027.060
26242.539
   Coarse Sand:8 Stone Agg.40 mm) in Foundation
3. Brick-work in Foundation in 1:8   0.000 Cum  809.300   829.532   0.000
   (1 Cement:8 Fine Sand) cement mortar
4. Brick-work in Foundation in 1:6    37.599 Cum  842.000   863.050
32449.723
   (1 Cement:6 Fine Sand) cement mortar
5. Providing/Laying DPC with Cement Conc.  1.156 Cum 1780.065   1824.567
2108.470
   1:2:4 (1Cem:2C.Sand:4S.AG10)
6. Brick-work in Superstructure in 1:8    0.000 Cum   846.700   867.867
0.000
```

```
  (1 Cement:8 Fine Sand) cement mortar
7. Brick-work in Superstructure in 1:6     156.090 Cum   879.400     901.385
140697.031
  (1 Cement:6 Fine Sand) cement mortar
8. Cement plaster 15 mm thick (rough side of    452.192 Sqm   36.067    36.969
16716.924
  wall) in 1:6 (1Cem.:6F.Sand) cement mortar
9. Cement plaster 12 mm thick in wall          452.192 Sqm   30.462    31.224
1499.027
  in 1:6 (1Cem.:6F.Sand) cement mortar
10. Cement plaster 6 mm thick (roof/ceiling) in   300.000 Sqm    24.127
24.730      7419.043
  1:4 (1Cem.:4F.Sand) cement mortar
9. Re-infrd.cem.conc. in roof/floor 1:3:5 (1Cem.   0.000 Cum   1404.820
1439.941      0.000
  :3C.Sand:5S.AG20) excldng.shutrng./re-infrcmnt.
12. Re-infrd.cem.conc. in roof/floor 1:2:4 (1Cem.   36.000 Cum   1682.725
1724.793    62092.465
  :2C.Sand:4S.AG20) excldng.shutrng./re-infrcmnt.
13. Mild Steel Reinforcement for RCC Work incl.1649.998 Kg   16.075   16.477
27186.805
  bending, binding, placing in position
14. Colour washing (green,blue,buff) on New Work 1204.383 Sqm    3.120
3.198      3851.616
  (2 or more coats) including a base coat of
  white washing with lime to give even shade
15. Cement Concrete Flooring 1:2:4 (1 Cement:2    18.000 Cum 2338.925
2397.398    43153.109
  Coarse Sand:4 Stone Aggregate 20 mm Nom.Size)
  finished with a floating coat neat cement
16. Cost per Dwelling Unit (As per Above Detailed Estimate for Building
Cost) =378285.594
17. Land Cost per Dwelling Unit (As per FAR, GPA & Land Price in Model)
=1049999.250
18. Cost per D.Unit (Difference between Least Cost (ac) & Land Cost/D.
Unit)  =378285.625
19. Total Cost per D.Unit (Bldg.Cost per Dwelling Unit + Land Cost/D.
Unit)  =1428284.875
20. Total Least Cost per Dwelling Unit (As Given by the LCO Model)
=1428284.875
21. Minimum Acceptable Built-Unit-Space (Given as Input to the LCO
Model) = 300.000 Sq.M.
```

9.12.4.3 Dimensional Diagram of Built-Unit-Design Pattern: E, Compatible with the Above Least-Cost Output Optimal Results in Numerical Terms Given by Model

As outlined in Sect. 9.11.4 above, the drafting and drawing task is carried out by the LISP program 'GMOD1' forming graphic component of MODULE-I of HudCAD. Typical *cpp output files given by the C++ programs CLC1.cpp and CMB1.cpp are shown, respectively, in Sects. 9.10.1 and 9.10.2 above. It is pointed out that while running the above C++ programs CLC1.cpp (for **Least-Cost**) and CMB1.cpp (for **Most-Benefit**), these programs also generate the output LISP files named as: dg1.lsp and dg2.lsp, incorporating the numerical data of the *optimal design solution*, corresponding to the output *cpp file generated. Adopting this procedure example **Most-Benefit** *optimal* built-unit design drawings instantly given by 'HudCAD', compatible with the above output *optimal* numerical results, are shown in Figs. 9.9 and 9.10 above. The current LISP program 'GMOD1', which is a demonstration program, presently incorporates only the 4 built-unit-design patterns (i.e. S2, S4, D2 and D4) included in Module- 1, Group-I of 'HudCAD', as shown in Fig. 9.8 above. Interested readers can modify or enlarge the current LISP program 'GMOD1', including in it built-unit-design pattern: F2, i.e. two dwelling units with five bed-rooms in each served by one staircase/elevator, as in the present case. However, the site-specific building design drawing of built-unit-design pattern: F2, compatible with above least-cost output optimal results in numerical terms, is shown diagram-matically (not to scale) in Fig. 9.13 above, just to demonstrate applicability of this technique in high-income countries case also.

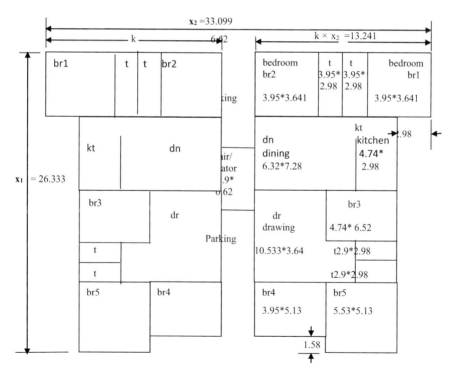

Fig. 9.13 Diagram of least-cost dwelling design module: E for high-income housing, showing optimum block/room dimensions (not to scale-Example: LCH1) given by the model

9.13 Integrated CAD Application Results: Some Major Differences with the Conventional Planning-Design Practices and Implications in Resource-Efficient Problem Solution in Building-Housing

There are some major differences between the conventional planning and design practices and the technique of integrated computer-aided design by optimization (Integrated CAD). Some of these are outlined below, stressing their implications in achieving a *resource-efficient* problem solution in building-housing.

9.13.1 Large Variations in QDEIs: Limitation of Conventional Practices and Utility of Integrated CAD in Achieving a Resource-Efficient Problem Solution

As discussed earlier, in any planning and design solution, there are a number of Quantitative Design Efficiency Indicators (QDEIs) such ass Built-Unit-Cost (BUC), Built-Unit-Space(BUS), Semi-Public Open-space (SPO), Built-Unit-Density (BUD), Ground Coverage Ratio (GCR), and so on. The Least-Cost Design **Examples: LCL1** and **LCL4** (**Context: Middle/Low-Income Countries**) presented in Sect. 9.10.1 above show that only due to change of BR from 4 to 2, the **Least-Cost** per Built-Unit (i.e. BUC) increases from 299,999.844 MU to 412,607.688 MU, i.e. 1.38 times, FAR reduces from 2.189 to 1.415 (i.e. by 35%), GCR increases from 0.547 to 0.707(i.e. by 29%), and BUD reduces from 301.819 to 195.065(i.e. by 35%), keeping BUS (i.e. Built-Unit-Space $= 72.53$ m^2) and other parameters unchanged in both cases. Similarly, Least-Cost Design Examples: LCH1 and LCH2, i.e. Dwelling Design Module-E: Five Bedrooms (**Context: High-Income Countries**) presented in Sect. 9.10.1 above, show that only due to change of BR from 4 to 2, the BUC, i.e. Least-Cost per Built-Unit increases from 966,863.625 MU to 162,5981.250 MU, i.e. **1.39 times**, FAR reduces from 2.222 to 1.429 (i.e. by about 36%), GCR increases from 0.556 to 0.714(i.e. by about 28%), and BUD reduces from 74.074 to 47.619 (i.e. by about 36%), keeping BUS (i.e. Built-Unit-Space $= 300$ m^2) and other parameters unchanged in both cases. Thus, only due to variation in only one variable, i.e. BR in the above cases, both in the context of low- and high-income countries, the least-cost per built-unit increases by 38–39% with similar variations in other *efficiency indicators* as mentioned above. Again, 'HudCAD' application results of most-benefit optimizing models presented in Sect. 9.9.5 above, also show very large variations in the BUS and other *efficiency indicators* such as increase of BUS by **2.52 times**, of optimal FAR by 3.38 times, and of BUD by 1.34 times, while the input unit land price and the investment/cost per built-unit are kept unchanged in both the designs presented; even due to interaction of only two variables, i.e. BR and SPO in the above case.

It may be seen from above optimal results that there are large cost and design *efficiency* differences in different results due to interaction of only one or two variables as cited above, *irrespective of the context of income category of housing or countries*. As part of the planning and design process, it is an *ethical responsibility* of the AEC professionals to analyse and bring on table in a *transparent* manner all such large cost and design *efficiency* differences for consideration by all *stakeholders* for appropriate decision. Moreover, such large variations cannot be ignored by AEC professionals, multiple *stakeholders,* and the *user citizens*, who would like to maximize benefits in terms of space and other design parameters for each unit of money spent or invested in a project, fulfilling the desired environmental standards. As the number of variables and their interactions are too numerous in real life, the conventional planning and design practices rarely permit consideration of all such interactions to help derive an *optimal* solution in a given context, as each case has to be examined on scientific basis in a given context to determine the *optimality*. However using a *computer-aided optimal design method* as presented above, it is possible to consider all such interactions in a *transparent* manner and bring such results on table for consideration by all *stakeholders* (including *user citizens*), and thus, help achieve instantly the *optimal* and a *resource-efficient* design solution with informed acceptance by all *stakeholders*, in a given context.

9.13.2 Achieving Instantly Exact Value of Designer Specified Built-Space Together with Least-Cost

As mentioned earlier, 19 input set options are available in this software module, where AC, LP, BR, and SPO are affordable cost or investment per Built-Unit (BU), market land price, building rise, and semi-public open space per BU, respectively. The input set options can be enlarged/modified as per the priority of the concerned interest group(s) including the professional groups, i.e. architects, engineers, and so on. The built-unit design coefficients are mostly related to the conceptual design pattern(s) and the relative proportion of different built-unit components decided by an architect. The cost coefficients are related to cost of labour, materials, and the specification chosen for building components such as foundation, wall, and floor/roof normally decided by an engineer. Using cost engineering principles, the software can derive these cost coefficients based on site-specific unit cost data on labour and materials and the respective construction specification chosen, or an engineer based on past data can choose specific values. During the iteration process, the appropriate values of these input parameters (which are in effect design-decisions and conditions) can be chosen by different interest groups and professional *disciplines* in an *interactive* way in response to the output solution given for each input set option, adopting the procedure outlined in Fig. 10.7 in Chap. 10. The software derives the optimizing model constants based on such input values and solves the nonlinear programming models giving the numerical optimal solution. Some input data could be preset in the computer to initialize the iteration process. In this technique, instead of prior preparation of drawing and time-consuming cost estimate to know the cost per BU, the desired cost itself (AC) is given as input, and the software gives instantly (in fact, time is taken only for pressing a few keystrokes for data input) the physical design maximizing the BUS within this cost per BU. Time thus saved could be used for more creative activity of preparing concept design patterns, refining the designs by giving different input sets and carrying out scientific sensitivity analysis for more informed decision-making in the planning and design process.

Using this software and MBO models, the *sensitivity* of BUS to the variation of AC and other variables such as LP, BR, SPO, etc. could be instantly ascertained (always real, as each case represents a physical design, which can be instantly drawn). Similarly, *sensitivity* of COST to the variation of BUS/other variables could be determined using the software and LCO models, which also gives the least-cost building design in graphic form *complying with any exact value of built-space per built-unit specified* by the designer, say 50 m²/BU, as shown in Fig. 9.14.

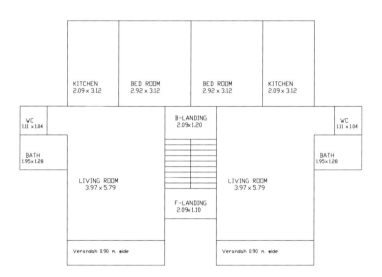

SOME INPUT PARAMETER VALUES GIVEN AS DESIGN CONSTRAINTS TO THE SOFTWARE:-

NO. OF FLOORS ('BUILDING RISE' OR 'BR') = 4   OPEN SPACE ('SPO' : SQ.M./'BU') =   15.000

MINIMUM ACCEPTABLE SPACE PER 'BUILT-UNIT' OR 'BU' (IN SQ.M.)   =   50.000

MARKET LAND PRICE (MONETARY UNITS/SQ.M.) APPLICABLE IN THE ABOVE COST =   5000.00

SOME OUTPUT OPTIMAL PHYSICAL DESIGN VALUES GENERATED BY THE SOFTWARE:-

MINIMUM POSSIBLE HOUSING COST/INVESTMENT PER'BUILT-UNIT'(MONETARY UNITS)=   238630.00

GROUND COVERAGE RATIO ACHIEVED IN ANY LAYOUT OF THE BUILT-UNITS   =   0.455

OPTIMAL 'FAR' TO BE ALLOWED, AS IT ACHIEVES ABOVE 'SPO' IN ANY LAYOUT =   1.818

'BUILT-UNIT DENSITY'('BUD') ACHIEVED (NO OF BU'S/HECTARE) IN ANY LAYOUT =   363.636

Fig. 9.14 Least-cost optimal built-unit design drawing given by 'HudCAD', achieving exact value of built-space (here 50 m²) specified by a designer

9.13.3   Comparative Visual Analysis of Designs Along with QDEIs

In addition to comparative *analysis* of numerical results discussed above, running 'HudCAD' an architect-designer can obtain instantly different optimal built-unit design drawings as output side by side (if required in different layers) for easy *visual* comparison, with the values of different QDEIs (quantitative design efficiency indicators such as Built-Unit-Cost (BUC), BUS, SPO, BR, DEN, GCR, and so on) also depicted for each design, as shown in Fig. 9.15.

Values of some QDEIs for the respective Design- 1, 2, and 3 shown above are as follows:

```
Design-1- INPUT: nf=5, SPO = 15 m² /BU, AC= 300000 MU/BU, CL= 5000 MU/m²
          OUTPUT: BUS = 81.132 m², DEN= 320.242 BU/HA, FAR = 2.598,
GCR=0.520
Design-2- INPUT: nf =4, SPO = 15 m² /BU, AC= 300000 MU/BU, CL= 5000 MU/m²
          OUTPUT: BUS = 72.319 m², DEN= 302.3 BU/HA, FAR = 2.186, GCR=0.547
Design-3- INPUT: nf=2, SPO = 15 m² /BU, AC= 300000 MU/BU, CL= 5000 MU/m²
          OUTPUT: BUS = 45.69 m², DEN= 264.512 BU/HA, FAR = 1.206, GCR=0.603
```

The designs shown above can be seen in more readable details by going to the 'edit' mode. However, some QDEIs extracted for the respective design-1, 2, and 3, are presented above, from which it will be seen that the designer/user by changing only the value of one design parameter, i.e. number of floor or 'nf' from 2 to 5, the BUS (built-unit-space) increases from 45.69 to 81.132 m², values of the other INPUT parameters including AC (affordable cost or investment) per Built-Unit (BU) being kept unchanged by the designer/user citizen. As may be seen, there is similar variations in values of other QDEIs. The designer/user citizen can change values of other INPUT design parameters and get the corresponding OUTPUT physical design solutions with corresponding QDEIs for evaluation and decision.

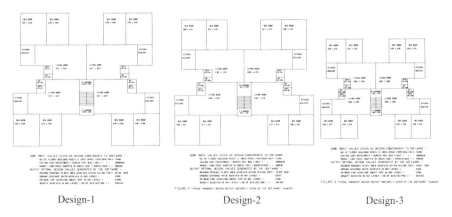

Design-1   Design-2   Design-3

Fig. 9.15 Running 'HudCAD' to obtain instantly the optimal built-unit design drawings as output side by side for easy comparison

In such input-output, design, analysis, *optimization,* and graphic presentation, multiple stakeholders including architect, planner, engineer, and *user citizens* can *participate* to give input and get the corresponding output in graphic form easily understood even by lay user, and thus, come to a planning and design solution acceptable to all stakeholders. Using this CAD and *optimization* technique, such *participatory planning and design process* can be applied in different components of urban development and management practice including layout planning and design presented in Chap. 10, with consequent benefits. This is rarely possible in conventional planning and design practice in AEC sector.

9.13.4 Synthesizing Quantitative and Qualitative Perspectives in Design

Fundamental to the optimizing computer modeling technique is the concept of a decision problem and a mathematical model for such a problem. This can be combined with the qualitative perspective obtained from AEC professional's personal experience, common sense, intuition, and *value judgement* of all stakeholders to achieve the *cost-effective* and most acceptable design solution.

As discussed below, even such qualitative perspectives could be indirectly quantified in terms of the '*value scale*' (quantitative measure) which may be adopted in a problem situation by all stakeholders. For example, from *qualitative perspective* a 'low rise' development (say BR = 2 in Design-1, Fig. 9.9) may be preferred compared to a 'medium rise' development (say BR = 4 in Design-2, Fig. 9.16). But, Design-2 adopting a 'medium rise' development may be more acceptable to the low-medium income households struggling for built-space within affordable cost, as it gives a BUS of 72.53 m^2 (see **Example: MBL1 Dwelling Module-D** in Sect. 9.10 above), i.e. **2.52 times** that in Design-1 ('low rise' design giving a BUS of 28.76 m^2 (see **Example: MBL2 Dwelling Module-D** in Sect. 9.10 above), even though it adopts a less open space standard (SPO = 15 m^2/BU) giving a higher ground coverage (GCR = 0.547 compared to GCR = 0.324 in Design-1). Thus, foregoing qualitative perspective of 'low rise' gives a higher BUS of **2.52 times**.

If the Design-1 in Fig. 9.9 is changed only by choosing BR = 4 (keeping other design-decisions unchanged including open space standard of SPO = 30 m^2/BU), the optimal design instantly obtained using the software '**HudCAD**', gives a BUS = 45.418 m^2 and BUD = 241.89 BU/Hectare (see **Example: MBL3** in section above). Between this case and the case shown in Design-1 (BR = 2) in Fig. 9.16, there is change of only one input parameter by the planner-designer, i.e. BR = 4 in the former case, and BR = 2 in the latter case (Fig. 9.17).

Comparison of the above two cases shows that an increase of BR from 2 (low-rise) to 4 (medium-rise) has *increased* the BUS from 28.76 to 45.418 m^2, i.e. by about *58%*, and the BUD is also increased from 225.327 to 241.89 BU/Hectare, i.e. by about 7%. This increase of 58% in BUS and of 7% in *land-use efficiency*, in terms of density (in case of 'medium rise' compared to 'low rise'), can be adopted as the *quantitative measure* of the *qualitative perspective* between the above two cases.

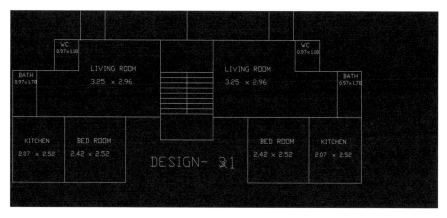

Fig. 9.16 Zoom view of optimal building/housing design-1 (BUS of 28.76 m²)

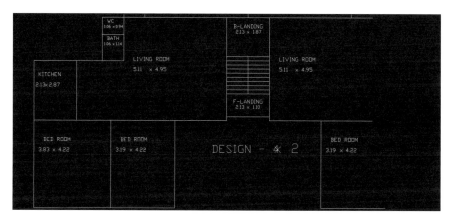

Fig. 9.17 Zoom view of optimal building/housing design-2 (BUS of 72.53 m²)

Such *value-analysis* will enable more *informed-decision-making* by the architect-planner-engineer-urban manager and the prospective user community to select building designs, type of development, and built-form synthesizing the quantitative and qualitative perspectives, depending on their preferences and value judgement in a specific context, rather than the frequent prescription of specific built-form on a predetermined notion. Large variations in above results also highlight the importance of synthesizing the quantitative and qualitative perspectives of *design-decisions* (instead of taking such decisions on ad-hoc basis) with the *participation of multiple stakeholders* including low-income households, to *maximize* benefits and the satisfaction level. Such an approach will also make the urban development and management process truly *participative* by the *multiple stakeholders* as per their *priority* within the national, local, and individual *resource-constraints*, and thus help achieve *resource-efficient* building, housing, and urban operations with *social equity*.

9.13.5 Participatory Planning and Design in Building-Housing to Promote Social Responsiveness

As cited above, it is an *ethical responsibility* of the AEC professionals to analyse and bring on table in a *transparent* manner all large variations in cost and other *efficiency indicators* for different planning and design solutions in a given context, for consideration by all *stakeholders* for appropriate decision. This is rarely possible in conventional practices. However, using the technique of integrated CAD by optimization, in addition to the instant display of *quantitative design efficiency indicators (QDEI)* as shown in above figures, it permits a *visual* comparative *analysis* to help synthesize the quantitative and qualitative perspectives in a design. Using this software and MBO models, the *sensitivity* of BUS to the variation of BUC and other variables such as LP, BR, SPO, etc. could be instantly ascertained (always real, as each case represents a physical design, which can be instantly drawn). Similarly, *sensitivity* of BUC to the variation of BUS/other variables could be determined using the software and LCO models, which also gives the least-cost building design in graphic form *complying with any exact value of built-space per built-unit specified* by the designer.

Thus, using this technique planner-designer can bring on table in a *transparent* manner, the *optimal* physical design of an urban development (here building-housing system) in a site-specific context, after a *participatory* and meaningful dialogue process between all *stakeholders* who can give their own input (and get corresponding output), availing the *opportunities* and respecting the *constraints* (say the '*affordability*') beyond control at that decision level. *Participation* in such a dialogue process by the urban local bodies (including 'Building' and 'Planning-Regulation' authorities) and the real estate developers—Builders, using such scientific tools, may lead to more realistic development control regulations, more *efficiency* and improved *social responsiveness* of various actors in the urban sector. Planner-designer can easily provide a wide selection of input set options in such a tool, covering concerns of the all interest groups, so as to have a meaningful dialogue process, and thus, not only promoting a *transparent* corruption-free urban planning and development process, but also help achieve a *resource-efficient* urban development with *social equity* satisfying multiple stakeholders. Possibility of using internet, for linkage of multiple *stakeholders* including *user citizens* with the above planning and design procedure, can also be explored which will lead to a comprehensive *participatory* planning and design process in the building-housing system enhancing *social responsiveness* while achieving a *resource-efficient* building, housing, and urban operations with *social equity*.

9.14 Universal Applicability of Quantitative Analysis, Optimization, and Integrated CAD in AEC Operations Including Building-Housing Planning and Design in all Countries

Today, information technology and computer application are everywhere altering the very nature of work in all sectors of society in almost all countries in the world irrespective of their income or development statuses. This is equally applicable in the knowledge disciplines of Architecture (including Urban Planning) Engineering-Civil-Building Engineering and Construction (usually classified as AEC sector), in various countries. Computer-Aided Design (CAD), quantitative analysis, and *optimization* are important application areas of IT. Generally, resources are always inadequate compared to the need, whether in low- or high-income countries. No doubt low-income countries are usually subjected to severe resource-constraints necessitating scaling-down needs in the context of available resources. But, even in affluent countries such resource-constraints are applicable because of scaling-up of needs to have higher standard of living consuming more resources. In view of large variations in Quantitative Design Efficiency Indicators (QDEI's) as shown above in building-housing irrespective of the context of income development statuses of countries, and the limitation of conventional practices (discussed above and also in earlier chapters), it is desirable that such variations are presented by AEC professionals in a *transparent* manner (by application of integrated CAD, quantitative analysis, and optimization techniques) to all *stakeholders* including *user citizens* in all countries, for appropriate planning and design-decisions, which will help achieve a **resource-efficient** solution with **social equity**. A general schematic optimum housing and urban development design procedure, which can also include optimal building-housing planning and design, is shown earlier in Fig. 10.7 in Chap. 10. In the same chapter, universal applicability of quantitative analysis and optimization in AEC operations including layout planning and design in all countries were highlighted. This is equally applicable in building-housing planning and design in all countries covering various income or development statuses, as discussed below citing some application examples.

It is sometimes argued that use of quantitative analysis and optimization techniques is applicable only in Third World Countries (i.e. low-income countries) which are subjected to severe resource-constraints and population pressure. But, we should not forget that all of us *live in a finite earth with finite resources*, and in this context, an ethical responsibility is enjoined on all of us to achieve *resource-efficiency* and *social equity* at each system level in all countries, and only through application of quantitative analysis and optimization techniques at each level, whether in low- or high-income countries, we can achieve this objective. As pointed out earlier in Chap. 10, there can be a large variation in QDEIs (e.g. a variation of 1.29 times in the maximum lot area within a given cost) depending on the layout module type and other design conditions chosen by planner-designer and other stakeholders. Similarly, as shown above, there can be very large variations in the

BUS and other *efficiency indicators* such as increase of BUS by **2.52 times** (due to interaction of only two variables, i.e. BR and SPO)**,** increase of BUC (Built-Unit-Cost) by **1.39 times** even in **High-Income Countries Context**, only due to change of BR (Building Rise). Likewise, as indicated in Chap. 7, there is complex interaction and sensitivity between various urban 'built-form-elements' causing large variations in planning and design *efficiency* indicators.

For example, in almost all countries (including high-income countries) a planning regulation: 'FAR' (floor area ratio, or floor space index—one of the urban-built-form elements) is used as a planning tool for urban planning and development control. However, in most of the cases, the value of FAR is prescribed intuitively (or even arbitrarily) without linking it with the prevailing land prices, land availability, demographic structure, and other relevant factors and constraints, in a scientific and *transparent* manner. This leads to many malpractices and even corruption and other urban problems in many countries including the high-income countries. A scientific analysis would show that there is a complex interaction between the land-price-FAR-building-rise-density-built-unit-cost, *and that, for a given planning and design condition and cost parameters including land prices, there is an optimum FAR which gives the least-cost per built-unit as shown in Chap. 7. It is also shown that there may be variations of 2-3 times in built-unit-costs, only due to planning regulations*, prescribed by the planning authorities. These variations may be even much higher at higher land prices, reaching to astronomical figures, and the consequent astronomical increase in the *built-unit-costs*, creating huge *inefficiency* and *social-inequity* and the consequent massive housing and urban problems. *Such interactions also exist in respect of other* urban built-form elements.

It is desirable that such variations are presented by AEC professionals in a *transparent* manner (by application of Quantitative Analysis and Optimization Techniques) to all stakeholders including user citizens in all countries, for appropriate planning and design-decisions to achieve *Efficiency* and *Social Equity*, resolving the conflicting interests, viewpoints, and demands of various stakeholders, and thus, reduce any scope of corruption and lessen many urban problems (e.g. slums, degeneration, and decay of urban areas) whether in low- or high-income countries. This is feasible only if scientific quantitative analysis and *optimization* are universally applied in all AEC operations. Urbanization, urban management, and computer-aided planning and design are discussed in Chap. 2 (see also ref city 2001), where a course design for *urban management* is presented to facilitate the above process. It may be noted that almost all countries in the world suffers many urban problems and achieving *efficiency* and *social equity* in the urbanization, urban development, urban-renewal, or urban re-development process, may be the key to solve such urban problems. This will not only help achieve a *resource-efficient* solution with *social equity*, but will also promote a corruption-free urban planning and development.

To make it easier for planners and designers to use such *optimization* techniques in building-housing designs, the author has developed many C++ programs and graphic programs and incorporated some of these in the software Module-I of '**HudCAD**' (presented in earlier sections), which covers all the five components of integrated CAD system including cost engineering discussed earlier. Out of these only two programs named as: CLC1.CPP and CMB1.CPP incorporated in the software Module-I as sub-models are elaborated above, showing that these are *Universal Optimization Models* and that the same model can be used, for the **Least-Cost Optimization** (CLC1.CPP) or the **Most-Benefit Optimization**(-CMB1.CPP) for any dwelling unit design module shown in Fig. 9.2, or any other design modules conceived by a creative designer, provided the corresponding design module and cost coefficients are derived and the model is calibrated accordingly. Complete computer code listing of these C++ programs is given in appendix/pen drive. Using CLC1.CPP and CMB1.CPP, the cited *optimization* problem can be solved in no time as also illustrated above.

The above application examples show universal applicability of quantitative analysis, *optimization,* and integrated CAD in AEC operations including building-housing planning and design in all countries. Building-housing designers-urban professionals (e.g. architects, planners, building/civil engineers) commit huge *resources* (e.g. money, materials, land, etc.) while taking planning and design-decisions, making them *ethically responsible-accountable* to the Society-*user-citizens* to achieve *productivity* (i.e. *efficiency* and *effectiveness*) and *social equity* in the use of *resources* (scarce-quantities with many claimants)so *committed* in their planning and designs to produce building-housing-urban services. Operations Research (**OR**) and Computer-Aided Design (**CAD**) are essential integrative techniques to improve *productivity* (Koontz & Weihrich, 1990). Hence, it is vital that above professionals apply such techniques in planning and design and achieve *Optimization* to discharge their above *responsibility-accountability* to the society-people for a **resource-efficient** solution of our mounting housing and urban problems with social equity, in the context of affordability and resource-constraints, and the conflicting interests-viewpoints-demands of multiple stakeholders in the building-housing-urban sector. Even the ancients applied optimization, which is rarely applied in our modern building-housing-urban operations creating many inefficiencies and inequities and consequent urban problems. With the increase in population and dwindling resources, applying such techniques in all AEC operations both in low- and high-income countries, to attain a resource-efficient and equitable urban problem solutions in all countries has become very crucial (also facilitated by modern computers), as (1) we live in a finite earth with finite resources, (2) no rational decision-making process is really complete without optimization (Beightler et al., 1979) and, (3) there can be very large variations in the resource-efficiency indicators depending on the planning and design-decisions in building-housing-urban operations, creating a large scope for resource-optimization.

9.15 Closure

In this chapter, building and housing design is considered as a problem-solving process highlighting the useful role of computers, and of optimization and CAD in AEC operations, i.e. building, housing, urban development, and management system. Module concept in building and housing development system design optimization is outlined, stressing the relevance of cost engineering; and a technique of computer-aided cost estimating in optimization is presented as cost is paramount in this process.

A typology of models, i.e. Least-Cost Optimizing Model (LCO model) and Most-Benefit Optimizing Model (MBO model), their calibration and illustrative numerical application examples of computer-aided building and housing design by optimization, are presented. These are *Universal Models* for building and housing development system design optimization, for the given objective function and constraint set. Model calibration and solution, using geometric programming technique, is computerized using C++ programs CLC1.CPP and CMB1.CPP for LCO and MBO model, respectively. Same *Universal Model* is applied both in low- and high-income countries context, using unit design module coefficients and cost coefficients corresponding to each module type. Thus, user planner-designer is to give only the desired input values prompted by the user-friendly C++ program. Based on the optimal design procedure presented, a planner-designer can develop such *Universal* optimizing model for any rectilinear module for any desired constraint set.

Application Software:'HudCAD' is presented for integrated computer-aided building-housing design by optimization and drafting-drawing, having both numeric (CLC1.CPP or CMB1.CPP) and graphic components (GMOD1.LSP). The numeric component solves in no time the optimization model using geometric programming techniques, and generates a LISP Output file incorporating the numeric optimal results. This LISP output file is loaded by the graphic component of 'HudCAD' and converts *instantly* the output numeric optimal results into drafting-drawing in soft or hard format. A number of application example results are presented showing a large variation in QDEIs highlighting the limitation of conventional practices (where search of numerous alternatives to find the optimal solution is rarely possible) and the *utility* of integrated CAD in achieving a *Resource-Efficient* problem solution. Application results show compatibility between all five components of integrated CAD by optimization including the site-specific building component unit costs and the output detailed estimate per dwelling unit INSTANTLY given by cost engineering model and the optimal results given by the LCO/MBO model both for the low-income countries case and the high-income countries case.

Comparative design efficiency indicators of housing designs for given design-decisions shown in Fig. 9.18 for low-medium income housing indicate that the BUS within the same BUC may increase by **2.52 times** depending on the value of other built-form-elements chosen by the planner-designer-user-citizen. Similarly, in the **Context: High-Income Countries**, the least-cost design examples: LCH1 and LCH2, i.e. Dwelling Design Module-E: Five Bedrooms, presented in Sect. 9.10.1 above, show that only due to reduction of BR from 4 to 2, the BUC, i.e. least-cost per built-unit increases from 966,863.625 MU to 1,625,981.250 MU, i.e. **1.39 times**, FAR reduces from 2.222 to 1.429 (i.e. by about 36%), GCR increases from 0.556 to 0.714 (i.e. by about 28%), and BUD reduces from 74.074 to 47.619 (i.e. by about 36%), keeping BUS (i.e. Built-Unit-Space $= 300$ m^2) and other parameters unchanged in both cases. Thus, only due to variation in only one variable i.e. BR

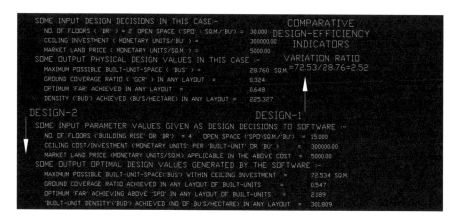

Fig. 9.18 Comparative design efficiency indicators of housing designs for given design-decisions

in the above cases, in the context of high-income countries, the Least-Cost per Built-Unit increases by 38-39% with similar variations in other *efficiency indicators* as mentioned above. Such large variations in Quantitative Design Efficiency Indicators (QDEIs) such as Built-Unit-Space(BUS), Built-Unit-Cost (BUC), Semi-Public Open-space (SPO), Built-Unit-Density (BUD), Ground Coverage Ratio (GCR) and so on, highlights the crucial need to use such quantitative analysis and *optimization* techniques in AEC operations like building and housing, if achieving *resource-efficiency* and *social equity* in planning and design are taken as important planning-design objective(s).

To help readers of this book to use such techniques to achieve their own problem solutions, complete listings of the respective C++ and AutoLISP programs are included in an online space connected to this book.

Chapter 10
Integrated Computer-Aided Layout Planning and Design by Optimization

10.1 Introduction

Layout planning and design is a multidisciplinary process involving physical planning, engineering, and architecture and so on. Many of these planning and design aspects are interlinked with each other, although in general, decisions on each of these aspects are taken by the respective professional discipline mentioned above, without considering the interlinkages and the combined effects on costs for example. It is desirable that planning and design decisions made by each of the above professional disciplines should take into account the combined effect of the decisions on design variables, belonging to each discipline in 'quantitative terms' with reference to the 'value scale' (say costs) chosen for measuring the overall planning and design objective achievement in an integrated manner in a given context. Such interdisciplinary communication choosing a common value scale such as costs will help avoid frequent conflicting planning and design decisions creating *inefficiency* and other problems. For example, a physical planner, while designing the subdivision of land for producing lots (lots) for a single-family *dwelling-development-system*, may provide different standards of public and semi-public open spaces, road widths, and lot widths. If the width of lots is increased, the subdivision plan will consume larger length of utility networks such as road pavement, water supply lines, sewer lines, and so on, with consequent effect on costs and other parameters. On the other hand, if the width of the lots is reduced, lot depths will increase (assuming same design lot area), affecting the length of utility service connection lines for the lots, again, affecting costs and other parameters. Similarly, the architect, while designing the dwelling may locate the 'service core' or 'wet core' (i.e. water closet, bathroom, and garage) at the rear, middle, or front of the lot. Again, the engineer may locate the off-lot utility lines at the rear or front of the lot. It is clear that all the above design-decisions regarding *dwelling-development-system* by the

© The Author(s), under exclusive license to Springer Nature Switzerland AG 2022
B. K. Chakrabarty, *Integrated CAD by Optimization*,
https://doi.org/10.1007/978-3-030-99306-1_10

respective professional discipline are interlinked and will have different cost implications for the ultimate product that is the 'serviced lot or plot'.

Resources (capital, land, materials, and so on) consumption and costs and should
be of paramount consideration in layout planning and design related to a *dwelling-
development-system*, in the context of dwindling *resources*, soaring prices, and
increasing populations in many countries. Therefore, one of the primary objectives
of layout planning and design should be to attain a cost-effective and *optimal* design
solution. To achieve such objectives, it is necessary to recognize the systems nature
of the problem and an attempt should be made to produce a system with the lowest
possible cost for a set level of effectiveness, or, a system with the maximum possible
effectiveness (say lot area) for a set level of cost, adopting an integrated systems
approach.

10.2   Layout Planning and Design Process: Systems Nature of the Problem

A system is defined as a collection of entities or components interrelated in some
ways, defining the structure of the system, to accomplish a specific purpose, fulfilling the constraint(s) that restrict the behaviour of the system(). Layout planning and
design involving land subdivision design incorporates all the characteristics of a
system, and therefore, it should be designed as a system to fulfill the given objective
(s). For this purpose, at the outset, the system will have to be defined in the form of a
descriptive problem model identifying the purpose or objective(s), the components,
structure, and constraints of the system, and thus giving a representation of the real
situation and reflecting the important factors that can be evaluated, manipulated, and
prescribed so that the systems engineer or designer could propose a specific system
solution, that met with the approval of the society in terms of the 'value scale' as may
be prescribed by the decision maker. The descriptive problem model then could be
converted into a mathematical model presenting the situation in mathematical
symbolic terms, thus permitting the decision makers to adopt an analytical procedure
to produce a system that will optimally satisfy the stated purpose or goal, fulfilling
the given constraints. A city or new township development will have a hierarchical
systems structure for different functions in which a residential neighbourhood could
be considered as a lower level system compared to the city as a whole. In many
cases, a residential neighbourhood may be planned adopting some repeatable planning units, which may be called 'planning modules' (further elaborated later) and are
lower level systems as compared to the neighbourhood system. Thus, the residential
land planning as a whole can be considered as a system of systems. The layout
planning and design process should consider this systems nature of the problem to
attain *efficiency* and *social equity* at each system level and in the urban development
process as a whole.

10.2.1 Large- and Small-Scale Urban Models

As highlighted in Chap. 1, post-war advances in information technology generated great interest in the application of systems analysis, mathematical models, computers, and data processing systems, in various disciplines. Systems analysis got us into space. Using such techniques, much planning has been achieved in the area of water resources, electrical engineering, and other 'hard' sciences, where systems are well defined, highly structured and can be quantified comparatively more easily, permitting development of large-scale models from well used small-scale models. However, this is not the case with the 'soft' sciences like city planning, housing, urban and regional planning. Based on the experience of applying large-scale urban models and their limitations, it is often opined that urban planning theory is not yet advanced enough to build large-scale models; and lacking a coherent body of theory that could explain land-use dynamics, the urban modellers often turn to analogies and descriptive regularities without establishing the validity of either the individual relationships or the combined structure (Lee). Many planners felt that formal models fitted urban systems only very superficially, and even applying simple techniques to immediate and partial problems was not easy. It is pointed out that gravity model (a large-scale urban model) while may be valid at the scale of metropolitan area, it has no statistical explanatory power at the neighbourhood level, and there are many other faults such as too comprehensive to give realistic results, data needs unmanageable, model does not operate the way it should, model too complicated and accumulated error so great that the model falls apart, and makes the model so expensive that it does not pay for itself (Lee). Large-scale urban land-use planning models generally emphasize on explaining and predicting human behaviour in the urban context. Such an approach for large-scale urban models conceives of the urban complex as a phenomenon to be explained scientifically as a changing configuration that can be predicted in the same way that, for example, the solar system can be predicted from the theories of physics. This is hardly the case in respect of urban complex, and as a result, such large-scale urban model solutions are remotely related to the synthesis of better urban and regional plans with the context. Simulation models cover a spectrum of possibilities from 'black box' to 'white box'. White box models are based on clear mathematical relationships. Black box models are based on unknown relationships. Formulas representing a system may be constructed from assumed relationships whose coefficients may be found by regression analysis, i.e. relationships are unclear or black. Most of the large-scale urban models reported in literature are black box models.

10.3   Design Viewpoint of Large- and Small-Scale Urban Models

A contrasting viewpoint can be to conceive urban complex as a subject for design. In this approach, the plan is a conscious synthesis of urban form to meet human needs. Simply, it is much easier to use a model to tell people what they should do instead of explaining what they are doing. Rather than serving as a negative restraint on undesirable aspects of human behaviour, the plan should serve as a positive force for the directed development of the community. This provoked the basis for architectural and engineering achievements for centuries. But the possibility of using recent advances in applied mathematics and electronic computation for plan design has opened up tremendous opportunities for more satisfying design solutions including optimization. In this process, the design, and not explanation and prediction, becomes the primary problem for solution.

10.3.1   Large-Scale Urban Modelling for Land-Use Design

Adopting the above concepts, a large-scale land-use design model, giving reorientation of land-use model development towards design, has been attempted assuming that the best plan for an area was that which minimized the cost of development using the linear programming technique (Schleger). Trips generated as a function of land use could be an input to this transportation 'structure plan' design. Trip distribution and traffic assignment models may be used to test the plan intuitively designed by the transportation planner, and based on this test intuitively designed transportation plan network may be revised until a satisfactory system is developed (Schleger).

Land development cost functions and many constraint relationships reflecting design requirements are nonlinear. In view of above, application of linear programming technique leads to error and may not satisfy the chosen design standards, and thus losing its usefulness as a design model. Moreover, it has been shown earlier in Chap. 7 that there can be very large variations in unit costs of land development (in terms of dollars per acre given as input in this large-scale model) depending on the geometrical design of layout plans, engineering specifications, and other factors, and therefore, incompatibility of such cost data, with the actual physical design layout cannot be ruled out. Thus, the objective of minimizing land development cost, which is the primary objective of this large-scale design model, may not be achieved in the real world situation.

Again, in this large-scale model, estimation of cost for a long planning period covering a few decades may be quite uncertain affecting the usefulness of the model for cost estimation. Moreover, actual decisions in the field are not made according to grand plans but, rather, in response to incremental problems which are within the immediate sphere of influence of the decision maker. An urban model should help such decision-making process, and building small-scale design models of 'white box' type, which will be free from many of the faults discussed above, could be more helpful in such decision-making process.

As mentioned above, an urban area is multisectoral and complex. To avoid the faults of large-scale urban models, small-scale models for the urban sectors or even functions within the sector could be developed, making model goals more modest, adopting a direct causal relationship between the design variables making the results more realistic and the benefits immediately applicable. Thus, a city can model its service system by attacking the problem sector by sector, say, transportation system model, education system model, housing development system models, and so on. Helweg has developed small-scale models to describe some simple service sectors and even functions within the sector such as education system, and suggests that running the model in annual segments enables the planner to programme various futures and thus can account for sociological changes and keep an annual update on the model so that the projections are always improved (Helweg, 1979). Similar model for education system is developed by other researchers (Eastman et al., 1970). Based on experience of applying such small-scale models accounting for sociological changes, it may be possible to combine these small-scale models into a larger scale model to derive more benefits.

As earlier mentioned, a city or new township development will have a hierarchical system structure for different functions in which a 'residential neighbourhood' (and its layout planning and design) could be considered as a lower level system as compared to the city as a whole. Thus, housing development could be assumed as a sector in the urban system, and synthesis of better housing development plans with the context could be achieved by developing small-scale design models in the housing development sector. In real world situations, housing development takes place in phases. Actual design-decisions are taken in response to incremental problems which were within the immediate sphere of influence of the designer of the housing development system. It is necessary to identify such incremental design phases and develop small-scale design model for each phase, linked with the city level service system models discussed above. Thus, small-scale models for transportation system could provide the guidance for developing the transportation plan or 'structure plan' of the urban planning area. Use of location models as well as analysis of facility locations on plane or on network could be helpful for this purpose.

The 'structure plan' will define the alignment of main transportation arteries, which are in any case built-in phases in the real world housing development projects. In the design and development of this 'structure plan', *optimization* could be attempted taking into account the socio-economic operating costs of transportation systems, using urban operations research techniques, which could be important inputs. Even this 'structure plan' should be flexible to permit adjustment during the planning period of long duration (for example, the right of way of transportation arteries could be for ultimate long duration stage, providing for flexible development of number of traffic lanes which could be in phases depending on the traffic demand at each phase), and in this operation use of small-scale accounting models of different service sectors as discussed above could be very useful input. In this process, the management *principles of flexibility and navigational change*, providing for plan review from time to time and redrawing them if required, as discussed in Chap. 2, could be implemented, and this process would be greatly facilitated if information technology and *integrated computer-aided planning and design by optimization* are adopted. Based on the above 'structure plan', housing development of different types could be designed in the form of 'design modules' discussed underneath using the small-scale residential development system models for *optimal* physical design of housing development system at each phase, as outlined below.

10.3.2   Small-Scale Residential Development Design Models

In view of a number of faults of large-scale urban models as discussed above and the utility of small-scale design models to serve as a positive force for the directed development of the community, it is desirable to develop small-scale residential development design models. Such models should be 'white box' type based on clear mathematical relationships between the design variables and the planning and design standards to permit such directed development of the community for better quality of life in a resource-efficient and equitable manner. Thus, model goals should be more modest, adopting a direct causal relationship between the design variables making the results more realistic and the benefits immediately applicable. To help this process such design models should allow numerical quantitative analysis to achieve *efficiency*, *effectiveness,* and *optimization* in design. Scientists frequently adopt the notion of modules or subsystems for analytical measurements. A spatial system is generally composed of such planning modules or units of design, which represent the structure and the relationships between the parts and the whole system. Such planning modules are frequently used by designers in actual physical layout designs, which can be considered as spatial systems or module concept in layout designs, and are elaborated below in more details.

10.3.2.1   Planning Module Concept in Layout Planning and Design

A representative layout of a real life mass housing project for single-family dwelling development is shown in Fig. 10.1 below. It may be seen that the layout is an aggregation of certain units of specific design which have been repeated to form the whole layout. Such repeatable basic unit design could be designated as 'Planning Module' for a residential layout. Such planning modules can be for different planning levels, and also, to meet the specific need of the planner. Some representative selections of layout planning modules, in the form of a typology of planning modules, are shown in Fig. 10.2 (HUDCO, 1982). These are introduced only for the

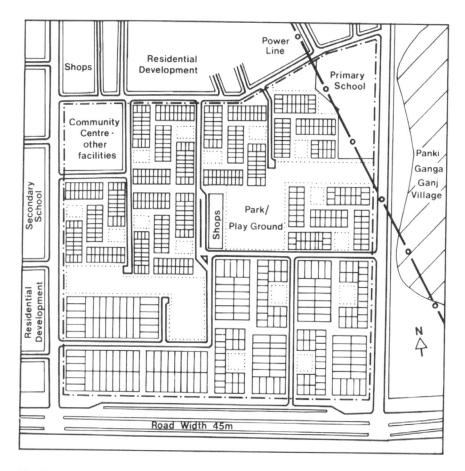

Fig. 10.1 Module pattern and structure in a real life layout: a HUDCO project, 1977

purpose of illustration because there can be a multiple diversity of such unit design modules depending on the ingenuity and creativity of the planner or designer, who can develop his/her own typology of layout planning modules which can be used in layout planning and design based on the site-specific requirements. Development of such dimensionless schematic conceptual design of planning modules is very helpful to study design economics and cost-effectiveness of layout plans by quantitative analysis and also to achieve optimization in layout planning and designs, as discussed below.

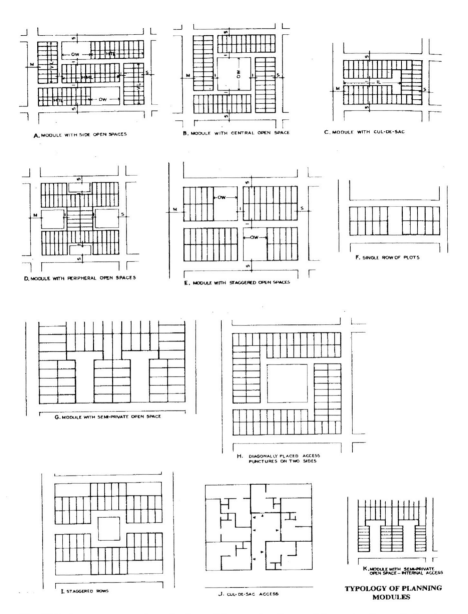

Fig. 10.2 Some representative selections of layout planning modules or typology of planning modules

10.4   Design Economics and Cost-Effectiveness of Residential Layout Plans: Limitations of Conventional Planning Practice

In conventional layout planning practice, the cost-effectiveness of design alternatives is judged by preparing individual site layouts, infrastructure designs, and cost estimates dealt with by each specialized discipline independently without considering the combined effect of decisions by each discipline. Thus, the physical planner prepares the layout plan in the form of a drawing with all dimensions, and thereafter, different ratios of land utilization such as marketable land ratio, circulation area ratio, open space ratio, lot density are calculated from this drawn layout. Based on the layout prepared by the physical planner, the engineer prepares the design and layout of the utility networks, roadways, and pavements. The dwelling units, which are constructed simultaneously with the residential land development in many housing development projects, are the final products of layout planning and design. Their designs, including the service connections, are dealt with by the architecture and engineering disciplines, taking into account the physical layout plan prepared by the physical planner, the utility networks, and roadways designs prepared by the engineer. Thus, in the conventional planning and design practice, all the above planning and design activities take place in series, dealt with by different disciplines, frequently without any intercommunication between them. This may involve differing and often conflicting interests, viewpoints, and demands, and may remain unresolved leading to increased costs in the overall planning and design solution of the ultimate product, i.e. the housing development system produced.

As earlier mentioned in any planning activity, it is desirable to choose a common 'value scale' for measuring the overall planning and design objective achievement in an integrated manner in a given context. Such interdisciplinary communication choosing a common 'value scale' such as costs will help avoid frequent conflicting planning and design-decisions creating *inefficiency* and other problems. It is desirable that planning and design-decisions made by each of the above professional disciplines should take into account the combined effect of the decisions on design variables, belonging to each discipline in 'quantitative terms' with reference to the 'value scale' (say costs) chosen. However, this is rarely done in conventional layout planning practice.

The efficiency of residential layout design can be measured by various efficiency parameters in physical terms such as marketable land ratio, circulation area ratio, density; and in cost terms such as cost of infrastructure per gross hectare and per lot and so on. Generally, a planner desires to maximize (say marketable land ratio) or minimize (say cost per lot) such land utilization and cost ratios. However, in conventional layout planning and design practice, a layout is first drawn with all dimensions already chosen. Thereafter, the above efficiency ratios are calculated from this drawn layout. If these derived values are not acceptable, another layout is drawn and the efficiency ratios are re-calculated from this redrawn layout. This process of trial and error may continue for considerable time, and still the desired results may not be attained. This is because there can be numerous planning and design solutions with very large variations in above efficiency indicators as a result of the interactions of many factors and planning and design variables ().

It is desirable to reverse the above layout planning and design process. Thus, a planner may first develop or choose a specific planning module in dimensionless form, and also decide the design objective(s) to be achieved in terms of the specific land utilization ratios like density, marketable land percentage, circulation space percentage, and so on; and cost ratios such as cost per hectare, cost per lot, and so on. In this reversed layout planning and design process, a layout could be designed directly to fit with the above design objective(s), if these are feasible, by adopting a quantitative analysis technique (). Similarly, mathematical programming technique, i.e. linear and nonlinear programming techniques can be used to achieve optimal design solutions, resolving the conflicting viewpoints and design-decisions mentioned above.

10.5   Quantitative Analysis and Optimization in Residential Layout Design: A Literature Review

In the literature, a number of studies are available on the economics of residential land subdivision and land development, land utilization, and quantitative analysis. One of the earliest studies considered a block of lots and assumed a given set of unit costs and improvements (ASCE, 1939). The variation in production costs of a lot, due to the change in any of the elements such as raw land prices, street width, block length, lot depth, and lot width, was analysed. A standard block, having specific dimensions and improvements was studied taking it as a yardstick for comparison. The costs of water mains, street lighting, gas and electric lines were omitted. However, the costs of house connections were included in the study. It was concluded that the combined effect of varying several of the elements did not equate with the effects of altering each element separately, highlighting the need for more comprehensive systems approach considering all the elements together. It is also necessary to analyse the effects of change in standards of improvements, taking the grouping of blocks in the form of a planning module as the planning level.

A study of the effect of varying urban development densities on the cost of municipal improvements, e.g. sanitary sewers and water distribution systems, was carried out to develop a population density-unit cost relationship, in respect of such improvements (Anderson, 1973). The population density was computed on the basis of lots assuming an average number of persons per household as determined by the Bureau of the Census. The unit costs of improvements were expressed as dollars per gross acre of land, dollars per lot of single-family residential layout, and so on. A number of curves, showing the variation of the unit costs of improvements with the variation in population densities, were developed in the study. It is indicated that the density and lot sizes are interrelated and that an absolute maximum density exists, beyond which exclusively single-family residential structures are not feasible. Further research was suggested to find the *optimum* arrangement of dwelling units with equal population densities (). One of the approaches for such arrangement of dwelling units can be in the form of planning modules for dwelling designs as outlined in Chap. 11 and as outlined in Chap. 9.

A number of economic indices—mainly land utilization and cost ratios such as percentage of land utilization for various categories, lengths of networks per hectare, costs per hectare for circulation and utilities—have been proposed as *optimum* indices for quantitative evaluation of urban layouts (Caminos & Goethert, 1978). Even in the limited number of model layouts studied in this case, large variations in the percentages of land used for private lots, circulation, and so on were observed. A variation of as high as 100 percent was observed in respect of cost ratio and costs per hectare of land for improvements between similar layouts of comparable size. This emphasizes the need for more comprehensive quantitative analysis of layouts adopting a systems approach, to achieve the optimal layout design or the optimum indices for quantitative evaluation of urban layouts.

A method for calculating land utilization and cost ratios, such as the circulation space percentage and the on-site infrastructure costs per square metre, has been presented, which are based on a block of lots having specified block length, lot size, lot ratio (between lot depth and width), and street widths; and also on a site plan module having specified length, lot sizes, lot widths, street widths, and so on. The above derived land utilization and cost ratios are used to calculate the lot size within a given affordable investment per lot. The results presented in Chap. 7 and in HUDCO model show that both circulation space and the cost of on-site infrastructure can vary as the lot sizes change due to changes in affordable investment, which itself can also vary. All these factors may affect the compatibility of the physical layout design with the above calculated results. It is also to be noted that there can be large variations in the cost and land utilization ratios from layout to layout, depending on the various interacting factors as outlined in Chap. 7 and in HUDCO model. This again highlights the need for development of mathematical models for quantitative analysis, representing directly the physical layout designs with the cost and land utilization ratios, and other desired design parameters, so that both are compatible with each other, and thus become realistic permitting realistic assessment of cost-effectiveness and land utilization efficiency between different layouts.

The above literature review outlines attempts towards *optimization* of urban layouts, and also emphasizes the need to develop quantitative analysis and optimization models representing directly the physical layout designs with the cost and land utilization ratios, and other desired design parameters, so that both are compatible with each other, and thus, become realistic. Although *optimum* arrangement of dwelling units or development of *optimum* indices for quantitative evaluation of urban layouts was attempted in the studies discussed above, no optimization techniques were applied for such optimization. It is essential that optimization technique such as linear and nonlinear programming techniques should be used for such optimization. It is possible to develop small-scale mathematical models for quantitative analysis and optimization for different system planning levels, representing directly the physical layout designs with the cost and land utilization ratios, and other desired design parameters, so that both are compatible with each other, and thus, become realistic. This is attempted below.

10.6   Models for Quantitative Analysis and Optimization

In order to illustrate the concept of small-scale modeling and systems design in residential planning, we can consider the representative selections of layout planning modules shown in Fig. 10.2. Each of the above layout planning modules is amenable to mathematical modeling, linking costs with the physical design of the module for different planning levels, and thus ensuring compatibility between the physical designs and costs, which is rarely possible in large-scale urban models discussed earlier. Each of the layout planning modules will have a number of physical parameters such as lot area, lot ratio, lot width, lot depth, street width, and open space standards and so on. There are also a number of cost parameters such as unit costs of street pavements, unit costs of utility lines like water supply lines and sewer lines and so on.

Quantitative design efficiency indicators (QDEIs) of the layout planning module could be the lot area, percentage of marketable or saleable land, circulation area ratio, lot density, cost of infrastructure per unit area (e.g. square metre, hectare), cost per lot, and so on, depending on the design condition in a given context. Physical planning, architectural and engineering design-decisions are independent decision variables within the purview of the respective discipline, but the value of specific QDEI is dependent on these decision variables, although many of these dependent and independent variables might be interchangeable depending on the design condition imposed by the decision maker at the specific system level. Adopting the above process, the HUDCO model for shelter design and analysis was developed to help a direct planning and design process and quantitative analysis incorporating a number of handheld calculator-based programmes (HUDCO, 1982). These programmes are now converted to C++ computer programmes enabling easy application by any planner and designer using any computer. In view of the interests in quantitative analysis and optimization in residential land planning by many researchers, as indicated in the literature review presented in Sect. 10.4 above (including the research study by ASCE), the C++ computer codes for HUDCO model are presented here for use by any interested researcher, planner, and designer for quantitative analysis and design.

10.6.1   Developing a Mathematical Model for Quantitative Analysis in Layout Planning and Design

As earlier mentioned, mathematical models for quantitative analysis and optimization can be developed for any of the layout planning modules shown in Fig. 10.2, or for any specific layout planning module conceived by a planner or designer. In fact, there is a hierarchy of layout planning modules for different system planning levels; starting from a 'lot or lot' at the lowest system planning level, to a higher level in the form of a unit 'layout planning module' consisting of a number of blocks of lots

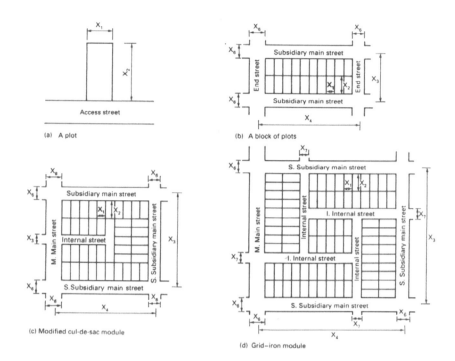

Fig. 10.3 Hierarchies of dimensionless schematic conceptual design of layout planning modules

arranged in a pattern such as cul-de-sac module or, a grid-iron module as shown in Fig. 10.3. Similarly, the layout plan shown in Fig. 10.1 can be conceived to be the next higher system planning level, as it is an aggregation of a number of 'layout planning modules' designed for different system planning levels. Again, a planned city as a whole may consist of a number of layout plans for different sectors, and therefore can be visualized to be the highest system planning level, in the above system hierarchy. Thus, small-scale mathematical models for quantitative analysis and *optimization* can be developed for each system planning level mentioned above, but the criteria for quantitative analysis and optimization will be different for different system planning levels.

As for example, the criteria for quantitative analysis and optimization for lower system planning levels, say, for the 'layout planning module' system level, may be only the capital cost of development within the immediate sphere of influence of the planner-designer; but this criteria may change to the *dynamic* 'socio-economic operations costs' (e.g. for transportation, health, education, and other social facilities), generally, not within the immediate sphere of influence of the planner-designer; requiring many studies including probability studies to arrive at the system solution at a specified point of time, and this system solution should be flexible and subject to periodical navigational change(s) as discussed in Chap. 2, to make the system solution realistic at that point. Depending on the phase of development, the adjacency conditions of such 'planning modules' may be flexible or constrained. Both conditions permit optimization of such 'planning modules'.

In order to help evaluation of alternative system design solution with reference to the desired objective, the first step is to develop mathematical models in the form of a set of algebraic equations, linking the design variables with the objective value scale, say, the cost per lot. Thereafter, a combinatorial approach of optimization could be adopted by identifying the most probable range of values for each design variable, analysing the maximum possible combinations of the values of these design variables and selecting the one considered best (Meredith et al., 1973). Although the number of system solutions obtained by this process will be numerous, many of these combinations may not be acceptable on technical and other relevant reasons. This may leave a smaller number of cases to be evaluated. Thus, in some situations, the adoption of such a combinatorial approach may be feasible to obtain a near optimal system solution as shown in a publication (HUDCO, 1982), but still the unique optimal system solution may not be reached by this method. Therefore, it is desirable to apply mathematical programming techniques such as linear and nonlinear programming technique to attain a resource-efficient optimal system solution, which is presented later.

It is possible to develop mathematical models for quantitative analysis for any of the layout module types shown in Fig. 10.2. Similarly, mathematical models for quantitative analysis can be developed for all the three layout module types shown in Fig. 10.3 or for any new layout module type conceived by a planner-designer. To illustrate the method of development of mathematical models for quantitative analysis, the grid-iron module type shown in Fig. 10.3(d) above (with suitably renaming design variables for convenience as shown in Fig. 10.6 below, which is later converted to an *optimization model* as shown in Sect. 10.6.4 below) is taken as example. Some of the design variables defining the layout planning module are described below, and accordingly, the expressions for the cost per lot and other physical design parameters such as percentage of marketable or saleable land, circulation area ratio, lot density, cost of infrastructure per unit area, and so on, can be developed as shown below (Chakrabarty, 1988a, 1988b):

x_1 = width of the layout planning module between the centre lines of two East-West access streets (for block of lots module) or subsidiary main streets (for other modules)

x_2 = length of the layout planning module between the centre lines of the end streets (for block of lots module) or between the main street and the subsidiary main street (for other modules) in North-South direction

x_3 = gross width of the lot including the allocated width of lot for the reservation of open space per lot where applicable

x_4 or PLD = depth of the lot

x_5 = a dependent variable related to the number of lots in a layout planning module

w or PLW = net width of the lot excluding the allocated width of lot for the reservation of open space per lot where applicable

PLA = a variable defining the net area per lot

I = right of way width of the internal street (also can be designated as x_7)

S = right of way width of the subsidiary main street (also can be designated as x_8) or end street

M = right of way width of the main street (also can be designated as x_9)

O = allocated open space per lot in the layout planning module

R = module ratio or the ratio between the length and the width of the layout planning module

A = area of the layout planning module

NP = number of lots in the layout planning module (model may sometime give the number of lots in fractions, which can be rounded off to the nearest integer value giving negligible error. However, if desired model can also be developed to give exact integer value of number of lots, as illustrated in Sect. 10.6.1 below)

SLR = saleable land ratio, i.e. ratio between the lotted/plotted area and the layout module area

OSR = open space ratio, i.e. ratio between the open space area and the layout module area

CAR = circulation area ratio, i.e. ratio between the circulation area and the layout module area

$PDEN$ = lot density, i.e. number of lots per unit area of the layout module

CI = combined unit cost of the internal street considering all utility lines

CS = combined unit cost of the subsidiary main street considering all utility lines

CM = combined unit cost of the main street considering all utility lines

CL = cost or price of prepared land per unit area of layout planning module

CSC = combined unit cost of the service connection lines to a lot (it is assumed that each lot will have service connection lines linking the utility networks in layout module with the service core in the lot/dwelling)

SLC = service-core location coefficient describing the relative location of the service core in the lot with reference to the lot depth (i.e. x_4 here)

SPC = serviced lot cost, i.e. cost per lot including cost of land, utility networks, service connection lines to a lot

$$x_1 = 6x_4 + 2I + S; \quad x_2 = x_1 R; \quad A = R\, x_1^2; \quad PLA = w \times x_4; \quad SLR = NP \times PA/A;$$
$$OSR = NP \times O/A; \tag{10.1}$$

$$NP = x_4 \times (6 \times x_2 - 4 \times I - 3 \times (M + S))/(w \times x_4 + O); \quad PDEN = NP/A;$$
$$CAR = 1 - SLR\text{-}OSR; \tag{10.2}$$

$$Q = (2 \times x_2 + 4 \times x_4 - M\text{-}S) \times CI + x_2 \times CS + 0.5 \times x_1 \times (CS + CM)$$
$$+ x_1 \times x_2 \times CL; \tag{10.3}$$

$$SPC = Q/NP$$
$$= \frac{\begin{aligned}((2 \times x_2 + 4 \times x_4 - M - S) \times CI + x_2 \times CS + 0.5 \times x_1 \times (CS + CM) \\ + x_1 \times x_2 \times CL) \times (w \times x_4 + O)\end{aligned}}{x_4 \times (6 \times x_2 - 4 \times I - 3 \times (M + S))}$$
$$+ CSC \times SLC \times x_4.$$

$$(10.4)$$

The above mathematical model is developed for grid-iron module type shown in Fig. 10.3(d), and adopting similar procedure the mathematical models for other module types can be developed. It may be noted that there are an infinite number of solutions to the equation set (10.1)–(10.4) that yield different values of cost per lot, or, the same value of cost per lot. Hence, a planner or designer has to assume the value of one or more of the variables x_1, x_2, x_3, and x_4 and so on; in a trial solution, and start an iterative design process to evaluate its suitability; and if not satisfied, a new solution is adopted and the process is repeated. But, only a few trial selections can be evaluated by such numerical analysis even by using IT tools and computers, and still, the *least-cost* solution to achieve *efficiency* and *effectiveness* may not be achieved, as shown by some results of computer-aided iterations presented in Tables 10.1, 10.2, 10.3, 10.4, and 10.5 below.

Table 10.1 Sensitivity of lot cost to the lot area-given land price and infrastructure level (low lot size case)

PLA (SQ.M.)	SLR (PL/HA)	PDEN	CAR	OSR	x_1 (M)	x_2 (M)	SPC (MU)	INFC (MU/Sq.M.)
10.0	0.14	139.72	0.65	0.21	39.50	39.50	5217.86	56.37
15.0	0.23	152.03	0.54	0.23	49.50	49.50	4206.64	46.44
20.0	0.31	152.53	0.47	0.23	59.50	59.50	3789.47	39.45
25.0	0.37	148.15	0.41	0.22	69.50	69.50	3600.08	34.26
30.0	0.43	141.87	0.36	0.21	79.50	79.50	3521.28	30.28
35.0	0.47	135.01	0.32	0.20	89.50	89.50	**3504.29**	27.11
40.0	0.51	128.19	0.29	0.19	99.50	99.50	3525.29	24.55
45.0	0.55	121.66	0.27	0.18	109.50	109.50	3571.27	22.43
50.0	0.58	115.54	0.25	0.17	119.50	119.50	3634.55	20.64
55.0	0.60	109.87	0.23	0.16	129.50	129.50	3710.30	19.12
60.0	0.63	104.62	0.22	0.16	139.50	139.50	3795.32	17.80

INPUT VALUES Chosen in the above CASE are:
$I = 6.0$ M; $S = 7.5$ M; $M = 9.0$ M; $O = 15$ Sq.M; $R = 1$; $CL = 15.00$ MU/SQ.M
PLW $= 3.0$ M; CI $= 480.00$ MU; CS $= 633.0$ MU; CM $= 711.0$ MU; CSC $= 100.0$ MU; SLC $= 0.33$

Table 10.2 Sensitivity of lot cost to the lot area—given land price and infrastructure level (medium lot size case)

PLA (Sq.M.)	PLD (M)	SLR	CAR	OSR	PDEN (PL/HA)	x_1 (M)	x_2 (M)	SPC (MU)	INFC (MU/Sq.M.)
80.0	8.00	0.29	0.64	0.07	36.47	88.00	88.00	53,196.43	159.74
90.0	9.00	0.32	0.60	0.07	36.02	94.00	94.00	52,080.93	151.20
100.0	10.00	0.35	0.58	0.07	35.42	100.00	100.00	51,352.94	143.50
110.0	11.00	0.38	0.55	0.07	34.72	106.00	106.00	50,906.26	136.53
120.0	12.00	0.41	0.52	0.07	33.96	112.00	112.00	50,670.23	130.18
130.0	13.00	0.43	0.50	0.07	33.18	118.00	118.00	**50,596.02**	124.39
140.0	14.00	0.45	0.48	0.06	32.38	124.00	124.00	50,648.79	119.08
150.0	15.00	0.47	0.46	0.06	31.59	130.00	130.00	50,803.03	114.20
160.0	16.00	0.49	0.45	0.06	30.81	136.00	136.00	51,039.64	109.70
170.0	17.00	0.51	0.43	0.06	30.04	142.00	142.00	51,344.02	105.53
180.0	18.00	0.53	0.41	0.06	29.30	148.00	148.00	51,704.84	101.67
190.0	19.00	0.54	0.40	0.06	28.57	154.00	154.00	52,113.14	98.08
200.0	20.00	0.56	0.39	0.06	27.88	160.00	160.00	52,561.78	94.73
210.0	21.00	0.57	0.37	0.05	27.20	166.00	166.00	53,044.98	91.60
220.0	22.00	0.58	0.36	0.05	26.55	172.00	172.00	53,558.00	88.66
230.0	23.00	0.60	0.35	0.05	25.93	178.00	178.00	54,096.94	85.91
240.0	24.00	0.61	0.34	0.05	25.33	184.00	184.00	54,658.55	83.32
250.0	25.00	0.62	0.33	0.05	24.75	190.00	190.00	55,240.10	80.89
260.0	26.00	0.63	0.32	0.05	24.20	196.00	196.00	55,839.29	78.59
270.0	27.00	0.64	0.31	0.05	23.66	202.00	202.00	56,454.14	76.41
280.0	28.00	0.65	0.31	0.05	23.15	208.00	208.00	57,082.98	74.36

INPUT VALUES (i.e. some design conditions) Chosen in the above CASE are:
$I = 10.0$ M; $S = 20.0$ M; $M = 25.0$ M; $O = 20$ Sq.M./Lot; $R = 1$; CL $= 15.00$ MU/Sq.M
PLW $= 10.0$ M; CI $= 3000.0$ MU/M; CS $= 4000.0$ MU/M; CM $= 5000.0$ MU/M; CSC $= 2000.0$ MU/M; SLC $= 0.33$

10.6.2   Complex Interaction between Cost and Physical Parameters in a Layout System: Need for Application of Quantitative Analysis and Optimization in AEC Sector Both in Low- and High-Income Countries

AEC professionals commit considerable resources in terms of land, materials, and other items in their planning-design-decisions related to a layout system covering all relevant AEC disciplines, and not only the planning discipline. There is a complex interaction between the land price, utility network cost, service connection cost, and layout physical parameters; and the lot cost value is determined by the combined effect of all these factors, indicating the *efficiency* achieved in the use of resources committed in the layout planning and design. It is necessary to examine the complex interaction between the land price, utility network cost, service connection cost, and layout physical parameters, at a layout system level, considering all involved

Table 10.3 Sensitivity of lot cost to the lot size-different land prices (low lot size case)

Lot area	SPC1	SPC3	SPC5	SPC6	SPC7
(square meter)	(CL = 0) (MU)	(CL = 15) (MU)	(CL = 50) (MU)	(CL = 100) (MU)	(CL = 1000) (MU)
10.0	4144.29	5217.86	7722.84	11,301.40	75,715.39
15.0	3219.97	4206.64	6508.89	9797.82	68,998.49
20.0	2806.06	3789.47	6084.07	9362.08	68,366.25
25.0	2587.63	3600.08	5962.48	9337.32	70,084.57
30.0	2463.99	3521.28	5988.29	9512.59	72,950.05
35.0	2393.30	3504.29	6096.60	9799.91	76,459.41
40.0	2355.13	3525.29	6255.66	10,156.19	80,365.70
45.0	2338.34	3571.27	6448.11	10,557.88	84,533.71
50.0	2336.35	3634.55	6663.70	10,991.05	88,883.32
55.0	2345.01	3710.30	6895.96	11,446.91	93,364.01
60.0	2361.61	3795.32	7140.65	11,919.69	97,942.40

INPUT VALUES Chosen in the above CASE are:

$I = 6.0$ M; $S = 7.5$ M; $M = 9.0$ M; $O = 15$ Sq.M./Lot; $R = 1$

PLW $= 3.0$ M; CI $= 480.0$ MU/M; CS $= 633.0$ MU/M; CM $= 711.0$ MU/M; CSC $= 100.0$MU/M; SLC $= 0.33$

INPUT-Land Price i.e. CL in MU/Sq.M.is shown above for each case

Table 10.4 Sensitivity of lot cost to the lot area-different land prices (medium lot size case)

Lot area	SPC1	SPC3	SPC5	SPC6	SPC7
(square meter)	(CL = 0) (MU)	(CL = 15) (MU)	(CL = 50) (MU)	(CL = 100) (MU)	(CL = 1000) (MU)
60.0	50,500.12	54,611.15	64,203.56	77,907.00	324,568.92
70.0	47,555.93	51,646.69	61,191.80	74,827.67	320,273.37
80.0	45,403.12	49,516.43	59,114.16	72,825.21	**319,624.08**
90.0	43,776.58	47,940.93	57,657.75	71,538.93	321,400.11
100.0	42,517.65	46,752.94	56,635.29	70,752.94	324,870.59
110.0	41,525.58	45,846.26	55,927.86	70,330.15	329,571.33
120.0	40,733.33	45,150.23	55,456.34	**70,179.34**	335,193.43
130.0	40,094.59	44,616.02	55,166.03	70,237.47	341,523.37
140.0	39,576.30	44,208.79	55,017.93	70,459.55	348,408.84
150.0	39,154.27	43,903.03	**54,983.47**	70,812.67	355,738.29
160.0	38,810.37	43,679.64	55,041.26	71,272.15	363,428.16
170.0	38,530.77	43,524.02	55,174.96	71,819.15	371,414.60
180.0	38,304.69	43,424.84	55,371.86	72,439.02	379,648.00
190.0	38,123.66	43,373.14	55,621.94	73,120.22	388,089.23
200.0	37,980.89	**43,361.78**	55,917.20	73,853.50	396,707.01
210.0	37,870.90	43,384.98	56,251.16	74,631.41	405,476.03
220.0	37,789.20	43,438.00	56,618.52	75,447.84	414,375.60
230.0	37,732.09	43,516.94	57,014.92	76,297.75	423,388.65
240.0	37,696.48	43,618.55	57,436.71	77,176.93	432,500.93

(continued)

Table 10.4   (continued)

Lot area	SPC1	SPC3	SPC5	SPC6	SPC7
250.0	**37,679.79**	43,740.10	57,880.83	78,081.87	441,700.52
260.0	37,679.83	43,879.29	58,344.70	79,009.58	450,977.30
270.0	37,694.72	44,034.14	58,826.12	79,957.52	460,322.68
280.0	37,722.89	44,202.98	59,323.21	80,923.53	469,729.28

INPUT VALUES Chosen in the above CASE are:
$I = 10.0$ M; $S = 20.0$ M; $M = 25.0$ M; $O = 20$ Sq.M./Lot; $R = 1$
PLW = 10.0 M; CI = 3000.00 MU/M; CS = 4000.0 MU/M; CM = 5000.0 MU/M; CSC = 2000.0
MU/M; SLC = 0.10
INPUT-Land Price i.e. CL in MU/Sq.M.is shown above for each case

disciplines, to derive the lot cost. It may be noted that almost all countries in the world suffers many urban problems and achieving *efficiency* and *social equity* in the urbanization, urban development, urban-renewal or urban re-development process, may be the key to solve such urban problems. Moreover, all of us *live in a finite earth with finite resources*, and in this context, an *ethical responsibility* is enjoined on all of us to achieve *resource-efficiency* and *social equity* at each system level in all countries, and only through application of quantitative analysis and optimization techniques at each level, whether in low- or high-income countries, we can achieve this objective.

As pointed out in Chap. 1, Operations Research (**OR**) and Computer-Aided Design (**CAD**) are essential integrative techniques to improve *efficiency & effectiveness*, and since, AEC professionals commit huge resources in their planning-design-decisions, it is basic that they use such basic tools to discharge their accountability to the society to achieve *efficiency* and *social equity* in the use of huge (but inadequate compared to the needs in all countries) *resources* committed by them in their plans and designs to produce building and urban services. *In view of above, it is desirable that such urban planning and development problems are seen as part of integrated urban management and AEC professionals uses management techniques like OR/CAD including quantitative analysis and optimization, to attain a resource-efficient and equitable urban problem solutions in all countries.*

To permit computer-aided iterations and optimization of layout planning modules a C++ program named as: LAY.CPP is developed, which incorporates five iteration-models including three sub-models or iteration-models for quantitative analysis and optimization of layout module: **3(d)** shown and discussed above. The software sub-model-1 or ITMODEL-1 provides for a series of lot sizes, incorporating desired lot sizes of low, high and incremental interval size given as input. Application examples of quantitative analysis and optimization in layout planning and design are presented below, using the mathematical model for quantitative analysis given above incorporated in the C++ computer program LAY.CPP. Thus, a series of values of different lot parameters are generated to determine the sensitivity of lot cost to the lot size for given land price.

Table 10.5 Sensitivity of lot cost to the lot area for different land prices (high/medium lot size case-applicable in high-income countries)

Lot area	PC1	PC2	PC3	PC4	PC5	PC6	PC7
(square meter)	(CL = 0) (MU)	(CL = 5) (MU)	(CL = 15) (MU)	(CL = 30) (MU)	(CL = 50) (MU)	(CL = 100) (MU)	(CL = 1000) (MU)
100.0	24,073.5	28,055.1	36,018.4	47,963.2	63,889.7	103,705.9	820,397.1
125.0	21,767.0	25,587.5	33,228.4	44,689.8	59,971.6	98,176.1	785,858.0
150.0	20,220.0	23,971.9	31,475.8	42,731.6	57,739.3	95,258.7	770,607.6
175.0	19,114.2	22,851.2	30,325.3	41,536.4	56,484.5	93,854.8	**766,520.2**
200.0	18,288.5	22,044.5	29,556.5	40,824.5	55,848.6	**93,408.7**	769,490.4
225.0	17,652.0	21,449.8	29,045.4	40,438.9	**55,630.1**	93,608.2	777,214.4
250.0	17,149.6	21,005.3	28,716.8	**40,284.0**	55,707.0	94,264.3	788,297.1
275.0	16,745.8	20,671.3	28,522.3	40,298.7	56,000.7	95,255.5	801,842.9
300.0	16,416.7	20,420.8	28,429.2	40,441.7	56,458.3	96,500.0	817,250.0
325.0	16,145.5	20,235.3	**28,414.8**	40,684.1	57,043.1	97,940.8	834,098.5
350.0	15,920.2	20,101.0	28,462.7	41,005.2	57,728.5	99,536.8	852,086.6
400.0	15,573.9	19,949.2	28,699.9	41,826.0	59,327.4	103,081.0	890,644.9
450.0	15,328.9	**19,910.8**	29,074.6	42,820.3	61,147.9	106,966.9	931,709.2
500.0	15,155.7	19,952.6	29,546.5	43,937.3	63,125.0	111,094.3	974,542.5
550.0	15,035.1	20,053.3	30,089.6	45,144.2	65,216.9	115,398.7	101,8671.4
600.0	14,954.3	20,198.4	30,686.7	46,419.0	67,395.5	119,836.7	106,3778.2
650.0	14,904.4	20,378.0	31,325.4	47,746.5	69,641.2	124,378.0	110,9641.2
700.0	14,878.8	20,584.9	31,997.1	49,115.4	71,939.9	129,001.0	115,6101.1
750.0	**14,872.8**	20,813.6	32,695.3	50,517.8	74,281.2	133,689.6	120,3040.6
800.0	14,882.8	21,060.3	33,415.1	51,947.5	76,657.2	138,431.6	125,0371.1

INPUT VALUES Chosen in the above CASE are:

I = 20.0 M; S = 25.0 M; M = 25.0 M; O = 50 Sq.M; R = 1

PLW = 20.0 M; CI = 500.00 MU; CS = 1000.0 MU; CM = 1000.0 MU; CSC = 500.0 MU; SLC = 0.10

INPUT-Land Price i.e. CL in MU/SQ.M.is shown for each case as above

Table 10.1, prepared for the low lot size case and usually applicable in low-income countries, is presented above. It will be seen that in the low lot size case (Table 10.1), the value of SPC (serviced lot cost, i.e. cost per lot including cost of land, utility networks, service connection lines to a lot) for a lot size of 10 sq. m. = 5217.86 MU (monetary units) which reduces to **3504.29** MU for a lot size of 35 sq.m. It should be noted here that for the given design conditions, a higher lot size reduces the value of SPC, which is quite contrary to the usual notion that higher the lot size higher should be the lot cost value, i.e. SPC. This is because there is a complex interaction between the land price, utility network cost, service connection cost, and layout physical parameters, and the lot cost value is determined by the combined effect of all these factors. Thus, while reduction in lot size reduces the land cost component, it increases the utility network cost component because of provision of utility networks at closer interval necessitated by less depth of lot, consequent to adoption of lesser lot size as the width of the lot is kept unchanged.

Moreover, Table 10.1 shows that lesser lot sizes reduce the value of saleable land ratio (SLR) and increase the circulation area ratio (CAR); which also causes higher lot costs for lesser lot sizes. It may be seen that after a lot size of 15 sq.m. the value of lot density (PDEN) goes on decreasing, the value of saleable land ratio (SLR) goes on increasing, the value of circulation area ratio (CAR) goes on decreasing, and the value of SPC (serviced lot cost) goes on decreasing, reaching a minimum value of **3504.29** MU at a lot size of **35.0** sq.m. However, on increase of lot size over **35.0** sq. m. the lot cost values, i.e. SPC goes on increasing and reaching a value of 3795.32 MU for a lot size of 60 sq.m. as shown in Table 10.1. These results indicate that the lot size cost relationship is not linear, and that, it is a case of *optimization*. Thus, in these given design conditions, a lot size of approximately **35.0** sq.m. appears to be the *optimum* lot size giving the least-cost per lot. However, using the optimization techniques as discussed in Chap. 8, the exact value of optimum lot size giving the least-cost per lot, and the exact least-cost value of SPC can be determined.

As done for the low lot size case presented above, a series of values of different lot parameters are generated for the medium lot size case (usually applicable in medium- or high-income countries), using the C++ computer program LAY.CPP. These are presented in Table 10.2 given below to determine the sensitivity of lot cost to the lot size for given land price. It may be seen from Table 10.2 that the serviced cost per lot (i.e. SPC) starts at a value of 53,196.43 MU (i.e. monetary units) for a lot area of 80 m^2, reduces to a value of 50,596.02 MU (i.e. monetary units), which is approximately the least value of SPC in the present chosen case of design condition. This value of SPC is for a lot area of 130 m^2, which is approximately the *optimum lot size* giving the **Least-Cost** in the chosen design condition here. Again, SPC increases to a value of 57,082.98 MU (i.e. monetary units) for a lot area of 280 m^2. Similar to low lot size case presented in Table 10.1 above, the Table 10.2 prepared for the medium lot size case also shows that lesser lot sizes reduce the value of saleable land ratio (SLR) and increase the circulation area ratio (CAR). All these factors cause higher lot costs for lesser lot sizes, in this medium lot size case also.

It is also shown that even a lower lot size such as 80 m^2 size in this medium lot size case, gives a higher lot cost, i.e. 53,196.43 MU compared to a lower lot cost of

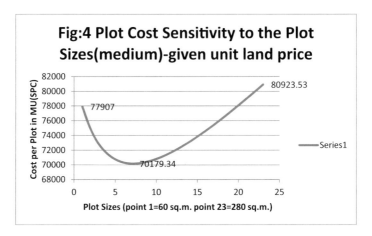

Fig. 10.4 Plot cost sensitivity to the plot sizes (medium)-given unit land price

50,596.02 MU for a higher lot size of 130 m², which appears to be the *optimum* lot size in the given design condition chosen here. Any other lot size chosen by a planner, whether lower or higher than 130 m², will increase the cost per lot in this case. In Tables 10.1 and 10.2, a low land price of 15.00 MU/Sq.M. with SLC (service-core location coefficient) = 0.33 are taken. In case of comparatively higher land prices, say, CL = 100 MU/Sq.M. with SLC = 0.10, and all other INPUT VALUES remaining unchanged as above, and using the C++ computer program LAY.CPP, a series of values of different lot parameters including the values of SPC can be generated and converted to a graphic form. Such a graphic form is shown in Fig. 10.4 below, taking only the single variable of lot size ranging from 60 Sq.M. to 280 Sq.M., with land price of 100 MU, CSC = 2000 MU, SLC = 0.1, CI = 3000 MU, CS = 4000 MU, CM = 5000 MU, I = 10 M, S = 20 M, M = 25 M; O = 20 (as shown in note below Table 10.4) giving the corresponding SPC values also shown in Table 10.4.

10.6.3 *Layout System Planning and Design: A Case of Optimization*

The series of values of different lot parameters to determine the sensitivity of lot cost to the lot sizes are generated, both for the low lot size case (generally applicable in low-income countries) in Table 10.1, and for the medium lot size case (applicable in medium- or high-income countries) in Table 10.2. Both Tables 10.1 and 10.2 indicate a nonlinear relationship between the lot cost and the lot size, where only one land price (i.e. CL = 15.00 MU/Sq.M.) is adopted. As already pointed out, there is a complex interaction between the land price, utility network cost, service connection cost, and layout physical parameters, and the lot cost value is determined by the combined effect of all these factors. Thus, while reduction in lot size reduces the land cost component, it increases the utility network cost component because of provision of utility networks at closer interval necessitated by less depth of lot, consequent to adoption of lesser lot size as the width of lot is kept unchanged.

10.6.3.1   Lot Cost Sensitivity to the Lot Areas-Different Land Prices with a Given Infrastructure Level

Instead of only one land price, Table 10.3 is generated to give SPC values for different land prices for the low lot size case. Figure 10.5 uses these SPC values showing the sensitivity of lot cost to the lot sizes for the low lot size case, adopting different land prices (i.e. CL = 0, 15, 50, and 100 MU/Sq.M.). It may be seen from the nonlinear curves generated for each land price, there is a clear indication of optimum lot size for each of the four cases, and this indication is more and more pronounced for higher land prices, and that, the curves become flatter for lower land prices.

As shown in Table 10.3, the approximate values of the **Least-Cost** lot sizes (for low lot size case) in square metres are 20, 25, 25, 35, and 50 for the land prices in MU/Sq.M. of 1000, 100, 50, 15, and 0, respectively. Similarly, Table 10.4 is generated to give SPC values for different land prices for the medium lot size case, indicating the sensitivity of lot cost to the lot area for this lot case, when different land prices are adopted. It may be seen from the Table 10.4, that the approximate values of the **Least-Cost** lot size (for medium lot size case) in square metres are 80, 120, 150, 200, and 250 for the land prices in MU/Sq.M. of 1000, 100, 50, 15, and 0, respectively. Again, Table 10.5 is generated to give SPC values

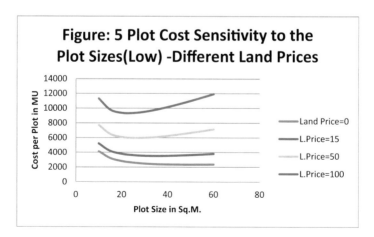

Fig. 10.5 Plot cost sensitivity to the plot sizes (low)-different land prices

INPUT VALUES (i.e. some design conditions) Chosen in the above CASE are:

$I = 10.0$ M; $S = 20.0$ M; $M = 25.0$ M; $O = 20$ Sq.M./Lot; $R = 1$; CL = 100 MU/Sq.M.
PLW = 10.0 M; CI = 3000.0 MU/M; CS = 4000.0 MU/M; CM = 5000.0 MU/M; CSC = 2000.0 MU/M; SLC = 0.10

for different land prices for the high lot size cases applicable in high-income countries, indicating the sensitivity of lot cost to the lot area for this lot case, when different land prices are adopted. It may be seen from the Table 10.5 that the approximate values of the **Least-Cost** lot size (for high/medium lot size case) in square metres are **175** (taking an interval of 25 sq.m.), 200, 225 (taking an interval of 25 sq.m.), 250, 325 (taking an interval of 25 sq.,m.), 450, and **750** for the land prices in MU/Sq.M. of 1000,100, 50, 30, 15, 5, and 0, respectively. Thus, in high lot size high-income-country case, the **Least-Cost** lot size may be as low as: 175 sq.m. and as high as: 750 sq.m., (i.e. a variation of **4.29 times**) for the same infrastructure and open space standards adopted, only depending on the unit land price applicable in a site-specific case.

It may be noted from the above tables that lower the land prices, higher is the optimum lot size value giving the **Least-Cost** per lot. One reason for this is that the land cost component of lot cost gets reduced in all lot size cases, i.e. low, medium, and high lot size cases, due to less unit land prices, while unit costs of utility and service connection lines are kept unchanged. Assuming a land price of 100 MU/Sq. M and using the SPC values given in Table 10.4 for this land price for medium lot size case, sensitivity of lot cost to the lot sizes for this case is shown graphically in Fig. 10.4, which clearly gives an indication of *optimum* lot size and the Least-Cost per lot. This property of the *optimum lot size value* giving the **Least-Cost** per lot is applicable for all cases, i.e. low, medium, and high lot size cases (with appropriate minimum acceptable width of lot for each case) applicable in low-, medium-, and high-income countries, respectively. This clearly highlights the need to use *quantitative analysis and optimization as universal planning and design tools in such AEC operations both in low- and high-income countries, to attain a resource-efficient and equitable problem solutions in all countries.*

10.6.3.2   Lot Cost Sensitivity to the Lot Area-Different Infrastructure Levels with a Given Land Price

Instead of only one infrastructure level, Table 10.6 presented below is generated to give SPC values for different infrastructure levels with a given land price. Here, one unit land price, i.e. $CL = 100$ MU/Sq.M., and two infrastructure levels (low-indicated by $INFL = 1$, and high- indicated by $INFL = 2$) are considered with corresponding unit costs of utility networks as shown in the note below Table 10.6. SPC and other parameter values for each lot area are shown in the table for the respective infrastructure level (i.e. for $INFL = 1$ and $INFL = 2$, respectively). It may be seen from the table that the indicated *optimum* lot area giving the **Least-Cost** per lot is 30 Sq.M. for $INFL = 1$ with SPC value of 12,699.7 MU per lot; and that the indicated *optimum* lot area giving the **Least-Cost** per lot is 70 Sq.M. for $INFL = 2$ with SPC value of 36,084.1 MU per lot. Thus, in this case a higher infrastructure level (i.e. INFL) gives a higher value of *optimum* lot area giving the **Least-Cost** per lot, while the unit land price is kept unchanged at $CL = 100$ MU/Sq.M., whereas in the case of changing unit land prices a higher unit land price provides a lower *optimum* lot area giving the **Least-Cost** per lot as shown in Sect. 10.6.3.1 above.

Table 10.6 Sensitivity of lot cost to the lot area for different infrastructure levels (given land price)

INFL	PLA (Sq.M.)	PLD (M)	SLR	CAR	OSR	PDEN (PL/HA)	N	x_1 (M)	x_2 (M)	SPC (MU)	INFC (MU/Sq.M.)
1	20.0	4.00	0.23	0.60	0.17	113.24	21	43.5	43.5	13,457.3	51.9
2	20.0	4.00	0.12	0.80	0.09	58.31	24	64.0	64.0	52,397.7	205.3
1	**30.0**	6.00	0.34	0.49	0.17	112.33	35	55.5	55.5	**12,699.7**	42.0
2	30.0	6.00	0.19	0.71	0.10	64.87	37	76.0	76.0	43,207.7	179.9
1	40.0	8.00	0.42	0.42	0.16	105.83	48	67.5	67.5	12,854.3	35.2
2	40.0	8.00	0.27	0.64	0.10	66.30	51	88.0	88.0	39,253.9	159.7
1	50.0	10.00	0.49	0.36	0.15	98.22	62	79.5	79.5	13,363.7	30.3
2	50.0	10.00	0.33	0.57	0.10	65.38	65	100.0	100.0	37,341.2	143.5
1	60.0	12.00	0.55	0.32	0.14	90.87	76	91.5	91.5	14,047.3	26.6
2	60.0	12.00	0.38	0.52	0.10	63.39	80	112.0	112.0	36,430.4	130.2
1	70.0	14.00	0.59	0.28	0.13	84.18	90	103.5	103.5	14,829.1	23.7
2	**70.0**	14.00	0.43	0.48	0.09	60.95	94	124.0	124.0	**36,084.1**	119.1
1	80.0	16.00	0.63	0.26	0.12	78.21	104	115.5	115.5	15,671.6	21.3
2	80.0	16.00	0.47	0.45	0.09	58.37	108	136.0	136.0	36,087.0	109.7
1	90.0	18.00	0.66	0.23	0.11	72.92	119	127.5	127.5	16,554.3	19.4
2	90.0	18.00	0.50	0.41	0.08	55.80	122	148.0	148.0	36,320.5	101.7
1	100.0	20.00	0.68	0.22	0.10	68.23	133	139.5	139.5	17,464.9	17.8
2	100.0	20.00	0.53	0.39	0.08	53.33	137	160.0	160.0	36,714.3	94.7
1	110.0	22.00	0.70	0.20	0.10	64.07	147	151.5	151.5	18,395.6	16.4
2	110.0	22.00	0.56	0.36	0.08	50.98	151	172.0	172.0	37,224.1	88.7

INPUT VALUES Chosen in the above CASE are:

INFL-1:

$I1 = 6.0$ M; $S1 = 7.5$ M; $M1 = 9.0$ M; $O = 15$ Sq.M; $R = 1$; $CL = 100.00$ MU/Sq.M

$PLW = 5.0$ M; $CI1 = 480.00$ MU; $CS1 = 633.0$ MU; $CM1 = 711.0$ MU; $CSC = 100.0$ MU; $SLC = 0.10$

INFL-2:

$I2 = 10.0$ M; $S2 = 20.0$ M; $M2 = 25.0$ M; $O = 15$ Sq.M; $R = 1$; $CL = 100.00$ MU/Sq.M

$PLW = 5.0$ M; $CI2 = 3000.00$ MU; $CS2 = 4000.0$ MU; $CM2 = 5000.0$ MU; $CSC = 100.0$ MU; $SLC = 0.10$

10.6.3.3   Quantitative Analysis and Optimization: Universal Applicability in Layout Planning and Design

Results shown in Tables 10.1, 10.2, 10.3, 10.4, 10.5, and 10.6, Fig. 10.4, and in Fig. 10.5 clearly indicate that the relationship between the lot cost and lot area is nonlinear, and that there can be a large variation in the values of QDEIs (Quantitative Design Efficiency Indicators) such as lot cost, land utilization ratios, cost ratios, lot density, and so on. This is because of complex interactions between the various design parameters chosen by planners and designers in respect of a layout plan including the physical design pattern of a layout planning module, planning standards adopted, and also the external design conditions such as unit land price, unit cost of utility network, and so on. The results presented above are in respect of low and medium lot size cases which also clearly furnish an indication of *optimum* lot area for **Least-Cost** per lot. Even in respect of comparatively high lot size case (usually applicable in high-income countries), it is shown earlier (Chakrabarty, 1991) that the maximum plot (i.e. lot) area within a given cost can vary from 426.62 to 548.7 m^2 (i.e. a variation of 1.29 times) depending on the layout module type and other design conditions chosen by planner/designer and other stakeholders.

Moreover, results shown in Tables 10.1, 10.2, 10.3, 10.4, 10.5, and 10.6, Fig. 10.4, and in Fig. 10.5 are quite contrary to the usual notion that reducing the lot size means automatic reduction in lot cost, which is not always true. In fact, in many cases, a lower lot size may increase the cost per lot and cause other inefficiencies due to the complex interactions between the land prices, utility network and service connection line costs, and the geometry of the layout planning module, which cannot be gauged by intuition and needs to be assessed by using the quantitative analysis models and optimization techniques. Thus, it is clear that the layout planning and design, for any lot size range (say low, medium, or high lot size cases) is a case of *optimization*, where each design condition, i.e. land price, utility network cost, service connection cost, and physical parameters such as road widths, open space standards, module pattern, and so on, produces an optimum lot size giving the *least-cost* per lot, which cannot be determined by intuition and requires use of quantitative analysis and optimization techniques if *efficiency* is the objective.

It is sometimes argued that use of quantitative analysis and optimization techniques is applicable only in Third World Countries (i.e. low-income countries) which are subjected to severe resource-constraints and population pressure (Chakrabarty, 1987a, 1987b TWPR). Results in Table 10.5 for high/medium lot size case show that the values of the **Least-Cost** lot size (for high/medium lot size case) in square metres may be as high as **750.0** and **450.0** for a land price of 0 or 5 MU per Sq.M. respectively, clearly highlighting the need to carry out such quantitative analysis and optimization studies even in high-income countries. In view of results presented above any notion that use of quantitative analysis and optimization techniques is applicable only in Third World Countries, i.e. low-income countries should be discarded, and such techniques, as available in literature and presented in Chap. 8, should be *universally applied* in all countries (all low-, medium-, and high-income countries). It is also desirable that any large variations in the values of QDEIs for different planning and design solutions are presented by AEC professionals in a *transparent* manner (by application of quantitative analysis and optimization techniques) to all *stakeholders* including user citizens in all countries, for appropriate planning and design-decisions, which will not only help achieve a *resource-efficient* solution with *social equity*, but will also promote a *corruption-free* urban planning, development, and management.

Taking a single variable hypothetical function $y = f(x)$, some of its critical points (i.e. stationary points) such as global maximum, valley, local maximum, local minimum, and global minimum, are explained in Chap. 8. A stationary point is any point where the gradient vanishes, i.e. $dy/dx = 0$ in case of a single variable function; in which case, it can be shown that when the first derivative vanishes, it is both *necessary and sufficient conditions* for a *global minimum*. Figure 10.4 above clearly shows a stationery point where gradient vanishes i.e. $dy/dx = 0$, which is near about the SPC value $= 70,179.34$ MU as shown, which is near about the *global minimum* point of the lot cost function, i.e. least-cost per lot. It is difficult to find graphically the exact value of this *global minimum* point. However, using mathematical programming and the C++ computer program LAY.CPP the exact value of this *global minimum* point can be instantly determined. This *global minimum* point is **70,175.88**MU for an *optimum* lot size of **121.85** Sq.M. as shown for 'Medium Lot Size Case' in Sect. 10.6.5 below.

Figure 10.4 clearly shows that lowering the lot size to 60 Sq.M. (i.e. to 50% of approximate optimum lot size of **120.0** Sq.M. indicated in Table 10.4) increases the cost per lot (i.e. SPC $= 77,907$ MU) by 11% compared to the least-cost of about 70,179.34 MU per lot for the optimum lot size of 120 Sq.M. in the present case. Thus, results presented in Tables 10.1, 10.2, 10.3, and 10.4, in Fig. 10.4 (medium lot size case—usually applicable in medium/high-income countries), and in Fig. 10.5 with different land prices (for low lot size case—usually applicable in low-income countries) clearly indicating *optimum* points, show that layout planning and design, at the planning module system level in all countries, is an obvious case of *Optimization*. This should be recognized by all urban planners and other AEC professionals in all countries, who should apply quantitative analysis and optimization technique in the AEC sector as a whole and in all urban operations, to achieve *efficiency* and *social equity* which are very important planning and design objectives, as also highlighted in Chaps. 1, 2, 6, 8, and 9.

Using Tables 10.1, 10.2, 10.3, 10.4, 10.5, and 10.6, Figs. 10.4 and 10.5, we can determine only approximately, the *optimum lot size* giving the *Least-Cost* per lot. However, by using *optimization models,* the exact *least-cost* solution can be achieved in no time, and by developing suitable expressions for module parameters and design parameters for each module type, the same *optimization model* can be used for different module types with many variables as against single variable adopted above. These are illustrated in later sections. The above quantitative analysis model can be easily converted to a simple optimization model, using only the lot depth as a single variable in a nonlinear lot cost function or equation, and thus, find the optimum lot depth and consequently the optimum lot size (assuming a given lot width) giving the **Least-Cost** per lot. In fact, in many cases, a lower lot size may increase the cost per lot due to complex interaction between the land prices, utility network costs, and the geometry of the layout planning module, which needs to be assessed by using the optimization model developed below, based on the quantitative analysis model presented above.

10.6.4 *Converting Quantitative Analysis Model to an Optimization Model: Assessing Land Price Utility-Networks-Cost Interaction in Land Subdivision to find the exact Least-Cost Lot Size*

The quantitative analysis model represented by the equation set (10.1)–(10.4) above can be easily converted to an optimization model. In order to simplify the computation, we can assume a fixed value of net width of lot, i.e. PLW as 'w', and put value of lot depth as x, and the value of serviced lot cost as y, and thus, express y in terms of only one variable x, using the equation set (10.1)–(10.4) above. Thereafter, we can differentiate y with respect to x and derive the expression for dy/dx and put its value as zero, and thus, derive the value x (i.e. lot depth) to give the minimum value of y, i.e. the *least-cost* per lot. Using this derived value of x, the values of all other parameters of the layout module can be derived, and thus, solving the complete problem as shown below.

$$x_1 = 6x + 2I + S; x_2 = x_1 R; A = R x_1^2;$$
$$\text{NP} = x \times (6 \times x_2 - 4 \times I - 3 \times (M + S))/(w \times x + O)$$
$$= x \times (36Rx + 12IR + 6SR - 4I - 3S - 3M)/(w \times x + O);$$
$$Q = (2 \times x_2 + 4 \times x - M\!-\!S) \times \text{CI} + x_2 \times \text{CS} + 0.5 \times x_1 \times (\text{CS} + \text{CM}) + x_1$$
$$\times x_2 \times \text{CL};$$
$$= (12Rx + 4IR + 2SR + 4x\!-\!M\!-\!S) \times \text{CI} + (6Rx + 2IR + SR) \times \text{CS}$$
$$+ 0.5 \times (6x + 2I + S) \times (\text{CS} + \text{CM}) + R \times (6x + 2I + S)^2 \times \text{CL};$$
$$\text{SPC} = y = Q/N + \text{CSC} \times \text{SLC} \times x;$$

To simplify the above equations, we can express the constants: k_1, k_2, k_3, k_4, k_5, and k_6 as follows:

$$k_1 = 36RCL$$
$$k_2 = 6R(2CI + CS + 2CL(21 + S)) + 4CI + 3(CS + CM)$$
$$k_3 = (2I + S)(R(2CI + CS) + 0.5(CS + CM) + RCL(2I + S))–CI(S + M)$$
$$k_4 = 36R$$
$$k_5 = 6R(2I + S)–4I–3(S + M)$$
$$k_6 = CSC \times SLC.$$

Substituting these constants, the expressions for NP and Q become as follows:

$$NP = x \times (36Rx + 12IR + 6SR - 4I - 3S - 3M)/(w \times x + O)$$
$$= x \times (k_4x + 6R(2I + S)–4I–3(S + M))/(w \times x + O)$$
$$= x \times (k_4x + k_5)/(w \times x + O)$$

$$Q = (12Rx + 4IR + 2SR + 4x–M–S) \times CI + (6Rx + 2IR + SR) \times CS + 0.5$$
$$\times (6x + 2I + S) \times (CS + CM) + R \times (6x + 2I + S)^2 \times CL$$
$$= (12Rx + 2R(2I + S) + 4x–S – M)CI + (6Rx + R(2I + S))CS$$
$$+ 3(CS + CM)x + 0.5(CS + CM)(2I + S)$$
$$+ RCL\left(36x^2 + (2I + S)^2 + 12x(2I + S)\right)$$
$$= 36R \ CLx^2 + (12RCL(2I + S) + 3(CS + CM) + 4CI + 6RCS + 12RCI)x$$
$$+ 2RCI(2I + S) - (S + M)CI + RCS(2I + S) + 0.5(CS + CM)(2I + S)$$
$$+ RCL(2I + S)^2$$
$$= 36RCLx^2 + (6R(2CI + CS + 2CL(2I + S)) + 4CI + 3(CS + CM))x$$
$$+ (2I + S)(R(2CI + CS) + 0.5(CS + CM) + RCL(2I + S))–CI(S + M)$$
$$= k_1x^2 + k_2x + k_3$$

Substituting the above expressions, we get:

$$y = Q/NP + k_6x$$
$$= \{k_1x^2 + k_2x + k_3\}(w \times x + O)\}/\{x \times (k_4x + k_5)\} + k_6x$$

Differentiating y with respect to x and putting it equal to 0, we get:

$$\frac{dy}{dx} = \{(w(k_1 x^2 + k_2x + k_3) + (w \times x + O)(2k_1x + k_2)\}\{x \times (k_4x + k_5)\}^{-1} +$$
$$\{k_1x^2 + k_2x + k_3)(w \times x + O)\}(-1)\{x \times (k_4x + k_5)\}^{-2}(2k_4x + k_5) + k_6 = 0$$

Multiplying both sides by $\{x \times (k_4 x + k_5)\}^2$, we get:

$$\{(w(k_1x^2 + k_2x + k_3) + (w \times x + O)(2k_1x + k_2))\{x \times (k_4x + k_5)\} -$$
$$\{k_1x^2 + k_2x + k_3)(w \times x + O)\}(2k_4x + k_5) + k_6\{x \times (k_4x + k_5)\}^2 = 0 \quad (10.5)$$

Expression (10.5) is a single variable nonlinear equation and can be solved easily by the methods presented in Chap. 3. Solution of the above optimization problem is also possible using the C++ program named as: LAY.CPP mentioned above, which incorporates one sub-model or iteration-model for determining the *optimum lot depth* of layout module: Fig. 10.3(d) shown and discussed above, and thus, find the optimum lot size to give the Least-Cost per lot, using Eq. (10.5) above.

```
SOME OPTIMUM RESULTS GIVEN BY ITRN.MODEL ARE:
OPT.PLA= 491.57 SQ.M. x1=197.47 M x2= 197.47 M
OPT.SLR= 0.58 OPT.CAR= 0.38 OPT.OSR= 0.05M OPT.PDEN=11.74 PLOTS/HEC.
OPTIMUM PLOT DEPTH=24.58 M LEAST-PLOT-COST=20558.45MU Number of
plots=  46
SOME INPUT-VALUES Chosen in the above CASE are:
 I= 15.0 M. S= 20.0 M. M= 25.0 M. O= 40 Sq.M. R=1.00 SLC== 0.05 PLW= 20.00 M
CI=500.00 MU/M CS=1000.00 MU/M CM=2000.00 MU/M CL= 5.00 MU/SQ.M.
CSC=500.00 MU/M

OUTPUT File Name:'optmpld.cpp' of LAY.CPP 29/2/20

OPT.PLA= 491.57 SQ.M. Applicable for high income housing when CL is very
low
```

10.6.5 Comparing Results of Quantitative Analysis Model and Optimization Model

As earlier stated, the results shown in Table 10.1 (**Low Lot Size Case**), Table 10.2 (**Medium Lot Size Case**), Fig. 10.4 (**Medium Lot Size**)**,** and Table 10.5 (**High/ Medium Lot Size Case**) above give only an indicative value of the **Least-Cost** lot size. However, running the C++ program: LAY.CPP (and using sub-model or iteration-model for determining the *optimum lot depth*) incorporating the Eq. (10.5) above, the exact value of the **Least-Cost** lot size with all other optimal parameters, can be determined for each of the above cases, as shown below.

　　Low Lot Size Case—Table 10.1

```
SOME OPTIMUM RESULTS GIVEN BY ITRN.MODEL ARE:
OPT.PLA=34.37 SQ.M. x1=88.24 M x2= 88.24 M
OPT.SLR= 0.47 OPT.CAR= 0.33 OPT.OSR= 0.20M OPT.PDEN=135.88 LOTS/HEC.
OPTIMUM LOT DEPTH=11.46 M LEAST-LOT-COST=3503.98MU Number of lots= 106
SOME INPUT-VALUES Chosen in the above CASE are:
 I= 6.0 M. S= 7.5 M. M= 9.0 M. O= 15 Sq.M. R= 1.00 SLC== 0.33 PLW=  3.00 M
CI=480.00 MU/M CS=633.00 MU/M CM=711.00 MU/M CL=15.00 MU/SQ.M.
CSC=100.00 MU/M
OUTPUT File Name:'optmpld.cpp'
```

It may be seen from above results of optimization model, that the exact solution of optimum lot area (giving the Least-Cost per lot) is = **34.37** Sq.M. (with optimum lot depth = 11.46 M not shown) as against the value of **35** Sq.M.in Table 10.1. Similarly, the exact value of **Least-Cost** per lot is **3503.98** MU as against **3504.29** MU shown in Table 10.1. Likewise, there are only marginal differences between the values of other parameters like x_1 and x_2.

Medium Lot Size Case—Table 10.2

```
SOME OPTIMUM RESULTS GIVEN BY ITRN.MODEL ARE:
OPT.PLA=130.46 SQ.M. x₁=118.28 M x₂= 118.28 M
OPT.SLR= 0.43 OPT.CAR= 0.50 OPT.OSR= 0.07M OPT.PDEN=33.14 LOTS/HEC.
OPTIMUM LOT DEPTH=13.05 M LEAST-LOT-COST=50595.89MU Number of lots=  46
SOME INPUT-VALUES Chosen in the above CASE are:
 I= 10.0 M. S= 20.0 M. M= 25.0 M. O= 20 Sq.M. R= 1.00 SLC== 0.33 PLW= 10.00 M
CI=3000.00 MU/M CS=4000.00 MU/M CM=5000.00 MU/M CL=15.00 MU/SQ.M.
CSC=2000.00 MU/M
OUTPUT File Name: 'optmpld.cpp'
```

It may be seen from the above results of optimization model that the exact solution of optimum lot depth is 13.05 (M) as against 13.0(M) in Table 10.2. Similarly, the exact solution of optimum lot area (giving the Least-Cost per lot) is = **130.46** Sq.M. as against the value of **130** Sq.M. shown in Table 10.2, while the exact value of **Least-Cost** per lot is **50,595.89** MU as against **50,596.02** MU in Table 10.2. Likewise, there are only marginal differences between the values of other parameters like x_1 and x_2.

Medium Lot Size Case—Figure 10.5

```
SOME OPTIMUM RESULTS GIVEN BY ITRN.MODEL ARE:
OPT.PLA=121.85 Sq.M. x1=113.11 M x2= 113.11 M
OPT.SLR= 0.41 OPT.CAR= 0.52 OPT.OSR= 0.07M OPT.PDEN=33.82 LOTS/HEC.
OPTIMUM LOT/PLOT DEPTH=12.19 M LEAST-LOT/PLOT-COST=70175.88MU Number
of lots= 43
SOME INPUT-VALUES Chosen in the above CASE are:
 I= 10.0 M. S= 20.0 M. M= 25.0 M. O= 20 Sq.M. R= 1.00 SLC== 0.10 PLW= 10.00 M
CI=3000.00 MU/M CS=4000.00 MU/M CM=5000.00 MU/M CL=100.00 MU/SQ.M.
CSC=2000.00 MU/M
OUTPUT File Name:'optmpld.cpp'
```

It may be seen from the above results of optimization model that the exact solution of **Least-Cost** per lot is **70,175.88** MU as against an approximate value of 70,179.**34** MU indicated in Fig. 10.5.

Large Lot Size Case Applicable in High Income Countries—Table 10.5

```
SOME OPTIMUM RESULTS GIVEN BY ITRN.MODEL ARE:
OPT.PLA=742.15 SQ.M. x1=287.64 M x2= 287.64 M
```

```
OPT.SLR= 0.63 OPT.CAR= 0.33 OPT.OSR= 0.04M OPT.PDEN= 8.47 PLOTS/HEC.
OPTIMUM LOT/PLOT DEPTH=37.11 M LEAST-LOT/PLOT-COST=14872.59MU Number
of plots= 70
SOME INPUT-VALUES Chosen in the above CASE are:
 I= 20.0 M. S= 25.0 M. M= 25.0 M. O= 50 Sq.M. R= 1.00 SLC== 0.10 PLW= 20.00 M
CI=500.00 MU/M CS=1000.00 MU/M CM=1000.00 MU/M CL= 0.00 MU/SQ.M.
CSC=500.00 MU/M
OUTPUT File Name:'optmpld.cpp'
```

It may be seen from the above results of optimization model that the exact solution of optimum lot area (giving the **Least-Cost** per lot) is = **742.15** Sq.M. as against the value of **750** Sq.M. shown in Table 10.5, while the exact value of **Least-Cost** per lot is **14,872.59** MU as against **14,872.8** MU shown in Table 10.5.

All Tables 10.1, 10.2, 10.3, 10.4, 10.5, and 10.6 results indicate that there is a complex interaction between the land prices, utility network or infrastructure costs, and the geometry of a layout planning module, including the location of the service-core in a lot. The cost per lot is the combined effect of the above factors in a specific site. Therefore, the physical planning and the architectural and engineering design-decisions should take into account the above factors, for achieving the design objective(s), such as minimizing the cost per lot of given size or, maximizing the size of a lot within a given cost.

Minimum width of a residential lot is prescribed in many national building codes which is also necessary to comply with the adequate width of habitable rooms in a dwelling. An earlier study indicated that a narrow deep lot reduces the cost (ASCE, 1939). Therefore, a minimum lot width on the above considerations is assumed in both Tables 10.1 and 10.2 and also in the above optimum results. Thereafter, assuming different land prices, utility networks costs including service networks and house connections, the costs per lot have been calculated for different lot areas; where the cost parameters other than land price and all physical parameters, including location of service core and standard of open space per lot, are assumed same. The input values of physical parameters and cost parameters/infrastructure cost chosen in each case are shown below the optimal results for the respective case.

10.7   Quantitative Analysis of Layout Planning Modules Using Computers and Graphics

Special purpose application packages such as AutoCAD are widely used in AEC sector. Although both the tasks of geometric modelling and drafting/drawing can be carried out using such a graphic software system, but it is difficult to address the problem of integrating these tasks with the design analysis, *design optimization*, and data management, storage and transfer, using only a graphics programming language, for which a high-level computer programming language such as C++ is required. Fortunately, AutoCAD supports AutoLISP, which is implementation of the LISP programming language. Simple planning and design analysis using quantitative analysis models in terms of simple algebraic expressions, and the corresponding graphics, i.e. drafting and drawing can be integrated using only the AutoLISP.

This is illustrated in Chap. 5 with an example AutoLISP CAD Application Software: '**HUDCMOD**' (Chakrabarty, 2012). Application of graphics programming using AutoLISP is demonstrated taking the layout planning module-A, one of the many layout modules included in the HUDCO MODEL (18). The complete listing of this graphics programme, i.e. CAD Application Software: '**HUDCMOD**' using AutoLISP is given in Chap. 5, which can be run by any user planner and designer who can modify it as per specific requirement. A planner and designer can also develop such graphic programs for all layout modules and dwelling option incorporated in the HUDCO MODEL or for layout modules conceived. Typical layout design drawing given instantly by '**HUDCMOD**', integrating graphic designs with planning parameters and costs is shown in Chap. 5, where some input-design-decisions given to the software by a stakeholder, planner, or designer, and the corresponding output-design-parameter values instantly generated by the software, are also displayed. This Application Software: '**HUDCMOD**' does not incorporate any *optimization* feature, as it involves complex mathematical programming not easily amenable by the graphics programming language, but can easily be done using C++ language as presented below. However, using AutoLISP in combination with a high-level computer programming language like C++, it is possible to integrate geometric modeling and drafting/drawing tasks with the tasks of design analysis, *design optimization*, and data management, storage, and transfer. This approach is adopted in developing the integrated CAD Application Software: '**HudCAD**', presented later.

10.8 Optimization of Layout Planning Modules Using Mathematical Programming Techniques, Computers, and Graphics

As in case of quantitative analysis models, even for optimization models using mathematical programming technique, the first step is to develop mathematical models in the form of a set of algebraic equations for each planning module, linking the design variables with the objective 'value scale' such as costs.

10.8.1 Universal Optimization Models for any Rectilinear Shape Layout Planning Module

It is interesting to note that the same *optimization model* can be used for different types of layout planning modules, as the optimal model formulation is independent of the type of layout planning module chosen by a planner-designer, provided these are, of course, rectilinear in shape. However, variation in the module type is ensured by selecting appropriate input values of model constants as per the different expressions for module parameters and cost coefficients for each layout module type as shown in Table 10.7 below.

The modules can be conceived to have a hierarchy of street systems such as main street, subsidiary main street, and internal street and so on, which also provides access to lots. The parameters k_1 and k_2 in Table 10.7 above are the street-width hierarchy ratio for the three levels of street widths which have been assumed in Fig. 10.3 (thesis p. 177, 183). Thus, the value of k_1 and k_2 will be given by:

$k_1 = I/S$ and $k_2 = M/S$, where I, S, and M are widths of internal, subsidiary main, and main street, respectively.

Specifying the desired values of street-width hierarchy ratio for the three levels of street widths, a planner-designer can find the optimum widths of internal, subsidiary main, and the main street maintaining this specified street-width hierarchy ratio. In case, the widths of internal, subsidiary main, and the main street are predetermined, the above expressions of module parameters incorporating k values (i.e. m_2 and m_4) are to be multiplied by the predetermined width of subsidiary main street to derive the actual values of module parameters: m_2 and m_4 to be used as input to the *optimization model*. Each street may contain all the utility lines such as water line, sewer line, and electric line besides the road pavement. The parameters CI, CS, and CM designate the unit costs of internal, subsidiary main, and the main street, respectively (Chapter-III, Thesis). The parameters CL, CSC, and SLC are the unit price of prepared land, unit cost of on-lot service connection lines, and the on-lot service-core location coefficients.

Accordingly, expressions for module parameters (designated by m_1, m_2, m_3, and m_4) and cost coefficients (designated by c_1 for the East-West streets, c_2 for the North-South streets, c_3 related to land price, and c_4 related to on-lot service connection) can be developed as shown in Table 10.7 above for different module types. Here, the modules presented in Fig. 10.3 are taken as examples.

10.8.1.1 Adjacency Conditions and Optimality of Layout Planning Modules

Optimization is possible only if there is flexibility in choice in a planning and design problem. A large-scale layout plan usually consists of a number of layout planning modules with different adjacency conditions permitting varying flexibility in design. Accordingly, the models for 'layout planning module' for quantitative analysis and optimization can be developed for different design conditions taking into account the adjacency conditions in a site-specific case. To illustrate the method, we can take as examples a selection of the planning modules for different planning levels shown in Fig. 10.3. It may be noted that optimality of such rectilinear planning modules will depend on the adjacency conditions of the module such as flexibility on all four sides (which is assumed in most of the optimization models presented below), on three

Table 10.7 Module types, module parameters, and cost coefficients

Module type	Module parameters				Cost coefficients			
	m_1	m_2	m_3	m_4	c_1	c_2	c_3	c_4
A block of lots	½	$(1+k_2) \times 1/2$	2	1	CS	$(CS+CM) \times 1/2$	CL	CSC × SLC
Modified Cul-De-Sac	¼	$(1+k_2) \times 1/2$	4	$(1+k_1)$	CI + CS	$(CS+CM) \times 1/2$	CL	CSC × SLC
Grid-iron	1/6	$(4k_1 + 3k_2 + 3)/6$	6	$(1+2k_1)$	2CI + CS	$(CS+CM) \times 1/2$	CL	CSC × SLC
Module with loop circulation	1/6	$(1+k_2) \times 1/2$	6	$(1+2k_1)$	2CI + CS	$(CS+CM) \times 1/2$	CL	CSC × SLC

sides, on two sides, and on one side; and accordingly, expressions of adjacency-condition-constraints can be formulated in the optimization model. It may be noted that in case, there is no flexibility in all four sides of the rectilinear planning module, the length and width of the planning module will be of fixed dimensions and no optimization is possible.

Cost is paramount in all optimization problems. However, cost varies with time (it will vary considerably in the usually long planning horizon of the order of 20 years or more, in conventional master plan approach) and *optimization* will be realistic if cost is realistic taking into account the time period of implementation of a project. Therefore, a planner-designer should attempt optimization of a layout plan which is within his/her *immediate sphere of influence*, in terms of planning and design decisions and their implementation, so that costs are realistic. Based on this consideration, a layout planning module, such as those shown in Fig. 10.2 (giving some representative selections of layout planning modules or typology of planning modules) can be the most suitable 'design unit' for optimal design in a site-specific case giving considerable flexibility. However, the alignment of the major arterial roads, enclosing such planning modules, can be fixed with adequate width of the right of ways, to accommodate suitable number of carriage ways which may be developed in a phased manner matching with the traffic demand from time to time.

A planner-designer can conceive any planning module as per her/his creativity, ingenuity, and requirement and develop module parameters and cost coefficients for such a planning module, and also develop mathematical models for quantitative analysis and optimization of such layout modules. This is illustrated below with examples. Hopefully, this will encourage planners and designers to adopt such quantitative analysis and optimization techniques in their layout planning and design activities, and thus help achieve a *resource-efficient* problem solution attaining *social equity* in the urban development and management process in various countries.

In order to demonstrate the use of the same *optimization model* for two types of layout planning modules, we can choose the layout planning module shown in Fig. 10.3(c) and (d) above, with suitably renaming modules and design variables for convenience as shown in Fig. 10.6 above.

Based on the above concepts and assumptions, the planning modules shown in Fig. 10.6 can be defined by a number of design variables as indicated below:

$x_1 =$ width of the planning module between the centre lines of two East-West access streets (for block of lots module) or subsidiary main streets (for other modules)

$x_2 =$ length of the planning module between the centre lines of the end streets (for block of lots module) or between the main street and the subsidiary main street (for other modules) in North-South direction

$x_3 =$ gross width of the lot allowing for the reservation of open space per lot where applicable

$x_4 =$ depth of the lot

$x_5 =$ a dependent variable related to the number of lots in a planning module

$x_6 =$ a dependent variable defining the gross area per lot

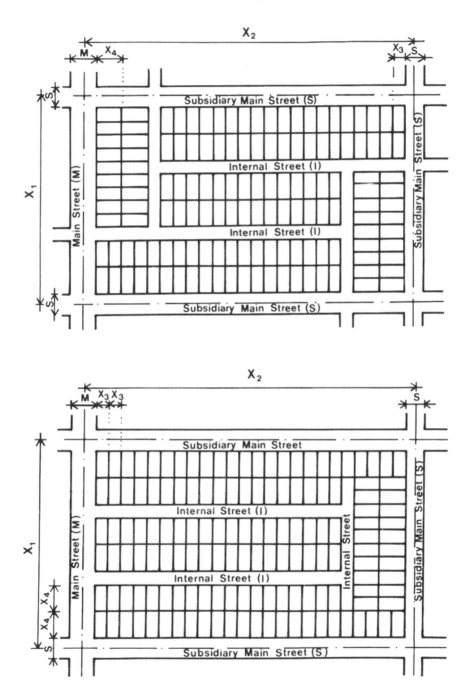

Fig. 10.6 Dimensionless schematic diagrams of two layout planning modules designated as: Module: A and Module: B

x_7 = right of way width of the internal street (also can be designated as I)
x_8 = right of way width of the subsidiary main street (also can be designated as S) or
end street
x_9 = right of way width of the main street (also can be designated as M)
x_{10} = a design variable defining the maximum permissible number of lots in the
layout module, which can also be specified as a constant design parameter
value by the planner-designer, and the model will generate a dimensioned
physical layout plan of the planning module fulfilling this requirement
x_{11} = cost of house service connection lines (which may also include the cost of
access pavement for garage inside a lot) per unit length
x_{12} = house service connection cost factor depending on the location of the service/
wet core (i.e. bath, water closet, garage) in the lot and related to the depth of
the lot decided by the architect planner-designer
x_{13} = cost of prepared land per unit area
x_{14} = open space reservation including the non-remunerative social facility space per
lot
T_j = total cost per unit length of the jth street including all components = $\sum t_{ij}$, where
t_{ij} represents unit cost of the ith component (such as road pavement, water line,
sewer line, electric cable line, and so on) in the j_{th} street (such as internal street,
subsidiary main street, main street, and so on)

Even if we consider three types of streets as in case of the planning module-A and
the planning module-B shown in Fig. 10.6; and unit cost of four components in each
street, there will be 12 variables defining the unit cost streets. Thus, taking 12 design
variables for other components of the planning module as described above, there will
be 24 design variables which will determine the cost per lot, out of which 23 are
independent design variable values which can be decided by a planner and designer.
Even if each design variable has the possibility of having, say, 4 discrete values, it
will give 4^{23} system design solutions which have to be evaluated before selecting a
particular system solution (Meredith et al., 1973). Obviously, evaluation of such
numerous system solutions by a manual method will be very tedious and time
consuming, but still the *optimal unique least-cost* or *most-benefit* design (explained
later) may not be reached. However, a computer-aided layout planning and design by
optimization will give the *optimal unique least-cost* or *most-benefit* design *instantly*
in numeric terms, which can be converted to graphic form, i.e. design drawing in no
time as presented below.

As mentioned above, in many situations the number of cases of system solutions
to be evaluated may be very large and as a result application of combinatorial
approach may not be practicable. Even if there are a smaller number of cases to be
evaluated, the adoption of a combinatorial approach may be feasible only to obtain a
near optimal system solution; but still the *unique optimal system solution* may not be
reached by this method. In such situations, it is imperative to apply mathematical
programming techniques like linear and nonlinear programming techniques. However, implementation of such an analytical procedure for systems design depends on
the degree of order that can be established in the problem, i.e. how well the system

structure and behaviour characteristics are known and how well the system constraints and criteria for optimization can be quantified. In the case of layout planning and design at the 'planning module' level, it is possible to quantify the objective function, which may be the 'minimisation of cost per lot for a given minimum acceptable lot area' or 'maximisation of area per lot for a given maximum acceptable cost per lot' in the present case. Similarly, it is possible to quantify a variety of design constraints depending on the emphasis or priorities accorded by the planner and stakeholders to particular aspects of design principles, and select the design variables and constraints accordingly. In view of above, the set of design variables mentioned above should be considered only as an illustration, which may be enlarged or modified depending on the priority and importance placed on the particular aspects of layout planning and design by the planners and stakeholders.

10.9   Optimization Models for Layout Planning and Design

A number of mathematical programming techniques are available in literature. Layout planning and design at the 'planning module' level usually involves nonlinear terms, and therefore, it is easier to apply nonlinear programming techniques in such problem for more efficient solution, although by linearization of nonlinear terms, the linear programming techniques can also be applied in certain situations (Philips et al). In such planning and design problems, the objective function is usually a summation of product functions and the constraints also contain the product functions. The geometric programming technique is an efficient tool for solving such problems as discussed in Chap. 8. A number of such optimization models are developed and presented elsewhere (Chakrabarty, 1986a, 1986b, 1987a, 1987b, 1988a, 1988b, 1990, 1991, 1992a, 1992b, 1996, 1997a, 1997b, 1998, 2001, 2007, 2012). To illustrate the method including applications of integrated computer-aided layout planning and design by optimization using the geometric programming technique, two optimization models—one for **Least-Cost** optimal layout design and the other for **Most-Benefit** optimal layout design—are presented below.

10.9.1   A Least-Cost Optimal Layout Design Model

Layout planning modules shown in Fig. 10.6 are defined in above section by a large number of design variables. As earlier stated, the optimal design of layout planning modules involves nonlinear programming. Therefore, it will be helpful to order and group the variables properly so that the number of variables in each cycle of calculation is less and optimal design model computation is easier. This can be done by assigning specific values for some design variables shown above, making them as constant input parameters while analysing the system response at each stage of computation. Thus, the total unit costs of various streets defined by the variable T_j

could be constant input parameters in each cycle of computation, adopting particular values of design variables constituting these parameters, i.e. defining a particular infrastructure standard and specification chosen by planner-designer for these components and get the system response, thereafter, vary and iterate with new standards and specifications until a satisfactory system solution acceptable to all stakeholders, is obtained (HUDCO, 1982). Similarly, specific values could be assigned to the right of way width values designated by the design variables x_6, x_7, x_8 which are normally specified in building codes or local planning regulations depending on the expected traffic volume in a site-specific situation. Accordingly, these can be used as constant input parameters in the optimization model for each cycle of computation.

A layout planning module can be designed for a given number of lots, and accordingly the design variable x_{10} will be constrained to have a specific value chosen by the planner-designer, which can be used as a constant input parameter. Similarly, the design variables x_{11} and x_{12} relating to costs will have specific values depending on the site-specific situations. The design variable x_{13} will have a specific value depending on the architectural design of the dwelling, and thus, will be a constant input parameter to the model. Depending on the desirable environmental standards, the design variable x_{14} could be constrained to have specific value chosen by the planner stakeholder, which can be used as a constant input parameter to the model, and accordingly, the net area per lot will be determined. Adopting this procedure, the number of basic independent design variables in the system design problem reduces to only five for each cycle of computation, but gives a *unique optimal solution*. However, different values could be assigned to the design variables, constituting each of the constant input parameter values mentioned above, and a comprehensive analysis of the system design could be done, covering all possible combinations, before arriving at a specific system design solution considered most optimal. This procedure will also provide the needed flexibility to planner-designer-stakeholder so that a planner-designer could use her/his ingenuity for maximizing the benefit for specified objective achievement. Taking the layout planning modules shown in Fig. 10.6 above, and properly ordering and grouping the design variables as mentioned above, the total cost per lot designated by $y(x)$, including the cost of house connection or service connection, could be expressed in terms of the design variables of the above layout planning module as follows:

$$y(x) = c_1 x_2 x_3 x_5^{-1} + c_2 x_1 x_3 x_5^{-1} + c_3 x_1 x_2 x_3 x_5^{-1} + c_4 x_4 \qquad (10.6)$$

Here $y(x)$ is the cost (in terms of Monetary Units or MU) per lot, and c_1, c_2, c_3, and c_4 are cost coefficients defined as follows:

c_1 = a cost coefficient defining the cost per unit length of East-West streets (in terms of Monetary Units or MU)

c_2 = a cost coefficient defining the cost per unit length of North-South streets (in terms of Monetary Units or MU)

c_3 = a cost coefficient defining the prepared land cost per unit of the gross area of the layout planning modules (in terms of Monetary Units or MU)

c_4 = a cost coefficient defining the cost of house connection or service connection per unit length across the depth of the lot in the layout planning modules (in terms of Monetary Units or MU)

Accordingly, the optimization model giving the objective function and the constraints for the above design conditions could be developed as follows:

Objective: *Minimize* the Total Cost per Lot $= y(x) = c_1 x_2 x_3 x_5^{-1} + c_2 x_1 x_3 x_5^{-1} + c_3 x_1 x_2 x_3 x_5^{-1} + c_4 x_4$.

Subject to:

$$a_1 x_3^{-1} x_4^{-1} \qquad\qquad\qquad \leq 1 \qquad\qquad\qquad (10.7)$$

(*Constraint* for Minimum Acceptable Lot Area)

$$a_2 x_2^{-1} x_5^{-1} + a_3 x_2^{-1} \qquad\qquad \leq 1 \qquad\qquad\qquad (10.8)$$

(*Constraint* for geometric compatibility—in E-W direction of layout module)

$$a_4 x_3^{-1} x_5 \qquad\qquad\qquad\qquad \leq 1 \qquad\qquad\qquad (10.9)$$

(*Constraint* for design number of lots in the layout module)

$$a_5 x_4 x_1^{-1} + a_6 x_1^{-1} \qquad\qquad\quad \leq 1 \qquad\qquad\qquad (10.10)$$

(*Constraint* for geometric compatibility—in N-S direction of layout module)

$$x_1, x_2, x_3, x_4, x_5 \qquad\qquad\qquad > 0 \qquad\qquad\qquad (10.11)$$

(Non-negativity *constraint*)

The above optimization problem is a posynomial geometric programming problem with inequality constraints as explained in Sect. 8.12.5 of Chap. 8. As per terminology used in that section, the subscript o for c should be used to designate the objective function coefficients, and the coefficient of each term in constraints should be double subscripted one subscript designating the constraint equation and the other for specific term of constraint. Similarly, each exponent should be triple subscripted identifying the equation, variable, and the posynomial term to which each belong. However, since the above problem is comparatively simpler with less number of variables and constraints; coefficients of all terms in objective function and constraint are single subscripted as indicated above, where 'c' with appropriate subscript indicates objective function coefficient and 'a' with appropriate subscript indicates constraint coefficient.

Here, c_1, c_2, c_3, and c_4, are the model constants related to the cost parameters as described below:

c_1 = a cost coefficient defining the cost per unit length of the East-West streets,
c_2 = a cost coefficient defining the cost per unit length of the North-South streets,
c_3 = a cost coefficient defining the prepared land cost per unit of gross area of the layout planning module,
c_4 = a cost coefficient defining the cost of house connection or service connection per unit length, across the depth of the lot in the layout planning module.

Similarly, the model constants a_1, a_2, a_3, a_4, a_5, and a_6 are related to the design and module parameters as described below:

a_1 = a constant input design parameter, defining the minimum acceptable lot area and can also designated as d_1 to high light this aspect;
a_2 = a constant input module parameter related to the number of blocks in the East-West direction of the layout planning module, i.e. m_1 in Table 10.7 above;
a_3 = a constant input module parameter related to the width of the streets in the North-South direction of the layout planning module, i.e. m_2 in Table 10.7 above;
a_4 = a constant input design parameter, defining the designed maximum number of lots in the layout planning module and can also designated as d_2 to high light this aspect;
a_5 = a constant input module parameter related to the number of rows of lots in the East-West direction of the layout planning module, i.e. m_3 in Table 10.7 above;
a_6 = a constant input module parameter related to the width of the streets in the East-West direction of the layout planning module, i.e. m_4 in Table 10.7 above.

Here, the model *variables* are as follows:

x_1 = A variable giving the optimum width of the layout module between the centre lines of the two East-West subsidiary main streets of planning module as shown in Fig. 10.6 above;
x_2 = A variable giving the optimum length of the layout module between the centre lines of the two East-West subsidiary main streets of planning module as shown in Fig. 10.6 above;
x_3 = A variable giving the optimum gross width of the lot allowing for the reservation of open space per lot;
x_4 = A variable giving the optimum depth of the lot;
x_5 = A variable defining the optimum number of lots in a planning module.

It may be noted that the value of the variables x_1, x_2, x_3, x_4, and x_5 not only affect the cost of the lot but also determine the optimum size of the layout module. Therefore, the solution of an optimization model should give the optimum value of each of the variables: x_1, x_2, x_3, x_4, and x_5 so as to minimize the cost of the lot, i.e. the value of the objective function $y(x)$, and simultaneously, achieve the geometric compatibility between the geometrical design of layout plans including minimum acceptable lot size, engineering specifications, and other factors, so that any incompatibility of such cost and physical design data, with the actual physical design of layout is ruled out, giving a realistic system solution. Accordingly, expressions for the constraints 1, 2, 3, and 4 of the model are developed and shown above as expressions (10.7), (10.8), (10.9), and (10.10), respectively, indicating within bracket of each expression, the specific design condition to be fulfilled by the model such as minimum acceptable lot area, number of lots, and so on as shown. Since the model has to give only the realistic values of the variables, none of the model variables can be negative which design condition is specified by the non-negativity constraint-5, i.e. expression (10.11) above.

10.9.1.1   Model Solution

The above optimal design model is a typical geometric programming problem. Thus, it may be seen that the objective function in the above problem is a summation of product functions and the constraints also contain the product functions. The geometric programming technique is an efficient tool in solving such problems as explained in Chap. 8. In this technique, emphasis is placed on the relative magnitudes of the terms of the objective function rather than on the variables, and instead of finding the optimal values of the design variables directly, the optimal way to distribute the total cost, among the various terms or components in the objective function, is found first. This indicates how resources should be apportioned among the components contributing to the design objective, which is very useful in engineering and architectural design. Even in the presence of great degree of difficulty, the geometric programming formulation uncovers important invariant properties of the design problem which can be used as design rules for optimality as discussed below (soc86).

In this technique, if the primal geometric programming problem has N variables and T terms; in the dual geometric programming problem, there will be N orthogonality conditions-one for each variable x_n, one normality condition and T weights, one for each term. Thus, there will be $(N + 1)$ independent linear constraint equations as against T independent dual variables which are unknown. In case $T - (N + 1)$, which is the degree of difficulty in the problem. In the present optimal design problem, there are ten terms (T) and five variables (N) in the primal objective function and constraints. Thus, the degree of difficulty in the problem is $= T - (N + 1) = 10 - (5 + 1) = 4$, which is comparatively small, and therefore, a set of $(N + 1)$ dual weights could be expressed in terms of the remaining $T - (N + 1)$ dual weights, and substituted into the dual objective function, which is then maximized by setting the first derivative to zero with respect to each of the dual weights coming into the substituted dual objective function which is being maximized. Thus, a sufficient number of equations are available for solving the problem. Accordingly, using the geometric programming technique, and taking the terminology used in literature (Beightler et al., 1979) on geometric programming as explained in Sect. 8.12.5 of Chap. 8, the dual objective function and the associated normality and orthogonality constraints for the above primal problem are given by:

$$d(w) = \left(\frac{C_{o1}}{w_{o1}}\right)^{w_{01}} \left(\frac{C_{o2}}{w_{o2}}\right)^{w_{02}} \left(\frac{C_{o3}}{w_{o3}}\right)^{w_{03}} \left(\frac{C_{o4}}{w_{o4}}\right)^{w_{04}} (c_{11})^{w_{11}} \left(\frac{C_{21}(W_{21}+W_{22})}{w_{21}}\right)^{w_{21}}$$

$$\left(\frac{C_{22}(W_{21}+W_{22})}{w_{22}}\right)^{w_{22}}$$

$$\times (c_{31})^{w_{31}} \left(\frac{C_{41}(W_{41}+W_{42})}{w_{41}}\right)^{w_{41}} \left(\frac{C_{42}(W_{41}+W_{42})}{w_{42}}\right)^{w_{42}}$$

$$(10.12)$$

As discussed above, the coefficients for the objective function are simplified as:

$$c_{01} = c_1, c_{02} = c_2, c_{03} = c_3, c_{04} = c_4.$$

Similarly, the coefficients for the constraint functions are simplified as:

$$c_{11} = a_1, c_{21} = a_2, c_{22} = a_3, c_{31} = a_4, c_{41} = a_5, c_{42} = a_6.$$
$$c_{11} = a_1 = d_1, c_{21} = a_2 = m_1, c_{22} = a_3 = m_2, c_{31} = a_4 = d_2, c_{41} = a_5 = m_3, c_{42} = a_6 = m_4.$$

Accordingly, the dual objective function to be maximized can be expressed in a simplified form as follows:

$$\text{Maximize } d(w) = [c_1/w_{01}]^w{}_{01} [c_2/w_{02}]^w{}_{02} [c_3/w_{03}]^w{}_{03} [c_4/w_{04}]^w{}_{04}$$
$$\times [a_1]^w{}_{11} [a_2(w_{21}+w_{22})/w_{21}]^w{}_{21} [a_3(w_{21}+w_{22})/w_{22}]^w{}_{22}$$
$$\times [a_4]^w{}_{31} [a_5(w_{41}+w_{42})/w_{41}]^w{}_{41} [a_6(w_{41}+w_{42})/w_{42}]^w{}_{42}$$

$$(10.13)$$

Subject to:

$$w_{01} + w_{02} + w_{03} + w_{04} = 1 \quad \text{(normality condition)} \qquad (10.14)$$

$$w_{02} + w_{03} - w_{41} - w_{42} = 0 \quad \text{(orthogonality condition for } x_1) \quad (10.15)$$

$$w_{01} + w_{03} - w_{21} - w_{22} = 0 \quad \text{(orthogonality condition for } x_2) \quad (10.16)$$

$$w_{01} + w_{02} + w_{03} - w_{11} - w_{31} = 0 \quad \text{(orthogonality condition for } x_3) \quad (10.17)$$

$$w_{04} - w_{11} + w_{41} = 0 \quad \text{(orthogonality condition for } x_4) \quad (10.18)$$

$$-w_{01} - w_{02} - w_{03} + w_{21} + w_{31} = 0 \quad \text{(orthogonality condition for } x_5) \quad (10.19)$$

Here, w_{01}, w_{02}, w_{03}, and w_{04} represent the four cost weights related to the four terms in the objective function $y(x)$, i.e. expression (10.6) above. Similarly, w_{11}, w_{21}, w_{22}, w_{31}, w_{41}, and w_{42} represent the dual variables related to the single term in constraint-1, i.e. expression (10.7), to the first term in constraint-2, i.e. expression (10.8), to the second term in constraint-2, i.e. expression (10.8), to the single term in constraint-3, i.e. expression (10.9), to the first term in constraint-4, i.e. expression (10.10), and to the second term in constraint-4, i.e. expression (10.10) above, respectively.

There are ten terms (T) and five primal variables (N) in the primal objective function and constraints [expressions (10.6)–(10.10)], and therefore, the degree of difficulty in the problem is $= T - (N + 1) = 4$. Thus, any six of the dual variables can be expressed in terms of the remaining four dual variables as shown below, choosing w_{02}, w_{03}, w_{04}, and w_{41} as the four dual variables:

$$w_{01} = 1 - w_{02} - w_{03} - w_{04};\ w_{11} = w_{04} + w_{41} = w_{21};\ w_{22} = 1 - w_{02} - 2w_{04} - w_{41};$$
$$w_{31} = 1 - 2\,w_{04} - w_{41};\ w_{42} = w_{02} + w_{03} - w_{41}.$$

$$(10.20)$$

The cost weights w_{01}, w_{02}, w_{03}, and w_{04} represent the optimum proportion of total cost per lot, distributed between the East-West streets including the utility lines, the North-South streets including the utility lines, the prepared land cost, and the lot service connection costs, respectively. The set of Eq. (10.20) represent the relationship between these cost distributions among the above components, which are independent of the values of the cost coefficients c_1, c_2, c_3, and c_4. This invariant property of geometric programming could be used for development of suitable *design rules* for *optimality* even in the presence of a great degree of difficulty, having general application in any problem of this nature. If we substitute the expressions in equations set (10.20) into the dual objective function [Eq. (10.12)] and maximize this substituted dual objective function over the four variables w_{02}, w_{03}, w_{04}, and w_{41} (i.e. by setting the first derivative of the function with respect to each variable to zero), we will get four additional equations, thus providing sufficient conditions for optimality. The process is simplified by maximizing the natural logarithm of the dual function; that is, by substituting

$$\ln d(w_{02}, w_{03}, w_{04}, w_{41}) = z(w_{02}, w_{03}, w_{04}, w_{41}).$$

Thus, by above substitution and taking logarithm on both sides of Eq. (10.13), we get the following equations:

$$\begin{aligned}
z(w) = &\ (1 - w_{02} - w_{03} - w_{04})\ (\ln c_1 - \ln (1 - w_{02} - w_{03} - w_{04})) \\
&+ w_{02}(\ln c_2 - \ln w_{02}) + w_{03}(\ln c_3 - \ln w_{03}) \\
&+ w_{04}(\ln c_4 - \ln w_{04}) + (w_{04} + w_{41}) \ln a_1 + (w_{04} + w_{41}) \\
&\ (\ln a_2 + \ln (1 - w_{02} - w_{04}) - \ln (w_{04} + w_{41})) \\
&+ (1 - w_{02} - 2w_{04} - w_{41})\ (\ln a_3 + \ln (1 - w_{02} - w_{04}) \\
&- \ln (1 - w_{02} - 2w_{04} - w_{41})) \\
&+ (1 - 2w_{04} - w_{41}) \ln a_4 + w_{41}(\ln a_5 + \ln (w_{02} + w_{03}) - \ln w_{41})) \\
&+ (w_{02} + w_{03} - w_{41})(\ln a_6 + \ln (w_{02} + w_{03}) - \ln (w_{02} + w_{03} - w_{41})).
\end{aligned}$$

$$(10.21)$$

By setting the first derivative $dz/dw_{03} = 0$, we get:

$$c_3 a_6 (1 - w_{02} - w_{03} - w_{04}) (w_{02} + w_{03})$$
$$- c_1 (w_{02} + w_{03} - w_{41}) w_{03} = 0 \tag{10.22}$$

From (10.22), we get:

$$w_{41} = ((w_{02} + w_{03}) (c_1 w_{03} - c_3 a_6 (1 - w_{02} - w_{03} - w_{04})) / c_1 w_{03}) \tag{10.23}$$

By setting the first derivative $dz/dw_{02} = 0$, we get:

$$c_2 a_6 (1 - w_{02} - w_{03} - w_{04}) (1 - w_{02} - 2w_{04} - w_{41}) (w_{02} + w_{03})$$
$$- c_1 a_3 (1 - w_{02} - w_{04}) (w_{02} + w_{03} - w_{41}) w_{02} = 0 \tag{10.23}$$

Again, by setting the first derivative $dz/dw_{04} = 0$, we get:

$$c_4 a_1 a_2 (1 - w_{02} - w_{03} - w_{04}) (1 - w_{02} - 2w_{04} - w_{41})^2$$
$$- c_1 a_3^2 a_4^2 (1 - w_{02} - w_{04}) (w_{04} + w_{41}) w_{04} = 0 \tag{10.25}$$

Similarly, setting the first derivative $dz/dw_{41} = 0$, we get:

$$a_1 a_2 a_5 (1 - w_{02} - 2\, w_{04} - w_{41}) (w_{02} + w_{03} - w_{41})$$
$$- a_3 a_4 a_6 (w_{04} + w_{41}) w_{41} = 0 \tag{10.26}$$

To simplify, we may make the following substitution:

$$w_{02} = x, w_{03} = y, w_{04} = z;$$
$$k1 = c_2 a_6 / c_1 a_3, k2 = c_4 a_1 a_2 / c_1 a_3^2 a_4^2, k3 = a_1 a_2 a_5 / a_3 a_4 a_6.$$

Then, the expression (10.23) above becomes:

$$w_{41} = (x + y) \left(c_1 y - c_3 a_6 (1 - x - y - z) c_1^{-1} y^{-1} \right) \qquad dz/dw_{03} = 0 \tag{10.27}$$

The expression (10.24) above becomes: $dz/dw_{02} = 0$

$$k1(x + y) (1 - x - y - z)(1 - x - 2z - w_{41}) - (1 - x - z)(x + y - w_{41})x = 0 \tag{10.28}$$

The expression (10.25) above becomes: $dz/dw_{04} = 0$

$$k2(1 - x - y - z) (1 - x - 2z - w_{41}) (1 - x - 2z - w_{41}) - (z + w_{41})(1 - x - z)z = 0 \tag{10.29}$$

The expression (10.26) above becomes: $dz/dw_{41} = 0$

$$k3(1 - x - 2z - w_{41}) (x + y - w_{41}) - (z + w_{41}) w_{41} = 0 \tag{10.30}$$

It may be noted that the value of w_{41} given by expression (10.27) is in terms of three unknown variables x, y, and z. If we substitute this value of w_{41} in expressions (10.28), (10.29), and (10.30), we get three nonlinear simultaneous equations in terms of three unknown variables $x(= w_{02})$, $y(= w_{03})$, and $z(= w_{04})$ which can be solved by the Newton–Raphson method as explained and illustrated in Chaps. 3 and 8. The author has developed C++ computer programmes for solving such nonlinear simultaneous equations covering many variables and details are presented in Chap. 3. As outlined in Chap. 3, we can write each of the above three nonlinear simultaneous equations in a more simplified form as: f, g, and h as shown below, and also express the partial derivatives as: $f_x, f_y, f_z, g_x, g_y, g_z, h_x, h_y$, and h_z where x, y, and z subscripts to f, g, and h denote the partial derivative of each of the functions with respect to the x, y, and z variables, respectively. Similarly, we can make other substitutions as below:

$$f = f(x_n, y_n, z_n), \qquad g = g(x_n, y_n, z_n), \qquad h = h(x_n, y_n, z_n)$$
$$F = f^* \left(g_y{}^* h_z - h_y{}^* g_z \right) + g^* \left(h_y{}^* f_z - f_y{}^* h_z \right) + h^* \left(f_y{}^* g_z - g_y{}^* f_z \right);$$
$$G = g^* \left(h_z{}^* f_x - f_z{}^* h_x \right) + h^* \left(f_z{}^* g_x - g_z{}^* f_x \right) + f^* \left(g_z{}^* h_x - h_z{}^* g_x \right);$$
$$H = h^* \left(f_x{}^* g_y - g_x{}^* f_y \right) + f^* \left(g_x{}^* h_y - h_x{}^* g_y \right) + g^* \left(h_x{}^* f_y - f_x{}^* h_y \right);$$
$$R = f_x{}^* \left(g_y{}^* h_z - h_y{}^* g_z \right) + g_x{}^* \left(h_y{}^* f_z - f_y{}^* h_z \right) + h_x{}^* \left(f_y{}^* g_z - g_y{}^* f_z \right);$$

Thus the Newton–Raphson formula for solution of three simultaneous nonlinear equations in three variables can be written as:

$$x_{n+1} = x_n - (F/R); \quad y_{n+1} = y_n - (G/R); \quad z_{n+1} = z_n - (H/R); \qquad (10.31)$$

By this method, we get a sequence of approximate values $x_1, x_2 \ldots x_n$; $y_1, y_2 \ldots y_n$; and $z_1, z_2 \ldots z_n$; each successive term of which is closer, compared to its predecessor, to the unknown exact values of the root of the three nonlinear simultaneous equations. Accordingly, the above three functions, i.e. expressions (10.28), (10.29), and (10.30), and their partial derivatives: $f_x, f_y, f_z, g_x, g_y, g_z, h_x, h_y$, and h_z (substituting expression (10.27) for w_{41}, and making other substitutions as shown below) become:

$$k1 = (c_2{}^* a_6)/(c_1{}^* a_3);$$
$$k2 = (c_4{}^* a_1{}^* a_2)/(c_1{}^* a_3{}^* a_3{}^* a_4{}^* a_4);$$
$$k3 = a_1{}^* a_2{}^* a_5/(a_3{}^* a_4{}^* a_6);$$
$$k4 = (c_3{}^* a_6)/c_1;$$

$$r = 1 - x - z, m = x + y, u = 1 - x - y - z,$$
$$w = m^* (1 - (c3^* a6^* u)/(c1^* y)), p = r - z - w, q = w/m;$$
$$f = k1^* m^* u^* p - r^* (m - w) * x;$$
$$g = k2^* u^* p^* p - (z + w) * r^* z;$$
$$h = k3^* p^* (m - w) - (z + w) * w;$$

$$wx = q + (m^*k4)/y;$$
$$wy = q + (m^*k4)/y + (m^*k4^*u)/(y^*y);$$
$$wz = (m^*k4)/y;$$

$$f_x = k1^*(u - m) * p + k1^*m^*u^*(-1 - wx) - (r - x) * (m - w) - r^*(1 - wx) * x;$$
$$f_y = k1^*(u - m) * p + m^*k1^*(-wy) * u - r^*(1 - wy) * x;$$
$$f_z = -m^*k1^*p + k1^*m^*u^*(-2 - wz) + x^*(m - w) + x^*r^*wz;$$

$$g_x = -k2^*u^*2^*p^*(1 + wx) - p^*k2^*p - wx^*z^*r + (z + w) * z;$$
$$g_y = k2^*(-wy) * u^*p^*2 - k2^*p^*p - wy^*z^*r;$$
$$g_z = -k2^*u^*2^*p^*(wz + 2) - k2^*p^*p - ((1 + wz) * z + z + w) * r + (z + w) * z;$$

$$h_x = -k3^*(1 + wx) * (m - w) + k3^*p^*(1 - wx) - wx^*w - (z + w) * wx;$$
$$h_y = -k3^*wy^*(m - w) + k3^*p^*(1 - wy) - w^*wy - (z + w) * wy;$$
$$h_z = k3^*(-2 - wz) * (m - w) - p^*k3^*wz - (1 + wz) * w - (z + w) * wz;$$

Since all the functions and their partial derivatives are expressed in terms of the three unknown variables: x, y, and z as above, putting these values in equation set (10.31) we get a sequence of approximate values x_1, x_2...x_n; y_1, y_2...y_n; and z_1, z_2...z_n; each successive term of which is closer, compared to its predecessor, to the unknown exact values of the root of the three nonlinear simultaneous Eqs. (10.28), (10.29), and (10.30). Thus, after a few iterations, we get the unknown exact values of the root of the three nonlinear simultaneous Eqs. (10.28), (10.29), and (10.30), i.e. the value of x, y, and z, fulfilling the tolerance limit prescribed by the planner-designer.

To make it easier for planner-designer, the author has developed the C++ program named as: LAY.CPP mentioned above, which incorporates one sub-model or iteration-model for least-cost optimization of any layout of rectilinear shape. Complete computer code listing of this C++ program is given in appendix. Using LAY.CPP, the above optimization problem can be solved in no time as illustrated below, giving the *optimum* layout parameters for least-cost as output which can be integrated with graphic program to generate *optimum* layout design drawing also.

Application of the geometric programming technique gives the following simple expressions for determining the optimum values of the model variables, thus, fully solving the model.

$$x_2 = (c_4w_{02})/(c_2w_{04}), x_3 = (c_3w_{02})/(c_2w_{03}),$$
$$y(x) = (c_2x_2x_3)/w_{02}, x_1 = y(x) w_{01}/c_1, \qquad (10.32)$$
$$x_4 = a_2 + a_3x_2x_3, x_5 = x_1/(a_1x_3).$$

10.9.2   A Most-Benefit Optimal Layout Design Model

In many cases, a planner/designer may be required to provide the most beneficial planning and design solution within a given budget of investment. In such a situation, a most-benefit optimal design model can be a useful tool. A most-benefit optimal layout design model is presented below, taking the layout planning modules shown in Fig. 10.6 above and properly ordering and grouping the design variables as done in Sect. 10.8.1. Here, the objective function is taken as maximizing the gross lot area (which is considered as benefit) within a given maximum total cost, instead of objective of minimizing the total cost per lot designated by $y(x)$ for a given minimum acceptable lot area. Thus, in this case, the minimum acceptable lot area constraint (i.e. constraint function (10.5) above) is replaced as the objective function of $y(x)$ for maximizing the lot area (i.e. benefit), and similarly, the objective function (10.4) above is converted to a constraint of maximum cost within which the above objective is to be achieved. Accordingly, the most-benefit optimal layout design model, giving the objective function and the constraints for the above design conditions could be developed as follows:

$$\text{Objective : Maximise the gross lot area } y(x) = x_3 x_4 \qquad (10.33)$$

Here $y(x)$ is the grass lot area (in terms of Area Units, e.g. Sq.M.) per lot. As before c_1, c_2, c_3, and c_4 are cost coefficients defined as follows:

c_1 = a cost coefficient defining the cost per unit length of East-West streets (in terms of Monetary Units or MU)

c_2 = a cost coefficient defining the cost per unit length of North-South streets (in terms of Monetary Units or MU)

c_3 = a cost coefficient defining the prepared land cost per unit of the gross area of the layout planning modules (in terms of Monetary Units or MU)

c_4 = a cost coefficient defining the cost of house connection or service connection per unit length across the depth of the lot in the layout planning modules (in terms of Monetary Units or MU).

As cited above in this case, the objective function (10.4) is converted to a constraint function designated as: (10.34) below, indicating the maximum acceptable total cost per lot, within which the above objective of gross lot area is to be achieved. Therefore, here, each of the cost coefficients: c_1, c_2, c_3, and c_4 is divided by the above given or acceptable maximum cost, in the constraint expression: (10.34), giving respectively, the model constants: a_7, a_8, a_9, and a_{10} below. Accordingly, the most-benefit optimal layout design model giving the objective function and the constraints for the above design conditions could be developed as follows:

$$\text{Objective : } \textit{Maximize} \text{ the gross area per lot } y(x) = x_3 x_4 \qquad (10.33)$$

Subject to:

$$a_2 x_2^{-1} x_5 + a_3 x_2^{-1} \qquad\qquad\qquad\qquad \leq 1 \qquad\qquad\qquad (10.8)$$

(*Constraint* for geometric compatibility—in E-W direction of layout module)

$$a_4 x_3^{-1} x_5 \qquad\qquad\qquad\qquad \leq 1 \qquad\qquad\qquad (10.9)$$

(*Constraint* for design number of lots in the layout module)

$$a_5 x_4 x_1^{-1} + a_6 x_1^{-1} \qquad\qquad\qquad\qquad \leq 1 \qquad\qquad\qquad (10.10)$$

(*Constraint* for geometric compatibility—in N-S direction of layout module)

$$a_7 x_2 x_3 x_5^{-1} + a_8 x_1 x_3 x_5^{-1} + a_9 x_1 x_2 x_3 x_5^{-1} + a_{10} x_4 \qquad\qquad \leq 1 \qquad\qquad (10.34)$$

(*Constraint* for Maximum Acceptable Total Cost per Lot)

$$x_1, x_2, x_3, x_4, x_5 \qquad\qquad\qquad\qquad > 0 \qquad\qquad\qquad (10.11)$$

(Non-negativity *constraint*)

10.9.2.1   Model Solution

The above Objective: *Maximize* the gross area per lot $y(x) = x_3 x_4$ can also be written in standard geometric programming form as:

$$\textit{Minimize } y(x) = -x_3 x_4$$

Thus, this optimization model is a typical signomial geometric programming problem as outlined in Chap. 8, and can be solved accordingly using the procedure suggested in the above Chap. 8. To solve the problem, let us guess the signum function $\sigma_0 = -1.0$ for $y(x)$ the negative value of $y(x)$, since the original objective function is to *Maximize*. Similarly, the value of the signum function σ_{01} is guessed as $= -1.0$. The values of other signum functions in the problem are guessed as: $\sigma_m = +1.0$ and $\sigma_{mt} = +1.0$. To simplify the notations, we can denote the sum of dual variables w_{41}, w_{42}, w_{43}, and w_{44} by w_4. With these simplified notations and other simplifications outlined in Sect. 10.8.1.1 above, the dual objective function, and the associated normality and orthogonality constraints for the above primal problem can be expressed as:

$$\textbf{Optimise}: d(w) = (-1)\left(\left(\frac{a_2(w_{11} + w_{12})}{w_{11}}\right)^{w_{11}} \left(\frac{a_3(w_{11} + w_{12})}{w_{12}}\right)^{w_{12}}\right.$$

$$(a_4)^{w_{21}}\left(\frac{a_5(w_{31} + w_{32})}{w_{31}}\right)^{w_{31}}\right)$$

$$\times \left(\frac{a_6(w_{31} + w_{32})}{w_{32}}\right)^{w_{32}} \left(\frac{a_7(W_4)}{w_{41}}\right)^{w_{41}} \left(\frac{a_8(W_4)}{w_{42}}\right)^{w_{42}}$$

$$\left(\frac{9(W_4)}{w_{43}}\right)^{w_{43}} \left(\frac{a_{10}(W_4)}{w_{44}}\right)^{w_{44}}\right)^{-1}$$

$$(10.35)$$

Subject to:

$$w_{01} = (-1) \times (-1) = 1 \qquad \text{(normality condition)}$$

$$-w_{31} - w_{32} + w_{42} + w_{43} = 0 \qquad \text{(orthogonality condition for } x_1) \qquad (10.36)$$

$$-w_{11} - w_{12} + w_{41} + w_{43} = 0 \qquad \text{(orthogonality condition for } x_2) \qquad (10.37)$$

$$-w_{01} - w_{21} + w_{41} + w_{42} + w_{43} = 0 \qquad \text{(orthogonality condition for } x_3) \qquad (10.38)$$

$$-w_{01} + w_{31} + w_{44} = 0 \qquad \text{(orthogonality condition for } x_4) \qquad (10.39)$$

$$w_{11} + w_{21} - w_{41} - w_{42} - w_{43} = 0 \qquad \text{(orthogonality condition for } x_5) \qquad (10.40)$$

Here, w_{01} represent the dual variable related to the single term in the objective function $y(x)$, i.e. expression (10.33) above. Similarly, w_{11}, w_{12}, w_{21}, w_{31}, w_{32}, w_{41}, w_{42}, w_{43}, and w_{44} represent the dual variables related to the first term in constraint-1, i.e. expression (10.8), to the second term in constraint-1, i.e. expression (10.8), to the single term in constraint-2, i.e. expression (10.9), to the first term in constraint-3, i.e. expression (10.10), to the second term in constraint-3, i.e. expression (10.10), to the first term in constraint-4, i.e. expression (10.34), to the second term in constraint-4, i.e. expression (10.34), to the third term in constraint-4, i.e. expression (10.34), and to the fourth term in constraint-4, i.e. expression (10.34) above, respectively.

There are ten terms (T) and five primal variables (N) in the primal objective function and constraints [expressions (10.6)–(10.10)], and therefore, the degree of difficulty in the problem is $= T - (N + 1) = 4$. As shown above, the value of $w_{01} = 1$ is already determined leaving remaining nine variable values to be determined. Therefore, any five of these nine dual variables can be expressed in terms of the remaining four dual variables as shown below, choosing w_{41}, w_{42}, w_{43} and w_{44} as the four dual variables.

From Eqs.(10.36) and(10.38) we get : $w_{11} = w_{01} = 1$

From Eq.(10.35) we get : $w_{12} = w_{41} + w_{43} - 1$

From Eq.(10.36) we get : $w_{21} = w_{41} + w_{42} + w_{43} - 1$

From Eqs.(10.34) and(10.37) we get : $w_{32} = w_{42} + w_{43} + w_{44} - 1$

From Eq.(10.34) we get : $w_{31} = w_{42} + w_{43} - w_{32} = w_{42} + w_{43} - w_{42}$

$$- w_{43} - w_{44} + 1 = 1 - w_{44} w_{11} + w_{12} = w_{41} + w_{43}; w_{31} + w_{32} = w_{42} + w_{43};$$

$$(10.41)$$

If we substitute the expressions in equations set (9.39) into the dual objective function [expression (10.35)] and maximize this substituted dual objective function over the four variables w_{41}, w_{42}, w_{43}, and w_{44} (i.e. by setting the first derivative of the function with respect to each variable to zero), we will get four additional equations, thus providing sufficient conditions for optimality. The process is simplified by maximizing the natural logarithm of the dual function; that is, by substituting

$$\ln d(w_{41}, w_{42}, w_{43}, w_{44}) = z(w_{41}, w_{42}, w_{43}, w_{44}).$$

Thus, by the above substitutions and taking logarithm on both sides of eq. (10.35), we get the following equation:

$$
\begin{aligned}
z(w) =\ & 1 \times (\ln a_2 + \ln (w_{41} + w_{43}) - \ln 1) + (w_{41} + w_{43} - 1)(\ln a_3)+ \\
& (\ln (w_{41} + w_{43}) - \ln (w_{41} + w_{43} - 1)) + (w_{41} + w_{42} + w_{43} - 1) \ln a_4 \\
& +(1 - w_{44}) (\ln a_5 + \ln (w_{42} + w_{43}) - \ln (1 - w_{44})) + (w_{42} + w_{43}+ \\
& w_{44} - 1) (\ln a_6 + \ln (w_{42} + w_{43}) - \ln (w_{42} + w_{43} + w_{44} - 1)) \\
& +w_{41}(\ln a_7 + \ln (w_{41} + w_{42} + w_{43} + w_{44}) - \ln w_{41}) + w_{42} (\ln a_8+ \\
& \ln (w_{41} + w_{42} + w_{43} + w_{44}) - \ln w_{42}) + w_{43}(\ln a_9 + \ln (w_{41} + w_{42} + w_{43} \\
& +w_{44}) - \ln w_{43}) + w_{44}(\ln a_{10} + \ln (w_{41} + w_{42} + w_{43} + w_{44}) - \ln w_{44}).
\end{aligned}
$$

$$(10.42)$$

By setting the first derivative $dz/dw_{41} = 0$, we get:

$$
\begin{aligned}
& 1/(w_{41} + w_{43}) + \ln a_3 + \ln (w_{41} + w_{43}) + (w_{41} + w_{43} - 1)/(w_{41} + w_{43}) \\
& - \ln (w_{41} + w_{43} - 1) - 1 + \ln a_4 + \ln a_7 + \ln (w_{41} + w_{42} + w_{43} + w_{44}) \\
& +w_{41}/(w_{41} + w_{42} + w_{43} + w_{44}) - \ln w_{41} - 1 + w_{42}/(w_{41} + w_{42} + w_{43} + w_{44}) \\
& +w_{43}/(w_{41} + w_{42} + w_{43} + w_{44}) + w_{44}/(w_{41} + w_{42} + w_{43} + w_{44}) = 0
\end{aligned}
$$

Or,

$$a_3 a_4 a_7 (w_{41} + w_{43}) (w_{41} + w_{42} + w_{43} + w_{44}) - (w_{41} + w_{43} - 1) w_{41} = 0 \quad (10.43)$$

By setting the first derivative $dz/dw_{42} = 0$, we get:

$$\ln a_4 + (1 - w_{44} + w_{42} + w_{43} + w_{44} - 1)/(w_{42} + w_{43}) + \ln a_6 + \ln (w_{42} + w_{43})$$
$$- \ln (w_{42} + w_{43} + w_{44} - 1) - 1 + \ln a_8 + \ln (w_{41} + w_{42} + w_{43} + w_{44})$$
$$+ (w_{41} + w_{42} + w_{43} + w_{44})/(w_{41} + w_{42} + w_{43} + w_{44}) - \ln w_{42} = 0$$
$$a_4 a_6 a_8 (w_{42} + w_{43}) (w_{41} + w_{42} + w_{43} + w_{44}) - (w_{42} + w_{43} + w_{44} - 1) w_{42} = 0$$

$$(10.44)$$

By setting the first derivative $dz/dw_{43} = 0$, we get:

$$1/(w_{41} + w_{43}) + \ln a_3 + \ln (w_{41} + w_{43}) + (w_{41} + w_{43} - 1)/(w_{41} + w_{43})$$
$$- \ln (w_{41} + w_{43} - 1) - 1 + \ln a_4 + \ln a_6 + \ln (w_{42} + w_{43})$$
$$+ (1 - w_{44} + w_{42} + w_{43} + w_{44} - 1)/(w_{42} + w_{43}) - \ln (w_{42} + w_{43} + w_{44} - 1)$$
$$- 1 + (w_{41} + w_{42} + w_{43} + w_{44})/(w_{41} + w_{42} + w_{43} + w_{44}) + \ln a_9$$
$$+ \ln (w_{41} + w_{42} + w_{43} + w_{44}) - \ln w_{43} - 1 = 0$$

Or,

$$a_3 a_4 a_6 a_9 (w_{41} + w_{43}) (w_{42} + w_{43}) (w_{41} + w_{42} + w_{43} + w_{44})$$
$$- (w_{41} + w_{43} - 1) (w_{42} + w_{43} + w_{44} - 1) w_{43} = 0$$

$$(10.45)$$

By setting the first derivative $dz/dw_{44} = 0$, we get:

$$- \ln a_5 - \ln (w_{42} + w_{43}) + \ln (1 - w_{44}) + 1 + \ln a_6 + \ln (w_{42} + w_{43}) - \ln (w_{42}$$
$$+ w_{43} + w_{44} - 1) - 1 + (w_{41} + w_{42} + w_{43} + w_{44})/(w_{41} + w_{42} + w_{43} + w_{44})$$
$$+ \ln a_{10} + \ln (w_{41} + w_{42} + w_{43} + w_{44}) - \ln w_{44} - 1 = 0$$
$$(a_6 a_{10}/a_5) (1 - w_{44}) (w_{41} + w_{42} + w_{43} + w_{44}) - (w_{42} + w_{43} + w_{44} - 1) w_{44} = 0$$

$$(10.46)$$

By dividing (10.45) by (10.44), we get:

$$a_3 a_4 a_6 a_9 (w_{41} + w_{43}) (w_{42} + w_{43}) (w_{41} + w_{42} + w_{43} + w_{44})/a_4 a_6 a_8 (w_{42} + w_{43})$$
$$(w_{41} + w_{42} + w_{43} + w_{44}) - (w_{41} + w_{43} - 1) (w_{42} + w_{43} + w_{44} - 1) w_{43}/(w_{42}$$
$$+ w_{43} + w_{44} - 1) w_{42} = 0 (a_3 a_9/a_8) (w_{41} + w_{43}) - (w_{41} + w_{43} - 1) w_{43}/w_{42} = 0$$

$$w_{42} = (a_8/a_3 a_9) (w_{41} + w_{43} - 1) w_{43}/(w_{41} + w_{43}) \qquad (10.\text{or, })$$

To simplify, we may make the following substitution:

$$w_{41} = x, \ w_{43} = y; \ r = x + y - 1 = w_{12}, \ m = x + y,$$
$$k1 = a_8/a_9 a_3, \ k2 = a_3 a_4 a_7, \ k3 = a_4 a_6 a_8, \ k4 = a_{10} a_6/a_5.$$

Hence,we can express $w_{42} = k1(x + y - 1)y/(x + y) = k1\, ry/m = u$ (10.47)

from (10.43)

$$a_3\, a_4\, a_7(w_{41} + w_{43})\, (w_{41} + w_{42} + w_{43}) + a_3 a_4\, a_7(w_{41} + w_{43})\, w_{44}$$
$$- (w_{41} + w_{43} - 1)\, w_{41} = 0$$

Hence,

$$w_{44} = (w_{41} + w_{43} - 1)\, w_{41}/a_3 a_4 a_7(w_{41} + w_{43}) - a_3 a_4\, a_7(w_{41} + w_{43})\, (w_{41} + w_{42}$$
$$+ w_{43})/a_3 a_4 a_7(w_{41} + w_{43})$$
$$= (w_{41} + w_{43} - 1)\, w_{41}/a_3 a_4 a_7(w_{41} + w_{43}) - (w_{41} + w_{43}) - w_{42}$$
$$= rx/k2m{-}m - u = w$$

$$(10.48)$$

It may be noted that the value of w_{42} given by expression (10.47) is in terms of two unknown variables x and y. Similarly, the value of w_{44} given by expression (10.48) is also in terms of two unknown variables x and y. If we substitute these values of w_{42} and w_{44} in expressions (10.44) and (10.46) along with the above simplified expressions, we get two nonlinear simultaneous equations in terms of two unknown variables $x(= w_{41})$ and $y(= w_{43})$ as shown below as: f and g, respectively, which can be solved by the Newton–Raphson method as explained and illustrated in Chaps. 3 and 8. The author has developed C++ computer programmes for solving such nonlinear simultaneous equations covering many variables and details are presented in Chap. 3.

As outlined in Chap. 3, we can write each of the above two nonlinear simultaneous equations in a more simplified form as: f and g as shown below, and also express the partial derivatives as: f_x, f_y, g_x, g_y where x and y subscripts to f and g denote the partial derivative of each of the functions with respect to the x and y variables, respectively. Similarly, we can make other substitutions as below:

$$f = f(x_n, y_n); g = g(x_n, y_n)$$
$$F = f^* g_y {-} g^* f_y; G = g^* f_x {-} f^* g_x; R = f_x^{\,*} g_y - f_y^{\,*} g_x$$

Thus the Newton–Raphson formula for solution of two simultaneous nonlinear equations in two variables can be written as:

$$x_{n+1} = x_n - (F/R); \quad y_{n+1} = y_n - (G/R); \quad\quad (10.49)$$

By this method, we get a sequence of approximate values x_1, $x_2\ldots x_n$ and y_1, $y_2\ldots y_n$; each successive term of which is closer, compared to its predecessor, to the unknown exact values of the root of the two nonlinear simultaneous equations.

Accordingly, the above two functions, i.e. expressions (10.44) and (10.46), and their partial derivatives: f_x, f_y, g_x, and g_y, after making other substitutions for simplification as show below, become:

$$p = (w_{41} + w_{42} + w_{43} + w_{44}) = x + y + u + w, q = (w_{42} + w_{43} + w_{44} - 1) = y + u$$
$$+ w - 1 = w_{32}$$

$$rx/k2m - m - u = w$$

$$f = k3(u + y) \, p - qu = 0 \tag{10.50}$$

$$g = k4(1 - w)p - qw = 0 \tag{10.51}$$

To simplify, we can express the partial derivatives as: u_x, u_y, w_x, w_y, p_x, p_y, q_x, q_y, where x and y subscripts to u, w, p and q denote the partial derivative of each of these variables with respect to x and y respectively, as shown below.

$$u_x = k1y/m - k1ry/m^2, u_y = (k1r + k1y)/m - k1ry/m^2$$

$$w_x = (x + r)/k2m - k2rx/(k2m)^2 - u_x - 1$$

$$w_y = (x/k2m) - rx/(k2m)^2) - u_y - 1$$

$$p_x = 1 + u_x + w_x, p_y = 1 + u_y + w_y$$

$$q_x = u_x + w_x, q_y = 1 + u_y + w_y$$

$$f_x = k3 \, u_x p + k3(u + y) \, p_x - q_x.u - q.u_x$$

$$f_y = k3 \, (u_y + 1) \, p + k3(u + y)p_y - q_y.u - q.u_y$$

$$g_x = k4.(-w_x)p + k4(1 - w)p_x - q_x.w - q.w_x$$

$$g_y = k4.(-w_y)p + k4(1 - w)p_y - q_y.w - q.w_y$$

Since all the functions and their partial derivatives are expressed in terms of the two unknown variables: x and y as above, putting these values in equation set (10.49) we get a sequence of approximate values $x_1, x_2 \ldots x_n$; and $y_1, y_2 \ldots y_n$; each successive term of which is closer, compared to its predecessor, to the unknown exact values of the root of the two nonlinear simultaneous Eqs. (10.50) and (10.51). Thus, after a few iterations, we get the unknown exact values of the root of the two nonlinear simultaneous Eqs. (10.50) and (10.51). i.e. the value of x and y, fulfilling the tolerance limit prescribed by the planner-designer.

As discussed earlier, the C++ program named as: LAY.CPP developed by the author also incorporates one sub-model or iteration-model for most-benefit optimization of any layout of rectilinear shape. Complete computer code listing of this C++ program is given in appendix. Using LAY.CPP the above optimization problem can be solved in no time as illustrated below, giving the *optimum* layout parameters for most-benefit (i.e. maximum lot area within given cost) design as output, which can be integrated with graphic program to generate *optimum* layout design drawing also.

Application of the geometric programming technique gives the following simple expressions for determining the optimum values of the model variables, thus fully solving the model.

From expression (10.34) we get: $a_9 x_2/a_8 = a_9 x_1 x_2 x_3 x_5^{-1}/a_8 x_1 x_3 x_5^{-1} = w_{43}/w_{42}$

Or, $x_2 = a_8 \times w_{43}/a_9 \times w_{42} = a_8 \times y/a_9 \times u$

Similarly, $x_1 = a_7 \times w_{43}/a_9 \times w_{41} = a_7 \times y/a_9 \times x$, $x_4 = (x_1 - a_6)/a_5$, $x_5 = (x_2 - a_3)/a_2$, $x_3 = a_4 \times x_5$

Therefore, $y(x) = x_3 \times x_4$

From expression (10.41) we get: $w_{11} = w_{01} = 1$, $w_{12} = r$, $w_{21} = x + u + y - 1$, $w_{31} = 1 - w$, $w_{32} = q$.

10.9.3 Calibration of the Least-Cost and Most-Benefit Optimal Design Models

The above models need to be calibrated before being used for optimal design of a layout planning module. This is done by calculating the cost coefficients: c_1, c_2, c_3, and c_4 which are related to the site-specific cost parameters such as cost of labour, materials, and so on; and also to the geometric properties of the layout planning module for the cost coefficients: c_1 and c_2 as shown in Table 10.7 above. Similarly, the design parameters d_1 related to the minimum acceptable lot area, and d_2 related to the desired number of lots in module, both defining the desired design conditions, need to be calculated so that the optimal layout design obtained fulfills these design conditions. Again, the module parameters m_1, m_2, m_3, and m_4 need to be calculated for each layout module type depending on its geometric design. Table 10.7 gives the expressions for calculating the module parameters m_1, m_2, m_3, and m_4 for the four module types adopted here. A planner-designer can choose or develop any other layout planning module, and accordingly, calculate the cost coefficients and module parameters which are dependent on the specific geometric design of the layout planning module, and also on the site conditions for the relevant cost coefficients. Accordingly, the expressions for the values of c_1, c_2, c_3, c_4, m_1, m_2, m_3, and m_4, are developed and given in Table 10.7 above. However, with reference to the model constants in constraint set in expressions: (10.7)–(10.10), the following substitutions may be made:

$$a_1 = d_1, a_2 = m_1, a_3 = m_2, a_4 = d_2, a_5 = m_3, a_6 = m_4.$$

Here, the letter '*d*' indicates the design parameter value externally decided by the planer or designer such as lot area, number of lots, density, and so on; and the letter '*m*' indicates the module parameter to be calculated by the planer or designer by

developing suitable expression for each layout module pattern based on its geometry. All the design parameter and module parameter values are to be given as input to
the model, by the planner or designer. It may be noted that these optimization models
are *Universal Models* independent of the geometry of the layout module chosen,
provided it is of rectilinear shape, and the same model can be used for different
geometrical patterns of layout module, by developing expressions and calculating
values of module parameters for each module type, which are given as input to the
model. The C++ program named as: LAY.CPP developed by the author automatically calculate the module parameters and calibrate the model *instantly* if the
required site-specific inputs such as unit costs and width of streets (in case widths
are pre-fixed), unit cost of utility lines and service lines, and the design-specific
inputs such as chosen module type, desired design parameter values like number of
lots, minimum acceptable lot area to be achieved; and also the desired service-core
location are provided by the planner or designer as input, while running the C++
program as illustrated below. Even the width of streets, lot density, and so on can be
optimized if desired, by developing suitable models which are available elsewhere
(Chakrabarty, 1988a, 1988b), where 15 *optimization models* in two sets (one set for
least-cost design and the other for most-benefit design) adopting a variety of
constraint sets decided by the planner or designer, representing different design
conditions such as lot area, lot density, lot cost, number of lots, open space standard,
circulation interval ratio, module area are presented.

10.10   Application of the Least-Cost and Most-Benefit Optimal Design Models

Today, information technology and computer application are everywhere altering
the very nature of work in all sectors of society. This is equally applicable in the
knowledge disciplines of Architecture (including Urban Planning) Engineering-
Civil/Building Engineering and Construction (usually classified as AEC sector), in
various countries. CAD, quantitative analysis, and optimization are important application areas of IT. Generally, resources are always inadequate compared to the need
whether in low- or high-income countries. No doubt low-income countries are
usually subjected to severe resource-constraints necessitating scaling-down needs
in the context of available resources. But, even in affluent countries such resource-
constraints are applicable because of scaling-up of needs to have higher standard of
living consuming more resources. In view of large variations in QDEIs discussed
earlier, it is desirable that such variations are presented by AEC professionals in a
transparent manner (by application of quantitative analysis and optimization techniques) to all stakeholders including user citizens in all countries, for appropriate
planning and design decisions, which will help achieve a *resource-efficient* solution
with *social equity*. A general schematic optimum housing and urban development
design procedure, which includes optimal layout planning and design, is shown in
Fig. 10.7, below.

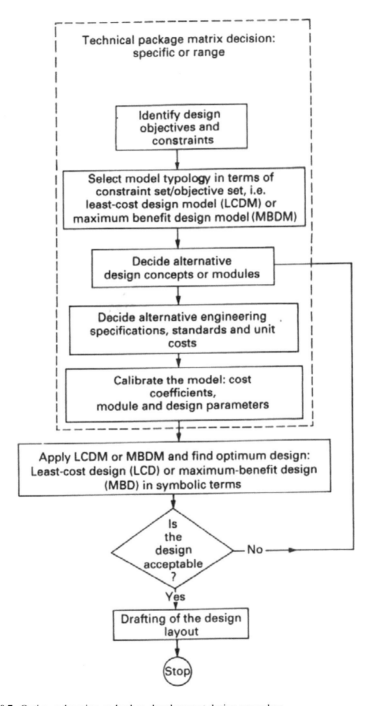

Fig. 10.7 Optimum housing and urban development design procedure

To help application of such quantitative analysis and optimization techniques in layout planning and design the author has developed a C++ program LAY.CPP, which incorporates five models as shown below:

1. QUANTITATIVE ANALYSIS MODEL-ONE CL VALUE LAYOUT MODULE-A (ITMODEL-1)
2. QUANTITATIVE ANALYSIS MODEL-MULTIPLE CL VALUES LAYOUT MODULE-A (ITMODEL-2)
3. MODEL TO OPTIMIZE LOT DEPTH (LOT AREA) FOR LEAST-COST LOT LAYOUT MODULE-A (ITMODEL-3)
4. LEAST-COST OPTIMIZATION MODEL-ANY RECTILINEAR SHAPE LAYOUT PLANNING MODULE (ITMODEL-4)
5. MOST-BENEFIT OPTIMIZATION MODEL-ANY RECTILINEAR SHAPE LAYOUT PLANNING MODULE (ITMODEL-5)

10.11 Illustrative Numerical Application Examples of Computer-Aided Layout Planning and Design by Optimization

Earlier computer-aided iterations for quantitative analysis are shown in Sect. 10.6 using the C++ program LAY.CPP to determine (1) the sensitivity of lot cost to the lot area with single land price (ITMODEL-1), (2) the sensitivity of lot cost to the lot size with multiple land prices (ITMODEL-2), and (3) the exact least-cost lot size for different land prices and infrastructure cost. Here, the use of the Least-Cost Optimization Model (ITMODEL-4) and the Most-Benefit Optimization Model (ITMODEL-5) in different contexts highlighting their relevance and utility in each context is illustrated below by some application results. As earlier mentioned, a layout is generally an aggregation of certain basic units of specific design designated as 'Planning Module' which are repeated to form the whole residential layout. Such planning modules can be for different planning levels, and also, to meet the specific need of the planners and designers who frequently conceive or develop a typology of 'planning modules'. Before using any such planning module in the planning and design solution, a planner-designer may be interested to know the comparative efficiency of each planning module type. Some representative selections of layout planning modules are shown earlier, out of which three planning module types are included in the application examples below, indicating their relative *efficiency*. Expressions for module parameters and cost coefficients for these three planning module types are included in Sect. 10.8.1, outlining the fact that both the **Least-Cost** and **Most-Benefit** Optimization Models presented here are *Universal Optimization Models* applicable for any rectilinear shape layout planning module using the corresponding expressions for module parameters and cost coefficients. A planner-designer can conceive or develop any other rectilinear shape layout planning module as per requirement and use the same optimization models to get optimal results;

provided expressions for module parameters and cost coefficients are also developed for the corresponding planning module and given as input to these optimization models.

10.11.1 Least-Cost Design

Numerical application example results of least-cost computer-aided layout design using the ITMODEL-4 of the C++ program LAY.CPP are shown below. On running the C++ program, it prompts the planner-designer to provide the desired values of input parameters, and thereafter, it solves the complex nonlinear programming problem in no time and gives instantly the output optimal results as shown below. It is necessary to recognize that there is some difference between low-income and middle/high-income countries in terms of availability of resources, affordability of the prospective beneficiaries of housing and urban development projects, planning and design standards, and in many other aspects. However, there is no difference in planning and design principles and objectives such as achieving *resource-utilization efficiency*, *social equity*, and so on, in the planning and design solutions. The example results are provided below both for the context of low-income countries and for the middle/high-income countries, indicating the relevance of the results with reference to the national, local, and individual contexts in each income group of countries.

10.11.1.1   Context: A—Low-Income Countries

In case of low-income countries, one of the prime considerations is the affordability of the prospective beneficiary, and therefore, the lot sizes and lot costs have to be low to make them affordable to the respective income category. Accordingly, infrastructure standards have to be low to reduce costs. Similarly, open space standards need to be low and lot densities need to be high so as to reduce costs and maximize utilization of land, which is often scarce because of low per capita land in many low-income countries. Here, Monetary Units (MU) such as INRs are assumed with representative unit costs, but a planner-designer can provide as input the unit costs as per the local situation and get the output results accordingly.

Example: LC1 Grid-Iron Module with Service-Core Location at Front

```
LEAST-COST OPTIMAL DESIGN OF RESIDENTIAL LAYOUT PLAN :'ITMODEL- 4'
GET ANY DESIRED NUMBER OF LOTS IN LAYOUT MODULE CHOSEN IS;'MODULE- 1'

SOME KEY INPUT VALUES CHOSEN(INDEPENDENT VARIABLES):
CI= 231.0 MU/m, CS= 281.0 MU/m, CM= 414.0 MU/m I= 6.0 m S= 7.5 m M= 9.0 m
CSC= 100.0 MU/m, SLC= 0.25, CL= 10.0 MU/sq.m O=15.00 sq.m MPA= 100.0 sq.m.
LAYOUT MODULE TYPE CHOSEN IS= 'GRID-IRON'
CALIBRATED MODEL CONSTANTS(for the Chosen DESIGN CONDITION)are:
```

a1=100.0000 a2=   0.1667 a3= 12.2500 a4= 0.0100000 a5=6.00 a6=19.50
c1=743.00000000 , c2= 347.50000000 , c3= 10.00000000 , c4= 25.00000000

SOME KEY OUTPUT OPTIMAL RESULTS BY MODEL INSTANTLY SOLVING 'NLP' MODEL:
The values of Optimal Objective & Constraint Weights are:
w01= 0.2705, w02=0.1414, w03=0.4455, w04=0.1425, t=1.0000,
w11=0.6359
w21= 0.6359, w22=0.0801, w31=0.2215, w42=0.0935, w41=0.4934
The Values of Some Optimal Land Utilisation Ratio's are:
Saleable Land Ratio= 0.63, Circulation Area Ratio= 0.25, Open Space
Ratio= 0.11
Optimal WIDTH of the LAYOUT MODULE- N-S Direction = 122.364 M
Optimal LENGTH of the LAYOUT MODULE-E-W Direction = 109.485 M
Optimal Gross Width(allowing for open space)of LOT in LAYOUT MODULE=
5.833 M
Optimal Depth of a LOT in the LAYOUT MODULE   = 17.144 M
Optimal NET Width(excluding reservation for open space)of LOT = 4.958 M
Optimal NET Lot Area (excludng.reservtn. of open space)of LOT = 85.0 Sq.M
Total LOT-FRONTAGE defining Total Number of LOTS in LAYOUT MODULE=
583.293 M
DESIGNED MINIMUM Gross AREA per LOT in LAYOUT MODULE = 100.000 SQ.M.
OPTIMAL LEAST-COST per LOT in the LAYOUT MODULE =3006.994 MU
OPTIMAL LOT DENSITY IN THE LAYOUT MODULE = 74.643 NUMBER OF LOTS/HECTARE
TOTAL NUMBER OF LOTS IN LAYOUT MODULE DESIGNED & ACHIEVED=   100
file:'lclaoutn.cpp' giving LAYOUT PLAN OPTIMAL DESIGN OUTPUT(Number of
Lots)
Lot Ratio = 17.144/4.958 = 3.458

Example: LC2 Cul-De-Sac Module with Service-Core Location at Front

LEAST-COST OPTIMAL DESIGN OF RESIDENTIAL LAYOUT PLAN :'ITMODEL- 4'
GET ANY DESIRED NUMBER OF LOTS IN LAYOUT MODULE CHOSEN IS;'MODULE- 2'

SOME KEY INPUT VALUES CHOSEN(INDEPENDENT VARIABLES):
CI= 231.0 MU/m, CS= 281.0 MU/m, CM= 414.0 MU/m I= 6.0 m S= 7.5 m M= 9.0 m
CSC= 100.0 MU/m, SLC= 0.25, CL= 10.0 MU/sq.m O=15.00 sq.m MPA= 100.0 sq.m.
LAYOUT MODULE TYPE CHOSEN IS= 'CUL-DE-SAC'
CALIBRATED MODEL CONSTANTS(for the Chosen DESIGN CONDITION)are:
a1=100.0000 a2=   0.2500 a3= 8.2500 a4= 0.0100000 a5=4.00 a6=13.50
c1=512.00000000 , c2= 347.50000000 , c3= 10.00000000 , c4= 25.00000000

SOME KEY OUTPUT OPTIMAL RESULTS BY MODEL INSTANTLY SOLVING 'NLP' MODEL:
The values of Optimal Objective & Constraint Weights are:
w01= 0.2529, w02=0.1159, w03=0.4534, w04=0.1778, t=1.0000,
w11=0.6634
w21= 0.6634, w22=0.0429, w31=0.1588, w42=0.0837, w41=0.4856
The Values of Some Optimal Land Utilisation Ratio's are:
Saleable Land Ratio= 0.68, Circulation Area Ratio= 0.20, Open Space
Ratio= 0.12
Optimal WIDTH of the LAYOUT MODULE- N-S Direction = 91.812 M
Optimal LENGTH of the LAYOUT MODULE-E-W Direction = 135.945 M
Optimal Gross Width(allowing for open space)of LOT in LAYOUT MODULE= 5.108 M
Optimal Depth of a LOT in the LAYOUT MODULE   = 19.578 M
Optimal NET Width(excluding reservation for open space)of LOT = 4.342 M

Optimal NET Lot Area (excludng.reservtn. of open space)of LOT = 85.0 Sq.M
Total LOT-FRONTAGE defining Total Number of LOTS in LAYOUT MODULE= 510.780 M
DESIGNED MINIMUM Gross AREA per LOT in LAYOUT MODULE = 100.000 SQ.M.
OPTIMAL LEAST-COST per LOT in the LAYOUT MODULE =2752.685 MU
OPTIMAL LOT DENSITY IN THE LAYOUT MODULE = 80.120 NUMBER OF LOTS/HECTARE
TOTAL NUMBER OF LOTS IN LAYOUT MODULE DESIGNED & ACHIEVED= 100
(Initial-Trial-Values of NLP Indpdnt.Variables are:x=0.1000,
y=0.4000 z=0.2000
file:'lclaoutn.cpp' giving LAYOUT PLAN OPTIMAL DESIGN OUTPUT(Number of
Lots)
Lot Ratio = 19.58/4.341 = 4.51

Example: LC3 Modified Cul-De-Sac (Loop) Module with Service-Core Location at Front

LEAST-COST OPTIMAL DESIGN OF RESIDENTIAL LAYOUT PLAN :'ITMODEL- 4'
GET ANY DESIRED NUMBER OF LOTS IN LAYOUT MODULE CHOSEN IS;'MODULE- 3'

SOME KEY INPUT VALUES CHOSEN(INDEPENDENT VARIABLES):
CI= 231.0 MU/m, CS= 281.0 MU/m, CM= 414.0 MU/m I= 6.0 m S= 7.5 m M= 9.0 m
CSC= 100.0 MU/m, SLC= 0.25, CL= 10.0 MU/sq.m O=15.00 sq.m MPA= 100.0 sq.m.
LAYOUT MODULE TYPE CHOSEN IS= 'LOOP'
CALIBRATED MODEL CONSTANTS(for the Chosen DESIGN CONDITION)are:
a1=100.0000 a2= 0.1667 a3= 8.2500 a4= 0.0100000 a5=6.00 a6=19.50
c1=743.00000000 , c2= 347.50000000 , c3= 10.00000000 , c4= 25.00000000

SOME KEY OUTPUT OPTIMAL RESULTS BY MODEL INSTANTLY SOLVING 'NLP' MODEL:
The values of Optimal Objective & Constraint Weights are:
w01= 0.2620, w02=0.1481, w03=0.4401, w04=0.1498, t=1.0000,
w11=0.6460
w21= 0.6460, w22=0.0561, w31=0.2042, w42=0.0919, w41=0.4963
The Values of Some Optimal Land Utilisation Ratio's are:
Saleable Land Ratio= 0.66, Circulation Area Ratio= 0.22, Open Space
Ratio= 0.12
Optimal WIDTH of the LAYOUT MODULE- N-S Direction = 124.783 M
Optimal LENGTH of the LAYOUT MODULE-E-W Direction = 103.252 M
Optimal Gross Width(allowing for open space)of LOT in LAYOUT MODULE=
5.699 M
Optimal Depth of a LOT in the LAYOUT MODULE = 17.547 M
Optimal NET Width(excluding reservation for open space)of LOT = 4.844 M
Optimal NET Lot Area (excludng.reservtn. of open space)of LOT = 85.0 Sq.M
Total LOT-FRONTAGE defining Total Number of LOTS in LAYOUT MODULE=
569.903 M
DESIGNED MINIMUM Gross AREA per LOT in LAYOUT MODULE = 100.000 SQ.M.
OPTIMAL LEAST-COST per LOT in the LAYOUT MODULE =2929.367 MU
OPTIMAL LOT DENSITY IN THE LAYOUT MODULE = 77.616 NUMBER OF LOTS/HECTARE
TOTAL NUMBER OF LOTS IN LAYOUT MODULE DESIGNED & ACHIEVED= 100
(Initial-Trial-Values of NLP Indpdnt.Variables are:x=0.1000,
y=0.4000 z=0.2000
file:'lclaoutn.cpp' giving LAYOUT PLAN OPTIMAL DESIGN OUTPUT(Number of
Lots)
Lot Ratio = 17.547/4.844 = 3.622

Example: LC4 Grid-Iron Module with Service-Core Location at Rear

```
LEAST-COST OPTIMAL DESIGN OF RESIDENTIAL LAYOUT PLAN :'ITMODEL- 4'
GET ANY DESIRED NUMBER OF LOTS IN LAYOUT MODULE CHOSEN IS;'MODULE- 1'

SOME KEY INPUT VALUES CHOSEN(INDEPENDENT VARIABLES):
CI= 231.0 MU/m, CS= 281.0 MU/m, CM= 414.0 MU/m I= 6.0 m S= 7.5 m M= 9.0 m
CSC=100.0 MU/m, SLC=0.75, CL= 10.0 MU/sq.m O=15.00 sq.m MPA= 100.0 sq.m.
LAYOUT MODULE TYPE CHOSEN IS= 'GRID-IRON'
CALIBRATED MODEL CONSTANTS(for the Chosen DESIGN CONDITION)are:
a1=100.0000 a2=  0.1667 a3= 12.2500 a4= 0.0100000 a5=6.00 a6=19.50
c1=743.00000000 , c2=347.50000000 , c3= 10.00000000 , c4= 75.00000000

SOME KEY OUTPUT OPTIMAL RESULTS BY MODEL INSTANTLY SOLVING 'NLP' MODEL:
The values of Optimal Objective & Constraint Weights are:
w01= 0.2947, w02=0.0871, w03=0.3703, w04=0.2479, t=1.0000,
w11=0.6098
w21= 0.6098, w22=0.0552, w31=0.1423, w42=0.0955, w41=0.3619
The Values of Some Optimal Land Utilisation Ratio's are:
Saleable Land Ratio= 0.62,Circulation Area Ratio= 0.27,Open Space
Ratio= 0.11
Optimal WIDTH of the LAYOUT MODULE- N-S Direction  = 93.360 M
Optimal LENGTH of the LAYOUT MODULE-E-W Direction = 147.679 M
Optimal Gross Width(allowing for open space)of LOT in LAYOUT MODULE=
8.124 M
Optimal Depth of a LOT in the LAYOUT MODULE   = 12.310 M
Optimal NET Width(excluding reservation for open space)of LOT = 6.905 M
Optimal NET Lot Area (excludng.reservtn. of open space)of LOT =  85.0 Sq.M
Total LOT-FRONTAGE defining Total Number of LOTS in LAYOUT MODULE=
812.409 M
DESIGNED MINIMUM Gross AREA per LOT in LAYOUT MODULE  = 100.000 SQ.M.
OPTIMAL LEAST-COST per LOT in the LAYOUT MODULE =3723.845 MU
OPTIMAL LOT DENSITY IN THE LAYOUT MODULE = 72.536 NUMBER OF LOTS/HECTARE
TOTAL NUMBER OF LOTS IN LAYOUT MODULE DESIGNED & ACHIEVED=   100
(Initial-Trial-Values of NLP Indpdnt.Variables are:x=0.1000,
y=0.4000 z=0.2000
file:'lclaoutn.cpp' giving LAYOUT PLAN OPTIMAL DESIGN OUTPUT(Number of
Lots)
Lot Ratio = 12.310/6.905 = 1.783. Cost Ratio =3723.845(Example: LC4)/
2752.685 (Ex.LC2)=1.353
```

10.11.1.2   Context: B—Middle/High-Income Countries

As mentioned above in low-income countries, the lot area and other planning standards are generally low to fit with the very low affordability of prospective beneficiaries of housing and urban development projects in these countries. But in case of middle/high-income countries, the lot area and other planning standards can be high to fit with the high affordability of prospective beneficiaries in these countries. But, even in this case, there can be large variation in costs, land utilization ratios and in optimal lot densities, although the scale in terms of absolute value may

be different as shown in examples below. Here, Monetary Units (MU) such as USDs
are assumed with representative unit costs, but a planner-designer can provide as
input the unit costs as per the local situation and get the output results accordingly.

Example: LC5 Cul-De-Sac Module-Service-Core Location at Front

```
LEAST-COST OPTIMAL DESIGN OF RESIDENTIAL LAYOUT PLAN :'ITMODEL- 4'
GET ANY DESIRED NUMBER OF LOTS IN LAYOUT MODULE CHOSEN IS;'MODULE- 2'

SOME KEY INPUT VALUES CHOSEN(INDEPENDENT VARIABLES):
CI= 108.6 MU/m, CS= 157.4 MU/m, CM= 170.6 MU/m I= 12.0 m S= 15.0 m M= 20.0 m
CSC=  58.6 MU/m, SLC= 0.25, CL=   4.0 MU/sq.m O=60.00 sq.m MPA= 515.5 sq.
m. N=   50
LAYOUT MODULE TYPE CHOSEN IS= 'CUL-DE-SAC'
CALIBRATED MODEL CONSTANTS(for the Chosen DESIGN CONDITION)are:
a1=515.5000 a2=   0.2500 a3=  17.5000 a4= 0.0200000 a5=4.00 a6=27.00
c1=266.08000000 , c2= 164.00000000 , c3=  4.00000000 , c4= 14.64000000

SOME KEY OUTPUT OPTIMAL RESULTS BY MODEL INSTANTLY SOLVING 'NLP' MODEL:
The values of Optimal Objective & Constraint Weights are:
w01= 0.2066, w02=0.1219, w03=0.5560, w04=0.1155, t=1.0000,
w11=0.6912
w21= 0.6912, w22=0.0714, w31=0.1933, w42=0.1022, w41=0.5757
The Values of Some Optimal Land Utilisation Ratio's are:
Saleable Land Ratio= 0.68, Circulation Area Ratio= 0.23, Open Space
Ratio= 0.09
Optimal WIDTH of the LAYOUT MODULE- N-S Direction  = 179.060 M
Optimal LENGTH of the LAYOUT MODULE-E-W Direction = 187.006 M
Optimal Gross Width(allowing for open space)of LOT in LAYOUT MODULE=
13.560 M
Optimal Depth of a LOT in the LAYOUT MODULE   = 38.015 M
Optimal NET Width(excluding reservation for open space)of LOT = 11.982 M
Optimal NET Lot Area (excludng.reservtn. of open space)of LOT = 455.5 Sq.M
Total LOT-FRONTAGE defining Total Number of LOTS in LAYOUT MODULE=
678.024 M
DESIGNED MINIMUM Gross AREA per LOT in LAYOUT MODULE  = 515.500 SQ.M.
OPTIMAL LEAST-COST per LOT in the LAYOUT MODULE = **4817.915** MU
OPTIMAL LOT DENSITY IN THE LAYOUT MODULE = 14.932 NUMBER OF LOTS/HECTARE
TOTAL NUMBER OF LOTS IN LAYOUT MODULE DESIGNED & ACHIEVED=   50
(Initial-Trial-Values of NLP Indpdnt.Variables are:x=0.1000,
y=0.4000 z=0.2000
file:'lclaoutn.cpp' giving LAYOUT PLAN OPTIMAL DESIGN OUTPUT(Number of
Lots)
**Lot Ratio**=37.439/11.752= 3.186
```

Example: LC6 Modified Cul-De-Sac (Loop) Module with Service-Core Location at Rear

```
LEAST-COST OPTIMAL DESIGN OF RESIDENTIAL LAYOUT PLAN :'ITMODEL- 4'
GET ANY DESIRED NUMBER OF LOTS IN LAYOUT MODULE CHOSEN IS;'MODULE- 3'

SOME KEY INPUT VALUES CHOSEN(INDEPENDENT VARIABLES):
```

CI= 108.6 MU/m, CS= 157.4 MU/m, CM= 170.6 MU/m I= 12.0 m S= 15.0 m M= 20.0 m
CSC= 58.6 MU/m, SLC= 0.75, CL= 4.0 MU/sq.m O=60.00 sq.m MPA= 515.5 sq.
m. N= 50
LAYOUT MODULE TYPE CHOSEN IS= 'LOOP'
CALIBRATED MODEL CONSTANTS(for the Chosen DESIGN CONDITION)are:
a1=515.5000 a2= 0.1667 a3= 17.5000 a4= 0.0200000 a5=6.00 a6=39.00
c1=374.72000000 , c2= 164.00000000 , c3= 4.00000000 , c4= 43.92000000

SOME KEY OUTPUT OPTIMAL RESULTS BY MODEL INSTANTLY SOLVING 'NLP' MODEL:
The values of Optimal Objective & Constraint Weights are:
w01= 0.2351, w02=0.1039, w03=0.4768, w04=0.1842, t=1.0000,
w11=0.6457
w21= 0.6457, w22=0.0662, w31=0.1701, w42=0.1192, w41=0.4615
The Values of Some Optimal Land Utilisation Ratio's are:
Saleable Land Ratio= 0.64, Circulation Area Ratio= 0.28, Open Space
Ratio= 0.08
Optimal WIDTH of the LAYOUT MODULE- N-S Direction = 190.009 M
Optimal LENGTH of the LAYOUT MODULE-E-W Direction = 188.221 M
Optimal Gross Width(allowing for open space)of LOT in LAYOUT MODULE=
20.482 M
Optimal Depth of a LOT in the LAYOUT MODULE = 25.168 M
Optimal NET Width(excluding reservation for open space)of LOT = 18.098 M
Optimal NET Lot Area (excludng.reservtn. of open space)of LOT = 455.5 Sq.M
Total LOT-FRONTAGE defining Total Number of LOTS in LAYOUT MODULE=
1024.118 M
DESIGNED MINIMUM Gross AREA per LOT in LAYOUT MODULE = 515.500 SQ.M.
OPTIMAL LEAST-COST per LOT in the LAYOUT MODULE =6000.331 MU
OPTIMAL LOT DENSITY IN THE LAYOUT MODULE = 13.981 NUMBER OF LOTS/HECTARE
TOTAL NUMBER OF LOTS IN LAYOUT MODULE DESIGNED & ACHIEVED= 50
(Initial-Trial-Values of NLP Indpdnt.Variables are:x=0.1000,
y=0.4000 z=0.2000
file:'lclaoutn.cpp' giving LAYOUT PLAN OPTIMAL DESIGN OUTPUT(Number of
Lots)

Example: LC7 Grid-Iron Module-Service-Core at Rear with Less Number of Lots and Wider Streets

LEAST-COST OPTIMAL DESIGN OF RESIDENTIAL LAYOUT PLAN :'ITMODEL- 4'
GET ANY DESIRED NUMBER OF LOTS IN LAYOUT MODULE CHOSEN IS;'MODULE- 1'

SOME KEY INPUT VALUES CHOSEN(INDEPENDENT VARIABLES) :
CI= 108.6 MU/m, CS= 157.4 MU/m, CM= 170.6 MU/m I= 15.0 m S= 20.0 m M= 25.0 m
CSC= 58.6 MU/m, SLC= 0.75, CL= 4.0 MU/sq.m O=60.00 sq.m MPA= 515.5 sq.
m. N= 26
LAYOUT MODULE TYPE CHOSEN IS= 'GRID-IRON'
nl= 59.28 , ml= 2.25 , vl= 114.06
CALIBRATED MODEL CONSTANTS(for the Chosen DESIGN CONDITION)are:
a1=515.5000 a2= 0.1667 a3= 32.5000 a4= 0.0384615 a5=6.00 a6=50.00
c1=374.60000000 , c2= 164.00000000 , c3= 4.00000000 , c4= 43.95000000

SOME KEY OUTPUT OPTIMAL RESULTS BY MODEL INSTANTLY SOLVING 'NLP' MODEL:
The values of Optimal Objective & Constraint Weights are:
w01= 0.2566, w02=0.1425, w03=0.4824, w04=0.1185, t=1.0000,

```
w11=0.5660
w21= 0.5660, w22=0.1730, w31=0.3155, w42=0.1774, w41=0.4475
The Values of Some Optimal Land Utilisation Ratio's are:
Saleable Land Ratio= 0.48,Circulation Area Ratio= 0.45,Open Space
Ratio= 0.06
Optimal WIDTH of the LAYOUT MODULE- N-S Direction  = 176.097 M
Optimal LENGTH of the LAYOUT MODULE-E-W Direction = 138.813 M
Optimal Gross Width(allowing for open space)of LOT in LAYOUT MODULE=
24.529 M
Optimal Depth of a LOT in the LAYOUT MODULE   = 21.016 M
Optimal NET Width(excluding reservation for open space)of LOT = 21.674 M
Optimal NET Plot Area (excludng.reservtn. of open space)of LOT =  455.5
Sq.M
Total LOT-FRONTAGE defining Total Number of LOTS in LAYOUT MODULE=
637.750 M
DESIGNED MINIMUM Gross AREA per LOT in LAYOUT MODULE  = 515.500 SQ.M.
OPTIMAL LEAST-COST per LOT in the LAYOUT MODULE =7795.132 MU
OPTIMAL PLOT DENSITY IN THE LAYOUT MODULE = 10.636 NUMBER OF PLOTS/
HECTARE
TOTAL NUMBER OF LOTS IN LAYOUT MODULE DESIGNED & ACHIEVED=  26
(Initial-Trial-Values of NLP Indpdnt.Variables are:x=0.1000,
y=0.4000 z=0.2000
file:'lclaoutn.cpp'/lch26n giving LAYOUT PLAN OPTIMAL DESIGN OUTPUT
(Number of Lots)
Lot Ratio =21.016/21.674= 0.97
```

Cost Ratio =7795.132/4817.915 (Example: LC5)=**1.62**

10.11.2   Most-Benefit Design

As mentioned above in many cases, a planner/designer may be required to provide the most beneficial planning and design solutions within a given budget of investment of prospective beneficiary (AFC). In such situations, a most-benefit optimal design model needs to be used. To illustrate this process, a number of examples of application of the most-benefit optimal layout design model are presented below.

10.11.2.1   Context: A—Low-Income Countries

Adopting grid-iron module and the same unit cost matrix, and other design parameters as in example LC1 of LC Model, and choosing the input affordable cost AFC same as the least-cost, i.e. 3006.990 MU given as output by the LC model in above example, and running the above MB model, we get the following output results for example: MB1. It may be seen that because of the above assumptions, the optimum layout module design and the lot area in MB1 case are exactly same as the example LC1 of LC model, which had the same unit cost matrix and location of service core in the lot. Similarly results of Example MB2, with output least-cost (LC) value of example LC2, i.e. 2752.685 MU as input value of affordable cost, i.e. AFC value, having the same unit cost matrix, location of service core in the lot, and the same number of lots in layout and module type, i.e. cul-de-sac module as in example LC2, are exactly same as in the example LC2 of LC model. Thus, MB model is the reverse of the LC model with the interchange of output least-cost (LC) value in LC model with the input affordable cost (AFC) value in the MB model, other parameter values remaining unchanged in both cases.

Example: MB1 Grid-Iron Module with Service-Core Location at Front

```
MOST-BENIFIT OPTIMAL DESIGN OF RESIDENTIAL LAYOUT PLAN :'ITMODEL- 5'
GET ANY DESIRED NUMBER OF LOTS IN LAYOUT MODULE CHOSEN IS;'MODULE- 1'

SOME KEY INPUT VALUES CHOSEN(INDEPENDENT VARIABLES):
CI= 231.0 MU/m, CS= 281.0 MU/m, CM= 414.0 MU/m I= 6.0 m S= 7.5 m M= 9.0 m
CSC= 100.0 MU/m, SLC= 0.25, CL= 10.0 MU/sq.m O=15.00 sq.m N= 100
DESIGNED MAXIMUM AFFORDABLE-COST per LOT in the LAYOUT MODULE =3006.990 MU
LAYOUT MODULE TYPE CHOSEN IS= 'GRID-IRON'
CALIBRATED MODEL CONSTANTS(for the Chosen DESIGN CONDITION)are:
a7=0.2471 a2= 0.1667 a3= 12.2500 a4= 0.0100000 a5=6.00 a6=19.50
a8= 0.11556407 , a9= 0.00332558 , a10= 0.00831396

SOME OUTPUT OPTIMAL RESULTS AFTER INSTANTLY SOLVING THE 'NLP' MODEL:
The values of Optimal Objective & Constraint Weights are:
w01= 1.0000, w11=1.0000, w12=0.1260, w21=0.3483, w44=0.2241
w31= 0.7759, w32=0.1471, w41=0.4254, w42=0.2224, w43=0.7006
The Values of Some Optimal Land Utilisation Ratio's are:
Saleable Land Ratio= 0.63,Circulation Area Ratio= 0.25,Open Space
Ratio= 0.11
Optimal WIDTH of the LAYOUT MODULE- N-S Direction = 122.364 M
Optimal LENGTH of the LAYOUT MODULE-E-W Direction = 109.485 M
Optimal Gross Width(allowing for open space)of LOT in LAYOUT MODULE=
5.833 M
Optimal Depth of a LOT in the LAYOUT MODULE = 17.144 M
Optimal NET Width(excluding reservation for open space)of LOT = 4.958 M
Optimal NET Lot Area (excluding reservation of open space)of LOT = 85.0 Sq.M
Total LOT-FRONTAGE defining Total Number of LOTS in LAYOUT MODULE=
583.292 M
OPTIMAL MAXIMUM POSSIBLE Gross AREA per LOT in LAYOUT MODULE = 100.0 SQ.M.
TOTAL NUMBER OF LOTS IN LAYOUT MODULE DESIGNED & ACHIEVED= 100
OPTIMAL LOT DENSITY IN THE LAYOUT MODULE = 74.643 NUMBER OF LOTS/HECTARE
(Initial-Trial-Values of NLP Indpdnt.Variables are:x=0.4000,
y=0.9000
file:'mblaoutn.cpp' giving LAYOUT PLAN OPTIMAL DESIGN OUTPUT(Number of
Lots)
```
Lot Ratio = 17.144/4.958 = 3.458

Example: MB2 Cul-De-Sac Module with Service-Core Location at Front

```
MOST-BENIFIT OPTIMAL DESIGN OF RESIDENTIAL LAYOUT PLAN :'ITMODEL- 5'
GET ANY DESIRED NUMBER OF LOTS IN LAYOUT MODULE CHOSEN IS;'MODULE- 2'
```

```
SOME KEY INPUT VALUES CHOSEN(INDEPENDENT VARIABLES):
CI= 231.0 MU/m, CS= 281.0 MU/m, CM= 414.0 MU/m I= 6.0 m S= 7.5 m M= 9.0 m
CSC= 100.0 MU/m, SLC= 0.25, CL=  10.0 MU/sq.m O=15.00 sq.m N=  100
DESIGNED MAXIMUM AFFORDABLE-COST per LOT in the LAYOUT MODULE =2752.685 MU

LAYOUT MODULE TYPE CHOSEN IS= 'CUL-DE-SAC'
CALIBRATED MODEL CONSTANTS(for the Chosen DESIGN CONDITION)are:
a7=0.1860 a2=   0.2500 a3=  8.2500 a4= 0.0100000 a5=4.00 a6=13.50
a8=  0.12624038 , a9=  0.00363282 , a10=  0.00908204

SOME OUTPUT OPTIMAL RESULTS AFTER INSTANTLY SOLVING THE 'NLP' MODEL:
The values of Optimal Objective & Constraint Weights are:
w01= 1.0000, w11=1.0000, w12=0.0646, w21=0.2393, w44=0.2680
w31= 0.7320, w32=0.1262, w41=0.3811, w42=0.1747, w43=0.6835
The Values of Some Optimal Land Utilisation Ratio's are:
Saleable Land Ratio= 0.68, Circulation Area Ratio= 0.20, Open Space
Ratio= 0.12
Optimal WIDTH of the LAYOUT MODULE- N-S Direction  = 91.812 M
Optimal LENGTH of the LAYOUT MODULE-E-W Direction = 135.946 M
Optimal Gross Width(allowing for open space)of LOT in LAYOUT MODULE=
5.108 M
Optimal Depth of a LOT in the LAYOUT MODULE   = 19.578 M
Optimal NET Width(excluding reservation for open space)of LOT = 4.342 M
Optimal NET Lot Area (excluding reservation of open space)of LOT =  85.0
Sq.M
Total LOT-FRONTAGE defining Total Number of LOTS in LAYOUT MODULE=
510.784 M
OPTIMAL MAXIMUM POSSIBLE Gross AREA per LOT in LAYOUT MODULE  = 100.0 SQ.M.
TOTAL NUMBER OF LOTS IN LAYOUT MODULE DESIGNED & ACHIEVED= 100
OPTIMAL LOT DENSITY IN THE LAYOUT MODULE = 80.119 NUMBER OF LOTS/HECTARE
(Initial-Trial-Values of NLP Indpdnt.Variables are:x=0.4000,
y=0.9000
file:'mblaoutn.cpp' giving LAYOUT PLAN OPTIMAL DESIGN OUTPUT(Number of
Lots)
Lot Ratio = 19.578/4.342 = 4.51
```

10.11.2.2   Context: B—Middle/High-Income Countries

In this case, five application examples are presented below. Examples **MB3**, **MB4**, and **MB5** are presented to show the comparative efficiency of different layout planning modules. It may be seen that the maximum possible gross area per lot is 491.6 SQ.M. (Example: **MB5**), 515.5 SQ.M. (Example: **MB4**), and 548.7 SQ.M. (Example: **MB3**) for the 'GRID-IRON', 'LOOP' and 'CUL-DE-SAC' planning module, respectively, where all other parameter values are kept unchanged, including the location of service core which is located at rear of the lot for all three cases. Thus, in this case also 'CUL-DE-SAC' module is most efficient, while 'LOOP' and 'GRID-IRON' modules, respectively, give 6.1 and 10.4% less lot area compared to the 'CUL-DE-SAC' module within the same AFC value of 6000 MU in this case.

Example: MB3 Cul-De-Sac Module with Service-Core Location at Rear

```
MOST-BENIFIT OPTIMAL DESIGN OF RESIDENTIAL LAYOUT PLAN :'ITMODEL- 5'
GET ANY DESIRED NUMBER OF LOTS IN LAYOUT MODULE CHOSEN IS;'MODULE- 2'

Some Key Input Values Chosen(Independent Variables):
CI= 108.6 MU/m, CS= 157.4 MU/m, CM= 170.6 MU/m I= 12.0 m S= 15.0 m M= 20.0 m
CSC= 58.6 MU/m, SLC= 0.75, CL=   4.0 MU/sq.m O=60.00 sq.m N=   50
Designed Maximum Affordable-Cost Per Lot in the Layout Module =6000.000 MU
Layout Module Type Chosen is= 'CUL-DE-SAC'
Calibrated Model Constants(for the Chosen Design Condition) are:
a7=0.0443 a2=   0.2500 a3=  17.5000 a4= 0.0200000 a5=4.00 a6=27.00
a8= 0.02733333 , a9= 0.00066667 , a10= 0.00732000


Some Output Optimal Results after Instantly Solving the 'NLP' MODEL:
The values of Optimal Objective & Constraint Weights are:
w01= 1.0000, w11=1.0000, w12=0.0730, w21=0.1894, w44=0.3152
w31= 0.6848, w32=0.1615, w41=0.3432, w42=0.1164, w43=0.7299
The Values of Some Optimal Land Utilisation Ratio's are:
Saleable Land Ratio= 0.67,Circulation Area Ratio= 0.25,Open Space
Ratio= 0.08
Optimal WIDTH of the LAYOUT MODULE- N-S Direction  = 141.486 M
Optimal LENGTH of the LAYOUT MODULE-E-W Direction = 257.136 M
Optimal Gross Width(allowing for open space) of LOT in LAYOUT MODULE=
19.171 M
Optimal Depth of a LOT in the LAYOUT MODULE  = 28.621 M
Optimal NET Width(excluding reservation for open space) of LOT = 17.075 M
Optimal NET Lot Area (excluding reservation of open space) of LOT = 488.7
Sq.M
Total LOT-FRONTAGE defining Total Number of LOTS in LAYOUT MODULE=
958.544 M
OPTIMAL MAXIMUM POSSIBLE Gross AREA per LOT LOT in LAYOUT MODULE  = 548.7
SQ.M.
TOTAL NUMBER OF LOTS IN LAYOUT MODULE DESIGNED & ACHIEVED=   50
OPTIMAL LOT DENSITY IN THE LAYOUT MODULE = 13.743 NUMBER OF LOTS/HECTARE
(Initial-Trial-Values of NLP Indpdnt.Variables are:x=0.4000,
y=0.9000
file:'mblaoutn.cpp' giving LAYOUT PLAN OPTIMAL DESIGN OUTPUT(Number of
Lots)
```

Example: MB4 Modified Cul-De-Sac (Loop) Module with Service-Core Location at Rear

```
MOST-BENIFIT OPTIMAL DESIGN OF RESIDENTIAL LAYOUT PLAN :'ITMODEL- 5'
GET ANY DESIRED NUMBER OF LOTS IN LAYOUT MODULE CHOSEN IS;'MODULE- 3'

SOME KEY INPUT VALUES CHOSEN (INDEPENDENT VARIABLES):
CI= 108.6 MU/m, CS= 157.4 MU/m, CM= 170.6 MU/m I= 12.0 m S= 15.0 m M= 20.0 m
CSC= 58.6 MU/m, SLC= 0.75, CL=   4.0 MU/sq.m O=60.00 sq.m N=   50
DESIGNED MAXIMUM AFFORDABLE-COST per LOT in the LAYOUT MODULE =6000.000 MU
LAYOUT MODULE TYPE CHOSEN IS= 'LOOP'
CALIBRATED MODEL CONSTANTS(for the Chosen DESIGN CONDITION) are:
a7=0.0625 a2=   0.1667 a3=  17.5000 a4= 0.0200000 a5=6.00 a6=39.00
a8= 0.02733333 , a9= 0.00066667 , a10= 0.00732000
```

SOME OUTPUT OPTIMAL RESULTS AFTER INSTANTLY SOLVING THE 'NLP' MODEL:
The values of Optimal Objective & Constraint Weights are:
w01= 1.0000, w11=1.0000, w12=0.1025, w21=0.2634, w44=0.2853
w31= 0.7147, w32=0.1846, w41=0.3641, w42=0.1609, w43=0.7384
The Values of Some Optimal Land Utilisation Ratio's are:
Saleable Land Ratio= 0.64,Circulation Area Ratio= 0.28,Open Space
Ratio= 0.08
Optimal WIDTH of the LAYOUT MODULE- N-S Direction = 190.002 M
Optimal LENGTH of the LAYOUT MODULE-E-W Direction = 188.215 M
Optimal Gross Width(allowing for open space)of LOT in LAYOUT MODULE=
20.482 M
Optimal Depth of a LOT in the LAYOUT MODULE = 25.167 M
Optimal NET Width(excluding reservation for open space)of LOT = 18.098 M
Optimal NET Lot Area (excluding reservation of open space)of LOT = 455.5
Sq.M
Total LOT-FRONTAGE defining Total Number of LOTS in LAYOUT MODULE=
1024.084 M
OPTIMAL MAXIMUM POSSIBLE Gross AREA per LOT in LAYOUT MODULE = 515.5 SQ.M.
TOTAL NUMBER OF LOTS IN LAYOUT MODULE DESIGNED & ACHIEVED= 50
OPTIMAL LOT DENSITY IN THE LAYOUT MODULE = 13.982 NUMBER OF LOTS/HECTARE
(Initial-Trial-Values of NLP Indpdnt.Variables are:x=0.4000,
y=0.9000
file:'mblaoutn.cpp' giving LAYOUT PLAN OPTIMAL DESIGN OUTPUT(Number of
Lots)

Example: MB5 Grid-Iron Module with Service-Core Location at Rear

MOST-BENIFIT OPTIMAL DESIGN OF RESIDENTIAL LAYOUT PLAN :'ITMODEL- 5'
GET ANY DESIRED NUMBER OF LOTS IN LAYOUT MODULE CHOSEN IS;'MODULE- 1'

SOME KEY INPUT VALUES CHOSEN(INDEPENDENT VARIABLES):
CI= 108.6 MU/m, CS= 157.4 MU/m, CM= 170.6 MU/m I= 12.0 m S= 15.0 m M= 20.0 m
CSC= 58.6 MU/m, SLC= 0.75, CL= 4.0 MU/sq.m O=60.00 sq.m N=   50
DESIGNED MAXIMUM AFFORDABLE-COST per LOT in the LAYOUT MODULE =6000.000 MU
LAYOUT MODULE TYPE CHOSEN IS= 'GRID-IRON'
CALIBRATED MODEL CONSTANTS(for the Chosen DESIGN CONDITION)are:
a7=0.0625 a2=   0.1667 a3= 25.5000 a4= 0.0200000 a5=6.00 a6=39.00
a8= 0.02733333 , a9= 0.00066667 , a10= 0.00732000

SOME OUTPUT OPTIMAL RESULTS AFTER INSTANTLY SOLVING THE 'NLP' MODEL:
The values of Optimal Objective & Constraint Weights are:
w01= 1.0000, w11=1.0000, w12=0.1488, w21=0.3070, w44=0.2782
w31= 0.7218, w32=0.1955, w41=0.3897, w42=0.1581, w43=0.7591
The Values of Some Optimal Land Utilisation Ratio's are:
Saleable Land Ratio= 0.60,Circulation Area Ratio= 0.32,Open Space
Ratio= 0.08
Optimal WIDTH of the LAYOUT MODULE- N-S Direction = 182.491 M
Optimal LENGTH of the LAYOUT MODULE-E-W Direction = 196.819 M
Optimal Gross Width(allowing for open space)of LOT in LAYOUT MODULE=
20.554 M
Optimal Depth of a LOT in the LAYOUT MODULE = 23.915 M
Optimal NET Width(excluding reservation for open space)of LOT = 18.045 M
Optimal NET Lot Area (excluding reservation of open space)of LOT = 431.6

Sq.M
Total LOT-FRONTAGE defining Total Number of LOTS in LAYOUT MODULE=
1027.710 M
OPTIMAL MAXIMUM POSSIBLE Gross AREA per LOT LOT in LAYOUT MODULE = 491.6
SQ.M.
TOTAL NUMBER OF LOTS IN LAYOUT MODULE DESIGNED & ACHIEVED= 50
OPTIMAL LOT DENSITY IN THE LAYOUT MODULE = 13.921 NUMBER OF LOTS/HECTARE
(Initial-Trial-Values of NLP Indpdnt.Variables are:x=0.4000,
y=0.9000
file:'mblaoutn.cpp' giving LAYOUT PLAN OPTIMAL DESIGN OUTPUT(Number of
Lots)

Example: MB6 Grid-Iron Module-Service-Core at Rear with Wider Streets

MOST-BENIFIT OPTIMAL DESIGN OF RESIDENTIAL LAYOUT PLAN :'ITMODEL- 5'
GET ANY DESIRED NUMBER OF LOTS IN LAYOUT MODULE CHOSEN IS;'MODULE- 1'

SOME KEY INPUT VALUES CHOSEN(INDEPENDENT VARIABLES):
CI=108.6 MU/m, CS=157.4 MU/m, CM=170.6 MU/m I=15.0 m S=20.0 m M=25.0 m
CSC= 58.6 MU/m, SLC=0.75, CL= 4.0 MU/sq.m O=60.00 sq.m N= 50
DESIGNED MAXIMUM AFFORDABLE-COST per LOT in the LAYOUT MODULE =6000.000 MU
LAYOUT MODULE TYPE CHOSEN IS= 'GRID-IRON'
CALIBRATED MODEL CONSTANTS(for the Chosen DESIGN CONDITION) are:
a7=0.0625 a2= 0.1667 a3= 32.5000 a4= 0.0200000 a5=6.00 a6=50.00
a8= 0.02733333 , a9= 0.00066667 , a10= 0.00732000

SOME OUTPUT OPTIMAL RESULTS AFTER INSTANTLY SOLVING THE 'NLP' MODEL:
The values of Optimal Objective & Constraint Weights are:
w01= 1.0000, w11=1.0000, w12=0.2031, w21=0.3745, w44=0.2815
w31= 0.7185, w32=0.2578, w41=0.3982, w42=0.1714, w43=0.8049
The Values of Some Optimal Land Utilisation Ratio's are:
Saleable Land Ratio= 0.53,Circulation Area Ratio= 0.39,Open Space
Ratio= 0.08
Optimal WIDTH of the LAYOUT MODULE- N-S Direction = 189.342 M
Optimal LENGTH of the LAYOUT MODULE-E-W Direction = 192.519 M
Optimal Gross Width(allowing for open space)of LOT in LAYOUT MODULE=
19.198 M
Optimal Depth of a LOT in the LAYOUT MODULE = 23.224 M
Optimal NET Width(excluding reservation for open space)of LOT = 16.615 M
Optimal NET Lot Area (excluding reservation of open space)of LOT = 385.9
Sq.M
Total LOT-FRONTAGE defining Total Number of LOTS in LAYOUT MODULE=
959.921 M
OPTIMAL MAXIMUM POSSIBLE Gross AREA per LOT LOT in LAYOUT MODULE = 445.9
SQ.M.
TOTAL NUMBER OF LOTS IN LAYOUT MODULE DESIGNED & ACHIEVED= 50
OPTIMAL LOT DENSITY IN THE LAYOUT MODULE = 13.717 NUMBER OF LOTS/HECTARE
(Initial-Trial-Values of NLP Indpdnt.Variables are:x=0.4000,
y=0.9000
file:'mblaoutn.cpp' giving LAYOUT PLAN OPTIMAL DESIGN OUTPUT(Number of
Lots)

Example: MB7 Grid-Iron Module-Service-Core at Rear with Wider Streets and Less Number of Lots

```
MOST-BENIFIT OPTIMAL DESIGN OF RESIDENTIAL LAYOUT PLAN :'ITMODEL-5'
GET ANY DESIRED NUMBER OF LOTS IN LAYOUT MODULE CHOSEN IS;'MODULE-1'

SOME KEY INPUT VALUES CHOSEN(INDEPENDENT VARIABLES):
CI=108.6 MU/m, CS=157.4 MU/m, CM=170.6 MU/m I=15.0 m S=20.0 m M=25.0 m
CSC= 58.6 MU/m, SLC=0.75, CL=   4.0 MU/sq.m O=60.00 sq.m N=   28
DESIGNED MAXIMUM AFFORDABLE-COST per LOT in the LAYOUT MODULE =6000.000 MU
LAYOUT MODULE TYPE CHOSEN IS= 'GRID-IRON'
nl=  50.59 , ml=   0.98 , vl=  99.23
CALIBRATED MODEL CONSTANTS(for the Chosen DESIGN CONDITION)are:
a7=0.0624 a2=   0.1667 a3=  32.5000 a4=  0.0357143 a5=6.00 a6=50.00
a8=  0.02733333 , a9=  0.00066667 , a10=  0.00732500

SOME OUTPUT OPTIMAL RESULTS AFTER INSTANTLY SOLVING THE 'NLP' MODEL:
The values of Optimal Objective & Constraint Weights are:
w01= 1.0000, w11=1.0000, w12=0.3603, w21=0.6428, w44=0.2385
w31= 0.7615, w32=0.3666, w41=0.5147, w42=0.2825, w43=0.8456
The Values of Some Optimal Land Utilisation Ratio's are:
Saleable Land Ratio= 0.41, Circulation Area Ratio= 0.50, Open Space
Ratio= 0.09
Optimal WIDTH of the LAYOUT MODULE- N-S Direction  = 153.843 M
Optimal LENGTH of the LAYOUT MODULE-E-W Direction = 122.705 M
Optimal Gross Width(allowing for open space)of LOT in LAYOUT MODULE=
19.326 M
Optimal Depth of a LOT in the LAYOUT MODULE   = 17.307 M
Optimal NET Width(excluding reservation for open space)of LOT = 15.859 M
Optimal NET Plot Area (excluding reservation of open space)of LOT =
274.5 Sq.M
Total LOT-FRONTAGE defining Total Number of LOTS in LAYOUT MODULE=
541.122 M
OPTIMAL MAXIMUM POSSIBLE Gross AREA per LOT LOT in LAYOUT MODULE  = 334.5
SQ.M.
TOTAL NUMBER OF LOTS IN LAYOUT MODULE DESIGNED & ACHIEVED=   28
OPTIMAL PLOT DENSITY IN THE LAYOUT MODULE = 14.833 NUMBER OF PLOTS/
HECTARE
(Initial-Trial-Values of NLP Indpdnt.Variables are:x=0.5600,
y=0.8600
file:'mblaoutn.cpp'(layh7.cpp) giving LAYOUT PLAN OPTIMAL DESIGN
OUTPUT(Number of Lots)
```
lot ratio=17.307/15.859=1.09

Example: MB8 Cul-De-Sac Module-Service-Core at Front, Less Open Space and More Number of Lots

```
MOST-BENIFIT OPTIMAL DESIGN OF RESIDENTIAL LAYOUT PLAN :'ITMODEL-5'
GET ANY DESIRED NUMBER OF LOTS IN LAYOUT MODULE CHOSEN IS;'MODULE-2'

SOME KEY INPUT VALUES CHOSEN(INDEPENDENT VARIABLES):
CI=108.6 MU/m, CS=157.4 MU/m, CM=170.6 MU/m I=12.0 m S=15.0 m M=20.0 m
```

CSC= 58.6 MU/m, SLC= 0.25, CL= 4.0 MU/sq.m O=40.00 sq.m N= 100
DESIGNED MAXIMUM AFFORDABLE-COST per LOT in the LAYOUT MODULE =6000.000 MU
LAYOUT MODULE TYPE CHOSEN IS= 'CUL-DE-SAC'
CALIBRATED MODEL CONSTANTS(for the Chosen DESIGN CONDITION)are:
a7=0.0443 a2= 0.2500 a3= 17.5000 a4= 0.0100000 a5=4.00 a6=27.00
a8= 0.02733333 , a9= 0.00066667 , a10= 0.00244000

SOME OUTPUT OPTIMAL RESULTS AFTER INSTANTLY SOLVING THE 'NLP' MODEL:
The values of Optimal Objective & Constraint Weights are:
w01= 1.0000, w11=1.0000, w12=0.0488, w21=0.1391, w44=0.1799
w31= 0.8201, w32=0.0990, w41=0.2200, w42=0.0903, w43=0.8288
The Values of Some Optimal Land Utilisation Ratio's are:
Saleable Land Ratio= 0.81,Circulation Area Ratio= 0.15,Open Space
Ratio= 0.04
Optimal WIDTH of the LAYOUT MODULE- N-S Direction = 250.580 M
Optimal LENGTH of the LAYOUT MODULE-E-W Direction = 376.119 M
Optimal Gross Width(allowing for open space)of LOT in LAYOUT MODULE=
14.345 M
Optimal Depth of a LOT in the LAYOUT MODULE   = 55.895 M
Optimal NET Width(excluding reservation for open space)of LOT = 13.629 M
Optimal Maximum NET Lot Area (excluding reservation of open space)of LOT
= **761.8** Sq.M
Total LOT-FRONTAGE defining Total Number of LOTS in LAYOUT MODULE=
1434.477 M
OPTIMAL MAXIMUM POSSIBLE Gross AREA per LOT LOT in LAYOUT MODULE = 801.8
SQ.M.
TOTAL NUMBER OF LOTS IN LAYOUT MODULE DESIGNED & ACHIEVED= 100
OPTIMAL LOT DENSITY IN THE LAYOUT MODULE = 10.610 NUMBER OF LOTS/HECTARE
(Initial-Trial-Values of NLP Indpdnt.Variables are:x=0.2000,
y=0.9000
file:'mblaoutn.cpp' giving LAYOUT PLAN OPTIMAL DESIGN OUTPUT(Number of
Lots)
Area Ratio =761.8/240.6 =3.17   Lot Ratio = 55.895/13.629 = 4.1

10.11.3   Discussion of Applications Example Results

A large number of illustrative numerical application examples of computer-aided layout planning and design by optimization are deliberately presented above, to show the likely pitfall or drawback in terms of *efficiency* and *social equity* objective achievement, in the conventional planning practice, where rarely quantitative analysis and optimization techniques are applied. In view of scarcity of resources compared to needs and the low affordability of substantial section of population in all countries, although the level of scale may be different between the low-income and middle/high-income countries, it is imperative that the planners and designers in all countries discharge their *ethical responsibility* to achieve a *resource-efficient* urban development with *social equity*.

The application example results of least-cost design, in the context of low-income countries presented above, show that the LEAST-COST per LOT is 3006.994 MU

(**Example: LC1**), 2929.367 MU (**Example: LC2**) and 2752.685 MU (**Example: LC3**) for 'GRID-IRON', 'LOOP' and 'CUL-DE-SAC' planning module, respectively, where all other parameter values are kept unchanged, including the location of service core which is located in front of the lot for all three cases. Thus, 'CUL-DE-SAC' module is most efficient, while 'LOOP' and 'GRID-IRON' modules, respectively, are 6.4% and 9.2% costlier than the 'CUL-DE-SAC' module in this case. In case, the service core is located in rear of the lot the LEAST-COST per LOT is increased to 3723.845 MU (**Example: LC4**) for 'GRID-IRON' module, where all other parameter values are kept unchanged. Thus, only because of change in module pattern and location of service core the LEAST-COST per LOT is increased from 2752.685 MU to 3723.845 MU, i.e. by **35%** for the same net lot size of 85.0 Sq.M. There are similar differences in the optimal land utilization ratios and in the optimal lot densities achieved in the respective layout module. Thus, saleable land ratio increases from 0.62 to 0.68, i.e. by 11%; and lot density increases from 72.536 to 80.120 lots per hectare, i.e. by 10.5% between the Cul-De-Sac Module with service-core location at front (Example: LC2), and the grid-iron module with service-core location at rear (**Example: LC4**), while open space standard in terms of open space per lot is same in both cases.

Such a large variation in costs and in land utilization ratios and optimal lot densities, not only influence the planning design objectives of *efficiency* and *social equity*, but also, seriously affect the financial viability of a housing development project. It may also be seen that the optimal lot ratios in the four cases are **3.458**, **4.51**, **3.622**, and **1.783** for the examples: LC1, LC2, LC3, and LC4, respectively. Thus, the optimal lot ratio depends on many interacting factors which should be considered in a combined manner in a specific case; and no generalization can be made in this regard, as frequently attempted in conventional planning practice, affecting *efficiency*. All the above factors should be very important concerns of the planner and designer to discharge their *ethical responsibility* to achieve a *resource-efficient* urban development with *social equity*. It is extremely difficult and time consuming to attain these objectives by conventional planning practice, but by using quantitative analysis and *optimization* techniques such as the C++ program LAY. CPP these objectives are achieved in no time, and therefore, need to be important part of planning and design practice.

The application example results of least-cost design presented above also emphasize the fact that even in the context of middle/high-income countries adopting high planning standards, there can be large differences in cost and other efficiency indicator values for the same lot area, say, 515.5 Sq.M. as shown above, where the cost per lot, for the grid-iron module with service-core at rear with less number of lots and wider streets (**Example: LC7**), increases by **62%** compared to the Cul-De-Sac Module with service-core location at front (**Example: LC5**). Similarly, saleable land ratio increases from 0.48 to 0.68, i.e. by 42%, and the optimal lot densities increases from 10.636 lots/hectare to 14.932 lots/hectare, i.e. by **40%** between the two cases. There are similar differences in cost per lot, saleable land ratio, and in lot densities between grid-iron module and loop module as shown above. Planners and designers have an *ethical responsibility* to achieve a *resource-efficient* urban development with *social equity*, and therefore, it is incumbent on the planners and designers to reiterate their plans in terms of cost-effectiveness and other quantitative design *efficiency* indicators; so that their plans are free from such *inefficiencies* or the user/beneficiary stakeholders are informed about such parameters and *inefficiencies* if any, and thus, planning and design decisions are taken in a *transparent* manner by all stakeholders. Such an approach will require use of *quantitative analysis* and CAD techniques as discussed above, which will also promote a corruption-free planning and design practice and process in built-environment, as also highlighted in Chap. 7.

It may be seen that the results given by the C++ program LAY.CPP in application **Example: LC6** and **MB4**, for Modified Cul-De-Sac (Loop) Module with Service-Core Location at Rear, are same (with minor decimal error) as reported earlier (Chakrabarty, 1991) before developing the above C++ program, with same values of input parameters. These two application examples also show that, as in the context of low-income countries mentioned above, in the context of middle/high-income countries also, the MB model is the reverse of the LC model with the interchange of Output Least-Cost (LC) value in LC model with the Input Affordable Cost (AFC) value in the MB model, other constraint sets and parameter values remaining unchanged in both cases. Thus, the planning solution in Example: **MB4** is the mirror image (with very minor decimal error) of planning solution **Example: LC6**, with the above interchange of LC and AFC values, other parameter values remaining unchanged in both cases.

Comparative effect of wider streets on the gross lot area is indicated by **Example: MB5** and **Example: MB6**, in which all other parameters, including the module type, number of lots, and cost per lot, remain same in both cases. It may be seen that the gross lot area in **MB6** reduces to 445.9 Sq.M from 491.6 Sq.M. in **MB5,** i.e. by 9.3%, due to the provision of wider streets in **MB6**.

Again, as shown in case of least-cost design model, in the most-benefit design case also, there can be large variations in costs and in land utilization ratios and optimal lot densities, although the scale in terms of absolute value may be different in low and middle/high countries contexts, as also revealed by the application examples above. Thus, while application Example: **MB7**, grid-iron module-service-core at rear with wider streets and less number of lots, gives maximum NET lot area (excluding reservation of open space) of only 274.5 Sq.M, it increases to 761.8 **Sq. M.,** i.e. by **2.78** times in application Example: **MB8,** adopting Cul-De-Sac module with service-core at front, less open space, and more number of lots. Similarly, the saleable land ratio increases from 0.41 to 0.81, i.e. by **1.98** times, although lot density reduces from 14.833 to 10.610, i.e. by 28% between the above two cases. Planners and designers cannot ignore such large variations if *efficiency* and *social equity* are the planning and design objectives. These factors not only influence the planning and design objectives of *efficiency* and *social equity*, but also, seriously affect the financial viability and sustainability of housing and urban development projects, which are very important concerns for all stakeholders in the process, including the prospective beneficiaries who should have very important say in the matter. Therefore, planners and designers in all countries need to apply quantitative analysis and *optimization* techniques and reiterate their planning and design solutions to the prospective beneficiaries in terms of the above objective (s) achievement, and thus, discharge their *ethical responsibility* in these regard, using tools such as the C++ program LAY.CPP which necessitates being important part of planning and designing practice to achieve the above objectives.

10.11.4   Typology of Optimization Models in Layout Planning and Design by Optimization

As earlier pointed out in conventional layout planning and design practice, a layout is first drawn with all dimensions already chosen. Thereafter, the efficiency ratios are calculated from this drawn layout. If these derived values are not acceptable, another layout is drawn and the efficiency ratios are re-calculated from this redrawn layout. This process of trial and error may continue for considerable time, and still the desired results may not be attained. This is because there can be numerous planning and design solutions with very large variations in various efficiency indicators as a result of the interactions of many factors and planning and design variables, as also outlined above. It is desirable to reverse the above layout planning and design process. Thus, a planner may first develop or choose a specific planning module in dimensionless form, and also decide the design objective(s) to be achieved in terms of the specific land utilization ratios like density, marketable land percentage, circulation space percentage, and so on; and cost ratios such as cost per hectare, cost per lot, and so on. In this reversed direct layout planning and design process, a layout could be designed directly to fit with the above design objective(s), if these are feasible, by using direct quantitative analysis and design models. Similarly, intricate mathematical programming technique, i.e. linear and nonlinear programming techniques can be used to achieve *optimal* design solutions, resolving the conflicting viewpoints, demands, and design-decisions earlier discussed.

In case only direct quantitative design is intended, avoiding intricate mathematical programming techniques involved in optimization, a planner/designer can develop simple algebraic equations for this purpose. As for example, a planner/designer can develop the following simple algebraic direct design models for layout design:

1. Designing a layout module for a given lot area or combination of lot areas and finding the cost per lot
2. Designing a layout module for a given lot density
3. Designing a layout module for a given marketable or saleable land ratio
4. Designing a layout module for a given affordable cost per lot and finding the lot area within that cost
5. Designing a layout module for a given infrastructure cost per unit area of module
6. Designing a layout module for a given number of plots or lots

The method of development of algebraic direct design models for quantitative analysis, taking the grid-iron module type shown in Fig. 10.3(d) as example, is illustrated in Sect. 10.6.1 above. The above mathematical model is developed for grid-iron module type shown in Fig. 10.3(d), to design a layout module for a given lot area and finding the SPC value, i.e. serviced plot or lot cost, i.e. cost per lot including cost of land, utility networks, and service connection lines to a lot. Adopting similar procedure direct design models can be developed for lot density, for marketable or saleable land ratio, for a given affordable cost (given as input) and finding the lot area (output) within that cost, and so on (Chakrabarty, 1988a, 1988b). Similarly, mathematical models for other module types can also be developed.

It may be noted that the algebraic direct design models of the above types give infinite solutions to the algebraic equation sets that yield different output values of cost per lot, or, the same value of cost per lot, depending on values of different input variables which could also be numerous. Hence, a planner or designer has to assume the value of one or more of the input variables in a trial solution, and start an iterative design process to evaluate its suitability; and if not satisfied a new solution is adopted and the process is repeated, and still, optimal solution may not be achieved. However, in case of *optimization* models, say, with the objective function of Least-Cost (LC) or Most-Benefit (MB), the *Global Optimal Unique* LC or MB values are achieved in no time fulfilling the given *constraint set* prescribed by the planner and designer. In the past, *optimization* was rarely applied mainly because it involves intricate and complex mathematical equations difficult to solve manually. But, with the availability of high-power computers with advanced graphics capability, it has become very easy to apply *optimization* in the computer-aided planning and design process, as demonstrated in a large number of application example problems presented above. As already stated, to permit computer-aided iterations and optimization of layout planning modules, the author has developed a C++ program named as: LAY.CPP. It incorporates five models including three *Optimization-Models* (one for lot depth *optimization* and two *universal optimization models* for layout planning and design) and two iteration-models for quantitative analysis and direct design of layout module, which are used in the above application example problems.

Two *universal optimization models* for layout planning and design presented above, one with **Least-Cost** as the *objective function* and the other with **Most-Benefit** as the objective function, have the number of lots as one of the *constraint set*. A planner/designer can develop similar models with other *objective functions* and *constraint sets* in the form of a *typology* of *optimization models* to meet specific requirements. As earlier mentioned, to help planners/designers 15 *optimization models* in the form of a typology with two sets of *objective functions* (one set for **Least-Cost** design and the other for **Most-Benefit** design) adopting a variety of constraint sets decided by the planner or designer, representing different design conditions, are developed, solved, and obtainable elsewhere (Chakrabarty, 1988a, 1988b). The *typology* of *optimization models* with typology name, indicating the nature of *objective function* and *constraint set* for the above 15 *optimization models*, is presented below (for ready reference):

1. Model Typology CPCV—Model-I
2. Model Typology CPCF—Model-II
3. Model Typology CPNF—Model-III
4. Model Typology CPCDV—Model-IV
5. Model Typology CPCDF—Model-V
6. Model Typology CPCMV—Model-VI
7. Model Typology CDCDV—Model-VII
8. Model Typology CPCV (Curvilinear Shape Layout Module)—Model-VIII
9. Model Typology BPCV—Model-IX
10. Model Typology BPCF—Model-X
11. Model Typology BPNF—Model-XI
12. Model Typology BPCDV—Model-XII
13. Model Typology BPCDF—Model-XIII
14. Model Typology BDCF—Model-XIV
15. Model Typology BDCDV—Model-XV

The above typology of models is only representative and not exhaustive, and it could incorporate any other model (s) developed by interested planners/designers to meet specific requirements in any specific situation. Out of the above 15 models, fourteen models (excluding Model-VIII which is for curvilinear shape layout module, and presented only to show that even in curvilinear shape layout module *optimization* is applicable) are *universal optimization models* in the sense that each model can be applied in any layout planning module of rectilinear shape (including those presented above) developed by a planner/designer, provided the corresponding expressions for the module parameters and cost coefficients are also developed, as done in Table 10.7 for the four module types presented above, and given as input to the model chosen by the planner/designer for application.

Each model typology name above starts with 'C' or 'B' depending on whether, the model is for minimizing the costs denoted by the letter 'C', or, for maximizing the benefit denoted by 'B'. The model may incorporate only layout or the dwelling-layout system for optimization. Accordingly, the next letter in typology name is 'P' (for plot or lot), or, 'D' (for dwelling-layout system). Thus, the model typology BDCF—Model-XIV and BDCDV—Model-XV are for *optimization* of single-family dwelling-layout system. Intervals, or, spacing between the lines of circulation is a basic factor in urban layouts. These are determined as a compromise between the opposing requirements of intercommunication between the community elements, e.g. dwellings, social facilities, and so on; and the cost as well as the land utilization limitations (Caminos & Goethert, 1978). In case of layout planning module, the spacing between peripheral streets, meant for both pedestrian and vehicular traffic, could be considered as the circulation interval. However, this spacing should be related to the width (or right of way) of the corresponding street, for equivalent performance in respect intercommunication between the community elements and ease of traffic flow (both pedestrian and vehicular). Thus, the circulation interval ratio (CIR), or, the ratio between the length of subsidiary main street and the width of such streets (CIR), could be considered as a quantitative design efficiency indicator, and can be specified as a design constraint, or, design parameter in the optimization model. Therefore, the next letter in model typology name is 'C', where circulation interval ratio is a design constraint. The next letter in the model name is 'N', or, 'D', or, 'M', depending whether the number plots (or lots), or, density, or, the layout module area, is a design constraint. However, in case none of these design constraints is included; these letters will not appear in the model names. The width of the streets could be fixed or variable in a model. Keeping in view the adjacency conditions and engineering requirements, streets may have fixed widths and predetermined, or, variable widths—in which case optimal street widths can be derived by the optimization model. Accordingly, the last letter in model name is 'V' (variable width of streets), or, 'F' (fixed width of streets), depending on the design condition chosen by the planner and designer.

As already stated above, solutions of each of the above 15 *optimization models* are available elsewhere (Chakrabarty, 1988a, 1988b), out of which *automated computer-aided* solutions are offered, for the Model Typology: CPNF—Model-III and BPNF—Model-XI, in the C++ program: LAY.CPP, application of which is demonstrated in a large number of application example problems presented above. In Chap. 12, the *automated computer-aided* solutions are presented for the Model Typology BDCF—Model-XIV and BDCDV—Model-XV, which are for the *optimization* of single-family dwelling-layout systems, as also stated above. Similar *automated computerized* solutions could be derived by interested planners/designers for all the remaining models, who could also incorporate their own models in this package. This will help in developing a more comprehensive system for *integrated computer-aided planning and design by optimization*, not only for small-scale layout planning, but also for a large-scale urban modelling and development system as a whole in future, based on *Information Technology*, Urban Operations Research, CAD, and *Optimization*.

10.12 Integration of Numerical Analysis and Graphics in Layout Planning and Design by Optimization

As cited above C++ program named as: **lay.cpp** gives optimal design solutions in numerical terms, which have to be converted to graphic form, i.e. drawing, by using a graphic program. Integration of numerical analysis and graphics, in layout planning and design by optimization, can be done by developing the Integrated Computer-Aided Design Drafting (CADD) Software: LAY**CAD** (under development), to carry out simultaneously, numerical analysis including *optimization* by the C++ program: lay.cpp (already fully developed), which also generates a LISP output file incorporating the numerical optimal output data, which is used as input by the LISP program: laydg.lsp (under development) to carry out graphic and drafting tasks including developing the working drawings of the optimal layout plan design as numerically analysed and optimized by lay.cpp. Thus, the C++

program: lay.cpp and the LISP program: Laydg.lsp can form the *numerical part* and *graphic part,* respectively, of the integrated CADD software: LAYCAD (which is under development and interested readers may fully develop it) for layout planning and design, in line with the already developed software: RCCAD presented in Chap. 8, and the software: HudCAD presented in Chap. 9.

Application of the optimization model is illustrated using two examples—one for least-cost design and the other for most-benefit design as outlined below.

10.12.1   *Illustrative Application Example: Integrated Computer-Aided Layout Design by Optimization for Least-Cost*

In this application example for integrating numerical analysis and graphics in design, a grid-iron layout module with service-core location at front is selected while running the C++ program: lay.cpp which carries out **instantly** the numerical analysis including *optimization* and gives the output results as shown in the numerical **Example: LC1** grid-iron module with service-core location at front**,** presented above. The C++ program: lay.cpp also generates a LISP output file: **lcn.lsp** including the numerical optimal output data as shown below:

```
(setq dlis
(list 100 122.36 109.48   5.83 17.14 583.29
60.95 20.66 74.58 1500.00  0.00  0.00
1 100.00   0.87   4.96  6.00   7.50   9.00  15.00 4 ) )
```

It may be seen that a list of 21 values have been generated, which designates the values of the 21 numerical model parameters:N, x_1, x_2, x_3, x_4, x_5, nl, ml, vl, TOA, il, tl, module, MPA, OW, NW, I, S, M, O, and $itrn$, respectively. The above model parameters are explained in the model description in Sect. 10.8 above and are also given below for ready reference:

$x_1 =$ width of the planning module between the centre lines of two East-West access streets (for block of lots module) or subsidiary main streets (for other modules)

$x_2 =$ length of the planning module between the centre lines of the end streets (for block of lots module) or between the main street and the subsidiary main street (for other modules) in North-South direction

$x_3 =$ gross width of the lot allowing for the reservation of open space per lot where applicable

$x_4 =$ depth of the lot

$x_5 =$ a dependent variable related to the number of lots in a planning module

$x_6 =$ a dependent variable defining the gross area per lot

$x_7 =$ right of way width of the internal street (also can be designated as I)

x_8 = right of way width of the subsidiary main street (also can be designated as S) or end street

x_9 = right of way width of the main street (also can be designated as M)

The LISP program: laydg.lsp (under development) when run loads the file: **lcn. lsp** (generated by the C++ program: lay.cpp, already fully developed) incorporating the numerical output data as above, and converts them as data input for the graphic program: laydg.lsp (partly developed), and thus, generates **instantly** the drawing of the optimal layout plan accordingly as shown below (presently showing only the street layout).

10.13   Universal Applicability of Quantitative Analysis and Optimization in AEC Operations including Layout Planning and Design in all Countries

Today, information technology and computer application are everywhere altering the very nature of work in all sectors of society. This is equally applicable in the knowledge disciplines of Architecture (including Urban Planning) Engineering-Civil/Building Engineering and Construction (usually classified as AEC sector), in various countries. CAD, quantitative analysis, and optimization are important application areas of IT. It is sometimes argued that use of quantitative analysis and optimization techniques is applicable only in Third World Countries (i.e. low-income countries) which are subjected to severe resource-constraints and population pressure. But, we should not forget that all of us *live in a finite earth with finite resources*, and in this context, an ethical responsibility is enjoined on all of us to achieve *resource-efficiency* and *social equity* at each system level in all countries, and only through application of quantitative analysis and optimization techniques at each level, whether in low- or high-income countries, we can achieve this objective. In fact it is shown earlier (Chakrabarty, 1991) that in the high-income countries context, the maximum lot area within a given cost can vary from 426.62 to 548.7 m^2 (i.e. a variation of 1.29 times) depending on the layout module type and other design conditions chosen by planner/designer and other stakeholders. It is desirable that such variations are presented by AEC professionals in a *transparent* manner (by application of quantitative analysis and optimization techniques) to all *stakeholders* including user citizens in all countries, for appropriate planning and design decisions, which will not only help achieve a *resource-efficient* solution with *social equity*, but will also promote a corruption-free urban planning and development.

Moreover, there is a complex interaction between the land price, utility network cost, service connection cost, and layout physical parameters; and the lot cost value is determined by the combined effect of all these factors. Results of such interactions are shown both for 'low lot size case' (i.e. Table 10.1, generally applicable in low-income countries) and for 'medium lot size case' (i.e. Table 10.2, generally applicable in medium- and high-income countries). In both cases, it is shown that there is no simple linear relationship between the lot size and the serviced lot cost; in other words, reducing the lot size does not necessarily mean reduction in the serviced lot cost which is erroneously assumed in conventional planning practices. In fact, in many cases, lot size reduction may be the cause for increase in the serviced lot cost, due to a complex interaction between the land price, utility network cost, service connection cost, and layout physical parameters; and the lot cost value is determined by the combined effect of all these factors, as shown in above tables and also discussed in reference (Chakrabarty, 1987a, 1987b TWPR). Each of the quantitative analysis, Tables 10.1 and 10.2, indicates an optimum lot size giving the least-cost per lot in each case, which needs to be determined by using *optimization* techniques as illustrated in Sects. 10.6.4 and 10.6.5 above.

Similarly, as indicated in Chap. 6, there is complex interaction and sensitivity between various urban 'built-form-elements' causing large variations in efficiency indicators. Therefore, it is desirable to apply traditional management principles in urban development planning and management (i.e. in AEC sector as a whole) to achieve *Efficiency* and *Social Equity*, resolving the conflicting interests, viewpoints, and demands of various stakeholders in a *transparent* manner, and thus, reduce any scope of corruption and avoid many urban problems whether in low- or high-income countries. This is feasible only if scientific quantitative analysis and optimization are universally applied in all AEC operations. Urbanization, urban management, and computer-aided planning and design are discussed in Chap. 2, where a course design for *urban management* is presented to facilitate the above process. It may be noted that almost all countries in the world suffer many urban problems and achieving *efficiency* and *social equity* in the urbanization, urban development, urban-renewal or urban re-development process including layout planning and design with appropriate planning and regulations (e.g. FAR discussed earlier in Chap. 6 and in Chap. 9) may be the key to solve such urban problems.

10.14   Closure

Layout planning and design is a multidisciplinary process involving several base-disciplines such as physical planning, engineering, architecture, and so on. Many of these planning and design aspects are interlinked. Therefore, it is desirable that planning and design decisions made by each base-discipline take into account the combined effect of the decisions on design variables, belonging to each discipline in 'quantitative terms' with reference to a common 'value scale' (say costs) chosen for measuring the overall planning and design objective achievement in an integrated manner in a given context, and thus, avoid frequent conflicting planning and design decisions creating inefficiencies and other problems.

In this chapter, the layout planning and design process, involving large- and small-scale urban models, the design viewpoint of such models, and the planning module concept for small-scale urban models, are outlined. In view of *resource-constraints* applicable in all countries, one of the primary objectives of layout planning and design should be to attain a *resource-efficient* and cost-effective design solution, recognizing the systems nature of the problem. Complex interaction between cost and physical parameters in a layout system is outlined using mathematical models, establishing that it is a case of *optimization* with *universal applicability* in layout planning and design in all countries.

Using quantitative analysis and optimization models developed, the lot-cost *sensitivity* to lot areas is presented; indicating that the lot area reduction does not automatically mean lot-cost reduction; and also the land-price-utility-networks-cost interaction in land subdivision is assessed giving the exact *least-cost* lot size in a given context. To help application of such quantitative analysis and optimization techniques in layout planning and design a C++ program LAY.CPP is developed and presented with source code, which incorporates five models as shown below:

1. Quantitative Analysis Model-one land price (Itmodel-1)
2. Quantitative Analysis Model-Multiple land price (Itmodel-2)
3. Model to Optimize Lot Depth for Least-Cost per Lot of given area (Itmodel-3)
4. Universal Least-Cost Optimization Model-Any Rectilinear Shape Planning Module (Itmodel-4)
5. Universal Most-Benefit Optimization Model-Any Rectilinear Shape Planning Module (Itmodel-5)

It is pointed out that the optimization models cited above are *universal models,* i.e. the same optimization model can be used for different types of layout planning modules, as the optimal model formulation is independent of the type of layout planning module chosen by a planner-designer, provided these are, of course, rectilinear in shape. However, variation in the module type is ensured by selecting appropriate input values of model constants as per the different expressions for module parameters and cost coefficients for each layout module type as shown in Table 10.7 in Sect. 10.8.1 above for some modules.

Even a few application example results of C++ program LAY.CPP presented reveal a variation of 35% and 62% in the values of *Least-Cost* per lot for the same net lot size in the low-income and high-income country context, respectively. There are similar differences in the optimal land utilization ratios and in the optimal lot densities achieved in the respective cases. Application examples of C++ program LAY.CPP, in most-benefit design case, show that the maximum net lot area within same value of maximum affordable cost per lot may vary by 2.78 times in high-income country context, with similar variations in the optimal land utilization ratios and in the optimal lot densities achieved.

Planners and designers cannot ignore such large variations if *efficiency* and *social equity* are also the planning and design objectives. These factors not only influence the planning and design objectives of *efficiency* and *social equity*, but also, seriously affect the financial viability and sustainability of housing and urban development projects, which are very important concerns for all *stakeholders* in the process, including the prospective beneficiaries who should have very important say in the matter. Therefore, planners and designers in all countries need to apply quantitative analysis and *optimization* techniques and reiterate their planning and design solutions to the prospective beneficiaries in terms of the above objective(s) achievement, and thus, discharge their *ethical responsibility* in these regard, using tools such as the C++ program LAY.CPP which necessitates being important part of planning and designing practice to achieve the above objectives.

As earlier pointed out in conventional layout planning and design practice, a layout is first drawn with all dimensions already chosen. Thereafter, the efficiency ratios are calculated from this drawn layout. If these derived values are not acceptable, another layout is drawn and the efficiency ratios are re-calculated from this redrawn layout. This process of trial and error may continue for considerable time, and still the desired results may not be attained. This is because there can be numerous planning and design solutions with very large variations in various efficiency indicators as a result of the interactions of many factors and planning and design variables, as also outlined above. It is desirable to reverse the above layout planning and design process. Thus, a planner may first develop or choose a specific planning module in dimensionless form, and also decide the design objective(s) to be achieved in terms of the specific land utilization ratios like density, marketable land percentage, circulation space percentage, and so on; and cost ratios such as cost per hectare, cost per lot, and so on. In this reversed direct layout planning and design process, a layout could be designed directly to fit with the above design objective(s), if these are feasible, by using direct quantitative analysis and design models presented. Similarly, intricate mathematical programming techniques such as nonlinear programming techniques, i.e. geometric programming technique as presented here, can be used to achieve *optimal* design solutions, resolving the conflicting viewpoints, demands, and design-decisions earlier discussed.

Integration of numerical analysis and graphics, in layout planning and design by optimization, can be done by developing the integrated Computer-Aided Design Drafting (CADD) Software: LAYCAD, in line with the already developed Software: RCCAD presented in Chap. 8, and the Software: HudCAD presented in Chap. 9. However, a drawing of optimal layout plan related to numerical example: LC1 (only street layout), generated by the partly developed CADD Software: LAYCAD, integrating numerical analysis-optimization and graphics is shown in Fig. 10.8 above.

There are a number of C++ and AutoLISP programs presented in this chapter along with application examples of optimization models. To help readers of this book to use such techniques to achieve their own problem solutions, complete listings of the above C++ and AutoLISP Programs are included in an online space connected to this book.

As pointed out in Chap. 1, Operations Research (**OR**) and Computer-Aided Design (**CAD**) are essential integrative techniques to improve *efficiency and effectiveness*, and since, AEC professionals commit huge resources in their planning-design-decisions, it is basic that they use such basic tools to discharge their *ethical responsibility* and *accountability* to the society to achieve *efficiency* and *social equity* in the use of huge (but inadequate compared to the needs in all countries) *resources*

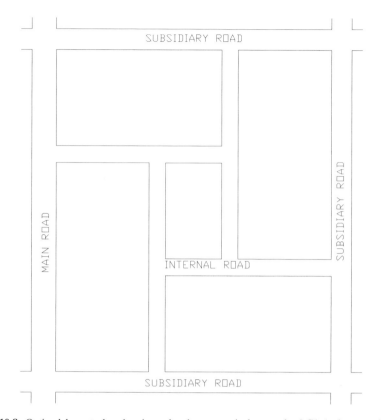

Fig. 10.8 Optimal layout plan drawing related to numerical example: LC1 (only street layout shown)

committed by them in their plans and designs to produce building and urban services. *In view of above, it is desirable that such urban planning and development problems including layout planning and design are seen as part of integrated urban management and AEC professionals uses management techniques like OR/CAD including quantitative analysis and optimization as universal planning and design tools in all AEC operations both in low- and high-income countries, to attain a resource-efficient and equitable urban problem solutions in all countries.*

Chapter 11
Integrated Computer-Aided Dwelling-Layout Systems Design by Optimization

11.1 Introduction

Small-scale modeling and system design in residential land planning, layout, building, and housing development have been presented in earlier chapters. The combination of the layout and 'dwelling unit design modules', with the utility lines, road pavement, dwelling structure, building-housing service connection lines along with the semi-public open-to-sky spaces, could be considered as the dwelling-layout systems and designed accordingly covering all the components. Layout planning and design process and the systems nature of the problem are outlined in Chap. 10, where it is emphasized that the layout planning and design process should consider this systems nature of the problem to help attain *efficiency* and *social equity* at each system level and in the urban development process as a whole. This is equally applicable in dwelling-layout systems planning and design process. Dwelling-layout systems may be of two types—one for single-family dwelling and the other for multi-family dwelling. In this chapter, both types of dwelling-layout systems are considered.

11.2 Dwelling-Layout Systems Design Process

As earlier stated in many cases, a residential neighbourhood may be planned adopting some repeatable planning units, which may be called 'planning modules' or the 'unit design modules'. The combination of the layout and the dwelling 'unit design modules', with the utility lines, road pavement, dwelling structure, house service connections, and public open-spaces, could be considered as the 'dwelling layout system'. In view of a number of faults of large-scale urban models and the utility of small-scale design models to serve as a positive force for the directed development of the community, it is desirable to develop small-scale residential

B. K. Chakrabarty, *Integrated CAD by Optimization*,
https://doi.org/10.1007/978-3-030-99306-1_11

development design models including 'dwelling layout system' design models. Such models should be 'white box' type based on clear mathematical relationships between the design variables and the planning and design standards to permit such directed development of the community for better quality of life in a *resource-efficient* manner achieving *social equity*. Thus, model goals should be more modest, adopting a direct causal relationship between the design variables making the results more realistic and the benefits immediately applicable. To help this process, such design models should allow numerical quantitative analysis to achieve *efficiency*, *effectiveness,* and *optimization* in design. Scientists frequently adopt the notion of modules or subsystems for analytical measurements. In this chapter module concept in Dwelling-Layout Systems Design is outlined.

11.3   Module Concept in Dwelling-Layout Systems Design

The first step in the optimization of dwelling-layout systems is the module concept in dwelling-layout systems design since outlined below. As mentioned earlier, a spatial system is generally composed of planning modules or 'unit design modules', which represent the structure and the relationships between the parts and the whole of a system. Such planning modules are frequently used by designers in actual physical layout and dwelling designs, which can be considered as spatial systems. The module concept in layout planning and in building and housing development system design are outlined in Chaps. 10 and 11, respectively. Such concepts are equally applicable in dwelling-layout system design both for single-family and multi-family dwelling-layout system. As outlined in Chap. 10, there can be a variety of layout planning modules or typology of planning modules for residential layout or dwelling development with a number of physical parameters such as lot area, lot ratio, lot width, lot depth, street-width, and open space standards, and so on. There are also a number of cost parameters such as unit costs of street pavements, unit costs of utility lines like water supply lines and sewer lines, and so on. Similarly, there can be an assortment of 'dwelling unit design modules' in building and housing development system design, both for single-family and multi-family dwelling development as outlined in Chap. 11. Such module concept is equally applicable in dwelling-layout systems design both for single-family and multi-family dwelling development, where the layout and the dwelling 'unit design modules' along with the utility lines, services, road pavement, dwelling structure, and public/semi-public open-to-sky spaces are combined.

In many cases, building, housing, and urban development take place in phases in accordance with the basic 'structure plan', in response to incremental effective demand from time to time, as also discussed in earlier chapters. Design and development of building, housing, and dwelling-layout system at the level of 'unit design modules' are generally within the immediate sphere of influence of the planner-designer, in response to the immediate or existing effective demand. Depending on the phase of development of building, housing, and dwelling-layout system, the adjacency conditions of such 'unit design modules' may be completely flexible or somewhat constrained. Both conditions permit optimization of such 'unit design modules', but the benefit of optimization is generally more if the adjacency conditions of such 'unit design modules' are more flexible. Hence, it is desirable to strive for such flexibility to derive maximum benefit from optimization.

11.4 Typology of Optimizing Models for Dwelling-Layout Systems Design

Generally, based on the physical designs, dwelling-layout systems can be categorized into two types of, namely: (a) single-family dwelling-layout systems, where single-family dwellings are located in individual lots in a layout module of lots described in Chap. 10 and a typical single-family dwelling-layout system is also shown below and (b) multi-family dwelling-layout systems, where multi-family-dwellings are located in different building-blocks which are again, sited in a layout pattern as shown below. Again, for each category of dwelling-layout systems design mentioned above, there can be two types of *optimizing models,* namely: (a) least-cost optimizing model and the (b) most-benefit optimizing model. The details of such *optimizing models* are presented in Chap. 8, which are equally applicable in single-family as well as multi-family dwelling-layout systems, as outlined later.

11.4.1 Single-Family Dwelling-Layout System

Some representative selections of 'unit design modules' or typology of planning modules for the layout are presented in Chap. 10. Two typical dimensionless unit design modules or schematic layout planning modules suitable for single-family dwelling-layout system are presented in Fig. 11.1. Similarly, schematic models of dwelling unit design modules for single-family dwelling-layout system, showing different room-arrangement matrices are presented in Fig. 11.2. There can be a variety of room arrangements for different types of dwelling development, such as row-housing, semi-detached housing, and detached housing, in a single-family dwelling-layout system. The number of rooms provided along the width and the depth of a lot, which can be called as room-arrangement matrix, is one aspect of such room arrangement, which is very relevant for cost optimization.

In Fig. 11.2, two modules of a dwelling-shell are shown, having same number of rooms along the depth of a lot (i.e. 2) but with one room along the width of the lot which can be designated as (1×2 matrix), and with two rooms along the width of the lot which can be designated as (2×2 matrix). These dwelling modules could be developed as: row-housing, semi-detached housing, and detached housing, in a single-family dwelling-layout system, having considerable effect on costs.

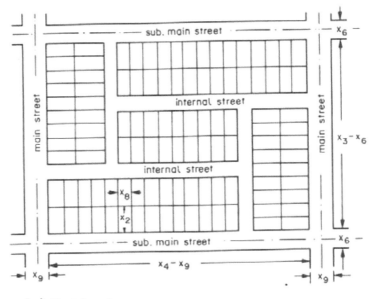

(a) Module-A

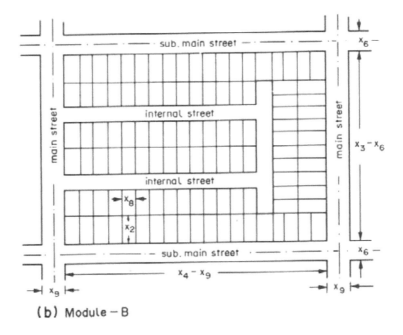

(b) Module-B

Fig. 11.1 Single-family dwelling-layout system-typical schematic layout planning modules

Fig. 11.2 Single-family
dwelling-layout system-
typical schematic dwelling
unit design modules

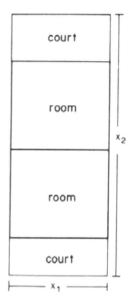

(a) Room arrangement 1 x 2 Matrix

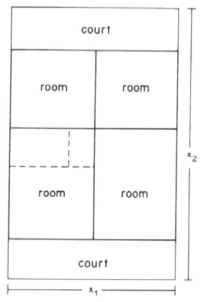

(b) Room arrangement 2 x 2 Matrix

The 'Layout Planning Modules' and the 'Dwelling Unit Design Modules' shown here are only for the purpose illustration. Many other 'unit design modules' for the layout and dwelling of different patterns and interesting shapes can be conceptualized by a creative designer, depending on the lifestyle of people, the terrain, climate, site-constraints, and so on. Similarly, the concept of room-arrangement matrix can be enlarged to a 'dwelling unit design module coefficients' based on proportion coefficients for various rooms giving complete optimal design of building and housing as outlined in Chap. 9.

11.4.2 Multi-family Dwelling-Layout System

As in case of single-family dwelling development, a designer generally adopts some repeatable basic units of designs or dwelling unit design modules with different combination or typology of grouping patterns in a multi-family dwelling development system also. Figure 11.3 below shows such dimensionless schematic diagrams of a set of layout planning modules for a multi-family dwelling development. Such module concept can also be applied in planning and design of public buildings like schools, hospitals and also for the commercial buildings like shopping centre and so on.

Some representative selections of 'unit design modules' or typology of planning modules for the layout are presented in Chap. 10. Two typical dimensionless unit design modules or schematic layout planning modules suitable for multi-family dwelling-layout system are presented in Fig. 11.3. Similarly, schematic models of dwelling unit design modules for multi-family dwelling-layout system, showing different room-staircase-arrangements are presented in Fig. 11.4. We can consider the dimensionless schematic layout, building, or dwelling unit design modules as shown in Figs. 11.3 and 11.4, along with their layout and the corresponding streets, utility networks, and other services, as the building or dwelling-layout-systems. A schematic multi-family dwelling unit design module for high-income housing is also shown in Fig. 11.5 just for illustration. Such multi-family dwelling unit design module for high-income housing can also be accommodated in a suitable layout planning module (as done in Fig. 11.3) conceived by a planner-designer, giving the multi-family dwelling-layout system for high-income housing which can be optimized.

The building or dwelling-layout systems can be optimized to achieve the desired objective such as minimizing the costs per dwelling unit fulfilling the given constraints such as the desired built-space, in a site-specific situation as presented later. The above schematic diagrams of unit design modules are presented only for the purpose of illustration. A creative designer generally develops and adopts a variety of such unit design modules in building and housing projects depending on the site-specific situation and other considerations.

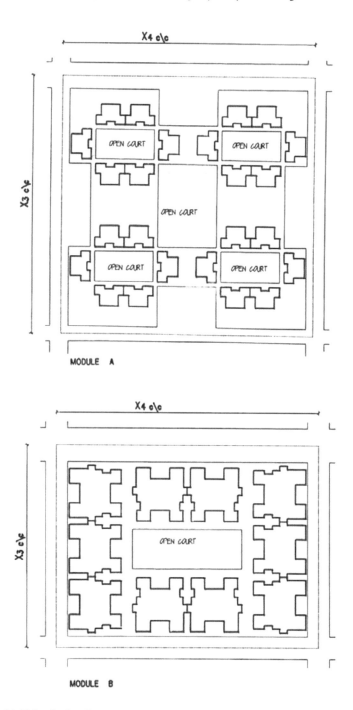

Fig. 11.3 Multi-family dwelling-layout system-typical schematic layout planning modules

Fig. 11.4 Multi-family dwelling-layout system-typical schematic dwelling unit design modules

11.5   Developing Model(s) for Optimal Design of Dwelling-Layout Systems

After the unit design modulesfor layout and dwelling designs forming the dwelling-layout system are decided, the first step in developing models to obtain the optimal design solution, i.e. least-cost or most-benefit design of dwelling-layout systems, is the derivation of unit design module coefficients for different unit layout and dwelling design modules, conceived by a planner/designer keeping in view the different contexts, one of which can be the intended target income group of beneficiaries. Again, it is necessary to formulate the building or dwelling cost coefficients based on the respective unit design module coefficients derived as above, which can be used in the development of cost function for the optimization model for least-cost or most-benefit design of the dwelling-layout systems, as presented later. These models need to incorporate various components and steps as outlined above, and also, some basic parameters such as design module coefficients, cost coefficients,

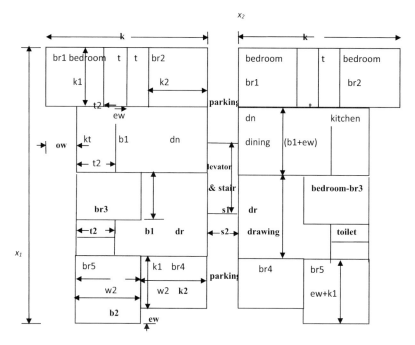

Fig. 11.5 Schematic multi-family dwelling unit design module: E for high-income housing showing the built-unit design proportion coefficients

cost functions, and constraints depending on the specific optimal design problem to be solved, as discussed below. Development of unit design module coefficients, cost coefficients, cost functions, and an optimization model for the representative selection of 'unit design modules' discussed above, is presented below, separately for single-family and multi-family dwelling-layout systems.

Ideally, the cost coefficients and cost functions in layouts and dwelling-layout systems should be derived based on site-specific-input-data in terms of labour and material costs. In respect of layouts, the unit costs and the method of deriving the combined unit costs of the internal streets (designated as CI), subsidiary main streets (designated as CS), and the main street (designated as CM) considering all the utility lines and adopting different standards are presented earlier (HUDCO, 1982). Adopting the same procedure and with suitable increase in unit costs, to allow for escalation in cost of labour and materials, the cost coefficients for layout component can be derived. In case of 'single-family dwelling-layout systems', the dwelling units are located in individual lots with individual on-lot service connections, forming part of layout modules of different patterns as shown in Chap. 10. However, in case of 'multi-family dwelling-layout systems', the dwelling units are located in different building blocks with low, medium, and high building rise (with consequent effect on costs and land-use patterns), which are located in some patterns without any individual lot identification, and therefore, layout modules in this case can be much simpler. These differences have to be considered while formulating optimization models for single- and multi-family dwelling-layout systems.

Cost coefficients and cost functions in building and housing development systems can be derived using the building component unit costs, which again, depend on the site-specific-input-data in terms of labour and material costs. A cost engineering model is presented in Chap. 4 which can generate item wise and building component wise unit costs. The model provides for change of values of all or some 'Input Cost-Engineering Co-Efficient' and also 'Input Labour-Material-Unit-Cost (MKT. RATES) Co-Efficient' as per the site-specific requirement in a particular location in a particular country. Thus, it permits use of realistic unit costs in the design process generating the optimal designs more realistic in any country. An example of Schedule of Rates (SOR) and also the building component unit costs instantly given by the cost engineering model is also presented in the above chapter. Using the above concepts, the process of developing **Cost Coefficients**, for building-housing and layout are illustrated in Chaps. 9 and 10, respectively.

In the process of developing **Cost Coefficients**, different building components such as foundation, wall, roof including flooring have to be considered. The unit costs of these building components are designated as: 'cf' in MU (Monetary Units) per meter, 'cw' in MU per meter, 'cr' in MU per square metre, and 'cfl' in MU per square metre; for the components: foundation, wall, roof, and floor, respectively. In case of multi-family dwelling-layout systems, a factor designated as: 'fic' is included to allow for the increase in unit costs due to building-rise. Suitable value can also be assigned to this factor to incorporate the cost of elevators in a building depending on its rise and size. The number of floors in a multi-family building is designated as: 'nf'. Thus, the expressions for the **Cost Coefficients** for different components in a multi-family dwelling-layout system can be developed using the above designated terms and the dwelling unit design module coefficients: cx_1, cx_2, and cba developed earlier for various dwelling unit design modules, as illustrated below.

Development of models for optimal design of dwelling-layout systems, both for single-family and multi-family considering the above factors are outlined below.

11.5.1 Single-Family Dwelling-Layout Systems

Derivation of unit design module coefficients for layout and dwelling designs forming the single-family dwelling-layout systems is outlined below. These derivations take into account suitably the design context for low- and high-income countries as may be applicable and relevant in a site-specific case.

11.5.1.1   Unit Design Module Cost Coefficients for Layout Forming Part of the Dwelling-Layout System

Typical schematic layout planning modules and dwelling unit design modules for single-family dwelling-layout system are shown above in Fig. 11.1 for layout and Fig. 11.2 for dwelling, respectively. One can easily derive the layout module cost coefficients for different patterns of layout modules selected by a designer, using these typical schematic layout planning modules and Table 10.6 in Chap. 10, for the corresponding single-family dwelling-layout system. The first step in developing models to obtain the optimal design solution, i.e. least-cost or most-benefit design of dwelling-layout systems, is the derivation of unit design module coefficients for different unit layout and dwelling design modules, conceived by a planner/designer keeping in view the different contexts, one of which can be the intended target income group of beneficiaries. Again, it is necessary to formulate the building or dwelling cost coefficients based on the respective unit design module coefficients derived as above, which can be used in the development of cost function for the optimization model for least-cost or most-benefit design of the dwelling-layout systems, as presented later. These models need to incorporate some basic parameters such as design module coefficients, cost coefficients, cost functions, and constraints depending on the specific optimal design problem to be solved.

As outlined in Chap. 10, the models for 'layout planning module' for quantitative analysis and optimization can be for different design conditions taking into account the adjacency conditions in a site-specific case. It is also pointed out that such models are universal optimization models for any rectilinear shape layout planning module, and that the same optimization model can be used for different types of layout planning modules, as the optimal model formulation is independent of the type of layout planning module chosen by a planner-designer, provided these are, of course, rectilinear in shape. However, variation in the module type is ensured by selecting appropriate input values of model constants as per the different expressions for module parameters and cost coefficients for each layout module type as shown in Table 10.7 in Chap. 10 which is reproduced below.

The modules can be conceived to have a hierarchy of street systems such as main street, subsidiary main street, internal street, and so on, which also provides access to lots. The parameters k_1 and k_2 in Table 11.1 above are the street-width hierarchy ratio for the three levels of street-widths which have been assumed (Chakrabarty, 1988a, 1988b, thesis p. 177, 183). Thus, the value of k_1 and k_2 will be given by:

$$k_1 = I/S \text{ and } k_2 = M/S, \text{ where } I, S, \text{ and } M \text{ are widths of internal,}$$
$$\text{subsidiary main, and main street, respectively.}$$

Specifying the desired values of street-width hierarchy ratio for the three levels of street-widths, a planner-designer can find the optimum widths of internal, subsidiary main, and the main street maintaining this specified street-width hierarchy ratio. In case the widths of internal, subsidiary main, and the main street are predetermined, the above expressions of *Module Parameters* incorporating k values (i.e. m_2 and m_4) are to be multiplied by the predetermined width of subsidiary main street to derive

Table 11.1 Module types, module parameters, and unit design module cost coefficients

Module type	Module parameters				Cost coefficients			
	m_1	m_2	m_3	m_4	c_1	c_2	c_3	c_4
A block of lots	½	$(1+k_2) \times 1/2$	2	1	CS	$(CS + CM) \times 1/2$	CL	CSC × SLC
Modified Cul-De-Sac	¼	$(1+k_2) \times 1/2$	4	$(1+k_1)$	CI + CS	$(CS + CM) \times 1/2$	CL	CSC × SLC
Grid iron	1/6	$(4k_1 + 3k_2 + 3)/6$	6	$(1+2k_1)$	2CI + CS	$(CS + CM) \times 1/2$	CL	CSC × SLC
Module with loop circulation	1/6	$(1+k_2) \times 1/2$	6	$(1+2k_1)$	2CI + CS	$(CS + CM) \times 1/2$	CL	CSC × SLC

the actual values of *Module Parameters: m_2 and m_4* to be used as input to the optimization model. Each street may contain all the utility lines such as water line, sewer line, and electric line besides the road pavement. The parameters CI, CS, and CM designate the unit costs of internal, subsidiary main, and the main street, respectively (Chapter-III, Thesis). The parameters CL, CSC, and SLC are the unit price of prepared land, unit cost of on-lot service connection lines, and the on-lot service-core location coefficients.

Accordingly, expressions for module parameters (designated by m_1, m_2, m_3, and m_4) and cost coefficients (designated by c_1 for the East-West streets, c_2 for the North-South streets, c_3 related to land price, and c_4 related to on-lot service connection) can be developed as shown in Table 11.1 above for different module types. A planner-designer can conceive any planning module as per her/his creativity, ingenuity, and requirement and develop module parameters and cost coefficients for such a planning module, and also develop mathematical models for quantitative analysis and optimization of such layout modules.

11.5.1.2 Unit Design Module Cost Coefficients for Dwelling Forming Part of the Dwelling-Layout System

As in case of layout unit design modules mentioned above, a planner-designer can conceive a variety of dwelling unit design modules for single-family dwelling-layout system, keeping in view the different contexts. One important aspect of such dwelling unit design modules for the purpose of optimization related to costs is the availability of wall sharing options, which can be easily represented by room-arrangement matrices such as 1×2, 2×2, 2×3, 2×4 and so on, as shown below:

In Fig. 11.1, two layout modules of single-family dwelling-layout system are shown in Fig. 1(a) and (b) which are adopted here. In Fig. 11.2, two modules of a dwelling-shell are shown, having same number of rooms along the depth of a lot (i.e. 2) but with one room along the width of the lot which can be designated as (1×2 matrix), and with two rooms along the width of the lot which can be designated as (2×2 matrix). In case dwelling-shell consists of only one room, its room-arrangementmatrix will be 1×1, and if it has 2 rooms its room-arrangement matrix will be 1×2. A variety of schematic room-arrangement matrices applicable for single-family dwelling-layout system are shown in Fig. 11.6. These dwelling modules could be developed as: detached housing, semi-detached housing and row-housing, in a single-family dwelling-layout system, having considerable effect on costs. Referring to above Fig. 11.6, one can easily note the value of wall sharing coefficients such as WCD (i.e. wall sharing coefficients along the lot depth) and WCW (i.e. wall sharing coefficients along the lot width) for each case as shown in Table 11.2 below. In Fig. 11.6, room-arrangement matrix for one room dwelling-shell which is 1×1 is not shown but its wall sharing coefficients for different cases are given in Table 11.2 below.

It may be noted that such wall sharing coefficients can be developed for any number of rooms. In fact, for multi-room multi-family dwelling development

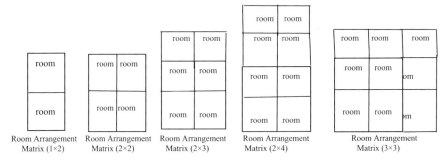

Room Arrangement Matrix (1×2)   Room Arrangement Matrix (2×2)   Room Arrangement Matrix (2×3)   Room Arrangement Matrix (2×4)   Room Arrangement Matrix (3×3)

Fig. 11.6 Schematic room-arrangement matrices in single-family dwelling-layout system

Table 11.2 Some wall sharing coefficients for different cases in single-family dwelling-layout system

Number of rooms	Detached housing		Semi-detached housing		Row-housing	
	WCD	WCW	WCD	WCW	WCD	WCW
One room (1 × 1 matrix)	2	2	1.5	2	1	2
Two rooms (1 × 2 matrix)	2	3	1.5	3	1	3
Four rooms (2 × 2 matrix)	3	3	2.5	3	2	3
Six rooms (2 × 3 matrix)	3	4	2.5	4	2	4
Eight rooms (2 × 4 matrix)	3	5	2.5	5	2	5
Nine rooms (3 × 3 matrix)	4	4	3.5	4	3	4

systems, the dwelling module coefficients can be developed incorporating all such wall sharing options in a combined manner as done in Chap. 9 for optimization of building and housing. It is also to be noted that an architect has full freedom to move the wall-segments horizontally or vertically in the dwelling design as per his/her requirement, without affecting the costs and *optimization model* provided the same room-arrangement matrix is maintained. Since the bathrooms and toilets are generally of fixed dimensions based on the anthropometric data and not subject to optimization, these are not considered in the above room-arrangement matrices, but can be added in the dwelling design as per requirements. But the location of such wet-core in the single-family dwelling-layout-system with individual lots affects the optimization of layout module component, and can be considered with reference to the lot depth by assigning suitable effective location coefficients such as 0.25 for front location, 0.5 for middle location, and 0.75 for rear location and so on, or calculating the effective location coefficient in a specific case.

11.5.1.3 Combined Unit Design Module Cost Coefficients for the Single-Family Dwelling-Layout System

Taking into account the unit design module coefficients for layout and dwelling, the combined cost coefficients for the single-family dwelling-layout system can be easily developed as follows using the GRID and LOOP layout as illustration.

GRID LAYOUT

$$c_1 = 2CI + CS, \quad c_2 = (CS + CM)1/2, \quad c_3 = CL,$$
$$c_4 = WCD^*(cw + cf) + SLC^*CSC, \quad c_5 = WCW^*(cw + cf).$$

LOOP LAYOUT

$$c_1 = 2CI + CS, \quad c_2 = (CS + CM)1/2, \quad c_3 = CL,$$
$$c_4 = WCD^*(cw + cf) + SLC^*CSC, \quad c_5 = WCW^*(cw + cf).$$

Here, the cost coefficients c_1 and c_2 define the cost per unit length of East-West streets and of North-South streets of layout module as explained above. Similarly, cost coefficient c_3 is related to land price, c_4 is related to the dwelling wall costs (including foundation) along the lot depth and the on-lot service connection, and c_5 is related to the dwelling wall costs (including foundation) along the lot width as explained above. Parameters CI, CS, CM, CL, CSC, SLC, WCD, and WCW are also defined above. The parameters *cf* and *cw* designate the unit costs of foundation and super-structure walls, respectively. It may be noted that the parameter *cw* includes joinery items such as doors and windows which are not shown separately in *various* **Built-Unit-Design Patterns and** *Schematic Diagrams of different Dwelling Unit Design Modules* shown in this chapter and in room-arrangement matrices given in Fig. 11.6 above. To simplify, it is assumed that the unit cost of joinery items and wall material is same, and accordingly, the value of 'CW' is derived adopting the procedure presented in Chap. 4, and incorporated in cost functions and *optimization models*. However, considering the desired ratio of joinery area and room area and the unit cost of wall and joinery separately, as presented in reference HUDCO, 1982, one can develop more refined cost function and optimization models. Interested readers can consider developing such cost function and optimization models. The above unit design module cost coefficients are for the least-cost-optimizing model for single-family dwelling-layout system. In case of the most-benefit-optimizing models, the above unit design module cost coefficients are to be divided by the AFC, i.e. affordable cost or investment per BU, i.e. Built-Unit.

To take into account, the effect of wall sharing coefficients (WCD and WCW) as well as the built-space coverage of individual lot, we can designate three factors, i.e. Lot Area Coverage = AC, Lot Depth Coverage = DC, Lot Width Coverage = WC. Thus, the value of c_4 and c_5 will then be modified as:

$$c_4 = \mathrm{DC} \times \mathrm{WCD} \times (cf + cw) + \mathrm{CSC} \times \mathrm{SC}, \quad c_5 = \mathrm{WC} \times \mathrm{WCW} \times (cf + cw)$$

It may be noted that $\mathrm{AC} = \mathrm{DC} \times \mathrm{WC}$.

Using the above formulations, the values of the unit design module cost coefficients for each case can be found and included in the **Cost Functions** in the respective optimization model as illustrated in examples below.

11.5.2 Multi-family Dwelling-Layout Systems

Typical schematic dwelling unit design modules for multi-family dwelling-layout system shown in Fig. 11.4 above include Module–D, which is same as the Alternative-I in Fig. 9.3(a) schematic diagrams of dwelling unit design Module–D, shown in Chap. 9.

11.5.2.1 Deriving Dwelling Unit Design Module Coefficients: Multi-family Dwelling-Layout System

The expressions for the dwelling unit design module coefficients: cx_1, cx_2, and cba, for different 'dwelling design modules' are given in Chap. 9. If these dwelling design modules form part of the multi-family dwelling-layout system, the same dwelling unit design module coefficients can be used in the multi-family dwelling-layout system design optimization model. Thus, referring to the Fig. 9.3(a) in Chap. 9, the expressions for 'cx_1', 'cx_2', and 'cba' are reproduced below for dwelling unit design module-D (which is applicable in Context-A—low and medium-income countries):

$$cx_1 = 4 + 6.3 \times w1 + 3 \times k1 \quad \text{or}$$
$$cx1 = 4 + 6.3 \times w1 + 2.5 \times k1 \text{ (if one } k1 \text{ wall is shared)}$$
$$cx_2 = 5 + 11 \times w2 - 3 \times s2 - 4 \times k2 - 4 \times ew$$
$$cba = 1 - (s2 + 4 \times ew)(0.5 - k1 - 2.1 \times w1) - 4(k2 - w2)(0.5 - k1) - 0.5 \times s2.$$

In the above expressions, the proportion coefficients $k1$, $k2$, $w1$, $w2$, $s2$, and ew are independent design variables, and a designer can assign a value to each of them based on his/her experience, desire, and feasibility.

Similarly, using the schematic unit design module: E, the expressions for the dwelling unit design module coefficients: cx_1, cx_2, and cba, for the multi-family dwelling unit design Module: E (shown in Fig. 9.5 in Chap. 9 and also using the proportion *coefficients* indicated in the diagram (see Sect. 9.6.2.2 of Chap. 9) could be reproduced as follows (which is applicable in Context-B—High-income countries):

$$k1 = (1 - 2(ew + b1 + t1))/2, \qquad k2 = (k - 2 \times t2)/2;$$
$$w2 = 0.5 \times (k - ow), \qquad s2 = (1 - 2 \times k);$$
$$cba = 1 - 2 \times ow \times (1 - k1) - s2 \times 1 - ew \times w2 \times 2 + s1 \times s2;$$
$$cx_1 = 4 + 8 \times k1 + 4 \times b1 + 4 \times t1;$$
$$cx_2 = 2 + 8 \times (k - ow) + 2 \times (b2 + t2).$$

In the above expressions, the proportion coefficients k, $b1$, $b2$, $t1$, $t2$, ow, ew, and $s1$ are independent design variables, and a designer can assign a value to each of them based on his/her experience, desire, and feasibility. In fact, the suitability and feasibility of such values chosen can be verified in each cycle of iteration as per the optimum housing and urban development design procedure shown in Fig. 10.7 in Chap. 10. Representative input and output results of optimization given by the **C++** **program: clc1.cpp** and **cmb1.cpp** using the above expressions for the dwelling design module coefficients cx_1, cx_2, and cba for different dwelling design modules including the Module: E are shown and discussed in Sect. 9.11.3 giving some applications example results.

11.5.2.2   Cost Coefficients and Cost Functions in Multi-family Dwelling-Layout Systems Optimization

Typical schematic layout planning modules and dwelling unit design modules for multi-family dwelling-layout system are shown above in Fig. 11.3 and Fig. 11.4, respectively. The layout module cost coefficients can be derived, using the typical schematic layout planning modules and Table 10.7 in Chap. 10, as discussed above in case of single-family dwelling-layout systems. Similarly, using the above dwelling unit design module coefficients: cx_1, cx_2, and cba, the dwelling unit design module cost coefficients could be derived as follows:

c_1 = a cost coefficient defining the cost per unit length of walls and foundation for the multi-family dwelling unit block in the x_1 direction = $cx_1((nf(1 + (nf - 1)$ $fic))cw + (1 + (nf - 1)fic)cf)$,

c_2 = a cost coefficient defining the cost per unit length of walls and foundation for the multi-family dwelling unit block in the x_1 direction = $cx_2 \times c_1/cx_1$,

c_3 = a cost coefficient defining the cost per unit area of roof/floors of the multi-family dwelling unit block = $(cba \times cr \times cif \times nf + cba \times cfl \times cif \times 1)/nf$,

$$cif = 1 + (nf - 1)\,fic.$$

Here,
nf = number of floors in the multi-family dwelling unit block.
cf = cost per unit length of foundation of multi-family dwelling unit block.
cw = cost per unit length of walls of multi-family dwelling unit block.

cr = cost per unit area of roof/floors of the multi-family dwelling unit block.

fic = a factor to allow for the increase in unit costs due to building-rise. Suitable value can also be assigned to this factor to incorporate the cost of elevators in a building depending on its rise and size.

cif = a cost increase factor covering total number of floors.

To illustrate the process of developing **Cost Coefficients**, four building components are considered, namely: foundation, wall, roof, and flooring. The unit costs of these building components are designated as 'cf' in MU (Monetary Units) per meter, 'cw' in MU per meter, 'cr' in MU per square metre, and 'cfl' in MU per square metre; for the components: foundation, wall, roof, and floor, respectively. A factor designated as 'fic' is included to allow for the increase in unit costs due to building-rise. Suitable value can also be assigned to this factor to incorporate the cost of elevators in a building depending on its rise and size. The number of floors in a building is designated as 'nf'. Thus, using the above designated terms and the *dwelling unit design module coefficients*: cx_1, cx_2, and cba developed earlier for various dwelling unit design modules, the expressions for the **Cost Coefficients** representing the total unit costs of walls in the direction of x_1 and x_2, which are chosen as: c_1 and c_2, respectively, can be developed as follows:

$$c_1 = cx_1 \times ((nf \times (1 + (nf - 1) \times fic)) \times cw + (1 + (nf - 1) \times fic) \times cf); \quad (11.1)$$

$$c_2 = cx_2 \times c_1/cx_1. \quad (11.2)$$

Similarly, using the *dwelling unit design module coefficients*: 'cba' developed earlier for various dwelling unit design modules, the expression for the cost coefficients representing the total unit cost of floors and roof designated as: c_3, can be developed as follows:

$$c_3 = (cba \times cr \times cif \times nf + cba \times cfl \times cif \times 1)/nf; \quad (11.3)$$

Thus, the total cost of walls (including foundation) in the direction of x_1 will be = c_1x_1.

Similarly, the total cost of walls (including foundation) in the direction of x_2 will be = c_2x_2.

Again, using the above coefficients the total cost of roof and floors will be = $c_3x_1x_2$.

Thus, three terms of the **Cost Function** in multi-family dwelling-layout systems will incorporate c_1x_1, c_2x_2, and $c_3\ x_1x_2$ representing the total cost of walls in the direction of x_1, the total cost of walls in the direction of x_2, and the total cost of floors and roof, respectively.

In multi-family dwelling-layout system, layout unit design modules are generally simple rectilinear shape in which building blocks are located. Accordingly, the expression for and c_5 representing the cost of road pavement/utility lines in layout module in North-South direction and in the East-West direction can be developed somewhat similar to given in Table 11.1 above.

The relationship between Built-Unit-Space (BUS), Building Rise (BR), Semi-Public Open space (SPO), and Floor Area Ratio (FAR) is shown in Chap. 6. Using this relationship, the allocated land area per Built-Unit (BU) can be calculated, wherefrom the expression for the cost coefficients: c_6 representing the cost of developed land per BU can be developed.

Thus, the complete **Cost Function** *of* the multi-family dwelling-layout system incorporating all the terms mentioned above, giving the total cost per built-unit can be developed as done in the LCO model and MBO models for optimal design of single-family and multi-family dwelling-layout systems as presented below.

11.6 Single-Family Dwelling-Layout System Design: An Example Least-Cost Optimizing Model (LCO)

Here, a representative **Least-Cost Optimizing Model (LCO)** for single-family dwelling-layout system design with dwelling (or lot for a single family) density, open space standard, and circulation interval ratio standard as some of the design objectives to be achieved along with the least-cost. Therefore, these are put as constraints in the optimization model so that these objectives are attained in the optimal solution. A planner or designer can select any other set of design objectives (provided these are feasible) as per requirement and put these as constraint in the model and develop it accordingly.

$$\textit{Minimize the Cost per Built-Unit} = y(x) = c_1 x_1 x_4 x_5^{-1} + c_2 x_1 x_3 x_5^{-1} + \\ c_3 x_1 x_3 x_4 x_5^{-1} + c_4 x_2 + c_5 x_7 \tag{11.4}$$

Subject to:

1. $a_1 x_2^{-1} x_7^{-1}$ ≤ 1 (11.5)
   (Minimum acceptable lot area *constraint*)
2. $a_2 x_4^{-1} x_5 + a_3 x_4^{-1} x_6$ ≤ 1 (11.6)
   (*Constraint* for geometric compatibility—in E-W direction of layout module)
3. $a_5 x_2 x_3^{-1} + a_6 x_6 x_3^{-1}$ ≤ 1 (11.7)
   (*Constraint* for geometric compatibility—in N-S direction of layout module)
4. $x_7 x_1^{-1} + x_8 x_1^{-1}$ ≤ 1 (11.8)
   (*Constraint* for geometric compatibility-lot width and allocated width of open-space in module)
5. $a_4 x_1^{-1} x_3^{-1} x_4^{-1} x_5$ ≤ 1 (11.9)
   (*Constraint* for designed lot density in the layout module)
6. $a_7 x_4 x_6^{-1}$ ≤ 1 (11.10)
   (*Constraint* for designed circulation interval ratio in the layout module)

7. $a_8 x_2^{-1} x_8^{-1}$ ≤ 1 (11.11)

   (*Constraint* for designed minimum acceptable open-space per built-unit)

8. $x_1, x_2, x_3, x_4, x_5, x_6, x_7, x_8$ > 0 (11.12)

   (Non-negativity *constraint*)

Here, the model constants are as follows:

c_1 = Model cost coefficient related to cost of road pavement/utility lines in layout in East-West direction,

c_2 = Model cost coefficient related to cost of road pavement/utility lines in layout in North-South direction,

c_3 = Model cost coefficient related to cost of raw land including site grading per unit area,

c_4 = Model cost coefficient related to cost of service connection, wall/foundation of dwelling along x_2 direction,

c_5 = Model cost coefficient related to cost of wall/foundation of dwelling along x_1 direction,

a_1 = a module parameter related to the desired net lot area per dwelling in layout module,

a_2 = a module parameter depending on the type of layout module chosen,

a_3 = a module parameter depending on the type of layout module chosen,

a_4 = a design parameter related to the designed lot or population density per unit area,

a_5 = a module parameter depending on the type of layout module chosen,

a_6 = a module parameter depending on the type of layout module chosen,

a_7 = design parameter regarding desired circulation interval ratio between main and subsidiary main street-widths,

a_8 = a design parameter related to the designed semi-public open space standard in m^2 per lot.

It may be noted that the model constants, related to the cost parameters (giving unit cost coefficients), dimensionless physical design patterns (giving built-unit design coefficients) and the design standards, mostly dependent on the input design conditions, decided by an architect, engineer, urban planner/urban manager as per requirements, and on the constraints beyond control (say market price of land) as may be applicable. It is imperative that AEC professionals decide values of such design parameters to ensure *resource-efficiency* and *social equity* while maintaining environmental standards.

Here, model variables: x_1, x_2, x_3, x_4, x_5, x_6, x_7, and x_8 are primal variables related to lot width, lot depth, layout module width, layout module length, variable related to number of lots in layout module, width of subsidiary main street, allocated lot width for open space, and gross width of lot, respectively.

11.6.1   Model Solution

The above optimal design model is a typical geometric programming problem. Thus, *it* may be seen that the objective function in the above problem is a summation of product functions and the constraints also contain the product functions. The geometric programming technique is an efficient tool in solving such problems as explained in Chap. 7. In this technique, emphasis is placed on the relative magnitudes of the terms of the objective function rather than on the variables, and instead of finding the optimal values of the design variables directly, the optimal way to distribute the total cost, among the various terms or components in the objective function, is found first. This indicates how resources should be apportioned among the components contributing to the design objective, which is very useful in engineering and architectural design. Even in the presence of great degree of difficulty, the geometric programming formulation uncovers important invariant properties of the design problem which can be used as design rules for optimality as discussed above (Chakrabarty, 1986a, 1986b, 1996).

In this technique, if the primal geometric programming problem has N variables and T terms; in the dual geometric programming problem, there will be N orthogonality conditions-one for each variable x_n, one normality condition and T weights, one for each term. Thus, there will be $(N + 1)$ independent linear constraint equations as against T independent dual variables which are unknown. In this case $T - (N + 1)$ is the degree of difficulty in the problem.

In the above problem, there are 15 terms and 8 variables. Therefore, the degree of difficulty in the problem is 6. using the geometric programming technique, and taking the terminology used in literature (Beightler et al., 1979) on geometric programming as explained in Sect. 7.11.5 of Chap. 7, the dual objective function and the associated normality and orthogonality constraints for the above primal problem (with simplified nomenclature of coefficients) are given by:

$$d(w) = \left(\frac{c_1}{w_{o1}}\right)^{w_{o1}} \left(\frac{c_2}{w_{o2}}\right)^{w_{o2}} \left(\frac{c_3}{w_{o3}}\right)^{w_{o3}} \left(\frac{c_4}{w_{o4}}\right)^{w_{o4}} \left(\frac{c_5}{w_{o5}}\right)^{w_{o5}}$$

$$\times (a_1)^{w_{11}} \left(\frac{a_2(W_{21} + W_{22})}{w_{21}}\right)^{w_{21}} \left(\frac{a_3(W_{21} + W_{22})}{w_{22}}\right)^{w_{22}} \left(\frac{a_5(W_{31} + W_{32})}{w_{31}}\right)^{w_{31}}$$

$$\times \left(\frac{a_6(W_{31} + W_{32})}{w_{32}}\right)^{w_{32}} \left(\frac{(W_{41} + W_{42})}{w_{41}}\right)^{w_{41}} \left(\frac{(W_{41} + W_{42})}{w_{42}}\right)^{w_{42}} (a_4)^{w_{51}}$$

$$\times (a_7)^{w_{61}} (a_8)^{w_{71}}$$

$$(11.13)$$

Subject to:

$$w_{01} + w_{02} + w_{03} + w_{04} + w_{05} = 1 \quad \text{(normality condition)} \qquad (11.14)$$

$$w_{01} + w_{02} + w_{03} - w_{41} - w_{42} - w_{51} = 0 \quad \text{(orthogonality condition for } x_1)$$

$$(11.15)$$

$$w_{04} - w_{11} + w_{31} - w_{71} = 0 \quad \text{(orthogonality condition for } x_2) \quad (11.16)$$

$$w_{02} + w_{03} - w_{31} - w_{32} - w_{51} = 0 \quad \text{(orthogonality condition for } x_3) \quad (11.17)$$

$$w_{01} + w_{03} - w_{21} - w_{22} - w_{51} + w_{61} = 0 \quad \text{(orthogonality condition for } x_4)$$

$$(11.18)$$

$$-w_{01} - w_{02} - w_{03} + w_{21} + w_{51} = 0 \quad \text{(orthogonality condition for } x_5) \quad (11.19)$$

$$w_{22} + w_{32} - w_{61} = 0 \quad \text{(orthogonality condition for } x_6) \quad (11.20)$$

$$w_{05} - w_{11} + w_{41} = 0 \quad \text{(orthogonality condition for } x_7) \quad (11.21)$$

$$w_{42} - w_{71} = 0 \quad \text{(orthogonality condition for } x_8) \quad (11.22)$$

Here, c_1, c_2, c_3, c_4, c_5, a_1, a_2, a_3, a_4, a_5, a_6, a_7, and a_8 are the model constants, related to the cost parameters (giving unit cost coefficients), *dimensionless* physical design patterns (giving built-unit design coefficients) and the design standards, mostly dependent on the input design conditions, decided by an architect, engineer, urban planner/urban manager as per requirements, and on the *constraints* beyond control (say market price of land) as may be applicable.

Here, w_{01}, w_{02}, w_{03}, w_{04}, and w_{05} represent the five cost weights related to the five terms in the objective function $y(x)$, i.e. expression (11.4) above. Similarly, w_{11}, w_{21}, w_{22}, w_{31}, w_{32}, w_{41}, w_{42}, w_{51}, w_{61}, and w_{71} represent the dual variables related to the single term in constraint-1, i.e. expression (11.5), to the first and second term in constraint-2, i.e. expression (11.6), to the first and second term in constraint-3, i.e. expression (11.7), to the first and second term in constraint-4, i.e. expression (11.8), to the single term in constraint-5, i.e. expression (11.9), and to the single term in constraint-6, i.e. expression (11.10) above, and to the single term in constraint-7, i.e. expression (11.11) above, respectively.

The degree of difficulty in the problem is $= T - (N + 1) = 15 - (8 + 1) = 6$, which is comparatively small, therefore, a set of $(N + 1)$ dual weights could be expressed in terms of the remaining $T - (N + 1)$ dual weights, and substituted into the dual objective function, which is then maximized by setting the first derivative to zero with respect to each of the dual weights coming into the substituted dual objective function, which is being maximized. Thus, a sufficient number of equations are available for solving the problem. Therefore, any nine of the dual variables can be expressed in terms of the remaining six dual variables as shown below, choosing w_{02}, w_{04}, w_{05}, w_{11}, w_{22}, and w_{51} as the six dual variables:

$$w_{01} = w_{04} - w_{02} - w_{05}; w_{03} = 1 - 2w_{04};$$

$$w_{21} = 1 - w_{04} - w_{05} - w_{51}; w_{31} = 1 - 2w_{04} - w_{51}; w_{32} = w_{02}; w_{41} = w_{11} - w_{05};$$

$$w_{42} = 1 - w_{04} - w_{11} - w_{51}; w_{61} = w_{02} + w_{22}; w_{71} = 1 - w_{04} - w_{11} - w_{51};$$

$$w_{21} + w_{22} = 1 - w_{04} - w_{05} - w_{51} + w_{22}; w_{31} + w_{32} = 1 - 2w_{04} - w_{51} + w_{02};$$

$$w_{41} + w_{42} = 1 - w_{04} - w_{05} - w_{51};$$

$$(11.23)$$

If we substitute the expressions in equations set (11.23) into the dual objective function [Eq. 11.13)] and maximize this substituted dual objective function over the six variables, i.e. w_{02}, w_{04}, w_{05}, w_{11}, w_{22}, and w_{51}, we will get six equations for the six variables to give the unique solution. The process is simplified by maximizing the natural logarithm of the dual function; that is, by substituting:

$$\ln d(w_{02}, w_{04}, w_{05}, w_{11}, w_{22}, w_{51}) = z(w_{02}, w_{04}, w_{05}, w_{11}, w_{22}, w_{51}).$$

Thus, by above substitution and taking logarithm on both sides of Eq. (11.13), we get the following equations:

$$
\begin{aligned}
z(w) = &(w_{04} - w_{02} - w_{05})\,(\ln c_1 - \ln(w_{04} - w_{02} - w_{05})) + w_{02}(\ln c_2 - \ln w_{02}) \\
&+ (1 - 2w_{04})(\ln c_3 - \ln(1 - 2w_{04})) + w_{04}(\ln c_4 - \ln w_{04}) \\
&+ w_{05}(\ln c_5 - \ln w_{05}) + w_{11}\ln a_1 + (1 - w_{04} - w_{05} - w_{51}) \\
&\quad(\ln a_2 + \ln(1 - w_{04} - w_{05-51} + w_{22}) - \ln(1 - w_{04} - w_{05} - w_{51})) \\
&+ w_{22}(\ln a_3 + \ln(1 - w_{04} - w_{05} - w_{51} + w_{22}) - \ln w_{22}) + (1 - 2w_{04} - w_{51}) \\
&\quad(\ln a_5 + \ln(1 - 2w_{04} - w_{51} + w_{02}) - \ln(1 - 2w_{04} - w_{51})) \\
&+ w_{02}(\ln a_6 + \ln(1 - 2w_{04} - w_{51} + w_{02}) - \ln w_{02})) \\
&+ (w_{11} - w_{05})(\ln(1 - w_{04} - w_{05} - w_{51}) - \ln(w11 - w_{05})) \\
&+ (1 - w_{04} - w_{11} - w_{51})(\ln(1 - w_{04} - w_{05} - w_{51}) \\
&- \ln(1 - w_{04} - w_{11} - w_{51})) + w_{51}\ln a_4 + (w_{22} + w_{02})\ln a_7 \\
&+ (1 - w_{04} - w_{11} - w_{51})\ln a_8.
\end{aligned}
$$

$$(11.24)$$

Using Eq. (11.24) and setting the first derivative $dz/dw_{22} = 0$, we get:

$$\ln a_3 + \ln a_7 + \ln(1 - w_{04} - w_{05} - w_{51} + w_{22}) - \ln w_{22} = 0$$

$$\text{or,} \quad a_3 a_7(1 - w_{04} - w_{05} - w_{51} + w_{22}) - w_{22} = 0 \qquad (11.25)$$

$$\text{or,} \quad w_{22} = (a_3 a_7/(1 - a_3 a_7))(1 - w_{04} - w_{05} - w_{51}).$$

Again, by setting the first derivative $dz/dw_{11} = 0$, we get:

$$\ln a_1 + \ln (1 - w_{04} - w_{05} - w_{51}) - \ln (w_{11} - w_{05}) - \ln a_8 = 0$$
$$\text{or,} \quad w_{11} = (a_1/(a_1 + a_8))(1 - w_{04} - w_{51}) + (a_8/(a_1 + a_8))w_{05}.$$

(11.26)

Similarly, setting the first derivative $dz/dw_{51} = 0$, we get:

$$- \ln a_2 - \ln (1 - w_{04} - w_{05-51} + w_{22}) - \ln a_5 - \ln (1 - 2w_{04} - w_{51} + w_{02})$$
$$+ \ln (1 - 2w_{04} - w_{51}) + \ln (1 - w_{04} - w_{11} - w_{51})$$
$$+ \ln a_4 - \ln a_8 = 0$$

Or, $(a_4/(a_2 a_5 a_8))(1 - 2w_{04} - w_{51})(1 - w_{04} - w_{11} - w_{51})$
$$- (1 - w_{04} - w_{05-51} + w_{22})(1 - 2w_{04} - w_{51} + w_{02})$$
$$= 0$$

(11.27)

From Eqs. (11.25), (11.26), and (11.27), we can get the expression for 'w_{02}' as follows:

$$w_{02} = \left[\frac{(1 - a_3 a_7)a_4 - a_2 a_5(a_1 + a_8)}{a_2 a_5(a_1 + a_8)} \right] (1 - 2w_{04} - w_{51})$$

(11.28)

To simplify, we can make the following substitutions:

$w_{04} = x, w_{05} = y, w_{51} = z,$ and
$$k1 = (a_3 a_7/(1 - a_3 a_7)), \quad k2 = (a_1 + a_8), \quad k3 = \frac{c_2 a_6 a_7}{c_1},$$
$$k4 = \left[\frac{(1 - a_3 a_7)a_4 - a_2 a_5(a_1 + a_8)}{a_2 a_5(a_1 + a_8)} \right], \quad k5 = \frac{c_1 c_4}{c_3^2 a_2 a_5^2 a_8}, \quad k6 = \frac{c_5}{c_1 a_2},$$
$$k = k3(1 + k4), \quad k7 = \frac{(1 - a_3 a_7)a_5 a_8}{((1 + k4)(1 + k4)k2)}, \quad k8 = k4(k + k4)/k, \quad (11.29)$$
$$k9 = k6(1 - a_3 a_7)a_1/k2, \quad k10 = [k6a_1(1 - a_3 a_7)/(a_1 + a_8)],$$
$$k11 = k10(k8 - k4) + k8, \quad k12 = (1 - k8/k11),$$
$$k13 = k5[a_8/(a_1 + a_8)](1 - a_3 a_7)/(1 + k4)^2,$$
$$k14 = (1 - k12 - k4/k11).$$

Hence,

$$w_{02} = k4(1 - 2w_{04} - w_{51})$$

Again, setting the first derivative $dz/dw_{02} = 0$, and using above substitutions, we finally get:

$$k3(w_{04} - w_{02} - w_{05})(1 - 2w_{04} - w_{51} + w_{02}) - w_{02}^2 = 0$$

or $k3(w_{04} - w_{05}-k4(1 - 2w_{04} - w_{51}))(1 - 2w_{04} - w_{51} + k4(1 - 2w_{04} - w_{51})$

$-k4^2(1 - 2w_{04} - w_{51})^2 = 0$

or $k3(w_{04} - w_{05}-k4(1 - 2w_{04} - w_{51}))(1 + k4)(1 - 2w_{04} - w_{51})$

$-k4^2(1 - 2w_{04} - w_{51})^2 = 0$

or $k3(1 + k4)(x - y - k4(1 - 2x - z)) - k4^2(1-2x - z) = 0$

or, $y = x - k8(1 - 2x - z).$

$$(11.30)$$

Similarly, setting the first derivative $dz/dw_{05} = 0$, and using above substitutions, we get:

$$K6(w_{04} - w_{02} - w_{05})(w_{11} - w_{05}) - (1 - w_{04} - w_{05} - w_{51} + w_{22})w_{05} = 0$$

Or $k6(w_{04} - w_{05}-k4(1 - 2w_{04} - w_{51}))(1 - w_{04} - w_{05} - w_{51})a_1/(a_1 + a_8)$

$- (1 - w_{04} - w_{05} - w_{51})w_{05}/(1 - a_3a_7) = 0$

Or $[k6a_1(1 - a_3a_7)/(a_1 + a_8)](w_{04} - w_{05}-k4(1 - 2w_{04} - w_{51})) - w_{05} = 0,$

Or $k10(x - y - k4(1 - 2x - z)) - y = 0$

or $k10(x - x + k8(1 - 2x - z)-k4(1 - 2x - z)) + k8(1 - 2x - z) - x = 0$

Or $k10(k8 - k4)(1 - 2x - z)-x + k8(1 - 2x - z) = 0$

or, $1 - 2x - z = x/(k10(k8 - k4)) + k8) = x/k11$

or, $z = 1 - 2x - x/k11,$ and $y = x - k8x/k11 = (k11 - k8)/k11x.$

$$(11.31)$$

Again, setting the first derivative $dz/dw_{04} = 0$, and using above substitutions, we get:

$$k5((1 - 2w_{04})^2(1 - w_{04} - w11 - w_{51})(1 - 2w_{04} - w_{51})^2 - (w_{04} - w_{02} - w_{05})$$

$(1 - w_{04} - w_{05-51} + w_{22})(1 - 2w_{04} - w_{51} + w_{02})^2w_{04} = 0$

or,

$k5(1 - 2x)^2[a_8/(a_1 + a_8)] - (x - y - k4(1 - 2x - z))[1/(1 - a_3a_7)](1 + k4)^2x$

$= 0$

Using the earlier substitutions, the above equation finally reduces to a single variable nonlinear equation as:

$$k13(1 - 2x)^2 - k14x^2 = 0 \qquad\qquad (11.32)$$

Equation (11.32) can be easily solved using N-R method and using the value of 'x', i.e. w_{04} so determined all other values of dual variables: $w_{01}, w_{02}, w_{03}, w_{05}, w_{11},$

w_{21}, w_{22}, w_{31}, w_{32}, w_{41}, w_{42}, w_{51}, w_{61}, and w_{71} can be determined, and thus, completely solving the model.

As earlier mentioned, using *optimization concepts*, one can discern general *decision rules* appropriate to problems, and develop an *insight* regarding proper form of an optimal solution even when a problem is not completely formulated mathematically. This is illustrated below, noting that the cost function in the above expression (11.1) literally means that the total cost per built-unit $=$ (a) cost contributed by streets and utility lines in the East-West direction (i.e. in x_4 direction in Fig. 11.1) $+$ (b) cost contributed by streets and utility lines in the North-South direction (i.e. in x_3 direction in Fig. 11.1) $+$ (c) cost contributed by raw land $+$ (d) cost contributed by all walls (including foundation and finishes) in x_2 direction including service connection (see Fig. 11.2) $+$ (e) cost contributed by all walls (including foundation and finishes) in x_1 direction (see Fig. 11.2). Thus, the cost weights w_{01}, w_{02}, w_{03}, w_{04} and w_{05} represent the optimum proportion of total cost per built-unit, distributed between the cost contributed by streets and utility lines in the East-West direction, cost contributed by streets and utility lines in the North-South direction, cost contributed by raw land, cost contributed by all walls in x_2 direction including service connection, and cost contributed by all walls in x_1 direction, respectively.

Thus, by a cursory look at the equation set (11.23), one can discern the following *decision* or *design rules* if one has to operate *optimally*:

1. no term of the cost function can contribute more than 100% of the total cost nor less than 0% at the *optimality*,
2. cost contributed by streets and utility lines in the East-West direction (w_{01}) should be equal to the cost contributed by all walls in the x_2 direction (w_{04}) minus the cost contributed by the streets and utility lines in the North-South direction (w_{02}) minus cost contributed by all walls in x_1 direction (w_{05}), at the *optimality*,
3. cost contributed by raw land (w_{03}) must be equal to the total cost per built-unit minus twice the cost contributed by all walls in the x_2 direction ($2w_{04}$), at the *optimality*,
4. the literal meaning of the cost function explained above also gives an *insight* to the designer regarding general *decision rules* appropriate to problems, and to develop a proper form of an optimal solution even when a problem is not completely formulated mathematically.

11.7   Single-Family Dwelling-Layout System Design: An Example Most-Benefit Optimizing Model (MBO)

As stated earlier in many real world problems, there may be terms with negative coefficients or there may be mixed constraints requiring the application of signomial geometric programming (as outlined in Chap. 8) to achieve optimal design. As, for example, in case of *Design Optimization* of **Dwelling-Layout System**, the AEC professionals may have to achieve the optimal design within the constraint of investment or cost which intended beneficiaries and stakeholders can afford, and at the same time maximize the *dwelling space* which is of high priority particularly for low-income households. In this context, optimal design model for affordable dwelling-layout-system is developed as follows.

The *dwelling space* can be expressed as $= x_1 x_2$, where these variables x_1 and x_2 represent the width and depth, respectively, of a residential lot. Since, the objective is to maximize the *dwelling space*, the objective function can be written as:

$$\text{Maximize } y_o(x) = x_1 x_2$$

This objective function can be expressed in standard geometric form as:

$$\text{Minimize } y_o(x) = -x_1 x_2$$

Since the objective function has negative coefficient, it is a signomial geometric programming problem and the entire design optimization model can be developed as follows:

$$\text{Minimize } y_o(x) = \sigma_0 x_1 x_2$$

Subject to:

$$\left.\begin{array}{c} \sigma_9 c_9 x_2 x_3^{-1} + \sigma_{11} c_{11} x_6 x_3^{-1} \le 1 \\ \sigma_{21} c_{21} x_5 x_4^{-1} + \sigma_{22} c_{22} x_6 x_4^{-1} \le 1 \\ \sigma_{31} x_1 x_8^{-1} + \sigma_{32} x_7 x_8^{-1} \le 1 \end{array}\right\} \text{Geometric Compatibility Constraints}$$

$$\sigma_{41} c_{41} x_3 x_4 x_5^{-1} x_8 \le 1 \qquad\qquad\qquad \text{Density Constraint}$$

$$\sigma_{51} c_{51} x_2^{-1} x_7^{-1} \le 1 \qquad\qquad\qquad\qquad \text{Open Space Constraint}$$

$$\qquad\qquad\qquad\qquad\qquad\qquad\qquad \text{Circulation Interval Ratio}$$

$$\sigma_{61} c_{61} x_4 x_6^{-1} \le 1 \qquad\qquad\qquad\qquad\quad \text{Constraint}$$

$$\sigma_{71} c_{71} x_4 x_5^{-1} x_8 + \sigma_{72} c_{72} x_3 x_5^{-1} x_8 + \sigma_{73} c_{73} x_3 x_4 x_5^{-1} x_8$$

$$\qquad\qquad\qquad\qquad\qquad\qquad\qquad\qquad \text{Affordable Cost Constraint}$$

$$+ \sigma_{74} c_{74} x_2 + \sigma_{75} c_{75} x_1 \le 1$$

$$x_1, x_2, x_3, x_4, x_5, x_6, x_7, x_8 > 0 \qquad\qquad \text{Non-negativity Constraint}$$

Here,

x_1, x_2, x_3, x_4, x_5, x_6, x_7, and x_8 are primal variables related to lot width, lot depth, layout module width, layout module length, variable related to number of lots in layout module, width of subsidiary main street, allocated lot width for open space, and gross width of lot, respectively. Coefficients c_9, c_{11}, and so on are model constants related to various design parameters regarding number of rows of lots, width of streets and their hierarchy, dwelling or population density, unit costs of streets, unit cost of land, unit costs of building walls, and service connection lines.

11.7.1   Model Solution

Simplifying the notations and other simplifications outlined in Chap. 7, the primal objective function and constraints for this problem can be expressed as follows:

$$\text{Maximize } y(x) = x_1 x_2 \tag{11.33}$$

Subject to:

$$
\left.
\begin{array}{l}
a_1 x_2 x_3^{-1} + a_2 x_6 x_3^{-1} \leq 1 \\
a_3 x_5 x_3^{-1} + a_4 x_6 x_4^{-1} \leq 1 \\
x_1 x_8^{-1} + x_7 x_8^{-1} \leq 1
\end{array}
\right\} \text{Geometric Compatibility Constraints} \tag{11.34}
$$

$$a_5 x_3 x_4 x_5^{-1} x_8 \leq 1 \qquad\qquad\qquad \text{Density Constraint}$$

$$a_6 x_2^{-1} x_7^{-1} \leq 1 \qquad\qquad\qquad\quad \text{Open Space Constraint}$$

$$a_7 x_4 x_6^{-1} \leq 1 \qquad\qquad\qquad\quad\ \text{Circulation Interval Ratio Constraint}$$

$$a_8 x_4 x_5^{-1} x_8 + a_9 x_3 x_5^{-1} x_8 + a_{10} x_3 x_4 x_5^{-1} x_8$$

$$\qquad\qquad\qquad\qquad\qquad\qquad\qquad \text{Affordable Cost Constraint}$$

$$+ a_{11} x_2 + a_{12} x_1 \leq 1$$

$$x_1, x_2, x_3, x_4, x_5, x_6, x_7, x_8 > 0 \qquad\quad \text{Non-negativity Constraint}$$

Here, the model constants are as follows:

a_1 = a module parameter related to the number rows of lots in layout module,

a_2 = a module parameter related to the width of streets and their hierarchy in layout module,

a_3 = a module parameter related to the number rows of lots in layout module,

a_4 = a module parameter related to the width of streets and their hierarchy in layout module,

a_5 = a design parameter related to the designed dwelling or population density per unit area,

a_6 = a design parameter related to the designed semi-public open space standard in m^2 per dwelling,

a_7 = a design parameter related to the width of subsidiary main street and the designed interval of main streets,

a_8 = a cost ratio related to the cost per unit length of the E-W streets in layout module,

a_9 = a cost ratio related to the cost per unit length of the N-S streets in layout module,

a_{10} = a cost ratio related to the cost per unit area of prepared land in the layout module,

a_{11} = a cost ratio related to the cost per unit length of the walls and service connection lines across the lot depth,

a_{12} = a cost ratio related to the cost per unit length of the walls along the width of the lot.

The variables in the model are as follows:

x_1 = width of the lot,
x_2 = depth of the lot,
x_3 = width of the layout module in the N-S direction between the centre lines of two subsidiary main streets,
x_4 = length of the layout module in the E-W direction between the centre lines of two main streets,
x_5 = a layout module parameter related to the number of lots in the module,
x_6 = width of the subsidiary main street,
x_7 = allocated width of the lot, to allow for the reservation of semi-public open-to-sky space per lot,
x_8 = gross width of the lot to allow for the reservation of semi-public open-to-sky space per lot.

As earlier stated, the objective function can also be expressed in standard geometric form as:

$$\text{Minimize } y(x) = -x_1 x_2 \qquad (11.35)$$

Thus, this optimization model is a typical signomial geometric programming problem as outlined in Chap. 7, and can be solved accordingly using the procedure suggested in the above Chap. 7. To solve the problem, let us guess the signum function $\sigma_0 = -1.0$ for $y(x)$ the negative value of $y(x)$, since the original objective function is to maximize. Similarly, the value of the signum function σ_{01} is guessed as $= -1.0$. The values of other signum functions in the problem are guessed as: $\sigma_m = +1.0$ and $\sigma_{mt} = +1.0$.

Accordingly, the dual objective function and the associated normality and orthogonality constraints for the above primal problem can be expressed as:

$$\text{Optimize} : d(w) = (-1)\left(\left(\frac{a_1(w_{12} + w_{11})}{w_{11}}\right)^{w_{11}} \left(\frac{a_2(w_{12} + w_{11})}{w_{12}}\right)^{w_{12}} \right.$$

$$\times \left(\frac{a_3(w_{21} + w_{22})}{w_{21}}\right)^{w_{21}} \left(\frac{a_4(w_{21} + w_{22})}{w_{22}}\right)^{w_{22}} \left(\frac{(w_{31} + w_{32})}{w_{31}}\right)^{w_{31}}$$

$$\times \left(\frac{(w_{31} + w_{32})}{w_{32}}\right)^{w_{32}} (a_5)^{w_{41}} (a_6)^{w_{51}} (a_7)^{w_{61}} \left(\frac{a_8\, W_7}{w_{71}}\right)^{w_{71}} \left(\frac{a_9 W_7}{w_{72}}\right)^{w_{72}}$$

$$\left. \times \left(\frac{a_{10} W_7}{w_{73}}\right)^{w_{73}} \left(\frac{a_{11} W_7}{w_{74}}\right)^{w_{74}} \left(\frac{a_{12} W_7}{w_{75}}\right)^{w_{75}}\right)$$

$$(11.36)$$

Subject to:

$$w_{01} = (-1) \times (-1) = 1 \qquad \text{(normality condition)}$$
$$-w_{01} + w_{31} + w_{75} = 0 \qquad \text{(orthogonality condition for } x_1)$$
$$-w_{01} + w_{11} - w_{51} + w_{74} = 0 \qquad \text{(orthogonality condition for } x_2)$$
$$-w_{11} - w_{12} + w_{41} + w_{72} + w_{73} = 0 \qquad \text{(orthogonality condition for } x_3)$$
$$-w_{21} - w_{22} + w_{41} + w_{61} + w_{71} + w_{73} = 0 \quad \text{(orthogonality condition for } x_4)$$
$$w_{21} - w_{41} - w_{71} - w_{72} - w_{73} = 0 \qquad \text{(orthogonality condition for } x_5)$$
$$w_{12} + w_{22} - w_{61} = 0 \qquad \text{(orthogonality condition for } x_6)$$
$$w_{32} - w_{51} = 0 \qquad \text{(orthogonality condition for } x_7)$$
$$-w_{31} - w_{32} + w_{41} + w_{71} + w_{72} + w_{73} = 0 \quad \text{(orthogonality condition for } x_8)$$

$$(11.37)$$

Here, w_{01} represent the dual variable related to the single term in the objective function $y(x)$, i.e. expression set (11.37) above. Similarly, w_{11}, w_{12}, w_{21}, w_{22}, w_{31}, w_{32}, w_{41}, w_{51}, w_{61}, w_{71}, w_{72}, w_{73}, w_{74}, and w_{75} represent the dual variables related to the respective term in the respective constraint above, as explained in earlier chapters. The term $w_7 = w_{71} + w_{72} + w_{73} + w_{74} + w_{75}$.

The problem has $T = 15$ terms and $N = 8$ variables giving degrees of difficulty $= T - (N + 1) = 15 - 9 = 6$. As shown above, the value of $w_{01} = 1$ is already determined leaving values of remaining 14 variables to be determined. Therefore, any eight of these dual variables can be expressed in terms of the remaining six dual variables as shown below, choosing w_{11}, w_{12}, w_{22}, w_{31}, w_{32}, and w_{41} as the six dual variables.

$$w_{01} = 1; w_{21} = w_{31} + w_{32}; w_{51} = w_{32}; w_{61} = w_{12} + w_{22};$$
$$w_{71} = w_{31} + w_{32} - w_{11} - w_{12}; w_{72} = w_{12}; w_{73} = w_{11} - w_{41}; \qquad (11.38)$$
$$w_{74} = 1 - w_{11} + w_{32}; w_{75} = 1 - w_{31}; w_7 = 2 + 2w_{32} - w_{11} - w_{41}.$$

Thus, the dual objective function $d(w)$ could be expressed in terms of the above six dual variables. By setting to zero, the first derivatives of the substituted dual objective function, with respect to each of the above six variables, i.e. w_{11}, w_{12}, w_{22}, w_{31}, w_{32}, and w_{41}, we will get six equations for the six variables to give the unique solution. Adopting this procedure, we will finally get the following equations:

$$w_{22} = \frac{a_4 a_7}{(1 - a_4 a_7)} \left[w_{31} + \frac{a_1 a_5 a_6 a_{12}}{a_2 a_7 a_9} \frac{w_{31} w_{12}^2}{(1 - w_{31}) w_{11}} \right]$$
$$w_{32} = \frac{a_1 a_5 a_6 a_{12}}{a_2 a_7 a_9} \frac{w_{31} w_{12}^2}{(1 - w_{31}) w_{11}}$$

$$(11.39)$$

$$w_{41} = \frac{2a_1a_5a_6a_{10}\,a_{12}\,w_{31}w_{12}^2}{a_2a_7a_9(a_{10}-a_5)(1-w_{31})w_{11}} - \frac{a_{10}+a_5}{a_{10}-a_5}w_{11} + \frac{2a_{10}}{a_{10}-a_5}$$

$$(1-w_{31})\left[w_{11}^2 + w_{11}w_{12} + \frac{a_3a_8}{a_{12}(1-a_4a_7)}(1-w_{31})w_{11} \right. \tag{11.40}$$

$$\left. + \frac{a_1a_3a_5a_6a_8}{a_2a_7a_9(1-a_4a_7)}w_{12}^2 - w_{11}w_{31} \right] - \frac{a_1a_5a_6a_{12}}{a_2a_7a_9}w_{31}w_{12}^2 = 0$$

$$\frac{a_2a_7a_9}{a_8}(w_{11}+w_{12})\left[w_{31} - w_{11} - w_{12} + \frac{a_1a_5a_6a_{12}}{a_2a_7a_9}\frac{w_{31}w_{12}^2}{(1-w_{31})w_{11}} \right] - w_{12}^2 = 0 \tag{11.41}$$

$$\left[\frac{2a_5}{a_{10}-a_5} \right]^2 \frac{a_3a_6a_8a_{11}}{(1-a_4a_7)}\left[1 + \frac{a_1a_5a_6a_{12}}{a_2a_7a_9}\frac{w_{12}^2}{(1-w_{31})w_{11}} \right]\left[1 - w_{11} + \frac{a_1a_5a_6a_{12}}{a_2a_7a_9}\frac{w_{31}w_{12}^2}{(1-w_{31})w_{11}} \right]$$

$$- \left[w_{31} - w_{11} - w_{12} + \frac{a_1a_5a_6a_{12}}{a_2a_7a_9}\frac{w_{31}w_{12}^2}{(1-w_{31})w_{11}} \right]\frac{a_1a_5a_6a_{12}}{a_2a_7a_9}\frac{w_{12}^2}{(1-w_{31})w_{11}} = 0 \tag{11.42}$$

Equations (11.40), (11.41), and (11.42) are three nonlinear equations in terms of three unknown variables: w_{11}, w_{12}, and w_{31}. Using the values of w_{11}, w_{12}, and w_{31} so determined in the equation set (11.39) the values of w_{22}, w_{32}, and w_{41} can be found out. Thus, the value of dual objective function and those of the other dual variables can be determined. Using these values of the dual variables, the optimal values of the primal variables x_1, x_2, x_3, x_4, x_5, x_6, x_7, and x_8 could be found out and, thus, completely solving the optimization model.

The above model would appear impossible to crack by conventional manual means, but can be easily solved by the N-R Method using computers as presented in earlier chapters.

To help this process the **C++ program: dlsf.cpp** to solve **instantly** both the **LCO** and **MBO Models** for the single-family dwelling-layout systems, respectively, is developed and presented in an online space attached to the book. Illustrative numerical application results of the **C++ programs: dlsf.cpp** for single-family **Most-Benefit Design** and for **Least-Cost Design** is presented below separately.

11.8   Multi-family Dwelling-Layout System Design: An Example Least-Cost Optimizing Model (LCO)

Here, a representative **Least-Cost Optimizing Model (LCO)** for multi-family dwelling-layout system design with built-unit space (per family), density, building rise, i.e. number of floors in a block, number of building blocks as some of the design objectives to be achieved along with the least-cost. Therefore, these are put as constraints in the optimization model so that these objectives are attained in the optimal solution. A planner or designer can select any other set of design objectives (provided these are feasible) as per requirement and put these as constraint in the model and develop it accordingly.

$Minimize$ the Cost per Built $-$ Unit

$$= y(x)$$

$$= c_1x_1x_5^{-1}x_7^{-1} + c_2x_2x_5^{-1}x_7^{-1} + c_3x_1x_2x_7^{-1} + c_4x_3x_5^{-1}x_6^{-1}x_7^{-1} \qquad (11.43)$$

$$+ c_5x_4x_5^{-1}x_6^{-1}x_7^{-1} + c_6x_3x_4x_5^{-1}x_6^{-1}x_7^{-1}$$

Subject to:

$$a_1x_1^{-1}x_2^{-1}x_7 \le 1 \qquad (11.44)$$

(*Constraint* for minimum acceptable built-unit space)

$$a_2x_3^{-1}x_4^{-1}x_5x_6x_7x_8^{-1} \le 1 \qquad (11.45)$$

(*Constraint* for designed built-unit density and layout area compatibility)

$$a_3x_5 \le 1 \qquad (11.46)$$

(*Constraint* for building rise, i.e. number of floors in building blocks)

$$a_4x_6 \le 1 \qquad (11.47)$$

(*Constraint* for building blocks, i.e. number of building blocks in layout module)

$$a_5x_7 \le 1 \qquad (11.48)$$

(*Constraint* for number of built-units or dwellings per floor of building blocks)

$$a_6x_8 \le 1 \qquad (11.49)$$

(*Constraint* for designed built-unit density in layout module)

$$x_1, x_2, x_3, x_4, x_5, x_6, x_7, x_8 > 0$$

(Non-negativity *constraint*)

Here, c_1, c_2, c_3, c_4, c_5, c_6, a_1, a_2, a_3, a_4, a_5, and a_6 are the model constants, related to physical design and the cost parameters of the buildings (i.e. dwellings) and layout modules; and the design standards mostly dependent on the input design conditions, decided by an architect, engineer, urban planner-urban manager as per requirements, and as may be applicable. These are explained in more details below:

c_1 = Model cost coefficient related to cost of wall/foundation of dwelling along x_1 direction,

c_2 = Model cost coefficient related to cost of wall/foundation of dwelling along x_2 direction,

c_3 = Model cost coefficient related to cost of roof/floor of dwelling,

c_4 = Model cost coefficient related to cost of road pavement/utility lines in layout in North-South direction,

c_5 = Model cost coefficient related to cost of road pavement/utility lines in layout in East-West direction,

c_6 = Model cost coefficient related to cost of land,

a_1 = Model constant related to BUS constraint, (BUS = $cba * x_1 * x_2/nd$ or $x_1 * x_2/$ $x_7 = a_1 = BUS/cba$),

a_2 = Model constant related to density-layout area compatibility constraint = 1,

a_3 = Model constant related to BR constraint,

a_4 = Model constant related to NB constraint,

a_5 = Model constant related to number of BU/floor constraint,

a_6 = Model constant related to BUD constraint.

Here, the model *variables* are as follows:

x_1 = Optimum depth of the built-unit block measured along the direction of staircase as shown in figure,

x_2 = Optimum width of the built-unit block measured perpendicular to the direction of staircase,

x_3 = A variable giving the optimum width of the layout module in the North-South direction,

x_4 = A variable giving the optimum length of the layout module in the East-West direction,

x_5 = A variable giving the optimum building rise (BR) or number of floors in the building or built-unit block,

x_6 = A variable giving the optimum number of building blocks in the layout module,

x_7 = A variable giving the optimum number of built-units or dwellings per floor of building blocks,

x_8 = A variable giving the optimum built-unit density (BUD) in layout module.

Choosing those values of the system *variables* (i.e. $x_1, x_2, x_3, x_4, x_5, x_6, x_7,$ and x_8) that yield the *optimum* effectiveness, which is the *third-step* of the *three-step rational decision-making process* (i.e. mathematical *optimization*) discussed earlier, and to complete this *third-step,* it is necessary to solve the above *Nonlinear Programming Problem,* which is a typical geometric programming problem and can be solved accordingly (Beightler et al., 1979). As discussed in earlier chapters, in this technique, if the primal geometric programming problem has N variables and T terms; in the dual geometric programming problem, there will be N orthogonality conditions-one for each variable x_n, one normality condition, and T weights, one for each term. Thus, there will be $(N + 1)$ independent linear constraint equations as against T independent dual variables which are unknown.

In case $T - (N + 1)$, which is the degree of difficulty in the problem. Accordingly, using the geometric programming technique, the dual objective function and the associated normality and orthogonality constraints for the above primal problem are given by:

$$d(w) = \left(\frac{c_1}{w_{o1}}\right)^{w_{01}} \left(\frac{c_2}{w_{o2}}\right)^{w_{02}} \left(\frac{c_3}{w_{o3}}\right)^{w_{03}} \left(\frac{c_4}{w_{o4}}\right)^{w_{04}} \left(\frac{c_5}{w_{o5}}\right)^{w_{05}} \left(\frac{c_6}{w_{o6}}\right)^{w_{06}}$$
$$\times (a_1)^{w_{11}} (a_2)^{w_{21}} (a_3)^{w_{31}} (a_4)^{w_{41}} (a_5)^{w_{51}} (a_6)^{w_{61}} \qquad (11.50)$$

Accordingly, the dual objective function to be maximized can be expressed in a simplified form as follows:

Maximize $d(w) = [c_1/w_{01}]^{w_{01}} [c_2/w_{02}]^{w_{02}} [c_3/w_{03}]^{w_{03}} [c_4/w_{04}]^{w_{04}} [c_5/w_{05}]^{w_{05}} [c_6/w_{06}]^{w_{06}}$
$$\times [a_1]^{w_{11}} [a_2]^{w_{21}} [a_3]^{w_{31}} [a_4]^{w_{41}} [a_5]^{w_{51}} [a_6]^{w_{61}}$$

$$(11.51)$$

Subject to:

$$w_{01} + w_{02} + w_{03} + w_{04} + w_{05} + w_{06} = 1 \quad \text{(normality condition)} \qquad (11.52)$$

$$w_{01} + w_{03} - w_{11} = 0 \quad \text{(orthogonality condition for } x_1) \qquad (11.53)$$

$$w_{02} + w_{03} - w_{11} = 0 \quad \text{(orthogonality condition for } x_2) \qquad (11.54)$$

$$w_{04} + w_{06} - w_{21} = 0 \quad \text{(orthogonality condition for } x_3) \qquad (11:55)$$

$$w_{05} + w_{06} - w_{21} = 0 \quad \text{(orthogonality condition for } x_4) \qquad (11.56)$$

$$- w_{01} - w_{02} - w_{04} - w_{05} - w_{06} + w_{11} + w_{21} + w_{31}$$
$$= 0 \text{(orthogonality condition for } x_5) \qquad (11.57)$$

$$- w_{04} - w_{05} - w_{06} + w_{21} + w_{41} = 0 \quad \text{(orthogonality condition for } x_6) \qquad (11.58)$$

$$- w_{01} - w_{02} - w_{03} - w_{04} - w_{05} - w_{06} + w_{11} + w_{21} + w_{61}$$
$$= 0 \text{(orthogonality condition for } x_7) \qquad (11.59)$$

$$w_{51} - w_{21} = 0 \quad \text{(orthogonality condition for } x_8) \qquad (11.60)$$

Here, $w_{01}, w_{02}, w_{03}, w_{04}, w_{05},$ and w_{06} represent the six cost weights related to the six terms in the objective function $y(x)$, i.e. expression (11.43) above. Similarly, $w_{11},$ $w_{21}, w_{31}, w_{41}, w_{51},$ and w_{61} represent the dual variables related to the respective term in the respective constraint above, as explained in earlier chapters.

In the present optimal design problem, there are 12 terms (T) and eight variables (N) in the primal objective function and constraints [expressions (11.43)–(11.49)]. Thus, the degree of difficulty in the problem is $= T - (N + 1) = 11 - (8 + 1) = 3$, which is comparatively small, and therefore, a set of $(N + 1)$ dual weights could be expressed in terms of the remaining $T - (N + 1)$ dual weights, and substituted into the dual objective function, which is then maximized by setting the first derivative to zero with respect to each of the dual weights coming into the substituted dual objective function, which is being maximized. Thus, a sufficient number of equations are available for solving the problem. Therefore, any nine of the dual variables can be expressed in terms of the remaining three dual variables as shown below, choosing w_{01}, w_{03}, and w_{05} as the three dual variables:

$$w_{02} = w_{01}; w_{04} = w_{05}; w_{06} = 1 - 2w_{01} - 2w_{05} - w_{03}; w_{11} = w_{01} + w_{03};$$

$$w_{21} = 1 - 2w_{01} - w_{03} - w_{05};$$

$$w_{31} = 2w_{01} + w_{05}; w_{41} = w_{05}; w_{51} = 1 - 2w_{01} - w_{03} - w_{05}; w_{61} = w_{01} + w_{05}.$$

$$(11.61)$$

As earlier mentioned, using *optimization concepts*, one can discern general *decision rules* or *design rules* for *optimization* which are appropriate to problems, and develop an *insight* regarding proper form of an optimal solution even when a problem is not completely formulated mathematically. This is illustrated below, noting that the cost function in the above expression (9.5) literally means that the total cost per built-unit = Cost contributed by all walls (including foundation and finishes) in x_1 direction (say, along the staircase) + Cost contributed by all walls (including foundation and finishes) in x_2 direction (say, perpendicular to the staircase) + Cost contributed by floors/roof (including finishes) + Cost contributed by services, i.e. road pavement/utility lines in Layout in North-South direction + Cost contributed by services, i.e. road pavement/utility lines in Layout in East-West direction + Cost contributed by land.

Thus, the cost weights w_{01}, w_{02}, w_{03}, w_{04}, w_{05}, and w_{06} represent the optimum proportion of total cost per built-unit, distributed between the cost contributed by all walls (including foundation and finishes) in x_1 direction (say, along the staircase), cost contributed by all walls (including foundation and finishes) in x_2 direction (say, perpendicular to the staircase), cost contributed by floors/roof (including finishes), cost contributed by services, i.e. road pavement/utility lines (N-S), cost contributed by services, i.e. road pavement/utility lines (E-W), and the cost contributed by land, respectively.

In this case by developing only the exponent matrix, without solving the model completely, one can discern (from equation set (11.61)), the following *decision* or *design rules* if one has to operate *optimally*:

1. no term of the cost function can contribute more than 100% of the total cost nor less than 0% at the *optimality*,
2. the cost contributed by all walls in the x_1 direction (w_{01}) must be equal to the cost contributed by all walls in the x_2 direction (w_{02}), at the *optimality*,
3. 3the cost contributed by services, i.e. road pavement/utility lines (N-S) (w_{04}), must be equal to the cost contributed by services, i.e. road pavement/utility lines (E-W) (w_{05}), at the *optimality*,

4. total cost *minus* twice the cost contributed by all walls in the x_1 direction, *minus* twice the cost contributed by services (E-W), *minus* the cost of roof/floor; is equal to the land cost, at the *optimality*.

The set of Eq. (11.61) represent the relationship between the cost distributions among the above components, which are independent of the values of the cost coefficients c_1, c_2, c_3, c_4, c_5, and c_6. This invariant property of geometric programming could be used for development of suitable *design rules* for *optimality* even in the presence of a great degree of difficulty, having general application in any problem of this nature. Thus, even without solving the mathematical model, and using the above design rules (particularly rules: 2, 3, and 4) a designer can move towards optimization. In fact, if his/her design violates any of the above *design rules*, obviously, it can be concluded that the design is inefficient compared to the optimal design and suitable corrections in design could be made to move towards optimization (say by choosing wall dimensions in x_1 and x_2 directions in such a way that costs in both directions are equal, and similar procedure can be followed for services in N-S and E-W direction). But, solving the optimization model will be a better course of action, particularly in these days of information technology and computer application revolution, when such models can be solved in no time as part of CAD as illustrated later.

If we substitute the expressions in equations set (11.61) into the dual objective function (Eq. 11.51) and maximize this substituted dual objective function over the three variables w_{01}, w_{03}, and w_{05} (i.e. by setting the first derivative of the function with respect to each variable to zero), we will get three additional equations, thus providing sufficient conditions for optimality. The process is simplified by maximizing the natural logarithm of the dual function; that is, by substituting:

$$\ln d(w_{01}, w_{03}, w_{05}) = z(w_{01}, w_{03}, w_{05}).$$

Thus, by above substitution and taking logarithm on both sides of Eq. (11.51), we get the following equations:

$$
\begin{aligned}
z(w) = {}& w_{01}(\ln c_1 - \ln w_{01}) + w_{01}(\ln c_2 - \ln w_{01}) + w_{03}(\ln c_3 - \ln w_{03}) \\
& + w_{05}(\ln c_4 - \ln w_{05}) + w_{05}(\ln c_5 - \ln w_{05}) \\
& + (1 - 2w_{01} - 2w_{05} - w_{03})(\ln c_6 - \ln(1 - 2w_{01} - 2w_{05} - w_{03})) \\
& + (w_{01} + w_{03})\ln a_1 + (1 - 2w_{01} - w_{03} - w_{05})\ln a_2 + (2w_{01} + w_{05})\ln a_3 \\
& + w_{05}\ln a_4 + (1 - 2w_{01} - w_{03} - w_{05})\ln a_5 + (w_{01} + w_{05})\ln a_6.
\end{aligned}
$$

$$(11.62)$$

By setting the first derivative $dz/dw_{03} = 0$, we get:

$\ln c_3 - \ln w_{03} - 1 - \ln c_6 + \ln (1 - 2w_{01} - 2w_{05} - w_{03}) + 1 + \ln a_1 - \ln a_2 - \ln a_5$
$= 0$

Or, $(c_3 a_1/c_6 a_2 a_5)(1 - 2w_{01} - 2w_{05}) - (c_3 a_1/c_6 a_2 a_5)w_{03} - w_{03} = 0$

Or, $w_{03} = \dfrac{c_3 a_1}{c_3 a_1 + c_6 a_2 a_5}(1 - 2w_{01} - 2w_{05})$

$$(11.63)$$

Using above equation, we can also express:

$$(1 - 2w_{01} - 2w_{05} - w_{03}) = \left(\dfrac{c_6 a_2 a_5}{c_3 a_1 + c_6 a_2 a_5}\right)(1 - 2w_{01} - 2w_{05}) \qquad (11.64)$$

By setting the first derivative $dz/dw_{01} = 0$, we get:

$\ln c_1 - \ln w_{01} - 1 + \ln c_2 - \ln w_{01} - 1 - 2 \ \ln c_6$
$\qquad + 2 \ln (1 - 2w_{01} - 2w_{05} - w_{03}) + 2 + \ln a_1 - 2 \ln a_2 + 2 \ln a_3 + \ln a_6 - 2 \ln a_5$
$\qquad = 0$

or, $c_1 c_2 a_1 a_3^2 a_6(1 - 2w_{01} - 2w_{05} - w_{03})^2 - c_6^2 a_2^2 a_5^2 w_{01}^2 = 0$

Substituting expression (11.64) in above equation, we get:

$\left(c_1 c_2 a_1 a_3^2 a_6/c_6^2 a_2^2 a_5^2\right)\left(\dfrac{c_6 a_2 a_5}{c_3 a_1 + c_6 a_2 a_5}\right)^2 (1 - 2w_{01} - 2w_{05})^2 - w_{01}^2 = 0$

$$(11.65)$$

or, $\left[(c_1 c_2 \, a_1 a_3^2 a_6)/(c_3 a_1 + c_6 a_2 a_5)^2\right] (1 - 2w_{01} - 2w_{05})^2 - w_{01}^2 = 0$

Again, by setting the first derivative $dz/dw_{05} = 0$, we get:

$\ln c_4 - \ln w_{05} - 1 + \ln c_5 - \ln w_{05} - 1 - 2 \ln c_6 + 2$
$\qquad + 2 \ln (1 - 2w_{01} - 2w_{05} - w_{03}) - \ln a_2 + \ln a_3 + \ln a_4 - \ln a_5 + \ln a_6 = 0$

Or, $c_4 c_5 a_3 a_4 a_6(1 - 2w_{01} - 2w_{05} - w_{03})^2 - c_6^2 a_2 a_5 w_{05}^2 = 0$

Substituting expression (11.64) in above equation, we get:

$\left(c_4 c_5 a_3 a_4 a_6/c_6^2 a_2 a_5\right)\left(\dfrac{c_6 a_2 a_5}{c_3 a_1 + c_6 a_2 a_5}\right)^2 (1 - 2w_{01} - 2w_{05})^2 - w_{05}^2 = 0$

or, $\left[(c_4 c_5 a_2 a_3 a_4 a_5 a_6)/(c_3 a_1 + c_6 a_2 a_5)^2\right] (1 - 2w_{01} - 2w_{05})^2 - w_{05}^2 = 0$

$$(11.66)$$

To simplify, we may make the following substitution:

$$k1 = \left[\left(c_1 c_2 a_1 a_3^2 a_6\right)/\left(c_3 a_1 + c_6 a_2 a_5\right)^2\right], k2 = \left(c_4 c_5 a_2 a_3 a_4 a_5 a_6\right)/\left(c_3 a_1 + c_6 a_2 a_5\right)^2],$$

$$k3 = \frac{c_3 a_1}{c_3 a_1 + c_6 a_2 a_5}, \quad w_{01} = x, \quad w_{05} = y.$$

Then, the Eqs. (11.65) and (11.66) can be expressed as:

$$\left. \begin{array}{ll} k1 \,(1 - 2x - 2y)^2 - x^2 = 0 & f(x, \ y), \\ k2 \,(1 - 2x - 2y)^2 - y^2 = 0 & g(x, \ y), \end{array} \right\} \tag{11.67}$$

and similarly, $w_{03} = k3(1 - 2x - 2y)$.

Thus, the Eq. (11.67) is a set of two simultaneous nonlinear equations and can be easily solved by the N-R method as presented in earlier chapters.

11.9   Multi-family Dwelling-Layout System: An Example of Most-Benefit Optimizing Model (MBO)

Here, an example of **Most-Benefit Optimizing Model (MBO)** for optimal design of multi-family dwelling-layout system is presented.

$$\text{Maximize } y_o(x) = a_1 x_1 x_2 x_7^{-1}$$

This objective function can be expressed in standard geometric form as:

$$\text{Minimize } y_o(x) = - a_1 x_1 x_2 x_7^{-1} \tag{11.68}$$

Subject to:

$$a_2 x_3^{-1} x_4^{-1} x_5 x_6 x_7 x_8^{-1} \leq 1 \tag{11.69}$$

(*Constraint* for designed built-unit density and layout area compatibility)

$$a_3 x_5 \leq 1 \tag{11.70}$$

(*Constraint* for building rise, i.e. number of floors in building blocks)

$$a_4 x_6 \leq 1 \tag{11.71}$$

(*Constraint* for building blocks, i.e. number of building blocks in layout module)

$$a_5 x_7 \leq 1 \tag{11.72}$$

(*Constraint* for number of built-units or dwellings per floor of a building block)

$$a_6 x_8 \leq 1 \tag{11.73}$$

(*Constraint* for designed built-unit density in layout module)

$$a_7 x_1 x_5^{-1} x_7^{-1} + a_8 x_2 x_5^{-1} x_7^{-1} + a_9 x_1 x_2 x_7^{-1} + a_{10} x_3 x_5^{-1} x_6^{-1} x_7^{-1}$$
$$+ a_{11} x_4 x_5^{-1} x_6^{-1} x_7^{-1} + a_{12} x_3 x_4 x_5^{-1} x_6^{-1} x_7^{-1} \leq 1 \tag{11.74}$$

(*Constraint* for affordable cost or maximum investment per built-unit)

$$x_1, x_2, x_3, x_4, x_5, x_6, x_7, x_8 > 0$$

(Non-negativity *constraint*)

Here, a_1, a_2, a_3, a_4, a_5, a_6, a_7, a_8, a_9, a_{10}, a_{11}, and a_{12} are the model constants, related to physical design and the cost parameters of the buildings (i.e. dwellings) and layout modules; and the design standards mostly dependent on the input design conditions, decided by an architect, engineer, urban planner-urban manager as per requirements, and as may be applicable. These are explained in more details below:

$a_1 =$ Model constant related to BUS constraint,
$a_2 =$ Model constant related to density-layout area compatibility constraint,
$a_3 =$ Model constant related to BR constraint,
$a_4 =$ Model constant related to NB constraint,
$a_5 =$ Model constant related to number of BU/floor constraint,
$a_6 =$ Model constant related to BUD constraint,
$a_7 =$ Model constant related to cost of wall/foundation of dwelling along x_1 direction,
$a_8 =$ Model constant related to cost of wall/foundation of dwelling along x_2 direction,
$a_9 =$ Model constant related to cost of roof/floor of dwelling,
$a_{10} =$ Model constant related to cost of road pavement/utility lines in layout in North-South direction,
$a_{11} =$ Model constant related to cost of road pavement/utility lines in layout in East-West direction,
$a_{12} =$ Model constant related to cost of land for layout area.

Here, the model *variables* are as follows:

$x_1 =$ Optimum depth of the built-unit block measured along the direction of staircase as shown in figure,
$x_2 =$ Optimum width of the built-unit block measured perpendicular to the direction of staircase,

x_3 = A variable giving the optimum width of the layout module in the North-South direction,

x_4 = A variable giving the optimum length of the layout module in the East-West direction,

x_5 = A variable giving the optimum building rise (BR) or number of floors in the building or built-unit block,

x_6 = A variable giving the optimum number of building blocks in the layout module,

x_7 = A variable giving the optimum number of built-units or dwellings per floor of building blocks,

x_8 = A variable giving the optimum built-unit density (BUD) in layout module.

It may be seen that the model constants a_1, a_2, a_3, a_4, a_5, and a_6 are same in both the least-cost and in the most-benefit models presented above. Using the expressions for model constants: c_1, c_2, c_3, c_4, c_5, and c_6 as given above, the model constants: a_7, a_8, a_9, a_{10}, a_{11} and a_{12} can be defined as follows:

$a_7 = c_1/ac$, $a_8 = c_2/ac$, $a_9 = c_3/ac$, $a_{10} = c_4/ac$, $a_{11} = c_5/ac$ and $a_{12} = c_6/ac$;

where ac = affordable or limit of investment per built-unit in monetary units (MU).

Here, the model *variables* (i.e. x_1, x_2, x_3, x_4, x_5, and x_6) are same as defined in Sect. 11.8 above. Thus, using these values of **Model Constants**, the above optimal built-unit design model for **Most-Benefit** can be easily calibrated for model solution in a site-specific case as outlined below.

11.9.1   Model Solution

The *Objective: Maximize the* built-unit-space $y(x) = a_1 x_1 x_2 x_7^{-1}$ can also be written in standard geometric programming form as:

$$Minimize \ y(x) = -a_1 x_1 x_2 x_7^{-1}$$

Thus, this optimization model is a typical signomial geometric programming problem as outlined in Chap. 8, and can be solved accordingly using the procedure suggested in the above Chap. 8. To solve the problem, let us guess the signum function $\sigma_0 = -1.0$ for $y(x)$ *the negative value of y(x), since the original objective function is to Maximize. Similarly, the value of the signum function* σ_{01} *is guessed as* $= -1.0$. *The values of other signum functions in the problem are guessed as:* $\sigma_m = +1.0$ and $\sigma_{mt} = +1.0$. To simplify the notations we can denote the sum of dual variables w_{51}, w_{52}, w_{53}, and w_{54} by w_5. With these simplified notations and other simplifications outlined in Chap. 8, the dual objective function and the associated normality and orthogonality constraints for the above primal problem can be expressed as:

Optimize :

$$d(w) = (-1)\left((a_2)^{w_{11}}(a_3)^{w_{21}}(a_4)^{w_{31}}(a_5)^{w_{41}}(a_6)^{w_{51}}\left(\frac{a_7 W_6}{w_{61}}\right)^{w_{61}}\left(\frac{a_8 W_6}{w_{62}}\right)^{w_{62}}\right.$$

$$\left.\times \left(\frac{a_9 W_6}{w_{63}}\right)^{w_{63}}\left(\frac{a_{10}W_6}{w_{64}}\right)^{w_{64}}\left(\frac{a_9 W_6}{w_{65}}\right)^{w_{65}}\left(\frac{a_{11}W_6}{w_{66}}\right)^{w_{66}}\right)^{-1}$$

$$(11.75)$$

Subject to:

$$w_{01} = (-1) \times (-1) = 1 \quad \text{(normality condition)}$$
$$-w_{01} + w_{61} + w_{63} = 0 \quad \text{(orthogonality condition for } x_1)$$
$$-w_{01} + w_{62} + w_{63} = 0 \quad \text{(orthogonality condition for } x_2)$$
$$-w_{11} + w_{64} + w_{66} = 0 \quad \text{(orthogonality condition for } x_3)$$
$$-w_{11} + w_{65} + w_{66} = 0 \quad \text{(orthogonality condition for } x_4)$$
$$w_{11} + w_{21} - w_{61} - w_{62} - w_{64} - w_{65}$$
$$-w_{66} = 0 \quad \text{(orthogonality condition for } x_5)$$
$$w_{11} + w_{31} - w_{64} - w_{65} - w_{66} = 0 \quad \text{(orthogonality condition for } x_6)$$
$$w_{01} + w_{11} + w_{41} - w_{61} - w_{62} - w_{63}$$
$$-w_{64} - w_{65} - w_{66} = 0 \quad \text{(orthogonality condition for } x_7)$$
$$-w_{11} + w_{51} = 0 \quad \text{(orthogonality condition for } x_8)$$

$$(11.76)$$

Here, w_{01} represent the dual variable related to the single term in the objective function $y(x)$, i.e. expression (11.68) above. Similarly, w_{11}, w_{21}, w_{31}, w_{41}, w_{51}, w_{61}, w_{62}, w_{63}, w_{64}, w_{65}, and w_{66} represent the dual variables related to the respective term in the respective constraint above, as explained in earlier chapters.

In the present optimal design problem, there are 12 terms (T) and eight variables (N) in the primal objective function and constraints [expressions (11.68)–(11.74)]. Thus, the degree of difficulty in the problem is $= T - (N + 1) = 11 - (8 + 1) = 3$, which is comparatively small, and therefore, a set of $(N + 1)$ dual weights could be expressed in terms of the remaining $T - (N + 1)$ dual weights, and substituted into the dual objective function, which is then maximized by setting the first derivative to zero with respect to each of the dual weights coming into the substituted dual objective function, which is being maximized. Thus, a sufficient number of equations are available for solving the problem. Therefore, any nine of the dual variables can be expressed in terms of the remaining three dual variables as shown below, choosing w_{61}, w_{65}, and w_{66} as the three dual variables:

$$w_{01} = 1; w_{64} = w_{65}; w_{62} = w_{61}; w_{11} = w_{65} + w_{66}; w_{21} = 2w_{61} + w_{65};$$

$$w_{31} = w_{65}; w_{41} = w_{61} + w_{65};$$ (11.77)

$$w_{51} = w_{65} + w_{66}; w_{63} = 1 - w_{61};$$

$$w = w_{61} + w_{62} + w_{63} + w_{64} + w_{65} + w_{66} = 1 + w_{61} + 2w_{65} + w_{66}.$$

If we substitute the expressions in equations set (11.77) into the dual objective function [Eq. (11.75)] and maximize this substituted dual objective function over the three variables $w_{61,}$ $w_{65,}$ and w_{66} (i.e. by setting the first derivative of the function with respect to each variable to zero), we will get three additional equations, thus providing sufficient conditions for optimality. The process is simplified by maximizing the natural logarithm of the dual function; that is, by substituting:

$$\ln d(w_{61}, w_{65}, w_{66}) = z(w_{61}, w_{65}, w_{66}).$$

Thus, by above substitution and taking logarithm on both sides of Eq. (11.75), we get the following equations:

$$
\begin{aligned}
\ln z(w) = {} & w_{01}(\ln a_1 - \ln w_{01}) + (w_{65} + w_{66})\ln a_2 + (2w_{61} + w_{65})\ln a_3 \\
& + w_{65}\ln a_4 + (w_{61} + w_{65})\ln a_5 + (w_{65} + w_{66})\ln a_6 \\
& + w_{61}(\ln a_7 + \ln(1 + w_{61} + 2w_{65} + w_{66}) - \ln w_{61}) \\
& + w_{61}(\ln a_8 + \ln(1 + w_{61} + 2w_{65} + w_{66}) - \ln w_{61}) \\
& + (1 - w_{61})(\ln a_9 + \ln(1 + w_{61} + 2w_{65} + w_{66}) - (1 - w_{61})) \\
& + w_{65}(\ln a_{10} + \ln(1 + w_{61} + 2w_{65} + w_{66}) - \ln w_{65}) + w_{65}(\ln a_9 \\
& + \ln(1 + w_{61} + 2w_{65} + w_{66}) - \ln w_{65}) \\
& + w_{66}(\ln a_{11} + \ln(1 + w_{61} + 2w_{65} + w_{66}) - \ln w_{66}).
\end{aligned}
$$

(11.78)

By setting the first derivative $dz/dw_{61} = 0$, we get:

$$
\begin{aligned}
& 2\ln a_3 + \ln a_6 + \ln a_7 + \ln(1 + w_{61} + 2w_{65} + w_{66}) - \ln w_{61} + \ln a_8 \\
& + \ln(1 + w_{61} + 2w_{65} + w_{66}) - \ln w_{61} - \ln a_7 \\
& - \ln(1 + w_{61} + 2w_{65} + w_{66}) - \ln(1 - w_{61}) = 0
\end{aligned}
$$ (11.79)

Or, $(a_3^2 a_6 a_7 a_8/a_9)(1 - w_{61})(1 + w_{61} + 2w_{65} + w_{66}) - w_{61}^2 = 0$

Similarly, by setting the first derivative $dz/dw_{65} = 0$, we get:

$\ln a_2 + \ln a_3 + \ln a_4 + \ln a_5 + \ln a_6 + \ln a_{10} + \ln a_9$
$\quad + \ln(1 + w_{61} + 2w_{65} + w_{66}) + \ln(1 + w_{61} + 2w_{65} + w_{66}) - w_{65} - w_{65}$
$\quad = 0$

Or, $(a_2a_3a_4a_5a_6a_{10}a_9)(1 + w_{61} + 2w_{65} + w_{66})^2 - w_{65}^2 = 0$

$$(11.80)$$

Again, by setting the first derivative $dz/dw_{66} = 0$, we get:

$$\ln a_2 + \ln a_5 + \ln a_{11} + \ln(1 + w_{61} + 2w_{65} + w_{66}) - w_{66} = 0$$
$$\text{Or,} \quad a_2a_5a_{11}(1 + w_{61} + 2w_{65} + w_{66}) - w_{66} = 0$$

From above, we get

$$w_{66} = (a_2a_5a_{11}/(1 - a_2a_5a_{11}))(1 + w_{61} + 2w_{65}) \qquad (11.81)$$

Substituting the above value of w_{66}, we get:
$1 + w_{61} + 2w_{65} + w_{66} = (1 + w_{61} + 2\,w_{65})/(1 - a_2a_5a_{11})$.
Thus, Eq. (11.79) becomes

$$\left[(a_3^2a_6a_7a_8)/(a_9(1 - a_2a_5a_{11}))\right](1 - w_{61})(1 + w_{61} + 2w_{65}) - w_{61}^2 = 0 \quad (11.82)$$

Similarly, Eq. (11.80) becomes

$$\left[(a_2a_3a_4a_5a_6a_{10}a_9)/(1 - a_2a_5a_{11})^2\right](1 + w_{61} + 2w_{65})^2 - w_{65}^2 = 0 \qquad (11.83)$$

Equations (11.82) and (11.83) are two simultaneous nonlinear equations in two variables w_{61} and w_{65} and can easily be solved using N-R Method. On finding the values of w_{61} and w_{65}, the optimum values of all the remaining geometric programming weights can be determined from Eq. (11.81) and equation set (11.77). Similarly, the GP weights the model design *variables*: x_1, x_2, x_3, x_4, x_5, x_6, x_7, and x_8 can be determined as follows:

$x_5 = 1/a_3$; $x_4 = a_{10}w_{66}/a_{11}w_{65}$; $x_3 = a_9w_{66}/a_{11}w_{65}$; $x_1 = a_8w_{63}/a_9w_{61}x_5$;
$x_2 = a_7w_{63}/a_9w_{61}x_5$;
$x_6 = 1/a_4$; $x_7 = 1/a_5$; $x_8 = 1/a_6$; MPA $= cba\ x_1x_2/x_7$;
SPO $= x_3x_4 - cba\ x_1x_2x_6/x_5x_7x_6$.

Thus, the model is completely solved.

11.10 Models Calibration and Computer-Aided Design of Dwelling-Layout Systems by Optimization, Using C++ Programs DLSF.CPP and DLMF.CPP: Illustrative Application Examples

All the above models would appear impossible to crack by conventional manual means which is one of the reasons for general reluctance and non-application of such optimization in the past by the AEC professionals; although even the *ancients* applied optimization by intuition (e.g. many ancient cities are circular to optimize length of protective peripheral walls of a city). Some such intuitive design rules for optimization are presented above in Sects. 11.6 and 11.8 above and in earlier chapters. However, with the ongoing information technology and computer science revolution, it has become extremely easy to develop and apply such optimization techniques and solve such complex problems in no time using computer. To help this process, the **C++ program: DLSF.CPP** covering both the **LCO Model** and **MBO Model** for the single-family dwelling-layout systems is developed and presented in an online space attached to this book. Similarly, the C++ **program: DLMF.CPP** covering both the **LCO Model** and **MBO Model** for the multi-family dwelling-layout systems is developed and presented as above. Illustrative numerical application results of the **C++ programs: DLSF.CPP** and **DLMF.CPP**, covering both the **Most-Benefit Design** and for **Least-Cost Design** are presented below separately.

As discussed in Chap. 9, low-income countries are usually subjected to severe *resource constraints* requiring scaling-down the needs in the context of their priority and available resources. Even in affluent countries such *resource constraints* are applicable because of scaling-up of needs to have higher standard of living consuming more resources. Therefore, results are presented below both for the middle/low-income countries designated as **Context-A**, and for the high-income countries designated as the **Context-B**. This also highlights the general applicability of the computer-aided layout, building, housing, and dwelling-layout-system planning and design by optimization technique in all countries for all income categories, to help achieve *resource-efficiency* and *social equity* in the layout, building, housing, and urban development sector as a whole.

In the optimum design procedure, after selecting the model typology, the alternative unit design modules, engineering specifications, planning and design standards, unit costs of various elements are to be determined based on the site-specific data as illustrated in Chap. 4, and accordingly the model is calibrated, i.e. model constants (as defined in various example models presented above) are determined. The optimum housing and urban development design procedure is shown diagrammatically in Fig. 10.7 in Chap. 10. Using this procedure, the results are presented separately for single-family and multi-family dwelling-layout systems design by optimization, using the **C++ program DLMF.CPP**. Again, the results are presented in two groups, namely most-benefit-design and least-cost-design, respectively, covering both single-family and multi-family dwelling-layout-system for Context-A (low- and middle-income countries) as well as Context-B (high-income countries). In the context of building, housing, and urban development projects, the *resource-productivity* improvement can be defined in three categories such as:

1. Increasing outputs (say, the number of built-units and the area of saleable built-up or lot-spaces produced) with the same inputs (say, capital investment, land, labour, materials),
2. Decreasing inputs (capital investment, land, labour, materials), while maintaining the same outputs (the number of built-units and the area of saleable built-up or lot-spaces produced),
3. Increasing outputs and decreasing inputs to change the output-input ratio favourably.

Thus, application of the MBO model in the optimal design of single-family and multi-family dwelling-layout system will help attain the resource-productivity improvement of category (a) above. Similarly, application of the LCO model in the optimal design of above dwelling-layout systems will help accomplish the resource-productivity improvement of category (b) above. Application of both MBO and LCO models in the optimal design of single-family and multi-family dwelling-layout systems in a site-specific context will help reach the resource-productivity improvement of category (c) above. Applications of both LCO and MBO models to maximize resource-productivity improvement with different input decisions, which can be considered as urban planning and management decision alternatives, are shown below.

The alternative urban planning and management decisions can be designated in the form of matrices. For example, even if we consider only four parameters such as land price (LP), built-unit-density (BUD), construction specifications (CS), and building rise (BR), there will be numerous input decision-matrices for each layout and dwelling module pattern selected by an urban planner-designer-manager, giving large variations in productivity indices of urban development designs. Considering only four parameters cited above the input urban planning and management decisions, alternatives can be designated in the form of decision-matrices such as LLLL, LMLL, LMML, LLML, HHLL, HHML, and so on, where first, second, third, and fourth letter represent the value of LP, BUD, CS, and BR, respectively, in terms of Low (indicated by L), Medium (M), and High (H) value, for say, Low, Medium, and High Land Price (LP) situations (which are beyond control of designers, and therefore, should be taken as a given design condition in a site-specific case). Thus, in the example problems design conditions and urban planning-management decisions adopted are: two levels of construction specifications, i.e. CS-low and CS-medium and accordingly unit costs are calculated as follows:

$$CS - low$$
$$CI = 1500 \text{ MU}/\text{M}, CS = 1770 \text{ MU}/\text{M}, CF = 280 \text{ MU}/\text{M},$$
$$CW = 656 \text{ MU}/\text{M}, CR = 382 \text{ MU}/\text{Sq.M.}$$

CS − medium

CI = 2211 MU/M,CS = 2528 MU/M,CF = 1076 MU/M,

CW = 837 MU/M,CR = 737 MU/Sq.M.

Similarly, three levels of land price (LP-low-87, medium-200, and high-1000), two levels of building rise (low-BR = 2, medium-BR = 4) and three levels of BU-Density (BUD:low-200, medium-400 high-800). Land prices and construction costs are determined by the market beyond control of planner-designer, although by choosing suitable construction specifications costs can be reduced. For example, the design conditions matrix: L-M-M-M low (L) land price, medium (M) 'den', i.e. BU-density, medium (M) construction specification, i.e. CS, and medium (M) Building Rise, i.e. BR. The above parameter values are only cited as examples, and the planner-designer can choose or derive these parameter values as per his/her own site-specific context.

These design conditions will give numerous decision-matrices where decisions on BUD, CS, and BR are entirely within the control of various professionals engaged in the urban development process, making it incumbent on them to take *informed-decisions* using appropriate techniques. It is shown earlier that there can be variation of even 4 times in the productivity indices of urban development designs only due to the urban management decision alternatives chosen by urban professionals, i.e. architects, planners, and engineers, highlighting the importance of application of such *Computer-Aided Planning and Design Techniques by Optimization* (Chakrabarty, 1998). Similar variations may be observed in the application examples presented below, again, stressing the need to apply such techniques by AEC professionals, e.g. architects, engineers, and planners, to maximize *resource-productivity* improvements in AEC operations.

11.11   Single-Family Dwelling-Layout Systems Design by Optimization

Here, a number of application examples of optimization of single-family dwelling-layout systems design are presented; both for least-cost design and most-benefit-design, and also covering both contexts of low- and high-income countries, showing general applicability of this *optimization* technique in all countries.

11.11.1   Application Examples of Least-Cost Design: Single-Family

The optimization model for **Least-Cost Design** of **Single-Family Dwelling-Layout System** developed and solved in Sect. 11.6 above is converted into a computer program in the form of Model-1 in the C++ program DLSF.CPP. Illustrative numerical examples of computer-aided single-family dwelling-layout systems design by optimization using the above C++ program DLSF.CPP is presented below.

As stated earlier, in this model, the factors c_1, c_2, c_3, c_4, c_5 are the model constants, related to the cost parameters giving the unit cost coefficients for layout and dwelling modules, i.e. cost coefficient related to the cost per unit length of utility lines and road pavement in E-W streets, in N-S streets, raw land per unit area, wall and service connection costs per unit length across the depth of a lot, and the wall cost per unit length along the width of a lot, respectively. The factor a_1 is the model constant related to the designed lot area per dwelling and the factors a_2, a_3, a_5, a_6 are the layout module constants (i.e. Module Parameters shown in Table 11.1). The factors a_4, a_7, and a_8 are the model constants, related to the designed lot density (i.e. number of lots per unit area of layout), chosen circulation interval ratio (indicating the spacing of peripheral streets related to the width of corresponding street for equivalent performance in respect of intercommunication and traffic flow—both pedestrian and vehicular) and the designed open-to-sky space standard per lot, respectively, which are dependent on the input design conditions, decided by an architect, engineer, urban planner-urban manager as per requirements, and as may be applicable. Accordingly, the model can be calibrated and applied in a site-specific case as illustrated below with application examples.

It is interesting to note that the above model formulations are independent of the dwelling and layout unit design module pattern chosen. The particular module pattern is taken care of by the appropriate values of module parameters and cost coefficients shown in Table 11.1. Thus, a designer has the full freedom to choose or create a particular module pattern, and still can use the same model for optimal design of the module so chosen or created, provided the model is calibrated accordingly.

Moreover, with the help of this model, a designer can achieve directly, not only the *least-cost*, but also, a number of other objectives like; a particular built-space, lot area, circulation interval ratio, lot density, and open space standard; all achieved directly, simultaneously and compatible with each other, which is rarely possible in conventional planning and design practice. Again, the design variables, objective function, and constraint-set could change depending on the emphasis given by the designer and other stakeholders on particular aspects of design and the model could be formulated accordingly. But, unlike the conventional practice, such an optimal design procedure always helps to achieve directly, the multiple design objectives simultaneously including *resource-efficiency* indicators, which are feasible and at the same time compatible with each other.

Using the model, the optimal designs of the single-family dwelling-layout-system, consisting of the 'layout unit design module' and the 'dwelling unit design module' could be obtained for different types of housing development like 'detached', 'semi-detached', and 'row' (Chakrabarty, 1988a, 1988b). Depending on the module pattern chosen, the constant values of the cost coefficients c_1 and c_2, could be derived in a site-specific case, from the unit costs of each street including utilities, assuming particular specification of road pavements, utility networks, and so on, based on field data (HUDCO, 1982) and cost engineering principles as outlined in Chap. 4. Similarly, the values of the other cost coefficients c_4 and c_5 could be derived from the unit costs of walls and house service connections, based on field data and cost engineering principles. The value of cost coefficient c_3 will depend on the unit market price of land in a site-specific case which is normally beyond control of the planner and designer. Accordingly, choosing the layout and dwelling module pattern as shown in Figs. 11.1 and 11.2, the values of module parameters and cost coefficients for the single-family dwelling-layout system are adopted in the application examples below.

11.11.1.1   Context-A: Low-Middle-Income Countries

Here, four application examples—three small and one medium size dwelling—are taken as examples with input parameters indicated in each application example problem.

1. **Application Example: LCSFL1 Dwelling-Layout System: Single-Family.**
   This is an example of small '**Detached Housing**' to achieve the *optimal design,* i.e. achieving the **Least-Cost** per dwelling unit for a given minimum acceptable *dwelling space*, with a desired open space per BU and a specified density in the layout. The results instantly given by the C++ program: DLSF.CPP is reproduced below, showing both input parameter values chosen by planner-designer and the output *optimal* results given by model.

```
LEAST-COST OPTIMAL DESIGN OF SINGLE-FAMILY DWELLING-LAYOUT SYSTEM
:'ITMODEL- 1'
GET ANY DESIRED LOT-DENSITY/OPEN-SPACE IN LAYOUT MODULE CHOSEN
IS;'MODULE- 1'

SOME KEY INPUT VALUES CHOSEN(INDEPENDENT VARIABLES):
Designed Minimum Acceptable Lot Space Per Built-Unit In The Layout
Module =  100.00 Sq.M.
LAYOUT MODULE TYPE CHOSEN IS= 'GRID-IRON'
SOME INPUT & CALIBRATED MODEL CONSTANTS(for the Chosen DESIGN
CONDITION)are:
a1=100.0000 a7= 0.0400 a2= 0.1667 a3= 1.6330 a4=151.5152
a5= 6.0000 a6= 2.6000
c1=743.0000 c2=347.5000 c3=10.0000 c4=447.0000 c5 =627.7500
a8=15.0000
CHOSEN construction specification:'LOW',and housing development
type is 'DETACHED'
lot-depth coverage ratio 'dc'= 0.667, lot-width coverage ratio
'wc= 0.750
wall-length coefficient along lot-depth wcd= 2.000
```

```
wall-length coefficient along lot-width wcw= 3.000
SOME OUTPUT OPTIMAL RESULTS AFTER INSTANTLY SOLVING THE 'NLP' MODEL:
The values of Optimal Objective & Constraint Weights are:
w01= 0.0826, w02=0.0214, w03=0.1112, w04=0.4444, w05=0.3404
w31= 0.0926, w32=0.0214, w41=0.1709, w51=0.0187, w61=0.0351
w71= 0.0256, w42=0.0256, w11=0.5113, w21=0.1966, w22=0.0137
The Values of Some Optimal Land Utilization Ratio's are:
Saleable Land Ratio= 0.66,Circulation Area Ratio= 0.24,Open Space
Ratio= 0.10
Optimal WIDTH of the LAYOUT MODULE- N-S Direction  =  100.0 M
Optimal LENGTH of the LAYOUT MODULE-E-W Direction =  180.6 M
Optimal Gross Width(allowing for open space)of LOT in LAYOUT
MODULE=  8.5 M
Optimal Depth of a LOT in the LAYOUT MODULE   =  13.5 M
Optimal NET Width(excluding reservation for open space)of
LOT =   7.4 M
Optimal NET Lot Area (excluding reservation of open space)of
LOT = 100.0 Sq.M
Optimal Least-Cost Possible Per Single Family Dwelling Unit in LAYOUT
MODULE= 13620.8 MU/BU.
Optimal Depth of the Dwelling Built-Unit-Space   =   9.0 M
Optimal Width of the Dwelling Built-Unit-Space   =   5.5 M
Total LOT-FRONTAGE defining Total Number of LOTS in LAYOUT
MODULE=  1012.8 M
Optimal Width of the Subsidiary-Main Street   =   7.2 M
Optimal Width of the Main Street   =   8.7 M
Optimal Width of the Internal Streets   =   5.8 M
TOTAL NUMBER OF LOTS IN LAYOUT MODULE =  119
 Optimal Lot Density In Layout Module Designed & Achieved= 66.000
 Number Of Lots/Hectare
 (Initial-Trial-Values of NLP Indpdnt.Variables are:x=0.2000
 file:'lsou15dg.cpp' giving Single-Family Dwelling-Layout System
 Optimal Design Output(Density Of Lots)
```

2. **Application Example: LCSFL2 Dwelling-Layout System—Single-Family.**
   This is an example of small **'Semi-Detached Housing'** to achieve the *optimal design,* i.e. achieving the least-cost per dwelling unit for a given minimum acceptable *dwelling space*, with a desired open space per BU and a specified density in the layout. The results instantly given by the C++ program: DLSF. CPP is reproduced below, showing both input parameter values chosen by planner-designer and the output *optimal* results given by model.

```
LEAST-COST OPTIMAL DESIGN OF SINGLE-FAMILY DWELLING-LAYOUT SYSTEM
:'ITMODEL- 1'
GET ANY DESIRED LOT-DENSITY/OPEN-SPACE IN LAYOUT MODULE CHOSEN
IS;'MODULE- 1'

SOME KEY INPUT VALUES CHOSEN(INDEPENDENT VARIABLES):
Designed Minimum Acceptable Lot Space Per Built-Unit In The Layout
Module = 100.00 Sq.M.
LAYOUT MODULE TYPE CHOSEN IS= 'GRID-IRON'
```

```
SOME INPUT & CALIBRATED MODEL CONSTANTS(for the Chosen DESIGN
CONDITION)are:
a1=100.0000 a7= 0.0400 a2= 0.1667 a3= 1.6330 a4=151.5152
a5= 6.0000 a6= 2.6000
c1=743.0000 c2=347.5000 c3=10.0000 c4=404.0000 c5 =627.7500
a8=15.0000
CHOSEN construction specification:'LOW',and housing development
type is 'SEMI-DETACHED'
lot-depth coverage ratio 'dc'= 0.667, lot-width coverage ratio
'wc= 0.750
wall-length coefficient along lot-depth wcd= 1.500
wall-length coefficient along lot-width wcw= 3.000
SOME OUTPUT OPTIMAL RESULTS AFTER INSTANTLY SOLVING THE 'NLP' MODEL:
The values of Optimal Objective & Constraint Weights are:
w01= 0.0822, w02=0.0213, w03=0.1163, w04=0.4418, w05=0.3384
w31= 0.0920, w32=0.0213, w41=0.1700, w51=0.0243, w61=0.0349
w71= 0.0255, w42=0.0255, w11=0.5084, w21=0.1955, w22=0.0137
The Values of Some Optimal Land Utilization Ratio's are:
Saleable Land Ratio= 0.66,Circulation Area Ratio= 0.24,Open Space
Ratio= 0.10
Optimal WIDTH of the LAYOUT MODULE- N-S Direction  =  105.2 M
Optimal LENGTH of the LAYOUT MODULE-E-W Direction =  190.0 M
Optimal Gross Width(allowing for open space)of LOT in LAYOUT
MODULE=  8.1 M
Optimal Depth of a LOT in the LAYOUT MODULE   =   14.2 M
Optimal NET Width(excluding reservation for open space)of
LOT =   7.0 M
Optimal NET Lot Area (excluding reservation of open space)of
LOT = 100.0 Sq.M
Optimal Least-Cost Possible Per Single Family Dwelling Unit In Layout
Module= 13023.8 MU/BU.
Optimal Depth of the Dwelling Built-Unit-Space   =   9.5 M
Optimal Width of the Dwelling Built-Unit-Space   =   5.3 M
Total LOT-FRONTAGE defining Total Number of LOTS in LAYOUT MODULE=
1065.3 M
Optimal Width of the Subsidiary-Main Street  =   7.6 M
Optimal Width of the Main Street  =   9.1 M
Optimal Width of the Internal Streets  =   6.1 M
TOTAL NUMBER OF LOTS IN LAYOUT MODULE =  132
Optimal Lot Density In Layout Module Designed & Achieved= 66.000
Number Of Lots/Hectare
(Initial-Trial-Values of NLP Indpdnt.Variables are:x=0.2000
file:'lsou15sg.cpp' giving Single-Family Dwelling-Layout System
Optimal Design Output(Density Of Lots)
```

3. **Application Example: LCSFL3 Dwelling-Layout System—Single-Family.**
   This is an example of small **'Row Housing'** to achieve the *optimal design,*
   i.e. achieving the least-cost per dwelling unit for a given minimum acceptable
   dwelling space, with a desired open space per BU and a specified density in the
   layout. The results instantly given by the C++ program: DLSF.CPP is
   reproduced below, showing both the input parameter values chosen by planner-
   designer and the output *optimal* results given by the model.

```
LEAST-COST OPTIMAL DESIGN OF SINGLE-FAMILY DWELLING-LAYOUT SYSTEM
:'ITMODEL- 1'
GET ANY DESIRED LOT-DENSITY/OPEN-SPACE IN LAYOUT MODULE CHOSEN
IS;'MODULE- 1'

SOME KEY INPUT VALUES CHOSEN(INDEPENDENT VARIABLES):
Designed Minimum Acceptable Lot Space Per Built-Unit In The Layout
Module = 100.00 Sq.M.
LAYOUT MODULE TYPE CHOSEN IS= 'GRID-IRON'
SOME INPUT & CALIBRATED MODEL CONSTANTS(for the Chosen DESIGN
CONDITION)are:
a1=100.0000 a7= 0.0400 a2= 0.1667 a3= 1.6330 a4=151.5152
a5= 6.0000 a6= 2.6000
c1=743.0000 c2=347.5000 c3=10.0000 c4=264.5000 c5 =837.0000
a8=15.0000
CHOSEN construction specification:'LOW', and housing development
type is 'ROW'
lot-depth coverage ratio 'dc'= 0.500, lot-width coverage ratio 'wc=
1.000
wall-length coefficient along lot-depth wcd= 1.000
wall-length coefficient along lot-width wcw= 3.000
SOME OUTPUT OPTIMAL RESULTS AFTER INSTANTLY SOLVING THE 'NLP' MODEL:
The values of Optimal Objective & Constraint Weights are:
w01= 0.0647, w02=0.0167, w03=0.1268, w04=0.4366, w05=0.3552
w31= 0.0724, w32=0.0167, w41=0.1338, w51=0.0544, w61=0.0275
w71= 0.0201, w42=0.0201, w11=0.4890, w21=0.1539, w22=0.0108
The Values of Some Optimal Land Utilization Ratio's are:
Saleable Land Ratio= 0.66,Circulation Area Ratio= 0.24,Open Space
Ratio= 0.10
Optimal WIDTH of the LAYOUT MODULE- N-S Direction  = 145.7 M
Optimal LENGTH of the LAYOUT MODULE-E-W Direction = 263.1 M
Optimal Gross Width(allowing for open space)of LOT in LAYOUT MODULE=
5.8 M
Optimal Depth of a LOT in the LAYOUT MODULE   = 19.7 M
Optimal NET Width(excluding reservation for open space)of LOT = 5.1
M
Optimal NET Lot Area (excluding reservation of open space)of LOT =
100.0 Sq.M
Optimal Least-Cost Possible Per Single Family Dwelling Unit In Layout
Module= 11948.4 MU/BU.
Optimal Depth of the Dwelling Built-Unit-Space   = 9.9 M
Optimal Width of the Dwelling Built-Unit-Space   = 5.1 M
Total LOT-FRONTAGE defining Total Number of LOTS in LAYOUT MODULE=
1475.1 M
Optimal Width of the Subsidiary-Main Street  = 10.5 M
Optimal Width of the Main Street  = 12.6 M
Optimal Width of the Internal Streets  = 8.4 M
TOTAL NUMBER OF LOTS IN LAYOUT MODULE = 253
Optimal Lot Density In Layout Module Designed & Achieved= 66.000
Number Of Lots/Hectare
(Initial-Trial-Values of NLP Indpdnt.Variables are:x=0.2000
file:'lsou15rg.cpp' giving Single-Family Dwelling-Layout System
Optimal Design Output(Density of Lots)
```

The above three examples are taken to demonstrate the variations in cost and in other quantitative design *efficiency* indicators, depending on the type of housing development chosen by planner-designers, while other factors such as size of dwelling in terms of dwelling space; construction specification standard; street-width hierarchy, circulation interval ratio and infrastructure standards; environmental standards in terms of open-space per Built-Unit (BU); and density standard in terms of number of single-family dwellings per unit area, are kept unchanged. It would be seen from the above computer results that the **Least-Cost** per BU are **13,681.7 MU/BU**, **13,023.8** MU/BU and **11,948.4 MU/BU.** the Housing Development Type: 'Detached' **(Example: LCSFL1)**, 'Semi-Detached' **(Example: LCSFL2)**, and 'Row Housing' **(Example: LCSFL3)**, respectively. Thus, the 'Detached' housing is **1.15 times** the cost of 'Row Housing' with the same size of dwelling, construction specification standard, street-width hierarchy, circulation interval ratio and infrastructure standards; environmental standards, and the density standard, as stated above. *Such an analysis is rarely possible in conventional planning and design practices, causing many inefficiencies and inequities and consequent housing and urban problems.* This highlights the importance of adopting the integrated *computer-aided design by optimization* in dwelling-layout systems planning and design, to achieve the *resource-efficiency* and *social equity* in urban operations. It may be noted that each of the above application example is real and represents a specific physical design which can be drawn, as shown in the next example, again, emphasizing the advantage of this technique compared to the conventional planning and design practice in AEC operations.

4. **Application Example: LCSFL4 Dwelling-Layout System—Single-Family.** This is an example of medium size 'Detached Housing' to achieve the *optimal design,* i.e. achieving the least-cost per dwelling unit for a given minimum acceptable *dwelling space*, with a desired open space per BU and a specified density in the layout. The results instantly given by the C++ program: DLSF. CPP is reproduced below, showing both the input parameter values chosen by the planner-designer and the output *optimal* results given by the model.

```
Least-Cost Optimal Design of Single-Family Dwelling-Layout System
:'ITMODEL- 1'
Get Any Desired Lot-Density/Open-Space in Layout Module Chosen
is;'Module- 1'
Some Key Input Values Chosen(Independent Variables):
Designed Minimum Acceptable Lot Space Per Built-Unit In The Layout
Module = 250.00 Sq.M.
Layout Module Type Chosen Is= 'Grid-Iron'
Some Input & Calibrated Model Constants(for the Chosen DESIGN
CONDITION)are:
a1=250.0000 a7= 0.0500 a2= 0.1667 a3= 1.6333 a4=384.6150
a5= 6.0000 a6= 2.6000
c1=1593.0000 c2=672.0000 c3=10.0000 c4=853.8759 c5 =837.4464
```

```
a8=20.0000
CHOSEN construction specification: 'medium',and housing development
type is 'detached'
lot-depth coverage ratio 'dc'= 0.660, lot-width coverage ratio
'wc= 0.670
wall-length coefficient along lot-depth wcd= 3.999
wall-length coefficient along lot-width wcw= 4.480
SOME OUTPUT OPTIMAL RESULTS AFTER INSTANTLY SOLVING THE 'NLP'
MODEL:
The values of Optimal Objective & Constraint Weights are:
w01= 0.941, w02=0.0266, w03=0.1064, w04=0.4468, w05=0.3061
w31= 0.0864, w32=0.0266, w41=0.2103, w51=0.0200, w61=0.0468
w71= 0.0168, w42=0.0168, w11=0.5163, w21=0.2271, w22=0.0202
The Values of Some Optimal Land Utilization Ratio's are:
Saleable Land Ratio= 0.65,Circulation Area Ratio= 0.30,Open Space
Ratio= 0.05
Optimal WIDTH of the LAYOUT MODULE- N-S Direction  = 148.5 M
Optimal LENGTH of the LAYOUT MODULE-E-W Direction = 268.9 M
Optimal Gross Width(allowing for open space)of LOT in LAYOUT
MODULE= 14.3 M
Optimal Depth of a LOT in the LAYOUT MODULE  = 18.9 M
Optimal NET Width(excluding reservation for open space)of LOT= 13.2
M
Optimal NET Lot Area (excluding reservation of open space)of
LOT = 250.0 Sq.M
Optimal Least-Cost Possible per Single Family Dwelling Unit in Layout
Module= 36155.0 MU/BU.
Optimal Depth of the Dwelling Built-Unit-Space   = 11.5 M
Optimal Width of the Dwelling Built-Unit-Space   = 8.9 M
Total LOT-FRONTAGE defining Total Number of LOTS in LAYOUT
MODULE= 1481.1 M
Optimal Width of the Subsidiary-Main Street   = 13.4 M
Optimal Width of the Main Street  = 16.1 M
Optimal Width of the Internal Streets  = 10.8 M
TOTAL NUMBER OF LOTS IN LAYOUT MODULE = 104
Optimal Lot Density In Layout Module Designed & Achieved= 26.000
Number of Lots/Hectare
(Initial-Trial-Values of NLP Indpdnt.Variables are:x=0.3000
file:'lcsfoutd.cpp' giving Single-Family Dwelling-Layout System
Optimal Design Output(Density of Lots)
```

The output optimal numerical values defining the physical design of the optimal dwelling-layout system design, can be easily drawn as revealed in Fig. 11.5 (Chakrabarty, 1988a, 1988b) showing the optimal physical design of single-family dwelling-layout system using the numerical results (with slight variation due to decimal error) instantly given by the C++ program: DLSF. CPP for the Case: LCSFL4 above. Again, by incorporating a suitable LISP file in the C++ program: DLSF.CPP the generation of output optimal design drawing, both soft copy and hard copy, can be completely automated as illustrated in earlier chapters including Chap. 9. Thus, planner-designers can use this computer-aided design optimal design technique including graphics, without being conversant with the complicated, tedious, and voluminous computations involved, in utilizing *Optimization* and operations research techniques and to solve and use the *Optimizing Models* like above. Interested planner-designers can even contribute for further development of this technique by going through and using the various C++ programs (including the C++ program: DLSF.CPP cited above) and Auto Lisp Programs, complete listing of which are supplied in an online space connected to this book.

11.11.1.2   Context-B: High-Income Countries

In single-family dwelling-layout system design, generally, a complete dwelling-layout system with a number of independent lots with dwelling in a repetitive pattern is constructed and supplied to the individual beneficiary, permitting *optimization* taking full advantage of flexibility in dimensions of various components of the dwelling-layout system. This is, usually acceptable to low- and middle-income families. However, for high-income families such dwelling design may be considered monotonous and not acceptable, who would like to get the developed lot but construct the dwelling with individual design according to their own choice. In such cases, the optimal layout may be developed to supply individual lots fixed size to the high-income families, who may be allowed to construct the individual dwelling according to their choice. Therefore, no application examples of single-family dwelling-layout systems design by *optimization* for high-income categories are included here. However, even in these cases, individual families can *optimize* various components of the dwelling, such as the wall lengths for different room-arrangement matrices chosen in individual dwelling design as outlined below.

11.11.1.3   Single-Family Dwelling with Fixed Lot
SIZE-Room-Arrangement Matrix and Optimization

Individual families belonging to any income group who construct their houses in a lot of fixed size can also apply *optimization* for various components of the dwelling such as wall length, reinforced concrete elements such as slab-floor, beams, and so on, and thus attain a *resource-efficient* dwelling design solution. In Sect. 11.5.2.2, it is stated that one important aspect of dwelling unit design modules for the purpose of optimization related to costs is the availability of wall sharing options, which can be easily represented by room-arrangement matrices as shown in Fig. 11.6. It is possible to *optimize* the dwelling wall length to minimize the total wall length in a single-family dwelling design. This can be done by suitably proportioning the dwelling coverage ratio along the width and depth of the lot of fixed size in a single-family dwelling design considering the prescribed built-space coverage ratio, using the optimization principles as presented below.

Earlier the room-arrangement matrices are shown in Fig. 11.6 and the corresponding wall number-sharing coefficients are also shown in Table 11.2.

Using the above parameters along with the other factors as defined below, we can express the total length of walls in a single-family dwelling with fixed lot-size in terms of these parameters as shown below.

NWCD = wall number-sharing coefficients along the depth of the lot.
NWCW = wall number-sharing coefficients along the width of the lot.
WCR = Lot width coverage ratio, i.e. dwelling built-space coverage ratio along the width of the lot.
DCR = Lot depth coverage ratio, i.e. dwelling built-space coverage ratio along the depth of the lot.
CVR = lot area coverage ratio prescribed by the planning authority.
x_1 = width of the lot,
x_2 = depth of the lot,
R = lot depth and width ratio, i.e. $R = x_2/x_1$.

Then, the total length of walls in the dwelling designated as 'L', will be given by

$$L = (\text{NWCD} + 1) \times \text{DCR} \times x_2 + (\text{NWCW} + 1) \times \text{WCR} \times x_1$$
$$= (\text{NWCD} + 1) \times \text{DCR} \times R \times x_1 + (\text{NWCW} + 1) \times x_1 \times \text{CVR}/\text{DCR}$$

To find the value of 'DCR' which will *minimize* the value of 'L', we can put $dL/\text{DCR} = 0$.

Thus, we get the equation:

$$-(\text{NWCW} + 1) \times \text{CVR} \times x_1/\text{DCR}^2 + (\text{NWCD} + 1) \times R \times x_1 = 0,$$
$$\text{or,} \quad \text{DCR}^2 \times (\text{NWCD} + 1) \times R \times x_1 = (\text{NWCW} + 1) \times \text{CVR} \times x_1$$

Or, the expression for the **Optimal Value** of 'DCR' will be:

$$\text{DCR} = \sqrt{\frac{((\text{NWCW} + 1) \times \text{CVR})}{(\text{NWCD} + 1) \times R}} \tag{11.84}$$

Thus, if an individual lot owner adopts room-arrangement matrix of (2 × 4, i.e. NWCD = 2 and NWCW = 4) as shown in Fig. 11.6, lots supplied by the housing agency with a lot ratio of 1:2 (i.e. $R = x_2/x_1 = 2$), and the planning authority prescribes a lot coverage ratio of 0.5 (i.e. CVR = 0.5), the *optimum* dwelling coverage ratio along the depth of the lot will be:

$$\text{DCR} = \sqrt{\frac{(5 \times 0.5)}{(3 \times 2)}} = 0.645,$$

and *optimum* dwelling coverage ratio along the width of the lot: WCR = 0.5/DCR =0.774.

We can now calculate the total length of wall along x_1 direction, i.e. $L1 = 5 \times x_1 \times WCR = 3.87 \times x_1$.

Similarly, the total length of wall along x_2 direction, i.e. $L2 = 3 \times R \times x_1 \times DCR = 3.87 \times x_1$.

Again, if only room-arrangement matrix is changed to (3×3) as shown in Fig. 11.6, the *optimum* dwelling coverage ratio along the depth of the lot will be:

$$DCR = \sqrt{\frac{(4 \times 0.5)}{(4 \times 2)}} = 0.5 \text{ and } WCR = 0.5/.5 = 1.$$

Here, $L1 = 4 \times x_1 \times WCR = 4 \times x_1$; and $L2 = 4 \times R \times x_1 \times DCR = 4 \times x_1 \times 2 \times x_1 \times 0.5 = 4 \times x_1$.

Thus, in this case, a 'row' housing with dwelling coverage ratio along the depth of the lot $= 0.5$, and dwelling coverage ratio along the width of the lot $= 1$, will be the *optimal* solution to *minimize* the total wall length in the dwelling.

On the other hand, if lots supplied by the housing agency with a lot ratio of 1:1, selection of room-arrangement matrix of (2×4) will give *optimal* solution in terms of the *optimum* dwelling coverage ratio along the depth of the lot as:

$$DCR = \sqrt{\frac{(5 \times 0.5)}{(3 \times 1)}} = 0.913, \text{ and } WCR = 0.5/DCR = 0.548$$

Here, $L1 = 5 \times x_1 \times WCR = 2.74 \times x_1$; and $L2 = 3 \times R \times x_1 \times DCR = 3 \times x_1 \times 1 \times 0.913 = 2.74 \times x_1$.

Thus, by a cursory look of the results of the Eq. (11.84) as presented above, one can discern the following simple *decision* or *design rules* if one has to operate *optimally*:

1. *The total length of walls along the depth of the lot should be equal to the total length of walls along the width of the lot.*
2. *The above design or decision rule is applicable for any room-arrangement matrix.*
3. *If the calculated value of DCR or WCR is more than '1', it signifies an infeasible solution for the chosen decision-set and indicating that a different room-arrangement matrix (which is entirely within the control of designer-individual dwelling owner) has to be chosen by the designer if value of other parameters of the decision-set, i.e. R and CVR cannot be changed.*

The above simple analysis shows that the 'lot width-depth ratio' and 'lot coverage ratio' both determined by the housing and planning authority, and the 'room-arrangement matrix' selected by the individual owner family, are interlinked for *optimization* of total wall length, which is a very important component of dwelling-building cost. Therefore, it is desirable that *optimization* principles should also form part of the *planning regulations* and the housing and planning authorities in all countries may consider all the above aspects while carrying out area development and formulating planning regulations, which will promote *optimization* and *resource-efficient* development of single-family dwelling-layout system even in context of high-income countries.

11.11.2   Application Examples of Most-Benefit-Design: Single-Family

The optimization model for **Most-Benefit-Design** of **Single-Family Dwelling-Layout System** developed and solved in Sect. 11.7 above is converted into a computer program in the form of Model-2 in the C++ program DLSF.CPP. Illustrative numerical application examples of computer-aided single-family dwelling-layout systems design by optimization using the above C++ program DLSF.CPP is presented below.

In this model, the factors a_1, a_2, a_3, a_4, a_5, a_6, a_7, a_8, a_9, a_{10}, a_9, and a_{11} are the model constants, related to the physical design and the cost parameters of the buildings (i.e. dwellings) and layout modules; and the design standards mostly dependent on the input design conditions, decided by an architect, engineer, urban planner-urban manager as per requirements, and as may be applicable. Accordingly, the model can be calibrated and applied in a site-specific case as illustrated below with application examples.

11.11.2.1   Context-A: Middle/Low-Income Countries—Single-Family

Here, three application examples are taken with input parameters indicated in each example problem:

1. **Application Example: MBSFL1 Dwelling-Layout System.**
   This is an example of small **'Detached Housing'** to achieve the *optimal design,* i.e. maximize the *dwelling space* (which is a high priority particularly for low-income households) within the constraint of investment or cost which intended beneficiaries and stakeholders can afford AFC = **30,000.0 MU** (Monetary Units) adopting **Layout Module: A,** with a desired open space per BU and a specified density in the layout. In Chap. 4, the procedure to determine the unit costs of foundation, wall, and so on, in a site-specific case, has been presented. Similarly, the procedure to derive unit design module coefficients for layout and dwelling designs forming the dwelling-layout system is also outlined in Sect. 11.5.1 above. The results instantly given by the **C++ program: DLSF.CPP** are reproduced below, showing both the input parameter values chosen by the planner-designer and the output *optimal* results given by the model.

Most-Benefit Optimal Design of Single-Family Dwelling-Layout System
:'ITMODEL- 2'
Get any Desired Lot-Density/Open-Space in Layout Module Chosen
is;'Module- 1'

SOME KEY INPUT VALUES CHOSEN(INDEPENDENT VARIABLES):
Designed Maximum Affordable-Cost per Built-Unit in the LAYOUT MODULE
= 30000.0 MU
Layout Module Type Chosen Is= **'GRID-IRON'**
Calibrated Model Constants (for the Chosen DESIGN CONDITION) are:
a7= 0.0400 a2= 2.6000 a3= 0.1667 a4= 1.6330 **a5= 0.0045 a6=15.0000**
a8= 0.079650, a9= 0.033600, a10= 0.001000, a11= 0.023693, a12=
0.020925, a1= 6.000000
CHOSEN construction specification:'medium',and housing development
type is **'DETACHED'**
SOME OUTPUT OPTIMAL RESULTS AFTER INSTANTLY SOLVING THE 'NLP' MODEL:
The values of Optimal Objective & Constraint Weights are:
w01= 1.0000, w11=0.3482, w12=0.0709, w21=0.6925, w22=0.0484
w31= 0.6323, w32=0.0602, w41=-0.0586, w51=0.0602, w61=0.1193
w71= 0.2734, w72=0.0709, w73=0.4068, w74=0.7120, w75=0.3677
The Values of Some Optimal Land Utilization Ratio's are:
Saleable Land Ratio= 0.71, Circulation Area Ratio= 0.22, Open Space
Ratio= 0.07
Optimal WIDTH of the LAYOUT MODULE- N-S Direction = 118.5 M
Optimal LENGTH of the LAYOUT MODULE-E-W Direction = 192.8 M
Optimal Gross Width(allowing for open space) of LOT in LAYOUT
MODULE= 10.5 M
Optimal Depth of a LOT in the LAYOUT MODULE   =   16.4 M
Optimal NET Width(excluding reservation for open space) of LOT= 9.6 M
Optimal NET Lot Area (excluding reservation of open space) of LOT =
157.5 Sq.M
Optimal Maximum Possible Built-Unit Space Per Single Family Dwelling
in Layout Module = **78.8 Sq.M.**
Optimal Depth of the Dwelling Built-Unit-Space   = 10.9 M
Optimal Width of the Dwelling Built-Unit-Space   =   7.2 M
Total LOT-FRONTAGE defining Total Number of LOTS in LAYOUT
MODULE= 1081.0 M
Optimal Width of the Subsidiary-Main Street   =   7.7 M
Optimal Width of the Main Street   =   9.3 M
Optimal Maximum Possible Gross Area Per Lot (Incldg.Open-Space) In
Layout Module = 172.5 Sq.M.
TOTAL NUMBER OF LOTS IN LAYOUT MODULE = 103
Optimal Lot Density In Layout Module Designed & Achieved= 45.000
Number of Lots/Hectare
(Initial-Trial-Values of NLP Indpdnt.Variables are:x=0.2000,
y=0.3000, z=0.3000
file:'bsou15d.cpp' giving Single-Family Dwelling-Layout System
Optimal Design Output(Density of Lots)

2. **Application Example: MBSFL2 Dwelling-Layout System.**
This is an example of **'Row Housing'** to achieve *optimal design,* i.e. maximize
the *dwelling space* within affordable cost (AFC) same as above = 30,000 MU

(Monetary Units), adopting same **Layout Module:A** as above. The results instantly given by MODEL-2 of C++ program: DLSF.CPP are reproduced below, showing both the input parameter values chosen by the planner-designer and the output *optimal* results given by the model.

```
Most-Benefit Optimal Design of Single-Family Dwelling-Layout System
:'ITMODEL- 2'
Get any Desired Lot-Density/Open-Space in Layout Module Chosen
is;'Module- 1'
SOME KEY INPUT VALUES CHOSEN(INDEPENDENT VARIABLES):
Designed Maximum Affordable-Cost per Built-Unit in the LAYOUT MODULE
= 30000.0 MU
LAYOUT MODULE TYPE CHOSEN IS= 'GRID-IRON'
CALIBRATED Model Constants(for the Chosen DESIGN CONDITION) are:
a7= 0.0400 a2= 2.6000 a3= 0.1667 a4= 1.6330 a5= 0.0045 a6=15.0000
a8=  0.079650 , a9=  0.033600 , a10=  0.001000 , a11=  0.017462 ,
a12=  0.020925, a1=  6.000000
CHOSEN construction specification:'medium',and housing development
type is 'ROW'
SOME OUTPUT OPTIMAL RESULTS AFTER INSTANTLY SOLVING THE 'NLP' MODEL:
The values of Optimal Objective & Constraint Weights are:
w01= 1.0000, w11=0.8577, w12=0.0528, w21=0.9802, w22=0.0685
w31= 0.9051, w32=0.0752, w41=0.7334, w51=0.0752, w61=0.1213
w71= 0.0698, w72=0.0528, w73=0.1243, w74=0.2175, w75=0.0949
The Values of Some Optimal Land Utilization Ratio's are:
Saleable Land Ratio= 0.81,Circulation Area Ratio= 0.12,Open Space
Ratio= 0.07
Optimal WIDTH of the LAYOUT MODULE- N-S Direction  =  141.8 M
Optimal LENGTH of the LAYOUT MODULE-E-W Direction =   79.1 M
Optimal Gross Width(allowing for open space) of LOT in LAYOUT MODULE=
8.8 M
Optimal Depth of a LOT in the LAYOUT MODULE    =   22.3 M
Optimal NET Width(excluding reservation for open space) of LOT=  8.1 M
Optimal NET Lot Area (excluding reservation of open space) of LOT =
180.6 Sq.M
Optimal Maximum Possible Built-Unit Space Per Single Family Dwelling
in Layout Module =  90.3 Sq.M.
Optimal Depth of the Dwelling Built-Unit-Space    =   11.1 M
Optimal Width of the Dwelling Built-Unit-Space    =    8.1 M
Total LOT-FRONTAGE defining Total Number of LOTS in LAYOUT MODULE=
443.5 M
Optimal Width of the Subsidiary-Main Street   =    3.2 M
Optimal Width of the Main Street   =    3.8 M
Optimal Maximum Possible Gross Area Per Lot (Incldg.Open-Space) In
Layout Module  =  195.6 Sq.M.
TOTAL NUMBER OF LOTS IN LAYOUT MODULE =   50
Optimal Lot Density In Layout Module Designed & Achieved= 45.000
Number Of Lots/Hectare
(Initial-Trial-Values of NLP Indpdnt.Variables are:x=0.2000,
y=0.3000, z=0.3000
 file:'bsou15r.cpp' giving Single-Family Dwelling-Layout System
Optimal Design Output(Density of Lots)
```

3. Application Example: MBSFL3 Dwelling-Layout System.

This is an example of **'Semi-Detached Housing'** to achieve the *optimal design,* i.e. maximize the *dwelling space* within affordable cost AFC same as above = 30,000 MU (Monetary Units), adopting the same **Layout Module: B as** above. The results instantly given by the C++ program: DLSF.CPP are reproduced below, showing both the input parameter values chosen by the planner-designer and the output *optimal* results given by the model.

```
Most-Benefit Optimal Design of Single-Family Dwelling-Layout System
:'ITMODEL-2'
Get any Desired Lot-Density/Open-Space in Layout Module Chosen
is;'Module-3'
Some Key Input Values Chosen(Independent Variables):
Designed Maximum Affordable-Cost per Built-Unit in the Layout Module
= 30000.0 MU
Layout Module Type Chosen is= 'Loop'
Calibrated Model Constants(For The Chosen Design Condition)Are:
a7=0.0400 a2= 2.6000 a3=0.1667 a4=1.2000 a5=0.0058 a6=25.0000
a8=  0.079600 , a9=  0.035600 , a10=  0.001000 , a11=  0.031200 ,
a12=  0.027900, a1=  6.000000
CHOSEN construction specification:'medium',and housing development
type is 'SEMI-DETACHED'
Some Output Optimal Results after Instantly Solving the 'NLP' Model:
The values of Optimal Objective & Constraint Weights are:
w01= 1.0000, w11=0.3455, w12=0.0723, w21=0.6870, w22=0.0346
w31= 0.5604, w32=0.1266, w41=0.0200, w51=0.1266, w61=0.1070
w71= 0.2692, w72=0.0723, w73=0.3255, w74=0.7811, w75=0.4396
The Values of Some Optimal Land Utilization Ratio's are:
Saleable Land Ratio= 0.64,Circulation Area Ratio= 0.21,Open Space
Ratio= 0.15
Optimal WIDTH of the LAYOUT MODULE- N-S Direction  =  96.2 M
Optimal LENGTH of the LAYOUT MODULE-E-W Direction =  160.2 M
Optimal Gross Width(allowing for open space)of LOT in LAYOUT MODULE=
10.2 M
Optimal Depth of a LOT in the LAYOUT MODULE   =  13.3 M
Optimal NET Width(excluding reservation for open space)of LOT=  8.3 M
Optimal NET Lot Area (excluding reservation of open space)of LOT =
110.7 Sq.M
Optimal Maximum Possible Built-Unit Space per Single Family Dwelling
in LAYOUT MODULE= 55.3 Sq.M.
Optimal Depth of the Dwelling Built-Unit-Space   =  9.9 M
Optimal Width of the Dwelling Built-Unit-Space   =  5.6 M
Total LOT-FRONTAGE defining Total Number of LOTS in LAYOUT MODULE=
914.8 M
Optimal Width of the Subsidiary-Main Street   =  6.4 M
Optimal Width of the Main Street   =  7.7 M
Optimal Maximum Possible Gross Area per LOT (incldg.open-space) in
LAYOUT MODULE  =  135.7 Sq.M.
Total Number of LOTS in LAYOUT MODULE =   89
Optimal Lot Density in Layout Module Designed & Achieved= 58.000
Number of LOTS/Hectare
```

```
(Initial-Trial-Values of NLP Indpdnt.Variables are:x=0.1000,
y=0.3000, z=0.6000
file:'mbsfoutd.cpp'-mbsfpprf.cpp giving Single-Family Dwelling-
Layout System Optimal Design Output (Density of Lots)
```

Application Examples: MBSFL1 and MBSFL2 above show that the optimal maximum possible built-unit space per single family dwelling within the same affordable cost are: **78.8 Sq.M.** and **90.3 Sq.M.** for **Detached** and **Row Housing,** respectively. Thus, if the row-type development is adopted, as against the detached type housing development, the built-space per dwelling increases by **1.15 times** within the same affordable cost of 30,000 MU/BU, although the same lot-density of **45 lots/hectare** and same open space standard of 15 **Sq.M. per lot** are specified and achieved by the planer-designer in both cases. In case of smaller size dwelling units, this increase ratio may be higher. Therefore, it is imperative to analyse and apply such *optimization* technique to maximize benefit, such as dwelling built-space, in which many user stakeholders would be interested. Density is an important indicator of land utilization *efficiency*. Using the *optimization* model presented a planner-designer can design a layout module for any desired and feasible dwelling density by assigning appropriate value to the design parameter a_5 in the model. Thus, it may be seen that in all the three cases presented above, the lot-density of **45 ($a5 = 0.0045$)**, **45 ($a5 = 0.0045$)**, and **58 ($a5 = 0.0058$)** lots/hectare are assigned and achieved in respective to layout unit-design module. Using this technique, a planner-designer can carry out a number of iterations to examine a range of lot-densities, and thus, derive the *optimal* lot-density which is feasible, and also, gives the maximum dwelling space in a site-specific case; simultaneously attaining the desired open space standard by assigning suitable value to the design parameter a_6 in the model. Putting the same values of input parameters (including affordable investment limit) except the layout module type chosen is = 'GRID-IRON' as against 'LOOP' in Example: MBSFL3, it can be shown that a 'LOOP' layout slightly increases the optimal maximum possible built-unit space per single family dwelling, indicating slightly more *efficiency* of 'LOOP' layout compared to 'GRID-IRON' layout.

Using the output optimal numerical values (with slight variation due to decimal error) instantly given by the C++ program: DLSF.CPP for the Case: **MBSFL3** above defining the physical design of the optimal dwelling-layout system design can be easily drawn as shown in Fig. 11.7 above. Again, by incorporating a suitable LISP file in the C++ program: DLSF.CPP the generation of output optimal design drawing, both soft copy and hard copy, can be completely automated as illustrated in earlier chapters including Chap. 9.

11.11.2.2   Context-B: High-Income Countries

As earlier stated in a single-family dwelling-layout system design, generally, a complete dwelling-layout system with a number of independent lot and dwelling

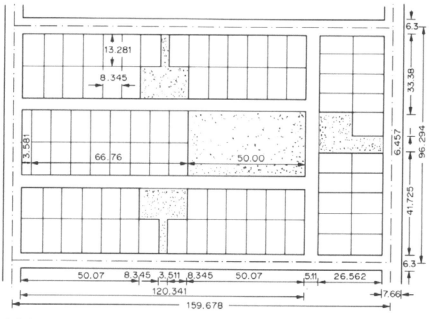

(a) Optimum physical design of layout

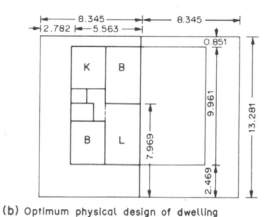

(b) Optimum physical design of dwelling

Fig. 11.7 Optimal physical design of single-family dwelling-layout system—case: MBSF

in a repetitive pattern is constructed and supplied to the individual beneficiary, which may be acceptable to low- and middle-income families, but not to high-income families who would like to get the developed lot but construct the dwelling with individual design according to their own choice. Therefore, no application examples of single-family dwelling-layout systems design by optimization for high-income categories are included here. However, even in these cases, individual families can *optimize* the wall lengths for different room-arrangement matrices chosen in individual dwelling design as outlined in Sect. 11.11.1.3 above.

11.12 Multi-family Dwelling-Layout Systems Design by Optimization

Here, a number of application examples of optimization of multi-family dwelling-layout systems design are presented; both for least-cost design and most-benefit-design, and also covering both contexts of low- and high-income countries, showing general and universal applicability of this *optimization* technique for all income categories and in all countries, i.e. low, middle as well as high-income countries.

11.12.1 Application Examples of Least-Cost Design: Multi-family

The optimization model for **Least-Cost Design** of Multi-**Family Dwelling-Layout System** developed and solved in Sect. 11.8 above is converted into a computer program in the form of Model-1 in the C++ program DLMF.CPP. Illustrative numerical examples of computer-aided multi-family dwelling-layout systems design by optimization using the above C++ DLMF.CPP is presented below. Here, c_1, c_2, c_3, c_4, c_5, c_6, a_1, a_2, a_3, a_4, a_5, and a_6 are the model constants, related to physical design and the cost parameters of the buildings (i.e. dwellings) and layout modules; and the design standards mostly dependent on the input design conditions, decided by an architect, engineer, urban planner-urban manager as per requirements, and as may be applicable. These are explained in more details in Sect. 11.8 above and also below. Accordingly, the model can be calibrated and applied in a site-specific case as illustrated below with application examples.

11.12.1.1 Context-A: Middle/Low-Income Countries—Multi-family

In this model, the factors a_1, a_2, a_3, a_4, a_5, a_6, a_7, a_8, a_9, a_{10}, a_9, and a_{11} are the model constants, related to the physical design and the cost parameters of the buildings (i.e. dwellings) and layout modules; and the design standards mostly dependent on the input design conditions, decided by an architect, engineer, urban planner-urban manager as per requirements, and as may be applicable. Accordingly, the model can be calibrated and applied in a site-specific case as illustrated below with application examples. Here, one application example is taken with input parameters indicated in the example problem:

1. Application Example: MFLCL1 Dwelling-Layout System.

```
SUMMARY OF ''outmflc.cpp'' RESULTS IN THE INSTANT CASE ARE:
OPTIMISATION MODEL CHOSEN IS = 1
INPUT MODEL CONSTANT VALUES CHOSEN BY USER ARE:
nf=    4 n=    4 nb=    8 den=400.0000 BUS=38.8230 I= 6.00 S= 9.00
c1=21868.1085 c2=20995.1980 c3=686.7049 c4=2211.0000 c5
=2370.0000 c6= 87.0000
a1=53.3870 a2= 1.0000 a3= 0.2500 a4= 0.1150 a5 =25.0000 a6= 0.2500
ix = 0.30000 iy= 0.20000
OUTPUT OPTIMAL VALUES GIVEN BY THE MODEL ARE:
Optimal Geometrical Programming Weights are:
w01= 0.2446 w02= 0.2446 w03= 0.4583 w04= 0.0116 w05 = 0.0116 w06=
0.0272
w11= 0.7029 w21= 0.0398 w31= 0.5019 w41= 0.0116 w51 = 0.0398 w61=
0.2573
Optimal Dimensions and Physical Parameter Values are:
Depth of Building-Blocki.e.Dwelling-Unit Design Module parallel to
staircase= 14.32 M
Width of Building-Blocki.e.Dwelling-Unit Design Module
perpendicular to staircase= 14.91 M
Width of Layout Unit Design Module in N-S direction(between road
center lines)= 58.55 M
Length of Layout Unit Design Module in E-W direction(between road
center lines)= 54.65 M
Optimal Building Rise i.e.number of floors in Dwelling-Unit Design
Module=    4
Optimal Number of Building Blocks in Layout Unit Design Module=    8
Optimal Number of Built-Units(BU)per floor in Dwelling-Unit Design
Module=    4
Optimal DENSITY in terms of Built-Units per unit area in Layout Unit
Design Module= 400.00BU/HA
Optimal Semi-Public Open space per BU in Layout Unit Design Module=
15.29 Sq.M./BU
Least-Cost per Built-Unit given by the LCO MODEL = 80000.08MU/BU
```

The output optimal numerical values given above by the C++ program: DLMF.CPP for the application example: **MFLCL1** defines the physical design of the optimal dwelling-layout system design in the above case and can be easily drawn as shown in Fig. 11.8 detailing both the layout and the dwelling components of the optimal physical design of multi-family dwelling-layout system using the numerical results instantly given above. Again, by incorporating a suitable LISP file in the C++ program: DLMF.CPP the generation of the output optimal physical design drawing, both soft copy and hard copy, can be completely automated, in the form of an integrated CAD by optimization software, as illustrated in earlier chapters including Chap. 9.

Thus, planner-designers can use this computer-aided optimal design technique including graphics, even without being conversant with the complicated, tedious, and voluminous computations involved, in utilizing *Optimization* and operations research techniques and to solve and use the *Optimizing Models* like above. Interested planner-designers can even contribute for further development of this

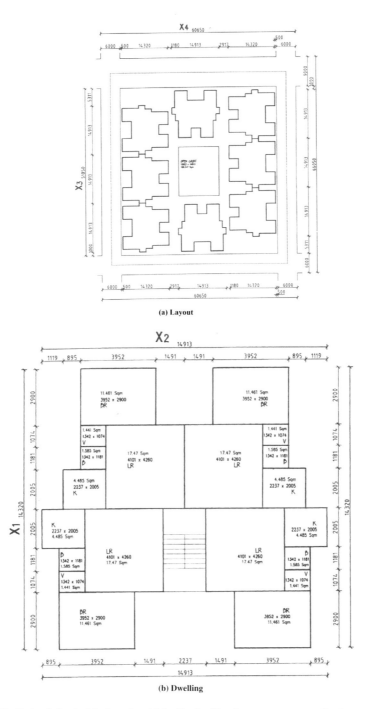

(a) Layout

(b) Dwelling

Fig. 11.8 Optimal physical design of multi-family dwelling-layout system—application example: MFLCL1

technique by going through and using the C++ programs and LISP programs, complete listing of which are supplied in an online space attached to this book.

11.12.1.2   Context-B: High-Income Countries—Multi-family

People belonging to high-income categories may not like to compromise on the desired built-space which may be an uncompromising rigid value for them. However, even they may like to reduce cost by optimization. Hence, three **Application Example**s are presented below to show how optimization can make large cost difference for the same built-space per built-unit. Here, three examples are taken, namely: (a) Low-Rise-Low-Density Case, (b) Low-Rise-Medium-Density Case, and (c) Medium-Rise comparatively High-Density Case.

1. **Application Example: MFLCH1 Dwelling-Layout System Design.**

    (a) **Low-Rise-Low-Density Case.**

    ```
    SUMMARY OF ''outmflc.cpp'' RESULTS IN THE INSTANT CASE ARE:
    OPTIMISATION MODEL CHOSEN IS = 1
    INPUT MODEL CONSTANT VALUES CHOSEN BY USER ARE:
    DWELLING MODULE TYPE CHOSEN IS== 'E'
    nf=   2 n=   5 nb=   4 den= 0.0025 I= 20.00 S= 20.00 BUS= 300.00
    c1=18364.5000 c2=14610.7500 c3=799.0775 c4=2212.0000 c5
    =2370.0000 c6= 1000.0000
    CI=2212.0000 CS=2528.0000 CM= 0.0000 CL=1000.0000 cf
    =1076.0000 cw= 837.0000 cr=737.0000
    cx1= 6.3600 cx2= 5.0600 cba= 0.6884
    a1=435.7931 a2= 1.0000 a3= 0.5000 a4= 0.2500 a5 =400.0000 a6=
    0.5000 ix = 0.20000 iy= 0.20000
    OUTPUT OPTIMAL VALUES GIVEN BY THE MODEL ARE:
    Optimal Geometrical Programming Weights are:
    w01= 0.1194 w02= 0.1194 w03= 0.3438 w04= 0.0113 w05 = 0.0113 w06=
    0.3949
    w11= 0.4631 w21= 0.4062 w31= 0.2500 w41= 0.0113 w51 = 0.4062 w61=
    0.1307
    Optimal Dimensions and Physical Parameter Values are:
     Depth of Building-Blocki.e.Dwelling-Unit Design Module parallel
    to staircase= 26.33 M
    Width of Building-Blocki.e.Dwelling-Unit Design Module
    perpendicular to staircase= 33.10 M
    Width of Layout Unit Design Module in N-S direction(between road
    center lines)= 82.81 M
    Length of Layout Unit Design Module in E-W direction(between road
    center lines)= 77.29 M
    Optimal Dimensions of Rooms(in Meters) are:
    wbr3=  6.62 wbr4=  5.13 wbr5=  5.13 lbr1=  3.95 lbr2=  3.95 lbr3=
    4.74 lbr5=  5.53
    ldn=  6.32 wdn=  7.28 ldr= 10.53 wdr=  3.64 wbr1=  3.64 wbr2=
    3.64 lbr4=  3.95
    Optimal Building Rise i.e.number of floors in Dwelling-Unit Design
    Module=   2
    ```

Optimal Number of Building Blocks in Layout Unit Design Module= 4
Optimal Number of Built-Units (BU) per floor in Dwelling-Unit Design Module= 2
Optimal DENSITY in terms of Built-Units per unit area in Layout Unit Design Module= 25.00BU/HA
Optimal Semi-Public Open space per BU in Layout Unit Design Module= 250.00 Sq.M./BU
Optimal Ground Coverage Ratio by Building Blocks in Layout Module= 0.375
Optimal Floor Area Ratio (FAR) Achieved in Layout Unit Design Module= 0.750
Least-Cost per Built-Unit given by the LCO MODEL =1012925.95MU/BU
mflce25.cpp

2. Application Example: MFLCH2 Dwelling-Layout System Design.

(b) Low-Rise-Medium-Density Case:

SUMMARY OF ''outmflc.cpp'' RESULTS IN THE INSTANT CASE ARE:
OPTIMISATION MODEL CHOSEN IS =1
INPUT MODEL CONSTANT VALUES CHOSEN BY USER ARE:
DWELLING MODULE TYPE CHOSEN IS== 'E'
nf= 2 n= 5 nb= 4 den= 0.0050 I= 20.00 S= 20.00 BUS=300.0000
c1=17927.2500 c2=14262.8750 c3=780.0519 c4=2212.0000 c5=2370.0000 c6= 1000.0000
CI=2212.0000 CS=2528.0000 CM= 0.0000 CL=1000.0000 cf=1076.0000 cw= 837.0000 cr=737.0000
cx1= 6.3600 cx2= 5.0600 cba= 0.6884
a1=435.7931 a2= 1.0000 a3= 0.5000 a4= 0.2500 a5 =200.0000 a6= 0.5000 ix = 0.30000 iy= 0.30000
OUTPUT OPTIMAL VALUES GIVEN BY THE MODEL ARE:
Optimal Geometrical Programming Weights are:
w01= 0.1490 w02= 0.1490 w03= 0.4291 w04= 0.0102 w05 = 0.0102 w06= 0.2525
w11= 0.5781 w21= 0.2627 w31= 0.3082 w41= 0.0102 w51 = 0.2627 w61= 0.1592
Optimal Dimensions and Physical Parameter Values are:
Depth of Building-Block i.e. Dwelling-Unit Design Module parallel to staircase= 26.33 M
Width of Building-Block i.e. Dwelling-Unit Design Module perpendicular to staircase= 33.10 M
Width of Layout Unit Design Module in N-S direction (between road center lines) = 58.54 M
Length of Layout Unit Design Module in E-W direction (between road center lines) = 54.64 M
Optimal Building Rise i.e. number of floors in Dwelling-Unit Design Module= 2
Optimal Number of Building Blocks in Layout Unit Design Module= 4
Optimal Number of Built-Units (BU) per floor in Dwelling-Unit Design Module= 2
Optimal DENSITY in terms of Built-Units per unit area in Layout Unit Design Module= 50.00BU/HA
Optimal Semi-Public Open space per BU in Layout Unit Design Module= 49.89 Sq.M./BU

```
Optimal Ground Coverage Ratio by Building Blocks in Layout
Module= 0.750
Optimal Floor Area Ratio (FAR) Achieved in Layout Unit Design
Module= 1.501
Least-Cost per Built-Unit given by the LCO MODEL = 792175.96MU/BU
mflce50.cpp
```

3. Application Example: MFLCH3 Dwelling-Layout System Design.

(c) Medium-Rise comparatively High-Density Case:

```
SUMMARY OF ''outmflc.cpp'' RESULTS IN THE INSTANT CASE ARE:
OPTIMISATION MODEL CHOSEN IS = 1
INPUT MODEL CONSTANT VALUES CHOSEN BY USER ARE:
DWELLING MODULE TYPE CHOSEN IS= = 'E'
nf=   4 n=   5 nb=   4 den=75.0000 I= 20.00 S= 20.00 BUS=300.0000
c1=28840.0560 c2=22945.0760 c3=650.0432 c4=2212.0000 c5
=2370.0000 c6= 1000.0000
CI=2212.0000 CS=2528.0000 CM= 0.7501 CL=1000.0000 cf
=1076.0000 cw= 837.0000 cr=737.0000
cx1= 6.3600 cx2= 5.0600 cba= 0.6884
a1=435.7931 a2= 1.0000 a3= 0.2500 a4= 0.2500 a5 =133.3333 a6=
0.5000 ix = 0.30000 iy= 0.30000
OUTPUT OPTIMAL VALUES GIVEN BY THE MODEL ARE:
Optimal Geometrical Programming Weights are:
w01= 0.1542 w02= 0.1542 w03= 0.4600 w04= 0.0076 w05 = 0.0076 w06=
0.2165
w11= 0.6142 w21= 0.2241 w31= 0.3159 w41= 0.0076 w51 = 0.2241 w61=
0.1617
Optimal Dimensions and Physical Parameter Values are:
Depth of Building-Blocki.e.Dwelling-Unit Design Module parallel
to staircase= 26.33 M
Width of Building-Blocki.e.Dwelling-Unit Design Module
perpendicular to staircase= 33.10 M
Width of Layout Unit Design Module in N-S direction(between road
center lines)= 67.61 M
Length of Layout Unit Design Module in E-W direction(between road
center lines)= 63.10 M
Optimal Building Rise i.e.number of floors in Dwelling-Unit Design
Module=   4
Optimal Number of Building Blocks in Layout Unit Design Module=   4
Optimal Number of Built-Units(BU)per floor in Dwelling-Unit Design
Module=   2
Optimal DENSITY in terms of Built-Units per unit area in Layout
Unit Design Module= 75.00BU/HA
Optimal Semi-Public Open space per BU in Layout Unit Design
Module= 58.33 Sq.M./BU
Optimal Ground Coverage Ratio by Building Blocks in Layout Module=
0.563
Optimal Floor Area Ratio (FAR) Achieved in Layout Unit Design
Module= 2.250
Least-Cost per Built-Unit given by the LCO MODEL = 615827.13MU/BU
mflce75.cpp
```

It may be seen from the above examples of high-income group housing **instantly** given by the **C++ program DLMF.CPP** as output, that the **Least-Cost** per Built-Unit are **1,012,925.95** MU**, 792,175.96** MU, and **615,827.13 MU** in application examples: MFLCH1 (**Low-Rise-Low-Density Case**), MFLCH2 (**Low-Rise-Medium-Density Case**), and MFLCH3 (**Medium-Rise comparatively High-Density**), respectively. Thus, for the same input BUS = 300 Sq.M. (Dwelling Unit Design Module Type: E—see Fig. 11.5 above) the **Least-Cost** per built-unit in low-rise-low-density case is **1.645 times** that of medium-rise and comparatively high-density case, while the unit costs of land, utility lines, foundation, and building component costs remain same in all cases. It may also be seen that the **Least-Cost** *optimal design* in the low-rise-low-density case, gives the optimal semi-public open space per BU in layout unit design module as high as 250.00 Sq.M./BU, with optimal ground coverage ratio by building blocks in layout module as: 0.375, and the optimal floor area ratio (FAR) in layout unit design module as 0.750 only. These factors have contributed to the increase of **Least-Cost** per Built-Unit by **1.645 times** that of medium-rise and comparatively high-density case, where the optimal semi-public open space per BU is 58.33 Sq.M./BU, the optimal Ground Coverage Ratio is 0.563, and the optimal Floor Area Ratio (FAR) achieved in layout unit design module is 2.250. It may be noted that even if the unit costs of utility lines, foundation, and building component costs are changed in a site-specific case, the relative cost difference per BU may be of the same order. However, if land price defined by the factor c_6 is very low, the relative cost difference per BU will be less. As, for example, increase in **Least-Cost** per built-unit will be 1.464 times (instead of **1.645** times cited above) if land price, i.e. c_6 = 100 MU/Sq.M. instead of 1000 MU/Sq.M. adopted in the above three example cases, other input parameter values remaining unchanged.

Developing an Integrated CAD by Optimization Software
for Dwelling-Layout System

To enable AEC professional to develop skills for **Integrated CAD by Optimization** in building and housing an application software: **'HUDCAD'**, integrating building design optimization and drafting along with complete computer code of **C++** and Lisp programmes, are presented in Chap. 9. The application results show that there can be very large variations in efficiency indicators depending on the planning and design decisions by AEC professionals, which are rarely considered in an integrated manner in conventional planning and design practice in building, housing, and dwelling-layout systems. The output optimal numerical values given above by the C++ program: DLMF.CPP for the application example: **MFLCL1** defines the physical design of the optimal dwelling-layout system design in the above case and can be easily drawn as shown in Fig. 11.8 detailing both the layout and the dwelling components of the optimal physical design of multi-family dwelling-layout system using the numerical results instantly given above. However, by incorporating a suitable LISP file in the C++ program: DLMF.CPP the generation of the output optimal physical design drawing, both soft copy and hard copy, can be completely automated, in the form of an integrated CAD by optimization software, as illustrated in Chap. 9.

Thus, planner-designers can use this computer-aided optimal design technique including graphics, even without being conversant with the complicated, tedious, and voluminous computations involved, in utilizing *Optimization* and operations research techniques and to solve and use the *Optimizing Models* like above. Interested planner-designers can even contribute for further development of this technique by going through and using the C++ programs and LISP Programs, complete listing of which are supplied in an online space attached to this book.

As for example, the copy of the LISP file: dg5.lsp, generated by the C++ program: DLMF.CPP incorporating the numerical results for the **Application Example: MFLCH1 Dwelling-Layout System Design,** is reproduced below:

```
(setq dlis1
(list 26.33 33.10 10.53  3.64  7.90        lb,wb,ldr,wdr,lsc
      7.28    6.32    2.98 300.00 2 4       wdn,ldn,wk,gpa,nf,nb
      3.64  3.64  6.62    5.13    5.13    2.90  wbr1,wbr2,wbr3,wbr4,
                                               wbr5,lbt
      3.95    3.95    4.74    3.95    5.53       lbr1,lbr2,lbr3,lbr4,lbr5
      2.98  6.62  1.58  4.74 20.00 20.00 ) )  wo,wsc,ewl,lk,I,S,
```

It may be seen that the dimensions of the dwelling components of the optimal physical design of multi-family dwelling-layout system, shown in Fig. 11.9 above, are same as those shown in the LISP file. Developing a suitable AutoLISP program and using these numerical results instantly given by the C++ program, as input to this AutoLISP program, the generation of the output optimal physical design drawing, both soft copy and hard copy, can be completely automated, in the form of an integrated CAD by optimization software.

Thus, planner-designers can use this computer-aided optimal design technique including graphics, even without being conversant with the complicated, tedious, and voluminous computations involved, in utilizing *Optimization* and operations research techniques and to solve and use the *Optimizing Models* like above. Interested planner-designers can even contribute for further development of this technique by going through and using the C++ programs and LISP Programs, complete listing of which are supplied in an online space attached to this book.

11.12.2 Application Examples of Most-Benefit-Design: Multi-family

Application of the MBO model to obtain the maximum built-space optimal design of multi-family dwelling-layout system, for a given maximum affordable investment per built-unit, with different other input decision-matrices, which can be considered as urban planning and management decision alternatives. The alternative urban

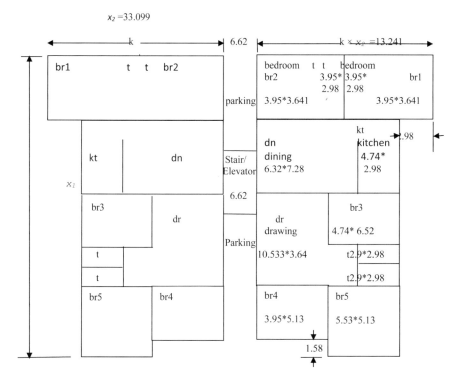

Fig. 11.9 Diagram of *Dwelling* unit design module: E *for High-Income Housing*, showing optimum block/room dimensions (not to scale-*Example: H1*) given by the model

planning and management decisions can be designated in the form of matrices. Even if we consider only four parameters such as land price (LP), built-unit-density (BUD), construction specifications (CS), and building rise (BR), there will be numerous input decision-matrices for each layout and dwelling module pattern selected by an urban planner-designer-manager, giving large variations in productivity ratios of urban development design. In the context of urban development projects, the resource-productivity improvement can be defined in three categories as cited in Sect. 11.10 above. Thus, application of the MBO model in the optimal design of multi-family dwelling-layout system will help attain the resource-productivity improvement of category (i) outlined in the above section.

It is shown earlier that there can be variation of even 4 times in the productivity of urban development designs only due to the urban management decision alternatives chosen by urban professionals, i.e. architects, planners, and engineers, and in view of large number of parameters involved, the need to develop suitable computer package was emphasized (Chakrabarty, 1998). Accordingly a number of C++ programs (including the C++ program: **DLMF.CPP** for the above MBO model to obtain the maximum built-space optimal design of multi-family dwelling-layout system in numeric terms) are developed and available in an online space attached to this book. Similarly, the corresponding LISP Programs to obtain maximum built-space optimal design of multi-family dwelling-layout system in graphic form can be developed, forming part of a comprehensive computer package, for *integrated computer-aided optimal design system* for urban development and management, in line with the procedure presented in Chap. 9. Interested planner-designers can carry out additional research and even contribute for further development of this technique by going through and using the source code of C++ programs and LISP Programs supplied in the book, and thus promote application of *integrated computer-aided design by optimization* in urban development and management, improving *resource-efficiency* and *social equity* in urban operations.

11.13   Sensitivity Analysis of Planning and Design Decisions to Derive an *Optimal* Solution in a *Site-Specific* Case

In the application examples of the above MBO model, using C++ program: **DLMF. CPP**, allowing for four parameters, i.e. **LP-BUD-CS-BR** cited above, the input urban planning and management decisions set can be expressed in the form of decision-matrices such as **LP-BUD-CS-BR** with value of Low (L), Medium (M), and High (H) for each of the above four parameters. These decision-matrices can be designated as: LLLL, LMLL, LMML, LLML, HHLL, HHML, and so on, as earlier explained in Sect. 11.10 above, to carry out a *sensitivity analysis* of various planning and design decisions, and thus arrive at an appropriate *optimal* solution in a *site-specific* case respecting the *constraints* beyond control of planner-designer such as LP which is almost entirely controlled by market; and availing the *opportunities* to maximize the *resource-productivity* (as earlier defined in Sect. 11.10 above) using the parameters such as BUD,CS, and BR which are almost entirely within the control of the planner-designer. Even though the number of alternatives to be considered in the above analysis will be numerous, use of the C++ program: **DLMF.CPP**, giving **instantly** the output results for each input urban planning and management decision-sets as above, will permit such *sensitivity analysis* and decision-making in no time, where all *stakeholders* can *participate*, particularly in selecting suitable input values for the parameters which are within the control of the planner-designer.

Here, the *resource-productivity* is indicated by *maximizing* the built-space per built-unit for each decision-set, given as input including the affordable cost and other parameter values provided by all *stakeholders*, while running the C++ program: **DLMF.CPP**. Incorporating suitable graphic features in such *integrated computer-aided design by optimization* for the urban development and management process (which is feasible as shown in Chap. 9) will not only improve *resource-efficiency* and *social equity* in urban operations, but will also make it a truly *informed* and *participatory* urban development involving all stakeholders, and thus, nipping in the bud many potential urban problems.

Some application examples of the above MBO model, using C++ program: **DLMF.CPP**, allowing for four parameters, i.e. **LP-BUD-CS-BR** cited above, is offered below to illustrate the above process. Input values of only Low (L) and Medium (M) for each of the above four parameters are taken as: 87 and 200 for LP, 200 and 400 for BUD, 2 and 4 for BR, and CS value as: Low and Medium described in Sect. 11.10 above.

11.13.1   Application Examples: Context-A— Middle/Low-Income Countries

Here, four application examples are taken with input parameters indicated in each example problem. Input parameter values in the **Application Example: MFMBL1** are same as in **Application Example: MFLCL1** above for least cost design with interchange of input BUS with output least-cost value in LCO model which is taken as input 'ahc' (also designated as 'afc') in above MBO model to derive the corresponding BUS value as output by this MBO model. It may be noted that the input parameter values related to costs in MBO model are actually cost ratios (i.e. respective cost parameter values such as c_1, c_2, and so on, are divided by 'ahc' value) and accordingly, suitable changes are made. The results **instantly** given by the MODEL-2 of C++ program: DLMF.CPP is reproduced below, showing both the input parameter values chosen by the planner-designer and the output *optimal* results given by the model for each application example.

1. **Application Example: MFMBL1 Dwelling-Layout System.**

```
SUMMARY OF ''outmfmb.cpp'' RESULTS IN THE INSTANT CASE ARE:
OPTIMISATION MODEL CHOSEN IS = 2
INPUT MODEL CONSTANT VALUES CHOSEN BY USER ARE:
nf=   4 n=    4 nb=   8 den=400.0000 ahc=80000.0000 I= 6.00 S= 9.00
CI=2212.00 CS=2528.0000 cf=1076.00 cw= 837.00 cr = 737.00 CL=
87.00
a7= 0.2734 a8= 0.2624 a9= 0.0086 a10= 0.0277 a11 = 0.0296 a12=
0.0011
a1= 0.1723 a2= 1.0000 a3= 0.2500 a4= 0.1250 a5 =25.0000 a6= 0.2500
ix = 0.30000 iy= 0.20000
OUTPUT OPTIMAL VALUES GIVEN BY THE MODEL ARE:
Optimal Geometrical Programming Weights are:
w01= 1.0000 w61= 0.3480 w62= 0.3480 w63= 0.6520 w64 = 0.0180 w65=
0.0180
w11= 0.0567 w21= 0.7141 w31= 0.0180 w41= 0.0567 w51 = 0.3660 w66=
0.0387
Optimal Dimensions and Physical Parameter Values are:
Depth of Building-Blocki.e.Dwelling-Unit Design Module parallel to
staircase= 14.32 M
Width of Building-Blocki.e.Dwelling-Unit Design Module
perpendicular to staircase= 14.91 M
```

```
Width of Layout Unit Design Module in N-S direction(between road
center lines)= 58.55 M
Length of Layout Unit Design Module in E-W direction(between road
center lines)= 54.65 M
Optimal Building Rise i.e.number of floors in Dwelling-Unit Design
Module=   4
Optimal Number of Building Blocks in Layout Unit Design Module=    8
Optimal Number of Built-Units(BU)per floor in Dwelling-Unit Design
Module=   4
Optimal DENSITY in terms of Built-Units per unit area in Layout Unit
Design Module= 400.00BU/HA
Optimal Semi-Public Open space per BU in Layout Unit Design Module=
15.29 Sq.M./BU
Maximum Possible Built-Unit-Space(BUS) given by the MBO MODEL =
38.82 Sq.M/BU
File:MFMBLMM.CPP
```

The above results show that the OUTPUT optimal dimensions and physical parameter values in above Application Example: MFMBL1 are same as those in Application Example: MFLCL1 for the LCO model with interchange of INPUT BUS in LCO model by the INPUT 'ahc' value = 80,000 MU (which is ≈ the least-cost per built-unit given as OUTPUT by the LCO MODEL) in the MBO model. It may also be seen that with this INPUT 'ahc' value the OUTPUT maximum possible built-unit-space (BUS) given by the MBO model = 38.82 Sq.M/BU, which is same as the INPUT BUS value in LCO model. Thus, the LCO and MBO models are interchangeable with the interchange of BUS and 'ahc' values as cited above.

2. Application Example: MFMBL2 Dwelling-Layout System.

```
SUMMARY OF ''outmfmb.cpp'' RESULTS IN THE INSTANT CASE ARE:
OPTIMISATION MODEL CHOSEN IS = 2
INPUT MODEL CONSTANT VALUES CHOSEN BY USER ARE:
nf=   2 n=   4 nb=   8 den=200.0000 ahc=80000.0000 I= 6.00 S= 9.00
a7= 0.1616 a8= 0.1552 a9= 0.0098 a10= 0.0277 a11 = 0.0296 a12=
0.0011
a1= 0.0000 a2= 1.0000 a3= 0.5000 a4= 0.1250 a5 =50.0000 a6=  0.2500
ix = 0.30000 iy= 0.20000
OUTPUT OPTIMAL VALUES GIVEN BY THE MODEL ARE:
Optimal Geometrical Programming Weights are:
w01= 1.0000 w61= 0.3894 w62= 0.3894 w63= 0.6106 w64 = 0.0393 w65=
0.0393
w11= 0.1237 w21= 0.8181 w31= 0.0393 w41= 0.1237 w51 = 0.4287 w66=
0.0844
Optimal Dimensions and Physical Parameter Values are:
Depth of Building-Blocki.e.Dwelling-Unit Design Module parallel to
staircase= 12.42 M
Width of Building-Blocki.e.Dwelling-Unit Design Module
perpendicular to staircase= 12.93 M
Width of Layout Unit Design Module in N-S direction(between road
center lines)= 58.55 M
Length of Layout Unit Design Module in E-W direction(between road
```

center lines)= 54.65 M
Optimal Building Rise i.e.number of floors in Dwelling-Unit Design
Module=   2
Optimal Number of Building Blocks in Layout Unit Design Module=   8
Optimal Number of Built-Units(BU)per floor in Dwelling-Unit Design
Module=   4
Optimal DENSITY in terms of Built-Units per unit area in Layout Unit
Design Module= 200.00BU/HA
Optimal Semi-Public Open space per BU in Layout Unit Design Module=
35.40 Sq.M./BU
Maximum Possible Built-Unit-Space(BUS) given by the MBO MODEL =
29.19 Sq.M/BU
File: MFMBLML.CPP   L-L-M-L   a5=50, BUD=200 i.e. L

3. Application Example: MFMBL3 Dwelling-Layout System.

SUMMARY OF ''outmfmb.cpp'' RESULTS IN THE INSTANT CASE ARE:
OPTIMISATION MODEL CHOSEN IS = 2
INPUT MODEL CONSTANT VALUES CHOSEN BY USER ARE:
nf=   4 n=   4 nb=   10 den=200.0000 ahc=80000.0000 I= 6.00 S= 9.00
a7= 0.1794 a8= 0.1723 a9= 0.0044 a10= 0.0187 a11 = 0.0204 a12=
0.0025
a1= 0.0000 a2= 1.0000 a3= 0.2500 a4= 0.1000 a5 =50.0000 a6= 0.2500
ix = 0.30000 iy= 0.20000
OUTPUT OPTIMAL VALUES GIVEN BY THE MODEL ARE:
Optimal Geometrical Programming Weights are:
w01= 1.0000 w61= 0.3360 w62= 0.3360 w63= 0.6640 w64 = 0.0171 w65=
0.0171
w11= 0.2129 w21= 0.6891 w31= 0.0171 w41= 0.2129 w51 = 0.3531 w66=
0.1958
Optimal Dimensions and Physical Parameter Values are:
Depth of Building-Blocki.e.Dwelling-Unit Design Module parallel to
staircase= 19.13 M
Width of Building-Blocki.e.Dwelling-Unit Design Module
perpendicular to staircase= 19.93 M
Width of Layout Unit Design Module in N-S direction(between road
center lines)= 93.38 M
Length of Layout Unit Design Module in E-W direction(between road
center lines)= 85.67 M
Optimal Building Rise i.e.number of floors in Dwelling-Unit Design
Module=   4
Optimal Number of Building Blocks in Layout Unit Design Module=   10
Optimal Number of Built-Units(BU)per floor in Dwelling-Unit Design
Module=   4
Optimal DENSITY in terms of Built-Units per unit area in Layout Unit
Design Module= 200.00BU/HA
Optimal Semi-Public Open space per BU in Layout Unit Design Module=
32.67 Sq.M./BU
Maximum Possible Built-Unit-Space(BUS) given by the MBO MODEL =
69.31 Sq.M/BU
File:MFMBLLM.CPP   M-L-L-M

4. Application Example: MFMBL4 Dwelling-Layout System.

```
SUMMARY OF ''outmfmb.cpp'' RESULTS IN THE INSTANT CASE ARE:
OPTIMISATION MODEL CHOSEN IS = 2
INPUT MODEL CONSTANT VALUES CHOSEN BY USER ARE:
nf=   4 n=   4 nb=   10 den=400.0000 ahc=80000.0000 I= 6.00 S= 9.00
CI=1500.00 CS=1770.0000 cf= 280.00 cw= 656.00 cr = 382.00 CL= 87.00
a7= 0.1794 a8= 0.1723 a9= 0.0044 a10= 0.0187 a11 = 0.0204 a12=
0.0011
a1= 0.0000 a2= 1.0000 a3= 0.2500 a4= 0.1000 a5 =25.0000 a6= 0.2500
ix = 0.30000 iy= 0.20000
OUTPUT OPTIMAL VALUES GIVEN BY THE MODEL ARE:
Optimal Geometrical Programming Weights are:
w01= 1.0000 w61= 0.3191 w62= 0.3191 w63= 0.6809 w64 = 0.0107 w65=
0.0107
w11= 0.0481 w21= 0.6489 w31= 0.0107 w41= 0.0481 w51 = 0.3298 w66=
0.0375
Optimal Dimensions and Physical Parameter Values are:
Depth of Building-Blocki.e.Dwelling-Unit Design Module parallel to
staircase= 20.65 M
Width of Building-Blocki.e.Dwelling-Unit Design Module
perpendicular to staircase= 21.51 M
Width of Layout Unit Design Module in N-S direction(between road
center lines)= 66.03 M
Length of Layout Unit Design Module in E-W direction(between road
center lines)= 60.58 M
Optimal Building Rise i.e.number of floors in Dwelling-Unit Design
Module=   4
Optimal Number of Building Blocks in Layout Unit Design Module=   10
Optimal Number of Built-Units(BU)per floor in Dwelling-Unit Design
Module=   4
Optimal DENSITY in terms of Built-Units per unit area in Layout Unit
Design Module= 400.00BU/HA
Optimal Semi-Public Open space per BU in Layout Unit Design Module=
4.81 Sq.M./BU
Maximum Possible Built-Unit-Space(BUS) given by the MBO MODEL =
80.77 Sq.M/BU
File:MFMBLLLM.CPP     L-M-L-M
```

To illustrate the above process of *sensitivity analysis* of planning and design decisions to derive an *optimal* solution, only three application examples with input *decision-matrices* of L-L-M-L, M-L-L-M, and **L-M-L-M** in Examples MFMBL2, MFMBL3, and MFMBL4, respectively, are presented above. To enlarge the scope of *sensitivity analysis*, interested planners-designers can consider more application examples, particularly because, possible application examples are numerous (for example, if each of the above four parameters can have 2 values, it will give 2^4 solutions which will have to be evaluated before choosing the most cost-effective solution). The parameters considered in these *decision-matrices* are LP-BUD-CS-BR with value of Low (L), Medium (M), and High (H) for each of the above four parameters, i.e. land price, built-unit-density, construction specification, and building-rise, respectively. It may be seen that in all the above examples, the same value of 'ahc' of 80,000 MU/BU is taken as INPUT. However, the *resource-productivity* indicated by the maximum possible built-unit-space (BUS) given by the MBO MODEL varies from 29.19 Sq.M/BU to 80.77 Sq.M/BU, i.e. by **2.767 times**.

It may be noted that both the Application Examples: MFMBL2 and MFMBL4 have same LP of 87 MU/Sq.M. (which is beyond the control of planner-designer), but the values chosen for other parameters, namely BUD, CS, and BR (which are entirely within the control of planner-designer) has resulted to the above large difference in the *resource-productivity* indicator value of **2.767 times.** Thus, the values of BUD, CS, and BR chosen as INPUT (entirely within the control of planner-designer and other *stakeholders*) are low (i.e. 200 built-unit per Hectare), medium (with corresponding medium unit rates of CI, CS, and so on, as shown under 'INPUT Model Constant Values' in the corresponding Application Example), and low (BR or nf $=2$), respectively, in Application Examples: MFMBL2. However, values of these parameters chosen as INPUT (again, entirely within the control of planner-designer and other *stakeholders*) are medium (i.e. 400 Built-Unit per Hectare), low (with corresponding low unit rates of CI, CS, and so on, as shown under 'Input Model Constant Values' in the corresponding Application Example), and medium (BR or nf $=4$), respectively, in Application Examples: MFMBL4.

If such 'what if' results **instantly** given by the **C++ program: DLMF.CPP** for various input *decision-matrices* mentioned above are brought to the table (along with corresponding graphic representation of each physical planning and design for each INPUT decision-set, which is feasible as shown in Chap. 9, to make it more meaningful) where *multiple stakeholders* including user citizens can *participate* (say, by giving suitable INPUT value to the above **C++ programs,** which are very user-friendly) will not only improve *resource-efficiency* and *social equity* in urban operations, but will also make it a truly *informed* and *participatory* urban development involving all *stakeholders* leading to more satisfying urban operations, and thus, nipping in the bud many potential urban problems.

11.14   Integrated Numeric-Analysis-Graphics in Dwelling-Layout System Design by Optimization for Resource-Efficient Problem Solution

As mentioned earlier, tedious and voluminous computations involved, in utilizing *Optimization* and operations research techniques and to solve and use *Optimizing Models* like above, have prevented their applications in building, housing, and urban *planning* and *design* in the past. The continued development of cheap and powerful computers with interactive graphics systems has made it possible to perform many planning and design functions integrating graphics with complex computations, design analysis, cost engineering, and *Optimization* techniques on a computer, revolutionizing the fields of design, engineering, and drafting. This is demonstrated with example problems in previous chapters and particularly in Chap. 9, where an Application Software: 'HudCAD', for integrated computer-aided building-housing design by optimization and drafting-drawing, is presented including all five components, and along with application results showing many advantages over the traditional manual methods, substantially improving **Productivity**. A module of 'HudCAD', for integrated computer-aided dwelling-layout system design by optimization and drafting-drawing, can be developed by interested readers adopting the same procedure as also indicated in Sect. 11.12.1.2.1 above, and thus, permitting adoption of an integrated numeric-analysis-graphics in dwelling-layout system design by optimization.

11.15   Closure

The combination of the layout and 'dwelling unit design modules', with the utility lines, road pavement, dwelling structure, building-housing service connection lines along with the semi-public open-to-sky spaces, could be considered as the dwelling-layout systems. In this chapter, the dwelling-layout systems design process based on module concept is presented. It is shown how using such concepts mathematical models both for quantitative analysis and for optimal design could be developed and used to attain *resource-efficiency* and promote *social equity* in dwelling-layout systems design, which may be of two types—one for single-family dwelling and the other for multi-family dwelling. Again, for each category of dwelling-layout systems design, there can be two types of *optimizing models,* namely: (a) least-cost optimizing model and the (b) most-benefit optimizing model, which are presented above along with application results.

Single-family dwelling-layout systems design by optimization of both types are presented in Sect. 11.11, covering both contexts of low- and high-income countries, showing general applicability of this *optimization* technique in all countries. Application examples of LCO models for least-cost design in Sect. 11.11.1 shows the variations in cost and in other quantitative design *efficiency* indicators, depending on the type of housing development chosen by planner-designers, while other factors such as size of dwelling in terms of dwelling space; construction specification standard; street-width hierarchy, circulation interval ratio and infrastructure standards; environmental standards in terms of open-space per built-unit (BU); and density standard in terms of number of single-family dwellings per unit area, are kept unchanged. It would be seen from the above computer results that the **Least-Cost** per BU are 13,681.7 MU/BU, 11,315.2 MU/BU, and 10,910.7 MU/BU, for the housing development type: 'Detached' (**Example: LCSFL1),** 'Semi-Detached' (**Example: LCSFL2**), and 'Row Housing' (**Example: LCSFL3**), respectively. Thus, the 'Detached' housing is **1.25 times** the cost of 'Row Housing' with the same size of dwelling, construction specification standard, street-width hierarchy, circulation interval ratio and infrastructure standards; environmental standards, and the density standard, as stated above. Application examples of MBO models also show similar large variation in the optimal maximum possible built-unit space per single family dwelling within the same affordable cost. In case of smaller size dwelling units, this increase ratio is higher.

In single-family dwelling-layout system design, generally, a complete dwelling-layout system with a number of independent lots with dwelling in a repetitive pattern is constructed and supplied to the individual beneficiary, permitting optimization taking full advantage of flexibility in dimensions of various components of the dwelling-layout system. This is, usually acceptable to low- and middle-income families. However, for high-income families, such complete single-family dwelling-layout system design may not be acceptable, who may like to get the developed lot but construct the dwelling with individual design according to their own choice. In such cases, the optimal layout may be developed to supply individual lots to the high-income families (as shown in Chap.10), who may like to construct the individual dwelling according to their choice. However, even individual families belonging to any income group who construct their houses in a lot of fixed size can also apply *optimization* for various components of the dwelling such as wall length, reinforced concrete elements such as slab-floor, beams, and so on, and thus attain a *resource-efficient* dwelling design solution. As shown in Sect. 11.11.1.3 above even in case of single-family dwelling with fixed lot size-room-arrangement matrix, one can optimize the wall lengths in the dwelling to minimize the cost by calculating the **Optimal Value** of 'DCR'(dwelling built-space coverage ratio along the depth of the lot) using the expression 11.28 presented above.

Application examples of MBO models also show a large variation in the optimal maximum possible built-unit space per single family dwelling within the same affordable cost. In case of smaller size dwelling units, this increase ratio is higher.

A number of application examples of optimization of multi-family dwelling-layout systems design are presented in Sect. 11.12; both for least-cost design and most-benefit-design, and also covering both contexts of low- and high-income countries, showing general and universal applicability of this *optimization* technique for all income categories and in all countries, i.e. low, middle as well as high-income countries. Application results show large variations in cost and in other quantitative design *efficiency* indicators. As, for example, in high-income group housing case **instantly** given by the **C++ program DLMF.CPP** as output, the **Least-Costs** per built-unit are **1,012,925.95 MU, 792,175.96 MU**, and **615,827.13 MU** in Application Examples: MFLCH1 (**Low-Rise-Low-Density Case**), MFLCH2 (**Low-Rise-Medium-Density Case**), and MFLCH3 (**Medium-Rise comparatively High-Density**), respectively. Thus, for the same input BUS (Built-Unit-Space) = 300 Sq.M. (Dwelling Unit Design Module Type: E—see Fig. 11.5 above) the **Least-Cost** per built-unit in low-rise-low-density case is **1.645 times** that of medium-rise and comparatively high-density case, while the unit costs of land, utility lines, foundation, and building component costs remain same in all cases. It may also be seen that the **Least-Cost** *optimal design* in the low-rise-low-density case, gives the optimal semi-public open space per BU in layout unit design module as high as 250.00 Sq. M./BU, with optimal Ground Coverage Ratio for building blocks in layout module as: 0.375, and the optimal Floor Area Ratio (FAR) in layout unit design module as 0.750 only. These factors have contributed to the increase of **Least-Cost** per Built-Unit by **1.645 times** that of medium-rise and comparatively high-density case, where the optimal semi-public open space per BU is 58.33 Sq.M./BU, the optimal Ground Coverage Ratio is 0.563, and the optimal Floor Area Ratio (FAR) achieved in layout unit design module is 2.250. It may be noted that even if the unit costs of utility lines, foundation, and building component costs are changed in a site-specific case, the relative cost difference per BU may be of the same order. However, if land price defined by the factor c_6 is very low, the relative cost difference per BU will be less.

Density is an important indicator of land utilization *efficiency*. Using the *optimization* model presented, a planner-designer can design a layout module for any desired and feasible dwelling density by assigning appropriate value to the design parameter a_5 in the model. Similarly, desired and feasible open space standard can be achieved by assigning suitable value to the design parameter a_6 in the selected model. Thus, using this technique, a planner-designer can carry out a number of iterations to examine a range of lot-densities, and thus, derive the *optimal* lot-density which is feasible, and also, gives the maximum dwelling space in a site-specific case; simultaneously attaining the desired open space standard by assigning suitable value to the design parameter a_5 in the model for optimal design for single-family dwelling-layout system.

Therefore, it is imperative to analyse and apply such *optimization* technique in all countries to maximize benefit, such as dwelling built-space, in which many user *stakeholders* would be interested. However, s*uch an analysis is rarely possible in conventional planning and design practices, causing many inefficiencies and inequities and consequent housing and urban problems*. This highlights the importance of adopting the integrated *computer-aided design by optimization* in dwelling-layout systems planning and design, to achieve the *resource-efficiency* and *social equity* in urban operations. It may be noted that each of the above application examples is real and represents a specific physical design which can be drawn, emphasizing the advantage of this technique compared to the conventional planning and design practice in AEC operations.

It may be noted that the output optimal numerical values defining the physical design of the optimal dwelling-layout system design can be easily drawn as revealed in Fig. 11.7 showing the optimal physical design of single-family dwelling-layout system, and in Fig. 11.8 showing the optimal physical design of multi-family dwelling-layout system, using the corresponding numerical results instantly given by the respective C++ program: DLSF.CPP or DLMF.CPP. Again, by incorporating a suitable LISP file in the C++ program: DLSF.CPP or DLMF.CPP the generation of output optimal design drawing, both soft copy and hard copy, can be completely automated as illustrated in earlier chapters including Chap. 9. Thus, planner-designers can use this computer-aided optimal design technique including graphics, without being conversant with the complicated, tedious and voluminous computations involved, in utilizing *Optimization* and operations research techniques and to solve and use the *Optimizing Models* like above. Interested planner-designers can even contribute for further development of this technique by going through and using the C++ programs and LISP programs, complete listing of which are supplied in an online space attached to this book.

References

Aguiler, R. J. (1973). *Systems analysis and design in engineering, architecture, construction and planning*. Prentice-Hall.

Alexander, C. (1964). *Notes on the synthesis of forms*. Harvard University Press.

Allen, T. J., & Scot Morton, M. S. (Eds.). (1994). *Information technology and the corporation of the 1990s—Research studies*. Oxford University Press.

American Society of Civil Engineers. (1939). *Land Subdivision—Manual of Engineering Practice No-16*.

Anderson M. L. (1973). Community improvements and services costs. *Journal of Urban Planning and Development Division, 99*(1).

Aoshima, N., & Kawakami, S. (1979). Weighting factors in environmental evaluation. *Journal of Urban Planning and Development Division, 105*, 2.

Atkins, R. S., & Krokosky, E. M. (1971). Optimum housing allocation model for urban areas. *Journal of Urban Planning and Development Division, American Society of Civil Engineers, 97*(1).

Autodesk Inc. (1985, 86, 87, 88, 89). *AutoLISP release 10-programmer's reference*, USA.

Ayres, F. (1974). *Schaum's outline of theory and problems of matrices*. McGraw-Hill Inc..

Barbara, W. (1976). *The home of man*. Penguin Books.

Beightler, C. S., Phillips, D. T., & Wilde, D. J. (1979). *Foundations of optimization*. Prentice-Hall.

Bellman, R. E., & Dryefus, S. E. (1962). *Applied dynamic programming*. Princeton University Press.

Bestor, G. C. (1967). Manual on urban planning, Chapter-III: Residential land planning. *Journal of Urban Planning and Development Division, Proceedings of the ASCE, 93*(2).

Branch, M. C. (1970). Delusions and diffusions of city planning in the United States. *Management Science, 16*(12).

Brotchie, J. F., Sharpe, R., & Toakley, A. R. (1973). Econometric model to aid urban planning decisions. *Journal of Urban Planning and Development Division, 99*(1).

Brown, L. (1978). *The Worldwide Loss of Cropland*. World Watch Institute.

Caminos, H., & Goethert, R. (1978). *Urbanisation primer*. MIT Press.

Cartwright, T. J. (1973). Problems, solutions and strategies: A contribution to the theory and practice of planning. *Journal of the American Institute of Planners, 39*, 179–187.

Catanese, A. J. (1972). *Scientific methods of urban analysis*. University of Illinois Press.

Chakrabarty, B. K. (1977a). Design and construction of brickwork structure. *Journal of the Institution of Engineers (India), 57*, 166–171.

© The Editor(s) (if applicable) and The Author(s), under exclusive license to Springer Nature Switzerland AG 2022
B. K. Chakrabarty, *Integrated CAD by Optimization*,
https://doi.org/10.1007/978-3-030-99306-1

Chakrabarty, B. K. (1977b). Ultimate strength design of reinforced concrete Slab-Hillerborg strip method. *Journal of the institution of Engineers (India), 59*, 119–126.

Chakrabarty, B. K. (1986a). Systems engineering approach to land subdivision design—An optimization model. *Journal of the Institution of Engineers (India), 67*, Pt. CI.1.

Chakrabarty, B. K. (1986b). Systems engineering approach in residential land planning. *Socio-Economic Planning Sciences, 20*, 33–39.

Chakrabarty, B. K. (1987a). Affordable dwelling-layout system design—A model for optimization. *Socio-Economic Planning Sciences, 21*, 283–289.

Chakrabarty, B. K. (1987b). Conventional planning practice and optimization in residential layout. *Third World Planning Review, 9*(2), 117.

Chakrabarty, B. K. (1988a). A model for optimal design of dwelling-layout system. *Building and Environment, 23*(1), 67–77.

Chakrabarty, B. K. (1988b). *Optimal design of housing development system.* Ph.D. Thesis, Indian Institute Of Technology, Delhi.

Chakrabarty, B. K. (1990). Models for the optimal design of housing development systems. *Environment and Planning B: Planning and Design, 17*(3), 331–340.

Chakrabarty, B. K. (1991). Optimization in residential land subdivisions. *Journal of Urban Planning and Development, American Society of Civil Engineers, 117*(1).

Chakrabarty, B. K. (1992a). Models for optimal design of reinforced concrete beams. *Computers & Structures, 42*(3), 447–451.

Chakrabarty, B. K. (1992b). A model for optimal design of reinforced concrete beam. *Journal of Structural Engineering, 118*(11).

Chakrabarty, B. K. (1996). Optimal design of multifamily dwelling development systems. *Building and Environment, 31*(1).

Chakrabarty, B. K. (1997a). Report on the programme design for education in urban management, A consultancy study sponsored by HUDCO, New Delhi.

Chakrabarty, B. K. (1997b). Urban management-a course design for education as a distinct discipline. *Spatio-Economic Development Record, 4*(1), 34–40.

Chakrabarty, B. K. (1998). Urban management and optimizing urban development models, *Habitat International, 22*(4).

Chakrabarty, B. K. (2001). Urban Management: concepts, principles, techniques and education. *Cities, 18*(5).

Chakrabarty, B. K. (2007). Computer-aided design in urban development and management—A software for integrated planning and design by optimization. *Building and Environment, 42*(1), 473–494.

Chakrabarty, B. K. (2012). *Chapter-5: Computer-aided design and computer graphics in Song Jia et al eds Computer aided design.* Nova Science Publishers Inc..

Cooper, W. W. (1970). An introduction to some papers on urban planning goals and implementation. *Management Science, 16*(12), 699–814.

Crosby, R. W. (1983). *Cities and regions as non-linear decision systems.* Westview Press Inc..

Danko, P. E., Popov, A. G., & Kozhevnikova, T. V. A. (1983). *Higher mathematics in problems and exercises: Part-1.* Mir Publishers.

Donald Hearn, M., & Baker, P. (1997). *Computer graphics C version.* Pearson Education Inc..

Eastman C., Johnson M., Kortanek K., (1970), A new approach to an urban information system. *Management Science, 16*(12).

Eastman, C., & Kortanek, K. (1970). Modelling school facility requirements in new communities. *Management Science, 16*(12).

Everard, N. J., & Tanner, J. L., III. (1987). *Theory and problems of reinforced concrete design.* McGraw-Hill.

Fenves, S. J. (1969). *Computer methods in civil engineering.* Prentice Hall of India.

Fintel, M. (1987). *Hand book of concrete engineering.* CBS Publishers.

Flachsbart, P. S. (1979). Residential site planning and perceived densities. *Journal of Urban Planning and Development Division, 105*(2).

Foot, D. (1981). *Operational urban models—An introduction*. Methuen.

Forrester, J. W. (1969). *Urban dynamics*. MIT Press.

Goetch, D. L. (1990). *MicroCADD: Computer-aided design and drafting on microcomputers*. Prentice-Hall of India.

Granfield, M. (1973). *An econometric model of residential location*. Bellinger Publishing Company.

Hadley, G. H. (1962). *Linear programming*. Addison-Wesley.

Harrington, S. (1987). *Computer graphics—A programming approach*. McGraw-Hill Publishing Company.

Harrison, B. (1973). The participation of ghetto residents in the model cities program, *Journal of the American Institute of Planners, 39*(1).

Harvey, J. (1987). *Urban land economics*. Macmillan Education Ltd..

Hasegawa, T., & Inoue, K. (1977). Urban, regional and national planning. *Proceedings of the International Federation of Automatic Control (IFAC) Workshop, Kyoto, Japan*, August, 1977, Pergamon Press.

Helma, B. P., & Clark, R. M. (1971). Locational model for solid waste management. *Journal of Urban Planning and Development Division, 97*, 1–13.

Helweg, O. J. (1979). Large scale urban modelling. *Journal of Urban Planning and Development Division*, UP-2.

Henry, S. (1992). *Integrating programming evaluation and participation in design*. Avebury.

Heuman, L. F. (1979). Low income housing planning: State-of-the-art. *Journal of Urban Planning and Development Division, 105*(2).

Hoch, I. (1969). The three-dimensional city: Contained urban space. In S. P. Harvey (Ed.), *The quality of built-environment*. Resources for the Future Inc..

Hsu, T., & Sinha, D. K. (1992). *Computer-aided design: An integrated approach*. West Publishing Company.

Hubbard, J. R. (1996). *Schaum's outline of theory and problems of PROGRAMMING WITH C++*. McGraw-Hill.

HUDCO. (1982). *Computer based design and analysis of affordable shelter*, HUDCO MODEL, New Delhi.

Hutchinson, B. (1973). Transport analysis of regional development plans. *Journal of Urban Planning and Development Division, 99*(1).

Jelen, F. C., & Black, J. H. (1983). *Cost and optimization engineering*. McGraw-Hill Book Company.

Koontz, H., & O'Donnel, C. (1976). *Management—A system and contingency analysis of managerial functions*. McGraw-Hill Publishing Company.

Koontz, H., & Weihrich, H. (1990). *Essentials of management*. McGraw-Hill Publishing Company.

Kuester, J. L., & Mize, J. H. (1972). *Optimization techniques with FORTRAN*. McGraw Hill Book Company.

Larson, R. C., & Odoni, A. R. (1981). *Urban Operations Research*. Prentice-Hall.

Lee, D. B. Jr. (1973). Requiem for large scale models. *Journal of the American Institute of Planners, 39*(3).

Mahapatra, P. B. (2000). *GRAPHICS PROGRAMMING IN C++*. Khanna Book Publishing Co..

Management Science—Urban Issues. (1970). The Institute of Management Science, Rhode Island, USA.

Mark, D. H., Revelle, C. S., & Liebman, J. C. (1970). Mathematical models of location: A review. *Journal of Urban Planning and Development Division, 99*(1).

McCormick, J. M., & Salvadori, M. G. (1968). *Numerical methods in fortran*. Prentice Hall of India.

McLoughlin, J. B. (1969). *Urban and regional planning—A systems approach*. Faber and Faber.

Meredith, D. D., Wong, K. W., Woodhead, R. W., & Wortman, R. H. (1973). *Design and planning of engineering systems*. Prentice Hall Inc..

Minai, A. T. (1984). *Architecture as environmental communication*. Mouton Publishers.

Mitchell, W. J. (1977). *Computer-aided architectural design*. Petrocelli/Charter.

Moore, D. W. (1971). Planning for a new town. *Journal of Urban Planning and Development Division, Proceedings of the ASCE*, UP-1.

NCU. (1988). *Report of the National Commission on Urbanisation* (Vol. II). Government of India Press.

Neufville, R. D., & Marks, D. (1979). *System planning and design: Case studies in modelling, optimization and evaluation*. Prentice Hall Inc..

Newman William, M., & Sproull Robert, F. (1981). *Principles of interactive computer graphics*. McGraw-Hill International Book Company.

Oliver, W. M. (1990). *Illustrated AUTOLISP*. BPB Publications.

Ozbekhan, H. (1967). The future of automation. *EKISTICS, 24*(142).

Penz, A. J. (1976). A resource constrained scheduling approach to land development projects. *Research Report No. 65*, Institute of Physical Planning, Carnegie-Mellon University, Pittsburgh, PA.

Phillips, D. T., & Beightler, C. S. (1976). *Applied geometric programming*. Wiley.

Piercy, N. (1984). Management and new information technology. In N. Piercy (Ed.), *The management implications of new information technology*. Croom Helm.

Pupescu, C., & Wilson, J. A. (1979). Development planning and scheduling model. *Journal of Urban Planning and Development Division, 105*(1).

Rasmussen, W. P. (1979). Forecasting housing requirements in a college town. *Journal of Urban Planning and Development Division, 105*(1).

Reynolds, C. E. (1961). *Reinforced concrete designer's hand-book*. Concrete Publications.

Rondinelli, D. A. (1973). Urban planning as policy analysis: Management of urban change. *Journal of the American Institute of Planners, 39*(1).

Rossman, L. A., & Graham, P. A. (1979). Distributing regional services costs. *Journal of Urban Planning and Development Division, 105*(1).

Ryan, D. L. (1994). *Computer-aided graphics and design*. Mercel Dekker Inc..

Sahlin, S. (1971). *Structural masonry*. Prentice Hall Inc..

Sanoff, H. (1992). *Integrating programming, evaluation and participation in design—A theory Z approach*. Avebury.

Scarborough, J. B. (1966). *Numerical mathematical analysis*. Oxford and IBH Publishing Company.

Schaefer, A. T., & Brittain, J. L. (1989). *The AutoCAD productivity book, Indian Reprint Edition*. Golgotia Publishers.

Schaefer, F. (1972). *The new town story*. Paladin.

Schlager, K. J. (1965). A land use plan design model. *Journal of the American Institute of Planners, 31*, 103–111.

Shapiro, M. H., & Rogers, P. R. (1979). Environmental model for screening land use plans. *Journal of Urban Planning and Development Division, 105*(2).

Shirley, P., Marschner, S., et al. (2011). *Fundamentals of computer graphics*. A K Peters/CRC Press, Taylor & Francis Group.

Smith, W. F. (1975). *Urban development—The process and the problems*. University of California Press.

Spillars, W. R. (1974). *Basic questions of design theory*. North Holland/American Elsevier.

Stark, R. M., & Nichols, R. L. (1972). *Mathematical foundations for design: Civil engineering systems*. McGraw Hill Book Company.

Stevens, R. T. (1993). *Graphics programming in C*. BPB Publications.

Stevens, R. T., & Watkins, C. D. (1993). *Advanced graphics programming in C and C++*. BPB Publications.

Teicholz, E., & Berry, B. J. L. (1983). *Computer graphics and environmental planning*. Prentice Hall Inc..

Terleckyj, M. E. (1970). Measuring progress towards social goals: Some possibilities at national and local levels. *Management Science, 16*(12), 765–778.

Touretzky, D. S. (2013). *Common LISP—A gentle introduction to symbolic computation*. Dover Publications, Inc..

Turner, A. (1980). *Cities of the poor*. Croom Helm.

United Nations. (1992). *World population trends*. Population Division, United Nations.

World Bank. (1990, 91). *World development reports*. The World Bank, Washington, DC.

Zecher, J. E. (1994). *Computer graphics for CAD/CAM systems*. Marcel Dekker Inc.

Zener, R. B., & Marans, R. W. (1973). Residential density, planning objectives and life in planned communities. *Journal of the American Institute of Planners, 39*(5), 337–345.

Index

Printed in the United States
by Baker & Taylor Publisher Services